Habib M. Ammari
Editor

The Art of Wireless Sensor Networks

Volume 1: Fundamentals

AF793429

Springer

Editor
Habib M. Ammari
WiSeMAN Research Lab
Department of Computer
and Information Science
College of Engineering
and Computer Science
University of Michigan-Dearborn
Dearborn, MI
USA

ISSN 1860-4862 ISSN 1860-4870 (electronic)
ISBN 978-3-662-50897-8 ISBN 978-3-642-40009-4 (eBook)
DOI 10.1007/978-3-642-40009-4
Springer Heidelberg New York Dordrecht London

© Springer-Verlag Berlin Heidelberg 2014
Softcover reprint of the hardcover 1st edition 2014
This work is subject to copyright. All rights are reserved by the Publisher, whether the whole or part of the material is concerned, specifically the rights of translation, reprinting, reuse of illustrations, recitation, broadcasting, reproduction on microfilms or in any other physical way, and transmission or information storage and retrieval, electronic adaptation, computer software, or by similar or dissimilar methodology now known or hereafter developed. Exempted from this legal reservation are brief excerpts in connection with reviews or scholarly analysis or material supplied specifically for the purpose of being entered and executed on a computer system, for exclusive use by the purchaser of the work. Duplication of this publication or parts thereof is permitted only under the provisions of the Copyright Law of the Publisher's location, in its current version, and permission for use must always be obtained from Springer. Permissions for use may be obtained through RightsLink at the Copyright Clearance Center. Violations are liable to prosecution under the respective Copyright Law.
The use of general descriptive names, registered names, trademarks, service marks, etc. in this publication does not imply, even in the absence of a specific statement, that such names are exempt from the relevant protective laws and regulations and therefore free for general use.
While the advice and information in this book are believed to be true and accurate at the date of publication, neither the authors nor the editors nor the publisher can accept any legal responsibility for any errors or omissions that may be made. The publisher makes no warranty, express or implied, with respect to the material contained herein.

Printed on acid-free paper

Springer is part of Springer Science+Business Media (www.springer.com)

Signals and Communication Technology

For further volumes:
http://www.springer.com/series/4748

To my first teachers: My mother,
Mbarka, and my father, Mokhtar
To my very best friends: My wife,
Fadhila, and my children, Leena,
Muath, Mohamed-Eyed, Lama,
and Maitham

To my first Dean: Dr. Bernard J. Firestone
Professor of Political Sciences and
Dean of the Hofstra College
of Liberal Arts and Sciences
at Hofstra University
for his wonderful friendship and
outstanding support to me during my
stay at Hofstra University
(September 2008–August 2011).

To my second Dean: Dr. Subrata Sengupta
Professor of Mechanical Engineering and
Dean of the College of Engineering
and Computer Science
at the University of Michigan-Dearborn
for his wonderful friendship and
outstanding support to me.

Foreword

This first volume in "The Art of Wireless Sensor Networks" series focuses on the "Fundamentals" behind wireless sensor networks, covering both the challenges and advances in networking and communications for these resource-constrained networks as well as the data management issues associated with the large volumes of sensor data produced by the individual and collective sensors. Wireless sensor networks continue to provide an important contribution to our information infrastructure, supporting a wide variety of industrial, medical, agricultural, research, and military applications. This book provides timely information for those desiring to understand some of the fundamental issues that make wireless sensor networks unique and challenging compared to more traditional wireless networks as well as proposed solutions to advance the use of sensor networks in theory and in practice. This volume will be an invaluable resource to those just beginning to explore this field as well as to those wanting to delve deeper into the issues involved in the design and implementation of this important class of networks.

Dr. Habib M. Ammari has assembled an impressive group of authors for the individual chapters, who are renowned for their work with wireless sensor networks and have provided both accessible surveys of their respective topics as well as theoretical, analytical, algorithmic, and protocol discussions, enabling the reader to benefit tremendously from their expertise. The first set of chapters touches on the networking aspects of wireless sensor networks, including the physical, medium access control, and routing layers. While these networking layers are found in virtually all networks, their design in wireless sensor networks must take into account the resource limitations of the individual sensor nodes as well as the application-specific purpose of these networks. These chapters describe how the protocols and algorithms for communication and networking can be optimized for such unique conditions in wireless sensor networks. Following this, the book explores the management of the network, in terms of topology, mobility, localization and task, and data management. Many of these are unique design issues for wireless sensor networks, and, once again, they must consider both the limited resources of the network as well as the application goals and quality of service requirements to appropriately manage the network. The final set of chapters delve into issues related with the data-driven nature of these networks, including gathering the data, securing the communications, enabling ease of use, and interoperability via middleware, and software design, all of which are vital for sensor networks deployed in real environments.

This volume entitled, "The Art of Wireless Sensor Networks": "Fundamentals," in conjunction with its sequel entitled, "The Art of Wireless Sensor Networks": "Advanced Topics and Applications," promises to be a staple for those wishing to learn the "art" as well as the science behind the design of wireless sensor networks!

February 10, 2013

Wendi Heinzelman
Dean of Graduate Studies for Arts, Sciences and Engineering
Professor of Electrical and Computer Engineering
Professor of Computer Science
University of Rochester
Rochester, NY
USA

Contents

Part VIII Security

Part IX Middleware

Part X Sensor Technology, Standards, and Operating Systems

Contributors

Hady S. AbdelSalam Microsoft Corporation, Redmond, USA

Ahmed E. A. A. Abdulla Tokyo University, Tokyo, Japan

Kemal Akkaya Southern Illinois University Carbondale, Carbondale, USA

Habib M. Ammari University of Michigan-Dearborn, Dearborn, USA

Nirwan Ansari New Jersey Institute of Technology, Newark, USA

He Ba University of Rochester, Rochester, USA

Carlo Alberto Boano University of Lübeck, Lübeck, Germany

Zhipeng Cai Georgia State University, Atlanta, USA

Qing Cao University of Tennessee, Knoxville, USA

Krishnendu Chakrabarty Duke University, Durham, USA

Marcus Chang Johns Hopkins University, Baltimore, USA

Chi-Yin Chow University of Hong Kong, Hong Kong, China

Flávia C. Delicato Federal University of Rio de Janeiro, Rio de Janeiro, Brazil

Ilker Demirkol Universitat Politecnica de Catalunya, Catalonia, Spain

Qi Han Colorado School of Mines, Golden, USA

Zygmunt J. Haas Cornell University, Ithaca, USA

Jing He Kennesaw State University, Kennesaw, USA

Tian He University of Minnesota, Minneapolis, USA

Wendi Heinzelman University of Rochester, Rochester, USA

Nishiyama Hiroki Tokyo University, Tokyo, Japan

Shouling Ji Georgia State University, Atlanta, USA

Nei Kato Tokyo University, Tokyo, Japan

Abdallah Khreishah New Jersey Institute of Technology, Newark, USA

Sookyoung Lee University of Maryland, Baltimore, USA

Fangxin Li Cornell University, Ithaca, USA

Qiao Li Cornell University, Ithaca, USA

Xiaoguang Li Temple University, Philadelphia, USA

Zhuo Li University of Texas at Arlington, Arlington, USA

Qilian Liang University of Texas at Arlington, Arlington, USA

Meng-Ling Liu Cornell University, Ithaca, USA

Mingyan Liu University of Michigan, Ann Arbor, USA

Sharad Mehrotra University of California, Irvine, USA

Mahta Moghaddam University of Southern California, Los Angeles, USA

Rajani Muraleedharan University of Rochester, Rochester, USA

Stephan Olariu Old Dominion University, Norfolk, USA

Felix Jonathan Oppermann University of Lübeck, Lübeck, Germany

Pouya Ostovari Temple University, Philadelphia, USA

Paulo F. Pires Federal University of Rio de Janeiro, Rio de Janeiro, Brazil

Christian Poellabauer University of Notre Dame, Notre Dame, USA

Zhijing Qin University of California, Irvine, USA

Surjya Ray University of Rochester, Rochester, USA

Kay Römer University of Lübeck, Lübeck, Germany

Rik Sarkar University of Edinburgh, Edinburgh, UK

Izzet Fatih Senturk Southern Illinois University Carbondale, Carbondale, USA

Agnelo Rocha da Silva University of Southern California, Los Angeles, USA

Andreas Terzis Johns Hopkins University, Baltimore, USA

Nalini Venkatasubramanian University of California, Irvine, USA

Lipeng Wan University of Tennessee, Knoxville, USA

Xin Wang University of Texas at Arlington, Arlington, USA

Jie Wu Temple University, Philadelphia, USA

Wenjian Xu University of Hong Kong, Hong Kong, China

Lin Yang Cornell University, Ithaca, USA

Ou Yang University of Rochester, Rochester, USA

Yanjun Yao University of Tennessee, Knoxville, USA

Mohamed Younis University of Maryland, Baltimore, USA

Albert Y. Zomaya University of Sydney, Sydney, Australia

Yi Zou Intel Corporation, Portland, USA

Part I
Introduction and Applications

Chapter 1
Introduction

Habib M. Ammari

> *"My original goal was to bring readers to the frontiers of knowledge in every subject that was treated. But it is extremely difficult to keep up with a field that is economically profitable, and the rapid rise of computer science has made such a dream impossible. The subject has become a vast tapestry with tens of thousands of subtle results contributed by tens of thousands of talented people all over the world. Therefore my new goal has been to concentrate on "classic" techniques that are likely to remain important for many more decades and to describe them as well as I can."*
>
> Donald E. Knuth
> *The Art of Computer Programming: Fundamental Algorithms (1997)*

1 The Art of Wireless Sensor Networks

Nowadays, the design and development of wireless sensor networks for various real-world applications, such as environmental monitoring, health monitoring, industrial process automation, battlefields surveillance, and seism monitoring, has become possible owing to the rapid advances in both of wireless communications and sensor technology. This type of network is cost-effective and appealing to a wide range of mission-critical situations. These two reasons helped them gain significant popularity compared to other types of networks. A wireless sensor network is a collection of low-powered, physically tiny devices, called *sensor nodes*, which are

H. M. Ammari (✉)
WiSeMAN Research Lab, Department of Computer and Information Science,
University of Michigan-Dearborn, Dearborn, MI 48128, USA
e-mail: hammari@umd.umich.edu

H. M. Ammari (ed.), *The Art of Wireless Sensor Networks*,
Signals and Communication Technology, DOI: 10.1007/978-3-642-40009-4_1,
© Springer-Verlag Berlin Heidelberg 2014

capable of sensing the physical environment, collecting and processing sensed data, and communicating with each other in order to accomplish certain common tasks. Furthermore, wireless sensor networks possess a central gathering point, called the *sink* (or *base station*), where all the collected data can be stored. The major challenge in the design and development of wireless sensor networks is mainly due to the severe constraints that are imposed on the sensing, storage, processing, and communication features of the sensor nodes. More precisely, the sensor nodes suffer from severely constrained power supplies, which shorten their lifetime and make them unreliable. It is worth noting that the sensor nodes may become faulty due to improper hardware functioning and/or low battery power (or energy). The latter is very crucial to be considered in the design and implementation of this type of network for their correct operation and longevity.

Since their inception in the late 1990s, wireless sensor networks have witnessed significant growth and tremendous development in both academia and industry. A large number of researchers, including computer scientists and engineers, have been interested in solving challenging problems that span all the layers of the protocol stack of sensor networking systems. Several venues, such as journals, conferences, and workshops, have been launched to cover innovative research and practice in this promising and rapidly advancing field. Because of these trends, I thought it would be beneficial to provide our sensor networks community with a comprehensive reference on as much of the findings as possible on a variety of topics in wireless sensor networks. As this area of research is in continuous progress, it does not seem to be a reasonable solution to keep delaying the publication of such reference any more.

This book series, titled "*The Art of Wireless Sensor Networks*," has two volumes that have been designed in a way to address challenging problems in traditional as well as new emerging areas of research in sensor networking. Moreover, all the book chapters in both volumes have been written as surveys of the state-of-the-art and state-of-the-practice of their corresponding topics. Our main goal is to help the readers understand the basic concepts of wireless sensor networks, and also be aware and knowledgeable of most of the underlying research topics although some of them are still in their infancy and not much work has been done to solve those new research problems. These two volumes are titled:

- *The Art of Wireless Sensor Networks: Fundamentals*
- *The Art of Wireless Sensor Networks: Advanced Topics and Applications*

This book relates to the first volume and focuses on the fundamentals concepts in the design, analysis, and implementation of wireless sensor networks. It covers the various layers of the lifecycle of this type of networks from the physical layer up to the application layer. Based on my fruitful discussion with all the contributing authors whom I invited, and, particularly, Drs. Wendi Heinzelman, Kay Römer, and Mohamed Younis, our rationale is that the first volume covers contemporary design issues, tools, and protocols for radio-based two-dimensional terrestrial sensor networks. Following Donald E. Knuth's above-quoted elegant strategy to focus on stable yet important "classic" techniques (*The Art of Computer*

Programming: Fundamental Algorithms, 1997), all the book chapters in this volume include up-to-date research work spanning various classic facets of the physical properties and functional behavior of wireless sensor networks, including physical layer, medium access control, data routing, topology management, mobility management, localization, task management, data management, data gathering, security, middleware, sensor technology, standards, and operating systems. This book will be an excellent source of information for both senior undergraduate and graduate students majoring in computer science, computer engineering, electrical engineering, or any related discipline. In addition, computer scientists, researchers, and practitioners in both academia and industry will find this book useful and interesting.

I would like to mention that I borrowed the title of this two-volume book series, "*The Art of Wireless Sensor Networks*," from Dr. Donald E. Knuth, computer scientist and Professor Emeritus at Stanford University, who is the author of the seminal multi-volume set of books, titled "*The Art of Computer Programming*." In fact, most of the problems being addressed in the area of wireless sensor networks are challenging and mathematical in nature. And, solving those problems requires an 'art' to find elegant yet efficient solutions in terms of time, space, and, especially, energy, which is a crucial resource in the design and implementation of algorithms and protocols for wireless sensor networking systems. I hope the readers will see the 'art' in this book and enjoy reading it as much as I enjoyed editing it.

2 Book Organization

This book has ten parts, each of which includes 2–3 chapters. Next, we briefly summarize the purpose of each part with a short description of its chapters.

In Part 1, titled "*Introduction and Applications*," Chap. 2 provides an overview of the most relevant applications of wireless sensor network, which have been deployed during the last one and a half decades. Furthermore, it gives a novel taxonomy of those applications with a goal to identify relevant programming constructs and run-time services. Chapter 3 investigates the problem of unattended deployment in a harsh environment and presents a discussion on the pros and cons of a specific wireless sensor network design.

In Part 2, titled "*Wireless Communications and Medium Access Control*," Chap. 4 investigates physical layer communications of wireless sensor networks. First, it presents an optimal power allocation scheme using the water-filling algorithm with Karush-Kuhn-Tucker conditions. Then, it describes two sensor selection schemes to enhance the parameter estimation in energy-constrained wireless sensor networks. Chapter 5 reviews various network coding techniques along with their assumptions and applications. It considers both general wireless networks and wireless sensor networks. Chapter 6 discusses several sensor sleeping techniques, which apply to either different layers of the protocol stack or multiple layers simultaneously. Also, it describes potential applications in each sleeping technique.

In Part 3, titled "*Routing*," Chap. 7 presents a classification of energy-aware routing algorithms and shows various issues with respect to data-aggregation, routing overhead, the energy hole phenomenon, and collisions/interferences. Chapter 8 discusses several utility-based routing protocols for wireless sensor networks and classifies them based on their utility properties, such as delay, cost, and packet delivery ratio. Moreover, it discusses the composition-based utility for wireless networks and its extensions in low duty-cycle wireless sensor networks.

In Part 4, titled "*Topology and Mobility Management*," Chap. 9 describes existing network topology management techniques for node failure tolerance. Also, it provides an analysis of the impact of node failure on network connectivity in wireless sensor networks, and proposes a classification of existing recovery schemes. Chapter 10 focuses on the problem of mobility in wireless sensor networks and its implications on sensing coverage, communication connectivity, and energy consumption. Precisely, it deals with target tracking in mobile wireless sensor networks using the Bayesian estimation theory. Also, it presents a purposeful and distributed mobility management scheme as a potential probabilistic solution to the problem of mobility management.

In Part 5, titled "*Localization and Task Management*," Chap. 11 surveys a variety of range-free localization techniques in wireless sensor networks, and provides a qualitative comparison of them. Also, it discusses current research directions in range-free localization. Chapter 12 considers two energy-aware task management protocols, which assign sensors to tasks based on their remaining energy while achieving balanced load among all the sensors. Also, it gives a comparison of these two protocols with an optimal task assignment protocol as well as energy-oblivious protocols with respect to the network lifetime.

In Part 6, titled "*Data Management*," Chap. 13 introduces a data management perspective on large-scale sensor environments applications whose goal is to meet non-functional requirements, such as timeliness, re-liability and accuracy, as well as functional needs of data collection. Chapter 14 presents geometric ideas to organize sensor data based on location information. It considers distributed methods for managing queries regarding isolated events, mobile objects, and general signal fields.

In Part 7, titled "*Data Gathering*," Chap. 15 considers two case studies of wireless sensor network monitoring systems, namely Life Under Your Feet and RACNet, in order to show the different components that constitute data collection networks. While the first system focuses on extreme duty-cycling and low data rate communications, the second one emphasizes high throughput and efficient channel utilization. Chapter 16 reviews existing techniques for data aggregation and presents their classification. In addition, it discusses a variety of tree-based and cell-based data collection algorithms. Also, it shows the dependency between those data aggregation and data collection techniques and potential applications.

In Part 8, titled "*Security*," Chap. 17 presents challenges for ensuring security in wireless sensor networks. It describes existing solutions, such as cryptography schemes, key management schemes, as well as some mechanisms for attack detection and prevention mechanisms. Also, it considers the problem of security in routing, localization, and data aggregation. Chapter 18 presents the technologies that are used for the protection of system privacy, data privacy and context privacy in wireless sensor networks, along with the threats in each of these three kinds of privacy. Moreover, it compares existing privacy-preserving techniques and discusses the strengths and weaknesses of each one of them.

In Part 9, titled "*Middleware*," Chap. 19 provides an overview of existing design approaches for middleware in wireless sensor networks. Also, it describes the most common middleware services and programming abstractions. Chapter 20 presents an approach to develop systems for wireless sensor networks, called Service Oriented Middleware, where the network is logically viewed as a service provider for consumer applications. This type of system provides abstractions of the network through a set of generic and/or application-specific services, such as data aggregation, adaptation, security, self-organization, resource management.

In Part 10, titled "*Sensor Technology, Standards, and Operating Systems*," Chap. 21 presents existing operating systems in wireless sensor networks and discusses their strengths and weaknesses. Chapter 22 describes various network simulators, such as NS-2, OMNET++, J-Sim, OPNET and TOSSIM, and network programming languages, such as NesC and Mate. Chapter 23 gives a review of wireless sensor network technologies, such as Zig-Bee, WirelessHART, 6LoWPAN, and ISA.100.11a, along with their network structure, protocol layers, and application areas.

3 Acknowledgments

This book of this complete two-volume series, titled "*The Art of Wireless Sensor Networks*," is a tribute to the fine work of the foremost leading authorities and scholars in their fields of research in the area of sensor networking. Frankly, it is not fair that I am the only one whose name appears on the book cover. And, it is a great pleasure and an honor for me to cordially recognize all of those who contributed a lot to this book and generously supported me throughout this project in order to make this two-volume series a reality. Without them, it would not be possible at all to finish this book and make it available to all the researchers and practitioners, who are interested in the fundamentals of wireless sensor networks.

First and foremost, I am sincerely and permanently grateful to Allah—the Most Gracious, the Most Merciful—for everything He has been providing me with. Particularly, I would very much love to thank Him for giving me the golden opportunity to work with such group of outstanding scientists and researchers to put together this book, and for helping me publish it within two years. I am very pleased to dedicate this modest book to Him and very much hope that He would kindly accept it and put His Blessing in it. His Saying "**And of knowledge, you (mankind) have been given only a *little***" has an endless, pleasant echo in my heart and always reminds me that our knowledge is much less than a drop in the ocean.

It is worth mentioning that all the contributing authors were invited to contribute to this book, and that no Call for Book Chapters had ever been sent through any mailing list. All of those authors whom I invited were chosen very selectively to cover most of the fundamental topics in wireless sensor networks. They have been contributing to the growth and development of the field of wireless sensor networks. This book would never have been written without their great contributions, support, and cooperation. Therefore, my cordial recognition is due to my colleagues—the ones whom I invited to contribute with their book chapters to this book—whose names are listed in the alphabetical order: Drs. Nirwan Ansari, Qing Cao, Krishnendu Chakrabarty, Xiuzhen Cheng, Flavia Delicato, Zygmunt Haas, Tian He, Wendi Heinzelman, Qilian Liang, Mingyan Liu, Sharad Mehrotra, Stephan Olariu, Christian Poellabauer, Kay Römer, Rik Sarkar, Andreas Terzis, Nalini Venkatasubramanian, Jie Wu, and Mohamed Younis. I am really honored to have worked with such an amazing crew of scientists. I learned a lot from them throughout this project, and it was an incredible experience for me in finishing this book.

Every book chapter has undergone two rounds of reviews. Moreover, in each round, every book chapter received 3–5 reviews by experts in the scope of the chapter. Our ultimate goal is to provide the readers with a high-quality reference on the fundamentals of wireless sensor networks. Precisely, all book chapters were carefully reviewed in both rounds by all the contributing authors. I would like to express my sincere gratitude to all the contributing authors for their constructive feedback to improve the organization and content of all book chapters. My special thanks go to Dr. Stephan Olariu for his generous offer to review all book chapters of both books of this two-volume series. Also, my original plan was to publish only one book, titled "*The Art of Wireless Sensor Networks*." But, I ended up with 40 book chapters. Therefore, I suggested to all the above-mentioned invited authors to split the book (i.e., 40 book chapters) into two volumes along with their book chapters and titles. Here, again, my special thanks go to all the invited authors for their very helpful feedback with regard to the content of each volume. Moreover, I am very grateful to Dr. Wendi Heinzelman, Professor of Electrical and Computer Engineering, and Dean of Graduate Studies for Arts, Sciences and Engineering at the University of Rochester, for her great foreword.

I started this project on Sunday, August 28, 2011 at 06:56 AM when I contacted the Publishing Editor, Dr. Thomas Ditzinger, who approved my proposal for an edited book. All book chapters for both volumes were uploaded on the website of Springer and made accessible to the Editorial Assistant, Mr. Holger Schaepe, on March 11, 2013. Hence, this project lasted over 18 months. During all this period of time, I exchanged 4, 840 emails with all contributing authors with regard to their book chapters. I would like to thank all the contributing authors for their invaluable time, flexibility, and wonderful patience in responding to all of my emails in a timely manner. Please forgive me for your time, and I hope that the readers will appreciate all of your great efforts and love all the materials in this book. We all have devoted a considerable time to finish this book and hope it will be paid off in the future.

I would like to acknowledge my family members who have provided me with excellent source of support and constant encouragement over the course of this project. In particular, I am most grateful to my best friend and beloved wife, Fadhila, for her genuine friendship and good sense of humor, and for being extremely supportive and unboundedly patient while I was working on this book. My special thanks and deep appreciation go to her for putting the *Art* into this book. In addition, I would like to express my hearty gratitude to my lovely and beautiful children, Leena, Muath, Mohamed-Eyed, Lama, and Maitham, for their endless support and encouragement. They have been one of my greatest joys, very patient, and understanding. I hope they will forgive me for spending several hours away from them while I was setting in front of my PC in my office or my laptop at home busy with this book. Several times, they all told me: "Daddy, your books and emails are always dragging you away from us!" My lovely wife and children have been a wonderful inspiration to me, and very patient throughout the life of this project. Without their warm love and care, this project would never even have been started. Also, I owe a lot to both of my first teachers, my mother, Mbarka, and my father, Mokhtar, for their sincere prayers, love, support, and encouragement, and for always teaching me and reminding me of the value of knowledge and the importance of family. Furthermore, I would like to thank my mother-in-law, Hania, and my father-in-law, Hedi, for their thoughtful prayers, concern, and valuable support. Besides, I would like to thank my sisters, sisters-in-law, brother, and brothers-in-law for their support and thoughts.

This project could not have been completed without the great support of the people around me who made this experience successful and more than enjoyable. I would like to thank all of my colleagues and friends at the University of Michigan-Dearborn and, particularly, the four departments of the College of Engineering and Computer Science (CECS), namely Department of Computer and Information Science, Department of Electrical and Computer Engineering, Department of Industrial and Manufacturing Systems Engineering, and Department of Mechanical Engineering, for the collegial and very friendly atmosphere they provided me with to finish this book. In particular, I am very grateful to Dr. Subrata Sengupta, Professor of Mechanical Engineering and CECS Dean at the University of Michigan-Dearborn, for his kindness, continuous encouragement, and outstanding support to WiSeMAN Research Lab since I joined the Department of Computer and Information Science at the University of Michigan-Dearborn on September 1, 2011. Also, I am very thankful to my colleagues at the University of Michigan-Dearborn, namely Dr. Kiumi Akingbehin, Professor of Computer and Information Science; Dr. Yubao Chen, Professor of Industrial and Manufacturing Systems Engineering, and China Programs Director; Dr. Bruce Elenbogen, Associate Professor of Computer and Information Science; Dr. Brahim Medjahed, Associate Professor of Computer and Information Science; Dr. Chris Mi, Professor of Electrical and Computer Engineering, and IEEE Fellow; Dr. Yi Lu Murphey, Professor and Chair of the Department of Electrical and Computer Engineering, and IEEE Fellow; Dr. Elsayed Orady, Professor of Industrial and Manufacturing Systems Engineering; Dr. Adnan Shaout, Professor of Electrical and Computer Engineering; Dr. Paul Watta, Associate Professor of Electrical and Computer Engineering; Dr. David Yoon, Associate Professor of Computer and Information Science; Dr. Armen Zakarian, Professor and Chair of the Department of Industrial and Manufacturing Systems Engineering; and Dr. Qiang Zhu, Professor of Computer and Information Science, and ACM Distinguished Scientist; for their wonderful friendship and for being so supportive and helpful along the way. In addition, I would like to express my special thanks and gratitude to my colleague,

Dr. Drew B. Buchanan, Director of the Office of Research and Sponsored Programs, for his excellent support to my research activities in several ways. This work is partially supported by the National Science Foundation (NSF) grants 0917089 and 1054935.

Last, but not the least, I would like to express my deep appreciation to Dr. Thomas Ditzinger, Publishing Editor, Dr. Dieter Merkle, Editorial Director, Mr. Holger Schaepe, Editorial Assistant, Springer-Verlag, Heidelberg (Germany) and New York (USA), Ms. Jacqueline Lenz, Springer Production Manager Book Production STM Heidelberg, Ms. Ramyakrishnan Murugesan, Springer Production Editor, Dr. S.A. Shine David, Project Manager, and Mr. Srinivas and his typesetting team, for their assistance throughout the lifecycle of this project, Varddhene V., Scientific Publishing Services, and Ms. Jessica Wengrzik, Springer DE. It was a great pleasure to work with all of them. I would like to acknowledge the publisher, Springer, for the professionalism and the high quality of their typesetting team as well as their timely publication of this book.

June 2013

Habib M. Ammari
WiSeMAN Research Lab

Chapter 2
A Decade of Wireless Sensing Applications: Survey and Taxonomy

Felix Jonathan Oppermann, Carlo Alberto Boano and Kay Römer

Abstract The popularity of low-power wireless sensors increased significantly in the last decade, triggering a golden era for wireless sensor network research and development. During the early years of the twenty-first century, wireless sensor network applications have evolved from small demonstrations with a lifetime of only a few hours to complete systems made up of hundreds of tiny wireless nodes deployed in a wide variety of settings, ranging from harsh and remote environments to residential buildings and clinical units. This survey gives an overview of the most relevant applications of wireless sensor network applications deployed during the last ten years, and classifies them using a novel taxonomy that aims to help identifying relevant programming constructs and run-time services. With more than 60 applications reviewed, ranging from military and civilian surveillance to tracking systems, from environmental and structural monitoring to home and building automation, from agriculture and industrial settings to health care, this survey will serve as a reference to guide researchers and system designers.

1 Introduction

Evolving from research at the University of California, Berkeley, CA, USA in the early years of the twenty-first century, wireless sensor networks (WSNs) have become an important research area with a high number of dedicated conferences and journals.

F. J. Oppermann (✉) · C. A. Boano · K. Römer
Institute of Computer Engineering, University of Lübeck, Lübeck, Germany
e-mail: oppermann@iti.uni-luebeck.de

C. A. Boano
e-mail: cboano@iti.uni-luebeck.de

K. Römer
e-mail: roemer@iti.uni-luebeck.de

H. M. Ammari (ed.), *The Art of Wireless Sensor Networks*,
Signals and Communication Technology, DOI: 10.1007/978-3-642-40009-4_2,
© Springer-Verlag Berlin Heidelberg 2014

The typical WSN consists of a number of tiny devices equipped with a micro-controller, a low-power radio, and a number of sensors to perceive their surrounding environment. These devices are usually networked in a multi-hop fashion, to enable cooperation among nodes and real-time delivery of sensed data to the user. The original vision of WSNs consisted of randomly dropping large quantities of tiny and low-cost embedded devices over a large area in order to enable ad hoc measurements. The resulting dense distribution of tiny sensor nodes would enable a better area coverage, an improved accuracy, and a greater fault tolerance compared to the use of traditional isolated sensors. However, this vision was beyond the technological capabilities at that time and the first prototypes of WSNs actually consisted of a small number of matchbox-sized devices, often called "motes." Still, their relatively small size allows a careful placement close to the phenomenon of interest, enabling unprecedented spatial and temporal resolution at rather low costs.

These properties, combined with the minimal need of human intervention, led to a great success of the WSN vision, and paved the way to the adoption in a wide range of applications, ranging from environmental monitoring and precision agriculture to industrial automation and personalized health care. Until today, most WSN deployments have a strong scientific background. Their main purpose is the demonstration of new technologies and the exploration of remaining limitations; the requirements of the actual application at hand are often secondary. Consequently, most deployments are carried out by computer scientists and not by the intended end-users.

One of the first application areas in the early years of WSN research was military surveillance, in which sensor nodes are seen as a tool to enable reliable and unobtrusive intrusion detection and tracking of enemy forces [9, 83]. In these scenarios, the sensor nodes are envisioned to be randomly dropped on the battlefield and to automatically self-organize into an operational WSN. A well-known early example is the 29 Palms project conducted in March 2001 by researchers from the University of California, Berkeley, CA, USA [83]. The demonstration employed a network of five sensor nodes that are dropped by an unmanned aerial vehicle (UAV) to monitor a road for passing vehicles. While environmental monitoring applications tend to be comparatively simple by only requiring a straightforward transmission of the sensed data to a single gateway station, early military scenarios like robust tracking of people and vehicles moving in the proximity of a WSN are more complex, as they require in-network processing of the sensed data. This explains why most of the early military deployments are rather small demo applications with a lifetime of only a few hours.

Another typical example of this kind of application is the deployment at Great Duck Island in the year 2002 [62, 72, 102], which is usually regarded as the first significant application deployment of a WSN. In this scenario, a sensor network was used to unobtrusively monitor the environmental conditions around the nests of storm petrels on a small island off the coast of Maine, USA. The goal was to provide a more detailed picture to the biologists examining the nesting behavior of birds. The deployment could not fully meet the expectations due to technical limitations, but it clearly highlighted the utility and usefulness of WSNs. Many deployments in the early years of sensor network research follow the example of Great Duck Island by focusing on environmental research as the primary application area. In these deployments, the WSN is typically used as a large-scale sensing instrument. Despite generating useful data for biological research, the primary aim is usually to demonstrate the usefulness and the advantages of low-power wireless systems compared to traditional approaches. Such advantages include higher spatial and temporal resolution of the measured data, greater flexibility, and lower costs.

Around the year 2004, the number of reported WSN deployments increased significantly. This increase was partly driven by the commercialization of the first WSN platforms, such as Mica2, Mica2Dot, and their later evolutions MICAz and TelosB, which became the de facto standard research platforms for WSNs [84]. Partly, this increase was also due to maturing software infrastructures (e.g., TinyOS, TinyDB), and the increasing robustness of networking protocols. As a consequence, WSNs started to cover a larger set of application areas, including more complex civilian scenarios such as structural monitoring [109], cold chain management [88], precision agriculture [20], emergency response [69], and health care [97, 106]. In the following years, the latter area started to evolve into an independent branch of WSN applications, often referred to as body sensor networks (BSN): networks of miniaturized and low-power noninvasive or invasive wireless biosensors used to monitor the vital signs of patients [114].

Simultaneously, the advent of wireless sensor and actor networks (WSANs) further broadened the application space of WSNs [4]. A traditional WSN is a pure measurement tool that only allows to observe the environment, and decision making processes typically happen outside of the network. WSANs, instead, also feature actuators and hence can exercise some control on the surrounding environment. In a WSAN, the control loop is usually closed within the network, and there is often support for the execution of decision processes. Consequently, the employed software tends to be more complex than in pure data collection applications. Despite their interesting features, the number of actual WSAN deployments is still rather low compared to the amount of WSN deployments. Only three out of over sixty deployments surveyed in this chapter feature actuators.

In the recent years, the number of WSN deployments largely increased and, driven by the overlap of neighboring research areas such as home automation and the Internet of Things, covers an even larger set of application areas. Furthermore, the increasing popularity of WSNs lead to an increase of unconventional application scenarios that combine sensor networks and other technologies, such as mobile robots [16], RFID [38, 39], cell phones, or smart cameras [79]. Such scenarios require more complex solutions than the ones employed in traditional WSN applications, and push the community away from simple low data-rate monitoring applications, which used to be the classical WSN deployments in the earlier years. Instead, classical WSN deployments experience a shift towards more economy-oriented scenarios and early real-world applications begin to appear. A good example is the SFpark project in San Francisco, CA, USA [92], in which sensor network technology is employed to monitor parking spots in a district of San Francisco. The collected data is used to enable demand-responsive pricing and a live search for empty parking spots while aiming to steer demand and to reduce congestion in the streets. In spite of such promising examples, the number of WSN applications outside the scientific community is still limited. Most deployments remain prototypical in character and are conducted by researchers working on sensor network technologies. Commercial applications tend to be conceptually simple and not to exploit the full potential of scientific innovations. For example, advanced multi-hop routing protocols are rarely used.

The following survey gives an overview of WSN applications deployed during the last ten years. It is based on a taxonomy that aims to help identifying relevant programming constructs and run-time services to support a broad range of existing and future WSN applications. Both the taxonomy and the survey are partially based on earlier work conducted in the context of the makeSense project [36]. The remainder of this chapter is structured as follows. In the next section we present a comprehensive taxonomy to classify WSN applications. In Sect. 3 we apply this taxonomy to a range of existing WSN applications and assign the applications to six basic categories based on the identified properties. For each category, exemplary applications are described in detail, and other representatives of the category with noteworthy distinctive characteristics are briefly highlighted. We conclude this chapter in Sect. 4 with an outlook on future WSN developments.

2 Taxonomy

Several taxonomies have been proposed in the literature to classify WSN applications. In 2002, Tilak, Abu-Ghazaleh and Heinzelman [104] defined an early taxonomy that allows the classification of WSNs according to different communication functions, data delivery models, and network dynamics.

Based on a discussion with European experts from academia and industry in 2004, Römer and Mattern [90] proposed an explicit definition of the WSN design space based on a taxonomy consisting of twelve categories. Their design space allows the characterization of WSN applications based on technical properties of the deployed network. It aims to cover the full spectrum of WSN application properties, but it is limited to sense-only applications without actuators. The design space is accompanied by a survey of WSN applications employing the newly defined taxonomy. Römer and Mattern's design space was later refined and simplified by Rocha and Gonçalves [89]. In addition to the simplification, the authors add categories to highlight application-specific needs independently of the employed technical solution.

Based on a survey of WSN taxonomies and the evaluation of existing WSN applications, Ruairí et al. [91] created a solution-neutral taxonomy to cover applications' primary requirements. In contrast to previous work, their taxonomy focuses on properties of the actual application and does not include technical properties of the deployed sensor network.

Bai et al. [11] proposed a WSN application taxonomy with eight dimensions that is used as a tool to identify typical application classes to fasciculate the creation of application-specific languages.

None of the proposed taxonomies seems to fully capture all aspects relevant to WSN design. We propose a refined taxonomy that aims to capture exactly these aspects while still staying concise enough to be useful. Our taxonomy is partially based on early work conducted by Mottola and Picco as part of a survey on WSN programming abstractions [77]. Their taxonomy is also intended as a tool to identify common application requirements that programming platforms must meet.

We now describe the dimensions of our taxonomy. For some dimensions, the possible values are not exclusive, and several of them may apply for a specific application.

Goal. Traditionally, WSN have been seen primarily as a *sense-only* tool to passively collect data. Such scenarios do not require control logic inside the network. For wireless sensor and actor networks (WSANs) this is not necessarily true, anymore [4]. The collected data could still be forwarded to a central location at which the control logic is implemented, but this is highly inefficient. Such applications follow a *sense and react* pattern.

$$goal \in \{\text{sense-only, sense-and-react}\}.$$

Sampling. Depending on the application scenario, WSNs follow different approaches for data collection and processing. In the *periodic* case, the nodes regularly read their sensors, process the resulting data, and possibly react accordingly. *Event-triggered* WSNs stay dormant for most of their lifetime and wait for some rare event. Each node monitors its sensors until a relevant event is detected. Following the successful detection of an event, the WSN becomes active and performs the required distributed processing.

$$sampling \in \{\text{periodic, event-triggered}\}.$$

Sensed phenomenon. Similarly to Ruairí et al. [91], we discriminate the sensed phenomena based on two orthogonal properties. The phenomenon sensed by a WSN can either be *discrete* or *distributed*. We consider a phenomenon to be discrete if it is located at a specific place and can usually be fully detected by a single sensor. Multiple sensors may be required if the phenomenon is mobile, for example to facilitate tracking, or they may allow a higher fidelity of the collected data. Distributed phenomena affect an area or volume and can only be fully captured by a larger number of sensors. WSNs may be used to monitor just a *single* phenomenon or *multiple* independent phenomena.

$$sensed\ phenomenon \in \{\text{single, multiple}\} \times \{\text{discrete, distributed}\}.$$

Data rate. The type of deployed sensors largely influences the capabilities and properties of a WSN. For this survey, we use an abstract classification of sensors based on the amount of generated data. We distinguish between two classes of sensors: *low* data-rate sensors, such as temperature and humidity sensors, that create a stream of simple numerical values and *high* data-rate sensors that produce large amounts of data per reading, like video cameras, or which require a high sampling rate, such as microphones and vibration sensors. The border between these classes is sometimes fuzzy. As a rule of thumb, we consider those sensors as high date-rate sensors that generate amounts of data that exceed the usual data-rate of WSN radio links. Most WSNs employ low data-rate sensors.

$$data\ rate \in \{\text{low, high}\}.$$

Heterogeneity. WSNs were originally envisioned to be largely *homogeneous*. In reality this is frequently not true: modern WSNs often consist of nodes that differ in the set of employed *sensors*. In addition, some WSNs employ multiple node *architectures* that offer different resources, for example in terms of storage space or processing power, and thus are bound to serve different purposes. Similarly, some nodes in the network may have a larger power supply available or they can harvest energy, which allows them to spend more time in an active state than the majority of the other nodes.

$$heterogenity \subset \{\text{sensors, architecture}\}.$$

Mobility. Most WSNs are *static*, but there are applications that require mobility. In the latter case, all the nodes or a subset of the *nodes* in the network are mobile. A typical example is having the sensor nodes directly attached to some animals. Active movement of the nodes is rare and thus not considered in this survey. The *base station* of the WSN is also not necessarily situated at a fixed location. It may move within the WSN or even occasionally leave the communication range of the network.

$$mobility \subset \{\text{mobile nodes, mobile base-station}\}.$$

Connectivity. The connectivity of a WSN depends on the communication range of the deployed radio, its surrounding environment, and the degree of network dynamics. Mobility influences the connectivity of a WSN, but disruption of communication can also be caused by an changing radio environment, even if the nodes stay at fixed locations. If at least one (multi-hop) communication path between each pair of nodes in the network is constantly expected, we consider the network to be *connected*. Note that sporadic unintended packet losses may also occur in the *connected* case. If the network is occasionally partitioned as part of normal operation, the network is considered to have *intermittent* connectivity. In some networks the nodes are isolated most of the time and enter the communication range of other nodes only *sporadically*.

$$connectivity \in \{\text{connected, intermittent, sporadic}\}.$$

Processing. In early WSN applications, the majority of data processing is performed outside of the network. To reduce the amount of data to be transmitted and to leverage the nodes' processing capabilities, it is sometimes beneficial to move parts of the data processing directly into the network. In-network processing can occur in different forms ranging from simple filtering to sophisticated control logic. We discriminate the following types of processing: *filtering*, *compression*, *aggregation*, *tracking*, *event detection*, *classification*, *interpretation*, and *decision making*. *Filtering* and *compression* allow to reduce the amount of data to be transmitted or stored. *Aggregation* further reduces the data by fusing values several sensors while passing it trough the network. If the network has the ability to autonomously scan for the occurrence of predefined events, we specify this as *event detection*. Further data processing can either take the form of *classification* or *decision making*. The result of the later is an immediately useful result suitable to act upon.

Orthogonally, processing can happen at different locations in the network. We distinguish local processing at a single *node*, distributed processing in the *network*, processing at a *gateway*, and processing at some *server* outside of the WSN system.

$$\begin{aligned} \mathit{processing} \subset \{&\text{filtering, compression, aggregation, tracking,} \\ &\text{event detection, classification, decision making}\} \\ \times \{&\text{node, network, gateway, server}\}. \end{aligned}$$

Storage. To allow later analysis of the gathered data, most applications require some form of *persistent* data storage. In addition, if the network is not constantly connected, some form of *caching* is required to permit a delayed transmission. Storage can happen either directly at the *node* that created the data, somewhere in the *network*, at the *gateway*, or at some *server* outside of the WSN system.

$$\mathit{storage} \subset \{\text{caching, persistent}\} \times \{\text{node, network, gateway, server}\}.$$

Services. Besides the processing of the data, further support services may be required for successful WSN operation. For many application scenarios it is necessary to spatially and temporally correlate the measurements of different sensors. This usually requires some form of *time synchronization* among the nodes. If node positions are not precisely know in advance there is also a need for some kind of *localization*.

If the sensed data is of privacy-critical nature, like vital patient data in a hospital, or its integrity is important, then the data needs to be protected against overhearing or tampering. This can be realized by implementing *encryption* and *authentication*.

Long-term deployments with changing environments or requirements raise a need for remote maintenance functions. *Reprogramming* or *reconfiguration* systems allow to meet new requirements by remotely adapting the nodes' software.

$$\begin{aligned} \mathit{service} \subset \{&\text{localization, time synchronization, authentication, encryption,} \\ &\text{reprogramming, reconfiguration}\}. \end{aligned}$$

Communication primitives. Deployed WSN employ a range of diverse communication primitives. Simple sense-only networks usually use only some form of *collection* to relay the gathered data towards the sink. More complex networks may employ a broad range of different processing primitives.

Especially heterogeneous networks may be organized in *clusters*. All nodes in a cluster communicate solely with a designated cluster head. This cluster head serves as a gateway to the outer world by communicating with other cluster heads and the base station of the WSN. In heterogeneous networks the nodes serving as cluster heads usually possess excessive resources in terms of energy supply, storage capacity, and computational power.

$$\mathit{communication\ primitives} \subset \{\text{single-hop unicast, multi-hop unicast, single-hop broadcast, flooding, collection, cluster}\}.$$

3 Survey

In this section, we present a comprehensive survey of well-documented WSN applications based on a systematic review of the leading WSN publication venues. The survey only includes applications whose feasibility has been demonstrated by either a prototype or a real-world deployment. We explicitly exclude pure testbeds and similar applications. A complete list of the applications covered in this survey can be found in Table 1. For each deployment, we report the approximate time frame, network size, and overall lifetime. If the year of the actual deployment is not explicitly stated, we assume the date of the earliest publication. We further map all the surveyed systems to the taxonomy presented in Sect. 2, and we summarize their properties in Tables 2 and 3. We finally categorize the surveyed applications into six classes based on the type of output returned to the user and on the complexity of the employed technical solution. For each class, we examine representative applications in detail and we highlight further deployments with distinguishing properties.

Most applications are using the WSN technology as a pure measurement instrument, and the collected data is typically transferred to a central server for post-processing and analysis. Hence, the network simply delivers the raw data, and its analysis is up to the end-user. This is typically the case for scientific deployments in which all the collected data is relevant. The indiscriminate collection of the complete raw data is also useful in prototypical deployments in order to be able to verify the correct operation of each single node. We divide this class of applications into three sub-classes: *Low-Rate Data Collection*, characterized by a periodic collection using low data-rate sensors; *High-Rate Data Collection*, characterized by periodic collection in which the data-rate generated by the sensors is particularly high; and *On-Demand Data Collection*, in which the user triggers the collection of data on-demand. Please note that in deployments that fall into one of these categories any processing or analysis of the data is left to the user: the WSN itself does not process the data. In another notable set of applications, the sensor network performs on-node processing and event detection or it even classifies the observed data within the network. As a result, the user does not get a raw collection of data anymore, but instead the notification about the occurrence of a given event or an instance of a class. We group these applications in the *Event Detection and Classification* class. Some other WSNs build upon the in-network event detection to localize or even track their position. The

Table 1 Basic properties of the presented WSN applications[1]

Application	Year	Size	Lifetime	Class
① Environment monitoring: PODS [19]	2002	*Tens*	Weeks	Low-rate data collection
② Habitat monitoring: Great Duck Island [62, 72, 102]	2002	Tens (100)	Months	Low-rate data collection
③ Glacier monitoring: GlacsWeb [74]	2004	Tens (9)	Months	Low-rate data collection
④ Power monitoring [53]	2004	*Tens*	Years	Low-rate data collection
⑤ Soil and Moisture monitoring [23, 24]	2004	Tens (9)	Weeks	Low-rate data collection
⑥ Vineyard monitoring: Unwired wine [17, 20]	2004	Tens (65)	Months	Low-rate data collection
⑦ Wildfire monitoring [35]	2004	Tens	Hours	Low-rate data collection
⑧ Environment monitoring: Redwood Eco-Physiology [32, 105]	2005	*Tens*	Days	Low-rate data collection
⑨ Forest fire detection: FireWxNet [37]	2005	*Tens*	Days	Low-rate data collection
⑩ Irrigation [57]	2005	Tens (5)	*Days*	Low-rate data collection
⑪ Landslide detection: SenSlide [94]	2005	*Tens*	Months	Low-rate data collection
⑫ Tunnel monitoring [27]	2005	Tens (18)	Weeks	Low-rate data collection
⑬ Water monitoring [81]	2005	Tens (5)	*Months*	Low-rate data collection
⑭ LOFAR-agro [63]	2006	*Tens*	Months	Low-rate data collection
⑮ Environment monitoring: SensorScope [15]	2007	Tens (16)	Months	Low-rate data collection
⑯ Irrigation: FLOW-AID [13]	2007	Tens (10)	Months	Low-rate data collection
⑰ Tunnel control and monitoring [31, 26, 78]	2007	*Tens*	Months	Low-rate data collection
⑱ Fire detection and tracking [8]	2008	Tens (12)	Hours	Low-rate data collection
⑲ Greenhouse Monitoring [3]	2008	Tens (4)	Days	Low-rate data collection
⑳ AC metering [49, 50]	2009	Tens (49)	Weeks	Low-rate data collection
㉑ Environment monitoring: GreenOrbs [75]	2009	Tens (120)	Weeks	Low-rate data collection
㉒ Environment monitoring: PermaSense [18]	2009	Tens (25)	Years	Low-rate data collection
㉓ Reliable Clinical Monitoring [29]	2009	Tens	Days	Low-rate data collection
㉔ Soil monitoring: Suelo [86]	2009	Tens (13)	Months	Low-rate data collection
㉕ Vineyard monitoring [7]	2009	Tens (27)	*Months*	Low-rate data collection
㉖ Wildlife and environmental monitoring [38, 39]	2009	Tens (36)	Months	Low-rate data collection
㉗ Duty cycling building: HVAC [2]	2010	Tens	*Days*	Low-rate data collection
㉘ MEDiSN [59]	2010	Tens (55)	Days	Low-rate data collection

(continued)

Table 1 (continued)

Application	Year	Size	Lifetime	Class
㉙ Electronic Shepherd [103]	2004	*Tens*	Months	Low-rate data collection
(56) Pipe monitoring: PIPENET [101]	2004	*Tens*	Months	Low-rate data collection
㉛ Relic protection in the forbidden city [65, 66]	2007	Tens (34)	*Weeks*	Low-rate data collection
㉜ Substation monitoring [80]	2008	Tens (45)	Weeks	Low-rate data collection
㉝ Structure monitoring: Four Seasons [109]	2004	*Tens*	Days	High-rate data collection
㉞ Industrial plant monitoring: Oil tanker [61]	2005	*Tens*	Months	High-rate data collection
㉟ Volcano monitoring [108]	2006	*Tens*	Days	High-rate data collection
㊱ Structure monitoring: Golden Gate Bridge [56]	2007	Tens (64)	Months	High-rate data collection
㊲ Structural monitoring [110]	2009	Tens (21)	*Days*	High-rate data collection
㊳ Structure monitoring: Torre Aquila [25]	2009	Tens (16)	Months	High-rate data collection
㊴ Underground animal tracking [73]	2010	Tens (14)	Months	High-rate data collection
㊵ Zebra monitoring: ZebraNet [51, 115]	2002	*Tens*	Months	On-demand data collection
㊶ Antelope monitoring: wildCENSE [87]	2006	*Tens*	Months	On-demand data collection
㊷ Environment Monitoring: LUSTER [93]	2007	Tens (20)	Weeks	On-demand data collection
㊸ High-fidelity Motion Analysis [70]	2009	Tens (9)	Days	On-demand data collection
㊹ Cane-toad monitoring [46, 47, 98]	2004	*Tens*	*Weeks*	Event detection and classification
㊺ Fence monitoring [111]	2007	Tens (10)	*Days*	Event detection and classification
㊻ Acoustic monitoring: VoxNet [6]	2008	Tens (8)	Days	Event detection and classification
㊼ Human monitoring: BehaviorScope [71, 14]	2008	*Tens*	*Days*	Event detection and classification
㊽ Coal mine monitoring [67]	2009	Tens (27)	*Days*	Event detection and classification
㊾ Fall detection: WeCare [5]	2010	*Tens*	Days	Event detection and classification
㊿ Activity recognition: PBN [54]	2011	Tens (5)	Hours	Event detection and classification
(51) Intrusion detection: 29 Palms [83]	2001	Tens (5)	Weeks	Localization and tracking
(52) Cold chain management [88]	2004	Tens (55)	Years	Localization and tracking

(continued)

Table 1 (continued)

Application	Year	Size	Lifetime	Class
⑬ Intrusion detection: A Line in the Sand [9]	2004	Tens (90)	Days	Localization and tracking
⑭ Robot navigation [16]	2004	Tens (9)	Hours	Localization and tracking
⑮ Sniper localization: PinPtr [99]	2004	Tens (60)	Months	Localization and tracking
⑯ Tracking: EnviroTrack [1, 44]	2004	Tens (70)	Days	Localization and tracking
⑰ Intrusion detection: ExScal [37]	2005	Thousands (1200)	Days	Localization and tracking
⑱ Parking lot surveillance [79]	2009	*Tens*	Days	Localization and tracking
⑲ Radio-based localization [113]	2012	Tens (16)	Days	Localization and tracking
⑳ Animal control: Networked Cows [22]	2004	Tens (10)	Weeks	Actuation
㉑ HVAC [34]	2005	*Tens*	*Weeks*	Actuation
㉒ Animal control [107]	2007	Tens (13)	Days	Actuation

[1]Values written in italics could not be determined with absolute certainty based on the literature available: these values have been estimated to the best of our knowledge

latter can be as simple as detecting a moving sensor or may require complex signal processing as in [99]. We group these applications in the *Localization and Tracking* class. Finally, applications that involve not only sensing but also actuation and actively manipulate the monitored environment are grouped in the *Actuation* class.

3.1 Low-Rate Data Collection

It is probably not surprising that low-rate data collection was the first application scenario for WSNs, and that it still represents the majority of existing deployments. These applications are typically characterized by periodic monitoring with low-data-rate sensors, such as simple temperature or infrared sensors that usually produce a single scalar value per measurement. Furthermore, they usually support an extensive lifetime of the network up to several years [21]. As it can be observed in Table 3, these applications rarely employ sophisticated in-network processing. The low data-rate makes it feasible to communicate the collected raw data without filtering, compression, or aggregation.

3.1.1 Environmental Monitoring

The most prominent example of this application class is probably a series of deployments between 2002 and 2004 at Great Duck Island in Maine,USA [72, 102] ②. The goal of these deployments was the long-term observation of the breeding behavior and nesting conditions of Leach's Storm Petrels. The involved biologists were especially interested in the usage patterns of the nesting burrows and how these are affected

Table 2 Mapping of WSN applications to the taxonomy presented in Sect. 2 (1/2)[2]

Application	Goal	Sampling	Phenomenon	Data rate	Heterogeneity	Mobility	Connectivity
① Environment monitoring: PODS [19]	SO	Periodic	Single distributed	Low	Homogeneous	Static	Connected
② Habitat monitoring: Great Duck Island [62, 72, 102]	SO	Periodic	Multiple discrete	Low	Sensors	Static	Connected
③ Glacier monitoring: GlacsWeb [74]	SO	Periodic	Single distributed	Low	Homogeneous	Nodes	Intermittent
④ Power monitoring [53]	SO	Periodic	Multiple discrete	Low	Architecture	Static	Connected
⑤ Soil and Moisture monitoring [23, 24]	SO	Periodic	Multiple distributed	Low	Sensors, architecture	Static	Connected
⑥ Vineyard monitoring: Unwired wine [17, 20]	SO	Periodic	Single distributed	Low	Homogeneous	Static	Connected
⑦ Wildfire monitoring [35]	SO	Periodic	Multiple distributed	Low	Homogeneous	Static	Connected
⑧ Environment monitoring: Redwood Eco-Physiology [32, 105]	SO	Periodic	Single distributed	Low	Homogeneous	Static	Connected
⑨ Forest fire detection: FireWxNet [37]	SO	Periodic	Single distributed	Low	Sensors	Static	Connected
⑩ Irrigation [57]	SO	Periodic	Multiple distributed	Low	Homogeneous	Static	Connected
⑪ Landslide detection: SenSlide [94]	SO	Periodic	Single distributed	Low	Architecture	Static	Connected
⑫ Tunnel monitoring [27]	SO	Periodic	Single distributed	Low	Homogeneous	Static	Connected
⑬ Water monitoring [81]	SO	Periodic	Multiple distributed	Low	Sensors, architecture	Static	Connected
⑭ LOFAR-agro [63]	SO	Periodic	Single distributed	Low	Homogeneous	Static	Connected
⑮ Environment monitoring: SensorScope [15]	SO	Periodic	Multiple distributed	Low	*Homogeneous*	Static	Connected
⑯ Irrigation: FLOW-AID [13]	SO	Periodic	Multiple distributed	Low	Sensors	Static	Connected
⑰ Tunnel control and monitoring [26, 31, 78]	SO	Periodic	Single distributed	Low	Homogeneous	Static	*Connected*

(continued)

Table 2 (continued)

Application	Goal	Sampling	Phenomenon	Data rate	Heterogeneity	Mobility	Connectivity
⑱ Fire detection and tracking [8]	SO	Periodic	Multiple distributed	Low	Homogeneous	Static	Connected
⑲ Greenhouse Monitoring [3]	SO	Periodic	Multiple distributed	Low	Homogeneous	Static	Connected
⑳ AC metering [49, 50]	SO	Periodic	Multiple discrete	Low	Sensors	Static	Connected
㉑ Environment monitoring: GreenOrbs [75]	SO	Periodic	Single distributed	Low	Homogeneous	Static	Connected
㉒ Environment monitoring: PermaSense [18]	SO	Periodic	Multiple distributed	Low	Homogeneous	Static	*Intermittent*
㉓ Reliable Clinical Monitoring [29]	SO	Periodic	Multiple discrete	Low	Sensors	Nodes	Intermittent
㉔ Soil monitoring: Suelo [86]	SO	Periodic	Multiple distributed	Low	Homogeneous	Static	Connected
㉕ Vineyard monitoring [7]	SO	Periodic	Multiple discrete	Low	Sensors, architecture	Static	Connected
㉖ Wildlife and environmental monitoring [38, 39]	SO	Event	Single discrete, Multiple distributed	Low	Sensors	Base sta-tion	Intermittent
㉗ Duty cycling building: HVAC [2]	SO	*Periodic*	Single distributed	Low	Sensors	Static	Connected
㉘ MEDiSN [59]	SO	Periodic	Multiple discrete	Low	Sensors	Nodes	Connected
㉙ Electronic Shepherd [103]	SO	Periodic	Multiple discrete	Low	Architecture	Nodes	*Sporadic*
(56) Pipe monitoring: PIPENET [101]	SO	Periodic	Multiple discrete	Low	Sensors	Static	*Connected*
㉛ Relic protection in the forbidden city [65, 66]	SO	Event	Multiple discrete	*Low*	Architecture	Static	Connected
㉜ Substation monitoring [80]	SO	Periodic	Multiple discrete	Low	Sensors	Static	Connected
㉝ Structure monitoring: Four Seasons [109]	SO	Periodic	Single distributed	High	Homogeneous	Static	Connected

(continued)

Table 2 (continued)

Application	Goal	Sampling	Phenomenon	Data rate	Heterogeneity	Mobility	Connectivity
㉞ Industrial plant monitoring: Oil tanker [61]	SO	Periodic	Multiple discrete	Low, high	Sensors	Static	Intermittent
㉟ Volcano monitoring [108]	SO	Periodic	Single distributed	High	Homogeneous	Static	Connected
㊱ Structure monitoring: Golden Gate Bridge [56]	SO	Periodic	Single distributed	High	Homogeneous	Static	Connected
㊲ Structural monitoring [110]	SO	Periodic	Single distributed	High, low	Homogeneous	Static	*Connected*
㊳ Structure monitoring: Torre Aquila [25]	SO	Periodic	Multiple distributed	Low, high	Sensors	Static	Connected
㊴ Underground animal tracking [73]	SO	Periodic	*Single discrete*	*Low*	Sensors, architecture	Nodes	Sporadic, connected
㊵ Zebra monitoring: ZebraNet [51, 115]	SO	Periodic	Multiple discrete	Low	Homogeneous	Nodes, base station	Sporadic
㊶ Antelope monitoring: wildCENSE [87]	SO	Periodic	Multiple discrete	Low	Homogeneous	Nodes, base station	*Sporadic*
㊷ Environment Monitoring: LUSTER [93]	SO	Periodic	Single distributed	Low	Sensors, architecture	Static	*Intermittent*
㊸ High-fidelity Motion Analysis [70]	SO	Periodic	Multiple distributed	*Low*	Homogeneous	Nodes	Intermittent
㊹ Cane-toad monitoring [46, 47, 98]	SO	Event	Multiple discrete	*High*	Architecture	Static	Connected
㊺ Fence monitoring [111]	SO	Event	Multiple discrete	High	Homogeneous	Static	Connected
㊻ Acoustic monitoring: VoxNet [6]	SO	Event	Multiple discrete	High	Homogeneous	Static	Connected
㊼ Human monitoring: BehaviorScope [14, 71]	SO	Periodic	Multiple discrete	High	Sensors	Static	Connected
㊽ Coal mine monitoring [67]	SO	Event	Multiple distributed	Low	Homogeneous	Nodes	Connected

(continued)

Table 2 (continued)

Application	Goal	Sampling	Phenomenon	Data rate	Heterogeneity	Mobility	Connectivity
㊾ Fall detection: WeCare [5]	SO	Event	Multiple discrete	Low, high	Sensors, architecture	Nodes	Connected
㊿ Activity recognition: PBN [54]	SO	Periodic	Single discrete, multiple distributed	High, low	Homogeneous	Nodes, base station	Connected
(51) Intrusion detection: 29 Palms [83]	SO	Event	Multiple discrete	*Low*	Homogeneous	Base station	Intermittent
(52) Cold chain management [88]	SO	*Periodic*	Multiple discrete	Low	Sensors, architecture	Nodes	Intermittent
(53) Intrusion detection: A Line in the Sand [9]	SO	Event	Multiple discrete	*Low*	Homogeneous	Static	Connected
(54) Robot navigation [16]	SO	Event	Single discrete	Low	Homogeneous	Base station	Connected
(55) Sniper localization: PinPtr [99]	SO	Event	Single discrete	*Low*	Homogeneous	Static	Connected
(56) Tracking: EnviroTrack [1, 44]	SO	Event	Multiple discrete	Low	Homogeneous	Static	Connected
(57) Intrusion detection: ExScal [37]	SO	Event	Multiple discrete	Low	Homogeneous	Static	Connected
(58) Parking lot surveillance [79]	SR	Event	Multiple discrete	*High*	Sensors, architecture	Static	Connected
(59) Radio-based localization [113]	SO	Peridic	*Single discrete*	*Low*	Sensors	Static	Connected
(60) Animal control: Networked Cows [22]	SR	Event	Multiple discrete	Low	Homogeneous	Nodes	Intermittent
(61) HVAC [34]	SR	Periodic	Multiple distributed	Low	*Homogeneous*	Static	*Connected*
(62) Animal control [107]	SR	Event	Multiple discrete	*Low*	Homogeneous	Nodes	Connected

[2]Values written in italics could not be determined with absolute certainty based on the literature available: these values have been estimated to the best of our knowledge

Table 3 Mapping of WSN applications to the taxonomy presented in Sect. 2 (2/2)[3]

Application	Processing	Storage	Services	Communication
① Environment monitoring: PODS [19]		Persistent[server]	Time synchronization	Collection, multi-hop unicast
② Habitat monitoring: Great Duck Island [102, 62, 72]		Persistent[server]		*Collection (cluster)*
③ Glacier monitoring: GlacsWeb [74]		Caching[node], persistent[gateway]	*Time synchronization*	*Multi-hop unicast*
④ Power monitoring [53]		Persistent[server]		*Collection*
⑤ Soil and Moisture monitoring [23, 24]		Persistent[node, server]		*Single-hop unicast*
⑥ Vineyard monitoring: Unwired wine [17, 20]		*Persistent[server]*	Synchronization	*Collection*
⑦ Wildfire monitoring [35]		Persistent[server]	Localization (GPS)	*Collection*
⑧ Environment monitoring: Redwood Eco-Physiology [105, 32]		Caching[node], persistent[gateway]		*Collection*
⑨ Forest fire detection: FireWxNet [37]		Persistent[server]	Time synchronization	Collection, multi-hop broadcast (cluster)
⑩ Irrigation [57]	Decision making [gateway]			*Collection*, Single-hop unicast
⑪ Landslide detection: SenSlide [94]	Filtering [node], aggregation [network], classification [server]	*Persistent[server]*		Collection, multi-hop unicast (cluster)
⑫ Tunnel monitoring [27]		Persistent [gateway]		*Collection*
⑬ Water monitoring [81]		Persistent[server], *Persistent[gateway]*		*Multi-hop unicast*
⑭ LOFAR-agro [63]	Compression[note]	Persistent [node]	Reprogramming	Collection
⑮ Environment monitoring: SensorScope [15]		*Persistent[server]*	Time synchronization	Multi-hop unicast
⑯ Irrigation: FLOW-AID [13]	Decision making [gateway]	Persistent[server]		*Multi-hop unicast*
⑰ Tunnel control and monitoring [26, 31, 78]				Collection, flooding

(continued)

Table 3 (continued)

Application	Processing	Storage	Services	Communication
⑱ Fire detection and tracking [8]	*Tracking[server]*	Persistent[server]		Collection
⑲ Greenhouse Monitoring [3]		Persistent[server]		*Collection*
⑳ AC metering [49, 50]		*Persistent[server]*		Single-hop broadcast
㉑ Environment monitoring: GreenOrbs [75]	*Filtering[node]*			Flooding
㉒ Environment monitoring: PermaSense [18]		*Caching[network]*		*Collection*
㉓ Reliable Clinical Monitoring [29]	Aggregation[network]	Persistent[server]		Collection
㉔ Soil monitoring: Suelo [86]	*Classification*			
㉕ Vineyard monitoring [7]		Persistent[server]		*Collection*
㉖ Wildlife and environmental monitoring [38, 39]	Compression[node]	Caching[network], Persistent[server]		Collection, multi-hop unicast
㉗ Duty cycling building: HVAC [2]	Decision making[server]	*Caching[server]*		Collection
㉘ MEDiSN [59]		Persistent[server]	Reconfiguration	Collection
㉙ Electronic Shepherd [103]		Persistent[server]	Localization (GPS)	Single-hop unicast (cluster)
(56) Pipe monitoring: PIPENET [101]	Aggregation[node], Event detection[node]	Caching[gateway]	Time synchronization	*Collection*
㉛ Relic protection in the forbidden city [65, 66]	*Event detection[server]*	*Caching[node], Persistend[server]*		*Collection*
㉜ Substation monitoring [80]				*Collection*
㉝ Structure monitoring: Four Seasons [109]	Compression[node]	Persistent[server]	Time synchronization	*Collection*
㉞ Industrial plant monitoring: Oil tanker [61]	Classification[server]	Persistent[server], caching[node]		Collection (cluster)
㉟ Volcano monitoring [108]	Event detection[node]	Caching[node], persistent[gateway]	Time synchronization, reprogramming	Collection, multi-hop unicast
㊱ Structure monitoring: Golden Gate Bridge [56]	*Filtering[node]*	Caching[node]	Time synchronization	*Collection*
㊲ Structural monitoring [110]		*Persistent[server]*	Time synchronization	*Collection*

(continued)

Table 3 (continued)

Application	Processing	Storage	Services	Communication
㊳ Structure monitoring: Torre Aquila [25]	Compression[node], classification[server]	Persistent[server]	Time synchronization, reconfiguration	Collection, flooding
㊴ Underground animal tracking [73]	Compression[node]	Caching[node], persistent[server]	Localization	Single-hop unicast, collection
㊵ Zebra monitoring: ZebraNet [51, 115]	Filtering[network]	Caching[network], persistent[server]	Localization (GPS)	Single-hop broadcast
㊶ Antelope monitoring: wildCENSE [87]	Compression[node]	Caching[node]	Localization (GPS), reconfiguration	*Flooding*
㊷ Environment Monitoring: LUSTER [93]	Aggregation[node]	*Persistent[network]*	Time synchronization	*Collection, single-hop unicast (cluster)*
㊸ High-fidelity Motion Analysis [70]	Filtering[node]	Persistent[node, server]	Reconfiguration, time synchronization	*Single-hop unicast*
㊹ Cane-toad monitoring [46, 47, 98]	Compression[node], filtering[node], classification[network], localization[network]	Caching[node], persistent[server]	Localization (GPS)	Single-hop unicast, collection (cluster)
㊺ Fence monitoring [111]	Filtering[node], event detection[node], event detection[network]		*Time synchronization*	Collection, neighborhood
㊻ Acoustic monitoring: VoxNet [6]	Event detection[network]	Caching[node], persistent[server]	Localization	*Multi-hop unicast*
㊼ Human monitoring: BehaviorScope [71, 14]	Classification[server]	Persistent[server]		
㊽ Coal mine monitoring [67]	Event detection[network], aggregation[network]		Reconfiguration	Collection, flooding
㊾ Fall detection: WeCare [5]	*Event detection[network]*			
㊿ Activity recognition: PBN [54]	Aggregation[node], *filtering[node]*, classification[gateway]	*Persistent[gateway]*		Collection

(continued)

Table 3 (continued)

Application	Processing	Storage	Services	Communication
⑤① Intrusion detection: 29 Palms [83]	Filtering[node], event detection[node]	*Persistent[network]*	Time synchronization, localization	*Collection*
⑤② Cold chain management [88]		*Persistent[server]*		*Collection*
⑤③ Intrusion detection: A Line in the Sand [9]	Event detection[node], classification[server], tracking[server]		Time synchronization	Single-hop broadcast, collection
⑤④ Robot navigation [16]	*Decision making[network]*			Flooding, single-hop broadcast
⑤⑤ Sniper localization: PinPtr [99]	Aggregation[network], classification[gateway], event detection[node]		Time synchronization, localization	Collection, single-hop unicast
⑤⑥ Tracking: EnviroTrack [1, 44]	Event detection[network], tracking[network], filtering[server]		Time synchronization, localization, reconfiguration	Single-hop broadcast, collection, flooding
⑤⑦ Intrusion detection: ExScal [37]	Classification[node], event detection[node]		Reprogramming	
⑤⑧ Parking lot surveillance [79]	Event detection[node], localization[network]			Collection, flooding
⑤⑨ Radio-based localization [113]	Decision making[gateway]			*Single-hop unicast*
⑥⓪ Animal control: Networked Cows [22]	*Decision making[node]*		Localization (GPS), reprogramming	Multi-hop unicast, single-hop broadcast, flooding
⑥① HVAC [34]	*Decision making[network]*			*Multi-hop unicast*
⑥② Animal control [107]	*Decision making[node]*		Localization (GPS)	Single-hop broadcast

[3] Values written in italics could not be determined with absolute certainty based on the literature available: these values have been estimated to the best of our knowledge

by environmental conditions. In detail, the Great Duck Island deployment consisted of several patches of sensing nodes, connected to a transit network via dedicated more powerful gateway nodes. A single base station provided Internet connectivity and database services for the whole deployment. The sensor patches consisted of two types of nodes: small sensor nodes monitor temperature and humidity in the nesting burrows, while infrared radiation sensors were used to detect the presence of a bird. A second type of nodes was used to monitor the weather conditions outside of the burrows. All nodes were carefully placed by hand and manually configured in advance, and no kind of self-organization or location detection was used [102]. The low data-rate of the employed sensors allowed to transfer all collected data to the base station without in-network aggregation or further processing. Another notable work very close in spirit to the Great Duck Island deployment is the long-term study of rare and endangered plant species by Biagioni et al. in the context of the PODS project [19] ①. Deployed in the Hawaii Volcanoes National Park (Hawaii, USA), the sensor network monitored several species of plants and their environment using temperature, humidity, rainfall, wind, and solar radiation sensors.

These two deployments were the first examples of long-term real-world deployment of WSNs and they became forerunners for a large number of similar deployments in the area of habitat and environmental monitoring. In the Redwood Eco-Physiology project [32, 105] ⑧, several redwood trees in a study area in Sonoma, CA, USA, were equipped with sensor nodes in order to allow a more fine-grained monitoring of the climate changes during the day than previously possible with conventional equipment. The involved quantities measured were air temperature, relative humidity, and photo-synthetically active solar radiation. In the context of the GreenOrbs project [75] ㉑, a WSN was used to observe the effect of different sunlight conditions in shrub thicket and to estimate canopy closure in a forest by collecting temperature, humidity, illumination, and carbon dioxide measurements. This application is especially notable for the high number of sensor nodes involved, with up to 330 nodes deployed in the forest.

In the GlacsWeb project [74] ③, a WSN was employed to generate insights on the conditions inside glaciers. The specific environment poses unusual challenges for the successful deployment of a WSN, as the glacier environment is especially hostile to sensor nodes and as radio communication through ice and water is known to be difficult and highly unreliable. In addition, due to its remote deployment location, the network had to reliably operate over long time intervals without direct interaction. A first prototype was deployed in the year 2003 at Briksdalsbreen in Norway, and an updated version was placed in the same area during 2005. Both networks were composed of a base station and eight sensor nodes. Each node was equipped with sensors to measure temperature, pressure, and the orientation in the ice. In order to survive in the harsh environment, the sensor nodes were encapsulated in robust and waterproof PVC capsules. The nodes were placed in previously drilled holes at predefined locations in the glacier, and data was sampled every four hours. Over time, however, the nodes slowly move with the ice, creating an additional challenge for radio communication. Once a day the collected sensor readings were transmitted to the more powerful base station situated on top of the glacier. The base station in turn relayed the collected data to a reference station with Internet access located at a nearby café via a long-range radio channel. Both prototype systems proved to be capable of gathering useful data and, in spite of the hostile environment, remained operational throughout the intended lifetime. Nevertheless, a high number of sensor nodes failed either because of hardware failure or because they lost radio connectivity with the base station.

GlacsWeb is not the only WSN successfully deployed in a harsh environment. In the context of the PermaSense project [18] ㉒, a WSN was deployed in a highly inaccessible terrain area in the Alps to support the creation of new temperature models. Another application similar in scope and execution is the SensorScope deployment at Le Génépi, Switzerland [15] ⑮. An embedded networked sensing system named Suelo was designed by Ramanathan et al. [86] ㉔ to collect high-resolution data on soil state. A distinctive feature of Suelo is to overcome problems with sensor calibration. If required, the system can automatically call for human verification and assistance. Another example of environmental monitoring application is the prediction of landslides through constant monitoring of ground stress [94, 95] ⑪. This deployment was part of the SenSlide project and showed how to use a WSN to protect human domiciles and infrastructures.

3.1.2 Wildfire Monitoring

A common application scenario with many similarities to environmental monitoring motivated by biological research is wildfire monitoring. Doolin et al. employed a WSN to monitor wildfire at the Pinole Point Regional Park (Contra Costa County, CA, USA) [35] ⑦. A similar low-rate data collection wildfire monitoring applications has also been developed by Antoine-Santoni et al. [8] ⑱. In a wildfire monitoring application by Hartung et al. [43] ⑨, a portable WSN called FireWxNet was used to monitor weather conditions in the proximity of wild-land fires. The collected data provides the firefighters with a more accurate picture of the local weather conditions and thus increases their efficiency and safety during fire suppression. In contrast to the previous examples, in the FireWxNet project the WSN is not permanently installed, but is intended to be deployed by the firefighters on demand.

3.1.3 Agricultural Settings

In addition to environmental and wildfire monitoring, low-rate data collection WSNs are also frequently used in agriculture. One example is the use of a WSN for monitoring a potato field in the LOFAR-agro project [63] ⑭. The main goal of the deployment was to generate new insights on climate conditions favoring Phytophthora, a fungal disease affecting potatoes and to enable more precise counteractive measures. The WSN was deployed at a remote field in Borger-Odoorn (Drentheand, The Netherlands) and should have supported a lifetime of one year in order to monitor the full growing cycle of the potatoes. It consisted of approximately one hundred nodes and a dedicated base station that was equipped with a Wi-Fi card to connect it to a backbone network. Each sensor node was attached to a combined temperature and humidity sensor. The collected data was relayed to a back-end system via the base station and it was stored for further processing. A second goal of the project was the evaluation of the suitability and reliability of sensor network technology under realistic environmental conditions. In this respect, the project is especially noteworthy for its overall failure. The deployed WSN never operated as intended and was hampered by a very high packet loss rate. According to the authors, only 2 % of the measurements made it to the back-end system. This deployment highlights many of the challenges faced in real-world WSN deployment that are still a major barrier to a more widespread use of sensor networks.

One specific agricultural application area is surprisingly popular among WSN researchers: vineyard monitoring. Several independent WSN deployments in vineyards exist [7, 17, 20, 78] ⑥ ㉕. In all these deployments, the sensor nodes were deployed at a vineyard in order to get a more fine-grained picture of the microclimate in the proximity of the plants. Anastasi et al. [7] further employed the WSN to monitor also humidity and temperature in the cellar used for wine storage and ripening. Other application areas in the agricultural field include greenhouse monitoring [3] ⑲, tracking of sheep [103] ㉙, irrigation control [13, 57, 81] ⑩ ⑬ ⑯ and soil moisture monitoring [23, 24] ⑤.

3.1.4 Industrial Settings

With an increase of reliability and the creation of more robust communication protocols, the use of WSNs has also been explored in industrial settings. Notable applications include monitoring of underground pipes in the PIPENET project [101] (56), monitoring of road tunnels [26, 27, 31, 78] ⑫ ⑰, and substation monitoring [80] ㉜. The increasing need to save energy in buildings opens a further area of application for low data-rate WSNs. Kappler et al. [53] ④ and Jiang et al. [49, 50] ⑳ have shown how WSNs allow a fine-grained monitoring of energy consumption of individual devices. Agarwal et al. [2] employed a WSN to provide information on room occupancy in order to operate a Heating, Ventilation and Air Conditioning (HVAC) system based on the actual demand ㉗. This allows to reduce the workload of the HVAC system and enables significant energy savings.

WSNs have also been used in museums or exhibitions to detect unsuitable climate conditions in the vicinity of exhibits. A prototype for relic protection has been deployed in the forbidden city in Beijing, China [65, 66] ㉛ .

3.1.5 Health Care

Low-Rate Data CollectionWSN applications have also been deployed in hospital environments to collect the vital signs of patients. Chipara et al. [29] have built a patient monitoring system and deployed it at the Barnes-Jewish Hospital (Saint

Louis, MO, USA) ㉓. The goal of the deployment was to monitor patients that do not require intensive care, but are at high risk. The patients wore TelosB-based wireless sensor nodes that measure pulse and blood oxygen saturation every 30 and 60 s, respectively. The data was forwarded to a base station through a number of static relay nodes carefully placed in the step-down hospital unit. This configuration also supported the mobility of patients that could hence be monitored even when they left the unit for diagnostic testing. A distinctive feature of this study is the thorough analysis of the system's reliability. On the one hand, the network performed pretty well and delivered more than 99 % of the data to the base station. On the other hand, the quality of the sensed data was affected by several factors, such as the mobility of the patient, the disconnection of the sensors, and the non-optimal placement of the pulse oximeters. Sensor disconnections typically cause long-term failures, and can hardly be noticed by the patients. In this specific deployment, sensor outages longer than 30 min were observed in more than 40 % of the patients, and lasted up to 14 h. Patient movements such as tapping or fidgeting, instead, only lead to short-term invalid sensor readings. In a first attempt to improve sensor reliability, the authors have discussed the impact of oversampling on sensing reliability and have developed an approach for early detection of sensor disconnection. Two other WSN pilot deployments in a clinical setting were carried out by Ko et al. [58, 59] at the Shock Trauma Center of the University of Maryland Medical Center, and in the Johns Hopkins Hospital Emergency Room (Baltimore, MD, USA) ㉘. Similarly to the work of Chipara et al. [29], the sensors employed measured blood oxygen levels and pulse rate, and a set of relay nodes forwarded the collected data to a central unit.

It is important to highlight that these two works are some of the few WSN applications designed for clinical monitoring in which the system was actually thoroughly tested on patients. The literature contains plenty of WSN architectures and prototypes specifically designed for medical applications and health care [12, 42, 96, 112], but they are rarely deployed in the real-world. Examples include the Health and Disaster Aid project [42], in which Gao et al. have proposed an architecture for medical WSN that collects real-time data in a mass casualty event. Similarly, in the context of the CodeBlue project [69], Lorincz et al. have proposed a WSN architecture for emergency response, which allows monitoring of the vital functions of a large number of patients during an emergency and tries to optimize the use of rescuers. Furthermore, in the ALARM-NET project [112], a heterogeneous WSN architecture for assisted-living and residential monitoring was developed using MICAz sensor nodes. We do not include such works in our survey due to the lack of real-world deployments.

3.1.6 Hybrid Systems

Finally, hybrid systems combining WSN and RFID technologies have also been implemented. An example is the wildlife monitoring deployment in Wytham Woods, Oxfordshire, UK by Dyo et al. [38, 39] ㉖. The system was supposed to provide the zoologists with a more detailed picture of the movement patterns of European badgers. The zoologists were especially interested in the social behavior of the animals, which is difficult to observe with traditional technology. Existing approaches, like VHF telemetry are labor-intensive and cannot be used on a large scale and for a prolonged time frame. The system consisted of three components; a number of RFID tags worn by a number of badgers; 26 detection nodes distributed at key locations, such as sets and latrines throughout the wood; and ten additional sensor nodes for micro-climate monitoring. RFID readers consume a significant amount of power when active. Hence, in order to extend the lifetime of the RFID detection, a two level adaptation process was employed. Short-term adaptation adjusts the detection interval of the reader based on recent detection events. This enables more accurate tracking if animals are present. Long-term adaptation adjusts the duty cycle of the detection nodes based on the observed activity pattern of the badgers. For example, if during the day activity is rarely detected, the intervals of time in which the system can be put into sleep mode can be increased. The system successfully operated for a one year period and is believed to be of great use to the involved zoologists [39].

3.2 High-Rate Data Collection

While it is usually feasible for low-rate data collection applications to transmit the collected raw data to a central server for processing, this is often not possible in high-rate data collection applications. In such scenarios, the data-rate generated by the sensors usually exceeds the available communication bandwidth or would quickly drain the limited energy budget if the raw sensor data is sent directly to a central unit. Hence, for such applications there is a need to either implement some form of data compression, filtering, or data processing into the network.

3.2.1 Structural Monitoring

A typical application area for high-rate data collection WSNs is structural monitoring, as demonstrated in the Four Seasons project, in which a WSN was deployed at an abandoned four-story building in Sherman Oaks, CA, USA, to monitor the health of the structure during earthquakes [28, 109] ㉝. The building was severely damaged during an earthquake in 1994 and consequently scheduled for demolition. Simultaneously to the sensor network experiment, a series of forced-vibration tests with conventional equipment were conducted. The nodes of the network were equipped with vibration sensors and accelerometers, which allowed to collect seismic structure response data in order to generate new insights on the cause of the damage in the building. The high amount of data generated by these sensors poses a major challenge as it is not possible to simultaneously transfer the data from all sensors. To limit the data to a maintainable rate, the system employed silence suppression and data compression. In addition, vibration analysis requires precise synchronization of the sensor readings. As a global time synchronization scheme would also require a significant amount of bandwidth, the system did not rely on a global synchronization of the nodes, but instead tracked the time it took a packet to travel through the network. This allowed to retrospectively correlate the measurements.

Similar WSN deployments for structural monitoring of bridges have been conducted at the Golden Gate Bridge in San Francisco, CA, USA [56] ㊱, and at a single-span bridge in St. Lawrence County, NY, USA [110] ㊲. The former deployment employed 64 nodes to measure ambient vibrations at a sampling rate of 1 kHz; the latter employed 20 sensor nodes with accelerometers and strain transducers.

A well-known deployment in the field of structural monitoring in heritage buildings is the "Torre Aquila" deployment in Trento, Italy [25] ㊳. The medieval tower Torre Aquila contains a precious and renowned medieval fresco called "Il ciclo dei mesi." The conservation of the tower and of the frescoes is endangered by the planned construction of a road tunnel below the building. Hence, the central goal of the sensor network deployment by Ceriotti et al. [25] was to generate a better insight into the structural behavior of the building and thus allow the assessment of how the construction work might affect the integrity of the tower. The actual deployment consisted of 16 nodes of different type and a dedicated base station. The captured data ranged from low-bandwidth measurements obtained using fiber optic sensors (FOSs) that detected deformations in the tower walls to high-bandwidth vibration measurements captured with a three-axis accelerometer. Additional nodes measure the temperature distribution in the building with the help of analog temperature sensors. The software employed in this deployment is especially noteworthy as it does not build directly on top of a WSN operating system, but it uses instead the TeenyLIME middleware [30], which provides basic network services such as routing and time synchronization.

3.2.2 Wildlife Monitoring

In a second deployment at Wytham Woods, Oxfordshire, UK, Markham et al. [73] employed a WSN for underground tracking of badgers ㊴. Limited radio propagation in the ground did not permit the use of radio-based localization. Instead, a number of magnetics coils was distributed over the set of interest. Each badger to be monitored wore a sensor node equipped with magneto-inductive sensors that periodically record the strength and properties of the magnetic field. All recordings were stored on the node until the badger left the set and moved into vicinity of a base station. In order to minimize the storage requirements, data compression was used. As soon as the badger reached the communication range of the base station, the data was uploaded and stored in an external database through a bulk transfer.

3.2.3 Environmental Monitoring

An example of environmental monitoring with high-rate data collection is an deployment at the active volcano Reventador in Ecuador by Werner-Allen et al. [108] ㉟.The goal of the deployment was the collection of high fidelity data on volcano activity to enable geologists to build a clearer picture of the seismic phenomena.

The deployment consisted of 16 sensor nodes equipped with seismic and acoustic sensors. High resolution seismoacoustic monitoring requires high data rates (up to 1200 bytes/s per node) that exceed the available communication bandwidth. Consequently, it is not possible to transmit the complete raw data. Werner-Allen et al. solved this challenge by only transmitting the data in case an interesting event is detected. Each node temporarily stored the collected data locally. As soon as a predefined pattern was detected, the node signaled a detection event to the base station. If a sufficient number of nodes reported an event, the base station triggered data collection and iteratively downloaded the last 60 s of recorded data from each sensor node. A second challenge is the precise synchronization of the logged events. To be useful to the geologists, the data needs to be correlated with a precision in the order of milliseconds. The deployment used global time synchronization based on the time signal of a single GPS receiver at the base station. In addition, a time rectification process was employed to further increase the accuracy of the recorded timestamps.

3.2.4 Industrial Settings

WSNs have also been used to monitor the status of in-field devices in industrial settings using high data-rate sensors. An ongoing deployment as part of the GINSENG project contemplates the replacement of the wired infrastructure at an oil refinery in Sines, Portugal [85]. To monitor the vibrations of industrial machinery and equipment, Krishnamurthy et al. have deployed a WSN in the engine of an oil tanker in the North Sea, and in a central utility support building at a semiconductor fabrication plant [61] ㉞. The system employed approximately 150 off-the-shelf accelerometers, and the data was stored persistently in a server located outside the sensor network.

3.2.5 Health Care

Acceleration measurements, as well as data obtained from cardiac or epilepsy care monitoring employing EKG, EEG also imply high data-rates. High-rate data collections from accelerometer sensors aimed for activity recognition and high-fidelity motor fluctuations monitoring have been carried out by Lombriser et al. [68] and Patel et al. [82]. Nevertheless, we do not include those works in our survey due to the lack of an actual real-world deployment.

3.3 On-Demand Data Collection

In on-demand data collection applications, the user triggers the collection of data on-demand. This class of applications typically involves a persistent data storage on the node or within the network in order to allow later retrieval of data.

3.3.1 Wildlife Monitoring

An example of on-demand data collection is ZebraNet [51, 115] ㊵. The main research goal of this project was to record data on migration patterns of zebras. A second goal was the exploration of the performance of a large-scale mobile WSN. A small amount of zebras were equipped with sensor nodes in the Sweetwaters Game Reserve (Nanyuki, Kenya) during January 2004. Each node was equipped with a Global Positioning System (GPS) receiver for localization in order to accurately log the position of each zebra for several months. The position of the animal was recorded once every hour, and more detailed information about the zebra's movement was recorded for three minutes of each hour. The network covered an area of 100 km^2 and was very sparse, hence the nodes could only sporadically communicate. Consequently, it was necessary to temporarily store the collected data on the nodes. To ensure a higher level of dependability, the data was replicated to other nodes in the vicinity, and the recorded data was collected by a mobile base station on a vehicle regularly driven by the end-user through the observed area. Very notable is the expected lifetime of the system. The application of the collars required the zebras to be tranquilized and put under high stress, hence it should be limited to once a year. Consequently, the network lifetime had to span at least this time frame. To achieve this goal, solar panels were used together with a rechargeable battery. Interesting features described by the authors are the inaccuracy of single GPS readings, and the design of the butyl belting that forms the collar. An application very close in spirit to Zebranet is the wildlife monitoring carried out by Ranjan et al. in the context of the WildCENSE project, in which a WSN was used to observe the movement patterns of antelopes [87] ㊶.

3.3.2 Environmental Monitoring

In the environmental monitoring project LUSTER [93], a WSN was used to monitor the light condition under shrub thickets ㊷. A network composed of 19 sensor nodes was deployed on Hog Island off the Eastern Shore of Virginia. As the remote location of the deployment did not permit a reliable continuous connection to an external data base, the network implemented distributed in-network storage for the collected data. The desired data could be fetched on demand either in situ or via a temporary directional long-range radio link.

3.3.3 Health Care

Another example of on-demand data collection is the deployment of Mercury at the Spaulding Rehabilitation Hospital in Boston, MA, USA [70] ㊸. Mercury is a software architecture running on Shimmer sensor nodes used to continuously sample and store sensor data in a MicroSD flash card for later retrieval. The system has been tested on patients undergoing treatments to measure accelerometer and electromyograph data for several days. Using a reliable transfer protocol based on acknowledgment messages, the end-user can extract selected raw data traces from each node and download them persistently to a server for later analysis. The transfer is triggered remotely by the end-user who needs to specify which specific set of data should be collected. A notable feature of Mercury is the local extraction of features from the collected data. To save considerable bandwidth, storage, and energy, Mercury provides a suite of custom feature-extraction algorithms such as maximum peak-to-peak amplitude, mean, and root mean square of the time series that are computed on the fly as sensor data is being acquired [70]. This implies that in addition to the raw data, the user can request on-demand a filtered dataset. In the next section, we will show a class of sensor network applications in which a filtered dataset is returned to the final user as soon as a given event has occurred.

3.4 Event Detection and Classification

In several WSN applications, the sensor network carries out on-node processing to detect user-defined events or to classify events according to a user-defined set of classes. In such applications, the end-user receives as output of the sensor network a notification of the occurrence of a given event or an instance of a class.

3.4.1 Structural Monitoring

An example of event detection and classification applications is the WSN designed by Li et al. [67] to detect collapses in coal mines ㊽. The goal of their "Structure-Aware Self-Adaptive Sensor System" is to quickly detect and report the collapse area in underground tunnels in order to ensure safer working conditions. The prototype was deployed on a tunnel wall 8 m wide and 4 m high, and 27 Mica2 motes preconfigured with their location coordinates were manually placed at carefully chosen points in the tunnel.

3.4.2 Wildlife Monitoring

The WSN designed by Hu et al. to detect the presence of Cane-toads in a specific area based on acoustic features [46, 47, 98] is a typical example of in-network classification ㊹. The authors deployed a large-scale WSN that incorporates in-network reasoning to autonomously classify toads. The goal of the sensor network was the monitoring of the increasing spread of cane toads in the North-East of Australia, due to its strong impact on Australian native fauna.

In the context of the VoxNet project, Allen et al. [6] employed a WSN to acoustically detect marmots at the Rocky Mountain Biological Laboratory (RMBL) in Gothic, CO, USA ㊻. The network consisted of eight rather powerful ARM-based nodes equipped with acoustic sensors that constantly monitored the environment for marmot alarm calls. The result could be used to notify a biologist on site in order to allow the gathering of further information. Although the network was designed to carry out in-network classification, in the actual deployment the nodes transfered the corresponding raw data to a gateway computer as soon as an interesting event was detected, and an external system employed the data from multiple nodes to deduce a position estimate for the call.

3.4.3 Civilian Surveillance

Wittenburg et al. [111] have demonstrated a rare example of civilian intrusion detection based on fence monitoring ㊺. In this example, the task of the WSN was to detect and report any incident occurring in the proximity of a fence, such as an intruder just probing the fence or actually climbing over it. In addition to the simple event detection and report, the network carried out also a classification of the activity, and has shown to be reliable even in a multi-hop scenario.

3.4.4 Health Care

The need for activity recognition has triggered a wide number of works in the body sensor network community. The PBN system [54] combined a five node BSN with an Android smartphone in order to enable reliable activity recognition ㊿. The system could detect and classify various daily activities, such as cleaning, eating, or watching television. Activity detection was primarily based on two-axis accelerometers attached to the sensor nodes. The necessary data processing was performed autonomously without relying on an external system. The detection and classification quality was improved by employing ensemble learning techniques based on user feedback provided through a smart phone. In BehaviorScope [14, 71], a BSN was used as part of a system to detect different activities of elderly people and to monitor for alarming deviations in their behavior ㊼. In contrast to PBN, this system did not employ sensor nodes worn on the body. Instead, nodes equipped with PIR sensors were distributed in the monitored apartment. WeCare [5] employed a combined BSN and WSN to detect falls ㊾. WeCare is more similar to PBN, but augmented the data from body-worn accelerometers with additional data sources. Fall detection was, for example, verified by video. Actual falls were automatically reported to caretakers or relatives using the cell-phone infrastructure.

3.5 *Localization and Tracking*

In many applications, not only an event has to be detected, but the location of that event has to be estimated or even tracked over time. Thus localization and tracking algorithms can be seen as a superset of event detection and classification applications.

Note that detecting an event that has to be localized can be as simple as receiving a message [88] or as complex as detecting a certain pattern in an acoustic signal [99].

3.5.1 Military Surveillance

Localization and tracking is frequently employed in military applications and surveillance systems, since the predominant goals are the detection, tracking, and classification of intruders in a given area. Data processing in these scenarios is especially challenging as it requires close co-operation of several nodes. Simon et al. have demonstrated in PinPtr how WSNs can be used to accurately estimate the position of snipers [99] (55). In their deployment, the sensor nodes were equipped with microphones to detect the muzzle blast of firearms, and they performed sound-based localization using distributed data processing. The network further performed a classification of the weapon generating the blast. In the 29 Palms Fixed/Mobile Experiment conducted at Marine Corps Air/Ground Combat Center (MCAGCC) in Twentynine Palms, CA, USA in March 2001, a WSN was dropped from an unmanned aerial vehicle (UAV) to monitor a road for vehicle movements [83] (51). Each of the nodes was equipped with a two-axis magnetometer, which allowed the detection of vehicles in a perimeter of 5–10 m. In addition, the WSN allowed to track vehicles once they were detected. The information on detected vehicles was temporarily stored in the network and later collected by a second flyover of the UAV. As the nodes were randomly dropped by a UAV, they needed to self-organize to allow collective monitoring of the environment. In the EnviroTrack project [1, 44], a similar but larger network was used to track intruders (56). "A Line in the Sand" [9] extended these capabilities by distinguishing between civilian or military vehicles and persons (53). In the ExScal project [37], a location detection of intruders based on proximity was carried out (57). The deployment is especially noteworthy for its unusual size of over 1000 nodes, all carefully placed in a preplanned layout.

3.5.2 Industrial Settings

On the industrial side, mobile networks are envisioned to be deployed for tracking of assets and goods, for instance, to ensure that certain climate conditions are constantly met while some goods travel through the cold chain [88] (52). These application scenarios are also especially challenging from a programming perspective, as they are highly dynamic and a high number of parties are involved.

Na et al. [79] have designed an application for parking lot surveillance that employs also traditional surveillance cameras (58). In this deployment, tracking information from the sensor network was used to control surveillance cameras.

3.5.3 Assisted Navigation

In [16], sensor nodes were employed to assist the navigation of an autonomous robot (54). The sensor nodes acted as signposts for the robot, that makes navigation decisions based on its closest node. In their setting, the robot did not have a predecided environment map, as the environment can be dynamically changing.

In their demonstration, Xu et al. [113] employed sensor nodes to track a single person based on received signal strength indication (RSSI) fingerprints (59). The system employed eight sender and eight receiver nodes. The senders periodically send beacon messages to each receiver. Based on previously recorded training data, the position of the person was inferred based on the effect that his or her presence had on the RSSI readings at the eight receivers.

3.6 Actuation

The addition of actuators to a sensor network allows not only to monitor the surrounding environment, but also to actively manipulate it. Such wireless sensor/actor networks (WSANs) raise additional challenges [4]. To allow the execution of control logic, it is necessary to implement control processes inside the network. Centralized decision making is usually not an option, as it would require excessive communication. The need to specify sophisticated control logic makes programming such WSANs especially difficult, which may explain why the number of WSAN applications is still comparatively low nowadays.

3.6.1 Building Automation

Modern buildings feature sophisticated heating, ventilation and air conditioning (HVAC) systems that can benefit from the use of a WSAN [34] (61). By replacing wired sensors and a centralized control system, WSANs promise to reduce costs and at the same time increase the flexibility of the solutions. The control logic of an HVAC system is usually based on the current climate in various parts of the building and a set of preferred temperature levels specified by the building users, and actuators can control heating or cooling devices [34].

3.6.2 Agricultural Settings

In the Animal Control project [107] a WSAN was used to control the behavior of bulls (62). Bull fights during the breeding season sometimes lead to serious injuries that significantly limit the value of the affected animal. As bulls are rather high-value animals, these injuries may lead to high losses for the farmer and are highly undesirable. Wark et al. [107] successfully employed a WSAN to separate bulls on a meadow and prevent them from fighting without a need for additional fences. The bulls were equipped with sensor/actor nodes that allowed to apply unpleasant but harmless stimuli to the animal. In addition, each sensor node was equipped with a GPS sensor that enabled precise localization of the animals. The equipment was worn by the animals in specially manufactured webbing collars. The WSAN constantly monitored the distance between the bulls and their aggressiveness level. If a bull was in the proximity of another bull and started moving in its direction, a small electrical shock was applied by the stimuli actuator. The network was also used to monitor the success of the control action and adjusts the stimuli accordingly. The efficiency of the approach was demonstrated by a 40 min controlled experiment in which all the relevant data was logged for later analysis. In the earlier Networked Cows project, Butler et al. [22] used a similar approach to keep cows within a limited area with the help of virtual fences [60]. Both scenarios combine the challenges of mobile WSNs and WSANs, such as the limited connectivity between sensor nodes. Currently, both projects rely on a central control instance and do not implement distributed in-network processing.

4 Summary and Outlook

The early years of the twenty-first century have seen a steep rise in the number and diversity of wireless sensor network applications. This survey examined over 60 applications spanning from scientific demonstrations to real-world deployments, and covered several application areas ranging from military and civilian surveillance to tracking systems, from environmental and structural monitoring to home and building automation, from agriculture and industrial settings to health care.

Triggered by the vision of Smart Dust [52], where thousands of tiny sensor nodes would be dispersed into the environment, researchers began to implement and deploy applications to drive and evaluate their research under realistic conditions. While the first of these systems addressed military applications, the focus quickly shifted towards environmental monitoring, and then agriculture, structural monitoring, home and building automation, health and sports. Due to the lack of real Smart Dust platforms, early systems used matchbox-sized motes assembled from off-the-shelf components [45]. As it turned out, however, these platforms were quite sufficient for many applications, also because the motes were carefully placed instead of dispersed. Yet, there are continuing efforts to build grain-sized motes [64], but so far they did not carry over to application deployments. However, as this technology matures, we may see applications in the future that truly require small size such as intra-body sensor networks, or swarms of tiny flying sensors [33] helping with pollination.

In fact, we can recently observe a general trend to broaden the field and move away from using homogeneous networks of mote-class devices with simple scalar sensors. New types of sensors and actuators such as cameras [10], RFID readers [39], or car controls [40] are being integrated, resulting also in new research challenges due to, for example, high data rates. Hence, high-performance microcontrollers [60] or even mobile phones are becoming interesting platforms that enable new types of applications. Especially smartphone-based participatory sensing applications are recently receiving substantial attention (e.g., [41, 55, 76]), mainly because mobile phones are already ubiquitously deployed and code can be easily distributed using app stores. Thus, very large-scale and redundant "sensor networks" as originally envisions are becoming feasible, but at the same time new challenges arise as placement and use of phones worn by people cannot be easily controlled, and collected data may expose sensitive information about people wearing the phones.

The recent vision of the Internet of Things takes scaling to an extreme in that all objects and places – respectively sensors and actuators embedded into them – shall be connected to the Internet. Thus, the state of the real world becomes accessible online and in real time and converges with the vast amount of information available on the Internet. In order to also connect motes to the Internet, TCP/IP stacks have been squeezed into 8-bit microcontrollers, enabling IP-based sensor networks [48]. Different from traditional sensor networks, there is typically less direct cooperation among sensor nodes in the Internet of Things, as each node monitors and controls the state of an object to which it is attached. One example are large deployments of parking spot occupancy sensors in Barcelona and San Francisco [92] where each node monitors a single parking spot. These deployments are the seeds for even-larger scale smart city projects [100] where many aspects of our urban environment will be monitored, and potentially even controlled and optimized.

This prospect raises serious questions about dependability, trustworthiness, specifically security and privacy aspects, but also ease of use. In fact, only a single application uses encryption to protect sensor data; and the large majority of the surveyed applications were deployed by scientists. While we could largely ignore the above issues in environmental monitoring applications where a system failure was annoying but not harmful and collected data revealed interesting but not privacy-violating insights, this is not the case any more for the applications that appear at the horizon.

Acknowledgments We would like to thank the anonymous reviewers for their helpful comments and suggestions for improving this manuscript. The research leading to these results has received funding from the European Union's Seventh Framework Programme (FP7/2007–2013) under grand agreement n° 258351 (makeSense: Easy Programming of Integrated Wireless Sensor Networks), n° 224053 (CONET, the Cooperating Objects Network of Excellence), and n° 317826 (RELYonIT: Research by Experimentation for Dependability on the Internet of Things).

References

1. T. Abdelzaher, B. Blum, Q. Cao, Y. Chen, D. Evans, J. George, S. George, L. Gu, T. He, S. Krishnamurthy, L. Luo, S. Son, J. Stankovic, R. Stoleru, A. Wood, EnviroTrack: towards an environmental computing paradigm for distributed sensor networks, in *Proceeding of the 24th International Conference on Distributed Computing Systems (ICDCS)*, pp. 582–589 (2004)

2. Y. Agarwal, B. Balaji, S. Dutta, R.K. Gupta, T. Weng, Duty-cycling buildings aggressively: the next frontier in HVAC control, in *Proceeding of the 10th International Conference on Information Processing in Sensor Networks (IPSN)*, pp. 246–257 (2011)
3. T. Ahonen, R. Virrankoski, M. Elmusrati, Greenhouse monitoring with wireless sensor network, in *Proceeding of the 4th International Conference on Mechatronic and Embedded Systems and Applications (MESA)*, pp. 403–408 (2008)
4. I.F. Akyildiz, I.H. Kasimoglu, Wireless sensor and actor networks: research challenges. Ad Hoc Netw **2**(4), 351–367 (2004)
5. H.O. Alemdar, G.R. Yavuz, M.O. Özen, Y.E. Kara, O.D. Incel, L. Akarun, C. Ersoy, Multimodal fall detection within the WeCare framework, in *Proceeding of the 9th International Conference on Information Processing in Sensor Networks (IPSN)*, demo session, pp. 436–437 (2010)
6. M. Allen, L. Girod, R. Newton, S. Madden, D.T. Blumstein, D. Estrin, VoxNet: an interactive, rapidly-deployable acoustic monitoring platform. in *Proceeding of the 7th International Conference on Information Processing in Sensor Networks (IPSN)*, pp. 371–382 (2008)
7. G. Anastasi, O. Farruggia, G. Lo Re, M. Ortolani, Monitoring high-quality wine production using wireless sensor networks, in *Proceeding of the 42nd International Conference on System Sciences (HICSS)*, pp. 1–7 (2009)
8. T. Antoine-Santoni, J.F. Santucci, E. De Gentili, X. Silvani, F. Morandini, Performance of a protected wireless sensor network in a fire: analysis of fire spread and data transmission. Sensors **9**(8), 5878–5893 (2009)
9. A. Arora, P. Dutta, S. Bapat, V. Kulathumani, H. Zhang, V. Naik, V. Mittal, H. Cao, M. Demirbas, M. Gouda, Y. Choi, T. Herman, S. Kulkarni, U. Arumugam, M. Nesterenko, A. Vora, M. Miyashita, A line in the sand: a wireless sensor network for target detection, classification, and tracking. Comput. Netw. **46**(5), 605–634 (2004)
10. R. Bagree, V.R. Jain, A. Kumar, P. Ranjan, Tigercense: wireless image sensor network to monitor tiger movement, in *Proceeding of the 4th International Conference on Real-World Wireless Sensor Networks (REALWSN)*, pp. 13–24 (2010)
11. L. Bai, R. Dick, P. Dinda, Archetype-based design: sensor network programming for application experts, not just programming experts, in *Proceeding of the 2009 International Conference on Information Processing in Sensor Networks (IPSN)*, pp. 85–96 (2009)
12. H. Baldus, K. Klabunde, G. Müsch, Reliable set-up of medical body-sensor networks, in *Proceeding of the 1st European Workshop on Wireless Sensor Networks (EWSN)*, pp 353–363 (2004)
13. J. Balendonck, J. Hemming, B. van Tuijl, L. Incrocci, A. Pardossi, P. Marzialetti, Sensors and wireless sensor networks for irrigation management under deficit conditions (FLOW-AID), in *Proceeding of the International Conference on Agricultural Engineering and Agricultural & Biosystems Engineering for a Sustainable World (AgEng)*, pp. 583–588 (2008)
14. A. Bamis, D. Lymberopoulos, T. Teixeira, A. Savvides, The behaviorscope framework for enabling ambient assisted living. Pers. Ubiquitous. Comput. **14**(6), 473–487 (2010)
15. G. Barrenetxea, F. Ingelrest, G. Schaefer, M. Vetterli, O. Couach, M. Parlange, Sensorscope: out-of-the-box environmental monitoring, in *Proceeding of the 7th International Conference on Information Processing in Sensor Networks (IPSN)*, pp 332–343 (2008)
16. M.A. Batalin, G.S. Sukhatme, M. Hattig, Mobile robot navigation using a sensor network, in *Proceeding of the IEEE International Conference on Robotics and Automation (ICRA)*, pp. 636–641 (2004)
17. R. Beckwith, D. Teibel, P. Bowen, Unwired wine: sensor networks in vineyards, in *Proceeding of IEEE Sensors*, pp. 561–564 (2004)
18. J. Beutel, S. Gruber, A. Hasler, R. Lim, A. Meier, C. Plessl, I. Talzi, L. Thiele, C. Tschudin, M. Woehrle, M. Yuecel, PermaDAQ: a scientific instrument for precision sensing and data recovery in environmental extremes, in *Proceeding of the 8th International Conference on Information Processing in Sensor Networks (IPSN)*, pp. 265–276 (2009)
19. E.S. Biagioni, K.W. Bridges, The application of remote sensor technology to assist the recovery of rare and endangered species. Int. J. High Perform. Comput. Appl. **16**(3), 315–324 (2002)

20. J. Burrell, T. Brooke, R. Beckwith, Vineyard computing: sensor networks in agricultural production. IEEE Pervasive Comput. **3**(1), 38–45 (2004)
21. N. Burri, P. von Rickenbach, R. Wattenhofer, Dozer: ultra-low power data gathering in sensor networks, in *Proceedings of the 6th International Conference on Information Processing in Sensor Networks (IPSN)*, pp. 450–459 (2007)
22. Z. Butler, P. Corke, R. Peterson, D. Rus, Networked cows: virtual fences for controlling cows, in *Proceeding of the MobiSys Workshop on Applications of Mobile Embedded Systems (WAMES)*, 2004
23. R. Cardell-Oliver, K. Smettem, M. Kranz, K. Mayer, Field testing a wireless sensor network for reactive environmental monitoring, in *Proceeding of the International Conference on Intelligent Sensors, Sensor Networks and Information Processing (ISSNIP)*, pp. 14–17 (2004)
24. R. Cardell-Oliver, M. Kranz, K. Smettem, K. Mayer, A reactive soil moisture sensor network: design and field evaluation. Int. J. Distrib. Sens. Netw. **1**(2), 149–162 (2005)
25. M. Ceriotti, L. Mottola, G.P. Picco, A.L. Murphy, C. Gună, M. Corra, M. Pozzi, D. Zonta, P. Zanon, Monitoring heritage buildings with wireless sensor networks: The Torre Aquila deployment, in *Proceeding of the 8th International Conference on Information Processing in Sensor Networks (IPSN)*, pp. 277–288 (2009)
26. M. Ceriotti, M. Corra, L. D'Orazio, R. Doriguzzi, D. Facchin, Ş. Gună, G.P. Jesi, A. Murphy, R.L. Cigno, L. Mottola, M. Pescalli, G.P. Picco, D. Prognolato, C. Torghele, Is there light at the ends of the tunnel? Wireless sensor networks for adaptive lighting in road tunnels, in *Proceeding of the 10th International Conference on Information Processing in Sensor Networks (IPSN)*, pp. 187–198 (2011)
27. S. Cheekiralla, Wireless sensor network-based tunnel monitoring, in *Proceeding of the 1st Workshop on Real-World Wireless Sensor Networks (REALWSN)*, poster session, 2005
28. K. Chintalapudi, T. Fu, J. Paek, N. Kothari, S. Rangwala, J. Caffrey, R. Govindan, E. Johnson, S. Masri, Monitoring civil structures with a wireless sensor network. IEEE Internet Comput. **10**(2), 26–34 (2006)
29. O. Chipara, C. Lu, T.C. Bailey, G.C. Roman, Reliable clinical monitoring using wireless sensor networks: experiences in a step-down hospital unit, in *Proceeding of the 8th International Conference on Embedded Networked Sensor Systems (SenSys)*, pp. 155–168 (2010)
30. P. Costa, L. Mottola, A.L. Murphy, G.P. Picco, Teeny Lime: Transiently shared tuple space middleware for wireless sensor networks, in *Proceeding of the 1st International Workshop on Middleware for Sensor Networks (MidSens)*, pp. 43–48 (2006)
31. P. Costa, G. Coulson, R. Gold, M. Lad , C. Mascolo, L. Mottola, G.P. Picco, T. Sivaharan, N. Weerasinghe, S. Zachariadis, The RUNES middleware for networked embedded systems and its application in a disaster management scenario, in *Proceeding of the 5th International Conference on Pervasive Computing and Communications (PERCOM)*, pp. 69–78 (2007)
32. D.E. Culler, Toward the sensor network macroscope, in *Proceeding of the 6th International Symposium on Mobile Ad Hoc Networking and Computing (MobiHoc)*, pp. 1–1 (2005)
33. K. Dantu, B. Kate, J. Waterman, P. Bailis, M. Welsh, Programming micro-aerial vehicle swarms with karma, in *Proceeding of the 9th ACM Conference on Embedded Networked Sensor Systems (SenSys), ACM*, pp. 121–134 (2011)
34. A. Deshpande, C. Guestrin, S.R. Madden, Resource-aware wireless sensor-actuator networks. IEEE Data Eng. **28**(1), 40–47 (2005)
35. D.M. Doolin, N. Sitar, Wireless sensors for wildfire monitoring, in *Proceeding of SPIE Symposium on Smart. Structures and Materials*, vol. 5765, pp. 477–484 (2005)
36. A. Dunkels, J. Eriksson, L. Mottola, T. Voigt, F.J. Oppermann, K. Römer, F. Casati, F. Daniel, G.P. Picco, S. Soi, S. Tranquillini, P. Valleri, S. Karnouskos, P. Spieß, P.M. Montero, D-1.1—application and programming survey. Technical report, makeSense (2010)
37. P. Dutta, M. Grimmer, A. Arora, S. Bibyk, D. Culler, Design of a wireless sensor network platform for detecting rare, random, and ephemeral events, in *Proceeding of the 4th International Symposium on Information Processing in Sensor Networks (IPSN)*, pp. 497–502 (2005)
38. V. Dyo, S.A. Ellwood, D.W. Macdonald, A. Markham, C. Mascolo, B. Pásztor, N. Trigoni, R. Wohlers, Wildlife and environmental monitoring using RFID and WSN technology, in

Proceeding of the 7th International Conference on Embedded Networked Sensor Systems (SenSys), poster session, pp. 371–372 (2009)
39. V. Dyo, S.A. Ellwood, D.W. Macdonald, A. Markham, C. Mascolo, B. Pásztor, S. Scellato, N. Trigoni, R. Wohlers, K. Yousef, Evolution and sustainability of a wildlife monitoring sensor network, in *Proceeding of the 8th International Conference on Embedded Networked Sensor Systems (SenSys)*, pp. 127–140 (2010)
40. T. Flach, N. Mishra, L. Pedrosa, C. Riesz, R. Govindan, Carma: towards personalized automotive tuning, in *Proceeding of the 9th ACM Conference on Embedded Networked Sensor Systems (SenSys), ACM*, pp. 135–148 (2011)
41. R.K. Ganti, N. Pham, H. Ahmadi, S. Nangia, T.F. Abdelzaher, GreenGPS: a participatory sensing fuel-efficient maps application, in *Proceeding of the 8th International Conference on Mobile Systems, Applications, and Services (MobiSys)*, pp. 151–164 (2010)
42. T. Gao, T. Massey, L. Selavo, D. Crawford, B. Chen, K. Lorincz, V. Shnayder, M. Welsh, The advanced health and disaster aid network: a light-weight wireless medical system for triage. IEEE Trans. Biomed. Circuits Syst. **1**, 203–216 (2007)
43. C. Hartung, R. Han, C. Seielstad, S. Holbrook, FireWxNet: a multi-tiered portable wireless system for monitoring weather conditions in wildland fire environments, in *Proceeding of the 4th International Conference on Mobile Systems, Applications and Services (MobiSys)*, pp. 28–41 (2006)
44. T. He, S. Krishnamurthy, J.A. Stankovic, T. Abdelzaher, L. Luo, R. Stoleru, T. Yan, L. Gu, J. Hui, B. Krogh, Energy-efficient surveillance system using wireless sensor networks, in *Proceeding of the 2nd International Conference on Mobile Systems, Applications, and Services (MobiSys)*, pp. 270–283 (2004)
45. J. Hill, R. Szewczyk, A. Woo, S. Hollar, D.E. Culler, K. Pister, System architecture directions for networked sensors. ACM SIGPLAN Not. **35**(11), 93–104 (2000)
46. W. Hu, V.N. Tran, N. Bulusu, C. tung Chou, S. Jha, A. Taylor, The design and evaluation of a hybrid sensor network for cane-toad monitoring, in *Proceeding of the 4th International Symposium on Information Processing in Sensor Networks (IPSN)*, pp. 503–508 (2005)
47. W. Hu, N. Bulusu, C.T. Chou, S. Jha, A. Taylor, V.N. Tran, Design and evaluation of a hybrid sensor network for cane toad monitoring. ACM Trans. Sens. Netw. (TOSN) **5**(1), 4:1–4:28 (2009)
48. J.W. Hui, D.E. Culler, IP is dead, long live IP for wireless sensor networks, in *Proceeding of the 6th ACM Conference on Embedded Networked Sensor Systems (SenSys)*, pp. 15–28 (2008)
49. X. Jiang, S. Dawson-Haggerty, P. Dutta, D. Culler, Design and implementation of a high-fidelity AC metering network, in *Proceeding of the 8th International Conference on Information Processing in Sensor Networks (IPSN)*, pp. 253–264 (2009)
50. X. Jiang, M. van Ly, J. Taneja, P. Dutta, D. Culler, Experiences with a high-fidelity wireless building energy auditing network, in *Proceeding of the 7th International Conference on Embedded Networked Sensor Systems (SenSys)*, pp. 113–126 (2009)
51. P. Juang, H. Oki, Y. Wang, M. Martonosi, L.S. Peh, D. Rubenstein, Energy-efficient computing for wildlife tracking: design tradeoffs and early experiences with ZebraNet, in *Proceeding of the 10th International Conference on Architectural Support for Programming Languages and Operating Systems (ASPLOS-X)*, pp. 96–107 (2002)
52. J.M. Kahn, R.H. Katz, K.S.J. Pister, Next century challenges: mobile networking for "smart dust", in *Proceeding of the 5th Annual ACM/IEEE International Conference on Mobile Computing and Networking (MOBICOM), ACM, New York, NY, USA*, pp. 271–278 (1999)
53. C. Kappler, G. Riegel, A real-world, simple wireless sensor network for monitoring electrical energy consumption, in *Proceeding of the 1st European Workshop on Wireless Sensor Networks (EWSN)*, pp. 339–352 (2004)
54. M. Keally, G. Zhou, G. Xing, J. Wu, A.J. Pyles, PBN: towards practical activity recognition using smartphone-based body sensor networks, in *Proceeding of the 9th International Conference on Embedded Networked Sensor Systems (SenSys)*, pp. 246–259 (2011)

55. D.H. Kim, Y. Kim, D. Estrin, M.B. Srivastava, Sensloc: sensing everyday places and paths using less energy, in *Proceeding of the 8th International Conference on Embedded Networked Sensor Systems (SenSys)*, pp. 43–56 (2010)
56. S. Kim, S. Pakzad, D. Culler, J. Demmel, G. Fenves, S. Glaser, M. Turon, Health monitoring of civil infrastructures using wireless sensor networks, in *Proceeding of the 6th International Conference on Information Processing in Sensor Networks (IPSN)*, pp. 254–263 (2007)
57. Y.J. Kim, R.G. Evans, W.M. Iversen, Remote sensing and control of an irrigation system using a distributed wireless sensor network. IEEE Trans. Instrum. Meas. **57**(7), 1379–1387 (2008)
58. J. Ko, R. Musăloiu-Elefteri, J.H. Lim, Y. Chen, A. Terzis, T. Gao, W. Destler, L. Selavo, Medisn: medical emergency detection in sensor networks, in *Proceeding of the 6th International Conference on Embedded Networked Sensor Systems (SenSys)*, demo session, pp. 361–362 (2008)
59. J. Ko, J.H. Lim, Y. Chen, R. Musvaloiu-E, A. Terzis, G.M. Masson, T. Gao, W. Destler, L. Selavo, R.P. Dutton, MEDiSN: medical emergency detection in sensor networks. ACM Trans. Embedded Comput. Syst. (TECS) **10**(1), 11:1–11:29 (2010)
60. J. Ko, K. Klues, C. Richter, W. Hofer, B. Kusy, M. Brünig, T. Schmid, Q. Wang, P. Dutta, A. Terzis, Low power or high performance? a tradeoff whose time has come (and nearly gone), in *Proceeding of the 9th European Conference on Wireless Sensor Networks (EWSN)*, pp. 98–114 (2012)
61. L. Krishnamurthy, R. Adler, P. Buonadonna, J. Chhabra, M. Flanigan, N. Kushalnagar, L. Nachman, M. Yarvis, Design and deployment of industrial sensor networks: experiences from a semiconductor plant and the north sea, in *Proceeding of the 3rd International Conference on Embedded Networked Sensor Systems (SenSys)*, pp. 64–75 (2005)
62. J. Kumagai, The secret life of birds. IEEE Spectr. **41**(4), 42–48 (2004)
63. K.G. Langendoen, A. Baggio, O.W. Visser, Murphy loves potatoes: experiences from a pilot sensor network deployment in precision agriculture, in *Proceeding of the 14th International Workshop on Parallel and Distributed Real-Time Systems (WPDRTS)*, 2006
64. Y. Lee, G. Kim, S. Bang, Y. Kim, I. Lee, P. Dutta, D. Sylvester, D. Blaauw, A modular 1mm3 die-stacked sensing platform with optical communication and multi-modal energy harvesting, in *Proceeding of the International Solid-State Circuits Conference (ISSCC)*, pp. 402–404 (2012)
65. D. Li, W. Liu, Z. Zhao, L. Cui, Demonstration of a WSN application in relic protection and an optimized system deployment tool, in *Proceedings of the 7th International Conference on Information Processing in Sensor Networks (IPSN)*, demo session, pp. 541–542 (2008)
66. D. Li, W. Liu, L. Cui, EasiDesign: an improved ant colony algorithm for sensor deployment in real sensor network system, in *Proceedings of the IEEE Global Telecommunications Conference (GLOBECOM)*, pp. 1–5 (2010)
67. M. Li, Y. Liu, Underground coal mine monitoring with wireless sensor networks. ACM Trans. Sens. Netw. (TOSN) **5**(2), 10:1–10:29 (2009)
68. C. Lombriser, N.B. Bharatula, D. Roggen, G. Tröster, On-body activity recognition in a dynamic sensor network, in *Proceedings of the 2nd International Conference on Body, Area Networks (BodyNets)* (2007)
69. K. Lorincz, D.J. Malan, T.R. Fulford-Jones, A. Nawoj, A. Clavel, V. Shnayder, G. Mainland, M. Welsh, Sensor networks for emergency response: challenges and opportunities. IEEE Pervasive Comput. **3**(4), 16–23 (2004)
70. K. Lorincz, B. rong Chen, G.W. Challen, A.R. Chowdhury, S. Patel, P. Bonato, M. Welsh, Mercury: a wearable sensor network platform for high-fidelity motion analysis, in *Proceedings of the 7th International Conference on Embedded Networked Sensor Systems (SenSys)*, pp. 183–196 (2009)
71. D. Lymberopoulos, A. Bamis, T. Teixeira, A. Savvides, BehaviorScope: real-time remote human monitoring using sensor networks, in *Proceedings of the 7th International Conference on Information Processing in Sensor Networks (IPSN)*, demo session, pp. 533–534 (2008)
72. A. Mainwaring, J. Polastre, R. Szewczyk, D. Culler, J. Anderson, Wireless sensor networks for habitat monitoring, in *Proceedings of the 1st International Workshop on Wireless Sensor Networks and Applications (WSNA)*, pp. 88–97 (2002)

73. A. Markham, N. Trigoni, S.A. Ellwood, D.W. Macdonald, Revealing the hidden lives of underground animals using magneto-inductive tracking, in *Proceedings of the 8th International Conference on Embedded Networked Sensor Systems (SenSys)*, pp. 281–294 (2010)
74. K. Martinez, R. Ong, J.K. Hart, GLACSWEB: a sensor network for hostile environments, in *Proceedings of the 1st IEEE Communications Society Conference on Sensor and Ad Hoc Communications and Networks (SECON)*, pp. 81–87 (2004)
75. L. Mo, Y. He, Y. Liu, J. Zhao, S.J. Tang, X.Y. Li, G. Dai, Canopy closure estimates with GreenOrbs: sustainable sensing in the forest, in *Proceedings of the 7th International Conference on Embedded Networked Sensor Systems (SenSys)*, pp. 99–112 (2009)
76. P. Mohan, V.N. Padmanabhan, R. Ramjee, Nericell: rich monitoring of road and traffic conditions using mobile smartphones, in *Proceedings of the 6th International Conference on Embedded Networked Sensor Systems (SenSys)*, pp. 323–336 (2008)
77. L. Mottola, G.P. Picco, Programming wireless sensor networks: fundamental concepts and state-of-the-art. ACM Comput. Surv. (CSUR) **43**(3), 19:1–19:51 (2011)
78. L. Mottola, G.P. Picco, M. Ceriotti, S. Gună, A.L Murphy, Not all wireless sensor networks are created equal: a comparative study on tunnels. ACM Trans. Sens. Netw. (TOSN) **7**(2), 15:1–15:33 (2010)
79. K. Na, Y. Kim, H. Cha, Acoustic sensor network-based parking lot surveillance system, in *Proceedings of the 6th European Conference on Wireless Sensor Networks (EWSN)*, pp. 247–262 (2009)
80. A. Nasipuri, R. Cox, H. Alasti, L.V. der Zel, B. Rodriguez, R. McKosky, J.A. Grazian, Wireless sensor network for substation monitoring: Design and deployment, in *Proceedings of the 6th International Conference on Embedded Networked Sensor Systems (SenSys)*, demo session, pp. 365–366 (2008)
81. J. Panchard, S. Rao, T. Prabhakar, H. Jamadagni, J.P. Hubaux, COMMON-sense net: improved water management for resource-poor farmers via sensor networks, in *Proceedings of the 1st International Conference on Communication and Information Technologies and Development (ICTD)*, pp. 22–33 (2006)
82. S. Patel, K. Lorincz, R. Hughes, N. Huggins, J. Growden, D. Standaert, M. Akay, J. Dy, M. Welsh, P. Bonato, Monitoring motor fluctuations in patients with parkinson's disease using wearable sensors. IEEE Trans. Inf. Technol. Biomed. **13**(6), 864–873 (2009)
83. K.S. Pister, Tracking vehicles with a UAV-delivered sensor network. (2001) Tech. rep., UC Berkeley and MLB, http://robotics.eecs.berkeley.edu/pister/29Palms0103/
84. J. Polastre, R. Szewczyk, D. Culler, Telos: enabling ultra-low power wireless research, in *Proceedings of the 4th International Symposium on Information Processing in Sensor Networks (IPSN)*, pp. 364–369 (2005)
85. W.B. Pöttner, L. Wolf, J. Cecílio, P. Furtado, R. Silva, J.S. Silva, A. Santos , P. Gil, A. Cardoso, Z. Zinonos, J.M. do Ó, B. McCarthy, J. Brown, U. Roedig, T. O'Donovan, C.J. Sreenan, Z. He, T. Voigt, A. Jugel, WSN evaluation in industrial environments first results and lessons learned, in *Proceedings of the 3rd International Workshop on Performance Control in Wireless Sensor Networks (PWSN)*, pp. 1–8 (2011)
86. N. Ramanathan, T. Schoellhammer, E. Kohler, K. Whitehouse, T. Harmon, D. Estrin, Suelo: human-assisted sensing for exploratory soil monitoring studies, in *Proceedings of the 7th International Conference on Embedded Networked Sensor Systems (SenSys)*, pp. 197–210 (2009)
87. P. Ranjan, P.K. Saraswat, A. Kumar, S. Polana, A. Singh, wildCENSE - sensor network for wildlife monitoring. Technical report, Dhirubhai Ambani Institute of Information and Communication Technology, Gandhinagar, Gujarat (2006)
88. R. Riem-Vis, Cold chain management using an ultra low power wireless sensor network, in *Proceedings of the MobiSys Workshop on Applications of Mobile Embedded Systems (WAMES)*, pp. 21–23 (2004)
89. V. Rocha, G. Goncalves, Sensing the world: challenges on WSNs, in *Proceedings of the IEEE International Conference on Automation, Quality and Testing, Robotics (AQTR)*, vol 1, pp. 54–59 (2008)

90. K. Römer, F. Mattern, The design space of wireless sensor networks. IEEE Wirel. Commun. **11**(6), 54–61 (2004)
91. R.M. Ruair, M.T. Keane, G. Coleman, A wireless sensor network application requirements taxonomy, in *Proceeedings of the 2nd International Conference on Sensor Technologies and Applications (SENSORCOMM)*, pp. 209–216 (2008)
92. San Francisco Municipal Transportation Agency (2011) SFpark: Putting theory into practice. Tech. rep., http://sfpark.org/wp-content/uploads/2011/09/sfpark_aug2011projsummary_print-2.pdf
93. L. Selavo, A. Wood, Q. Cao, T. Sookoor, H. Liu, A. Srinivasan, Y. Wu, W. Kang, J. Stankovic, D. Young, J. Porter, LUSTER: wireless sensor network for environmental research, in *Proceedings of the 5th International Conference on Embedded Networked Sensor Systems (SenSys)*, pp. 103–116 (2007)
94. A. Sheth, K. Tejaswi, P. Mehta, C. Parekh, R. Bansal, S. Merchant, T. Singh, U.B. Desai, C.A. Thekkath, K. Toyama, SenSlide: a sensor network based landslide prediction system, in *Proceedings of the 3rd International Conference on Embedded Networked Sensor Systems (SenSys)*, poster session, pp. 280–281 (2005)
95. A. Sheth, C.A. Thekkath, P. Mehta, K. Tejaswi, C. Parekh, T.N. Singh, U.B. Desai, Senslide: a distributed landslide prediction system. ACM SIGOPS Operating Syst. Rev. **41**(2), 75–87 (2007)
96. E.I. Shih, A.H. Shoeb, J.V Guttag, Sensor selection for energy-efficient ambulatory medical monitoring, in *Proceedings of the 7th International Conference on Mobile Systems, Applications, and Services (MobiSys)*, pp. 347–358 (2009)
97. V. Shnayder, B.R. Chen, K. Lorincz, T.R. Fulford-Jones, M. Welsh, Sensor networks for medical care. Technical report TR-08-05 (Harvard University, Cambridge, 2005)
98. S. Shukla, N. Bulusu, S. Jha, Cane-toad monitoring in kakadu national park using wireless sensor networks, in *Proceedings of the 18th Asia Pacific Advanced, Network Conference (APAN)* (2004)
99. G. Simon, M. Maróti, Á Lédeczi, G. Balogh, B. Kusy, A. Nádas, G. Pap, J. Sallai, K. Frampton, Sensor network-based countersniper system, in *Proceedings of the 2nd International Conference on Embedded networked Sensor Systems (SenSys)*, pp. 1–12 (2004)
100. SmartSantander Project (2012) SmartSantander project. http://www.smartsantander.eu
101. I. Stoianov, L. Nachman, S. Madden, T. Tokmouline, PIPENET: a wireless sensor network for pipeline monitoring, in *Proceedings of the 6th International Conference on Information Processing in Sensor Networks (IPSN)*, pp. 264–273 (2007)
102. R. Szewczyk, A. Mainwaring, J. Polastre, J. Anderson, D. Culler, An analysis of a large scale habitat monitoring application, in *Proceedings of the 2nd International Conference on Embedded networked Sensor Systems (SenSys)*, pp. 214–226 (2004)
103. B. Thorstensen, T. Syversen, T.A. Bjørnvold, T. Walseth, Electronic shepherd: A low-cost, low-bandwidth, wireless network system, in *Proceedings of the 2nd International Conference on Mobile Systems, Applications, and Services (MobiSys)*, pp. 245–255 (2004)
104. S. Tilak, N.B. Abu-Ghazaleh, W. Heinzelman, A taxonomy of wireless micro-sensor network models. SIGMOBILE Mobile Comput. Commun. Rev. **6**(2), 28–36 (2002)
105. G. Tolle, J. Polastre, R. Szewczyk, D. Culler, N. Turner, K. Tu, S. Burgess, T. Dawson, P. Buonadonna, D. Gay, W. Hong, A macroscope in the redwoods, in *Proceedings of the 3rd International Conference on Embedded Networked Sensor Systems (SenSys)*, pp. 51–63 (2005)
106. K. Van Laerhoven, B.P. Lo, J.W. Ng, S. Thiemjarus, R. King, S. Kwan, H.W. Gellersen , M. Sloman, O. Wells, P. Needham, N. Peters, A. Darzi, C. Toumazou, G.Z. Yang, Medical healthcare monitoring with wearable and implantable sensors, in *Proceedings of the 3rd International Workshop on Ubiquitous Computing for Pervasive Healthcare Applications (UbiHealth)*, pp. 115–123 (2004)
107. T. Wark, C. Crossman, W. Hu, Y. Guo, P. Valencia, P. Sikka, P. Corke, C. Lee, J. Henshall, K. Prayaga, J. O'Grady, M. Reed, A. Fisher, The design and evaluation of a mobile sensor/actuator network for autonomous animal control, in *Proceedings of the 6th International Conference on Information Processing in Sensor Networks (IPSN)*, pp. 206–215 (2007)

108. G. Werner-Allen, K. Lorincz, J. Johnson, J. Lees, M. Welsh, Fidelity and yield in a volcano monitoring sensor network, in *Proceedings of the 7th USENIX Symposium on Operating Systems Design and Implementation (OSDI)*, pp. 381–396 (2006)
109. D. Whang, N. Xu, S. Rangwala, K. Chintalapudi, R. Govindan, J. Wallace, Development of an embedded networked sensing system for structural health monitoring, in *Proceeedings of International Workshop on Smart Materials and Structures Technology* (2004)
110. M.J. Whelan, K.D. Janoyan, Design of a robust, high-rate wireless sensor network for static and dynamic structural monitoring. J. Intell. Mater. Syst. Struct. **20**(7), 849–864 (2009)
111. G. Wittenburg, K. Terfloth, F.L. Villafuerte, T. Naumowicz, H. Ritter, J. Schiller, Fence monitoring: experimental evaluation of a use case for wireless sensor networks, in *Proceedings of the 4th European Conference on Wireless Sensor Networks (EWSN)*, pp. 163–178 (2007)
112. A. Wood, J. Stankovic, G. Virone, L. Selavo, Z. He, Q. Cao, T. Doan, Y. Wu, L. Fang, R. Stoleru, Context-aware wireless sensor networks for assisted-living and residential monitoring. IEEE Network **22**(4), 26–33 (2008)
113. C. Xu, B. Firner, Y. Zhang, R. Howard, J. Li, X. Lin, Improving rf-based device-free passive localization in cluttered indoor environments through probabilistic classification methods, in *Proceedings of the 11th International Conference on Information Processing in Sensor Networks (IPSN)*, pp. 209–220 (2012)
114. G.Z. Yang (ed.), Body Sensor Networks, 1st edn. (Springer-Verlag, London, 2006)
115. P. Zhang, C. Sadler, S. Lyon, M. Martonosi, Hardware design experiences in ZebraNet, in *Proceedings of the 2nd International Conference on Embedded Networked Sensor Systems (SenSys)*, pp. 227–238 (2004)

Chapter 3
Design of Low Data-Rate Environmental Monitoring Applications

Agnelo Rocha da Silva, M. Moghaddam and M. Liu

Abstract The majority of low-cost and off-the-shelf Wireless Sensor Networks (WSNs) solutions cannot adequately address issues related to an unattended deployment in a harsh environment, especially if the network needs to scale and achieve high density or high coverage or both. This is usually the case in environmental applications. In this chapter, this problem is investigated and extensive discussion on the pros and cons of a specific WSN design is presented. However, before moving from generic and well-established WSN solutions to customization, a detailed analysis of the gains of having a tailored design is necessary. Accordingly, a case study involving sparse deployments in outdoors is used to illustrate the process.

The majority of existing WSN deployments are for low data-rate monitoring applications, and some of the first WSNs were designed for environmental monitoring. However, many challenges remain to be fully addressed before low-cost and large-scale outdoor WSNs can become a more common reality [13, 15, 23, 24]. In this chapter, the rationale supporting this statement is presented and some guidelines for the design of specialized WSNs for such a scenario are proposed. For this purpose, a real-world environmental monitoring application is used as a case study along with a specific WSN design. The taxonomy given in Chap. 2 is used to characterize the most relevant aspects of the design. More specifically, for each design dimension, the pros and cons of the proposed solution are discussed. We also highlight the pros and cons of designing a specialized WSN as opposed to adopting generic off-the-shelf WSN technologies.

A. R. da Silva · M. Moghaddam
Electrical Engineering–Electrophysics, University of Southern California,
Los Angeles, CA, USA
e-mail: agnelosi@usc.edu

M. Moghaddam
e-mail: mahta@usc.edu

M. Liu (✉)
EECS Department, University of Michigan, Ann Arbor, MI, USA
e-mail: mingyan@eecs.umich.edu

H. M. Ammari (ed.), *The Art of Wireless Sensor Networks*,
Signals and Communication Technology, DOI: 10.1007/978-3-642-40009-4_3,
© Springer-Verlag Berlin Heidelberg 2014

In the first section of the chapter, the challenges of environmental monitoring are discussed. We then present the requirements of an illustrative project called NatureMONITOR. A specialized WSN solution is proposed along with a discussion of its main advantages and limitations. In the third part of the chapter, the importance of properly applying taxonomy to a WSN project is highlighted. We also make the observation that this task is made easier when a solution is highly specialized. Each design option of the NatureMONITOR project is explained in more detail and alternative design options are also discussed. In the last section, lessons learned are highlighted with a conclusion that advocates WSN designs tailored to application-specific characteristics [33].

1 Challenges in Environmental Monitoring

One of the main challenges in typical environmental monitoring applications is actually the cost involved in deploying and maintaining a significant number of sensors in the area of interest. For instance, although commercial radio, cellular, and satellite telemetry solutions have been available for a long time [13], the cost associated with deploying and supporting such solutions is relatively high, even when only a dozen nodes are deployed. When larger areas (e.g., >1 km^2) are involved, the cost becomes a major factor in the feasibility of the project. About 10 years ago when WSNs were presented as a feasible low-cost solution to large-scale deployments, there was great expectation that these networks would be the key answer to indoor/outdoor environmental monitoring [1].

However, a careful investigation of recent WSN work in this area, such as the references provided in Chap. 2 shows that we have not achieved an *effective* solution. From a business (or project sponsor's) viewpoint, the effectiveness is measured by a good balance between the expected functionality of such a network and acceptable levels of cost, reliability, and scalability. We next discuss these 3 fundamental metrics, as illustrated in Fig. 1.

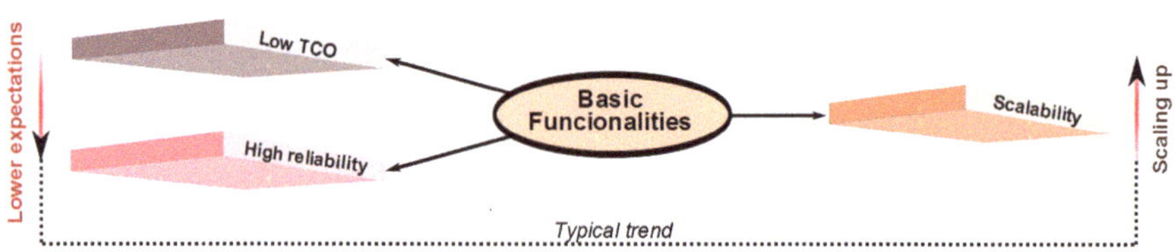

Fig. 1 The main challenge of designing a large outdoor WSN is to achieve a balanced solution involving TCO (effective cost), reliability, and scalability

1.1 Controlling the Total Cost of Ownership

While the terms *large-scale*, *long-term*, and *low-cost* are frequently used in the WSN literature, a certain proposed solution often falls short of meeting the requirement of an application. As an example, assume a sensor node carries a unitary cost of $50 and it employs a "perpetual" energy harvesting system using a solar panel. This does not automatically mean that the solution is truly low-cost. The true cost must be evaluated systemically, while sensors constitute only one part of the system. In particular, it is important to remember that the regular maintenance costs of a WSN are *not exclusively* related to the power modules. For instance, if periodic sensor calibrations are also necessary, then such costs must be included in the total cost of ownership (TCO) of the sensor node. Moreover, in an outdoor setting regular maintenance may be unavoidable due to a variety of reasons. Besides theft and vandalism, the following are some of the potential issues found in a typical outdoor deployment [27]:

- Negative effect of low-temperature on rechargeable batteries.
- Obstruction of the solar panel caused by dirt left by birds.
- Obstruction of the solar panel caused by shadow of trees (seasonal).
- Efficiency loss of the solar panel caused by long-term exposition to the environment (e.g., >1 year).
- Damage to the components of the sensor node, including the sensor itself, caused by insects and animals.
- Water infiltration.
- Condensing water inside waterproof enclosures.
- Communication issues due to new sources of radio frequency (RF) interference at the site.

Therefore, we argue that the key to controlling the total cost lies in reducing maintenance needs and operational cost. Best practices in the industry include preventive maintenance, which has long been used as a way to minimize the probability of failures of the system. During a *scheduled* maintenance, many components of the system are replaced simply because they are close to their expected lifetime.

Similarly, in the context of outdoor WSNs, it is important to determine whether there are tasks or situations that require regular human intervention; such needs may dictate other design aspects of the system. For instance, in a case where sensors must be physically inspected, cleaned, or calibrated every 6 months, the pros and cons of designing the lifetime of a sensor node for 2 years (in terms of its energy solution) instead of a value closer to 6 months must be re-evaluated. Observe that even a node with a perpetual energy lifetime will require such intervention. One way to minimize outdoor maintenance cost is to coordinate multiple maintenance tasks, e.g., scheduling them on the same, pre-defined dates. Knowing such a schedule affects the design consideration for the rest of the system. For instance, such maintenance cycle will ultimately define the expected lifetime of the node in terms of energy.

1.2 Reliability and Performance Metrics

In order to properly define maintenance dates for an outdoor WSN without compromising its functionality and costs, a *high degree of reliability* of system components along with a set of sound performance metrics are required. Observe that such statement does not follow the traditional assumption that WSNs comprise inexpensive nodes with a relative high failure probability. Component redundancy is a typical solution to achieve high availability, although it is not always feasible to have replicas of all components, not to mention that a complete self-healing solution is in general complex and expensive. An alternate approach is to develop smart strategies for the most vulnerable components. Observe that both approaches require a qualitative understanding of the reliability level of system components. Usually for mature solutions, this is known prior to implementation. However, early adopters of a new technology must use or develop measurement tools to acquire such knowledge. This is typically the case with large outdoor WSNs, where researchers and developers must determine what to measure to assess the reliability level of system components. The following is a list that exemplifies some of the expected performance metrics for this scenario:

1. Remaining energy level (real-time) of each node.
2. Ratio between the number of packets correctly received and that of packets transmitted.
3. Ratio between the number of measurements received by the sink node and the number actually taken by the sensor nodes.
4. Number of power shortages (or reset) on each node.
5. Average data-latency over the entire WSN.
6. Measurement health (or detection reliability) of each sensor module in each node.

We have now argued that low operating cost is key to making WSNs affordable for environmental monitoring applications, and that to lower maintenance cost it is important to identify the weakest components. However, to realize such analysis, at least basic network management functionalities must be available as briefly listed above. However, for the design of a large WSN, it is essential a clear understanding of how scalability affects both performance and cost, as highlighted in Fig. 1. Accordingly, in the next section, additional metrics related to the terms *large* and *sparse* are established.

1.3 Network Metrics for Large Network Design

The term *large-scale* in the WSN literature usually refers to a large number of sensor nodes over a relatively confined area (thus high node-density) [1, 23, 24]. However, we note that for outdoor WSNs sometimes large-scale is more related to large spatial coverage than the number of sensors. As an example, consider a 4 km-long beach that

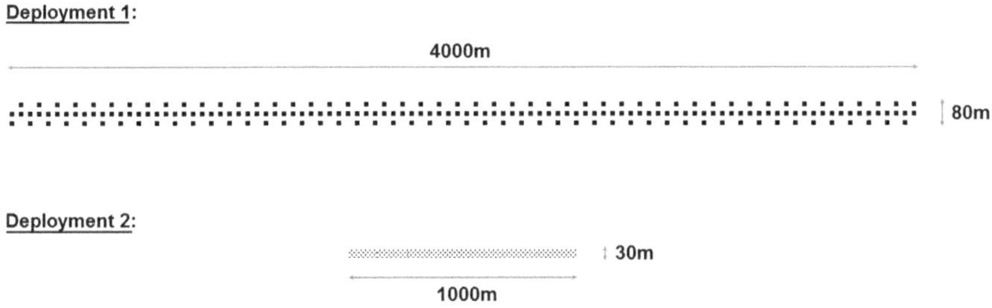

Fig. 2 *Two* possible deployment options for the beach example involving 200 sensor nodes

must be monitored for possible pollution, as shown in Fig. 2. If the budget constraint limits the number of nodes to 200, it may be more desirable to have these sensors covering all 4 km in a sparse deployment (Deployment 1) than to densely deploy these sensors at a few selected locations while ignoring the rest (Deployment 2). The technical issues underlying both approaches are distinctly different. While a solution may be scalable in terms of node-density, this does *not imply* that it is also scalable in terms of spatial coverage.

The above illustrates two types of scaling (at two extremes), one with increasing density within a fixed area (density scaling) and the other with increasing area with a fixed number of nodes (area scaling). In practice, a particular large-scale deployment may be some combination of the two with one being more dominant. Area scaling notably gives rise to the question of whether the network can remain connected as we stretch the area without adding more nodes. But the difference between the two also goes beyond the communication range. Depending on how the network scales, data-rate, duty-cycle, data-latency, communication errors, and energy consumption all vary in ways determined by the architecture, hardware, and protocols in use, and not all of this is well understood.

For the above example, if we assume a typical communication range of 50 m, the average number of one-hop neighbors for a node is 2 (Deployment 1) or 16.6 (Deployment 2). Therefore, the denser deployment (2) has the clear advantage of a plurality of data paths which is usually associated with a higher degree of data transfer reliability for the network. Similarly, the ratio between the communication range of the nodes and the average inter-node distance between a node and its 3 closest neighbors is 1.1 and 3.3 for deployments 1 and 2, respectively. Again, the probability of communication errors for the first case is significantly higher compared with the second one [43]. In short, for the same number of nodes, a sparser deployment is typically more critical.

1.3.1 Motivation for Metrics Toward Large Networks

The previous discussion highlights the importance of carefully considering the area scaling aspect when the scalability of a WSN solution is under analysis. Unfortunately, such information is typically ignored or not mentioned in WSN designs. For instance, many WSN papers discuss the scalability of networking protocols considering density scaling, but it is much less clear how the same architecture and protocols behave under area scaling. At the same time, cost concerns (both deployment and maintenance) suggest that area scaling is often the more relevant one, as shown in the preceding beach example. In this way, as the networks scales (area), it becomes more difficult to achieve the initial goals related to TCO and reliability, as illustrated in Fig. 1. In fact, rarely a large-scale deployment is reported as having a longer lifetime (e.g., >l year).

The majority of the reported large-scale WSN studies are for scenarios where the average inter-node distance among immediate neighbors is still relatively small compared to the communication range of the nodes, implying a high-density regime [32, 34]. Consequently, the emphasis has been on handling interference in dense deployments and on examining the performance of collaborative and multi-hopping protocols for this scenario. When typical WSN solutions suggested for wider physical coverage area are considered [1, 17], multi-hopping usually works due to a plurality of paths available. For instance, ZigBee specifications allow deep multi-hopping (e.g., 30 hops under ZigBee PRO [10, 14, 16]) and ZigBee-based networks are being successfully deployed especially indoors. This type of solution obviously requires the deployment of an increasing number of nodes to cover a larger area so as to maintain a similar degree of density and this is the case for deployments inside buildings and industrial plants.

We strongly believe that the success of using wireless sensors in environmental monitoring applications, especially outdoor WSN deployments, heavily lies in the proper support for *sparse* deployments. This is the case because the goal of any environmental monitoring application is to capture certain underlying phenomenon of interest. The quality of our observation is determined by both spatial resolution of data (node-density) as well as the spatial diversity of the data (coverage area), sometimes one more than the other. Ideally if we can have both then we have the best information quality. However, if both cannot be afforded simultaneously due to high cost, then one must prioritize. For some applications it makes sense to first focus on a few select, small areas and get high-resolution data limited to those areas. For many other applications it may be far more sensible to start with a wider coverage first and then gradually increase the node-density over time if necessary and when more resources become available. This observation is also relevant to many envisioned infrastructure monitoring applications, such as the detection of leakages in oil pipelines, structural failures of bridges, and the detection of landslides in roads. Precision agriculture is another example [39]. Note that in all these examples high spatial coverage is usually a requirement.

When the number of sensor nodes is limited, node placement becomes a critical issue in the design process. Consider a deployment with 100 sensor nodes over an area of 500 × 500 m. If we assume a square grid deployment, the average inter-node distance among immediate neighbors is around 56 m. Many low-power radio transceivers actually offer good communication performance for distances higher than this one. Therefore, we can realistically assume that one node in this network will have more than 6 neighbors within its communication range of 100 m. In this case, we expect multi-hop protocols to work properly, e.g., in [21] a routing protocol was shown via simulation to achieve satisfactory results for 100 nodes non-uniformly spread in a 100 × 100 m area. Moreover, because of the high density, this network is more robust to the actual placement of sensors, i.e., changes in nodes' locations (as long as not too drastic as to alter connectivity) do not severely impact the performance of the network.

Now consider a second scenario with a sparser deployment: 100 nodes over an area of 3000 × 3000 m. The average inter-node distance among immediate neighbors is higher than 330 m. Although some WSN transceivers at their maximum transmit power levels can operate for this distance, the potential number of neighbors of a node is drastically reduced. In fact, to the best of our knowledge, there is no published work reporting a satisfactory multi-hop WSN solution operating in this scenario.

The connectivity challenge becomes more complicated when non-uniform deployments, obstacles, and environmental issues are considered. In this case, well-known multi-hop and collaborative protocols may not work properly in even relatively small areas. It is also clear that, when the number of sensor nodes is limited, the quality of observation is highly affected by the placement of the nodes. Due to the resource constraint, a good placement may mean spatially varying node densities over the target area.

We will consider hereafter a deployment with 100 nodes and a coverage area of 1000 × 1000 m. While not a particularly very large network, this example will help illustrate the problem of finding WSN solutions for realistic sparse deployments. In Fig. 3a, b we show a node placement in a square grid and a more realistic shape for the 1000 × 1000 m target area, respectively. In Sect. 2, a WSN solution to placement in the area shown in Fig. 3b is presented. The proposed architecture can be applied in different scenarios that also require high coverage area, or high spatial scalability.

When comparing Fig. 3a with b, it is clear that the latter case has more associated challenges. However, we still have to find a way to quantitatively express the challenges for this scenario. Moreover, similar to the standard processing benchmarks used by the microprocessor and computer industries, there is a real need of a way to compare the performance of WSN protocols for distinct and strategic cases involving large networks. Such cases can be real-world examples and/or artificial scenarios of network placements involving different number of nodes and coverage areas. The first step toward this goal is the formalization of metrics that better translate the challenges involving communication range, number of neighbor nodes, and distance to these nodes. Accordingly, in the next section we propose one of such metrics, although it is not complete.

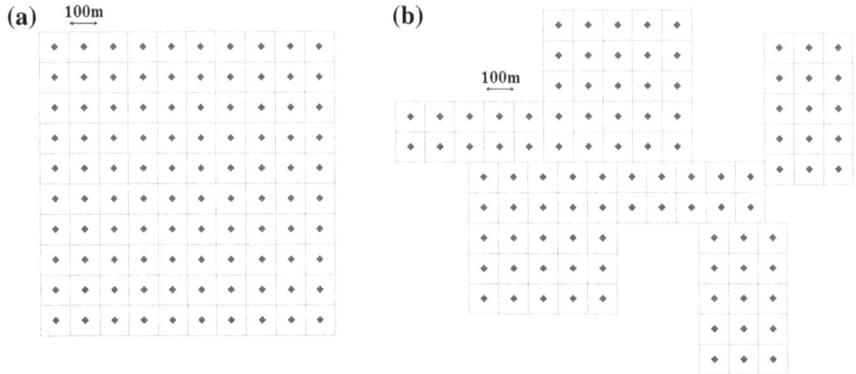

Fig. 3 **a** Grid-based node placement in a 1000×1000 m area. **b** An example of a realistic 1000×1000 m deployment area

1.3.2 AIND & ANON Metrics

In this section, it is introduced a novel network metrics, called AIND-ANON. AIND stands for Average Inter-Node Distance among immediate neighbors and ANON stands for Average Number of One-hop Neighbors. The metrics cannot be used in its current form as a network performance tool with the objective of comparing network solutions, although we acknowledge that such tool is potentially a first step toward this goal. The fact that the tool is not a complete one is revealed by the observation that the metrics does not capture the location and number of sinks (i.e., end-destinations) at the node topology. Without this information, it is not possible to calculate the average number of hops necessary to conclude an end-to-end communication transaction. Moreover, the sensing range of a node is assumed to be the same as the communication range, which cause distortions in the analysis of channel contention and interference [43]. Similarly, the tool does not capture aspects related to network congestion, data-latency, or network throughput.

Therefore, it is natural to question the effectiveness of the AIND-ANON metrics at the ongoing discussion in this chapter. The main compelling reason behind the development of this metrics is *simplicity*, that is, a simple way to formalize the concept of network sparsity. To this end, it is important to maintain the technical discussion in a higher level without involving too many details related to underlying protocols and deployment environment. Moreover, we are interested to understand how sparsity is related to network performance, at least in relation to the average success for delivering a packet. If we assume that such delivery success in a *multi-hop* network is strongly associated with the success of a node in communicating with its closest neighbors, we significantly reduce the complexity of the analysis. However, we will see that such assumption may not be realistic under certain circumstances.

The key input parameters of this high-level analysis are two: (a) the node topology in 2D (or 2D-equivalent from a 3D scenario) and (b) the expected communication range of the node. The rationale is simple: calculate the average distance of a node to its 3 closest one-hop neighbors and also calculate the total number of one-hop neighbors. By averaging these values for all nodes in the network, it is possible to have qualitative indicators related to the expected challenges for this specific topology when multi-hopping is used. This goal is achieved even without informing the location of the sink(s). If, on the average, the ratio between the communication range of a node and the distance to its closest neighbors are close to the unity, the probability of communication errors increase because the received signal strength is expected to be smaller. However, even for this worst scenario, if the number of one-hop neighbors of this node is relatively high, the likelihood of communication success increases. As expected, having multiple neighbors close to a node (dense network) is the ideal case for this analysis. At the contrary direction, having a scenario with few neighbors located at the limits of the communication range is the worst case (sparse network). Therefore, by using the AIND-ANON metrics, it is possible to get the sparsity degree of a network and infer about possible communication issues.

Observe that so far the use of the AIND-ANON metrics is always associated with the expressions *neighbors* and *multi-hopping*. It suggests that the analysis under this metrics will potentially degrade or fail when any form of segmentation or in-network aggregation is applied to the network thus decreasing the use of multi-hopping in that network. More specifically, assume that after analyzing a node topology using AIND-ANON we conclude that the network is very sparse and it has high probability of communication errors and low packet delivery rate. Although such deployment is potentially sparse from the physical point of view, it does not necessarily imply that the performance of the network is as critical as indicated by the metrics. The initial pessimistic conclusion about the network is potentially valid if the average number of hops used for the message delivery is high; otherwise, the conclusion can be invalid. The cases where the analysis with this metrics can be distorted are the ones where the average number of hops is close or equal to the unity and they are the following: (a) small networks and (b) any network that employs a form of segmentation or in-network aggregation.

Opportunistically, the above latter observation (b) is very important for our analysis. More specifically, if a multi-hopping network is analyzed under AIND-ANON as sparse (with high probability of network issues), any form of segmentation and aggregation can potentially mitigate the performance issues in that scenario. Although the AIND-ANON metrics cannot capture such improvement (there is no associated input parameter), it is not difficult to get the intuition behind the previous statement. While the metrics captures values on the average for the *entire* network, when aggregation and/or segmentation are used, the success of the message delivery is now constrained to a *smaller* physical region. For instance, in many cluster-based WSNs, the majority of nodes exchange their messages by means of a single hop to a cluster-head. Therefore, even if the average distance between a node and its immediate neighbors is high (compared to its communication range), the success of the solution is mainly associated with the average performance of the links involving nodes and their cluster-heads, not the links among regular nodes. Similarly, when one large

WSN employs a significant number of sinks, the same rationale applies because on the average a smaller number of hops is required.

We summarize this discussion as follows. The AIND-ANON metrics can be used as a simply tool to evaluate the sparsity level of a network with a specific node topology. Moreover, the metrics can be used as a qualitative tool to evaluate the expected challenges associated with the use of generic multi-hopping protocols assuming that no aggregation or segmentation effort is being used. If the metrics indicates that a network is very sparse, network performance issues are potentially expected. In this case, aggregation or segmentation efforts can be adopted to improve the performance of the network. Following this direction, it is possible to reach the extreme case where the communications at the entire network are based on one-hop links only. If these links have good performance, the overall network achieves good performance and easily scales. We will see that this rationale will be later applied for the design of the case study in this chapter.

Another alternative for enhancing performance in sparse networks without segmenting the network is the simple addition of extra nodes, mainly serving as relay-nodes. Typically, this is a very usual practice, although it comes with the penalty of a higher cost to install and maintain the solution. As expected, with more nodes, the network becomes denser and the existing WSN multi-hop protocols can potentially achieve better performance. If the AIND-ANON metrics are evaluated again for the new node topology, the values will indicate that the network density in fact increased, as expected.

Next, we will see how the values for the AIND and ANON metrics are produced by means of an algorithm. After that, we will study some cases of real-world deployments (large networks) and calculate the AIND and ANON values for each case in addition to brief discussions.

Before proceeding with the presentation of the AIND-ANON algorithm, two expressions must be properly explained. We define *immediate neighbors* as the closest one-hop neighbors of a node. In our proposed algorithm, we limited the number of these immediate neighbors to 3, meaning that AIND reflects the average inter-node distance in relation to the closest 3 one-hop neighbors. The reason to this constraint is simple: because AIND is ultimately related to the probability of communication errors due to high inter-node distances, we are interested to limit the analysis to the closest neighbors that have better potential to successfully perform the communication task. In relation to ANON, the term *neighbors* is related to all one-hop neighbors that are located at the communication range of a node. Each node is evaluated individually and the average values of AIND and ANON for the network are calculated.

Algorithm 1 AIND-ANON

Requires:	the network is, at least, weakly connected
Requires:	*n*: number of nodes, (n>1)
Requires:	*distance(a,b)*: inter-node distance between nodes *a* and *b*
Requires:	maximum communication range (MCR) of each node
Returns:	*AIND*: average inter-node distance among immediate neighbors
Returns:	*ANON*: average number of one-hop neighbors

```
For Each node i in the network
    AIND_i ← 0, ANON_i ← 0
    For Each node j in the network (j≠i)
        If node j is in the communication range (MCR) of node i
            AIND_i ← AIND_i + distance(i,j)
            ANON_i ← ANON_i + 1
        End If
    Next j
    If ANON_i >3
        For All node j in the network (j≠i)
            AIND_i ← first_smallest(distance(i,j))
            AIND_i ← AIND_i + second_smallest(distance(i,j))
            AIND_i ← AIND_i + third_smallest(distance(i,j))
        AIND_i ← AIND_i / 3
    Else If ANON_i > 0
            AIND_i ← AIND_i / ANON_i
    End if
Next i
```

$AIND = {}^{1}/_{n} \cdot \sum_{k=1}^{n} AIND_k$

$ANON = {}^{1}/_{n} \cdot \sum_{k=1}^{n} ANON_k$

In conjunction[1] with AIND and ANON, the maximum communication range of the nodes (MCR) will form the set of parameters that we want to consider. Note that the number of nodes is indirectly included when the node topology is analyzed and it is also used when the values for AIND and ANON are calculated. Such parameters (AIND, ANON, MCR) give an indication (although without accuracy) of how sparse the network is and the potentiality of reliability and connectivity issues when multi-hopping is intensively used. Using such metrics, we will eventually conclude that some scenarios involving distinct number of nodes can have the same *sparsity* degree and potentially share some of the network problems[2] associated with this fact. In other words, scalability issues are closely related to the sparsity level of the network. The AIND-ANON algorithm is presented in Algorithm 1.

The calculated ANON value represents the expected number of one-hop neighbors of a node in a network. A higher value for ANON typically is associated with a denser network and a smaller probability of issues related to the data transfer. In fact, with

---

[1] Contextual definition for a *weakly* connected network: considering the communication range of the nodes, there is at least one path that connects all nodes.

[2] The number of nodes, application/network duty-cycles, network topology, and selected WSN protocols will also be associated with potential bandwidth, contention, and similar issues. However, another metrics besides AIND-ANON must be developed to address these needs.

Table 1 AIND-ANON metrics: extreme cases interpretation

ANON value	MCR/AIND value	Extreme cases interpretation
Small	Small	Very sparse network (critical scenario): small number of data paths and high probability of communications errors
Small	High	Not a large network
High	Small	Rare case: an efficient deployment geometry (e.g., hexagonal tessellation) covering huge areas
High	High	Very dense network: issues related to the data transfer reliability and communication errors are significantly smaller

a higher number of neighbors, the potential existence of multiple paths for the same message increases the likelihood of having this message properly delivered.

The calculated value for AIND represents the expected average distance between a node and its closest one-hop neighbors. A smaller value for AIND in general is associated with a smaller probability of communication problems due to a higher signal-to-noise ratio (SNR) level. Naturally, the numerical analysis of AIND only makes sense if the maximum communication range (MCR) of a node is also included. Therefore, the ratio MCR/AIND is the ultimate metric of interest besides ANON. Although the Algorithm 1 supports heterogeneous MCRs, we will hereafter assume a single MCR for all nodes. An initial interpretation of the values of ANON and MCR/AIND is shown in Table 1.

In the next section, 13 real-world large[3] WSN cases will be investigated under the AIND-ANON metrics. Because only the node topology and the MCR is captured for each case study (no additional detail in relation to the protocols and location of the sink), the comparison between the cases are mainly to understand the network sparsity level for each case. Nonetheless, as already mentioned, the sparser the network is, the higher are the challenges to the deal with the performance issues if no segmentation or aggregation techniques are employed.

The focus of this chapter is related to the cases similar to the first scenario shown in Table 1, that is, large *and* sparse networks. In the next section, we will discover that few cases of the real-world deployments of large WSNs really lie on the first scenario. Nonetheless, important applications for outdoor WSNs require specific solutions for this scenario, as discussed in Sect. 2.

1.3.3 AIND & ANON Metrics Applied to Real-World Deployments

Chapter 2 listed 62 real-world deployments of WSNs and, from that list, only 10 are related to deployments of more than 30 nodes. On the other hand, there are deployments with fewer nodes that cover impressive large areas. In this section,

[3] The term large WSN in this context is associated to a high number of nodes, or to a large coverage area, or both.

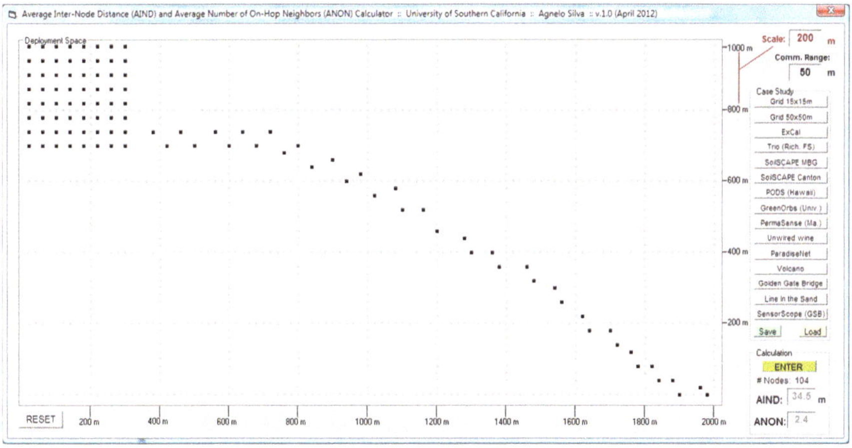

Fig. 4 AIND-ANON analysis for the PODS case [8]

some of these cases and additional ones are evaluated in relation to their AIND and ANON values. To this end, we developed a program (Microsoft Visual Basic) that implements the Algorithm 1. The interface of this program is shown in Fig. 4.

Due to the lack of precise topology information for some of the 13 selected cases, some of the distances are estimated based on additional reports and pictures provided by the authors or associated websites. Similarly, while in some cases the MCR value is explicitly provided in the related work, there are cases where we must derive the value for MCR based on the characteristics of the radio transceiver, antenna height, and related information. Usually, the adopted MCR value is significantly smaller than the communication range specified in the data sheets of the radio transceivers. It is explained by the fact that the MCR is associated with the *connected region* [43], that is, with the region related to good quality bi-communication. In particular for outdoor scenarios, the connected region is reduced due to the characteristics of the environment, such as vegetation and topography.

Next, the case studies are presented in chronological order and each figure shows the use of the AIND-ANON calculator software for each case.

Case 1: PODS [8]

This is a large deployment in both senses: number of nodes and coverage area. If the deployment was only restricted to the dense grid area (top-left of Fig. 4), the value of ANON would be higher. However, the network challenges significantly increase with the introduction of the non-regular part of the network. Consequently, the ratio MCR/AIND becomes significantly smaller. Therefore, the combination of a small ANON (2.4) with a small MCR/AIND (1.4) indicates that this network is sparse and critical network challenges are associated.

The details of the hardware are not extensively presented in [8] and, in particular network performance results are missing. However, based on the information about

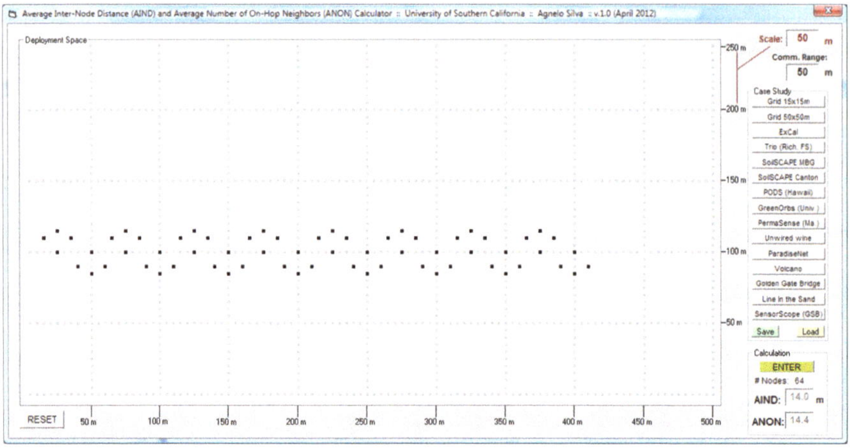

Fig. 5 AIND-ANON analysis for the Unwired Wine case [6]

the Tephranet node (900 MHz, +4.5 dBm transmit power), it is possible that the effective MCR (for a good quality communication) is higher than used in the calculation shown in Fig. 4. If this is the case, the network challenges can be potentially less critical than expected for this deployment. Again, the lack of network performance data in this case (the network operated for few weeks), impact a deeper analysis of the case.

Case 2: Unwired Wine [6]

The node topology for this case study was not provided in [6] and Fig. 5 represents the sparser configuration based on our interpretation based after reading the paper and researching associated work. The relatively high values for ANON (14.4) and MCR/AIND (3.6) indicate that generic WSN protocols would work properly for this scenario due to the high number of additional paths for the data transfer and potentially smaller communication error probability.

Case 3: A Line in the Sand [4]

This is a large network only in terms of the number of sensors. The Fig. 6 (observe the 2 m-scale) quickly reveals how dense this network is. Note that due to the lack of information for the antenna height, we adopted a conservative value for MCR: 20 m. Nonetheless, the very high values for ANON (77) and MCR/AIND (15.4) indicate that generic WSN protocols would potentially fit for this scenario. Nonetheless, there are other issues not revealed by the AIND-ANON metrics.

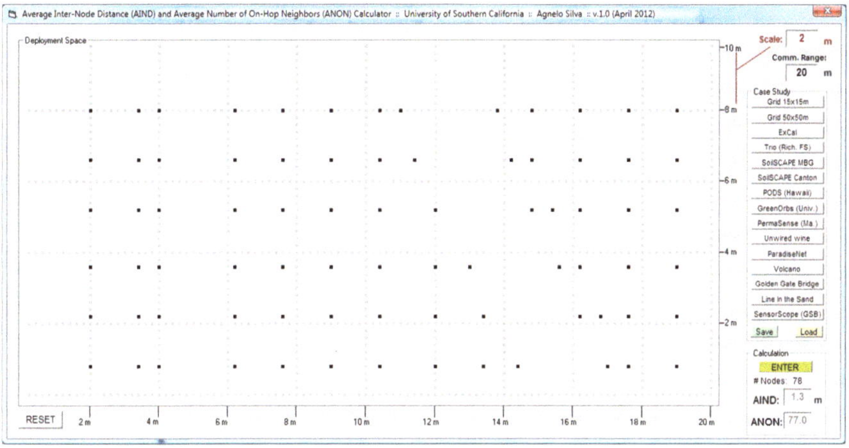

Fig. 6 AIND-ANON analysis for the A Line in the Sand case [4]

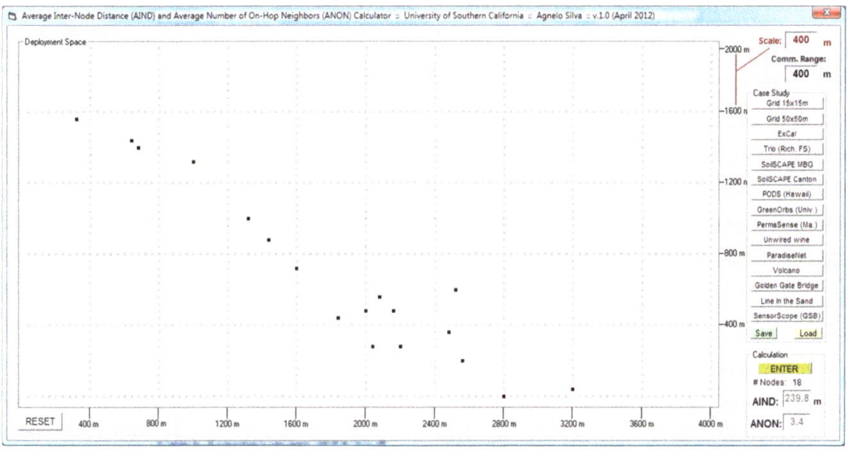

Fig. 7 AIND-ANON analysis for the Volcano (Ecuador) case [41]

Case 4: Volcano (Ecuador) [41]

This is a large network only in terms of coverage area. The proper choice for the antenna and radio transceivers allows the node to have a MCR of around 400 m. It is easy to observe how sparse this network is in the Fig. 7. Such degree of sparsity is also revealed by the small values in the AIND-ANON metrics: 3.4 for ANON and 1.7 for MCR/AIND. Similar to the first case (PODS), this is other case of a critical network with the potential of having high packet loss rate. In fact, the Fig. 4 in [41] indicates significant packet loss rate.

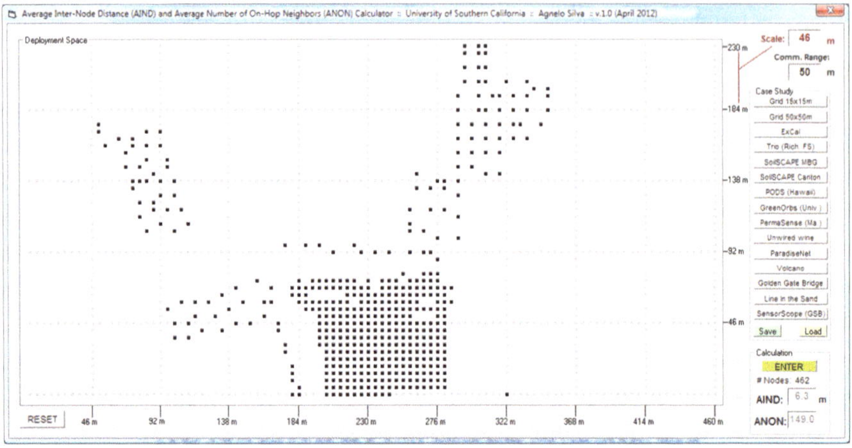

Fig. 8 AIND-ANON analysis for the Trio Testbed (Richmond Field Station) case [12]

Case 5: Trio Testbed [12]

The analysis of this case is partial: only 462 of the reported 557 nodes are actually shown in Fig. 8. Nonetheless, the additional nodes probably would not significantly change this metrics: 149 for ANON and 7.9 for MCR/AIND. Although the coverage area is not small, the network is mainly considered large due to the number of nodes. In fact, it represents one of the densest outdoor WSNs so far deployed. Similar to the Line in the Sand case, excluding contention and bandwidth issues, the network can potentially have good performance for the majority of the current WSN protocols, if energy and contention-related issues are disregarded.

Case 6: ExScal [3]

The AIND-ANON analysis in this case is also partial (now due to limitations of our program): only 824 of 1200+ nodes are actually shown in Fig. 9. Similar to the previous case, a high-dense network is identified: ANON = 35 and MCR/AIND = 3.4. Due to the existence of regions with sparser deployment, the expected network performance is a little bit more critical compared to the Line in the Sand and Trio cases.

Case 7: Golden Gate Bridge [19]

Due to the lack of 3D support of our program, the layout shown in Fig. 10 is an approximation of the real network. Nonetheless, the analysis of this layout must reveal some similarities with the Ecuador Volcano case, another very sparse network that has a large network section with a linear topology. In fact, ANON (2.3) and MCR/AIND (1.6) are small values and pretty similar to the values for the Volcano case. Unfortunately, there is no additional information about the network performance for this 3-week deployment, except the use of an efficient patch antenna in order to increase communication range.

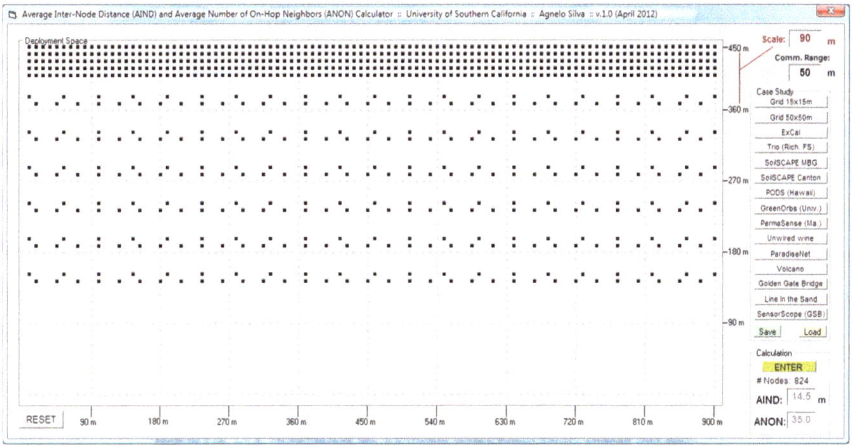

Fig. 9 AIND-ANON analysis for the ExScal case [3]

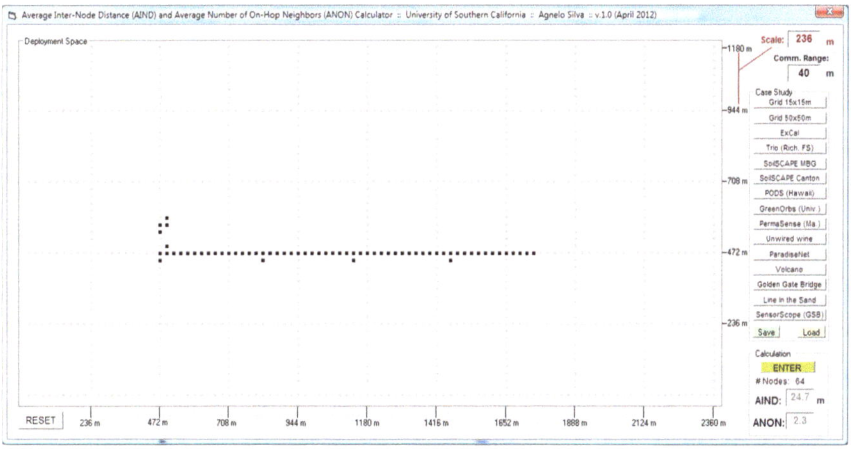

Fig. 10 AIND-ANON analysis for the Golden Gate Bridge case [19]

Case 8: SensorScope (Grand Saint Bernard) [5]

Due to the lack of precise information related to the node distances in [5] and also due to the lack of 3D support of our tool, the layout in Fig. 11 is an approximation of the actual network. Nonetheless, similar to the previous case, we still expect to have AIND-ANON metrics that reveal some degree of sparsity. In fact, MCR/AIND (2.2) is a relative small value. Although ANON (5.4) is not small, if we take into account that the each set of nodes is deployed in a distinct mountain, the real value of ANON can potentially be half of the calculated one. Therefore, the sparsity degree in this case is medium-to-large.

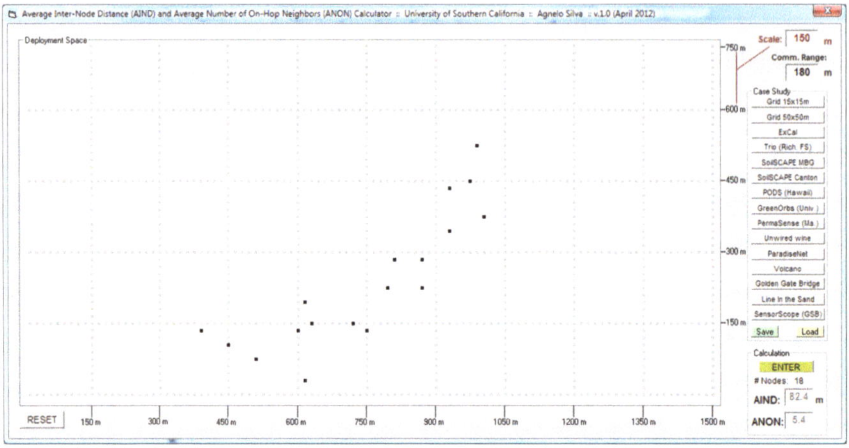

Fig. 11 AIND-ANON analysis for the SensorScope (Grand Saint Bernard) case [5]

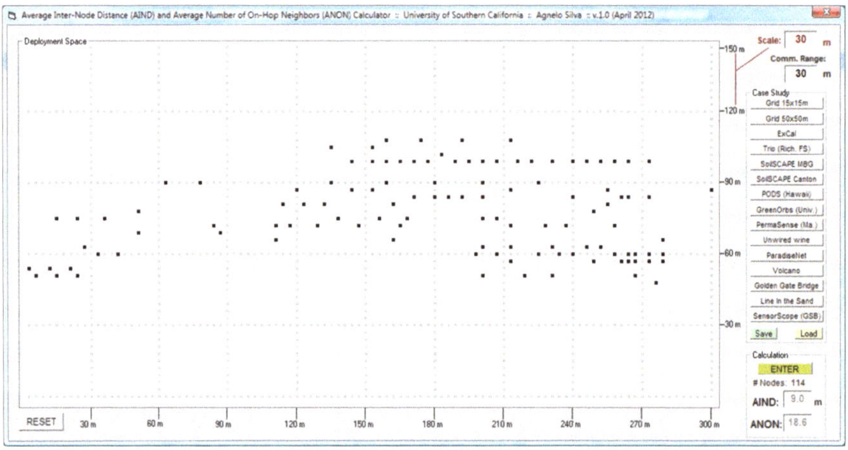

Fig. 12 AIND-ANON analysis for the ParadiseNet case [29, 30]

Case 9: ParadiseNet [29, 30]
The MCR in this case is strongly reduced due to the existence of large high-voltage transformers among the nodes. The Fig. 12 quickly reveals a very dense network. Accordingly, ANON is calculated as 18.6 and MCR/AIND as 3.3 which are relatively high values in our metrics.

Case 10: GreenOrbs (University Site) [24]
Another example of a dense network as shown in Fig. 13 : ANON = 9.5 and MCR/AIND = 2.1. Due to the relatively small MCR/AIND, this network can present a higher packet error rate. However, due to the possibility of multiple data paths, the

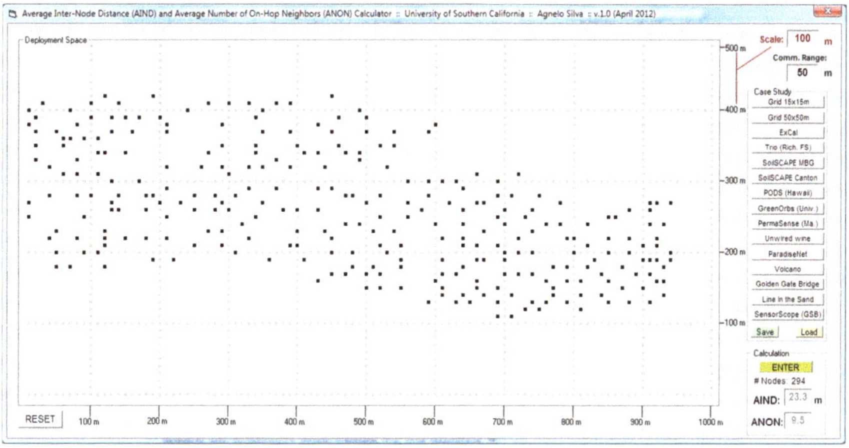

Fig. 13 AIND-ANON analysis for the GreenOrbs (University Site) case [24]

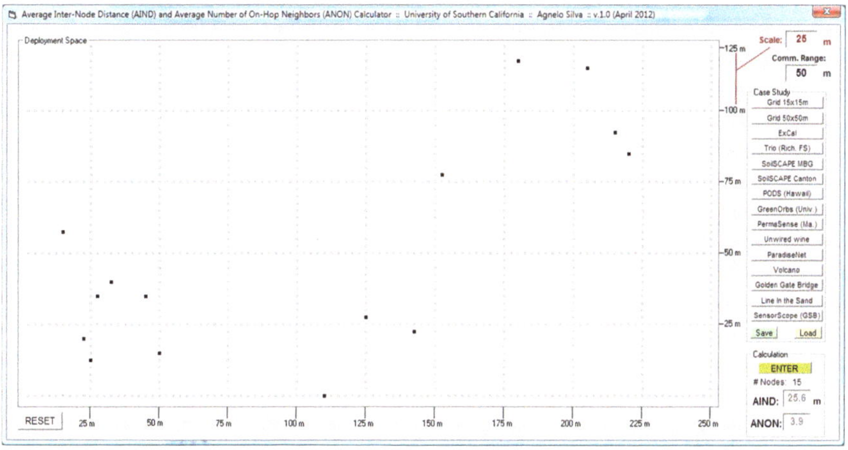

Fig. 14 AIND-ANON analysis for the PermaSense (Mattherhorn) case [7]

effective data transfer reliability for the network can be significantly superior. Therefore, such case study is another medium-to-large sparsity degree case.

Case 11: PermaSense (Mattherhorn) [7]
Although not involving a significant number of nodes, this is a large network in terms of coverage area. As shown in Fig. 14, this is definitely a sparse network: ANON = 3.9 and MCR/AIND = 1.9. The MCR was strongly reduced in this case due to existence of mountains in this scenario.

Case 12: SoilSCAPE I (Matthaei Botanical Gardens) [27, 28]
Involving an area of around $200 \times 300\,\text{m}^2$, this network is not very large but it is relatively sparse if we consider the number of nodes: only 27. With ANON = 6.4

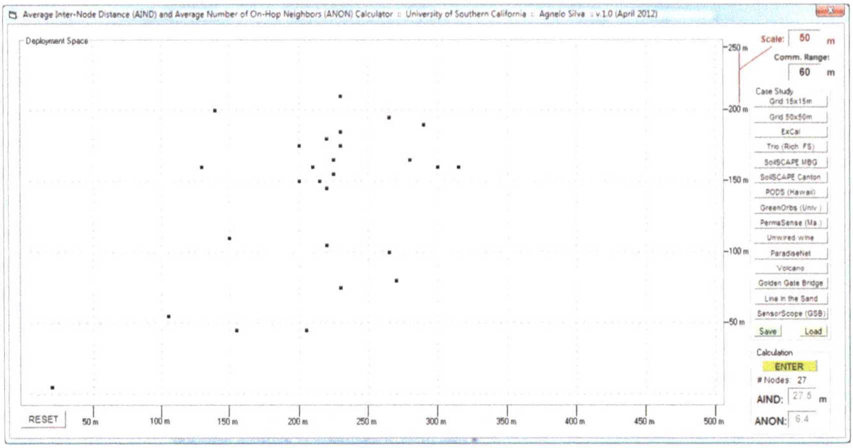

Fig. 15 AIND-ANON analysis for SoiSCAPE I case [27, 28]

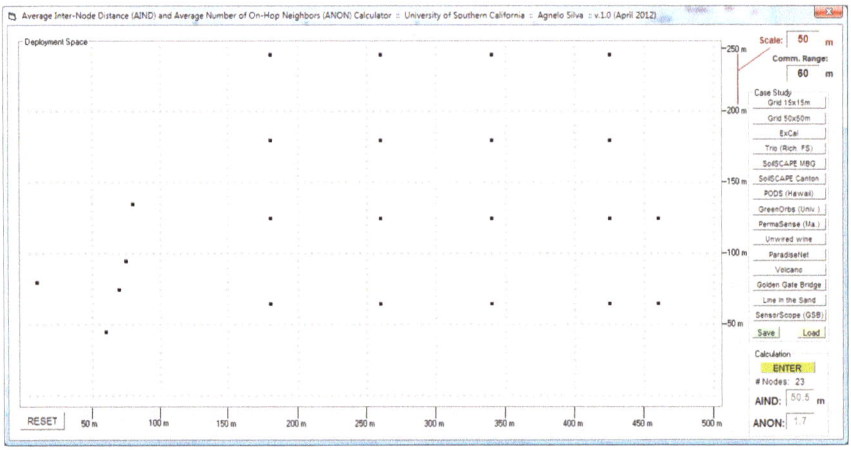

Fig. 16 AIND-ANON analysis for SoiSCAPE II case [28]

and MCR/AIND = 2.2, it has a medium-to-large sparsity degree. The majority of the WSN solutions will have good performance for this scenario.

Case 13: SoilSCAPE II (Canton Farm) [28]
Comparing Fig. 15 with Fig. 16, one can observe that SoilSCAPE II basically doubles the coverage area with a smaller number of nodes, i.e., a sparser network is achieved. In this particular scenario, the existence of obstacles (vegetation, trees, etc.) and differences at the topography make some nodes work better than others. Although MCR of around 400 m was achieved in ideal conditions, on the average 60m is the MCR for the nodes with their standard antennas [28]. The small values of ANON (1.7) and MCR/AIND (1.2) clearly indicate that this network is very sparse and the

Table 2 Estimated values of AIND and ANON for large-scale WSN deployments

Deployment case	#Nodes	MCR (m)	AIND (m)	MCR/AIND	ANON
Environment monitoring: PODS [8]	104	50	34.5	1.4	02.4
Vineyard monitoring: Unwired wine [6], approx. topology	64	50	14	3.6	14.4
Intrusion detection: A Line in the Sand [4]	78	20	01.3	15.4	77
Volcano monitoring (Ecuador) [41]	18	400	239.8	1.7	03.4
Outdoor Testbed: Trio (Richmond F. S.) [12], approx. topology	462	50	06.3	7.9	149
Intrusion detection: ExScal [3], based on a partial deployment	824	50	14.5	3.4	35
Structure monitoring: Golden Gate Bridge [19]	64	40	24.7	1.6	02.3
Environment monitoring: SensorScope (G. Saint Bernard) [5]	18	180	82.4	2.2	05.4
Substation monitoring: ParadiseNet [30, 29]	114	30	09	3.3	18.6
Environment. monitoring: GreenOrbs (Univ. woodland) [24]	294	50	23.3	2.1	09.5
Environment monitoring: PermaSense (Mattherhorn) [7]	15	50	25.6	1.9	03.9
Environment monitoring: SoilSCAPE I (Matthaei B. G.) [27, 28]	27	60	27.5	2.2	06.4
Environment monitoring: SoilSCAPE II (Canton farm) [28]	23	60	50.5	1.2	01.7

MCR: maximum communication range (on the average for all nodes). The values for MCR are approximated ones; they are based on the information provided by the related work and also on the characteristics of the environment, height of the antenna, and radio transceiver.

communication performance can be strongly impacted. Without a careful network design, this scenario is highly associated with high packet error loss and associated higher energy consumption. In fact, as reported in [28], the network only achieved certain degree of stability when 4 additional nodes (ZigBee routers) where added to the network.

The results of these 13 case studies are summarized in Table 2. Also, in Fig. 17, the values of MCR/AIND and ANON for these cases are plotted. It is clear that moving from the left-bottom part of Fig. 17 to the upper-right, the network becomes denser and many performance problems are potentially avoided in this way. However, due to the realistic budget constraints, more and more outdoor WSNs can be potentially located at the *sparser* area of the picture(left-bottom corner). In fact, this is the

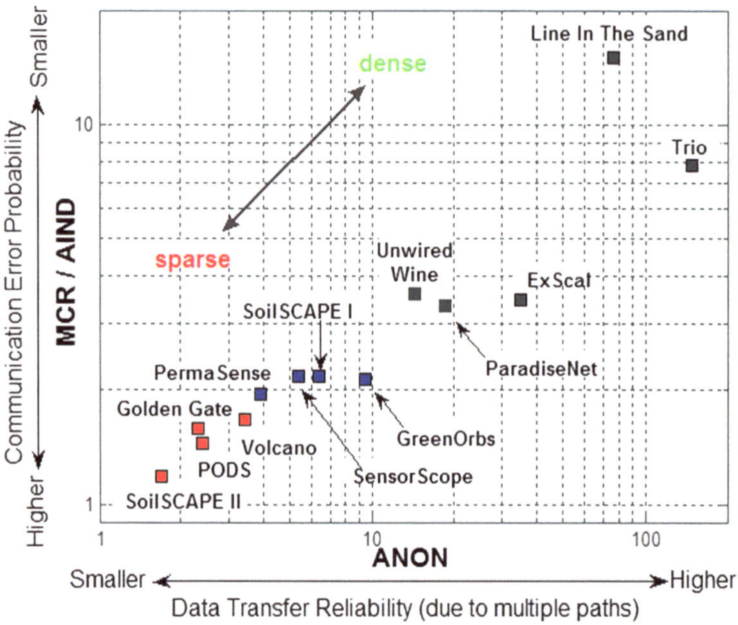

Fig. 17 AIND-ANON metrics applied to 13 large-scale WSN deployments

challenge behind the case study of this chapter, NatureMONITOR, as considered in the next section.

2 NatureMONITOR Case Study

In this section we introduce an illustrative project called NatureMONITOR to highlight the advantages of having a custom WSN design when off-the-shelf solutions cannot provide a complete answer. In this case study, the functional requirements are very strict (though realistic) such that the final solution is rather non-traditional compared to that typically found in the literature. Specifically, an application-centric approach is adopted in the design and many well-established principles are ignored. The main advantages and limitations of the proposed solution are presented. In the next sections more extensive discussion is provided where we also consider other, including more traditional, options. The final conclusions are not definitive ones but they stimulate discussions around the design of low data-rate environment monitoring and similar applications.

2.1 NatureMONITOR Project Specifications

Objective: 400 environmental sensors (humidity, solar radiance, soil moisture, wind spend, temperature, etc.) must be deployed and maintained over multiple years in an 1 km^2 area (hereafter called "cell") to provide in-situ measurements for a variety of environmental studies. Multiple cells will be deployed.

Business Requirements:

1. The deployment must follow the guidelines of the environmental scientists, the main users of the application, in that observations must be of sufficient spatial, topographical, and geological diversity across the 1 km^2 region. This could mean that some areas may need to have more sensors than others, so that the ensemble of data properly represents the region (e.g., its heterogeneity in soil makeup and so on).
2. At each node location, 4 environmental sensors are installed. Therefore, each cell has a maximum of 100 sensor nodes each one with up 4 sensors. While taking measurements, each environmental sensor consumes a maximum of 100 mW.
3. The default sampling rate is once every 15 min for all sensors. However, the system needs to be able to dynamically change this parameter for each location from a central web application.
4. The cell has roughly 1 km^2 of area but can have different geometrical shapes. It can be anywhere in the world and no assumption may be made about the topography and environmental parameters. In particular, support for extreme weather conditions must be provided.
5. The total material cost of a cell cannot exceed US\$25,000, excluding the environmental sensors.
6. The total deployment cost of a cell cannot exceed US\$15,000.
7. The annual operating cost of a cell (excluding the costs associated with the central data server and also the connection to it) must be smaller than \$15,000. This value assumes the default sampling rate of 15 min.
8. The additional maintenance costs due to the usage of a more frequent sampling rate at some nodes for a certain period of time must be known a priori before adopting a higher sampling rate.
9. The maximum allowed latency between the moment when the measurement is performed and the moment when the data is stored in an existing central database (with Internet connectivity) is 24 h.
10. The sensing data must be time-stamped with real local time and the maximum allowed timing error is 1 s. It must also be stamped with the location and depth/elevation of the measurement.
11. When the number of environmental sensors that simultaneously experience problems in a cell exceeds 30, the network cell is considered "unavailable." Such problems may be related to both communication and sensory devices, e.g., defective measurements. The availability of the network cell must be higher than 95 %

over a period of 1 year. If the network cell is unavailable due to maintenance, these periods of time effectively count *against* the network availability metric.

12. The maximum number of lost measurements over a period of 1 month in a cell cannot exceed 90 % of the expected value (i.e., the amount of measurements scheduled to take place).

2.2 Functional and Non-functional Requirements

Observe that at the above business requirements list there is no mention to the term WSN. However, the project budget quickly rules out more expensive connectivity solutions like satellite or cellular connections for each location. We will therefore seek a WSN solution. The first step for the network designer is to evaluate the feasibility of the project, and then to use incremental deployments to test parts of the design under real-world scenarios due to the lack of a more systematic guideline [13]. However, it is expected that outdoor WSN deployments will eventually achieve the same level of standardization and out-of-the-box solution that some indoor applications are starting to experiment in recent years [10, 14, 16].

One approach to the feasibility study of this project is to fill and analyze a list of 29 strategic technical aspects of WSNs, as shown in Table 3. Each technical aspect will be marked either "Required" or "Not Required." The term *Required* must be understood in this context as "must have" rather than "desired" (or "can be supported"). For instance, in our case study some form of time-synchronization is clearly required. However, "multi-hopping" or "reduced size of a node" are not explicit requirements and they are marked *Not Required* (even if desired). The main point here is to separate real functional requirements from features that are typically *expected* to be included in the design of a WSN solution. Each time a non-required feature is included in the design, the design becomes more generic and significant cost and performance penalties can be hidden in that decision.

Analyzing Table 3, we see for this case study that the majority of the *essential* characteristics of a typical WSN are marked as "not required." This suggests that a simpler WSN tailored to the application may be the right answer for this project; many checks in the "Required" column would be an indication that a generic WSN may be a better design direction. In short, the main constraints of the NatureMONITOR project are:

- High reliability;
- Spatial scalability with low-density: large and sparse network;
- Unattended operation at harsh environmental conditions;
- Dynamic sampling scheduling;
- Real-time-based time-stamp for the measurements;
- High degree of accuracy of energy prediction at the level of a single node;
- Critical budget for deployment and ongoing operation.

Table 3 High-level business/functional requirements

Feature	Required	Not required
Bi-directional communication	✓	
Unicast communication		✓
Multi-hopping communication		✓
Multi-cast communication		✓
Broadcast communication		✓
Heterogeneous network (nodes with different profiles)		✓
Real-time communication ($< 1\ s$ data-latency)		✓
Node-to-node communication		✓
Continuous network connectivity		✓
Node mobility		✓
Well-planned node location	✓	
Support for random node deployment		✓
High-accurate localization of mobile nodes		✓
High reliability	✓	
Spatial scalability without high node-density	✓	
Ability to withstand harsh environmental conditions	✓	
Dynamic sampling scheduling	✓	
Remote reconfiguration (exclude sampling scheduling)		✓
Remote reprogramming		✓
Small size of the node		✓
High data-rate		✓
In-network processing		✓
Persistent data storage at the nodes		✓
Localization and timestamp for the measurements	✓	
Time synchronization among nodes		✓
Authentication		✓
Data encryption		✓
Remaining available energy prediction	✓	
Unattended operation	✓	

Harsh environment in this chapter refers to the scenario where the nodes are exposed to extreme weather conditions (sun light, wind, humidity, temperature, etc.) and the action of insects and animals. In some cases, the area is also one with very difficult access.

Considering the availability and characteristics of off-the-shelf hardware and software solutions and the costs associated with the development of fully customized solutions, the design team in this case study finally decides on a balanced solution, a non-traditional mix of telemetry [13], short-range wireless solutions [10, 14, 16], WSN technologies, and some degree of customization. The project proposal (in fact, this is the initial formal feedback to the project sponsors) is described as follows, as an extension and/or adjustment to the original business requirements:

Overview:
To cover a network cell with 1 km^2 area, an open, asynchronous, and hybrid wireless sensor network (WSN) is proposed. The cell is divided into multiple physical

segments, each one with up to 30 sensor nodes. Four environmental sensors are attached to each sensor node. The nodes in a segment communicate with a special node called *master* of that segment. On the second layer of the hierarchy, master nodes communicate with the main gateway of the WSN. The wireless technologies used in these 2 layers are not necessarily the same. Therefore, this network solution is potentially a *hybrid* one.

Sensor nodes in the same segment are deployed up to 300 m away from the master provided that the communication performance is acceptable over that distance under different environmental conditions. No peer-to-peer communication or collaboration among nodes is provided. Communication between nodes and the master is based on commercial radio transceivers typically used in WSNs. The link(s) between the master(s) and the main gateway is (are) realized using point-to-point radio technologies capable of supporting distances of up to 5 km. However, short-range radios (including WSN-based links) can also be implemented depending on the distance.

The solution is considered *open* because both network tiers (sensor node-to-master, master-to-gateway) can adopt any current and future wireless technology that supports point-to-point connection. Possibilities include, but are not limited to, IEEE 802.15.4 [10, 14, 16], IEEE 802.11, Bluetooth [10, 16], Z-Wave [10], DASH7 [11], GPRS, VHF/UHF wireless modem, etc. The solution is also insensitive to the existence of low-level protocols that deal with medium access, reliability, and synchronization. All these features are actually implemented at the application layer (layover design), which facilitates the future change of wireless technologies if necessary.

All nodes, except the main gateway, are powered by non-rechargeable batteries. Assuming sensor measurements every 20 min, the expected lifetime of a sensor node (including the master) is 13 months. Thus every 12 months, human inspection is expected to replace all batteries and also to perform additional preventive maintenance tasks. This scheme is possible because all regular nodes are expected to have similar energy consumption. The nodes send data according to the sampling schedule provided by the gateway via the master. They follow a TDMA-like protocol implemented at the application layer in order to avoid medium contention [35]. Data collected by the master is stored in its persistent memory. Sometime later (e.g., minutes or hours), the master forwards the data to the gateway. These features outline an *asynchronous* behavior of the proposed network architecture.

In order to achieve the budget goals, the default sampling duration has been revised from 15 to 20 min. For many environmental monitoring applications we expect that such modification will not significantly impact the results, though this must be confirmed by the end user. Due to temporal and spatial correlation inherent in many environmental measurements, scientists routinely use subsampling and round-robin subsetting [23] techniques to make up for measurement losses (either in time or in space). However, to apply either technique, dynamic and individual measurement schedules must be supported by a sensor node. Incidentally, for this project such provision is also a business requirement. Therefore, it is possible to maintain the same energy budget (in terms of battery lifetime) and to have a default sampling schedule of 20 min while variations in this scheduling are supported for some of the nodes.

Besides the original functional requirements, the following functional and non-functional requirements are added to the project proposal:

- The maximum number of sensor nodes per network segment is 30, excluding the master itself which is also a sensor node (i.e., a cell has a maximum of 31 nodes).
- Adjacent network segments must use different wireless technologies or, alternatively, distinct radio channels.
- All nodes in the same segment must use the same wireless technology.
- The location of the master in a segment must be carefully chosen: each sensor node in that segment must communicate with (and only with) that master through a reliable communication link in order to satisfy the performance metrics previously mentioned.
- Every master has a persistent (non-volatile) memory and it must store at least 10 days worth of measurements from all sensors of the segment.
- The battery level of a node must be sent to the master along with sensor measurements. Such information must eventually reach the data server.
- Only non-rechargeable batteries are used on the sensor nodes (including the master) and such batteries must work from −40 to 70 °C. Many Li-$SOCL_2$ models support this range.
- Assuming that measurements from the 4 environmental sensors take place every 20 min, the lifetime of the non-rechargeable batteries must be at least 13 months considering the temperature range previously mentioned.
- A new sampling rate or schedule sent to a node must be applied within 1 h.
- The sensor node enclosure must be weatherproof and have IP67 or similar/superior rating.
- All nodes including masters and the gateway must have external watchdog circuitry in order to reset the device in case of a continuous (e.g., lasting more than 4 h) non-functional state.
- Nodes must synchronize their internal clocks with respect to the master at least every hour.
- The masters must synchronize their internal clocks with respect to the gateway at least every 24 h.
- The gateway must synchronize its internal clock with respect to the global real time clock at least every 24 h.
- The installation or presence of the sensor node (processor and radio modules) cannot interfere with the measurements taken by the environmental sensors at the same location.
- The system administrator or the end user must have a way to verify the health status (energy and communication performance) of each sensor node and the network as a whole.
- The system administrator or the end user must have a way to forecast the energy costs incurred by changes in the sampling scheduling of a node or a group of nodes.

As the design process progresses, some technical constraints emerges and some features that were not previously required become required. In the case of this project, the modifications are as follows:

- Unicast communication: **Required**.
- Heterogeneous network (nodes with different profiles): **Required**.
- Persistent data storage at the nodes : **Required for Masters**.
- Time synchronization among nodes: **Required**.

We next turn our attention to the issue of node placement. As already mentioned, the limited (and small) number of nodes over a large coverage area implies low node-density or areas with no nodes at all. Assume that, in this case study, the end user (i.e., the environmental scientists) indicate the areas that must be populated with sensors as shown in Fig. 18a. This placement is planned considering the topography, landscape, soil composition, and other science-related factors.

Using the Fig. 18a as a starting point, the WSN designer has now the task of considering available technologies and proposing the final node placement plan, an example of which is shown in Fig. 18b. In this case, it is clear that not all areas have an ideal representation due to the limited number of nodes, but all areas of *interest* are represented. The overall network architecture of the proposed solution is shown in Fig. 18c. As expected, this is a large and sparse network according to the AIND-ANON metrics, as shown in Fig. 19. In this case, assuming MCR = 100 m,[4] ANON is 2.9 and MCR/AIND = 1.4. We can easily anticipate that traditional WSN protocols will potentially face problems in this scenario.

The proposed system is a collection of 2-tier WSNs, each network segment with a star topology. For the lower tier (nodes-master), a customized overlay network solution is placed on top of a single-hop WSN. The master of each segment is usually close to the center of the segment. Note that the disc-shaped communication ranges are shown here for simplicity of illustration; it has been reported that such shapes are much more complex and irregular in practice [43]. We have used a smaller circle compared to the real communication range. While such circles include parts of the so called transitional region [43], the high-quality of the node-to-master links is achieved because the topology, transmit power level, antennas, hardware, data-rate, and protocols have been taken into account and verification tests are performed. In practice, only the most critical links of a network segment due to distance, topography, or existence of obstacles must be carefully tested and validated before the actual deployment.

Note that, in this case study, node placement is very carefully planned and some preliminary tests are even performed prior to the actual deployment. This is in stark contrast to a "random" placement that can be proposed for environment monitoring studies. As argued earlier, the cost constraint typically prohibits the adoption of a random or ad hoc placement. Accordingly, the domain knowledge that environmental

---

[4] Although the worst-case to be supported is 300 m, the AIND for this specific case is 69.2 m. Therefore, we want to consider a more realistic case where MCR = 100 m (rather than 300 m) considering the capabilities of existing WSN radios. The exceptions can be potentially solved with higher transmit power levels, special antennas, or the use of additional intermediate nodes as repeaters. However, for this preliminary analysis we want to see how critical would be the typical WSN solution (collaborative protocols) assuming the use of typical hardware and neglecting the worst scenarios.

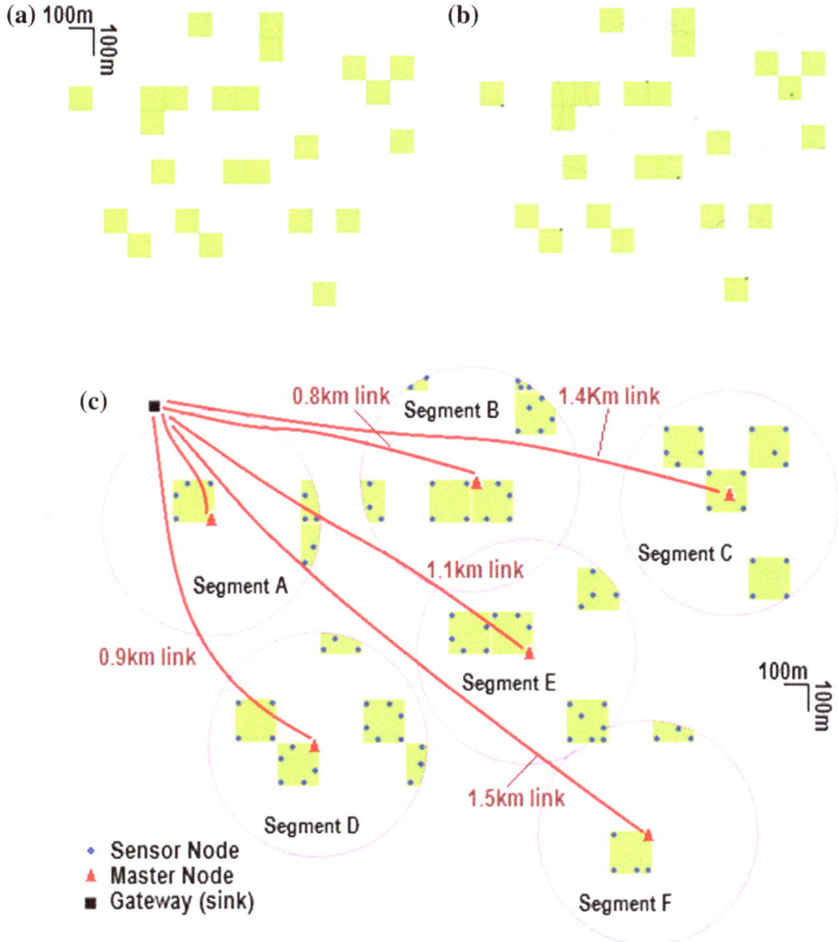

Fig. 18 **a** Areas to be covered by sensors as required by environmental scientists. **b** *Circles* representing the communication coverage of the sensor-to-master link; each *circle* is a segment. **c** Final placement plan for a cell (100 sensor nodes) based on a two-tier network

scientists have plays a fundamental role in the placement plan, as exemplified here.

The advantage of the 2-tier, non-collaborative, and asynchronous approach in the architecture for NatureMONITOR is the predictable network behavior which is highly deterministic. In addition, issues within one segment do not propagated to another. As a result, higher scalability is achieved both in terms of number of nodes and in terms of spatial coverage. It is clear that this architecture shares similarities with the existing LANs and WLANs: the scalability challenge is addressed by segmenting the network into small groups through the use of hubs/switches/access points and gateways/routers.

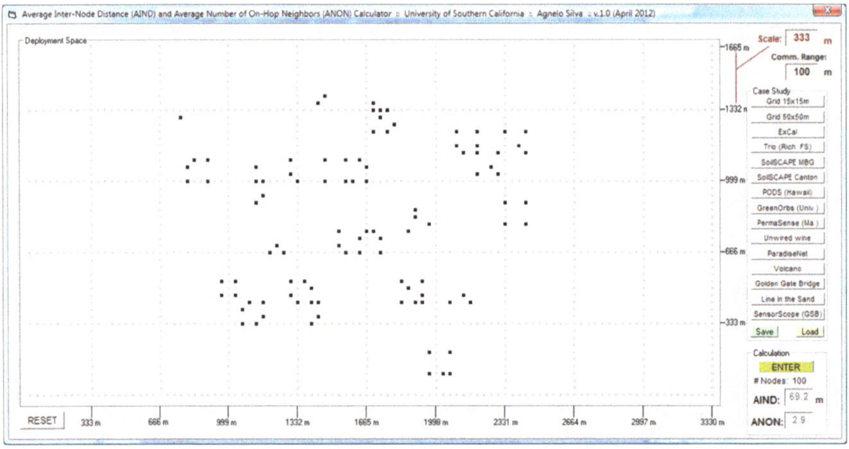

Fig. 19 AIND-ANON analysis for the NatureMONITOR case

In general, a planned deployment involving network segments or clusters has a special advantage rarely mentioned in the literature: the issues related to the communication of a regular node (end device, slave, etc.) and a cluster-head (coordinator, master, router, sink, etc.) can be solved case-by-case. For instance, a different antenna height or the adoption of a directional antenna can be investigated for the most critical cases involving a link of a regular node and its master. In ad hoc deployments and also in typical collaborative WSNs, such approach normally cannot be employed due to the plurality of neighbors of a given node. This fact is evidenced by the regular use of disc-shaped communication models for WSNs.

Another special feature of the proposed solution is the abstraction in relation to the underlying wireless technologies. There is no explicit tie between the architecture and a specific solution, such as IEEE 802.15.4, TinyOS-based, or similar low-range technology: any of these may be a proper answer. More importantly, a mix of these solutions is supported provided that a single segment shares the same technology. In other words, in a cell with 100 nodes, some segments may use 802.15.4 PHY 2.4 GHz standard, while one specific segment inside a dense forest adopts ISM 900 MHz links for the nodes. Such an approach is extremely flexible in controlling choices and costs. Moreover, the risks of having an outdated technology are significantly reduced. As an example, the design team may opt to use 802.15.4 transceivers as much as possible due to its high performance, current low cost, and high market availability. However, if after few years significant issues come to the scene due to over-utilization of the 2.4 GHz bands by devices nearby the deployment area, the radio transceivers can be simply changed by other solution (e.g., 900 MHz) that has similar or better performance.

In the next section we evaluate the proposed architecture underlying the NatureMONITOR project comparing it with more traditional WSN designs.

3 Why Applying Taxonomy to a WSN Project?

In this section we apply the taxonomy introduced in Chap. 2 to the NatureMONITOR project and its proposed architecture. More details about this architecture are provided and the advantages of applying such taxonomy are highlighted. By classifying our design, we have an opportunity to discuss different alternatives and be better prepared to identify the pros and cons of the solution [25].

Goal. NatureMONITOR is clearly a *sense-only* tool which is the case for the majority of WSN applications. Although the proposed architecture does provide some level of support for "sense-and-react," it is not a functional requirement of the project. Note that cost, energy, and data-latency constraints make the implementation of a sense-and-react tool relatively difficult to be achieved in this scenario.

Time. The solution which is being proposed is clearly a *periodic* data collection one. In order to take advantage of (a) the pre-defined static node placement, (b) the low data-rate, and (c) the limited number of nodes per segment (30 sensor nodes plus the master), a TDMA-like MAC scheme is adopted in NatureMONITOR for each segment. A sensor node wakes up according to its measurement schedule and the master node wakes up when one or more nodes of its segment become active. Contrast this approach with a typical ZigBee network configuration where a router device, in contrast with our master node, must be always active [10, 14, 16, 27]. Because each node in a segment has its own assigned slot-time, medium access contention is kept at a minimum. Moreover, the energy consumption of a node is almost entirely determined by its measurement scheduling, making it highly deterministic, homogeneous among the nodes, and can be accurately predicted.

On the other hand, the simplicity of this time-division scheme comes with the drawback of a weak support for *event-triggered/driven* measurements because neither the nodes nor the master are continuously active. This is a potential constraint for some applications and the architecture proposed for NatureMONITOR is thus tailored for applications with a ***sense-send-sleep*** data pattern. This solution is also not suited for critical real-time applications, such as security monitoring/surveillance, due to its significant high data-latency.

One may argue that many existing event-driven WSN solutions could also be adopted for this project. Indeed, many MAC protocols allow a sensor node to sleep more than 99 % of the time while quickly waking up for fractions of a second. During this small active time period, events can still be detected and measurements conducted. However, the tradeoff here is the energy consumption [20]. Considering a traditional WSN hardware platform, in order to achieve very quick wake-up times, many parts of the hardware cannot be simply turned off and on again. Instead, the modules are usually put into some form of sleeping or standby mode. With this approach the latency to wake up the hardware can be as low as a few microseconds. However, because the modules (e.g., radio transceivers and voltage regulators) cannot be completely shutdown, the required sleeping power consumption can be significant as discussed next.

Even with the recent advance in low-power technology, the difference between digitally switching off such hardware devices and putting them in standby mode is between 1 to 2 orders of magnitude. Again, for medium to high duty-cycle networks (e.g., >1 %), this difference is insignificant as the energy consumption is dominated by the active power, not the sleeping power. However, for low duty-cycle applications, such as NatureMONITOR, the sleeping power becomes critical and a barrier to achieve significant energy savings.

Since there is nothing at requirements of the NatureMONITOR project that points to a future need for critical real-time support and the average sampling rate is once every 20 min, many parts of the hardware can be simply turned off. This approach is possible because the activation delay when the module is completely shutdown (e.g., hundreds of milliseconds) is still very small compared to the 20 min-period. By doing so, the design achieves energy consumption far below current WSN solutions.

Sensed Phenomenon. NatureMONITOR deals with *multiple* | *distributed* sensed phenomenon, while uses a large number of sensors. In this regard the proposed architecture is highly scalable as discussed in the previous section. A detailed investigation of current software and hardware WSN solutions suggests that very few existing solutions can achieve a reliable and low-cost solution for a scenario similar to the one in Fig. 18a: 100 nodes non-uniformly deployed over an irregular area of 1 km^2.

To understand how difficult it is to achieve spatial scalability, consider a larger area of 9 km^2 instead of 1 km^2, e.g., 3000×3000 m. It is not difficult to see that the proposed architecture is ready to address this new scenario by the simply addition of more network segments. In essence, the segmentation and the two relatively independent communication layers (sensor node-master, master-gateway/data server) are the mechanisms that allow the architecture in NatureMONITOR to be highly scalable. By contrast, it is not clear how many large-scale solutions reported in the existing WSN literature would spatially scale without having to add infrastructure nodes, such as repeaters or relay-only nodes. Nonetheless, one notable solution for large and sparse WSNs is the use of mobile nodes [2, 36, 40]. Unfortunately, such infrastructure does not fit with the characteristics or requirements of NatureMONITOR.

A ZigBee-based solution has also been directly applied to a relative sparse network in [27], with 3 router nodes and 21 end device nodes. This is the scenario for the SoilSCAPE I case study. It is reported that when one or more of these routers failed, the network quickly became overwhelmed with excessive retransmissions and ZigBee-level traffic. Energy issues also emerged at the end devices due to damages to solar panels. Using a network analyzer, it was observed that the duty-cycle exceeded 10 %, far beyond the original application requirement ($\sim$0.3 %). This example shows the need of a significant network infrastructure (e.g., routers) to properly support the outdoor deployment based on ZigBee: 1 router per 7 nodes in this case [27]. Similarly, in [9] a large outdoor deployment with ZigBee uses 1 router per 8 nodes. Besides ZigBee, it is expected that many typical collaboration and multi-hopping WSN solutions also present network performance issues in the NatureMONITOR scenarios. A typical strategy in this case is the addition of *infrastructure nodes*, devices that are deployed only to maintain the network connectivity and better performance. As expected, this simple solution comes with a significant cost penalty.

Even if one artificially increases the network density (that is, extra nodes are not actually required by the application) and adopts a multi-hopping solution for the NatureMONITOR project without segmenting the cell, data-latency can be aggravated if a strong sleep policy for the nodes is maintained. In short, the higher is the number of hops in conjunction to longer sleeping periods of the nodes, the higher is the data-latency. For instance, for the deployment plan shown in Fig. 18c, assume the adoption of WSN nodes with transmit power between 0 and 17 dBm and frequency 2.4 GHz. In this case, the average number of hops required for one node to reach the sink of the cell is around 5. For the worst case, a node would have more than 10 hops toward the sink and the data-latency becomes significantly high.

Data-Rate. NatureMONITOR has *low data-rate* and the proposed architecture takes this as both an assumption and a feature to be exploited. If the data-rate requirement increases, even if each network segment has enough bandwidth to support this increase, two problems can potentially occur. First, the probability of network errors would significantly increase due to the combination of higher data-rates and high distances between the sensor nodes and the master. As a result, the proposed architecture may not achieve the expected reliability metric.

A second problem related to a higher data-rate is that long-distance links used to connect the master node to the gateway may not afford the bandwidth increase. Usually these links have smaller bandwidth compared to the short-range link used within a network segment. Therefore, some form of data aggregation and in-network processing would be necessary. In short, the lack of support for high data-rate applications is another drawback of the architecture proposed in NatureMONITOR.

Heterogeneity. NatureMONITOR uses an *architecture* that comprises 3 distinct: sensor node, master, and gateway, forming a heterogeneous WSN. Specifically, for each segment, the master node is in charge of collecting the data from all sensor nodes of that segment. In general, using another more powerful radio transceiver, the master node communicates with the gateway via a long-distance link. Due to the need of storing the measurements data in case of failure of this long-distance link, the master also requires some form of non-volatile memory. As a result, the communication, processing, storage, and power capabilities of the master are distinctly different from that of a sensor node.

The main weakness of this heterogeneous approach is the network unavailability in case of failure of the master node. If one regular sensor node stops working, the overall application does not suffer very much. However, if a master fails, its entire network segment becomes disconnected. If sensor nodes do not have any persistent storage, then all the sensing data of a segment will be lost in the meantime. One approach used in many WSNs that also employ special nodes is to apply redundancy for these nodes. Another approach is to increase the hardware/software reliability level of the special node (the master in this case) and to perform emergency maintenance for this specific node if necessary.

One of the resources commonly used to achieve high availability in embedded systems is the use of "watch-dog timers" (WDTs). Such optional device provides some form of initialization of the system when the latter freezes due to a software or hardware problem. For the master node used in this project, two independent WDTs are used. This solution obviously does not solve all the problems; thus long-term tests must be performed in order to verify the failure probability rate for the master node in particular.

Mobility. NatureMONITOR does not have mobility support, as it is not a functional requirement. It is also not hard to see that it would be quite difficult to modify the proposed architecture to support mobility: the proper operation and low energy consumption of the current solution heavily rely on the static nature of the system by means of a fixed node placement and a priori known sampling scheduling. On the other hand, the introduction of mobility brings randomness to the network topology and generic ad hoc WSN solutions surely are more effective. This observation highlights the fact that customized designs are adequate when the underlying fundamental assumptions are clearly understood and not broken. Otherwise, generic solutions are potentially in advantage.

For instance, one can argue that mobile data collectors can be a better solution for large and sparse networks [2, 36]. The design team of NatureMONITOR investigated this option, but because the sensor nodes are left unattended in harsh environment, no practical way to have a mobile data collector for this scenario was figured out. Nonetheless, there are scenarios, including environmental monitoring, that are suitable for mobile data collectors, such as the precision agriculture application in [39].

Connectivity. NatureMONITOR has an *intermittent* connectivity and this feature significantly simplifies the network design. Note that there is no peer-to-peer communication provision in this architecture and a fully connected network is not required. Moreover, the data over the most critical links (master-to-gateway) is protected by storage provision at the master side. The main weaknesses of having intermittent connectivity in NatureMONITOR are (a) lower reliability due to the lack of multiple data paths and (b) a higher data-latency. As already discussed, the sensor-to-master communication occurs over a one-hop link. Since a node is not logically connected to another in the same segment (even if both are within the communication range of each other), there is no other way for the data of one sensor node to reach the gateway or data server except passing through the master. Based on the requirements for NatureMONITOR, the total number of lost messages can be as high as 10 % of the total. For this particular application, due to the spatio-temporal correlation among measurements, such relaxed metrics are usually acceptable and the network architecture *exploits* this fact. However, other applications may require higher reliability on the data transfer.

NatureMONITOR adopts an acknowledge mechanism to guarantee the proper data delivery. However, the retransmissions are allowed in the same time slot. After the end of the assigned time slot, the sensor node simply discards the measurements for that cycle. Although more reliable transport protocols are available, it is clear that the simply approach under NatureMONITOR has a very small overhead while it still can satisfy the relatively relaxed reliability requirements of the project.

Message redundancy is another technique used to mitigate the reliability issue without compromising the deterministic behavior of the solution and this mechanism is also available for NatureMONITOR.

Besides weaker data transfer reliability, another drawback of having intermittent connectivity is the higher data-latency. In fact, the architecture behind NatureMONITOR does not properly support critical real-time and very time sensitivity applications, such as intruder detection. Two aspects of the architecture affect data-latency. First, the master collects the data from the nodes in its segment and only transmits that data to the gateway after receiving the data packet from the last node of a giving sampling cycle. Depending on the length of the node's time-slot and the number of active nodes, this delay can be on the order of seconds. A second source of data-latency is due to the transmission of data from the master to the gateway or data server. Although, it is possible that such data transfer occurs immediately after the conclusion of the sensor-master communication, the master-gateway data transfer can be delayed on the order of minutes or hours if energy or communication issues at the master node are considered.

Processing. The master nodes in the NatureMONITOR solution perform some form of *filtering and compression* in order to optimize the usage of time while transmitting data to the gateway. This is an important provision due to the high energy costs associated with the typical high-power transceivers and long-range links used for the master-gateway communication. However, due to the simplicity of this data collection application, there is no need for data processing at the sensor node itself, that is, a small message is enough for the transmission of the measurements.

On the other hand, some applications require significant in-network processing, particularly those associated with high data-rates. As previously mentioned, the architecture used in NatureMONITOR is not appropriate for these scenarios. In particular, typical in-network processing involves a high level of collaboration among nodes, which is not supported by this architecture.

Storage. The master node in NatureMONITOR has *persistent* data storage provision through the use of non-volatile memory and/or an SD Card. Thus, the measurements data are properly saved in case of energy issues at the master or communication problems in the master-gateway link. However, there is no similar provision for the sensor nodes. This is a decision aligned with the expected reliability metrics for the sensor nodes and the fact that communication range tests are conducted at the deployment site before the actual installation of the sensor nodes.

The main problem associated with storing data at the sensor node itself, an approach not used in NatureMONITOR, is the additional energy consumed by the operation. Moreover, under the current architecture, one time slot is only sufficient for the transmission of the measurements data collected by the 4 environmental sensors (also a single re-transmission in case of failure). The current slot structure does not allow transmission of more data, i.e., previous measurements that could not be transmitted. In short, there is no data queue at the sensor node's side.

Services. NatureMONITOR offers two network infrastructure services: ***time synchronization*** and a very basic form of node ***reconfiguration***. The former service is actually a requirement for the project because the measurements must be accompanied by a time stamp based on the real clock at the gateway. Even with cheap clock systems, the sensor and master nodes do not suffer significant clock skew effects in the proposed architecture. This is the case because every time a sensor node sends measurements to the master, it receives scheduling information that allows its clock to be automatically adjusted on the order of milliseconds. Therefore, only very large data sampling schedules, such as >15 h, may lead to time stamp errors beyond the project specification. Similarly, the master also adjusts its clock according to the gateway's clock every time it sends data to the gateway. Finally, the same process also occurs between the gateway and the central data server.

Node reconfiguration is a second requirement of the project. However, only the measurement scheduling is remotely configurable. In some cases, it may be desirable to implement full re-configurability of the sensor node without local intervention. The same is also true in relation to remotely upload a new version of the program that runs at the sensor node. NatureMONITOR does not fully support reconfiguration and reprogramming in order to maintain more deterministic network traffic, which is a key characteristic of the architecture. On the other hand, services that have regular and predictable behavior (fixed and regular bandwidth) are relatively easy to be implemented under the proposed architecture. Two examples are encryption and authentication services. They are not used in NatureMONITOR as they are not required, but their usage is feasible although resulting in additional network overhead. Finally, because NatureMONITOR is based on a static topology and a planned node placement, there is no need of localization techniques. A simple addressing scheme for the sensor nodes is adequate for this objective.

After applying the taxonomy to the NatureMONITOR project, we conclude that the proposed architecture is a highly specialized/customized WSN with the following characteristics: sense-only, periodic sampling, multiple sensors for a distributed phenomenon, low data-rate, heterogeneous and static nodes, intermittent connectivity, filtering and compression features at some special nodes, persistent storage at these nodes, time synchronization, and dynamic measurement scheduling.

We conclude this section with an important discussion related to the node size and battery choices. Note that small size is not a project requirement under NatureMONITOR. In fact, while size is critical for some WSN applications [32], it is usually not so important for outdoor scenarios compared to the cost of a sensor node. Also note that the antenna of a sensor node is usually placed more than 1m above the soil surface in order to increase the wireless channel quality. Therefore, cylindrical structures longer than 1m in length and 5–10 cm in diameter can be potentially used. There is no doubt that such form factor is far beyond the typical match-size of WSN nodes. However, for the NatureMONITOR project, there is no need to have a tiny sensor node. Also, the resulting gain in terms of available room at the node's enclosure can be exploited in favor of the adoption of **larger batteries**.

Despites the increasing popularity of energy-harvesting systems based on supercapacitors and/or rechargeable batteries [18], the design team in NatureMONITOR project opted for non-rechargeable batteries. This is because unknown weather conditions (recall that the system may be deployed anywhere) imply that the design must take into consideration extreme temperatures, which can significantly affect the performance of energy-harvesting solutions. Moreover, it was observed in [15] that the canopy would impact the efficient usage of solar panels. Snow, pollution, and dirt caused by animals are also potential sources of problems for these devices.

There are 3 additional factors that favor non-rechargeable batteries despite the fact that they must be replaced from time to time. First, non-rechargeable batteries have the highest energy-density in comparison with any other form of low-cost energy source for WSNs. Specifically, for the same physical volume the energy stored in a non-rechargeable battery can be 2 or 3 times that stored in a rechargeable battery. Second, more accurate methods to determine the end of life of these batteries are available in comparing with rechargeable cells. More specifically, when non-rechargeable batteries are used with a well-known discharging behavior, it is possible to predict the remaining lifetime of the node with a high degree of accuracy. Note that this aspect is actually a requirement for NatureMONITOR. Finally, non-rechargeable batteries have superior performance under extreme temperature variations. For instance, in [27] many problems with the nodes occurred when the rechargeable batteries stopped charging due to low temperatures ($< 0\,^{\circ}$C).

Therefore, in some outdoor scenarios it may be more economical to replace the batteries following a pre-determined schedule than having to deal with the uncertainty of maintenance of some energy-harvesting solutions. However, such a guideline is usually valid only when (a) the energy consumption is relatively small due to low duty-cycles and low data-rates and (b) the system has very stable network traffic. Because these aspects, in particular the latter one, are very difficult to be achieved in large-scale outdoors WSNs, the majority of these solutions employ rechargeable batteries usually associated to solar panels or other form of energy harvesting system.

4 Discussion and Conclusions

In this chapter, the advantages of designing a WSN according to specific application needs instead of going toward a generic WSN are highlighted. The filling practice involving the Table 3 allows the project manager and network designer to have a better understanding of the *actual* requirements that are behind a WSN project. It is highlighted that the fact that a certain feature is available and can be potentially supported in our design does not make it a *requirement* for the project. By removing such feature(s) from the design, it is possible to evaluate the option of having a customized WSN solution. In particular, low data-rate environmental monitoring applications can potentially follow this track.

However, before moving from generic and well-established WSN solutions to any form of customized solution, it is highly recommended to analyze the gains of having a tailored design. In fact, this project management principle must govern any project, in particular involving technologies that are quickly evolving. In this chapter, it is analyzed why sparse networks in outdoors deserve a special attention in their design. It is shown that, for some scenarios, the current WSN solutions may not provide the proper answer or, at least, the effective cost of the project may be significantly higher than initially planned. At the end of the chapter, a discussion about the pros and cons of adopting customized vs. typical WSN solutions is provided. The arguments are somewhat controversial because each WSN project has a significant number of design aspects to be considered and what is presented here as a proper solution for some particular scenarios may not satisfy the requirements of an ongoing project which a WSN designer is involved.

4.1 NatureMONITOR Project: Discussion

With the help of an illustrative project called NatureMONITOR, it is highlighted some of the advantages of tailoring the WSN design to the environmental monitoring application in comparison of simply adopting traditional WSN options. Nonetheless, there are tradeoffs to be considered. We summarize this high-level comparison in Table 4, where the architecture behind NatureMONITOR is compared with a commercial, ZigBee, and TinyOS-based solutions.

An interesting network metrics, called AIND-ANON, is presented in this chapter. The metrics can be used as a qualitative tool to evaluate the expected challenges associated with the use of multi-hopping protocols giving the topology of the nodes and their expected communication range. The proposed metrics are not complete but provide a formal way to define sparsity level in WSNs. Moreover, when a network is found to be sparser in this metrics, it is expected significant network issues. In the case of environmental monitoring systems, three solutions are suggested: (1) simply increase the network density by adding more nodes typically functioning as relay-nodes, (2) apply some sort of in-network aggregation in order to reduce the average number of hops required by the network, or (3) apply some of network segmentation. For the options (2) and (3), one can propose certain level of customization of the WSN design.

When a node topology for the NatureMONITOR project is analyzed under the AIND-ANON metrics, it is revealed that the network is very sparse and potential network issues are expected with traditional WSN multi-hopping solution are employed without a careful study. For this scenario, the network segmentation seems to be a good option. Moreover, considering the strict requirements of the project, a discussion about customization of the solution evolves. The idea of customizing a WSN design is not a novel approach. For instance, WSNs to be used in a human body are routinely tailored to the specific requirements of body-area networks (BANs). Additional examples are wireless underwater sensor networks [31] and wireless underground sensor network [39], both considered in this book.

For our illustrative large-scale and sparse monitoring application (NatureMONITOR), the proposed solution is highly flexible to future changes requested by environmental scientists or end users. For instance, based on data analysis from the previous year, an end user may decide to reduce the number of sensor nodes in certain areas with historically very similar results and spread them in areas where no

Table 4 Customized architecture (NatureMONITOR) compared to traditional WSN solutions

Aspect	NatureMONITOR	ZigBee PRO [9, 10, 16, 27]
Expected functionality	Appropriate for the project	Potential feasibility for multiple topologies assuming a significant number of routers (e.g., >1 per 10 nodes)
Cost	Appropriate for the project due to the network segmentation and open use of different wireless technologies	High deployment and operating costs. Just one network protocol is used and long-range communication must be achieved by multi-hopping and a high number of router devices
Reliability (and feasibility of accurate performance metrics)	Appropriate for the project assuming high-degree of reliability for the master nodes	*Uncertainty* due to inexistency of similar scenario in real-world
	The performance metrics control is achieved	Many of the requested performance metrics cannot be achieved with high accuracy
Scalability	Highly scalable due to the network segmentation and asynchronous behavior	Very good node-density scalability. *Reasonable* spatial scalability. Problems occur in very sparse networks
Aspect	TelosB+TinyOS [24, 43]	eKo solution [26]
Expected functionality	Potential *unfeasibility* for some topologies if protocols and design do not consider a sparser WSN. Potential use of multiple sink nodes	Potential *unfeasibility* for some topologies (maximum hop-distance is 5 for this solution). Multiple replicas of the solution are expected
Cost	Potential *unfeasibility* for some topologies if protocols and design do not consider a sparser WSN. Potential use of multiple sink nodes	Deployment cost far beyond the budget. Operating costs similar to TelosB case. In case of areas without sufficient solar radiation for extended periods of time, the rechargeable batteries must be changed frequently
Reliability (and feasibility of accurate performance metrics)	*Uncertainty* (same as the ZigBee case)	*Uncertainty* (same as the ZigBee case)
Scalability	Very good node-density scalability. *Uncertainty* related to the spatial scalability	Very good node-density scalability. *Uncertainty* related to the spatial scalability

sensor was previously deployed or in areas that require more resolution by means of a higher node-density. NatureMONITOR adapts and scales (number of nodes and space) nicely in this scenario. However, we still believe that this solution is not the unique solution. Different architectures can still satisfy the requirements while giving more emphasis on one of the three main design challenges discussed in this chapter: cost, reliability, or scalability.

It is not rare to see network projects (not only related to WSN) that start with a certain target related to the coverage area. Nonetheless, as the network grows, issues potentially rise and the deployment stops when the project reaches its budget. It is possible that the project never reaches the initial target in terms of scalability. In short, this is the problem with this design challenge (scalability): it is usually taken for granted. However, if the scalability aspect is analyzed since the beginning of the project, many constraints are properly imposed to the project. For instance, the designer can figure out that cheaper WSN platforms will not scale. Other more expensive platform claims to properly scale, but there is no evidence that it can achieve the level of expected reliability and tests must be realized. Even if the new solution scales and is reliable, its energy consumption can still be too high implying the need of frequent battery exchanges (higher cost). One can suggest the use of solar panels (or other form of energy harvester) but new problems can potentially come to the scene. Therefore, the real hidden challenge in designing large environmental monitoring systems is to achieve a balance involving functionality, cost, reliability, and scalability. In some cases, the solution comes in the form of a customized design as with the NatureMONITOR case study. However, other techniques can be employed as discussed next.

Among the architectural aspects of the environmental monitoring application discussed in this chapter, dynamic scheduling was mentioned as one of the project requirements of the case study. In fact, the potential advantages of using this mechanism can significantly improve the energy performance of large and sparse WSN networks. For instance, an adaptive scheduling has been proposed as a solution for similar scenario [15, 22, 27]. Similarly, aggregation and compressive sensing [42] can be properly exploited in the NatureMONITOR project. Therefore, instead of following the segmentation and customization approaches, one can propose some of the mentioned techniques to highly mitigate the expected issues in large and sparse deployments.

4.2 NatureMONITOR Project: Preliminary Results

Although the NatureMONITOR project was introduced in this chapter as a hypothetical one, it was actually implemented in real-world deployments by means of an architecture called Ripple-2 [38]. The requirements of Ripple-2 are, in fact, a subset of the NatureMONITOR project but with a particular focus on soil moisture and temperature. We are still working on this project to implement all the strict network management requirements mentioned in this chapter. Nonetheless, the majority of the discussions related to NatureMONITOR also correspond to Ripple-2. For instance, the numbers behind the discussion involving the lifetime of the nodes and application duty-cycle in Sect. 2.2 come from our work with Ripple-2 [38].

The networks for the cases SoilSCAPE I and SoilSCAPE II, analyzed at Sect. 1.3.2, were finally converted to the Ripple-2 architecture. It is important to highlight that SoilSCAPE II was initially deployed favoring the segmentation approach (ZigBee network). However, it still presented scalability issues even with few hops involved. Moreover, a significant number of ZigBee routers are necessary to cover the area. Because these special nodes cannot sleep, their energy requirements are a challenge. In addition, we had many issues related to solar panels and rechargeable batteries associated with extreme temperatures, in particular subzero temperatures [37].

The Ripple-2 architecture gives a strong emphasis on non-rechargeable batteries and, accordingly, we work on hardware and software solutions to enhance the lifetime of such batteries in one more folds in relation to the current technology [37]. Therefore, we definitely followed the customization track discussed in this chapter. The solution comes as a software and hardware overlay [38], meaning that it can work on top of many WSN platforms. Because environmental monitoring systems typically have a very low duty-cycle, the concept of hibernation can be exploited. During this state, the inactive modules in the node are not simply in sleep or standby mode, they are turned-off. However, to implement this solution, we had to develop a cross-layer protocol. The effort paid off when we confirmed that, on the average, the additional effective network overhead is smaller than 1 %. For low duty-cycled applications, this result is very important in terms of energy efficiency [38]. Moreover, the effective packet loss rate is consistently below 2 % in all sites.

Up to the date, we deployed 105 nodes in 5 different sites with an average of 1 node per 5,000 km^2. No *extra* nodes, such as repeaters or routers, were used in these sites and the master node (Ripple-2 Local Coordinator) is actually also a sensor node which can also hibernate. So far, all SoilSCAPE nodes are based on 802.15.4 (2.4 GHz) transceivers, 17 dBm transmit power. The energy and network performance associated with the solution are potentially above the average in comparison with current WSN protocols [38, 42]. We are currently working to integrate the solution to ISM 900 MHz radio modules and a well-known WSN platform. The drawbacks of the architecture are exactly the ones discussed in this chapter: high data-latency and low throughput. In short, the network performance is traded for excellent energy performance. However, the low-cost, relative high reliability, and high scalability goals are achieved.

4.3 Conclusions

Although the expressions ah-doc, collaboration, and multi-hopping are constantly mentioned in the WSN literature, we believe that these features are systematically overvalued. Despites their high value in military applications and academic researches, a significant number of real-world applications can be properly addressed with basic networking functionalities. The recent industrial trend toward 802.15.4 in star-topology and ZigBee standards is strong evidence that simple solutions can be enough for many scenarios.

The existence of a significant number of solutions related to WSN is an indicative that hybrid solutions mixing state-of-the-art technology with some level of customization can be a promising direction for many applications, in particular low data-rate environmental monitoring systems. Nonetheless, as time goes by, new technological options and standards allow us to return to a more conservative, flexible, and generic solution avoiding the extra cost (and risks) of customizing a solution. However, as new challenges eventually appear according to the current demand, the specialization of the solution becomes again an appealing option. And this cycle repeats and not only for WSNs.

References

1. I.F. Akyildiz et al., A survey on sensor networks. IEEE Commun. Mag. **40**(8), 102–114 (2002)
2. G. Anastasi et al., Reliable and energy-efficient data collection in sparse sensor networks with mobile elements. Perform. Eval. **66**(12), 791–810 (2009)
3. A. Arora et al., Exscal: Elements of an extreme scale wireless sensor network. in *Proceedings of IEEE RTCSA*, 2005, pp. 102–108
4. A. Arora et al., A line in the sand: a wireless sensor network for target detection, classification, and tracking. Comput. Netw. **46**(5), 605–634 (2004)
5. G. Barrenetxea et al., Sensorscope: Out-of-the-box environmental monitoring. in *Proceedings of IEEE IPSN*, 2008, pp. 332–343
6. R. Beckwith et al., Unwired wine: sensor networks in vineyards. in *Proceedings of IEEE Sensors*, 2004, pp. 561–564
7. J. Beutel et al., PermaDAQ: A scientific instrument for precision sensing and data recovery in environmental extremes. in *Proceedings of IEEE IPSN'09*, 2009, pp. 265–276
8. E.S. Biagioni, K.W. Bridges, The application of remote sensor technology to assist the recovery of rare and endangered species. Int. J. High Perform. Comput. Appl. **16**(3), 315–324(2002)
9. H.R. Bogena et al., Potential of wireless sensor networks for measuring soil water content variability. Vadose Zone J. **9**, 1002–1013 (2010)
10. C. Buratti et al., An overview on wireless sensor networks technology and evolution. Sensors **9**, 6869–6896 (2009)
11. DASH7 Alliance, http://www.dash7.org Accessed 23 April 2012
12. P. Dutta et al., Trio enabling sustainable and scalable outdoor wireless sensor network deployments. in *Proceedings of IEEE IPSN'06*, 2006, pp. 407–415
13. S.A. Ellwood et al., in *Key Topics in Conservation Biology*, ed. by D. Macdonald, K. Service. Technology in Conservation: A Boon but with Small Print (Blackwell Publishing, Oxford, 2007), pp. 105–119
14. S. Farahani, *ZigBee Wireless Networks and Transceivers* (Elsevier, Burlington, 2008)
15. D. Ganesan et al., Networking issues in wireless sensor networks. J. Parallel Distrib. Comput. **64**, 799–814 (2003)
16. N. Hunn, *Essentials of Short-Range Wireless* (The Cambridge University Press, Cambridge, UK, 2010)
17. J.M. Kahn et al., Next century challenges: mobile networking for smart dust. in *Proceedings of MobiCom '99*, (Seattle, WA, 1999)
18. A. Kansal et al., Power management in energy harvesting sensor networks. ACM Trans. Embedded Comput. Syst. **6**(4), 1–38 (2007)

19. S. Kim et al., Health monitoring of civil infrastructures using wireless sensor networks. in *Proceedings of IEEE IPSN*, 2007, pp. 254–263
20. R. Kuntz et al., Medium access control facing the reality of WSN deployments. ACM Comput. Commun. Rev. **39**(3) 2009
21. X.F. Li et al., A differential evolution-based routing algorithm for environmental monitoring wireless sensor networks. Sens. J. **10**(6), 5425–5442 (2010)
22. J.C. Lim, C. Bleakley, Adaptive WSN scheduling for lifetime extension in environmental monitoring applications. Int. J. Distrib. Sens. Netw. (2012). doi:10.1155/2012/286981
23. J.C. Lim, C.J. Bleakley, Adaptive WSN scheduling for lifetime extension in environmental monitoring applications. Int. J. Distrib. Sens. Netw. **2012** (2012) article ID 286981
24. Y. Liu et al., Does wireless sensor network scale? a measurement study on greenOrbs. in *Proceedings of IEEE INFOCOM '11*, Shanghai, China, 2011
25. K. Lu et al., Wireless sensor networks for environmental monitoring applications: a design framework. in *Proceedings of GLOBECOM*, 2007, pp. 1108–1112
26. Memsic Corp. ēKo Pro series system, http://www.memsic.com/products/wireless-sensor-networks/environmental-systems.html. Accessed 23 April 2012
27. M. Moghaddam et al., A wireless soil moisture smart sensor web using physics-based optimal control: concept and initial demonstration. IEEE-JSTARS **3**(4), 522–535 (2010)
28. M. Moghaddam et al., Ground network design and dynamic operation for validation of spaceborne soil moisture measurements: initial developments and results. in *Proceedings of ESTF-2010*, 2010
29. A. Nasipuri et al., Design considerations for a large-scale wireless sensor network for substation monitoring. in *SenseApp*, 2010, pp. 882–889
30. A. Nasipuri et al., Wireless sensor network for substation monitoring: Design and deployment. in *Proceedings of SenSys, Demo Session*, 2008, pp. 365–366
31. D. Pompili, T. Melodia, An architecture for ocean bottom underWater acoustic sensor networks. Poster Presentation, in *Proceedings of Mediterranean Ad Hoc Networking Workshop (Med-Hoc-Net)*, (Bodrum, Turkey, 2004)
32. V. Rajendran et al., Energy-efficient collision-free medium access control for wireless sensor networks. in *Proceedings of 1st International Conference on Embedded Networked Sensor Systems*, 2003, pp. 22–27. doi:10.1145/958491.958513
33. B. Raman, K. Chebrolu, Censor networks: a critique of sensor networks from a systems perspective. ACM SIGCOMM Comput. Commun. Rev. (2008), pp. 22–27. doi:10.1145/1384609.1384618
34. I. Rhee et al., Z-MAC: a hybrid MAC for wireless sensor networks. in *Proceedings of 3rd International Conference on Embedded Networked Sensor Systems*, (2004). doi:10.1145/1098918.1098929
35. L. Selavo et al., LUSTER: wireless sensor network for environmental research. in *Proceedings of 5th International Conference on Embedded Networked Sensor Systems*, 2007. doi:10.1145/1322263.1322274
36. R.C. Shah et al., Data MULEs: Modeling a three-tier architecture for sparse sensor networks. in *Proceedings of IEEE SNPA*, 2003, pp. 30–40
37. A. Silva, M. Liu, M. Moghaddam, Power management techniques for WSNs and similar low-power communication devices based on non-rechargeable batteries. J. Comput. Netw. Commun. 1–10 (2012)
38. A. Silva, M. Liu, M. Moghaddam, Ripple-2: a non-collaborative; asynchronous; and open architecture for highly-scalable and low duty-cycle WSNs. in *Proceedings of ACM MiSeNet*, (Istanbul, Turkey, 2012), pp. 39–44
39. A.R. Silva, M.C. Vuran, $(CPS)^2$: integration of center pivot systems with wireless underground sensor networks for autonomous precision agriculture. in *Proceedings of ACM/IEEE International Conference on Cyber-physical Systems (ICCPS '10)*, (Stockholm, Sweden, 2010)
40. A. Somasundara et al., Controllably mobile infrastructure for low energy embedded networks. IEEE Trans. Mob. Comput. **6**(8), 958–973 (2006)

41. G. Werner-Allen et al., Fidelity and yield in a volcano monitoring sensor network. in *Proceedings of the 7th USENIX Symposium on Operating Systems Design and Implementation (OSDI)*, 2006, pp. 381–396
42. X. Wu, M. Liu, In-Situ soil moisture sensing: measurement scheduling and estimation using compressive sensing, in *Proceedings of IEEE/ACM IPSN*, Beijing, China, April 2012
43. M. Zuniga, B. Krishnamachari, Analyzing the transitional region in low power wireless links. in *Proceedings of IEEE SECON '04*, vol. 2012 (Santa Clara, CA, 2004)

Part II
Wireless Communications and Medium Access Control

Chapter 4
Physical Layer Communications in Wireless Sensor Networks

Zhuo Li, Xin Wang and Qilian Liang

Abstract Wireless Sensor Networks (WSNs) are widely used in applications such as event detection, environment monitoring, target tracking, and home automation. Typically, manually deployed sensors measure environmental information, e.g., temperature, image, and wind speed, etc., and transmit noisy versions of these measurements over wireless fading channels to the local intelligent control center (ICC), in which an estimation of these state measurements is obtained and thus related control operations are conducted.

This chapter investigates physical layer communications of WSNs in two parts:

(1) Impulse Radio Ultra-Wideband (IR-UWB) capacity optimization under the coexistence with the OFDM-based wireless communication system—IEEE 802.11n. The optimal power allocation scheme is presented by using the water-filling algorithm with Karush-Kuhn-Tucker (KKT) conditions, which is also compared with a traditional equal power allocation scheme;

(2) sensor selection schemes for the parameter estimation in energy-constrained WSNs. To prolong the network lifetime and optimize the power consumption, only sensors experiencing favorable conditions will participate in the estimation process. And two sensor selection schemes are proposed to improve the estimation performance.

Partial work in this chapter was present in The 11th IEEE International Symposium on Communications and Information Technologies (ISCIT2011) as "capacity optimization of ultrawide band system under the coexistence with IEEE 802.11n system" and 2012 IEEE Innovative Smart Grid Technologies—Asia (ISGT Asia) as "Sensor selection schemes in smart grid."

Z. Li · X. Wang · Q. Liang (✉)
Department of Electrical Engineering, University of Texas at Arlington,
Arlington, TX 76019-0016, USA
e-mail: liang@uta.edu

Z. Li
e-mail: zhuo.li@mavs.uta.edu

X. Wang
e-mail: xin.wang51@mavs.uta.edu

H. M. Ammari (ed.), *The Art of Wireless Sensor Networks*,
Signals and Communication Technology, DOI: 10.1007/978-3-642-40009-4_4,
© Springer-Verlag Berlin Heidelberg 2014

1 Introduction

WSNs have been rapidly developed in recent years. Consisting of wireless interconnected sensor nodes, WSNs collaboratively collect, deliver and process information in various environmental monitoring, and commercial or human centric applications. The development of WSNs has attracted growing research on the relative technologies. An important issue of WSNs is energy efficiency. To optimize the throughput between the wireless sensors with stringent power constraint is of great significance to the robustness as well as the reliability of WSNs. At the mean time, if a sensor remains active continuously, its energy will be depleted quickly leading to death. To prolong the network lifetime, the sensors need alternate between being active and sleeping, which means an efficient sensor selection scheme is required of power-restricted WSNs.

In this chapter, we try to approach this from the physical layer design perspective in terms of the following three objectives:

(1) *Communication capacity*, while coexisting with other wireless users, the throughput of the WSN should be optimized with an efficient power allocation scheme;

(2) *Estimation accuracy*, estimation accuracy determines if a control center could perform reasonable operations based on the real-time measurements collected from the monitoring sensors;

(3) *Power efficiency*, to prolong the lifetime of the WSN, the total power consumption should be minimized.

The ultra wideband (UWB) radio is a promising technology that can be applied at very low power for short-range high data-rate communications. In light of its main advantages of high data rates, low cost, and low power consumption, UWB systems tend to be used for short-range and indoor applications, examples of which are communications between wireless monitors, digital camcorders, wireless printing, cell phones and personal computers [37]. Besides, owing to its precision capabilities combined with the extremely low power, the UWB radio is ideal for certain frequency sensitive environments such as hospitals and health care. It has also been used in military applications for covert communications. In this application, the military took advantage of the fact that UWB signals spread across a very wide bandwidth and could be made to appear as noise to most interception equipment [34]. UWB is part of "see-through-the-wall" precision radar imaging technology, precision locating and tracking. Moreover, UWB technology is an ideal candidate for future Wireless Personal Area Networks (WPAN) that requires processing information with low-power sources at very high speed across short distances [2]. Numerical studies indicate that UWB is one of the enabling technologies for sensor network applications. Short range communication systems could be applied to the fields of Smart Grid [39], which includes an intelligent monitoring system that would keep track of electrical power distribution systems [4].

In the existing literature, the coexistence issue of UWB systems and other wireless communication systems, such as cellular systems and NB wireless systems, is widely studied. The investigation is mainly categorized into two parts: one is to study the interference effect from other conventional systems on the UWB system, and the other one is to analyze how UWB signals could impact other coexisting systems. Particularly, previous literature focuses on the interference introduced by UWB signals to fixed wireless communications systems, for example, the global system for mobile communications (GSM), universal mobile telecommunication system (UMTS), and global positioning system (GPS) [15, 17]. Later on, the coexistence issue between UWB systems and NB communication systems attracted researchers' attention due to the wide application of NB communication systems. Giorgetti et al. Reference [14] evaluated the performance of wideband communication systems in the presence of narrowband interference (NBI). In [14] closed-form bit-error probability expressions for spread-spectrum UWB systems were derived under the additive white Gaussian noise (AWGN) channels, flat-fading channels, and frequency-selective multipath fading channels, respectively, where the NBI was considered as a tone interferer. Furthermore, coexistence between UWB and Orthogonal-Frequency-Division-Multiplexing (OFDM)-based systems was thoroughly analyzed in [8].

Considering the operating frequencies of IEEE 802.11n, an IEEE 802.11n system with either operating mode could be interfered by a UWB user, and the interference of an IEEE 802.11n system to a UWB user is also unavoidable. The coexistence issue of UWB and IEEE 802.11n systems has been investigated since the standardization of IEEE 802.11n. Reference [26] studied the maximum permissible emission power of a UWB system under the coexistence with IEEE 802.11n by setting up the physical layer models of these two systems. The spectrum sensing performance of UWB-based Cognitive Radio (CR) systems with the primary user of IEEE 802.11n was presented in [22]. Based on these studies, the topic of how to optimize the capacity of a UWB user while making sure that these two systems can still work simultaneously is seldom investigated. Therefore, a study on the proper spectrum management of UWB systems which optimizes its transmission rate, seems to be needed. Besides, this research will give a suggestion on the spectrum regulation of UWB system.

Throughput optimization for wireless communication systems has been widely studied. Some of the existing literatures focus on the coexistence situation of two systems. Reference [38] proposed a water-filling algorithm for a direct-sequence (DS) UWB cognitive radio (CR) network that maximizes the sum capacity while enabling each transmitter to fit its power spectral density into, and thus to make the most of, the spectrum void. Bansal et al. investigated an optimal power loading algorithm for an OFDM-based cognitive radio system [1]. The downlink transmission capacity of the CR user is thereby maximized, while the interference introduced to the primary user (PU) remains within a tolerable range. The non-convex NP-hard problem of weighted sum rate maximization in a multiuser Gaussian channel that models a cognitive wireless network with affine power constrains was studied in [31]. The key technique is the use of nonnegative matrix theory, in particular the Perron-Frobenius Theorem and the Friedland-Karlin inequalities. Reference [31] also extends to a multiuser-multiple carrier model, where a common spectrum is divided into K frequency tones. Reference [7] proposed a margin-based power allocation scheme that utilizes each UWB node's own position information, and an exclusive region-based scheduling scheme that takes into consideration the interaction among simultaneous transmission links.

The sensor selection problem already arises in various applications, including sensor placement for structures [24] and target tracking [33]. The work of [30] investigated the full sensing coverage of the field by identifying the appropriate sensors and turning off the redundant sensors. Sensor selection via convex optimization is discussed in [20] and a survey of sensor selection schemes in wireless sensor networks is summarized in [29].

Section 2 studies the coexistence issue between an WSN and a narrowband (NB) OFDM-based wireless communications system. The throughput optimization of wireless sensors within the coexisting operating bandwidth with an IEEE 802.11n user is explored, while making sure that its cumulative interference to the coexisting IEEE 802.11n user is below a certain permissible threshold. The optimal power allocation scheme is presented by using water-filling algorithm with Karush-Kuhn-Tucker (KKT) conditions, which is also compared with a traditional equal power allocation scheme.

In Sect. 3, we consider the sensor selection schemes to improve estimation accuracy and power efficiency in energy-constrained WSNs. To prolong the network lifetime and optimize power consumption, only sensors experiencing favorable conditions will participate in the estimation process. Two sensor selection schemes are proposed to improve the estimation performance. We first propose an opportunistic sensor selection scheme under equal power allocation and investigate the asymptotic behaviors. Then we address a sensor selection scheme under optimal power allocation and derive a reminiscent "water-filling" solution for this scenario. Finally we present numerical studies on improving the power efficiency using sensor selection. The theoretical analysis and proofs are instrumental to the sensor network design.

2 Coexistence of UWB and IEEE 802.11n: Throughput Optimization for IR-UWB

2.1 Study on Coexistence Between UWB and NB Systems

In principle, any wireless communication technology that produces signals with a bandwidth wider than 500 MHz or a fractional bandwidth greater than 0.2 can be considered as UWB [18], where the fractional bandwidth is defined as

$$\eta = 2 \times \frac{f_H - f_L}{f_H + f_L} \tag{1}$$

where f_L and f_H are the lower bound and upper bound of the spectral frequency of a UWB radio.

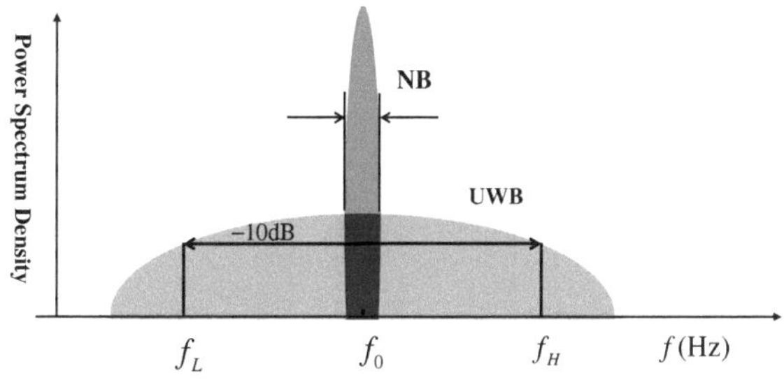

Fig. 1 Spectrum comparison of UWB and NB signals

The Federal Communication Commission (FCC) released a spectral mask with some restrictions for UWB. For indoor UWB systems, a maximum mean effective isotropic radiated power (EIRP) spectrum density of −41.3 dBm/MHz is established over the 3.1–10.6 GHz operating bandwidth [12]. Different from the NB wireless communications systems, i.e., Wireless Local Area Network (WLAN), Worldwide Interoperability for Microwave Access (WiMAX), and Long Time Evolution (LTE), UWB systems conventionally occupy a relatively large bandwidth in the frequency domain which is already allocated by these NB communication systems as illustrated in Fig. 1. Therefore, it is obvious that UWB systems would inevitably coexist with other NB wireless communication systems by sharing some operation bandwidth [26].

2.2 UWB Radio

Two types of UWB technologies are commonly used: impulse radio UWB (IR-UWB) and multiband OFDM UWB (MB-OFDM UWB). Due to the unique properties, such as low complexity and low cost, good resistance to severe multipath and jamming, and high time-domain resolution for location and tracking, IR-UWB is well suited to sensor network applications. In this subsection, we adopt IR-UWB as the objective. IR-UWB uses a short pulse in the time domain that occupies a large bandwidth in the frequency domain to modulate the information [3]. The most commonly used pulse is Gaussian pulse and its derivatives, i.e., first and second derivatives of Gaussian pulse. The normalized Gaussian pulse is represented as

$$p(t) = \frac{1}{\sqrt{2\pi}\sigma} \exp\left(-\frac{(t-\mu)^2}{(2\sigma)^2}\right), \tag{2}$$

where μ is the mean value of the Gaussian random variable, and σ is its standard deviation. Let $\sigma^2 = \alpha^2/(4\pi)$, where α is an index correlated to the width of pulse, indicating the shape of pulse, namely the shaping factor. Figure 2 shows a normalized Gaussian Impulse with $\mu = 0$, and $\sigma = 0.1$.

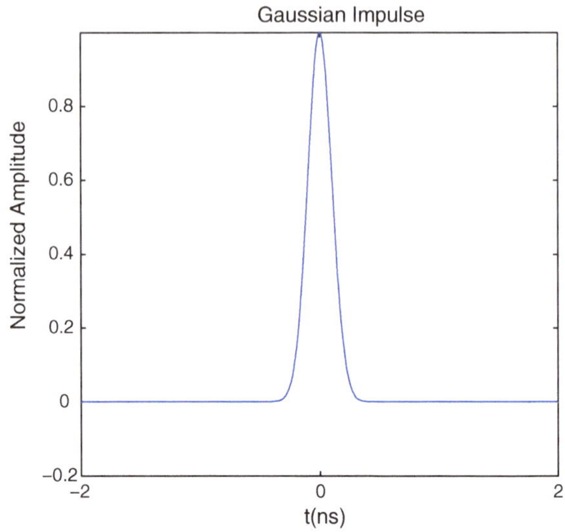

Fig. 2 Normalized Gaussian impulse

2.3 IEEE 802.11n

The Orthogonal Frequency Division Multiplexing (OFDM) technique is adopted in IEEE 802.11n systems, and it is essentially a multi-carrier modulation (MCM) method. The basic idea is to carry data in a large number of closely-spaced orthogonal sub-carriers at a relatively low symbol rate. OFDM is used to cope with severe channel conditions (for example, frequency selective fading due to multipath), to effectively eliminate inter-symbol interference (ISI), and also to reduce the overall amount of required spectrum due to the overlapping of the tones [28]. For the traditional FDM multicarrier modulation technique, as shown in Fig. 3a, each subcarrier in the frequency domain does not overlap with each other. At the same time, in order to reduce mutual interference between the various subcarriers, subcarriers need to retain sufficient frequency spacing, which results in low spectrum efficiency; But for the OFDM technique, due to the orthogonal overlapping between subcarriers, the protection bandwidth is greatly reduced and the spectrum efficiency is also highly improved as shown in Fig. 3b.

A typical OFDM system model is shown in Fig. 4. In the transmitter, the input serial data stream is shifted into a parallel format. The parallel data in each carrier is then separately modulated by traditional modulation methods, such as QPSK and QAM. After the required spectrum is worked out, an inverse fast fourier transform (IFFT) is performed, and the guard period, also called the cyclic prefix (CP) is added to the start of each symbol. The receiver basically does the reverse operation to the transmitter. CP is removed and fast fourier transform (FFT) of each OFDM symbol is then taken to obtain the original transmitted spectrum. Demodulation is performed in each carrier, which is followed by a parallel to serial conversion [25].

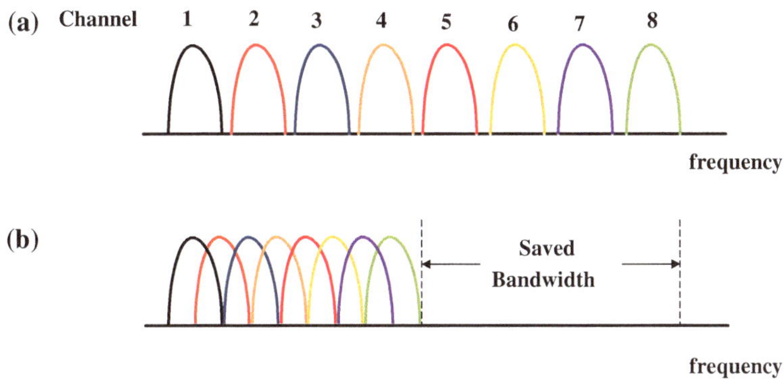

Fig. 3 Traditional FDM and OFDM techniques

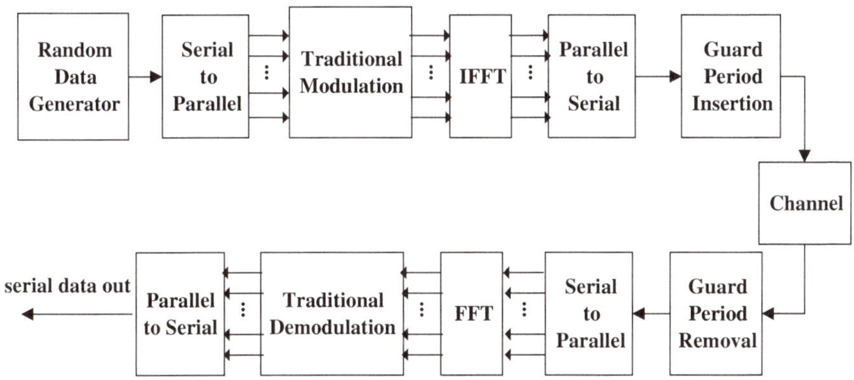

Fig. 4 A typical OFDM system model

The new Wireless Local Area Network (WLAN) standard, IEEE 802.11n, was finalized as a standard in 2009. This standard aims at improving network throughput over previous standards, such as 802.11b and 802.11g with the combination of OFDM and Multiple Input and Multiple Output (MIMO) techniques. Moreover, either the 2.4 GHz or the 5 GHz frequency band could be used for IEEE 802.11n. In the 5 GHz band, a high throughput (HT) OFDM system with 40 MHz bandwidth is specified [19]. Products with IEEE 802.11n technology, such as wireless LAN card, and wireless routers, have been widely used in personal computers, notebook computers and other digital terminals prior to the finalization of the official standard.

Since OFDM technique is used in IEEE 802.11n signals, based on an ideal Nyquist pulse, the power spectrum density (PSD) of the lth subcarrier in the IEEE 802.11n user is represented as

$$\Phi_l(f) = Q_l T_s \left(\frac{\sin \pi (f - f_l) T_s}{\pi (f - f_l) T_s} \right)^2, \tag{3}$$

where Q_l is the transmit power in the lth subcarrier, and T_s is the symbol duration.

2.4 Coexistence System Model

2.4.1 Distribution Model in the Spatial Domain

Before we study the throughput optimization issue of a UWB system under the coexistence with an IEEE 802.11n system, the coexistence model of these two systems in the spatial domain is set up. In the downlink transmission scenario, as shown in Fig. 5, the receiver of an IEEE 802.11n user could get the desired signal from its transmitter, and also get an interference signal from the coexisting UWB transmitter. Similarly, not only the useful information from the UWB transmitter, but also the jammer from the IEEE 802.11n transmitter is obtained at the receiving part of the UWB user. In this scenario, h_{ii} is denoted as the channel gain from the IEEE 802.11n transmitter to its receiver, h_{iu} is the channel gain from the IEEE 802.11n transmitter to the UWB receiver, h_{ui} represents the channel gain from the UWB transmitter to the IEEE 802.11n receiver, and h_{uu} is the channel gain from the UWB transmitter to its receiver. We assume these channel gains are perfectly known at the transmitters.

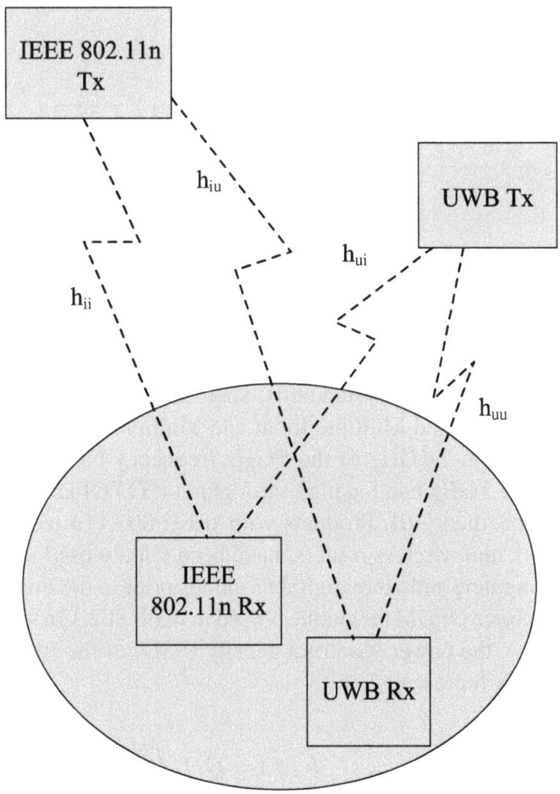

Fig. 5 Distribution of IEEE 802.11n and UWB users in the spatial domain

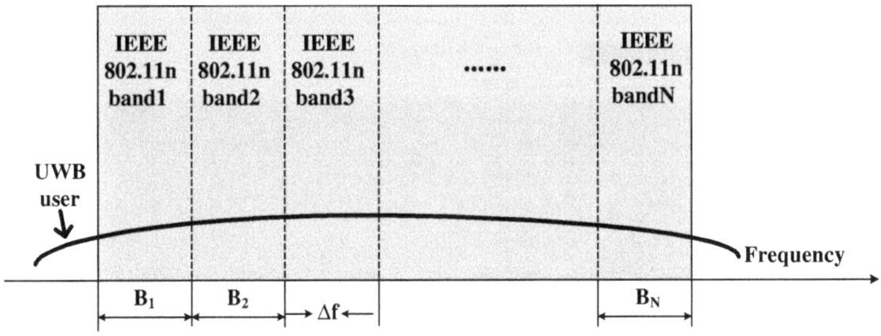

Fig. 6 Distribution of IEEE 802.11n and UWB users in the frequency domain

2.4.2 Distribution Model in the Frequency Domain

In the frequency domain, the available bandwidth of an IEEE 802.11n user can be divided into N subcarriers. It is assumed that the bandwidth of each subcarrier is Δf Hz. For band 1, band 2, ..., band N, the bandwidths are correspondingly $B_1, B_2, \ldots, B_N$. As shown in Fig. 6, a UWB user and an IEEE 802.11n user share the same spectrum in certain frequencies part. In this case, we can study the capacity issue of a UWB user within each subcarrier domain of an IEEE 802.11n user, and then sum them up to get the optimized capacity.

The frequency spectrum of a Time Domain Inc PulsON 220 UWB transceiver via an Agilent Spectrum Analyzer is shown in Fig. 7. It is obvious to see that the frequency spectrum of the UWB transceiver spans in a large range with center frequency 4.28 GHz. Additionally, the power is relatively low.

The frequency spectrum in the Max Hold mode of an OFDM-based NB signal is obtained from a WiFi-equipped cell phone. As shown in Fig. 8, the center frequency is 2.4 GHz with 20 MHz operating bandwidth.

2.5 Interference Model

Our objective is to optimize the capacity of the UWB user within the coexisting operating bandwidth of an IEEE 802.11n user, under the constraint that its interference to the IEEE 802.11n user is under an acceptable threshold. Meanwhile, studying the capacity of the UWB user, the interference at the receiver part of the UWB user should be considered. For simplicity, we study the scenario that the UWB system coexists with only one NB wireless communication system, which is the IEEE 802.11n system. In this way, the received interference of the UWB user includes the additive white Gaussian noise (AWGN) and interference from the coexisting IEEE 802.11n user.

We suppose $\mu = 0$ in (2), so the Power Spectrum Density (PSD) of the UWB user within the nth subcarrier of IEEE 802.11n is [21]:

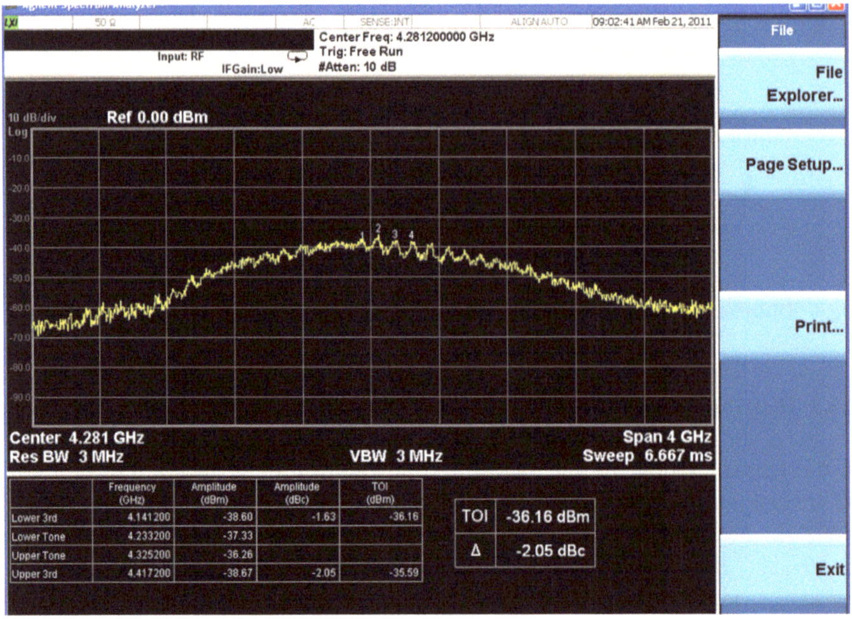

Fig. 7 Frequency spectrum of UWB transceiver

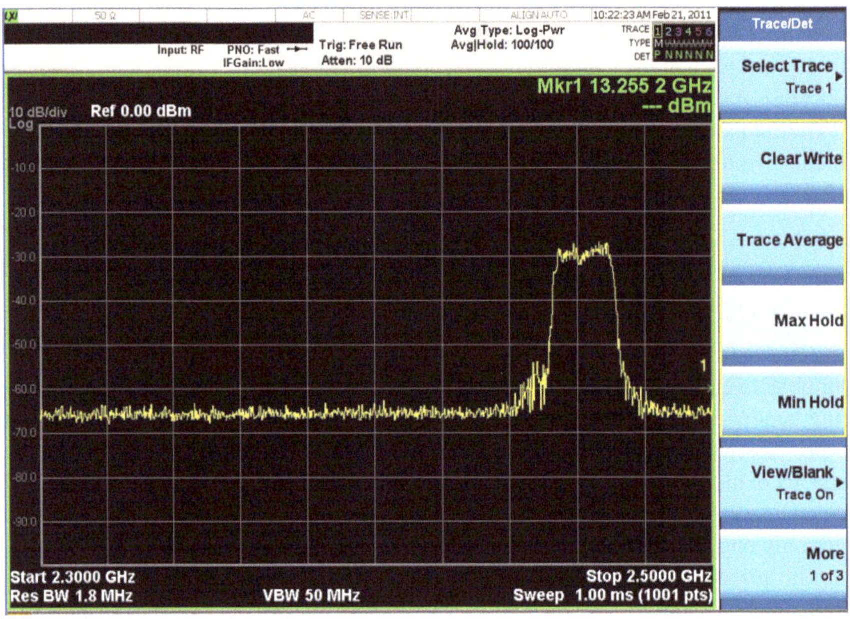

Fig. 8 Frequency spectrum of OFDM-based WiFi user

Table 1 Interference of IEEE 802.11n user $\sum_{l=1}^{N} J_n^{(l)}$

Operating frequency of IEEE 802.11n (GHz)	Bandwidth of IEEE 802.11n (MHz)	Range of transmit power (dBm)	Maximum $\sum_{l=1}^{N} J_n^{(l)}$ (dB)	Minimum $\sum_{l=1}^{N} J_n^{(l)}$ (dB)
2.417–2.437	20	18–21	−78.5308	−78.6660
2.417–2.437	20	15–18	−81.5167	−81.6516
5.170–5.190	20	18–21	−84.9494	−85.1046
5.170–5.190	20	15–18	−87.9597	−88.1149
5.160–5.200	40	18–21	−82.0331	−82.1809
5.160–5.200	40	15–18	−85.1469	−85.2966

$$\Psi_n(f) = P_n \exp\left(-(2\pi f \sigma)^2\right), \tag{4}$$

where P_n is the transmission power of the UWB user within the subcarrier B_n of IEEE 802.11n. In this way, the interference of the UWB user within the nth subcarrier to the lth subcarrier of the IEEE 802.11n user can be represented as

$$I_l^{(n)} = |h_{ui}|^2 \int_{d_{nl}-\Delta f/2}^{d_{nl}+\Delta f/2} \Psi_n(f)df, \tag{5}$$

where d_{nl} is the frequency separation between the nth subcarrier of and the lth subcarrier of the IEEE 802.11n user, and h_{ui} is the downlink channel gain from the UWB transmitter to the IEEE 802.11n receiver as mentioned in Sect. 2.4.

The interference to the UWB user within the nth subcarrier introduced by the lth subcarrier of the IEEE 802.11n can be written as

$$J_n^{(l)} = |h_{iu}|^2 \int_{d_{nl}-\Delta f/2}^{d_{nl}+\Delta f/2} \Phi_l(f)df, \tag{6}$$

where h_{iu} is the downlink channel gain from the IEEE 802.11n transmitter to the UWB receiver as mentioned in Sect. 2.4, and $\Phi_l(f)$ is the PSD of IEEE 802.11n signals within the lth subcarrier as represented in (3).

Table 1 lists the numerical values of IEEE 802.11n interference within the nth subcarrier bandwidth of a UWB user, namely $\sum_{l=1}^{N} J_n^{(l)}$, when the range of transmit power of the IEEE 802.11n user varies.

2.6 Interference Evaluation Based on I/N Criteria

The interference to noise ratio (I/N), which is defined as the power ratio of received interference and receiver noise floor, is a widely used interference evaluation method.

Table 2 Maximum permissible interference at the receiver part of IEEE 802.11n user

Operating frequency of IEEE 802.11n (GHz)	Bandwidth of IEEE 802.11n (MHz)	Ith (dBm)
2.417–2.437	20	−105.62
5.170–5.190	20	−105.62
5.160–5.200	40	−102.61

This method is originally adopted by FCC to regulate the UWB emission limits [12], and provides a simplified model for the calculation of the maximum permissible interference level I_{MAX} at the receiver input for an IEEE 802.11n user. The normalized I_{MAX} in *dBm* [11], can be represented as

$$I\mathrm{th} = I/N + N, \tag{7}$$

where I/N is the maximum permissible average or peak interference-to-noise ratio at the receiver's intermediate frequency (IF) output necessary to maintain an acceptable performance criteria, in *dB*; N is the receiver's inherent noise level at the receiver IF output referred to the receiver input, in *dBm*.

$$N = 10\log K + 10\log T + 10\log B + NF, \tag{8}$$

where K is Boltzmann's constant, normally 1.38×10^{-20}, in mW/K/Hz, T is the system noise temperature, in degrees Kelvin, B is the receiver IF bandwidth, in Hz, and NF is the noise coefficient, in dB.

According to (7) and (8), Table 2 lists the calculated maximum acceptable interference at the receiver part of the IEEE 802.11n user in three different operating modes.

It is obvious that the maximum acceptable interference at the receiver part of IEEE 802.11n is mainly determined by the bandwidth of the receiver, no matter on which center frequency it operates. Moreover, for the interfered system with larger operating bandwidth, higher interference is tolerable than that with smaller operating bandwidth.

2.7 Optimal Power Allocation Scheme

We firstly study the capacity of a UWB user within each operating bandwidth shared with an IEEE 802.11n subcarrier. The transmission power of UWB user within the subcarrier B_n is represented as P_n, h_{uu} is the channel gain of the nth subcarrier from the UWB transmitter to its receiver, $J_n^{(l)}$ denotes the interference introduced by the lth subcarrier of the IEEE 802.11n user to the UWB receiver within the nth subcarrier. According to the Shannon capacity formula, the transmission rate of the UWB user within the nth subcarrier can be represented as [9]:

$$R_n = \Delta f \log_2\left(1 + \frac{|h_{uu}|^2 P_n}{\sigma_{awgn}^2 + \sum_{l=1}^{N} J_n^{(l)}}\right), \tag{9}$$

where σ_{awgn}^2 is the power of the additive white Gaussian noise (AWGN).

As we know the capacity of the UWB user within each subcarrier, the cumulative capacity of the UWB user within the whole coexisting operating bandwidth with the IEEE 802.11n user is essentially the summation of these subcarrier-capacities. Our objective is to maximize the capacity of the UWB user while keeping its interference to the IEEE 802.11n user below a certain threshold.

$$C = \max_{P_n} \sum_{n=1}^{N} \Delta f \log_2\left(1 + \frac{|h_{uu}|^2 P_n}{\sigma_{awgn}^2 + \sum_{l=1}^{N} J_n^{(l)}}\right) \tag{10}$$

subject to

$$\sum_{l=1}^{N}\sum_{n=1}^{N} I_l^{(n)} \leq I\text{th} \tag{11}$$

and

$$P_n \geq 0, \forall n = 1, 2, \ldots, N. \tag{12}$$

Since $\log(\cdot)$ is a concave function, and the sum of the concave functions is still a concave function, to maximize a concave function is equivalent to minimizing the negative of the concave function, namely a convex function. Obviously this is a problem of convex optimization. In this case, by introducing a Lagrange multiplier ν to the inequality constraint in (11) and Lagrange multipliers λ_n to the inequality constraints in (12), we can get the following Karush-Kuhn-Tucker (KKT) condition [6]:

$$\begin{aligned}
&\sum_{l=1}^{N}\sum_{n=1}^{N} I_l^{(n)} \leq I\text{th} \\
&P_n^* \geq 0, \forall n = 1, 2, \ldots, N \\
&\lambda_n \geq 0, \forall n = 1, 2, \ldots, N \\
&\lambda_n P_n^* = 0, \forall n = 1, 2, \ldots, N \\
&-\frac{1}{\frac{\sigma_{awgn}^2 + \sum_{l=1}^{N} J_n^{(l)}}{|h_{uu}|^2} + P_n^*} - \lambda_n + \nu \sum_{l=1}^{N} \frac{\partial I_l^{(n)}}{\partial P_n^*} = 0, \forall n = 1, 2, \ldots, N.
\end{aligned} \tag{13}$$

From the last condition in (13), we know that

$$\lambda_n = \nu \sum_{l=1}^{N} \frac{\partial I_l^{(n)}}{\partial P_n^*} - \frac{1}{\frac{\sigma_{awgn}^2 + \sum_{l=1}^{N} J_n^{(l)}}{|h_{uu}|^2} + P_n^*}. \tag{14}$$

So we can eliminate the slack variable λ_n first and get:

$$
\begin{aligned}
&\sum_{l=1}^{N}\sum_{n=1}^{N} I_l^{(n)} \le I\text{th} \\
&P_n^* \ge 0, \forall n = 1, 2, \ldots, N \\
&\nu P_n^* \sum_{l=1}^{N} \frac{\partial I_l^{(n)}}{P_n^*} - \frac{P_n^*}{\frac{\sigma_{awgn}^2 + \sum_{l=1}^{N} J_n^{(l)}}{|h_{uu}|^2} + P_n^*} = 0, \forall n = 1, 2, \ldots, N \\
&\frac{1}{\left(\frac{\sigma_{awgn}^2 + \sum_{l=1}^{N} J_n^{(l)}}{|h_{uu}|^2} + P_n^*\right)\frac{\partial I_l^{(n)}}{P_n^*}} \le \upsilon, \forall n = 1, 2, \ldots, N.
\end{aligned}
\tag{15}
$$

We can denote $\frac{\partial J_n^{(l)}}{\partial P_n^*}$ as $K_l^{(n)}$ for simplicity, which is

$$K_l^{(n)} = |h_{ui}|^2 \int_{d_{nl}-\Delta f/2}^{d_{nl}+\Delta f/2} \exp\left(-(2\pi f \sigma)^2\right) df. \tag{16}$$

From the last condition in (15), we can see that if

$$\nu < \frac{1}{\left(\frac{\sigma_{awgn}^2 + \sum_{l=1}^{N} J_n^{(l)}}{|h_{uu}|^2}\right) \sum_{l=1}^{N} K_l^{(n)}} \tag{17}$$

then, $P_n^* > 0$. From the third condition in (15), we can get P_n^* as

$$P_n^* = \frac{1}{\nu \sum_{l=1}^{N} K_l^{(n)}} - \frac{\sigma_{awgn}^2 + \sum_{l=1}^{N} J_n^{(l)}}{|h_{uu}|^2}. \tag{18}$$

Otherwise, $P_n^* = 0$ if

$$\nu \ge \frac{1}{\left(\frac{\sigma_{awgn}^2 + \sum_{l=1}^{N} J_n^{(l)}}{|h_{uu}|^2}\right) \sum_{l=1}^{N} K_l^{(n)}}. \tag{19}$$

To summarize, the allocated power of UWB user can be written as

$$P_n^* = \max\left\{0, \frac{1}{\left(\frac{\sigma_{awgn}^2 + \sum_{l=1}^{N} J_n^{(l)}}{|h_{uu}|^2}\right) \sum_{l=1}^{N} K_l^{(n)}}\right\}. \tag{20}$$

Substituting (20) into (11), we can get

$$\sum_{l=1}^{N}\sum_{n=1}^{N} K_l^{(n)} \max\left\{0, \frac{1}{\left(\frac{\sigma_{awgn}^2+\sum_{l=1}^{N} J_n^{(l)}}{|h_{uu}|^2}\right)\sum_{l=1}^{N} K_l^{(n)}}\right\} \leq I\text{th}. \tag{21}$$

From (21), we can see that when the left side is equal to the right side, the left side can reach its maximum value, namely, the allocated power is maximized. In order to maximize the capacity of UWB user, (21) can be written as

$$\sum_{l=1}^{N}\sum_{n=1}^{N} K_l^{(n)} \max\left\{0, \frac{1}{\left(\frac{\sigma_{awgn}^2+\sum_{l=1}^{N} J_n^{(l)}}{|h_{uu}|^2}\right)\sum_{l=1}^{N} K_l^{(n)}}\right\} = I\text{th}. \tag{22}$$

We compare the proposed optimized capacity of the UWB user with the capacity by using the equal power allocation scheme [32] in three scenarios mentioned above. For the equal power allocation scheme, the allocated power for the UWB user within each subcarrier shared with subcarriers of IEEE 802.11n user P_{eq} can be represented as:

$$P_{eq} = \frac{I\text{th}}{\sum_{l=1}^{N}\sum_{n=1}^{N} K_l^{(n)}}. \tag{23}$$

The capacity of the UWB user with equal power allocation C_{eq} is:

$$C_{eq} = \sum_{n=1}^{N} \Delta f \log_2\left(1 + \frac{|h_{uu}|^2 P_{eq}}{\sigma_{awgn}^2 + \sum_{l=1}^{N} J_n^{(l)}}\right). \tag{24}$$

2.8 Numerical Simulation and Discussion

The maximum transmit power of an IEEE 802.11n user is 200 mW, namely 23 dBm both in 20 MHz and 40 MHz [19]. We can assume the transmit power of an IEEE 802.11n user Q is within the range of 64–128 mW (18–21 dBm) for indoor applications. There are 64 OFDM subcarriers in the mode of 20 MHz and 128 subcarriers with 40 MHz bandwidth. Dividing the number of subcarriers by the total bandwidth, we can get the subcarrier frequency spacing 312.5 KHz. The symbol duration of the IEEE 802.11n signal is 4 μs. According to the ITU-R requirements, the interference power due to unwanted emissions from sources sharing the same band on primary bases can be partitioned as the intraservice interference, the coprimary services interference, and other interfering aggregation. UWB systems are commonly classified as secondary services. For calculating the maximum permissible interference at the receiver of the IEEE 802.11n user, system noise temperature is 293 K, the noise

Table 3 Parameter settings

Parameter	Symbol	Value
Transmit power of IEEE 802.11n user	Q	64–128 mW (18–21 dBm)
Number of subcarriers	N	64/128
Subcarrier frequency spacing	Δf	312.5 KHz
Symbol duration of IEEE 802.11n signal	T_s	4×10^{-6} s
Power of AWGN	σ_{awgn}^2	4×1^{-3} s
Shaping factor of Gaussian pulse	α	0.2×1^{-9} s
Channel gains	h	10 dB
Boltzmann's constant	K	1.38×10^{-20} mW/K/Hz
System noise temperature	T	293 K
Noise coefficient	NF	5.3 dB
Interference to noise ratio	I/N	−10 dB

coefficient of an IEEE 802.11n receiver is 5.3 dB, and the interference to noise ratio is −10 dBm [15].

In this subsection, we select the operating frequency range of an IEEE 802.11n user as 2417–2437 MHz with 20 MHz bandwidth, 5170–5190 MHz with 20 MHz bandwidth, and 5160–5200 MHz with 40 MHz bandwidth. σ_{awgn}^2 is assumed to be 1×10^{-3}. The shaping factor of the Gaussian pulse is chosen as 0.2×10^{-9}. The channel gains between the IEEE 802.11n user and the UWB user h_{ii}, h_{iu}, h_{ui}, and h_{uu} are assumed to be Rayleigh flat fading with an average power gain of 10 dB. The detailed parameter settings are shown in Table 3.

Figure 9 is the interference introduced by the lth subcarrier of IEEE 802.11n to an UWB user within the nth subcarrier spectrum domain, where the operating frequency of IEEE 80.211n is 2.417–2.437 GHz and the operating bandwidth is 20 MHz. We can see that the interference to UWB changes periodically due to the OFDM characteristic for different distances between the lth subcarrier of IEEE 802.11n and the nth subcarrier of the UWB user in the spectrum domain.

The numerical results are shown in Table 4. From these results, we can see that: (1) the capacities of the UWB user within the coexisting frequencies shared with IEEE 802.11n user, using the proposed optimal power allocation scheme, are significantly larger than those with the equal power allocation scheme. (2) For the IEEE 802.11n user working on a certain frequency, i.e., in the 5 GHz band, the UWB user could achieve larger capacity when coexisting with higher-bandwidth IEEE 802.11n user. This could be explained according to (7), (8) and Table 2. Since the interfered system with larger operating bandwidth can tolerate higher interference, the transmitting power of the interfering system, namely the UWB user, will be larger, which means higher capacity is achievable. (3) Compared with a higher operating frequency of the IEEE 802.11n user, the capacity of the UWB user under the coexisting IEEE 802.11n user with lower operating frequency is larger. (4) When the range of transmit power of IEEE 802.11n varies, the capacity of the UWB user changes. When the range decreases, the corresponding interference introduced by the IEEE 802.11n user to the UWB user gets smaller, that means the UWB user could reach higher capacities.

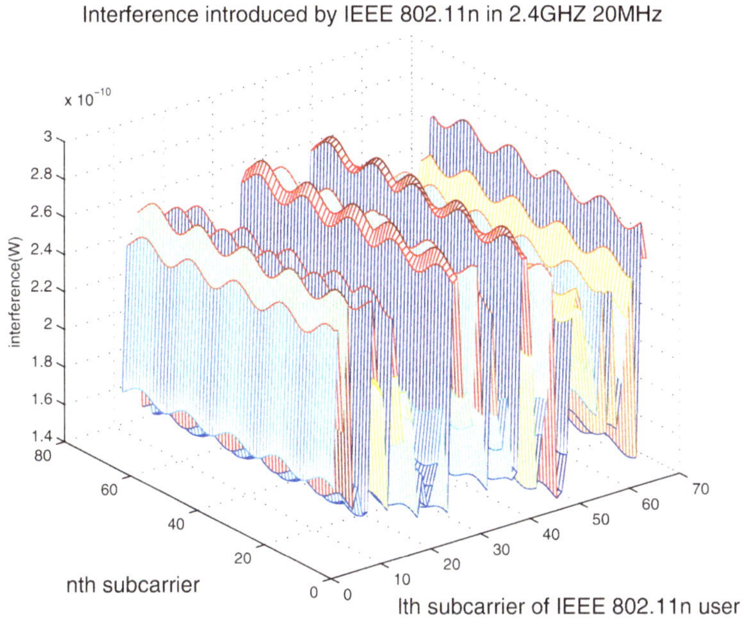

Fig. 9 Interference introduced by the lth subcarrier of IEEE 802.11n to UWB user (in 2.4 GHZ with 20 MHz bandwidth)

Table 4 Optimal capacity and capacity with equal power allocation scheme

Operating frequency of IEEE 802.11n (GHz)	Bandwidth of IEEE 802.11n (MHz)	Range of transmit power of IEEE 802.11n (dBm)	Optimal capacity (bps)	Capacity with equal power allocation scheme (bps)
2.417–2.437	20	18–21	1.22837×10^9	1.8420×10^8
2.417–2.437	20	15–18	1.22845×10^9	1.8427×10^8
5.170–5.190	20	18–21	1.13294×10^9	1.3815×10^8
5.170–5.190	20	15–18	1.13300×10^9	1.3819×10^8
5.160–5.200	40	18–21	1.86853×10^9	2.7631×10^8
5.160–5.200	40	15–18	1.86873×10^9	2.7648×10^8

3 Sensor Optimization and Selection in Wireless Sensor Networks Based on Physical Layer Design

The goal of this subsection is to consider the above points that have not yet been well-defined in previous works. The reminder of this subsection is organized as follows: The system model and problem formulation are given in Sect. 3.1. In Sect. 3.2, aiming at improving the estimation performance, we propose two sensor selection schemes: sensor selection under equal power allocation and optimal power allocation, respectively. Section 3.3 demonstrates the improvements in power efficiency using sensor selection. We present some concluding remarks in Sect. 4. Several Proofs omitted from the body of the subsection are presented in the Appendix.

3.1 System Model and Problem Formulation

3.1.1 System Model

Consider a WSN with m sensors $S_1, S_2, \cdots, S_m$ deployed to estimate an unknown parameter θ, such as the voltage, temperature level and so on. The system model is shown in Fig. 10.

Each sensor S_i is able to periodically measure the source θ. The measurement at each sensor is assumed to be a noise-corrupted version of θ:

$$x_i = \theta + \nu_i, \tag{25}$$

where ν_i is the noise introduced at each sensor. Here, we assume that θ and ν_i are independent and identically distributed (*i.i.d.*) random variables with zero mean and variance σ_θ^2 and σ_i^2, respectively.

After collecting the information about the source, each sensor transmits its observation directly to the control center without any coding. This strategy is known as *analog amplify and forward* [13]. Analog amplify and forward scheme with power allocation has been extensively studied in [10, 36]. Therefore, at sensor i, the transmitter can be simply modeled by a power amplifying factor a_i, and the average transmit power is given as

$$P_i = a_i P_{x_i} = a_i(\sigma_\theta^2 + \sigma_i^2), \tag{26}$$

where P_{x_i} represents the power of observation x_i.

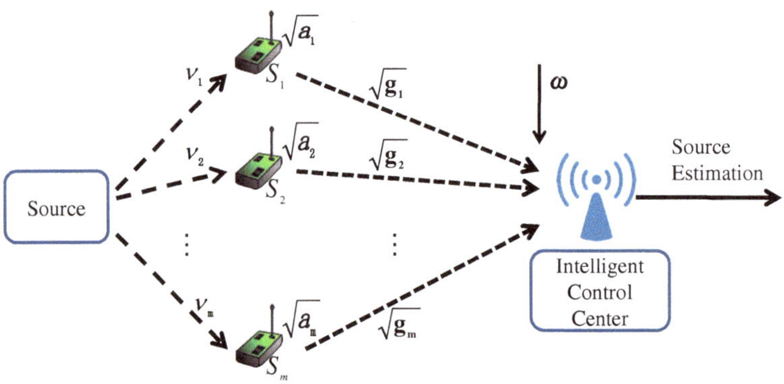

Fig. 10 System model

3.1.2 Multiple Access Scheme

We investigate two different multiple access schemes between the wireless sensors and the ICC: orthogonal and coherent.

A. Orthogonal Access Scheme
In the orthogonal access scheme, each sensor transmits its measurement to the ICC via orthogonal channels (e.g., using FDMA or CDMA) and the ICC receives

$$\begin{aligned} y_i &= \sqrt{g_i a_i}(\theta + v_i) + w_i \\ &= h_i \theta + n_i \qquad\qquad i = 1, 2, \ldots, m, \end{aligned} \tag{27}$$

where g_i is the power gain of the fading channel and w_i with variance ξ_i^2 is the noise introduced at the ICC. We assume that both the transmitter and receiver know the fading channel state.

Clearly, the received signal vector is

$$\mathbf{y} = \mathbf{h}\theta + \mathbf{n}, \tag{28}$$

where $\mathbf{h} = [\sqrt{a_1 g_1}, \ldots, \sqrt{a_m g_m}]^T$ and $\mathbf{n}$ stands for the AWGN with diagonal covariance matrix $\mathbf{R}$ given as $diag[\mathbf{R}] = [\sigma_1^2 a_1 g_1 + \xi_1^2, \ldots, \sigma_m^2 a_m g_m + \xi_m^2]^T$.

With the received signal, the ICC generates an estimation $\hat{\theta}$ to recover the source θ. Here, we adopt the maximum likelihood (ML) estimator [23]

$$\begin{aligned} \hat{\theta} &= [\mathbf{h}^T \mathbf{R}^{-1} \mathbf{h}]^{-1} \mathbf{h}^T \mathbf{R}^{-1} \mathbf{y} \\ &= \left(\sum_{i=1}^{m} \frac{a_i g_i}{\sigma_i^2 a_i g_i + \xi_i^2} \right)^{-1} \sum_{i=1}^{m} \frac{\sqrt{a_i g_i}\, y_i}{\sigma_i^2 a_i g_i + \xi_i^2}. \end{aligned} \tag{29}$$

The Mean Square Error (MSE) of this estimator is given as

$$D_{orth} = [\mathbf{h}^T \mathbf{R}^{-1} \mathbf{h}]^{-1} = \left(\sum_{i=1}^{m} \frac{a_i g_i}{\sigma_i^2 a_i g_i + \xi_i^2} \right)^{-1}. \tag{30}$$

B. Coherent MAC
Another case is that all sensors transmit simultaneously and the transmitted signals from all sensors reach the ICC as a coherent sum. Thus, the received signal at the control center can be expressed as:

$$\begin{aligned} y &= \sum_{i=1}^{m} h_i \theta + n \\ &= \sum_{i=1}^{m} \sqrt{a_i g_i}\, \theta + \sum_{i=1}^{m} \sqrt{a_i g_i}\, v_i + w. \end{aligned} \tag{31}$$

We assume the variance of noise w at the ICC is ξ^2.

Given the received y, the ICC generates an estimation of source θ. Accordingly, the ML estimator is

$$\hat{\theta} = (h^T R^{-1} h)^{-1} h^T R^{-1} y = \left(\sum_{i=1}^{m} \sqrt{a_i g_i} \right)^{-1} y. \tag{32}$$

This estimator is linear and achieves a MSE

$$\begin{aligned} D_{cmac} &= (h^T R^{-1} h)^{-1} \\ &= \left(\sum_{i=1}^{m} \sqrt{a_i g_i} \right)^{-2} \left(\sum_{i=1}^{m} \sigma_i^2 a_i g_i + \xi^2 \right). \end{aligned} \tag{33}$$

Remark 1: *Theoretical lower bound*

If all sensor measurements are directly available to the control center, we could get the following estimator

$$\theta_0 = \left(\sum_{i=1}^{m} \frac{1}{\sigma_i^2} \right)^{-1} \left(\sum_{i=1}^{m} \frac{x}{\sigma_i^2} \right), \tag{34}$$

which achieves the distortion

$$D_0 = \left(\sum_{i=1}^{m} \frac{1}{\sigma_i^2} \right)^{-1}. \tag{35}$$

This theoretical result could serve as a performance benchmark for later analysis.

Remark 2: $P \to \infty$

As the transmit power approaches infinity, we have

$$D_{cmac} \geq D_{orth} > D_0. \tag{36}$$

The detailed derivation is given in the Appendix. This inequality implies that as the transmit power increases, the distortion of the orthogonal access scheme could approach the lower bound D_0; however, even if the signal power approaches infinity, the coherent multiple access scheme cannot achieve the lower bound. One reasonable explanation is that the individual observations carry more information than a combination of these observations if the noise introduced at the ICC could be neglected.

Remark 3: $\xi^2 \to \infty$

When the noise variance ξ^2 is very large,

$$D_{cmac} \leq D_{orth}. \tag{37}$$

This is due to the fact that $D_{orth} - D_{cmac} \geq 0$ will stand if ξ^2 dominates. Thus, in this case, the coherent scheme will have smaller estimation distortion than the orthogonal scheme.

Since the coherent MAC scheme requires perfect carrier-level synchronization among sensors, in this subsection, we will adopt the orthogonal access scheme as the multiple access scheme between the sensors and the ICC.

3.1.3 Problem Formulation

Based on the above analysis, the power consumption of each sensor is given by (26) and the related estimation distortion with orthogonal access scheme is given in (30). Now we have two objectives: (1) minimizing the distortion to improve the estimation accuracy; (2) minimizing the power consumption to improve the sensor power efficiency. Several sensor selection schemes are proposed in the following sections to optimally accomplish these objectives.

3.2 Improving the Estimation Accuracy Using Sensor Selection

3.2.1 Sensor Selection Scheme Under Equal Power Allocation

This section aims to minimize the estimation error under total power constraint P_{tot}. Our strategy is first to select the K ($\leq m$) "*opportunistic*" sensors with the best channel conditions, which can be formulated as follows (without power consideration):

$$\begin{aligned} &min \quad \left(\sum_{i=1}^{m} z_i \frac{g_i}{\sigma_i^2 g_i + \xi_i^2} \right)^{-1} \\ &s.t. \quad \boldsymbol{1}^T \boldsymbol{z} = K \\ &z_i \in \{0, 1\}. \end{aligned} \tag{38}$$

The vector $\boldsymbol{1}$ is a vector with all entries one and the element z_i of $\boldsymbol{z}$ can be chosen from 0 or 1, which decides whether the ith sensor will be selected or not. Clearly, the above problem is an integer optimization problem which is nonconvex and hard to solve. Relaxing the nonconvex constraint $z_i \in \{0, 1\}$ with the convex constraint $0 \leq z_i \leq 1$, we reach the following relaxed sensor selection problem:

$$\begin{aligned} &max \quad \sum_{i=1}^{m} z_i \frac{g_i}{\sigma_i^2 g_i + \xi_i^2} + \gamma(log(z_i) + log\,(1 - z_i)) \\ &s.t. \quad \boldsymbol{1}^T \boldsymbol{z} = K, \end{aligned} \tag{39}$$

where γ is a positive parameter controlling the quality of approximation.

Next, solving the above convex optimization problem, we could obtain the target sensors. Here, our principle is to choose the K sensors with largest z_i weightness, which means only "strong" sensors with favorable channel condition will participate in the estimation process.

Finally, the power needs to be allocated in a reasonable way due to the total power constraint. In this subsection, we will apply equal power allocation; sensor selection under optimal power allocation will be discussed later. The equal power allocation scheme is:

$$a_i(\sigma_\theta^2 + \sigma_i^2) = \frac{P_{tot}}{K} \qquad 1 \le i \le K. \tag{40}$$

Applying the equal power allocation strategy (40) into (30), we could get the analytical estimation distortion under equal power allocation:

$$D_{orth} = \left(\sum_{i=1}^{m} t_i \frac{P_{tot} g_i}{\sigma_i^2 P_{tot} g_i + K \xi_i^2 (\sigma_\theta^2 + \sigma_i^2)} \right)^{-1}, \tag{41}$$

where t_i ($t_i \in \{0, 1\}$) represents the final sensor selection decision. The sensor selection scheme under equal power allocation is summarized by the following algorithm.

step 1 Formulate the relaxed sensor selection problem without power constraint (39);
step 2 Solve the optimization problem and obtain the target sensors t_i;
step 3 Activate the K selected sensors. Only active sensors will participate in the estimation process;
step 4 K active sensors adjust their transmit power accordingly (40) and send their measurements to ICC;
step 5 ICC adopts ML estimator to recover the source θ.

Let's discuss the following asymptotic scenarios:

Remark 4: $P_{tot} \to \infty$

Under equal power allocation, as total power P_{tot} approaches infinity, we have $D_{orth} \to D_0$. This result suggests that even if the power approaches infinity, the distortion could only achieve the performance benchmark D_0 rather than approaching zero. The reason is that the observation noise ν_i could not be eliminated even if power approach infinity. If the signal is amplified at each transmitter, the observation noise ν_i is inevitably amplified as well.

Remark 5: $K \to \infty$

As selected sensor number K approaches infinity, we obtain a similar asymptotic behavior as [10]:

$$\begin{aligned} D_{orth} &\sim \left[\frac{1}{K} \sum_{i=1}^{K} \frac{P_{tot} g_i}{\xi_i^2 (\sigma_\theta^2 + \sigma_i^2)} \right]^{-1} \\ &= \frac{E\left[\frac{\xi_i^2(\sigma_\theta^2 + \sigma_i^2)}{g_i} \right]}{P_{tot}}. \end{aligned} \tag{42}$$

Hence, with K approaching infinity, the estimation error could not decrease to zero either. One can attribute this limitation to the fact that, under the orthogonal access scheme, K different channel noises cannot be eliminated even if K approaches infinity.

3.2.2 Sensor Selection Scheme Under Optimal Power Allocation

In this subsection, we will address a sensor selection scheme under optimal power allocation. According to the Problem Formulation part in Sect. 3.1, sensor selection under optimal power allocation can be formulated as

$$\begin{aligned} &max \quad \sum_{i=1}^{m} \frac{a_i g_i}{\sigma_i^2 a_i g_i + \xi_i^2} \\ &s.t. \quad \sum_{i=1}^{m} a_i(\sigma_\theta^2 + \sigma_i^2) \leq P_{tot} \\ &a_i \geq 0. \end{aligned} \tag{43}$$

Unlike a preassigned sensor number K in the equal power allocation scenario, under optimal power allocation, the number of active sensors will not be fixed in advance. Obviously, this problem is convex and a_i is the variable to optimize. The Lagrangian G is given as

$$\begin{aligned} G = &-\sum_{i=1}^{m} \frac{a_i g_i}{\sigma_i^2 a_i g_i + \xi_i^2} \\ &+ \lambda \left[\sum_{i=1}^{m} a_i(\sigma_\theta^2 + \sigma_i^2) - P_{tot} \right] - \sum_{i=1}^{m} \mu_i a_i. \end{aligned} \tag{44}$$

From the Lagrangian function we can derive the following Karush-Kuhn-Tucker (KKT) conditions [5]:

$$\begin{aligned} \frac{\partial G}{\partial a_i} = -\frac{g_i \xi_i^2}{(\sigma_i^2 a_i g_i + \xi_i^2)^2} + \lambda(\sigma_\theta^2 + \sigma_i^2) - \mu_i &= 0 \\ \lambda \left(\sum_{i=1}^{m} a_i(\sigma_\theta^2 + \sigma_i^2) - P_{tot} \right) &= 0 \\ \mu_i a_i &= 0 \\ \lambda &\geq 0 \\ \mu_i &\geq 0 \\ a_i &\geq 0. \end{aligned}$$

Solving the KKT conditions, we obtain a result reminiscent of the "water-filling" solutions in wireless communications,

$$a_i = \frac{\xi_i^2}{\sigma_i^2 g_i}\left(\sqrt{\frac{g_i}{\lambda \xi_i^2(\sigma_\theta^2 + \sigma_i^2)}} - 1\right)^+, \tag{45}$$

where x^+ equals to 0 when x is less than zero, and otherwise equals to x. The solution is derived in the Appendix. For sensor i with $\eta_i = \frac{g_i}{\xi_i^2(\sigma_\theta^2+\sigma_i^2)}$, if $\eta_i > \lambda$, the corresponding sensor will be active; otherwise the related sensor will be switched off for power efficiency.

To practically implement the proposed sensor selection scheme, we need to split the transmission time τ into two sections. The first section consumes a fraction $\theta \in [0, 1]$ of τ and is used to select the target sensors and inform every selected sensor. The second section of $(1 - \theta)\tau$ is for the selected sensors to transmit the measurements to the control center.

3.2.3 Voltage Estimation Performance with Different Sensor Selection Schemes

The estimation performance of our proposed sensor selection schemes is evaluated. To simulate the practical communication environment, we assume the wireless channel follows Rayleigh fading [16, 27]. In the following simulations, we assume 100 voltage monitoring sensors are deployed and the power of the voltage source is set as $\sigma_\theta^2 = 1\,\text{mW}$. Figure 11 demonstrates the voltage estimation performance with different sensor selection schemes.

First, we examine the voltage estimation performance of the sensor selection scheme under equal power allocation. Note that our proposed sensor selection schemes will not be valid unless at least 10 voltage sensors are selected from the whole 100 available sensors, which implies if too few sensors are selected per time, the information collected at the selected sensors is not enough for the control center to make an accurate voltage estimation. Besides, if too many sensors ($K \geq 40$ in this case) are selected per time, the estimation performance will also degrade. The explanation is that if too many sensors are selected, under equal power allocation, voltage sensors with good channel conditions could not be assigned enough power, which will definitely impair the estimation performance. Hence, selecting the proper number of sensors could improve the voltage estimation performance. This conclusion is further illustrated in Fig. 12. Meanwhile, Fig. 12 shows that more power budget will always bring better voltage estimation performance.

Now let us compare the estimation performance of sensor selection schemes under equal power allocation and optimal power allocation. As shown in Fig. 11, sensor selection under optimal power allocation always outperforms the sensor selection scheme under equal power allocation. The improved estimation performance of the optimal power allocation scheme comes with the price of more complex

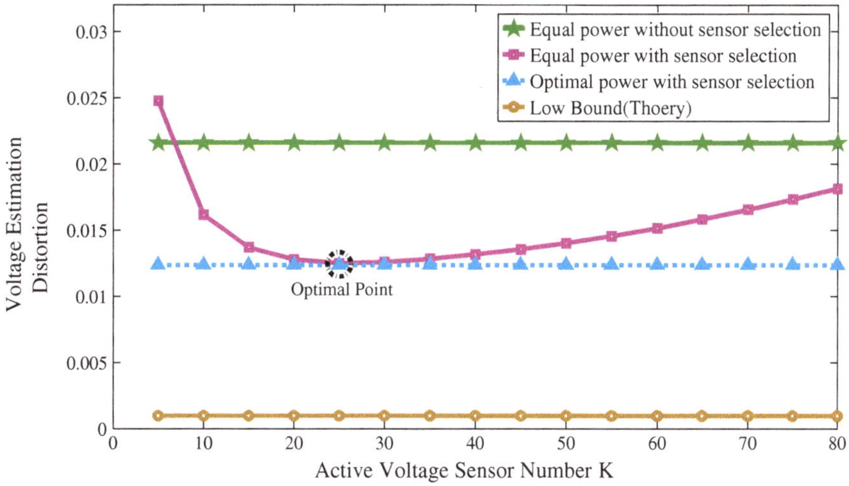

Fig. 11 Estimation performance with different sensor selection schemes

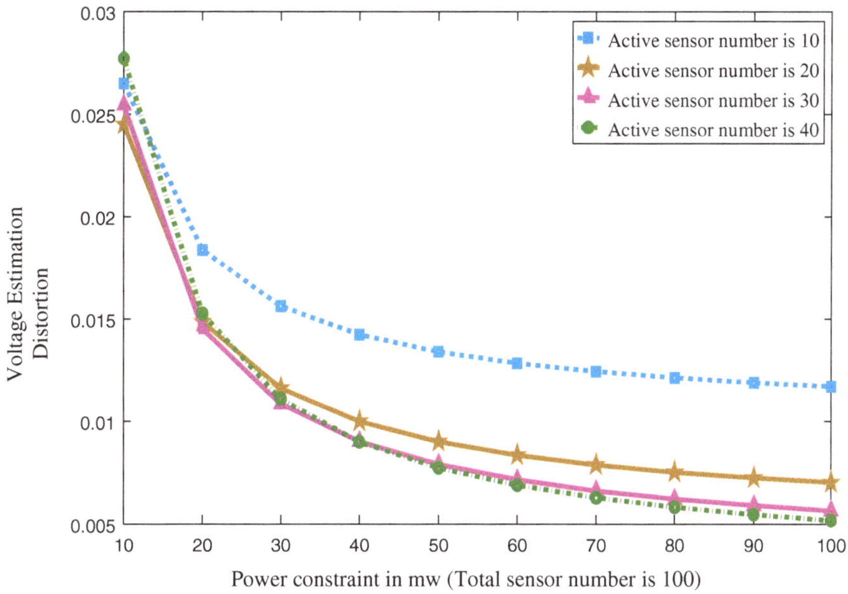

Fig. 12 Estimation performance with different power constraint

computation. Quite interestingly, if the proper number of voltage sensors ("optimal point") are selected, the equal power allocation scheme could even achieve the estimation performance of the optimal power allocation scheme. This conclusion verifies the merit and validity of our proposed opportunistic sensor selection algorithm. However, due to the power constraint and limited available sensors, the estimation performance could not approach the theoretical bound D_0.

3.3 Improving the Power Efficiency Using Sensor Selection

In this section, we will focus on another objective: minimizing the power consumption using sensor selection.

We first consider the scenario of equal power allocation without sensor selection, which means the sensor selection number K equals to m. Transforming the analytical distortion result (41), we could reach

$$\sum_{i=1}^{m} \frac{m\xi_i^2(\sigma_\theta^2 + \sigma_i^2)/\sigma_i^2}{\sigma_i^2 P_{tot} g_i + m\xi_i^2(\sigma_\theta^2 + \sigma_i^2)} = \sum_{i=1}^{m} \frac{1}{\sigma_i^2} - \frac{1}{D}. \tag{46}$$

Solving this equation, the power consumption P_{tot} could be obtained for any given estimation distortion D.

Then, in terms of the proposed sensor selection scheme under equal power allocation, the power consumption P_{tot} within a given distortion D is derived from

$$\sum_{i=1}^{m} \frac{K t_i \xi_i^2(\sigma_\theta^2 + \sigma_i^2)/\sigma_i^2}{\sigma_i^2 P_{tot} g_i + K\xi_i^2(\sigma_\theta^2 + \sigma_i^2)} = \sum_{i=1}^{m} \frac{t_i}{\sigma_i^2} - \frac{1}{D}, \tag{47}$$

where K represents the selected sensor number and t_i is the sensor selection decision as explained in Sect. 3.2.

Finally, to optimally allocate the power, we formulate the following optimization problem to minimize the total power consumption subject to a given estimation distortion D:

$$\begin{aligned} min \quad & \sum_{i=1}^{m} a_i(\sigma_\theta^2 + \sigma_i^2) \\ s.t. \quad & \left(\sum_{i=1}^{m} \frac{a_i g_i}{\sigma_i^2 a_i g_i + \xi_i^2} \right)^{-1} \le D \\ & a_i \ge 0. \end{aligned}$$

The first constraint is equivalent to

$$\sum_{i=1}^{m} \frac{\xi_i^2/\sigma_i^2}{\sigma_i^2 a_i g_i + \xi_i^2} \le \left(\sum_{i=1}^{m} \frac{1}{\sigma_i^2} - \frac{1}{D} \right). \tag{48}$$

For simplicity, let C represent the right-side part of (48) $\sum_{i=1}^{m} \frac{1}{\sigma_i^2} - \frac{1}{D}$. This problem is convex and a_i is the variable to optimize. The Lagrangian G is given by

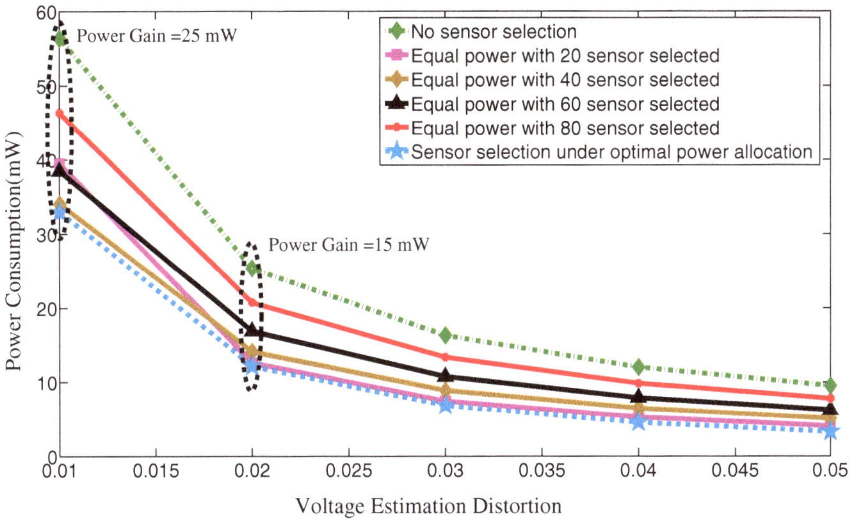

Fig. 13 Power efficiency with different sensor selection schemes

$$G = \sum_{i=1}^{m} a_i(\sigma_\theta^2 + \sigma_i^2) + \lambda \left[\sum_{i=1}^{m} \frac{\xi_i^2/\sigma_i^2}{\sigma_i^2 a_i g_i + \xi_i^2} - C \right] - \sum_{i=1}^{m} \mu_i a_i. \tag{49}$$

Implementing the KKT conditions, the optimal power allocation scheme for power efficiency is derived as

$$a_i = \frac{\xi_i^2}{\sigma_i^2 g_i} \left(\sqrt{\frac{\lambda g_i}{\xi_i^2(\sigma_\theta^2 + \sigma_i^2)}} - 1 \right)^+, \tag{50}$$

where λ is

$$\sqrt{\lambda} = \frac{\sum_{i=1}^{K''} \sqrt{\frac{\xi_i^2(\sigma_\theta^2+\sigma_i^2)}{g_i \sigma_i^4}}}{\sum_{i=1}^{K''} \frac{1}{\sigma_i^2} - \frac{1}{D}}. \tag{51}$$

The number of active sensors K'' can be solved if we substitute λ back to (50). Now the optimal strategy is that we will active the corresponding sensor if sensor index $1 \leq i \leq K''$, while we will switch off the sensor for all $i > K''$. Refer to the Appendix for the details about the solution of (50) which is similar to the derivation of (45).

Figure 13 illustrates the power efficiency with different sensor selection schemes. We use the same simulation setup as Sect. 3.2. We conclude that the sensor selection scheme under optimal power allocation always comes with the minimum power consumption. When the voltage estimation distortion is required as 0.01, the optimal power allocation scheme saves nearly 25 mW compared to the no-sensor-selection scheme; when the estimation distortion is set to 0.02, the optimal power allocation scheme saves more than 50 % of the power consumption (15 mW). Besides, we can observe that if proper voltage sensors are selected, the sensor selection scheme under equal power allocation could approach the performance of optimal power allocation. This conclusion further verifies the advantage of our proposed opportunistic sensor selection scheme.

4 Conclusion

In this chapter, the energy efficiency issue of WSNs from the physical layer design perspective is approached in two views: the capacity optimization of the IR-UWB WSN with energy constraints and coexistence of a NB communications system—IEEE 802.11n; an optimized sensor selection scheme to improve the estimation accuracy and power efficiency. A holistic investigation on the current literature is also conducted.

Appendix

Derivation of (36)

When the power approaches infinity, $D_{orth}^{-1} = \sum_{i=1}^{m} \frac{1}{\sigma_i^2}$ is readily derived. In term of the coherent multiple access case, as the power approaches infinity,

$$\begin{aligned} D_{cmac}^{-1} &= \frac{\left(\sum_{i=1}^{m} \sqrt{a_i g_i}\right)^2}{\sum_{i=1}^{m} \sigma_i^2 a_i g_i + \xi^2} \\ &\approx \frac{\left(\sum_{i=1}^{m} \sqrt{a_i g_i}\right)^2}{\sum_{i=1}^{m} \sigma_i^2 a_i g_i} \\ &\leq \sum_{i=1}^{m} \frac{a_i g_i}{\sigma_i^2 a_i g_i} = \sum_{i=1}^{m} \frac{1}{\sigma_i^2}, \end{aligned}$$

where the unequal step follows the *Cauchy-Schwarz* inequality. So the D_{cmac} could not approach D_0 unless the strict Cauchy-Schwarz equality condition $\sigma_i^2 a_i g_i = t\sqrt{a_i g_i}$ is satisfied for all i, where t is a constant. The proof is complete.

Derivation of (45)

Note that, if $\lambda = 0$, the first KKT condition implies that $\mu_i < 0$ for all sensor i. This will contradict with the fifth KKT condition. Thus, we must have $\lambda > 0$, which means

$$\sum_{i=1}^{m} a_i(\sigma_\theta^2 + \sigma_i^2) - P_{tot} = 0. \tag{52}$$

Transforming the first KKT condition, we have

$$a_i = \frac{\xi_i}{\sigma_i^2 g_i}\left(\sqrt{\frac{g_i}{\lambda(\sigma_\theta^2 + \sigma_i^2) - \mu_i}} - \xi_i\right). \tag{53}$$

For those sensors to be activated, a_i should satisfy $a_i > 0$. Then the third KKT condition tells us that if $a_i > 0$, then $\mu_i = 0$ holds. Thus the proof of (45) is complete.

To determine λ, let us assume that the sensors are ordered such that

$$\eta_1 \geq \eta_2 \ldots \geq \eta_m. \tag{54}$$

Clearly, this ranking favours the sensors with better channel conditions and higher observation quality.

Combining the first and second KKT conditions, we could get

$$\sqrt{\lambda} = \frac{\sum_{i=1}^{K'} \sqrt{\frac{\xi_i^2(\sigma_\theta^2+\sigma_i^2)}{g_i\sigma_i^4}}}{P_{tot} + \sum_{i=1}^{K'} \frac{\xi_i^2(\sigma_\theta^2+\sigma_i^2)}{g_i\sigma_i^2}}. \tag{55}$$

The number of active sensor number K' (which has been shown to be unique [35]) can be solved if we substitute λ back to (45).

References

1. G. Bansal, M.J. Hossain, V.K. Bhargava, Optimal and suboptimal power allocation schemes for OFDM-based cognitive radio systems. IEEE Trans. Wirel. Commun. **7**(11), 4710–4718 (2008)
2. J. Bellorado, S.S. Ghassemzadeh, L.J. Greenstein, T. Sveinsson, V. Tarokh, Coexistence of ultra-wideband systems with IEEE-802.11a wireless LANs, in *Proceedings of GLOBECOM '03*, p. 410 (2003)
3. M.D. Benedetto, G. Giancola, *Understanding Ultra Wide Band Radio Fundamentals* (Prentice Hall, 2004)
4. F. Berens, P. Jung, Wireless ultra-wide-band (UWB) communications: technology, regulation, standardization, and application areas. *2010 IEEE ICUWB*, vol. 1, pp. 12–19 (2010)

5. S. Boyd, L. Vandenberghe, *Convex Optimization* (Cambridge University Press, Cambridge, 2003)
6. S. Boyd, L. Vandenberghe, *Convex Optimization* (Cambridge University Press, Cambridge, 2004)
7. J. Cai, K.H. Liu, X. Shen, J.W. Mark, T.D. Todd, Power allocation and scheduling for ultra-wideband wireless networks. IEEE Trans. Veh. Technol. **57**(2), 1103–1112 (2008)
8. M. Chiani, A. Giorgetti, Coexistence between UWB and narrow-band wireless communication systems (invited Paper). Proc. IEEE. **97**(2), 231–254 (2009)
9. T.M. Cover, J.A. Thomas, *Elements of Information Theory*, 2nd edn. (Wiley, Hoboken, 2006)
10. S. Cui, J. Xiao, A. Goldsmith, Z. Luo, H. Poor, Estimation diversity and energy efficiency in distributed sensing. IEEE Trans. Sig. Process. **55**(9), 4683–4695 (2007)
11. Electronic Communications Committee, ECC Report 64: the protection requirements of radio communications systems below 10.6 GHz from generic UWB applications, pp. 5–7, 24–27, February 2005
12. Federal Communication Commission, Revision of part 15 of the commission's rules regarding ultra-wideband transmission systems, FIRST REPORT AND ORDER. ET Docket 98–153, FCC 02–48, 14 February 2002, pp. 2–94 (2002)
13. M. Gastpar, M. Vetterli, *Source-Channel Communication in Sensor Networks*, vol. 2634 (Springer, New York, 2003), pp. 162–177
14. A. Giorgetti, M. Chiani, M.Z. Win, The effect of narrowband interference on wideband wireless communication systems. IEEE Trans. Commun. **53**(12), 2139–2149 (2005)
15. R. Giuliano, F. Mazzenga, On the coexistence of power-controlled ultrawide-band systems with UMTS, GPS, DCS1800, and fixed wireless systems. IEEE Trans. Veh. Technol. **54**(1), 62–80 (2005)
16. A. Goldsmith, *Wireless Communications* (Cambridge University Press, Cambridge, 2005)
17. M. Hamalainen, V. Hovinen, R. Tesi, et al., On the UWB system coexistence with GSM900, UMTS/WCDMA, and GPS. IEEE J. Sel. Areas Commun. **2**(9), 1712–1721 (2002)
18. A. Huseyin, C. Zhining, D. Benedetto, *Ultra Wideband Wireless Communication* (Wiley-Interscience, Hoboken, 2006), pp. 10–11
19. IEEE 802.11n D.7.0, Part 11: Wireless LAN Medium Access Control (MAC) and Physical Layer (PHY) specifications (2008), p. 245
20. S. Joshi, S. Boyd, Sensor selection via convex optimization. IEEE Trans. Sig. Process. **57**(2), 451–462 (2009)
21. C.X. Juan, Q. Sheng, Spectrum analysis of Ultra-wide band signal based on Gaussian pulse, *Global Mobile Congress (GMC)*, pp. 1–4 (2010)
22. S. Kandeepan, G. Baldini, R. Piesiewicz, Preliminary experimental results on the spectrum sensing performances for UWB-cognitive radios for detecting, in *IEEE 802.11n Systems 6th International Symposium on Wireless Communication Systems (ISWCS)*, pp. 111–115 (2009)
23. S.M. Kay, *Fundamentals of Statistical Signal Processing: Estimation Theory* (Cambridge University Press, Cambridge, 1993)
24. K. Kincaid, S. Padula, D-optimal designs for sensor and actuator locations. Comput. Oper. Res. **29**(6), 701–713 (2002)
25. Z. Li, Q. Liang, Capacity optimization of ultra-wide band system under the coexistence with IEEE 802.11n, in *11th International Symposium on Communications and Information Technologies (ISCIT)*, pp. 553–557 (2011)
26. Z. Li, W. Zou, B. Li, Z. Zhou, X. Huang, Analysis on coexistence of ultra wideband with OFDM-based communication systems. IEEE Trans. Electromagn. Compat. **53**(3), 823–830 (2011)
27. J.G. Proakis, *Digital Communication*, 4th edn. (McGraw-Hill, New York, 2001)
28. A. Roca, Implementation of a WiMAX Simulator in Simulink. Diplomarbeit, Vienna, February 2007
29. H. Rowaihy, S. Eswaran, M. Johnson, D. Verma, A.B. Noy, T. Brown, T.L. Poota, A survey of sensor selection schemes in wireless sensor networks, in *Proceedings of SPIE* (2007)

30. K. Shih, Y. Chen, C. Chiang, B. Liu, A distributed active sensor selection scheme for wireless sensor networks, in *Proceedings of the IEEE Symposium on Computers and Communications* (2006)
31. C.W. Tan, S. Friedland, S.H. Low, Spectrum management in multiuser cognitive wireless networks: optimality and algorithm. IEEE J. Sel. Areas Commun. **29**(2), 421–430 (2011)
32. D. Tse, P. Viswanath, *Fundamentals of Wireless Communication* (Cambridge University Press, Cambridge, 2004)
33. H. Wang, K. Yao, G. Pottie, D. Estrin, Entropy-based sensor selection heuristic for target localization, in *Proceedings of the Third International Symposium on Information Processing in Sensor Networks* (2004)
34. S. Wood, R. Aiello, *Essentials of UWB* (Cambridge University Press, Cambridge, 2008)
35. J. Xiao, S. Cui, Z. Luo, A. Goldsmith, Power scheduling of universal decentralized estimation in sensor networks. IEEE Trans. Sig. Process. **54**(2), 413–422 (2006)
36. J. Xiao, S. Cui, Z. Luo, A. Goldsmith, Linear coherent decentralized estimation. IEEE Trans. Sig. Process. **56**(2), 757–770 (2008)
37. L. Yang, G.B. Giannakis, Ultra-wideband communications: an idea whose time has come. IEEE Sig. Process. Mag. **21**(6), 26–54 (2004)
38. R. Yang, K.S. Kwak, Z. Zhou, Distributed water-filling algorithm for direct-sequence ultra wideband cognitive radio network with limit on aggregate power emission. IET Commun. **4**(12), 1404–1414 (2010)
39. P. Yi, A. Iwayemi, C. Zhou, Developing Zigbee deployment guideline under WiFi interference for smart grid applications. IEEE Trans. Smart Grid **2**(1), 110–120 (2011)

Chapter 5
Network Coding Techniques for Wireless and Sensor Networks

Pouya Ostovari, Jie Wu and Abdallah Khreishah

Abstract Network coding is a technique where relay nodes mix packets using mathematical operations, which can increase the throughput. Network coding was first proposed for wired networks to solve the bottleneck in a single multicast session problem and to increase the throughput. However, the broadcast nature of wireless networks and the diversity of the links make network coding more attractive in wireless networks. Network coding can be classified as either inter or intra-session. Inter-session network coding allows the packets from different sessions (sources) to be mixed to increase the throughput. In contrast, intra-session network coding, which can be used to address the packet loss problem, uses the diversity of the wireless links and mixes packets from the same sessions. In this chapter, we survey the recent works on network coding in both general wireless networks and wireless sensor networks. We present various network coding techniques, their assumptions, applications, as well as an overview of the proposed methods.

1 Introduction

As the demand for communication services is growing, wireless solutions become more and more important. Due to their ease of deployment, wireless networks play a major role in our lives. They are also ideal to provide a convenient solution to the

P. Ostovari · J. Wu (✉)
Department of Computer and Information Sciences, Temple University,
Philadelphia, PA 19122, USA
e-mail: jiewu@temple.edu

P. Ostovari
e-mail: tuc71330@temple.edu

A. Khreishah
Department of Electrical and Computer Engineering, New Jersey Institute of Technology,
Newark, NJ 07102, USA
e-mail: abdallah.khreishah@njit.edu

H. M. Ammari (ed.), *The Art of Wireless Sensor Networks*,
Signals and Communication Technology, DOI: 10.1007/978-3-642-40009-4_5,

© Springer-Verlag Berlin Heidelberg 2014

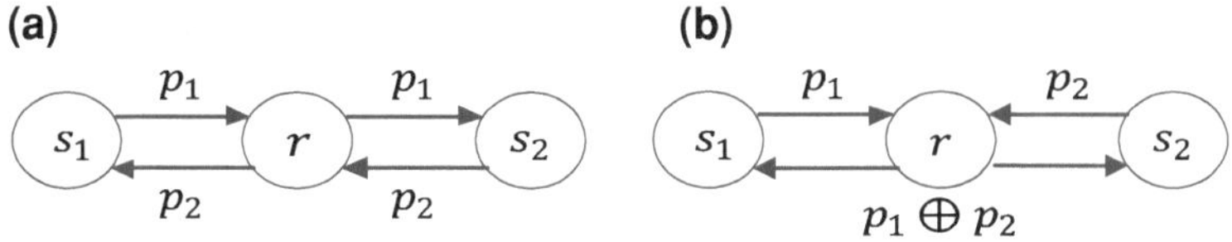

Fig. 1 Binary network coding

last mile problem [1, 2]. Wireless networks can be cellular networks that are used for mobile phones, or Wi-Fi networks that provide an Internet connection. Different multihop wireless network settings are used. Mesh networks can be used to provide Internet access and file sharing [3]. Wireless sensor networks (WSNs) [4] can be used for military applications, such as enemy detection in battlefields. They can also be used for disaster detection and monitoring applications.

Despite the diverse types of wireless networks and their applications, the common features of wireless networks create opportunities to be exploited and challenges to be addressed. These common features include the broadcast nature of wireless links, the interference among the links, the diversity of the links, and the lossy behavior of the links [5]. Also, the correlation between the links may affect the performance of the wireless communication protocols.

Inter-session Network Coding. The broadcast nature of wireless networks is considered a challenge, as it creates interference between the links and produces unnecessary multiple copies of the same packet. However, if we allow the intermediate wireless nodes to code the packets, the broadcast nature becomes an opportunity. Consider the example in Fig. 1a, where nodes s_1 and s_2 want to exchange their own packets, p_1 and p_2, respectively. Assuming that these nodes are out of range of each other, this communication incurs four transmissions; two transmissions for sending the packets to the relay node, and two transmissions for relaying the packets. However, the relay node can simply *XOR* the packets and send the coded packet $p_1 \oplus p_2$ [6], which is shown in Fig. 1b. The nodes s_1 and s_2 can retrieve each others' packets by XOR-ing $p_1 \oplus p_2$ with their own packets, p_1 and p_2, respectively. As a result, the number of transmissions has been reduced to three by using binary network coding. *Inter-session* network coding can reduce the number of transmissions, by allowing packets from different sessions (sources) to be coded together. By reducing the number of required transmissions, network coding increases the throughput and decreases the interference between the links in wireless networks. In the rest of the chapter, we use terms session and flow interchangeably.

Intra-Session Network Coding. Another important application of network coding is to provide reliability in wireless networks. The traditional way to provide reliability for both wired and wireless networks is to use feedback messages to report the received (or lost) packets. By using feedback messages, the sender node will know which packets need to be sent again. However, these feedback messages consume bandwidth. Consider the example in Fig. 2; the source node wants to deliver packets p_1 and p_2 to the node d. The reliability of the link $s_1 \rightarrow d$ is equal to $\frac{2}{3}$. In the case

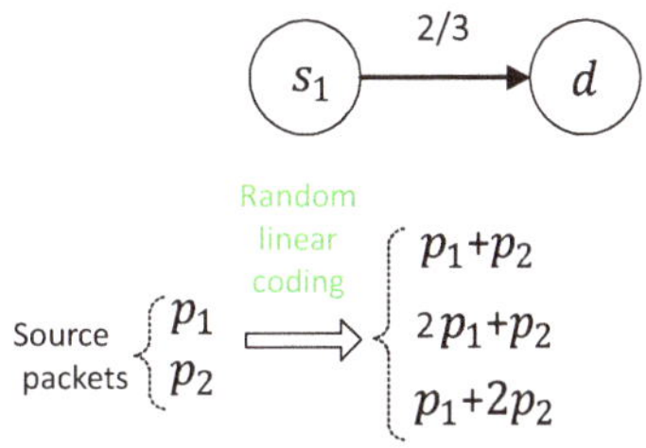

Fig. 2 Application of network coding to provide reliability

that the source node sends three coded packets, $p_1 + p_2$, $p_1 + 2p_2$, and $2p_1 + p_2$, on average, the destination node will receive two of the three coded packets. Therefore, the destination nodes will be able to retrieve the packets p_1 and p_2. However, without network coding, we need to use a feedback mechanism or else the source node needs to transmit each packet multiple times. As a result, communication schemes with network coding can provide reliability with a fewer transmissions than schemes without network coding.

Coding the packets from the same session (source) together is called *intra-session* network coding, which exploits the diversity of the links. In intra-session network coding, the packets from the same source are coded together (usually linearly), which makes the importance of the packets the same. Therefore, when k packets are coded together, a relay node does not need to know exactly which packets are received by the destination node; it is thereby enough to successfully deliver k coded packets out of the transmitted coded packets.

Opportunistic Routing. An efficient way to address packet loss in wireless networks without network coding is to use *opportunistic routing* approaches [7]. When a node broadcasts a packet, it is probable that the next-hop does not receive the packet. However, because of the broadcast nature of the wireless medium, and the diversity among the links, a neighbor of the sender can receive and forward the packet as the next-hop with high probability. In opportunistic routing, there is no specific path from the source to the destination, and any node that overhears the packet can relay it. Take Fig. 3 as an example, in which node s wants to send 4 packets to the destination d. The delivery rate of the links are shown beside the links. Assume that each relay node received the packets shown beside the nodes. If we use traditional shortest path routing, the route from s to d will be fixed. Assuming that the chosen route is $s \rightarrow r_1 \rightarrow d$, the source node needs to retransmit the packets p_3 and p_4. On the other hand, if we allow the other nodes that received the packets p_3 and p_4 to forward them, the source node will not need to retransmit any packet.

The main challenge in opportunistic routing is coordinating the intermediate nodes. To prevent redundant transmissions, the intermediate nodes need to send feedback or listen to the other nodes' transmissions to find out if there is a neighbor that has received the transmitted packet. For this purpose, the intermediate nodes need to be able to overhear each other, which might not be possible, as shown in Fig. 3. Network coding can solve this problem [8]. To this purpose, the source node divides the packets to be sent in batches of k packets. The source keeps sending

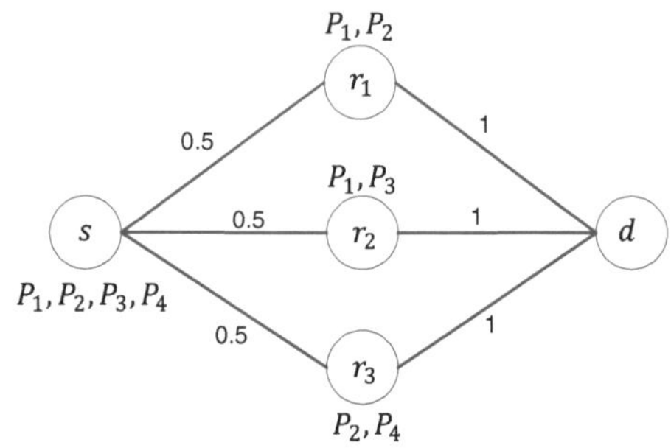

Fig. 3 Opportunistic routing

coded packets of the form $\sum_{i=1}^{k} \alpha_i p_i$, where α_i is a random coefficient chosen over a finite field. When an intermediate node receives a coded packet, the node checks if the coded packet is linearly independent to the previously received packets. If so, the node will add the packet to its buffer. Each intermediate node generates linear combinations of the packets in its buffer and sends the coded packets. The destination node can decode all of the packets of the batch when it receives k linearly independent packets. In this case, the destination node sends feedback to the source to stop sending the packets.

Cross-Layer Design. Using network coding methods in wireless protocols incurs new challenges. For example, previous routing protocols are unaware of network coding. However, the routing protocol affects the coding opportunity. If two flows pass through relay nodes that are far from each other, there will be no coding opportunity. On the other hand, flows that are close to each other result in more interference. Therefore, to increase the efficiency of the proposed protocols for wireless networks, cross-layer approaches are needed. In cross-layer approaches, the protocols of different layers are independent. However, they communicate with each other to make decisions and perform more efficiently.

Wireless Sensor Networks. Sensor networks differ from the general wireless networks in performance metrics, traffic patterns, and their amount of available memory and processing resources [9]. These differences make some of the network coding approaches proposed for general wireless networks inappropriate for WSNs. For example, in some of the network coding methods, the nodes should listen to their neighbors and store the overheard messages in their buffers. However, in sensor networks, because of the memory limitation, sensor nodes cannot cache overheard packets that might not be useful [10]. WSNs' protocols must be simple and easily implemented. Moreover, the links' quality between the sensor nodes vary over the time, and nodes can fail or disconnect. Therefore, the dynamic environment should be considered, and the algorithms should be adaptive to reflect this dynamic nature [10].

The rest of this chapter is organized as follows. We provide our classification methodology in Sect. 2. In Sect. 3, we describe some of the well-known proposed methods for unicast application, and we categorize them. A discussion about multicast and broadcast network coding approaches is provided in Sects. 4 and 5, respectively. Section 6 concludes the chapter. Note that fountain codes (also known as rateless erasure codes), such as online codes [11], LT codes [12], and raptor codes [13], are another type of coding. However, these codes are beyond the scope of this chapter.

2 Classification of Network Coding Approaches

From one perspective, network coding can be classified into *XOR* (binary) coding and *random linear* (RL) coding. In binary coding, XOR operations are performed between the packets. Take Fig. 4a as an example; we have two flows: one of them between nodes s_1 and d_1 and the other between nodes s_2 and d_2. Without network coding, the relay node needs two transmissions to send the packets, one for each flow. However, the relay node r can exploit the broadcast nature of its output links and reduce the number of transmissions to one by XOR-ing the two packets. The nodes d_1 and d_2 decode the coded packet by XOR-ing $p_1 \oplus p_2$ with the overheard packets, p_2 and p_1, respectively.

In random linear coding, the relay nodes create coded packets of the form $\sum_{i=1}^{k} \alpha_i p_i$, where α_i is a random coefficient chosen over a finite field, and p_i's can be coded or uncoded packets. Assume that the delivery rate of all of the links in Fig. 4b is 0.5. Each source node generates four random linearly coded packets and sends them. Two linearly independent packets from each session are received by the relay node r. Then, the relay node sends four random coded packets for each session (Fig. 4c). Each destination on average receives two linearly independent coded packets from the relay node, and is able to decode them. The decoding process is similar to solving a system of linear equations. Note that in the figures, we just show the received coded packets on the lossy links.

From another view, we can classify network coding as *local* or *global* coding. In local network coding, a relay node sends the coded packets such that the next hop nodes are able to decode the coded packets. Then, the next hop nodes decode the

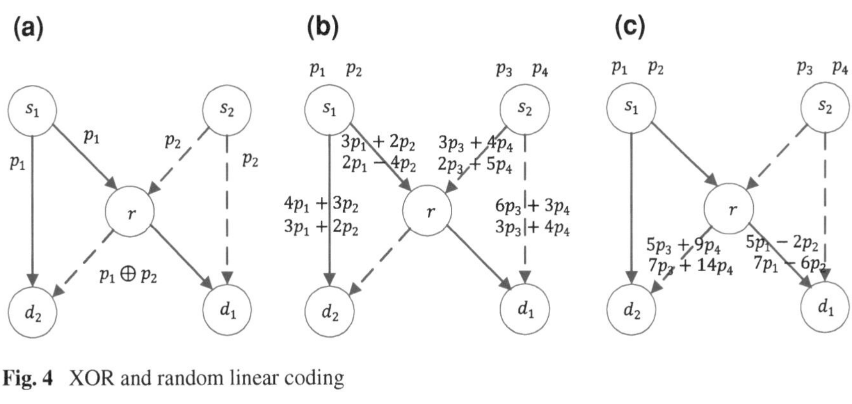

Fig. 4 XOR and random linear coding

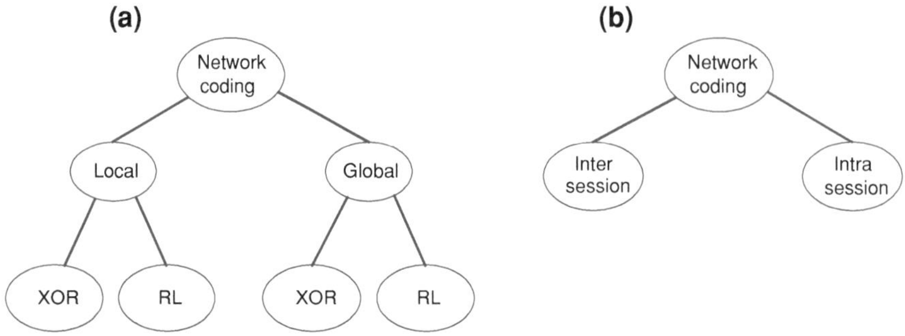

Fig. 5 Classification of network coding approaches

coded packets and use the same policy to code the packets. Therefore, in a multi-hop transmission, hop-by-hop coding and decoding is performed. In contrast, in global network coding, the intermediate nodes do not perform decoding; they just code the coded packets again. At the end, when the destination nodes receive enough packets, they will be able to decode them. Usually, local network coding protocols use XOR coding, and global protocols perform random linear coding. Figure 5 shows our classification of network coding methods.

As described in the introduction, network coding can be *inter-session* or *intra-session*. Inter-session network coding allows the relay nodes to code packets from the same session (source) to solve the bottleneck problem, and to reduce the number of transmissions (Fig. 4a). On the other hand, in intra-session network coding, the relay nodes code packets from the same session to make the importance of the packets the same. Intra-session network coding is a natural way to address the packet loss problem in wireless networks (Fig. 4b and c). Also, inter and intra-session coding can be combined together to increase the throughput. Instead of transmitting 4 coded packets over each session, the relay node can mix the two sessions together and send 4 coded packets over them. In this way, the number of transmissions by the relay node is reduced from 8 to 4. Each destination node receives two coded packets from the relay node and two from a source node. Using these 4 coded packets, each destination node can decode the packets.

3 Network Coding Methods for Unicast Applications

In this section, we describe some of the proposed network coding approaches for unicast application. We categorize the methods based on their methodologies, which are inter or intra-session network coding. Then, we compare the methods and summarize their advantages and drawbacks in the following sections.

3.1 Inter-session Network Coding

COPE. A practical forwarding architecture, called COPE, is proposed in [6] which increases the throughput of wireless networks. This paper addresses the case of unicast traffic: dynamic and potentially bursty flows. COPE incorporates three main techniques, opportunistic listening, opportunistic coding, and learning neighbors' states. In COPE, the nodes snoop on all communications and store the overheard packets for a limited period of time. The nodes broadcast reception reports to tell their neighbors which packets they have in their buffers. On the other hand, in the network coding phase, a node may have multiple choices for coding. However, the goal is to maximize the number of packets delivered in a single transmission, while making sure that all next-hops are able to decode the coded packet, so that they retrieve their respective packets. When it comes to learning a neighbor's state, COPE does not rely solely on the reception reports, since they may get lost or arrive late. For this reason, the delivery rate of the links are computed and broadcasted periodically. The authors show that in the absence of opportunistic listening, COPE's maximum coding gain is 2.

A forwarder node in COPE works as follows. First, it selects a packet at the head of the forwarding queue. Then, it sequentially selects another packet in the queue, and computes the decodability probability of the packets at the next-hops when the packets are coded together. If the decodability probabilities at all of the next-hop nodes are greater than a given threshold, the relay node will code the packets together. Assume that in Fig. 6, the next-hops for the packets from nodes s_1, s_2, and s_3 are nodes d_1, d_2, and d_3, respectively. Also, assume that the delivery rate of the shown links is 1, but the overhearing probability between the s nodes and d nodes is 0.8. The node r has received packets p_1, p_2, and p_3 from nodes s_1, s_2, and s_3, respectively. First, the relay node selects packet p_1. Then, it computes the decodability probability of the coded packet $p_1 \oplus p_2$ at the respective next-hops of packets p_1 and p_2, d_1 and d_2. This probability is equal to 0.8. Assuming that the coding threshold is equal to 0.8 (this is the default value in [6]), COPE allows these packets to be coded together. Then, the relay node checks the decodability probability when p_3 is coded with $p_1 \oplus p_2$. For all next-hops, this probability is equal to 0.64, which is less than the threshold. Thus, p_3 cannot be coded with the packet $p_1 \oplus p_2$.

With the current sensor nodes' technology, COPE might not be much appropriate for sensor networks. First, the nodes in COPE should snoop on all communications and store the overheard packets in their buffer, which is not practical in WSNs, because of the power and memory limitations. Second, the reception reports of the packets and the delivery rate of the links should be broadcasted in COPE periodically; this results in large amounts of power consumption in WSNs.

Centralized Approach. A network coding-aware routing method is proposed in [14] to achieve optimal throughput. In contrast with COPE, in which the routing and coding algorithms are separate, the proposed mechanism in [14] is a cross-layer approach. The authors argue that when the paths of two flows are far apart (the flows pass through nodes that are far from each other), the interference between them

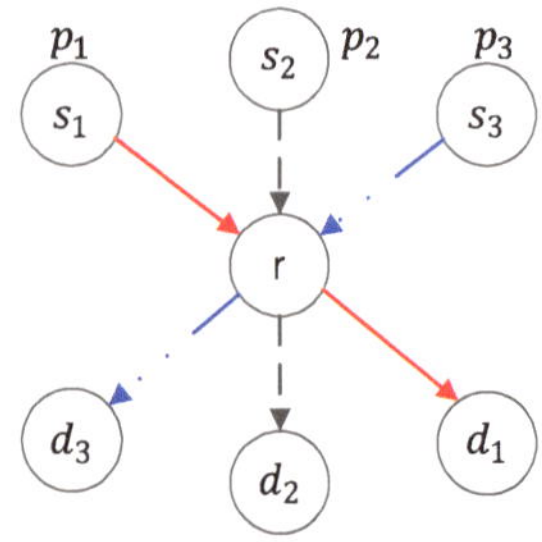

Fig. 6 COPE approach

is minimized. On the other hand, choosing close flows paths increases the coding opportunities. Therefore, a trade-off between coding opportunity and conflict should be performed. A conflict graph is used in this work to model the interference between the links, and linear programming is used to find the optimal solution for the joint routing and network coding problem. The main drawback of this work is that the authors do not consider all of the possible overhearing cases between the nodes. In the same way as the COPE approach, this approach is not suitable for sensor networks.

Distributed Approach. The problem of energy efficient opportunistic network coding for multiple unicast flows is addressed in [15]. The proposed inter-sessions network coding method, which is referred to as COPR, decomposes multiple unicast sessions into a superposition of multicast and unicast sessions, with coding within each session (note that these sessions are artificial, and they differ from the original sessions). The network is modeled as a directed hypergraph, and the achievable rate region of one-hop XOR network coding is determined under a primary interference model. To simplify the network operation, the authors propose a back pressure algorithm for dynamic scheduling that does not optimize overheard flows.

Network coding opportunities are not fully exploited in TCP flows over wireless network coding, due to the bursty behavior of the flows. Rate mismatches between the flows reduce the coding opportunities since the intermediate nodes may not have enough packets from different flows to code together. [16] addresses this problem by proposing coding-aware queue management for unicast flows. The authors formulate congestion control as a network utility maximization problem and solve it via a distributed scheme. Using the optimal solution, a network coding-aware queue management scheme at intermediate nodes (NCAQM) is proposed, which stores coded packets and drops packets based on network coding and congestion information. NCAQM does not change the TCP or MAC protocols, which makes the approach practical. The bursty flows are not usual in WSNs, so this method might not be very useful for their current applications.

Analysis. A formal analysis on the performance of COPE is provided in [17]. The authors use the encoding number as the performance measure. The encoding number is defined as the number of packets that can be coded together at a relay node in each transmission, and an upper bound on the encoding number at a single relay node is proposed. It is shown that, in the case of overhearing, the upper bounds of 2D and 3D networks are equal to 5 and 6, respectively. The authors also propose a methodology for computing the average coding number under a general class of a random access link-scheduling mechanism. They extend their analysis to general multi-hop wireless networks, and they formally prove the upper bound of the throughput gain for the practical XOR coding scheme.

Lossy Links. The CLONE approach, which is a loss-aware network coding method, is proposed for unicast sessions in [18]. The relay nodes use local XOR coding to code the packets from different sessions. However, in contrast with the relay nodes in COPE, which try to send the minimum number of transmissions, in order to achieve higher throughput, the relay nodes in CLONE use redundancy to increase the probability of delivering the packets. The idea can be motivated by the example in Fig. 1b. Assume that the links from nodes s_1 and s_2 to the relay node are loss free, and the loss probability of the links from the relay node to the source nodes is P'. Using COPE, the number of transmissions is equal to 3, and the number of received packets is equal to $(1 - P')$. Therefore, the throughput is equal to $(1 - P')/3$. On the other hand, if the relay node transmits the coded packet twice, the throughput will be equal to $(1 - P'^2)/4$; thus, for $P' = 0.5$, the throughput of the first and the second schemes are equal to 0.167 and 0.1875, respectively. In CLONE, the relay nodes construct redundant coded packets such that the delivery probabilities of the original packets to their next-hop achieve a given threshold. CLONE is not deployable in practice (especially in WSNs) due to its computational complexity. In addition, intra-session network coding provides a more efficient way to address the lossy behavior of the links, which is discussed in the following section.

Flow-Based Approach. The authors in [19, 20] use inter-session network coding to increase the throughput of multi unicast flows, while maintaining fairness between the flows. The optimal solution for lossy 2-hop relay networks is #P-complete when the packets are considered separately. For this reason, in this work, the authors consider flows instead of individual packets. Using this policy, they optimize the overhearing and characterize the capacity region in the form of linear equations when XOR network coding is used. Linear programming is used in this work to compute the capacity region.

SenseCode. The authors in [9] use network coding to provide a reliable and energy-efficient data gathering approach in WSNs. It is assumed that the sensing task is periodic, and during each round all of the nodes should send their sensed data to a sink node. They argue that the traditional tree-based methods, in which each intermediate node transmits the received packets from its children nodes to its parent, cannot provide reliability. The reason is that in the case of node or link failures, the data will not be able to reach the sink node. In order to solve this problem, in SenseCode, the sensed data from each sensor is transmitted through different paths. In this method, each node stores all of the messages it has generated by itself, and the packets it has received from its children nodes during the current round, in a queue. The node also stores the overheard packets to a separate queue. When a node has a new message to send to the sink node, the node creates a packet and marks it as uncodabe. Then, the message will be transmitted to the parent node. Moreover, the node sends $R - 1$ linear combinations of the packets from its queues and marks them as codable. Here, R is a configurable redundancy factor. Also, when a node receives a packet from its child node, and it is marked as uncodable, the node will relay the packet. If the packet is marked as codable, the node sends a linear combination of the received packet and the packets in its queues. In this way, some of the packets will be received by the sink node as uncoded packets, which provides reliability even in the case of high link loss rates.

Physical Layer Coding. A physical layer network coding scheme (PNC) is proposed in [21] for linear networks. In contrast to traditional network coding schemes, where coding is performed by the relay nodes on digital bits, PNC makes use of the additive nature of simultaneously-arriving electromagnetic waves for network coding. Take the example in Fig. 1, in which nodes s_1 and s_2 want to exchange their packets through relay node r. In binary network coding, the source nodes send their packets in different time slots, and the relay node XORs the packets after receiving them. In contrast, in PNC, the source nodes transmit their packets simultaneously, so the relay node receives a combined signal. Assume that the signals sent by nodes s_1 and s_2 are $a_1 \cos(\omega t) + b_1 \sin(\omega t)$ and $a_2 \cos(\omega t) + b_2 \sin(\omega t)$, respectively. Then, the signal received by the relay node will be in the form of $(a_1 + a_2)\cos(\omega t) + (b_1 + b_2)\sin(\omega t)$. The relay node maps the received signal, such that when the nodes s_1 and s_2 receive the mapped signal, they will be able to extract the signal sent by the other source node. Physical layer coding can be very useful and efficient for WSNs [22]. In these networks, the sensor nodes are placed in a line to monitor linear structures like roads, or long pipelines carrying oil, gas and water resources, etc.

3.2 Intra-session Network Coding

In the previous section, we reviewed some of the inter-session network coding approaches that have been proposed for lossy environments. However, the natural way to address the loss problem is to use intra-session network coding, which makes the importance of the transmitted packets the same. In this section, we will review some of these approaches.

MORE. An opportunistic routing method, called MORE, is proposed in [8]. This approach, which uses random linear network coding, can be used for unicast and multicast applications. In contrast with traditional routing methods, in which the path from the source to the destination node is predetermined, opportunistic routing allows any node that overhears the transmission and is closer to the destination node to participate in forwarding the packet [7]. However, opportunistic routing faces two challenges. Multiple nodes may overhear a packet and forward the packet. Also, the MAC protocol needs to be modified. MORE uses random linear network coding to address these problems.

Consider Figure 7. Traditional routing sends the packet along path $s \rightarrow r_1 \rightarrow r_2 \rightarrow d$. However, there is a chance that node r_2, which is closer to the destination node, will receive some of the packets. For example, assume that the source (s) sends two packets, p_1 and p_2. Both of them are received by the node r_1, and the

Fig. 7 Opportunistic routing

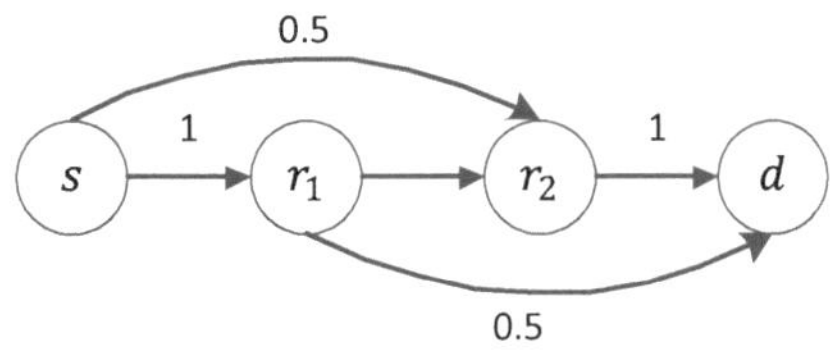

node r_2 received the packet p_1. Therefore, node r_1 does not need to forward packet p_1 since a node closer to the destination node (r_2) can forward the packet. In order to prevent this unnecessary transmission, the nodes need to be coordinated, which is hard for large networks. To solve this problem, node r_1 can forward a random linear combination of the packet, $c_1 p_1 + c_2 p_2$.

MORE works as follows. The source node breaks up the file into batches of k uncoded packets, called *native packets*. The source creates a random linear combination of the native packets in the current batch, and broadcasts the coded packet. A coded packet $\sum_{i=1}^{k} \alpha_i p_i$, where α_i is a random coefficient and p_i is the native packet of the current batch. The source node attaches a header to each packet, which contains the coefficients and the list of forwarder nodes. MORE uses ETX (expected number of transmissions) to compute the forwarder list. The source node includes the nodes which are closer to the destination node (in term of the ETX metric) in the forwarder list. When a forwarder node receives a packet, the node checks if the packet contains new information. In other words, the node checks whether the new received packet is linearly independent from the received packet in the node's buffer, in which case it is called an *innovative packet*. Non-innovative packets will be ignored. Otherwise, the node generates a linear combination of the received coded packets from the current batch and broadcasts it. When the destination node receives k linearly independent packets, it can decode the whole batch.

The remaining question in MORE is: how many packets does each forwarding node need to send when the node receives an innovative packet from an upstream node? The authors use the ETX metric to calculate the number of transmissions that should be done at a forwarder node upon receiving an innovative packet from an upstream node. They call this expected value TX_credit (transmission credit).

MORE is not suitable for WSNs. The reason is that in MORE, every node can be a potential forwarder to transmit the packets from the source node to the destination. Therefore, the nodes should remain in active mode to participate in opportunistic routing, which increases the energy consumption of the sensor nodes.

Extensions over MORE. MORE does not consider the possible congestion caused by multiple forwarders that have new packets to transmit. The problem arises when a large number of intermediate forwarders are involved in the unicast. A distributed optimization framework, called OMNC, is proposed in [23]; OMNC jointly optimizes rate control and multi-path routing. OMNC avoids network congestion through its rate control mechanism. Instead of determining the number of packets, OMNC assigns the encoding and broadcast rate to each node in a decentralized manner, and tries to optimize the bandwidth usage and congestion avoidance. OMNC is designed for long-lived unicast sessions in lossy wireless networks.

The authors in [24] address the problem of resolving conflicts of interest among multiple competing flows with wireless multi-path network coding. They use game theory to optimize resource allocation for network coding-based unicast protocols. In the proposed framework, called Dice, the problem is modeled as a network game, in which players share the bandwidth resource through negotiation or competition. For both cases, the players, which are the end users (destinations), perform a localized optimization of two subproblems: multi-path opportunistic routing, and the broadcast and coding rate allocation among competing players.

Dividing the packets into different batches (segments) and performing coding between the packets from the same segments is referred to as *segmented network coding*, which reduces the complexity of network coding. In MORE, the source node transmits only one segment at any time while waiting for acknowledgment from the destination node. This stop and wait policy degrades performance, as it leads to wasted wireless bandwidth. Also, the existence of just a single segment in the network may not be sufficient to saturate its delay-bandwidth product. This problem is addressed in [25] by allowing the coexistence of different segments. In the proposed method, called CodeOR, the source node transmits W (window size) concurrent segments. When the source node receives end-to-end feedback from the destination node, the node adds a new segment to the current window. In addition, each downstream node sends one-hop feedback after receiving a sufficient number of coded packets. The authors propose a heuristic to calculate the threshold for the sufficient packets at a given node. When a relay node (including the source node) receives an acknowledgment from all of its downstream nodes, it starts sending the packets of the next segment. The authors also adopt a similar algorithm to TCP Vegas [26], which uses increased queueing delays as congestion signals.

The authors in [27] propose an optimization framework for opportunistic routing based on network utility maximization (NUM), and they derive optimal scheduling, routing, flow control, and rate adaptation schemes. In this work, the links' rate constraints are defined per broadcast region instead of unicast links. The authors prove the optimality of their approach, and derive a primal-dual algorithm that is the basis of their practical protocol.

CCACK. The performance of MORE depends on the accuracy of the estimated loss rates. Loss rates change over time, but to reduce the overhead of calculating and collecting loss rates in MORE, the loss rates are collected only before the source node starts the transmission of the packets. The CCACK approach [28] solves this problem. In CCACK, nodes use cumulative coded acknowledgments, which allow nodes to acknowledge the coded received packets to their upstream nodes, using a single compressed feedback message, with almost zero cost. For this purpose, each node calculates the coefficients' null-space of the received coded packets, and the node adds the null-space to the forwarding messages. The null-space of a set of vectors V is a vector z, such that the inner products between z and each vector in V is zero. When an upstream node overhears a packet from a downstream node, the upstream node multiplies the coefficient of packets in its buffer with the received

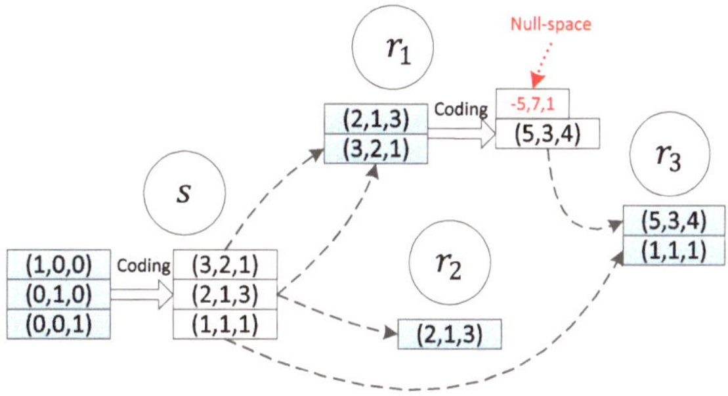

Fig. 8 CCACK approach

null-space. A nonzero result means that the packet in the buffer is innovative to the packet in the downstream node's buffer. CCACK is not applicable to WSNs, for the same reason that MORE is not applicable.

Let us consider Fig. 8. The source node s has three packets in its buffer. The node constructs three coded packets and broadcasts them. Assume that all nodes need to decode all of the packets. Node r_1 has only two packets, so it is not able to decode the received packets. The node r_1 will send a null-space of the received coefficient vectors, which can be any vector of form $(-5y, 7y, y)$. Suppose that z is chosen as $(-5, 7, 1)$. Since $(-5, 7, 1).(1, 1, 1) = 3 \neq 0$, node r_3 must transmit the packet $(1, 1, 1)$ to node r_1. On the other hand, $(-5, 7, 1).(2, 1, 3) = 0$, so node r_2 does not have any innovative packet for node r_1.

The use of null-space in opportunistic routing suffers from a problem called the collective space problem [28]. Suppose that nodes r_3, r_2, and r_1 are sorted in increasing order of their distance from the destination node. Nodes r_2 and r_3 collectively cover all of the three sent packets from the source node. As a result, the node r_1, which is farther from the destination node, does not need to transmit any more packets. However, the inner product of the packet $(3, 2, 1)$ in the buffer of r_1 and the null-space of r_2 and r_3 are not zero. The reason is that r_1 does not consider the collective covered space by nodes r_2 and r_3. In order to solve this problem, the authors in [28] use separate buffers at each node i for the coefficient of the received packets from upstream nodes (B_u), the coefficient of sent coded packets (B_w), and the received innovative packets (B_v). When the node i overhears a coded packet from the downstream nodes, the node marks the coefficients in B_u and B_v, if their inner product with the recited null-space is equal to zero. When the rank of the marked coefficient vectors in $B_u \cup B_w$ becomes equal to the rank of the packets in buffer (B_v), the downstream nodes (which are closer to the destination node) collectively cover all of the packets in the node i's buffer. Therefore, node i does not need to transmit more packets.

Most of the proposed methods for the networks with lossy links assume that the links are independent, and they do not consider the effect of the correlation between the links [29] on the performance. Take Fig. 9 for example. Assume that each node stops to transmit more packets when its next-hop nodes have collectively received the same number of linearly independent packets to what it has in its buffer. Assume

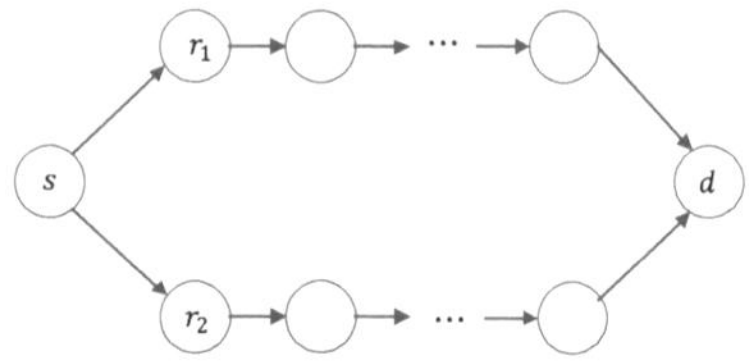

Fig. 9 The drawback of the CCACK approach when the links are highly correlated

that the delivery rate of the links between the source node s and the nodes r_1 and r_2 is 0.5, and the batch size is 6. In the case that the links $s \rightarrow d_1$ and $s \rightarrow d_2$ are independent, the source node needs to try 8 transmissions, as the probability of receiving a transmission by at least one of the nodes d_1 and d_2 is 0.75. In the case of highly correlated links, either both of the nodes will receive a transmission or none of them will. Therefore, the source node needs to transmit 12 packets. When the links are negatively correlated, exactly one of the nodes d_1 and d_2 will receive a transmitted packet. As a result, the number of required transmissions by the source node will be 6. It can be inferred that correlation between the links has a huge effect on the throughput of the methods. Now assume that the links are highly correlated, so the nodes r_1 and r_2 will receive all of the six packets. Since they are not aware of each other's received packets, both of them will send all of the packets, which results in unnecessary redundant transmissions. This problem can be solved by giving a credit, equal to 3, to nodes r_1 and r_2. The work in [30] considers correlation between the links and improves the performance of CCACK.

MIXIT. Symbol-level network coding for wireless mesh networks in introduced in [31]. The main idea behind the MIXIT approach is that even when no node receives a packet correctly, any given bit might be received by some node correctly. As a result, instead of insisting on forwarding only correct packets, the intermediate nodes can forward the correct received bits to the destination. For this purpose, the intermediate nodes in MIXIT use physical layer hints to guess which bits in a corrupted packet are likely correct. Unlike the previous work on network coding, the network code in MIXIT operates at the granularity of symbols, which is defined as a small sequence of bits, rather than packets. Take Fig. 10, in which the original symbols and the coded symbols are noted as s and S, respectively. In contrast with the packet-level network coding, each coded packet in MIXIT consists of multiple coded symbols. As a result, if some parts of a packet encounters with an error, the other symbols are still useful. In MIXIT, each router forwards random linear combinations of the high-confidence symbols belonging to different packets, and the destination node is able to decoded the symbols once it receives enough number of coded packets.

The first problem that MIXIT addresses is using an scalable coordination among the nodes in order to prevent duplicate transmissions of the same symbol. In contrast with node coordination-based approaches like ExOR, MIXIT uses the randomness from the network code and a dynamic programming algorithm to solve the cordination problem. The second issue is error recovery. The destination node needs to correct the errors that might exist in the received symbols. MIXIT uses symbol-level network coding along with an end-to-end maximum rank distance (MRD) codes [32]

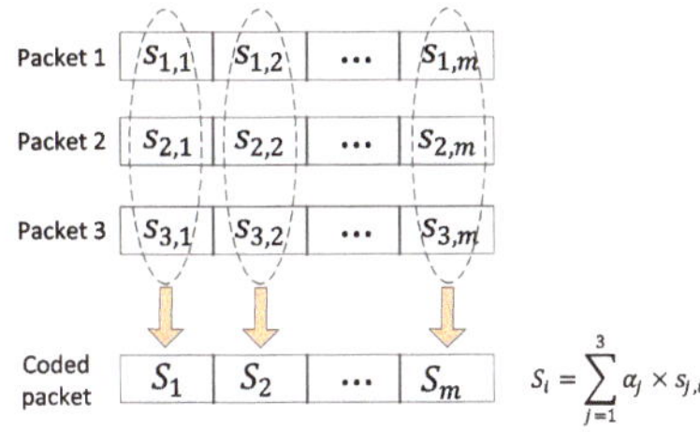

Fig. 10 Symbol-level network coding

for this purpose. The routers in MIXIT only forward random linear combinations of high-confidence symbols, and they do not perform any error correcting. MIXIT protocol benefits from a congestion-aware forwarding. It forwards coded symbols through paths that have small queues and high delivery probabilities. MIXIT may be applicable in WSNs to deliver data to sink nodes. In WSNs most traffic is from the sensors to the sink node, so data from different sensor nodes can be coded together to improve throughput. The MIXIT protocol can also be used for multicast applications in mesh networks. For this purpose, routers can keep transmitting coded packets until all destination nodes can decode them.

The authors in [33] show that the symbol-level network coding outperforms the packet level network coding for content distribution in vehicular ad hoc networks (VANETs). In [34], they later study the advantage of symbol level network coding for live media streaming in VANETs. As shown in Fig. 11, the goal in [34] is to designate live streaming multimedia to all of the nodes in a specific region of a road, called area of interest. The core part of the proposed method, called CodePlay, is a coordinated local push mechanism. In order to disseminate the content from sources to all the receivers smoothly and timely, a set of spatially separated relay nodes are selected in CodePlay distributively. The relay nodes are selected in such a way that their transmissions can bring most useful information to their nearby vehicles. For this purpose, CodePlay uses an objective function to calculate the contribution of each

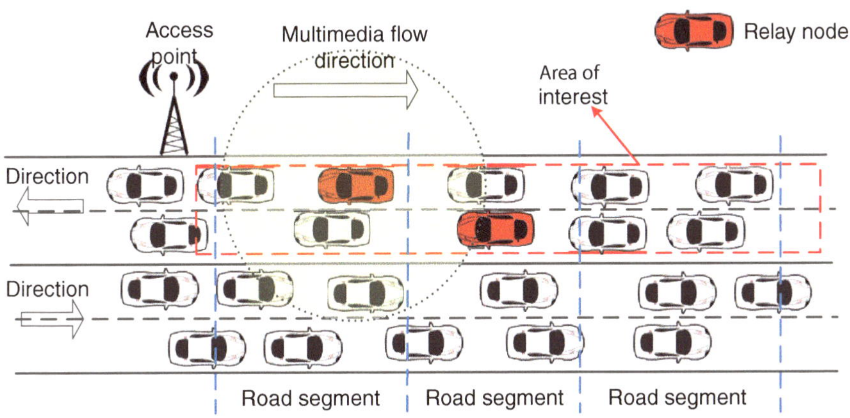

Fig. 11 Multimedia Streaming in VANETs

potential relay node. The proposed method segments the road during initialization so that the relay selection could be made locally within each segment. Each selected relay node actively pushes coded data to cover its neighborhood. Using symbol-level network coding CodePlay can better tolerance transmission interference, and concurrent transmissions of all relays could be optimally coordinated locally. In CodePlay, adjacent segments share the wireless channel resource in a round-robin fashion to reduce interference.

3.3 Joint Inter and Intra-session Network Coding

It is not desirable to use inter-session network coding alone in a lossy link environment, since intra-session network coding is an efficient way to deal with the lossy links. Thus, it is critical to have joint inter and intra-session network coding for wireless networks with lossy links.

The work in [35] proposes a heuristic to combine inter and intra-session network coding in lossy multi-hop wireless networks. This approach limits network coding to be within a hop and provides a limited performance gain in the range of 20 % to 30 %. Also, the proposed approach in [35] lacks theoretical analysis.

A joint inter and intra-session network coding scheme, called I^2NC, is proposed in [36]. This work is grounded in network utility maximization formulation of the problem. Assuming that the number of packets in each segment is k, each relay node constructs $k + k'$ linear combination of the packets instead of k coded packets. It is sufficient for the receiver nodes to receive k out of $k + k'$ packets. In other words, the k' additional packets work as parity packets. After adding redundancy to the packets, and coding them together, I^2NC uses inter-session network coding to mix the coded packets of different sessions. The authors propose two schemes: I^2NC-state and I^2NC-stateless. In the former scheme, each node listens to all transmissions in its neighborhood, stores the overheard packets in its buffer, and periodically informs its neighbors about the content of the buffer. In contrast, the I^2NC-stateless scheme only relies on the local loss-rates of the links.

A cross-layer optimization scheme for lossy 2-hop relay networks is proposed in [37] that uses joint inter and intra-session network coding. The work optimizes overhearing, considers flows instead of packets, and assumes limited feedback. Linear equations are used to characterize the capacity region for the problem of when the number of sessions is less than three. Also, a near-optimal coding scheme is proposed for the case with more than two sessions, and its performance is characterized using linear equations. However, the complexity of the near-optimal scheme is hyper-exponential.

A polynomial time coding method for the 2-hop relay network problem is proposed in [38, 39]. This scheme, which uses random linear network coding, makes a linear number of decisions. The authors characterize the performance of their scheme by using linear constraints in terms of link delivery rates. They use the proposed 2-hop relay scheme as a building block to extend the proposed scheme to multi-hop wireless networks. Based on this policy, a linear programming formulation of the achievable rate region is proposed.

3.4 Summary and Discussion

COPE is the first proposed practical inter-session network coding method. Its complexity is not high, and it works in networks with perfect links. However, COPE is not appropriate when the links have a moderate loss probability of 20 %, as it turns off coding in this case. CLONE solves this problem by sending different redundant coded packets such that a given level of reliability is provided. However, CLONE does not optimize the overhearing, and it limits the operation to XOR. As the optimal solution is #p-complete, approximation heuristics are proposed. In [19], the authors tackle the problem of optimal inter-session network coding from a different angle, as they consider flows instead of packets. They optimize overhearing and characterize the capacity region. The authors in [14] propose a cross-layer method that combines the routing and inter-session network coding. They model interference between the nodes as a conflict graph, and they find the optimal solution by using optimization techniques. The drawback of this work is that some overhearing cases are not considered during the formulation of the problem.

MORE is a practical opportunistic routing approach that uses intra-session network coding to provide reliability in lossy link environments. In MORE, there is no need to send feedback messages from the intermediate nodes, and only when the destination nodes receive all of the packets a feedback message is sent to stop the source node from sending more packets. The Dice method addresses the problem of a conflict of interests among multiple flows. The CodeOR protocol solves the stop and wait problem of MORE, which degrades performance. The CCACK method is proposed to solve vulnerability of MORE to links' quality changes. Instead of estimating the number of required transmissions, in CCACK, the intermediate nodes send the null-space of the received coded packets to help their neighbors discover when they should transmit more packets.

MIXIT proposes the idea of performing network coding in the granularity of symbols instead of packets to increase the transmission efficiency in lossy environments. CopePlay uses the idea of symbol-level network coding for live multimedia streaming in VANETs. Using symbol-level network coding for the highly mobile nodes in VANETs decreases the interference problem by enabeling using the received correct symbols even in the case that a packet is not received correctly.

Table 1 classifies the discussed methods for unicast application based on the used methodology. This table also shows the objective of the approaches, and whether they assume the existence of lossy links or perfect links.

Table 1 Classification of the network coding methods for unicasting

Approach	Methodology	Topology	Objective	XOR or RL	Local or Global	Links
COPE [6]	Inter-session	Multi-hop	Throughput	XOR	Local	Lossy
[14]	Inter-session	Multi-hop	Throughput	XOR	Local	Perfect
[15]	Inter-session	Multi-hop	Energy efficiency	XOR	Local	Lossy
NCAQM [16]	Inter-session	Multi-hop	Throughput	XOR	Local	Perfect
CLONE [18]	Inter-session	Multi-hop	Throughput	XOR	Local	Lossy
[19]	Inter-session	Multi-hop	Throughput and fairness	XOR	Local	Lossy
SensCode [21]	Inter-session	Multi-hop linear network	Throughput	Physical	Local	Perfect
MORE [8]	Intra-session	Multi-hop	Throughput	RL	Global	Lossy
OMNC [23]	Intra-session	Multi-hop	Throughput	RL	Global	Lossy
Dice [24]	Intra-session	Multi-hop	Throughput and fairness	RL	Global	Lossy
Dice [27]	Intra-session	Multi-hop	Throughput	RL	Global	Lossy
CCACK [28]	Intra-session	Multi-hop	Throughput	RL	Global	Lossy
[30]	Intra-session	Multi-hop	Throughput	RL	Global	Lossy
MIXIT [31]	Intra-session	Multi-hop	Throughput	RL (symbol-level)	Global	Lossy
CopePlay [31]	Intra-session	Multi-hop	Throughput	RL (symbol-level)	Global	Lossy
[35]	Joint Inter and Intra-session	Multi-hop	Throughput	XOR	Local	Lossy
[36]	Joint Inter and Intra-session	Multi-hop	Throughput	RL	Local	Lossy
[37]	Joint Inter and Intra-session	2-hop	Throughput and fairness	RL	Local	Lossy
[38]	Joint Inter and Intra-session	2-hop Multi-hop	Throughput and fairness	RL	Local	Lossy

4 Network Coding Methods for Multicast Applications

In this section, we look at network coding approaches that can be used for multicast applications. With simple modifications, some of the proposed approaches for unicast application can be applied for multicasting. To the best of our knowledge there is no inter-session network coding for multicasting in wireless networks. It should be noted that there are some works on inter-session network coding for multicasting in wired networks, which are beyond the scope of this chapter.

4.1 Intra-session Network Coding

In addition to unicast applications, MORE [8] can be used for multicasting. For this purpose, the authors make simple modifications to their unicast algorithm. The first reconciliation is that the source node does not proceed to the next batch until all of the destinations receive the packets in the current batch. Also, the list of forwarder nodes for multicast applications differs from the unicast. The source node computes the list of forwarder nodes for each unicast flow from itself to the destinations in the multicast group. The forwarder list of the multicast flow is the union of the forwarders of the unicast flows. Moreover, the TX_credit (transmissions credit) at each forwarder node is the maximum of the required transmissions for different unicast flows in the multicast group. The last modification is that when the source node receives a feedback message from a destination node, the source node recomputes the list of the forwarder nodes and their TX_credits for the remaining destinations. As discussed before, MORE is not applicable in WSNs.

The modified MORE for multicast applications suffers from two problems. First, it can lead to congestion since too many nodes may act as forwarder nodes, even for a single destination. This situation is worsened as the number of flows increases. Next, if one of the receivers has a poor connection, then trying to satisfy reliability for this receiver may result in throughput degradation for the other receivers. This problem is called the *crying baby* [40] problem and is unique to multicast. The Pacifier approach [41] proposes a multicast tree-based opportunistic routing design to solve these problems. Pacifier creates a multicast tree to connect the source node to the multicast receivers. The source node builds this shortest-ETX tree by taking the union of all of the shortest-ETX paths from the source to the receivers. The source node reconstructs this tree when a receiver node receives the complete batch. In contrast with MORE, in which every node with a greater ETX value can be the next-hop, Pacifier limits the forwarder nodes to the nodes in the multicast tree. The source node assigns a TX_credit to each forwarder node. TX_credit specifies how many packets that a forwarder node should transmit upon the reception of a packet from an upstream node (a node with a greater ETX).

To solve the crying baby problem, Pacifier changes the sending pattern of the batches. In MORE, the source node does not start transmitting the next batch until all of the destination nodes acknowledge reception of the current batch. However, the source node in the Pacifier approach transmits the packets in a Round-Robin pattern. In details, when one of the receivers sends the acknowledgment of receiving the current batch, the source node moves to the next batch. Forwarder nodes only buffer the packets belonging to the current batch, and the nodes delete their buffer upon reception of a packet from a new batch. The source node will continue sending the packets from the first batch when the other batches are received by at least one

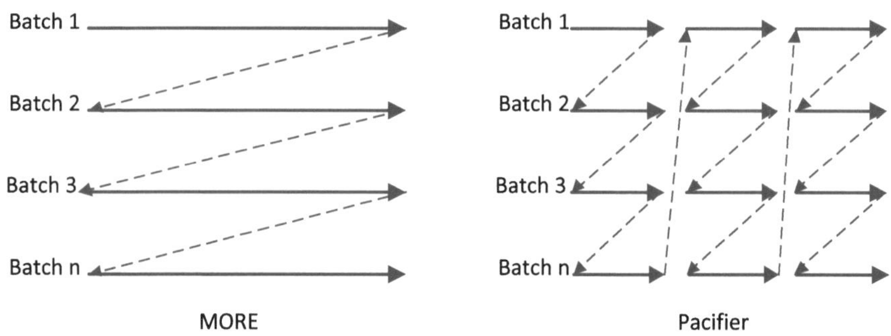

Fig. 12 The order of transmitting the batches of packets in the MORE and Pacifier

of the destination nodes. Figure 12 describes the order of transmitting the packets by the source node in the MORE and Pacifier approaches. In contrast to MORE, Pacifier is suitable for WSNs. The reason is that Pacifier limits the forwarder nodes to the nodes in the multicast tree. This method can be used in WSNs to send code updates or other data from the sink node to a group of sensor nodes.

In [42], the authors address the network coding-based opportunistic routing problem for multicast. They argue that the important factors that affect the performance of the multicast protocols are loss rate, the correlation among the links, and the reachability of the node. They formulate the optimal network coding-based opportunistic routing for multicast as an optimization optimization, and develop a distributed algorithm for the problem in which each node only requires local information. The proposed distributed algorithm consists of two phases. In the first phase, the proposed method uses ETX metric to construct the most reliable broadcasting tree. In the second phase, each node runs a credit assignment algorithm to calculate the number of coded packets that it has to send. The authors show that the proposed distributed algorithm adapts to the changes in the channel conditions, and converges to the optimal solution. Moreover, this approach does not need any explicit knowledge about the correlation among the links or the channel conditions. In addition to using coded packet, the authors use coded feedback messages to reduce the number of feedback messages, and to resolve the problem of delayed feedback. The simulation result in the paper show the effectiveness of the proposed method over the MORE approach. The distributed approach can be applied on WSNs in multicast applications such as software update of the sensor nodes.

Table 2 classifies the discussed methods for multicast application, based on their methodology, objective, and whether they assume the existence of lossy links or perfect links.

Table 2 Classification of the network coding methods for multicasting

Approach	Methodology	Topology	Objective	XOR or RL	Local or Global	Links
MORE [8]	Intra-session	Multi-hop	Throughput	RL	Global	Lossy
Pacifier [41]	Intra-session	Multi-hop	Throughput	RL	Global	Lossy
CCACK [28]	Intra-session	Multi-hop	Throughput	RL	Global	Lossy
[42]	Intra-session	Multi-hop	Throughput	RL	Global	Lossy

5 Network Coding Method for Broadcast Applications

In this section, we survey some of the works that address inter-session and intra-session network coding for broadcast applications. At the end, we compare the proposed methods and summarize the results.

5.1 Inter-session Network Coding

CODEB The problem of minimizing the number of transmissions in all-to-all broadcasting is addressed in [43]. All-to-all broadcasting is a special case of broadcasting, in which all node broadcast their respective packets to all other nodes. In this paper, the authors combine network coding with a deterministic forwarding approach, and they show that using network coding results in a significant reduction in the number of transmissions. They apply coding to the partial dominant pruning (PDP) [44] forwarding approach, which is a local forwarding method, but their coding algorithm can be applied to other localized deterministic approaches. The PDP approach is used to select a subset of the nodes as the relay nodes. In PDP, each relay node uses two-hop local information to select a subset of its neighbors such that they can cover all two-hop neighbors of the relay node. In the CODEB approach, each relay node maintains a neighbor reception table that shows the received packets by each neighbor. CODEB is inappropriate for WSNs, because of its overhead.

In the proposed XOR-based coding method, the relay nodes code the packets such that their neighbors can decode the received coded packets using the packets in their buffer. This means that the receiver nodes can decode the received XOR-ed coded packet without waiting for more packets to arrive. In more detail, each relay node with a set of packets in its output queue tries to find a subset of the packets to XOR, such that the number of native packets in the coded packet are maximized. In [43], it is shown that this problem is NP-hard. Therefore, a greedy algorithm is proposed. This greedy algorithm selects the first packet in the output queue and sequentially checks if other packets can be XOR-ed with this packet, such that all of the neighbors can decode the coded packet. In the case of delay-tolerant networks, when there is no coding opportunity at a relay node, the node will postpone the transmission of the packets for a random amount of time.

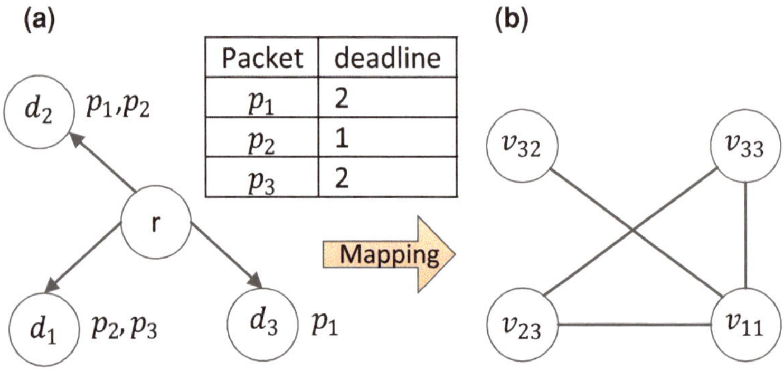

Fig. 13 Mapping delay-aware network coding problem to maximal clique problem

Directional Antenna. The problem of efficient broadcasting using network coding and directional antennas is studied in [45]. A node with directional antenna capabilities can divide the omnidirectional area into different sectors and turn a subset of them on for transmission. Therefore, in the proposed method, the forwarder nodes transmit the coded or uncoded broadcast messages to restricted sectors, which decreases the energy consumption. The authors assume that the links are perfect, and they use a directional connected dominating set (DCDS) [46] to construct a directional network backbone. A connected dominating set is a subset of the nodes such that all of the nodes in the set are connected together. In addition, each node of the network is either a member of this set or is connected to a member of this set. In the proposed method, each forwarder node performs the coding between the packets that should be sent in the same section. The proposed approach can be applied to WSNs that their sensor node are equipped with directional antenna; however, the method might not be realistic, as the links are assumed to be perfect.

Deadline-Aware. The problem of deadline-aware broadcast scheduling using network coding is considered in [47]. It is assumed that each packet has a deadline, by which it must be sent by a relay node. To solve the problem of minimizing the number of transmissions at a relay node subject to deadline constraints, the authors map the problem to a maximal clique problem. In the mapped problem, each vertex represents a packet needed by a node in the network. There is a link between two vertices if the vertices correspond to the same packet, or if the correspondent packet of each of the vertices is received by the correspondent node of the other vertex. This means that these two nodes have received the packet that the other node has missed. Therefore, if we code these packets together, the two destination nodes will be able to decode it.

Take Fig. 13a for example. The received packets of each node are shown beside it. Node v_{23} in the mapped problem (Fig. 13b) shows that node d_2 has not received packet p_3. Since both of the nodes d_2 and d_3 need packet p_3, there is a link between vertices v_{23} and v_{33}. Also, since node d_1 has packet p_2 and node d_3 has packet p_1, the vertices v_{11} and v_{32} are connected. After mapping the problem, a weight is assigned

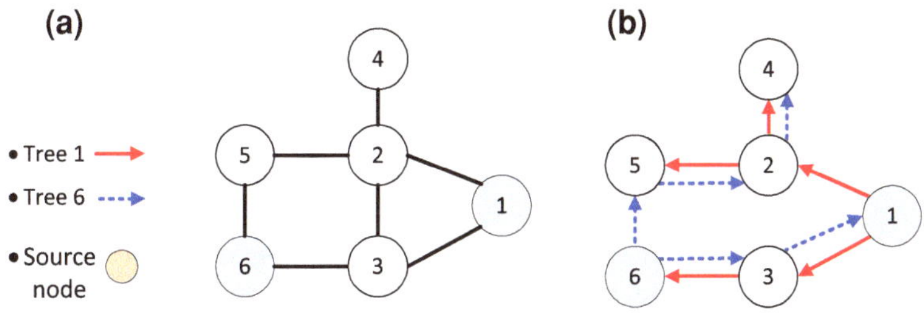

Fig. 14 **a** A given topology. **b** Two broadcasting trees routed at the source nodes 1 and 6

to each vertex. These weights are proportional to the deadline of the packets. As finding the maximal clique in a weighted graph is NP-complete, a greedy algorithm is used in [47] to find the maximal clique. After finding the maximal clique, the relay node codes the correspondent packets of the vertices in the clique together. The drawback of this work is that the deadline of the packets is considered for one-hop transmissions, but it is not clear how to calculate the one-hop delays to meet the global deadline. The authors study the effects of different weight functions in [48].

The problem of deadline-aware broadcasting using binary network coding is addressed in [49]. It is assumed that a subset of nodes are the source nodes, and each packet has a deadline to be received by all nodes, and the nodes have multi-channel multi-radio capability. Similar to [43], this work combines PDP, which is a deterministic forwarding approach with binary network coding. In [43], if there is a deadline constraint, the relay nodes will send the received packets immediately, which may decrease coding opportunities. In order to increase the coding opportunities, the authors in [49] propose three methods to compute the waiting time of the packets at the relay nodes, such that the packets meet the deadline constraints. The authors define the extra time as the remaining time to the deadline of the packet minus the maximum remaining hops to the farthest destination nodes (it is assumed that each transmission takes a unit of time to be received by the next hop). In the first method, which is velocity-based distribution of waiting time, the assigned waiting time to each relay node is equal to the extra time divided by the maximum remaining hops. Because of more coding opportunities at the nodes with more crossing flows, the second proposed method distributes the remaining time proportional to the number of crossing flows to the nodes, which increases coding opportunities. The last proposed method is a random distribution method, which randomly selects each node's waiting time from a specific range. All of the proposed methods in this work are very simple, and their computation complexity are low due to using XOR coding; thus, they can be applied in WSNs.

Authors in [50] study the problem of periodic broadcasting in wireless networks. In this work, a subset of the nodes are the source nodes and their packets should be broadcasted to all the nodes in the network. The authors use random linear network coding to reduce the number of required transmissions. In this work, a broadcasting tree is defined as a spanning tree routed at a source node. The authors propose the idea of using one broadcasting tree for disseminating each source packet, and preforming random linear network coding at the intermediate nodes that are relay nodes in more than one tree. The main idea behind using broadcasting trees is that it ensures decodability of the coded packets at every node, as every node receives enough linearly coded packets. Figure 14a show a given topology with two source nodes. The two broadcasting trees routed at the source nodes are shown in Fig. 14b. In this figure, nodes 2 and 3 are relay nodes in both of the trees. As a result, they can encode the received packets. Node 3 linearly combines the packets and transmits one coded packet. In contrast, node 2 needs to transmit two coded packets. This is because that there are two parallel edges from node 2 to node 4, which means node 2 should provide two packets to node 4. In order to minimize the number of parallel edges, which results in less number of transmissions, the broadcasting trees are constructed using a heuristic algorithm. In the next phase, in order to guarantee meeting the packets' deadlines, the authors propose a heuristic to partition the trees such that coding the packets of each partition does not result in any deadline misses. The proposed method can be used in WSNs for periodic broadcasting tasks. However, the drawback of this scheme for WSNs is that the unreliability of the links is not considered in this work.

Analysis. The problem of energy-efficient all-to-all broadcasting is studied in [51]. The presented theoretical analysis shows that network coding improves performance by a constant factor in fixed networks. The authors calculate this factor for some canonical networks, such as circular networks and square grid networks. They also propose a simple algorithm in which each node in the network sends a random linear combination of the received packets with a given probability, called the forwarding factor. To calculate the forwarding factors, two heuristics are proposed that use local two-hop local information. The first heuristic assigns forwarding factors to the nodes inversely proportional to the number of their 1-hop neighbors. The second heuristic sets the forwarding factors of the nodes inversely proportional to the minimum number of 1-hop neighbors of the node's neighbors. The authors extend their work in [52]. They show that, in networks where the topology dynamically changes and operations are restricted to a simple distributed algorithm, network coding offers improvements of factor $\log n$, where n represents the number of nodes in the network.

5.2 Intra-session Network Coding

One-hop. In [53], network coding is used to decrease the number of required retransmissions due to packet loss in one-hop broadcasting over packet-erasure channels. Firstly, the authors propose two NAK-based (negative acknowledgment) schemes without network coding to provide reliability for broadcasting. In the first proposed network coding-based broadcasting method, the source node receives a NAK message immediately after each message transmission. However, the source does not retransmit the lost packet immediately when it receives the NAK, and it maintains a list of lost packets and the receivers that lost each packet. The retransmission phase starts at a fixed interval of time. Then, the source node tries to code the maximum

Fig. 15 One-hop broadcasting

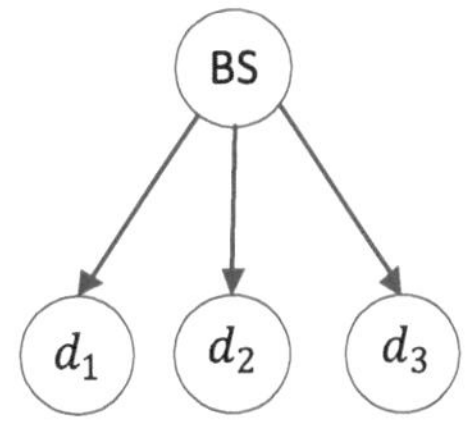

number of packets in a single coded packet. The source node retransmits the coded packet until all of the destination nodes that have a lost packet in the coded packet receive the packet. In an effort to improve the efficiency of the method, another method is proposed, in which the source node dynamically changes the coded packet based on the received feedback after each retransmission. This approach can be applied for applications such as driver or software updates of sensor nodes in single-hop WSNs. Base station sends the updates in one-hop transmissions to the sensor nodes, and receives NAK from the sensor nodes in the case of transmission errors.

The setting in [54] is the same as in [53], but here the base station (BS) broadcasts a fixed batch of packets. The proposed approach consists of two phases: information transmission phase and retransmission phase. In the first phase, the BS transmits the batch of N packets and receives a feedback message from each destination node. The BS uses the benefit of network coding in the retransmission phase to send the lost packets. The authors use XOR coding, and the constraint on coding packets together is that each destination should not have more than one lost packet in the coded packet. Firstly, the proposed algorithm finds the destination with the maximum number of lost packets, and adds each of its lost packets to a different coding set. Then, the algorithm sorts the remaining erased packets in increasing order, according to the number of coding sets, the lost packet can be allocated such that the coding constraint is satisfied. Starting from the packet with the minimum number of choices, the remaining lost packets are allocated to an eligible encoding set. If there is no eligible coding set for a packet, a new coding set will be generated. At the end, the BS node codes the packets of each coding set together and transmits them. This process is repeated until all of the destination nodes receive all of the packets. Similar to [53], this method can be used for driver updates of sensor nodes in WSNs.

Assume that in Fig. 15, node d_1 missed packets p_1, p_2, and p_3. Also, node d_2 missed packets p_1 and p_4, and node d_3 missed packet p_5. The node d_1 has the maximum number of lost packets. Therefore, the algorithm adds each of the lost packets by node d_1 as a separate coding set. Let $S_1 = \{p_1\}$, $S_2 = \{p_2\}$, and $S_3 = \{p_3\}$. Now, packets p_4 and p_5 are remaining. Packet p_4 can be added to sets S_2 and S_3, but packet p_5 can be added to S_1, S_2, or S_3. The packet p_4 has the smallest number of choices, so the algorithm will add it to one of the sets, S_2 or S_3. Assume that packet p_4 is added to set S_2. Packet p_5 can be added to all of the sets. The final result will be $S_1 = \{p_1, p_5\}$, $S_2 = \{p_2, p_4\}$, and $S_3 = \{p_3\}$.

In [55], the same problem as in [53] is addressed using a different approach. The authors map the problem to a graph coloring problem and introduce a greedy

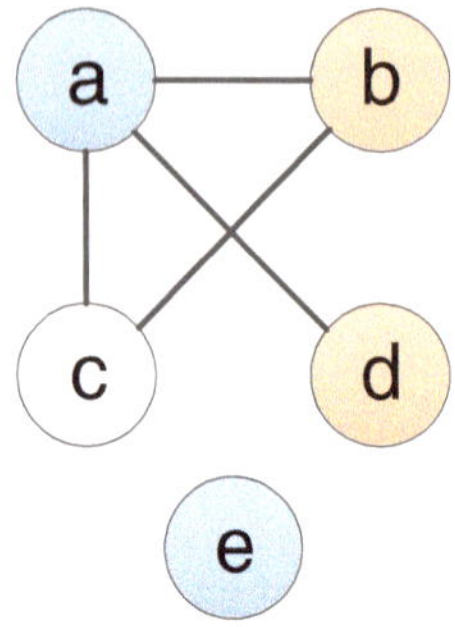

Fig. 16 Graph coloring in reliable broadcasting using network coding

heuristic to solve it. The mapping process is as follows. For each lost packet, a vertex is added to the graph. If two packets are missed by the same destination node, a link will be added between the corresponding vertices. A coding constraint implies that missed packets by the same destination nodes cannot be coded together since the destination node will not be able to decode the coded packet. This constraint is exactly the same as the coloring constraint, in which two neighbor nodes cannot be colored with the same color. Therefore, the vertices (packets) with the same color can be coded together, and the minimum number of required colors for coloring the correspondent graph is equal to the number of transmissions. The graph coloring problem is also NP-complete; the authors use the proposed greedy algorithm in [56] to address the mapped problem. This greedy algorithm sorts the vertices in descending order, according to their degree. Then, starting from the first node, the algorithm colors this node and all of the nodes that are not connected to this node with the same color. This process is repeated for the uncolored nodes. Figure 16 shows the mapping from the example in the previous paragraph to a graph coloring. The proposed approach is useful for broadcasting data from the base station to the sensor nodes in single-hop WSNs.

The problem of efficient one-hop broadcasting of layered-video is studied in [57]. In this problem, a server node broadcasts a layered-video to a set of users. Because of different channel conditions, the clients receive different number of transmissions from the server. A promising approach to overcome this problem is using multi-resolution coding (MRC) [58–60]. MRC is originally introduced for wired networks, and it divides a video into a base layer and multiple enhancement layers. In this scheme, the clients can independently decide how many layers to receive from the server according to their available bandwidth from the server. In contrast with the wired networks, in a wireless network all transmitted layers share the medium. As a result, sending higher layers reduces the available bandwidth for sending lower layers. The authors in [57] show that we can overcome the user diversity problem in broadcasting video over Wi-Fi by combining MRC with inter-layer network coding to increase the number of useful layers that can be retrieved by the users (It should be noted that inter-layer coding is different from inter-session coding. This work is

Fig. 17 **a** Original packets. **b** General form of random linear network coding. **c** Triangular network coding

(a)
$$\begin{cases} l_1 \\ l_2 \\ l_3 \end{cases}$$

(b)
$$\begin{cases} l_1,\ l_2\ ,\ l_3 \\ l_1 + l_2\ ,\ l_1 + l_3,\ \ l_2 + l_3 \\ l_1 + l_2 + l_3 \end{cases}$$

(c)
$$\begin{cases} l_1 \\ l_1 + l_2 \\ l_1 + l_2 + l_3 \end{cases}$$

in the category of intra-session network coding as the coding is done between the packets of the same session).

It is shown in [57] that inter-layer coding improves the number of decoded layers even for a single receiver. The reason is that it allows retrieving useful layers from more combinations of received transmissions. The authors show that even for a single receiver, the previously proposed even canonical triangular scheme [61, 62] for inter-layer network coding can perform poorly, and they propose two simple heuristics to enhance the gain. In triangular network coding, the encoded layers are in the form of $\sum_{j=1}^{k} \alpha_j l_j$, where $1 \leq k \leq h$ and α_j is a random coefficient. In other words, each coded layer is a combination of the first k original layers. The advantage of triangular network coding is that it reduces the number of possible coding strategies. Considering a video with n layers, the possible ways for coding the layers using inter-layer triangular coding and the general form of linear coding are equal to n and $2^n - 1$, respectively. Figures 17b and c show the possible coded layers of the original layers in Fig. 17a using the general form of network coding and triangular coding, respectively. We do not show the coefficients in the figures for simplicity. For example, $l_1 + l_2$ means $\alpha_1 l_1 + \alpha_2 l_2$, where α_1 and α_2 are two random coefficients. For the case of multiple receivers, the proposed method calculates the gain of all possible canonical triangular coding, and selects the best one. The authors propose three optimization technics that drastically reduce the complexity of scanning the gain of all the possible canonical triangular schemes.

Relay-Aided. The problem of efficient relay-aided one-hop broadcasting is addressed in [63]. This paper is an extension of [54], in an effort to use a relay node (Fig. 18). The authors assume that the links are lossy, and the base station (BS) to relay channel and the relay to users' (destinations) channels are better than the

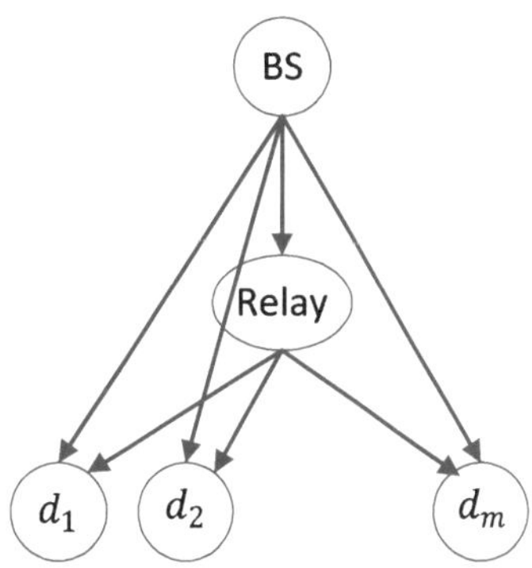

Fig. 18 Relay-aided broadcasting

BS to user channels. The proposed method has 3 phases. In the fist phase, the BS transmits N packets to the relay and user nodes. Then, the user nodes send feedback messages to notify the BS and the relay node about the received packets by the user nodes. Also, the relay node sends a feedback message to the BS node. In the second phase, the BS node retransmits the set of lost packets by the user nodes. The BS node uses network coding to increase the efficiency of this phase. After transmitting all of the lost packets, the relay and user nodes send new feedback messages. The BS node repeats this process until the user nodes receive all of the packets or the relay node receives all of the lost packets by the user nodes. The third phase is similar to the second phase, but the relay node performs retransmissions instead of the BS node. This is because the relay node has all of the lost packets by the user nodes, and the relay to user links are better than BS to user links. Much like as [53], this method is appropriate for data transmissions from the base station to the sensor nodes.

Multi-hop. A reliable data dissemination protocol using adaptive network coding, called AdapCode, is proposed in [10]. The authors use linear network coding to reduce the traffic in WSNs, which results in increasing the battery life of the sensors. They show that when nodes have more neighbors, we can increase the segment size (the number of packets that will be coded together). This allows us to encode more packets together without losing reliability, since nodes can get enough coded packets from their neighbors. Based on this observation and the fact that the network topology may change, an adaptive network coding protocol is proposed, where nodes dynamically change the segment size.

An extension over AdapCode method is proposed in [64]. In the AdapCode approach, throughout the code dissemination process, each sensor node dynamically decides how many packets should be coded together (decides about the segment size). The performance of the AdapCode method highly depends on the density of the sensor nodes. Therefore, it is important to calculate the number of the sensor nodes' neighbors correctly. However, in AdapCode the nodes can only find their full active neighbors, and in the case that their neighbors do not send any message, the number of neighbors cannot be calculated correctly. To solve this problem, the authors in [64] propose an energy-efficient neighbors discovery method. To make the discovery process efficient, they use network beacons. After running the discovery phase, a similar code dissemination phase to the AdapCode approach will be run to deliver the packets to the sensor nodes.

The R-Code approach, proposed in [65], uses network coding to provide reliable broadcast in wireless mesh networks with unreliable links. R-code uses ETX value of the links as their weights, and constructs a minimum spanning tree. In contrast with the AdapCode approach, in R-code only the non-leaf nodes in the spanning tree are the relay nodes. Each parent node is responsible for delivering a sufficient number of linearly coded packets to its children nodes. The parent node stops sending more packets after receiving acknowledgment messages from all of its children nodes. Similar to AdapCode, R-code can be used in WSNs to send code updates from sink node to the sensor nodes.

The DutyCode approach, which combines network coding with duty-cycling is proposed in [66]. Duty-cycling is a technique for saving energy in WSNs. In this scheme, the nodes turn off part or all of their systems for periods of time. Network coding and duty-cycling achieve energy efficiency through conflicting means. Network coding saves energy by exploiting the broadcast nature of the medium and overhearing, whereas duty-cycling saves energy by reducing idle listening, which reduces overhearing. The authors in [66] address the combination of these technics in flooding-based WSNs applications, such as code dissemination applications that require a non-negligible amount of time, possibly tens of minutes in large-scale WSNs.

The main idea in DutyCode is that due to the redundancy of coding, in some periods of time a sensor node does not benefit from overhearing coded packets. The goal of the authors is to determine these periods of time, and let sensor nodes that do not benefit from these useless packets, to go to the sleep mode. DutyCode is a cross layer method. In this approach, the MAC layer provides streaming, random sleeping and synchronization facilities. On the other hand, the proposed network coding-aware application layer uses information from the stream being transmitted to determine the time to sleep and its duration. In DytyCode approach, the network coding application specifies the sleep duration when it requests the node to go to the sleep mode. Then, the MAC protocol turns off the sensors radio for the requested duration if there is no pending transmission. The MAC protocol does not put the sensor node in the sleeping mode periodically. When requested, and if feasible, it shuts down the sensor's radio for the requested period.

5.3 Summary and Discussion

The proposed inter-sessions network coding approach in [43] is a distributed method that relies only on local 2-hop information. It is an appropriate scheme for the case of a delay-tolerant network. However, in the case of applications with deadline constraints, the relay nodes forward the received packets immediately, and the relay nodes do not postpone the transmission of the received packets to receive more packets. In order to increase the coding opportunities at the relay nodes, the authors in [49] proposed three methods to compute waiting time at the relay nodes; this is to assure that these waiting times do not result in any deadline misses. The main drawback of this work is that it is assumed that the nodes have multi-channel multi-radio capabilities. The work in [50] uses broadcasting trees to periodically disseminate the source packets to all the other nodes in the network, and performs random linear network coding at the intermediate nodes. In [48], relay nodes transmit the packets in an order which decreases the packet delay. It is assumed that there is a deadline to transmit a packet at each relay node. However, it is not clear how the deadlines can be calculated.

Table 3 Classification of the network coding methods for broadcasting

Approach	Methodology	Topology	Objective	XOR or RL	Local or Global	Links
CODEB	Inter-session	Multi-hop	Throughput	XOR	Local	Perfect
[45]	Inter-session	Multi-hop	Transmissions	XOR	Local	Perfect
[48]	Inter-session	Multi-hop	Throughput, Deadline	XOR	Local	Perfect
[49]	Inter-session	Multi-hop	Throughput, Deadline	XOR	Local	Perfect
[50]	Inter-session	Multi-hop	Throughput, Deadline	RL	Global	Perfect
[52]	Inter-session	Multi-hop	Energy efficiency	RL	Global	Lossy
[53]	Intra-session	One-hop	Transmissions	XOR	Local	Lossy
[54]	Intra-session	One-hop	Transmissions	XOR	Local	Lossy
[55]	Intra-session	One-hop	Transmissions	XOR	Local	Lossy
[57]	Intra-session	One-hop	Transmissions	RL (Triangular)	Local	Lossy
[63]	Intra-session	One-hop	Transmissions	XOR	Local	Lossy
[67]	Intra-session	One-hop	Transmissions, Delay	XOR	Local	Lossy
AdapCode [10]	Intra-session	Multi-hop	Transmissions, Reliability	RL	Global	Lossy
R-code [65]	Intra-session	Multi-hop	Transmissions, Reliability	RL	Global	Lossy
DutyCode [66]	Intra-session	Multi-hop	Transmissions, Reliability, and duty-cycle	RL	Global	Lossy

The work in [45] uses the advantage of directional antennas to reduce the energy consumption of relay nodes. The authors use a directional connected dominant set to find relay nodes. Inter-session network coding is used at relay nodes to reduce the number of required transmissions. In [52], random linear network coding is combined with a probabilistic forwarding approach. The performance of this approach is highly dependant on the computed forwarding factor of the relay nodes. In this scheme, overestimating the forwarding factor results in unnecessary redundant transmissions. On the other hand, by underestimating the forwarding factor, the nodes will not be able to decode the packets due to receiving insufficient number of coded packets.

The approaches in [53–55] are proposed for one-hop reliable broadcasting applications. In these methods, the source node uses intra-session network coding to retransmit the missed packets by different destination nodes. As the problem of efficient reliable broadcasting is NP-complete, all of the proposed approaches are heuristic algorithms. The work in [55] is extended in [63] to use the advantage of relay node for one-hop broadcasting applications. Triangular network coding is used in [57] in order to increase the efficiency of multi-layer video broadcasting to a set of client nodes over single-hop error-prone wireless networks.

The work in [10, 64–66] are proposed specifically for WSNs. In AdapCode, the sensor nodes use their neighbors density to adopt the number of packets coded together (segment size). In order to increase the efficiency of Adapcode, the authors in [64] use network beacons to propose an energy-efficient neighbors discovery method. The R-Code approach [65] uses ETX metric to construct a minimum spanning tree, and the non-leaf nodes in the tree keep sending linear coded packets until they receive acknowledgment messages from all of their children nodes. Duty-Code [66] combines network coding with duty-cycling to increase the battery conservation.

Table 3 classifies the discussed methods for broadcast application, based on the used methodology. This table also shows the objective of the approaches and whether they assume the existence of lossy links or perfect links.

6 Conclusion

In this chapter, we surveyed recently proposed network coding approaches for wireless networks and WSNs. In general, network coding methods can be classified as inter-session or intra-session network coding approaches. We surveyed some of the proposed inter-session network coding approaches, which allow mixing the packets from different sessions to solve the bottleneck problem. We also reviewed intra-session network coding methods, which use the diversity of the links and mix the packets from the same sessions to solve the packet loss problem. The network coding methods can be applied in unicast, multicast, or broadcast applications. Moreover, some of the network coding approaches have been proposed just for one-hop, two-hop, or multi-hop networks. Therefore, we classified the methods based on their objective, application, and network topology assumption. Some of the proposed approaches specific to WSNs are surveyed, and we argued which of the proposed network coding methods for general wireless networks are applicable and suitable for WSNs.

Acknowledgments This research was supported in part by NSF grants ECCS 1231461, ECCS 1128209, CNS 1138963, CNS 1065444, and CCF 1028167.

References

1. C. Cordeiro, H. Gossain, R. Ashok, D. Agrawal, The Last Mile: Wireless Technologies for Broadband and Home Networks, in *Tutorial Presented in the 21th Brazilian Symposium on, Computer Networks*, 2003
2. S. Cherry, The wireless last mile. IEEE Spectrum **40**(9), 18–22 (2003)

3. I. Akyildiz, X. Wang, A survey on wireless mesh networks. Commun. Mag. IEEE **43**(9), S23–S30 (2005)
4. I. Akyildiz, W. Su, Y. Sankarasubramaniam, E. Cayirci, Raptor codes. IEEE Commun. mag. **40**(8), 102–114 (2002)
5. D. Aguayo, J. Bicket, S. Biswas, G. Judd, R. Morris, Link-Level Measurements from an 802.11 b Mesh Network, in *ACM SIGCOMM*, 2004, pp. 121–132
6. S. Katti, H. Rahul, W. Hu, D. Katabi, M. Médard, J. Crowcroft, XORs in the air: Practical wireless network coding. ACM SIGCOMM Comput. Commun. Rev. **36**(4), 243–254 (2006)
7. S. Biswas, R. Morris, ExOR: Opportunistic multi-hop routing for wireless networks. ACM SIGCOMM Comput. Commun. Rev. **35**(4), 133–144 (2005)
8. S. Chachulski, M. Jennings, S. Katti, D. Katabi, Trading Structure for Randomness in Wireless Opportunistic Routing, in *ACM SIGCOMM*, 2007
9. L. Keller, E. Atsan, K. Argyraki, C. Fragouli, SenseCode: Network Coding for Reliable Sensor Networks, in *EPFL Technical, Report*, 2009
10. I. Hou, Y. Tsai, T. Abdelzaher, I. Gupta, AdapCode: Adaptive Network Coding for Code Updates in Wireless Sensor Networks, in *IEEE INFOCOM*, 2008
11. P. Maymounkov, Online Codes, in *Technical Report TR2002-833*, New York University, 2002
12. M. Luby, Lt Codes, in *The 43rd Annual IEEE Symposium on Foundations of Computer Science*, 2002, pp. 271–280
13. A. Shokrollahi, Raptor codes. IEEE Trans. Inform. Theor. **52**(6), 2551–2567 (2006)
14. S. Sengupta, S. Rayanchu, S. Banerjee, An Analysis of Wireless Network Coding for Unicast Sessions: The Case for Coding-Aware Routing, in *IEEE INFOCOM*, 2007, pp. 1028–1036
15. T. Cui, L. Chen, T. Ho, Energy Efficient Opportunistic Network Coding for Wireless Networks, in *IEEE INFOCOM*, 2008, pp. 361–365
16. H. Seferoglu and A. Markopoulou, Network Coding-Aware Queue Management for Unicast Flows Over Coded Wireless Networks, in *NetCod*, 2010, pp. 1–6
17. J. Le, J. Lui, D. Chiu, How Many Packets Can We Encode?-An Analysis of Practical Wireless Network Coding, in *IEEE INFOCOM*, 2008, pp. 371–375
18. S. Rayanchu, S. Sen, J. Wu, S. Banerjee, S. Sengupta, Loss-aware network coding for unicast wireless sessions: Design, implementation, and performance evaluation. SACM SIGMETRICS Perform. Eval. Rev. **36**(1), 85–96 (2008)
19. A. Khreishah, J. Wu, P. Ostovari, and I. Khalil, Flow Based Xor Network Coding for Lossy Wireless Networks, in *IEEE GLOBECOM*, 2011
20. A. Khreishah, I. Khalil, P. Ostovari, J. Wu, Flow-based xor network coding for lossy wireless networks. IEEE Trans. Wireless Commun. **11**(6), 2321–2329 (2012)
21. S. Zhang, S. Liew, P. Lam, Hot Topic: Physical-Layer Network Coding, in *MobiCom*, 2006, pp. 358–365
22. I. Jawhar, N. Mohamed, D.P. Agrawal, Linear wireless sensor networks: Classification and applications. J. Network Comput. Appl. **34**(5), 1671–1682 (2011)
23. X. Zhang, B. Li, Optimized multipath network coding in lossy wireless networks. IEEE J. Selected Areas Commun. **27**(5), 622–634 (2009)
24. X. Zhang, B. Li, Dice: A Game Theoretic Framework for Wireless Multipath Network Coding, in *ACM MobiHoc*, 2008
25. Y. Lin, B. Li, B. Liang, Codeor: Opportunistic Routing in Wireless Mesh Networks with Segmented Network Coding, in *IEEE ICNP*, 2008
26. L. Brakmo, L. Peterson, TCP vegas: End to end congestion avoidance on a global internet. IEEE J. Selected Areas Commun. **13**(8), 1465–1480 (1995)
27. B. Radunović, C. Gkantsidis, P. Key, P. Rodriguez, Toward practical opportunistic routing with intra-session network coding for mesh networks. IEEE/ACM Trans. Network. **18**(2), 420–433 (2010)
28. D. Koutsonikolas, C. Wang, Y. Hu, CCACK: Efficient Network Coding Based Opportunistic Routing Through Cumulative Coded Acknowledgments, in *IEEE INFOCOM*, 2010, pp. 1–9
29. K. Srinivasan, M. Jain, J. Choi, T. Azim, E. Kim, P. Levis, B. Krishnamachari, The κ Factor: Inferring Protocol Performance Using Inter-Link Reception Correlation, in *ACM MobiCom*, 2010, pp. 7317–328

30. A. Khreishah, I. Khalil, J. Wu, Universal Opportunistic Routing Scheme Using Network Coding, in *IEEE SECON*, 2012
31. S. Katti, D. Katabi, H. Balakrishnan, M. Medard, Symbol-level network coding for wireless mesh networks. ACM SIGCOMM Comput. Commun. Rev. **38**(4), 401–412 (2008)
32. È. Gabidulin, Theory of codes with maximum rank distance. Problemy Peredachi Informatsii **21**(1), 3–16 (1985)
33. M. Li, Z. Yang, W. Lou, Codeon: Cooperative popular content distribution for vehicular networks using symbol level network coding. IEEE J. Selected Areas Commun. **29**(1), 223–235 (2011)
34. Z. Yang, M. Li, W. Lou, CodePlay: Live multimedia streaming in vanets using symbol-level network coding. IEEE Trans. Wireless Commun. **11**(8), 3006–3013 (2012)
35. C. Qin, Y. Xian, C. Gray, N. Santhapuri, S. Nelakuditi, I^2MIX: Integration of Intra-Flow and Inter-Flow Wireless Network Coding, in *IEEE SECON Workshops*, 2008, pp. 1–6
36. H. Seferoglu, A. Markopoulou, K. Ramakrishnan, I2NC: Intra-and Inter-Session Network Coding for Unicast Flows in Wireless Networks, in *IEEE INFOCOM*, 2011, pp. 1035–1043
37. C. Wang, A. Khreishah, N. Shroff, Cross-Layer Optimizations for Intersession Network Coding on Practical 2-Hop Relay Networks, in *Asilomar*, vol. 41, 2009, pp. 771–775
38. A. Khreishah, I. Khalil, J. Wu, Polynomial Time and Provably Efficient Network Coding Scheme for Lossy Wireless Networks, in *IEEE MASS*, 2011
39. A. Khreishah, I. Khalil,, J. Wu, Low Complexity and Provably Efficient Algorithm for Joint Inter and Intrasession Network Coding in Wireless Networks, in *IEEE Transactions on Parallel and Distributed Systems*, 2012
40. H. Holbrook, S. Singhal, D. Cheriton, P. fan and c. zhi and c. wei and k. ben letaief. ACM SIGCOMM Comput. Commun. Rev. **25**(4), 328–341 (2005)
41. D. Koutsonikolas, Y. Hu, C. Wang, Pacifier: High-Throughput, Reliable Multicast without Crying Babies, in *Wireless Mesh Networks, in IEEE INFOCOM*, 2009, pp. 2473–2481
42. A. Khreishah, I. Khalil, J. Wu, Distributed Network Coding-Based Opportunistic Routing for Multicast, in *MobiHoc*, 2012, pp. 115–124
43. L. Li, R. Ramjee, M. Buddhikot, S. Miller, Network Coding-Based Broadcast in Mobile Ad-Hoc Networks, in *IEEE INFOCOM*, May 2007, pp. 1739–1747
44. W. Lou, J. Wu, On reducing broadcast redundancy in ad hoc wireless networks. IEEE Trans. Mobile Comput., **1**(2), 111–122 (2002)
45. S. Yang, J. Wu, Efficient broadcasting using network coding and directional antennas in MANETs. IEEE Trans. Parallel Distrib. Syst. **21**(2), 148–161 (Feb 2010)
46. S. Yang, J. Wu, F. Dai, Efficient Backbone Construction Methods in Manets Using Directional Antennas, in *ICDCS*, 2007
47. Z. Dong, C. Zhan, Y. Xu, Delay Aware Broadcast Scheduling in Wireless Networks Using Network Coding, in *IEEE NSWCTC*, 2010, pp. 214–217
48. C. Zhan, Y. Xu, Broadcast Scheduling Based on Network Coding in Time Critical Wireless Networks, in *IEEE International Symposium on Network Coding*, June 2010
49. P. Ostovari, J. Wu, A. Khreishah, Deadline-Aware Broadcasting in Wireless Networks with Local Network Coding, in *IEEE ICNC*, Jan 2012
50. P. Ostovari, A. Khreishahand, J. Wu, Deadline-Aware Broadcasting in Wireless Networks with Network Coding, in *IEEE GLOBECOM*, Dec 2012
51. C. Fragouli, J. Widmer, J.L. Boudec, A Network Coding Approach to Energy Efficient Broadcasting: From Theory to Practice, in *IEEE INFOCOM*, 2006, pp. 1–11
52. C. Fragouli, J. Widmer, J.L. Boudec, Efficient broadcasting using network coding. IEEE/ACM Trans. Network. **16**(2), 450–463 (2008)
53. D. Nguyen, T. Tran, T. Nguyen, B. Bose, Wireless broadcast using network coding. IEEE Trans. Vehicular Technol. **58**(2), 914–925 (2009)
54. L. Lu, M. Xiao, M. Skoglund, L. Rasmussen, G. Wu, S. Li, Efficient Network Coding for Wireless Broadcasting, in *IEEE WCNC*, 2010, pp. 1–6
55. W. Fang, F. Liu, Z. Liu, L. Shu, S. Nishio, Reliable Broadcast Transmission in Wireless Networks Based on Network Coding", in *IEEE INFOCOM Workshops (INFOCOM WKSHPS)*, 2011, pp. 555–559

56. D. Welsh, M. Powell, An upper bound for the chromatic number of a graph and its application to timetabling problems. Comput. J. **10**(1), 85–86 (1967)
57. D. Koutsonikolas, Y. Hu, C. Wang, M. Comer, A. Mohamed, Efficient Online WiFi Delivery of Layered-Coding Media Using Inter-Layer Network Coding, in *ICDCS, 2011*, pp. 237–247
58. U. Horn, K. Stuhlmüller, M. Link, B. Girod, Robust internet video transmission based on scalable coding and unequal error protection. Sig. Process. Image Commun. **15**(1), 77–94 (1999)
59. D. Wu, Y. Hou, Y. Zhang, Scalable video coding and transport over broadband wireless networks. Proc. IEEE **89**(1), 6–20 (2001)
60. A. Majumda, D. Sachs, I. Kozintsev, K. Ramchandran, M. Yeung, Multicast and unicast real-time video streaming over wireless lans. IEEE Trans. Circuits Syst. Video Technol. **12**(6), 524–534 (2002)
61. D. Koutsonikolas, Y. Hu, C. Wang, M. Comer, A. Mohamed, On the Performance of Network Coding in Multi-Resolution Wireless Video Streaming, in *NetCod*, 2010, pp. 1–6
62. M. Halloush, H. Radha, Practical Network Coding for Scalable Video in Error Prone Networks, in *PCS*, 2009, pp. 1–4
63. L. Lu, M. Xiao, L. Rasmussen, Relay-Aided Broadcasting with Instantaneously Decodable Binary Network Codes, in *ICCCN*, 2011, pp. 1–5
64. H. Shwe, F. Adachi, Power Efficient Adaptive Network Coding in Wireless Sensor Networks, in *IEEE ICC*, 2011, pp. 1–5
65. Z. Yang, M. Li, W. Lou, R-code: Network Coding Based Reliable Broadcast in Wireless Mesh Networks with Unreliable Links, in *IEEE GLOBECOM*, 2009
66. R. Chandanala, R. Stoleru, Network Coding in Duty-Cycled Sensor Networks, in *INSS*, 2010, pp. 203–210
67. S. Sorour, S. Valaee, Minimum Broadcast Decoding Delay for Generalized Instantly Decodable Network Coding, in *IEEE GLOBECOM*, 2010, pp. 1–5

Chapter 6
Sleeping Techniques for Reducing Energy Dissipation

Rajani Muraleedharan, Ilker Demirkol, Ou Yang, He Ba, Surjya Ray and Wendi Heinzelman

Abstract Sensors have limited resources such as energy, computational power and bandwidth, and thus they require protocols and techniques that are resource aware and energy efficient. As energy waste through idle listening, retransmissions and overhearing are some of the primary causes of reduced lifetime in wireless sensor networks, sensor sleeping is critically important. Sleeping techniques prolong the network lifetime by placing components of the sensor node into a sleep mode while aiming to minimize the impact on application performance. Sensor sleeping can be applied to different layers of the protocol stack, and a cross-layer sleep manager can orchestrate sleeping in multiple layers simultaneously. In this chapter, the importance of sensor sleeping, the various sleeping techniques proposed and the applications using these approaches are discussed.

R. Muraleedharan · O. Yang · H. Ba · S. Ray · W. Heinzelman (✉)
University of Rochester, Rochester, USA
e-mail: wendi.heinzelman@rochester.edu

R. Muraleedharan
Saginaw Valley State University, University Center, USA
e-mail: rmuralee@svsu.edu

O. Yang
e-mail: oyang@ece.rochester.edu

H. Ba
e-mail: ba@ece.rochester.edu

S. Ray
e-mail: ray@ece.rochester.edu

I. Demirkol
Universitat Politecnica de Catalunya, Barcelona, Spain
e-mail: ilker.demirkol@entel.upc.edu

H. M. Ammari (ed.), *The Art of Wireless Sensor Networks*,
Signals and Communication Technology, DOI: 10.1007/978-3-642-40009-4_6,

© Springer-Verlag Berlin Heidelberg 2014

1 Introduction

Wireless sensor networks (WSNs) have paved the way for a new era in ubiquitous computing, where information can be sensed, accessed, and distributed over a range of environments with minimal or no human interaction. Applications of wireless sensor networks range from military to commercial applications such as border security and surveillance, personal health monitoring, environmental monitoring, and structural monitoring [1–3]. As the sensor nodes in these networks are often small, inexpensive devices that are battery-operated and communicate wirelessly, they are constrained by limited resources such as bandwidth, processing power and energy. Therefore, the reliability, longevity, and correctness of these applications depend primarily on the effective exploitation of the sensors through intelligent protocol design. In particular, in this chapter we will focus on protocol techniques that support the efficient use of energy in enabling the dissemination of sensed data to the sink(s).

In order to efficiently use the limited energy available in a wireless sensor network while simultaneously supporting the application goals, we must determine where the energy is "wasted" and develop protocol techniques that specifically address this waste. Any action that directly relates to the sensor's required contribution to the application (whether sensing, transmitting or routing data) is considered useful energy. On the other hand, actions such as idle listening, retransmission, and overhearing as well as sensing, transmitting or routing data that are not useful to the application are considered wasteful, and these are the operations we aim to reduce through protocol design.

Such energy waste can be minimized through a variety of techniques such as in-network data processing, physical-layer optimization and sleeping techniques. *In-network data processing* can be used to reduce the amount of data that must be transmitted to the sink(s) through local processing of the data. As the energy expended in routing information from multiple source nodes to a sink node is high in comparison with processing the data locally using fusion [4] or data aggregation techniques [5], much energy can be saved via in-network data processing. *Physical-layer optimization* can reduce the energy needed for the successful transmission of data over a communication link. For instance, energy consumption can be reduced by choosing an appropriate modulation scheme based on current channel noise [6]. However, the most significant energy savings is attained by *sleeping techniques*, which enable the sensor nodes to enter sleep modes to save energy whenever they are not performing useful tasks for the application, including sensing and communicating data.

Sleeping techniques can reduce a sensor's idle listening, which is often a significant portion of the "wasted" energy in a sensor node. In sleep mode, the node's radio is turned off and the microcontroller is put into a deep sleep. In general, in this low power mode of the microcontroller, only the RAM, interrupt handler and timer are functional. It is common to draw only nA-level current using this power mode with current off-the-shelf sensor motes [7].

Once a node is in the sleep state, the node must be woken up by either an internal or external interrupt. While sleeping techniques can save tremendous amounts of energy in the sensor node, the disadvantage of using sleeping is that a node (sender) that wants to communicate with the sleeping node must wait until the receiver node is awake, which causes a delay in the communication. Hence the duration of sleeping represents a trade-off between the energy saved by turning the sensor off when not needed and the communication latency and reliability of the application.

In this chapter, we take an in-depth look at sleeping techniques. We begin with an exploration of different approaches to wake up a sensor node from the sleeping state. Following this, we explore sleeping techniques at the medium access control (MAC) layer, considering both synchronous and asynchronous duty cycling as well as time division multiple access (TDMA) protocols that enable nodes to sleep. Moving up a layer, we discuss sleeping techniques at the network (routing) layer, including the use of sleeping in multi-path routing. Finally, we explore cross-layered approaches for maximizing the effectiveness of using sleeping techniques at both the MAC and routing layers.

2 Wake-up Techniques

Sensor sleeping is an effective approach for reducing node energy dissipation, as nodes that are not currently transmitting or receiving data go to sleep instead of wasting energy in the idle state. However, once a node is in the sleep mode, it must be woken up in order to return to an active state, where the node can transmit or receive data.

A sensor can be awakened using either an internally-controlled wake-up or an externally-controlled wake-up. The internally-controlled wake-up is used in duty cycling, where nodes schedule their wake-up time when they go to sleep, which is referred to as *scheduled wake-up* in this chapter. For externally-controlled wake-up, an external factor is the wake-up trigger. An external trigger can be sent by the user through a radio (*radio controlled wake-up*), or it can be an event in the environment (*environmentally controlled wake-up*).

A detailed categorization of the different wake-up techniques is illustrated in Fig. 1. In this section, we introduce these wake-up techniques along with a description of state-of-the-art research under each category. In the subsequent sections, we will provide a discussion of sleeping techniques at different protocol stack levels and some approaches to cross-layer sleeping.

2.1 Scheduled (Internally Controlled) Wake-up

In scheduled wake-up, a sensor node sets an internal timer to fire after a specific duration, and turns the radio off and the microcontroller to a very low power mode [8]. A common use of this approach is through periodic wake-ups, where nodes sleep

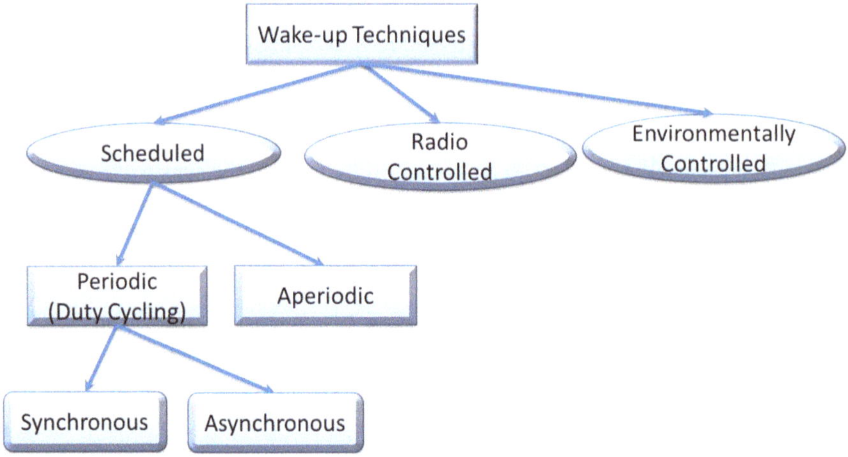

Fig. 1 Categorization of wake-up techniques for wireless sensor networks

and wake up according to a regular schedule. This is also called duty cycling, where the duty cycle of a node represents the percentage of time that the node is awake. Oftentimes the periods for sleeping and waking up are fixed a-priori, however in some cases this period can be adapted based on information learned by the sensors in the active periods [9].

A node may also set the timer such that it is awake for a specific event, i.e., with an aperiodic schedule. For example, if the node expects to transmit or receive a message at a certain future time (such as a time slot in a TDMA schedule), the timer is set to ensure that the sensor node is awake as needed. Both of these approaches for scheduled wake-ups trade off energy savings with the latency and the reliability of the application. Therefore, it is important to appropriately schedule the wake-up to meet application goals yet save as much energy as possible.

For periodic wake-up (duty cycling), nodes may be woken up either synchronously, where all the periodic wake-up times are synchronized among nodes, or asynchronously [10–13]. Synchronous duty cycling is beneficial for applications such as smart metering, environmental monitoring, and smart homes, as in these applications, the traffic load is often periodic and predictable, and nodes can sleep during the known inactive periods, avoiding idle listening and overhearing. Synchronous duty cycling, however, requires clock synchronization, which also means that frequent resynchronization is required to maintain schedule consistency among the nodes in the presence of clock drifts.

Synchronous duty cycling can be applied in groups or clusters of nodes [12], rather than to the entire network. The group synchronization approach is used by many MAC protocols, e.g., S-MAC [14], the operation of which is illustrated in Fig. 2. As seen in the figure, a SYNC period is reserved for synchronization purposes, where the nodes listen to the medium for the potential synchronization messages from their neighbors and send a synchronization message periodically. A node can therefore assume the

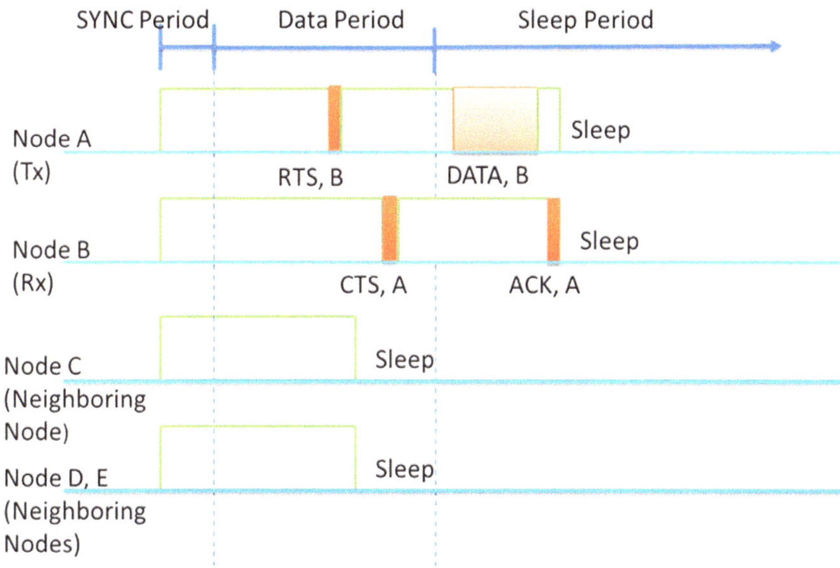

Fig. 2 Scheduled wake-up employed in S-MAC [16]

role of a synchronizer or follower based on whether it sends a SYNC message or follows a neighbor's schedule. Figure 2 illustrates a scheduled wake-up employed in S-MAC, where node A sends an RTS control signal to the follower node B, who in turn waits for the data exchange by forwarding a CTS control signal to node A. The neighboring nodes C, D, and E are synchronizers with independent sleep schedules. After receiving the data packet, node B sends an ACK to node A. Every node in a WSN is independent, and therefore issues such as collision can occur when multiple nodes attempt to be synchronizers, which is minimized using a random delay. As the overhead required to synchronize a network increases with the size of the network, group/cluster synchronization scales better than full network synchronization.

One issue with synchronous duty cycling is clock drift, whereby nodes become de-synchronized after a certain period of time. The SYNC period in S-MAC and related protocols helps to keep nodes synchronized even in the presence of clock drift, but for networks with extremely low duty cycles, there is still a chance that the nodes' active periods will not align due to de-synchronization of the clocks during the sleep period.

Asynchronous duty cycling, on the other hand, requires no prior knowledge about other nodes and no global timing of the network, as each node independently schedules its periodic wake-up [13]. While this presents a simple approach that requires no synchronization of all nodes in a network, and hence achieves minimal overhead traffic, the disadvantage of this technique is that nodes are not awake at the same time, and hence an awake node may need to remain awake for a long period of time in order to communicate with one of its neighbors. Sensor networks with low or bursty

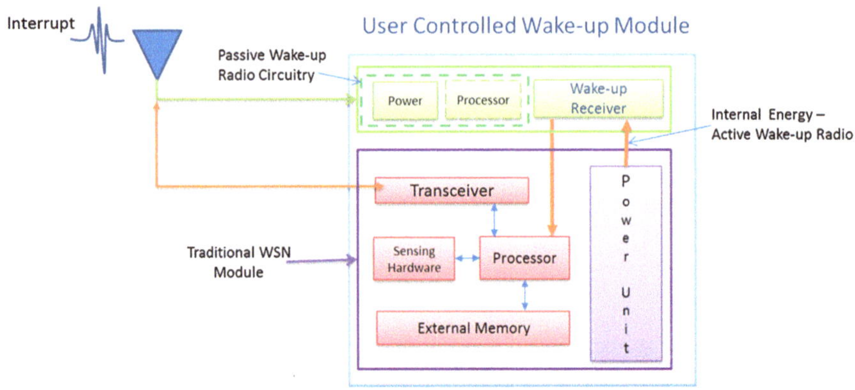

Fig. 3 Sensor node employing the radio controlled wake-up technique

traffic such as habitat and health monitoring can benefit from using asynchronous duty cycling, where nodes periodically wake up to check for traffic on the network and return to sleep until their next periodic wake-up if no traffic is detected [15].

2.2 Radio Controlled Wake-up

Rather than having the sensor node wake up periodically (duty cycling) or wake up for an expected event (such as a transmission slot), another approach is to have a remote user (through a sensor mote, actuator node, etc.) control the sensor wake-up via a radio signal. Radio controlled wake-up is enabled by the user sending a signal that triggers an interrupt that wakes up the sensor node [17].

Radio controlled wake-up is generally achieved with the use of a wake-up radio [18], which is a secondary, very low power radio for the wake-up functionality of the node, as illustrated in Fig. 3. Depending on the energy source of the wake-up radio receiver, the wake-up radio can be classified as active or passive [19]. Active wake-up radios use internal batteries to continuously provide power to the wake-up radio to search for a wake-up signal and to trigger an interrupt of the microcontroller to wake the node up. On the other hand, in passive wake-up radios, the energy from the wake-up signal triggers the wake-up circuitry, and results in the wake-up of the node.

Applications such as asset tracking, habitat monitoring, patient monitoring, and intelligent transportation systems [20] can greatly benefit from active and passive radio controlled wake-up. In these applications, replacing devices at remote locations is not feasible, or very cumbersome, and thus extending the lifetime of the sensor nodes as much as possible becomes the top priority. Radio controlled wake-up techniques that trigger the nodes on an as-needed basis rather than using predetermined schedules achieve higher lifetime when compared to the scheduled wake-up approach by eliminating the energy waste through idle listening and overhearing when the sensor is not actually needed.

The wake-up signal can be either broadcast to wake up all nodes within the range of the wake-up transmitter, or unicast, where upon receiving a wake-up signal, the wake-up receiver checks if the message is intended for the node before waking up the node [8]. While wake-up radios have the advantage of only waking up the nodes when another node or user is ready to receive data from the nodes, their use increases the complexity of the sensor nodes and the hardware costs [8], and their wake-up range may be limited, since the use of very low power or no amplifiers increases the received signal power requirement.

There have been many research advancements in hardware design for active and passive wake-up radios. For active wake-up radios, initial studies proposed the use of a super-regenerative architecture with a bulk acoustic wave (BAW) resonator. Theoretical receiver power consumptions of 400 and 62 μW were achieved in [21] and [22], respectively. Zero-bias Schottky diodes were used to design a 3-stage wake-up scheme in [23] to reduce the false alarms caused by environmental noise. The main transceiver is triggered only if an intermediate higher power stage verifies a wake-up signal received from the low power stage, which dissipates on the order of nWs and is always on searching for a potential wake-up signal.

Ansari et al. [24] achieved a further improvement in active wake-up radios through the design of a five-stage charge pump that enables energy to be stored for a period of time and, once accumulated to a certain level, this triggers a wake-up interrupt. The only active parts of this circuit are a digital comparator and a voltage divider, which consume 350 nA and 526 nA, respectively. Van der Doorn et al. [25] implemented a low cost active wake-up receiver in the T-node platform, which is quite similar to Mica2 motes using commercially available hardware. The signals received at the antenna are filtered and amplified to feed the microcontroller, which ensures accurate detection of the wake-up signal that interrupts the main processor. Their design can theoretically operate at 171 μW.

Recently, an idea based on using a commercially available low frequency (LF) wake-up receiver [26] with an envelope detector for an ultra high frequency (UHF) signal has shown incredible success compared to the former studies. In [27] and [28], for less than 1 μA current consumption at the receiver, 15 and 40 m ranges are achieved, respectively, in field tests for a transmit power of 10 mW.

In [29], the Sparse Topology and Energy Management (STEM) technique proposed the use of an active wake-up radio approach using dual radios with each operating in a different frequency band. The primary radio is woken up if the processor decides that information needs to be forwarded to the data sink. A paging signal is initiated by the sender with a target ID, therefore activating the intended receiver. STEM provides a benefit in energy dissipation by reducing idle listening, but the benefits are limited by the radio's wake-up range.

While these approaches show the feasibility of very low power active wake-up radios, passive wake-up radios have the advantage of not requiring any battery power from the sensor node. The trade-off, however, is often a reduced wake-up range compared with active wake-up radios. Gu et al. [18] proposed a passive radio wake-up circuit that theoretically could operate at a range of 10 feet with 5 ms latency. If a comparator and an amplifier are added, which respectively consume negligible currents of 350 and 880 nA, the radio could theoretically reach up to 100 feet with 55 ms latency. In [30, 31], Jurdak et al. proposed a passive radio-frequency identification (RFID) wake-up radio, RFIDImpulse, to accommodate varied network traffic. RFIDImpulse assumes that every sensor has an attached RFID reader and a passive RFID tag to trigger a radio wake-up signal. However, a sensor's ability to wake up one's neighbor is not feasible in reality, due to the large energy cost and size of the RFID reader. Additionally, the energy model used for the analysis in the RFIDImpulse work does not include the energy consumed by the nodes to wake-up.

An approach for passive wake-up radios was implemented by Ba et al. [8, 32], using an Intel WISP passive RFID tag as the wake-up radio for a sensor mote. The wake-up performance of this WISP-Mote was tested in open, closed and cluttered environments. The WISP RFID tag acts as a wake-up signal receiver, and can use broadcast and unicast wake-ups to activate a sensor node, requiring no extra energy from the node's battery. The experimental results show that the WISP-Mote can be woken up within 5 m using both broadcast and individually addressed wake-ups.

2.3 Environmentally Controlled Wake-up

In environmentally controlled wake-up, the sensor node is woken up by an environmental trigger. Mainly, two types of wake-ups can be defined in this category: sensory-based wake-up and energy-harvesting-based wake-up, as shown in Fig. 4.

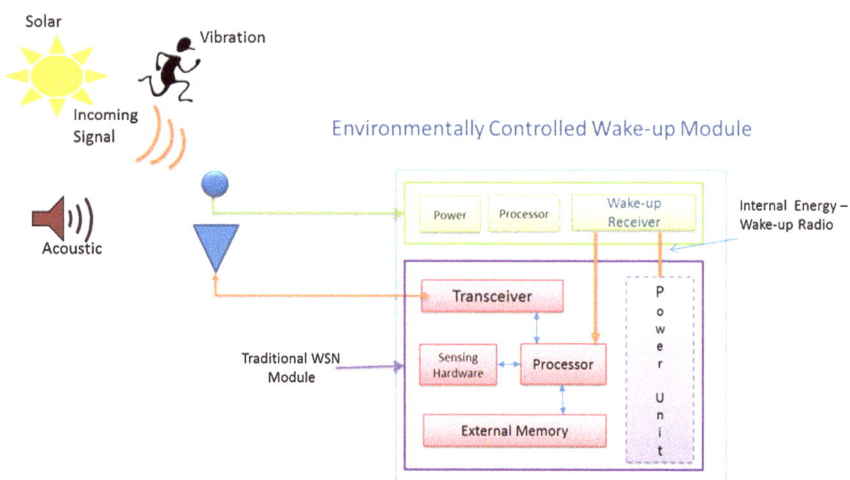

Fig. 4 Sensor node employing environmentally controlled wake-up technique

In sensory-based wake-up, a sensor triggers the microcontroller unit (MCU), based on the information sensed that may indicate the presence of an event of interest that requires the node to wake up. An example of sensory-based wake-up is described by Javaheri et al. [33], where a biological signal that is produced by a Bio-Mechanical Signal Interpreter (BMSI) triggers the node to wake up. The BMSI is an electrically passive decoder-amplifier, where weak electrical signals are converted into Bio-Mechanical signals based on the concentration of a specific ion. The signal amplification and decoding processes are performed in a biological environment, and hence are energy-efficient. Another example of environmentally controlled wake-up is the acoustic active wake-up radio with 835 nW consumption proposed by Goldberg et al. [34], where the level of sound triggers a wake-up interrupt. A multi-mode wake-up is designed by Malinowski et al. [35] for supply chain management and asset tracking applications using CargoNet, where sensors such as shock, vibration dosimeter, temperature, and humidity are used to trigger events at extremely low power. Applications such as intrusion detection and surveillance can benefit from using sensory-based wake-up systems.

In energy harvesting wake-up the sensor node is only woken up when "enough" energy is harvested to trigger a wake-up signal. There are several possible energy harvesting approaches, including solar, wind, vibration, and acoustic energy, with varied power density [36]. Although, these energy harvesting techniques can be effective wake-up triggers, waking up the sensor node when enough energy has been harvested to support data communication requires additional hardware and depends on environmentally available resources. The process of gathering signal strength to trigger a wake-up can delay the wake-up of the node, which affects the network performance, causing increased latency and reduced throughput. However, several studies target the efficient planning of the harvesting and consumption of the energy to increase the potential performance of such systems [33, 34, 36]. Among many, applications such as habitat monitoring [37] can benefit from using energy harvesting-based wake-up, because of the available outdoor harvestable energy resources.

2.4 Choosing the Wake-up Technique

The choice of wake-up technique, whether scheduled, radio controlled or environmentally controlled, depends mainly on the application goals, network traffic, and the desired *energy saving, latency, and reliability* trade-off. For radio controlled wake-up, factors such as the required wake-up range, the mechanism for powering the wake-up radio (active or passive), and the hardware cost, require additional considerations. Note that, combinations of these wake-up techniques are also possible. For example, energy harvesting wake-up can be combined with scheduled wake-up, in which case, if enough energy could not be harvested, the node will wake-up with a very low duty cycle timer.

As energy conservation is a growing concern, sleeping clearly will be an important technique to avoid energy waste in future wireless sensor networks, as well as

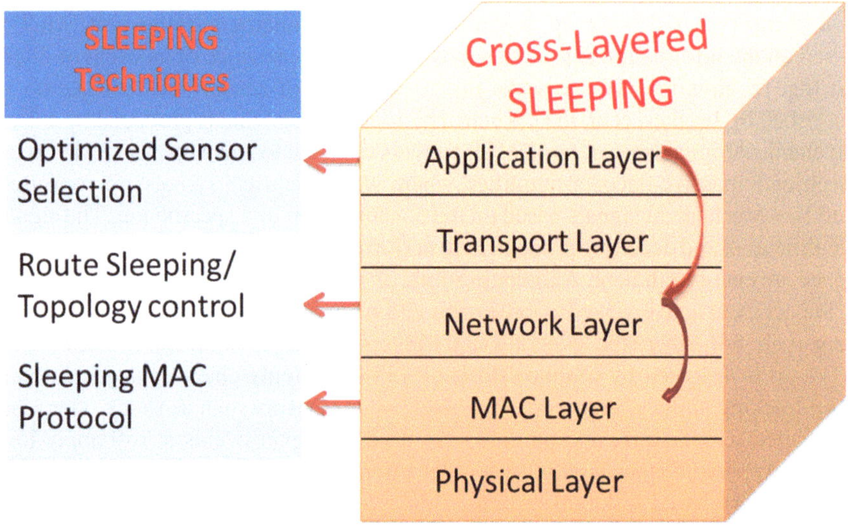

Fig. 5 Sleeping techniques for each layer in the protocol stack

in other wireless and wired networks. While the energy efficiency of the network depends on minimizing idle listening, overhearing, and data collisions, the reliability of the application depends on maximizing channel utilization and packet delivery and minimizing latency. Hence the sleeping and wake-up techniques should be selected carefully to provide the optimal trade-off.

The decision of which wake-up technique to use should be jointly done with the protocol design for maximizing the efficiency of the network by putting the sensors to sleep as much as possible to save energy. Sleeping can be exploited at different levels of the protocol stack, as shown in Fig. 5:

1. Application Layer Sleeping: Nodes can sleep when their sensed data is redundant and not required by the application. The wake-up technique to be used can be a decisive factor. For example, if energy harvesting wake-up is used, then the decision to send the data can involve the residual and predicted energy levels, along with the redundancy level of the data.
2. Routing Layer Sleeping: Nodes can be put to sleep when they have no data to relay. However, when selecting the routing approach to be used, one must consider the benefits and constraints of potential wake-up techniques. For instance, radio controlled wake-up will result in more relay nodes, while sleeping multipath routing will reduce the number of routing nodes. The potential benefit of putting more nodes to sleep for routing purposes is the reduction in energy waste. In addition, data collisions can also be reduced using sleeping multipath routing.
3. Medium Access Control (MAC) Layer Sleeping: Source and routing nodes that are involved in data delivery may sleep when it is not their turn to transmit or receive data. The MAC protocol to be used can be determined with the

wake-up technique. For instance, if unicast type radio controlled wake-up is to be employed, complicated coordination, time slot assignments, etc. will not be required.

Additionally, cross-layer techniques for managing sleeping are important when sensors are put to sleep in different layers of the protocol stack. Next, we discuss how protocols for the MAC and routing layer are designed to support sensor sleeping.

3 Medium Access Control Layer Sleeping Techniques

The MAC layer is responsible for ensuring that the sensor nodes access the medium efficiently and fairly. Because of the importance of energy efficiency at the MAC layer, there have been a host of different MAC protocols proposed to incorporate sleeping, using both fixed and variable schedules. In the following sections, protocols that exploit sleeping techniques at the MAC layer are discussed.

3.1 Taxonomy of Sleeping MAC Protocols

In traditional wireless networks, MAC protocols include fixed access protocols (e.g., TDMA, FDMA, CDMA) and random access protocols (e.g., contention-based protocols). Additionally, in traditional wireless networks, the performance metrics include fairness and throughput. For many protocols, the nodes are required to listen to the channel frequently to maintain channel state and thus avoid collisions. Since the idle power consumption can be of the same order as transmit and receive power consumption for sensor nodes [38], MAC protocols for wireless sensor networks are designed to conserve energy by incorporating sleeping techniques to reduce idle listening, while ensuring efficient medium access.

As shown in Fig. 6, most existing MAC protocols for wireless sensor networks are designed for scheduled wake-up. Among those, the protocols employing "periodic" scheduled wake-up (duty cycling) can be further categorized as synchronous, where nodes are time-synchronized to carry out rendezvous-based medium access and asynchronous, where nodes carry out the communications without rendezvous decided a-priori.

In synchronous duty cycling, since all nodes are awake at the same time, it is possible to use any traditional MAC approach during the awake period, such as TDMA, random access or hybrid (a combination of TDMA and random access) approaches. The drawback of synchronous wake-up MAC protocols is the need for control messages among the nodes to synchronize their clocks and hence their wake-up schedules.

In asynchronous duty cycling, nodes set their own wake-up times independent of other nodes in the network. Thus, nodes utilizing asynchronous wake-up can operate

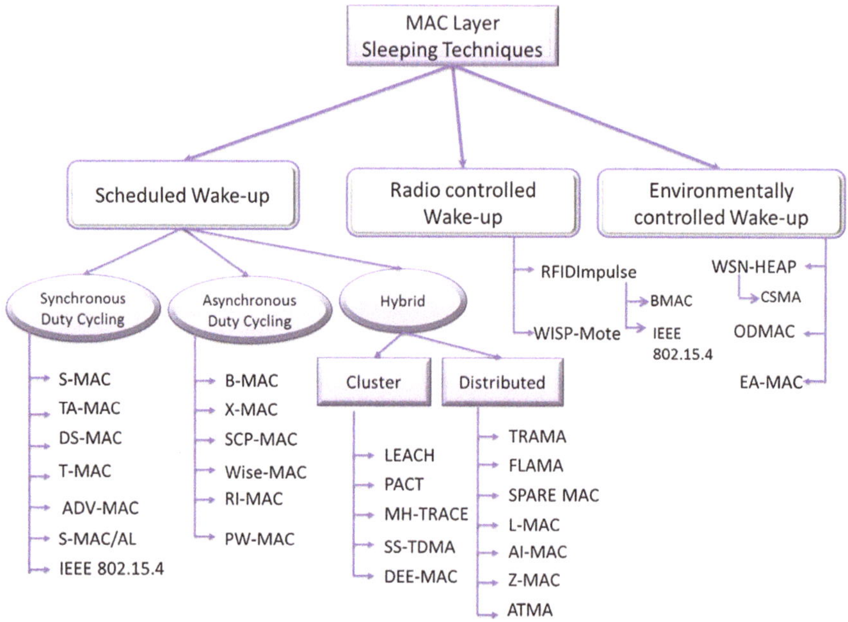

Fig. 6 Taxonomy of sleeping MAC protocols

with simple hardware, and this approach is robust and scalable to network changes. Protocols proposed for asynchronous duty-cycling ensure that the nodes listen to the channel periodically, and enable the nodes to sleep if there is no relevant activity. However, these approaches have the drawback that the transmitter is required to stay awake for a potentially long period of time to guarantee the intended receiver is awake to hear the transmission. This may cause high energy consumption and long delays, and hence, this approach requires careful planning that should consider the network characteristics.

Non-periodic scheduled wake-up is mainly employed by TDMA-based protocols, where nodes sleep until their assigned time slot in a known TDMA schedule. This may be a slot assigned to the node to transmit or to receive. In this approach, synchronization is required among all the nodes so that the slots align, but this approach is very energy-efficient as nodes are only awake when they are transmitting or receiving data. However, TDMA protocols require either some controller node to create the TDMA schedule such that collisions are avoided, or nodes must select a slot in a distributed manner using information from other nodes, which potentially incurs a large amount of overhead.

Additionally, MAC protocols can be designed for radio controlled wake-up and for environmentally-controlled wake-up. These MAC protocols must address the lack of state information at each node when they wake up.

3.2 MAC Protocols Employing Scheduled Wake-up

Based on the categorization that is introduced in Sect. 2 and depicted in Fig.2, MAC protocols that employ *scheduled wake-up* can be categorized as *synchronous duty cycling*, where all nodes are synchronized and wake up at the same time; *asynchronous duty cycling*, where the nodes are not synchronized and hence wake up at potentially different times yet have a common duty cycle; and *employing hybrid wake-up*, where the nodes employ both periodic (duty cycling) and aperiodic type of wake-up techniques. The following sections provide brief descriptions of the different MAC protocols in each category and their benefits and limitations.

3.2.1 Synchronous Duty Cycling MAC Protocols

In synchronous periodic sleeping MAC protocols, all nodes have synchronized clocks and wake up at the same time for communication. During the awake period, different protocols employ contention-based, TDMA based, and hybrid medium access and nodes may go to sleep at different times. Yet all of the nodes will wake up again at the same periodic interval for continued communication.

Contention-based MAC protocols that employ synchronous periodic sleeping are designed to handle nodes that have no coordination when accessing the channel. Using concepts such as carrier sensing and random back-offs, the collision probabilities can be reduced in contention-based protocols. While contention-based protocols for traditional wireless networks, such as IEEE 802.11's DCF [39] and PAMAS [40], are successful in achieving high throughput and fairness, these approaches still require a large amount of idle listening in networks with low traffic. Hence, the use of fixed sleep-listen schedules was introduced in S-MAC [14], where nodes use synchronization messages to ensure synchronized duty cycles in groups. In the active part of the duty cycle, the nodes communicate using RTS/CTS/DATA/ACK messages, as in IEEE 802.11 DCF. S-MAC ensures nodes do not waste energy in idle listening by forcing the nodes to sleep after a data message exchange. Due to the fixed sleep-listen schedule and only allowing a single data transmission in each active period of S-MAC, the energy savings is at the expense of reduced throughput and high latency. In addition, static sleep-listen schedules are best suited for known or constant traffic patterns.

To handle variable traffic patterns, techniques such as adaptive contention windows, latency-based duty cycles and time-outs were introduced as extensions to S-MAC in TA-MAC [41], DSMAC [42] and T-MAC [43], respectively. In all of these approaches, which still employ a synchronous, periodic wake-up of the nodes, the goal is to keep nodes awake to support variable traffic yet allow nodes to sleep as much as possible. TA-MAC utilizes an adaptive contention window that incorporates the back-off algorithm of IEEE 802.11 and the fast collision resolution algorithm from [44] to determine the contention window for the current traffic state. In DSMAC, the duty cycle is adjusted based on the perceived latency of the data transmissions. The timer introduced in T-MAC is used to help sensors sleep if no activation occurs after a certain period of time. This timer can be renewed based on the traffic load. S-MAC was further extended by S-MAC/AL [45] to reduce multi-hop latency by incorporating adaptive listening as in T-MAC.

Although these variations of S-MAC improved end-to-end delay and packet delivery ratio (PDR), unfortunately idle listening and the early sleep problem (where nodes go to sleep when there is still data pending for them) still exist. A data-centric MAC protocol called ADV-MAC [16] reduces the early sleep problem, by advertising for upcoming data to alert intended receivers that they have pending data transfers. In addition, a variable duty cycle that consists of fixed synchronization and advertisement periods and a variable data period is used in ADV-MAC. Unlike S-MAC or T-MAC, ADV-MAC handles variable traffic loads without compromising latency and throughput. The limitation of ADV-MAC is the need to do the contention twice, once for the advertisement of data and then for the transmission of the data. The duration of the advertisement period is based on the traffic model, which affects the energy consumption due to idle listening.

IEEE 802.15.4 [46], the IEEE standard for low power wireless networks, can use both synchronous and asynchronous duty cycling, but the protocol is more efficient using synchronous duty cycling in terms of packet delivery ratio and throughput [11]. The synchronous duty cycling is used with a cluster topology, where the nodes are synchronized by a beacon packet from the cluster-head. IEEE 802.15.4 has superframes that consist of a Contention Access Period (CAP) and a Contention Free Period (CFP). Within the CAP, medium access is acquired via standard contention access. Nodes that have a stream of data to transmit can also reserve slots in the superframe by sending requests to the coordinator during the CAP. If such requests are granted, the coordinator includes the relevant information, such as the number of slots allocated and the beginning slot, in the next frame. Nodes sleep following the CAP and CFP until the start of the next synchronized wake-up period. One drawback of this mode is that the nodes have to communicate through cluster-heads to other nodes. This incurs delay, and high energy consumption at the cluster-head, while reducing the achievable throughput.

3.2.2 Asynchronous Duty Cycling MAC Protocols

Asynchronous duty cycling MAC protocols do not require any locally or globally synchronized clocks, which reduces synchronization overhead. However, all nodes use a common duty cycle such that the duration of their sleep-awake cycle is the same. These MAC protocols can be either sender-initiated such as B-MAC [47], X-MAC [48], SCP-MAC [49] and WiseMAC [50] or receiver-initiated such as RI-MAC [51] and PW-MAC [52]. In sender-initiated protocols, the sender transmits a preamble until it receives a response from the receiver. On the other hand, in receiver-initiated protocols, the receiver sends a beacon as soon as it is awake to let all potential transmitters know that it is available for transmissions [52].

Sender-initiated protocols in this category are also known as low power listening (LPL) protocols. In LPL protocols, a node with data to send transmits an extended preamble. Once a node wakes up during its regular periodic wake-up, it detects the preamble and will stay awake until it determines if it is the receiving node. Since the wake-up schedules are not synchronized, the preamble must be transmitted for a duration long enough to ensure that the receiver wakes up at some time during the transmission of the preamble. This can require a large amount of energy from the transmitting node. Therefore LPL protocols are not feasible for networks with heavy data traffic.

In B-MAC, asynchronous duty cycles and long preambles are used to wake up nodes. Although, B-MAC does not require synchronization messages, the transmitter expends energy due to the long preamble, while the other nodes expend energy due to overhearing a preamble that may not be intended for them. X-MAC [48] reduces the energy expenditure by replacing the long preamble of B-MAC with a series of short preamble messages that include the receiver address. Thus, the nodes that are not the intended receiver waste only a small amount of energy before determining that the data is not for them and returning to sleep until the next wake-up period.

A similar approach is used in WiseMAC, which combines non-persistent CSMA with dynamic preamble sampling based on the neighbors' sleep schedules to reduce idle listening. The sender can predict the next wake-up time of a receiver based on sampling the medium with the same constant period independent of the actual traffic. If the medium is found busy, the receiver continues to listen until a data packet is received or until the medium becomes idle again. Every node has an updated table with the direct neighbor's sampling schedule offset obtained from the acknowledgement packets. Since WiseMAC uses only clock drift rate to adjust predicted receiver wake-up times, this may lead to prediction error caused by variable hardware and operating system latency. Also, disregarding the state of the receivers, senders in WiseMAC retransmit packets repeatedly until all packets are acknowledged, potentially causing collisions and larger sender duty cycle.

In receiver-initiated transmission approaches, such as RI-MAC [51], the sender is active but silent, whereas the receiver sends inviting beacon frames at regular periods of time, when the node wakes up. The data transmission begins when the sender receives a beacon frame from the targeted receiver. Due to the lack of a long transmitter preamble, as in transmitter-initiated approaches, energy is saved and the use of the medium is reduced. Upon detecting a collision, the receiver sends out a new beacon using a contention window, the size of which is increased based on the number of retransmissions. Similarly, in PW-MAC [52], independently generated pseudo random sequences are used to control each node's wake-up time, which allows senders to accurately predict receivers' wake-up times. However, PW-MAC considers the potential hardware and operating system latency that would corrupt the synchrony, so nodes wake up earlier by a specific amount of time, which is defined as a protocol parameter. In addition, since PW-MAC uses pseudorandom wake-up schedules, even if multiple nodes wake up at the same time for reception that might cause collisions, they will have different wake-up times in the following cycles.

3.2.3 MAC Protocols Employing Hybrid Wake-up

There are a wide range of MAC protocols that utilize both periodic and aperiodic wake up techniques. Mainly, TDMA-based wireless sensor network MAC protocols fall into this category, since these protocols generally employ periodic wake-up to coordinate for the time slot assignments, whereas they employ aperiodic wake-up for the data slot access.

In TDMA-based MAC protocols the channel is divided into time-slots that are shared by the nodes. Each node is allocated a dedicated time slot to communicate (send/receive). This approach allows the nodes to communicate without collisions in their allocated slot, thereby minimizing the transmit time. There are a number of factors that influence TDMA-based MAC protocols, including synchronization, creating distributed schedules, maximizing channel reuse, incorporating node mobility and ensuring energy efficiency [53].

There are two methods for using TDMA-based medium access in wireless sensor networks: (1) use of a hierarchical clustering structure where the cluster head creates and disseminates a TDMA schedule to the nodes in the cluster, and (2) the nodes themselves pick slots in a distributed manner to reduce collisions with other transmissions in their two-hop neighborhood.

3.2.4 Cluster-Based Protocols with Hybrid Wake-up

Protocols that use a hierarchical, cluster-based topology include LEACH [54], PACT [55], MH-TRACE [56], SS-TDMA [57] and DEE-MAC [58]. In these protocols, cluster-heads create the TDMA transmission schedules and broadcast these schedules to their cluster members. The nodes within each cluster turn their radios off at all times except during their own transmission time slots, thereby conserving energy. For example, in LEACH, nodes elect themselves as cluster-heads in a distributed manner, and nodes join nearby clusters for channel access. The cluster-head learns of all cluster members and then creates a TDMA schedule that allocates each cluster member a slot. This schedule is then sent to the cluster members, enabling the cluster members to sleep with an event-scheduled wake-up at their given transmission slot. In LEACH, all cluster members transmit directly to the cluster-head, so no other cluster members need to be awake.

PACT and MH-TRACE are proposed for networks with a clustered multi-hop topology. In order to achieve contention-free communication, these protocols use superframes. On the other hand, SS-TDMA handles traffic in a fixed sequence of rounds using directions such as North, South, East, and West to achieve collision free transmission by checking interference from its neighbors. This approach is best suited for grid-topologies. DEE-MAC also uses cluster-based TDMA, where each cluster is dynamically formed based on the remaining power of the nodes, and the cluster-head builds the TDMA schedule based on the interest received from nodes in its cluster. In addition, DEE-MAC employs an inter-cluster communication via nodes rather than just through the cluster-heads.

In general, the overhead involved in forming clusters, communicating within a cluster, and maintaining clusters adds complexity to cluster-based TDMA protocols. It is important in cluster-based TDMA protocols to perform resource-tracking (e.g., monitoring the energy of the cluster-head nodes) in order to maintain the stability of the clusters. The energy required to maintain the time slot assignments, synchronized clocks, and frame lengths for a dynamically changing topology heavily depends on the protocol. In addition, the nodes' resources, the data traffic and the environment may change over time, which would require dynamic slot assignments.

3.2.5 Distributed Protocols with Hybrid Wake-up

Creating a distributed TDMA schedule is difficult, requiring a large amount of overhead to converge on a feasible schedule such that collisions are minimized. Such an approach is used in TRAMA [59], FLAMA [60], SPAREMAC [61], LMAC [62], AI-LMAC [63], Z-MAC [66], and ATMA [67]. In these protocols, slots are allocated either based on a random priority or winning slot contention within a two-hop neighborhood [64], but in either case, when a node does not have any data to send, its allocated slot is wasted. In addition, the nodes contending for the slot waste energy during the contention process.

In TRAMA, all nodes compute priorities on the upcoming time slots with their 2-hop neighborhood information. Moreover, nodes regularly announce their data schedules to inform the intended receivers of the data slots for which they have priority, yet will not be used. This can lead to collisions when more than one node tries to access a particular data slot. FLAMA is an improved version of TRAMA, where instead of broadcasting information to the neighboring nodes, a pull-based mechanism that shares information only based on an explicit request is used, thus reducing overhead. SPAREMAC is a receiver oriented MAC protocol, where nodes announce their reception schedules (RSs), which designate the time slot(s) they will be awake and waiting for incoming data. The RS is propagated to neighboring nodes, such that the transmitting node can consequently become active in the RS of its intended receiver only. Therefore, SPAREMAC limits the impact of idle listening and traffic overhearing. This approach to determine a distributed TDMA schedule solves the hidden node problem, however it does not eliminate collisions and the control overhead is very high, leading to high energy consumption and data delay [65].

Self-organizing TDMA based communication without using a central manager or base station is also achieved by LMAC, where nodes can choose time slots on their own, and they share their selected slots with their one-hop neighbors. Nodes with conflicting time slots choose an ID-dependent back-off time to enable collision-free communication. The drawback of LMAC is that idle listening overhead is substantial, as nodes must listen to the control sections of all slots in a frame to learn about the neighbor schedules and to allow nodes to join the network. In a given time period there might be nodes that are forced to be in the awake state, leading to idle listening and energy dissipation. AI-LMAC is an improved version of LMAC where each node can choose multiple slots based on the expected traffic flow, which aids in adapting the operations based on the application requirements.

The Z-MAC protocol [66] combines CSMA and TDMA-based approaches, aiming to exploit the better performance of TDMA for high contention situations. The setup phase of Z-MAC includes neighbor discovery, local frame exchange of neighbors' lists and slots assignment. All the nodes are synchronized with a global time synchronization feature. Each node is assigned a slot but it is not fixed, and any node can contend for the channel within any slot for data transmission, although the assigned node will get the highest priority. Nodes can sleep until they find a time slot that is either not owned within their two-hop neighborhood or is owned by themselves.

Another protocol employing both periodic and aperiodic wake-up techniques is ATMA [67], where nodes advertise for their data transmissions in an advertisement period, as in ADV-MAC [16, 68], but they also select a TDMA slot at which the data transfer will occur. If the receiver node is free during the selected data slot, it acknowledges the advertisement, and then the transmitter and receiver both know that they can sleep until the assigned slot. After the data transfer in the assigned slot, both the transmitter and receiver go back to sleep until the start of the next frame. The efficiency of ATMA stems from the continued use of a selected data slot for a given number of frames, thus reducing contention in subsequent frames.

3.3 MAC Protocols Employing Radio Controlled Wake-up

MAC protocols that employ radio controlled wake-up techniques help in reducing the energy and latency involved in the frequent transition of nodes from the sleep to the active modes. In [69], an RFID-based wake-up mechanism called RFIDImpulse is evaluated for different sleep modes of the sensor devices, where the sleep modes observe a trade-off between wake-up delay, wake-up energy and power consumed during sleep. The authors implement RFIDImpulse both on the 802.15.4 MAC and B-MAC, where nodes are woken up through the use of an RFID signal. It is shown that if the network traffic is high, it is more energy efficient to use a light sleep mode, and with low network traffic, nodes should be put in deep sleep mode. RFIDImpulse is shown to work independent of the underlying MAC protocol, and hence can be applied to many MAC protocols to employ radio controlled wake-up. The applicability of RFIDImpulse is limited because of its assumption that every node has the ability to wake up its neighbors. However, battery-powered sensor nodes cannot afford the large amount of energy consumed by the RFID reader.

In [8, 32], an RFID-based wake-up scheme was proposed, where it is assumed that only a collector (i.e., a data MULE) with unlimited power and an RFID reader can wake up the sensor nodes. In the data MULE scenario, an ALOHA-based MAC protocol is applied and the simulation results in [32] show the energy efficiency benefits of using an RFID-based wake-up scheme over a traditional duty-cycling scheme. In order to improve the performance of RFID wake-up in a dense network,

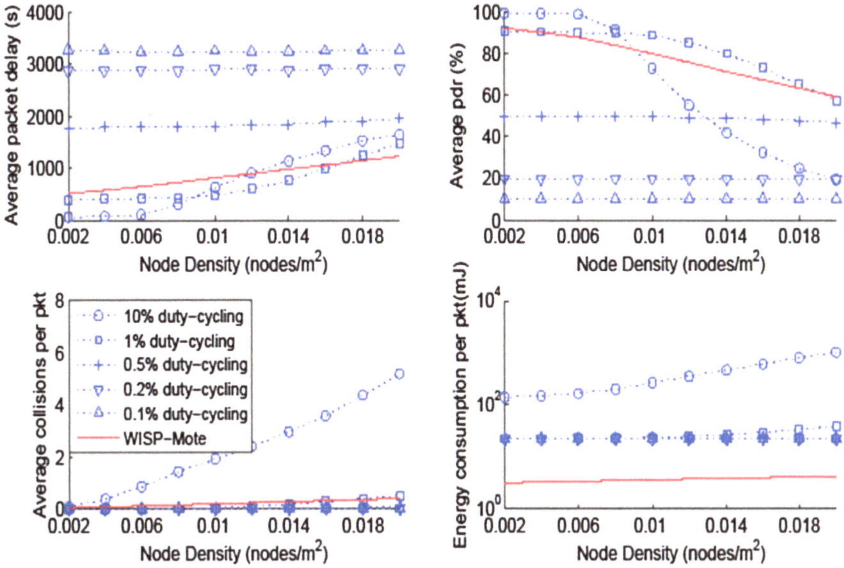

Fig. 7 Performance of radio-controlled wake-up (WISP-Mote) compared with scheduled wake-up (duty cycling) as node density varies [32]

a collision avoidance mechanism is incorporated in [8]. Figure 7 shows the benefit of using radio-controlled wake-up (denoted as WISP-Mote) over scheduled wake-up (duty cycling). As seen in the figure, the average packet delay and the number of collisions in 10 % duty cycling increase significantly as the node density varies. Meanwhile, the performance of the WISP-Mote is more stable. Additionally, the energy wasted due to unnecessary wake-ups and idle listening is reduced using the radio-controlled wake-up compared with 0.1, 0.25, 2 and 10 % duty cycling.

3.4 MAC Protocols Employing Environmentally Controlled Wake-up

MAC protocols that employ environmentally controlled wake-up use energy harvesting techniques [70] to wake up the nodes. This approach conserves energy and reduces the need to replace the node batteries, which is critical for applications that are deployed in remote or hostile regions, or where battery replacement or recharging is not feasible. Eu et al. [71] performed a study on single-hop MAC protocols based on slotted and unslotted CSMA, as well as ID-based polling and probabilistic polling techniques for wireless sensor networks powered by ambient energy harvesting. The authors utilize a charge-and-spend strategy, where nodes go into the receive mode when enough energy has been accumulated. They state that this strategy is employed instead of methods such as duty cycling due to its simplicity, the limited capacity of the energy storage, to reduce leakage, and to achieve lower delay. Through simulations they show that probabilistic polling achieves better fairness and is highly scalable, while achieving good throughput compared to the CSMA approach.

In [72], Fafoutis et al. proposed the ODMAC protocol, which targets the operation of sensor nodes in a state where the energy consumed is equal to the energy harvested by adjusting the duty cycle accordingly [70]. Due to the unpredictable nature of the energy harvesting, the authors proposed the use of a receiver-based opportunistic forwarding approach, where the transmitter opportunistically forwards frames to the first node from the potential forwarders list that sends a beacon.

Kim et al. [73] proposed the energy adaptive MAC protocol (EA-MAC) for wireless sensor networks powered by an RF energy source. In EA-MAC, the master node is assumed to be always in the awake state, and the slave nodes transit from the sleep to the active state depending on their remaining energy levels. Also, the slave node's energy harvesting level is inversely proportional to its distance from the master node. EA-MAC uses CSMA/CA with an energy adaptive contention algorithm, where the backoff time of each slave node depends on its energy harvesting rates. Therefore, EA-MAC adaptively manages both the duty cycle and the contention time of the slave nodes using the nodes' energy harvesting rates to achieve fairness among the nodes.

4 Routing Layer Sleeping Techniques

The goal of routing in wireless sensor networks is to get data from the source node(s) to the sink node(s) with high reliability, low latency, and using minimum energy. In wireless sensor networks, routing of the packets is generally done by the regular sensor nodes, and hence energy efficiency of the routing mechanism employed is paramount in order to extend the network lifetime.

In the previous section, we saw how sleeping can be used to save energy at the medium access control level. In this section, we explore how sleeping can save energy at the routing layer, by placing potential nodes that are not needed as routers into the sleep state. As shown in Fig. 8, there are two ways sleeping in the routing layer can be achieved: (1) *topology control*, where only a subset of the sensor nodes are selected as routers, enabling all other sensor nodes to go to sleep when idle, and (2) *sleeping routing*, where possible router nodes that are not needed are put to sleep. We discuss both approaches in the following sections.

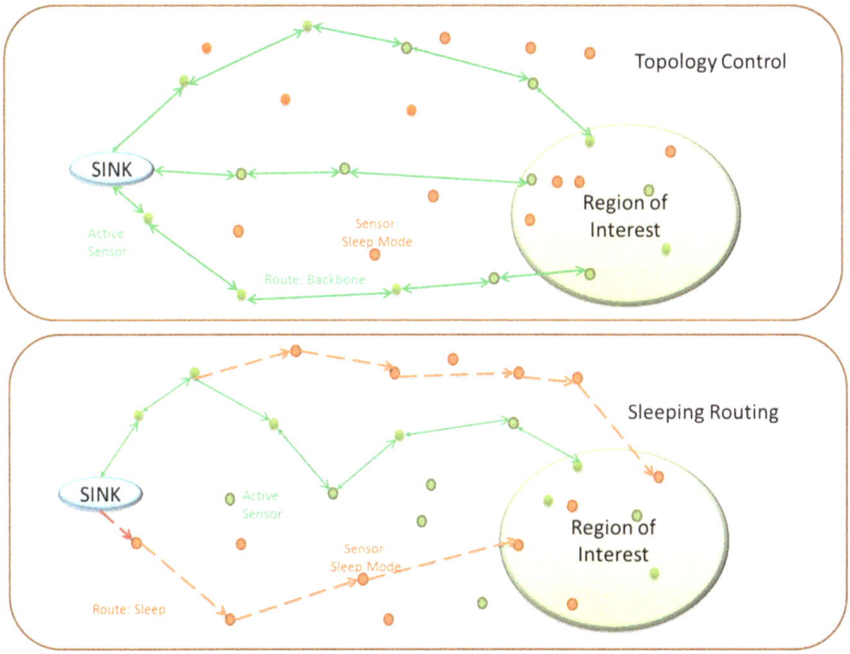

Fig. 8 Routing layer sleeping techniques. Topology Control (*top*), where nodes that form the backbone are awake. Sleeping Routing (*bottom*), where nodes on the selected routes are awake

4.1 Topology Control

Topology control is used to select a subset of available sensor nodes to route data in the network.[1] Traditionally, the sensors are selected to create a backbone that ensures network connectivity while maintaining network throughput. An additional goal for topology control in wireless sensor networks is to select the backbone in such a way as to reduce energy dissipation, and to select nodes that have enough remaining energy to support the additional task of routing. The benefit of topology control is that all nodes that are not selected as the backbone routers can go to sleep when they do not have sensed data to transmit. This technique ensures that non-backbone nodes save significant amounts of energy. However, given that the backbone nodes are still energy-limited sensor nodes, many topology control protocols rotate the backbone nodes such that the energy drain of performing the routing duties is shared among all the nodes in the network. Topology control is most effective in large, dense networks where the large number of nodes results in many redundant routes.

---

[1] There can be scenarios where there are sensors (actuators, relay nodes) with different communication and energy capacities, namely heterogeneous wireless sensor networks. In this chapter we will concentrate on homogeneous topology control, where all the sensors have the same capabilities. For details on topology control in heterogeneous wireless sensor networks, readers can refer to [74].

One such topology control protocol is GAF [75], a localized, distributed protocol that divides the network into virtual grids. Only one sensor node in each grid is activated at any given time. The virtual grid is set up in such a way that full connectivity is guaranteed. GAF requires location information so that the nodes know their grid ID to ensure that only one node per grid is activated as the backbone node. To prevent energy drain of a node, the activated node in each grid is rotated among the nodes in the grid.

While GAF is an effective approach for topology control, it does require that nodes have location information and know about the virtual grid. On the other hand, in the SPAN [76] protocol, the nodes make local decisions on whether to sleep or join the backbone as a "coordinator node." This decision is based on the connectivity information provided by the routing protocol. If the connectivity between any two nodes is poor and can be improved by the coordinator node joining the backbone network, then the coordinator node will be activated. The coordinator node will go to sleep when routing is no longer needed. Unlike GAF, SPAN does not require its nodes to know any geographical location information, instead SPAN broadcasts messages and utilizes overhearing to determine when to activate nodes and to discover topology changes in the network.

A similar approach is employed in Adaptive Self-Configuring sEnsor Networks Topology (ASCENT) [77], where nodes are turned on or off depending on the node's assessment of the operating conditions. When a node identifies high data loss in its neighborhood, it requests neighboring nodes to join the network as additional backbone relay nodes. Nodes that receive 'join' request messages probe the network, rather than joining immediately. This approach ensures that multiple nodes do not simultaneously activate and join the backbone. The sleeping nodes check periodically to see whether additional backbone nodes are needed in their neighborhood.

In [78] a geographical cluster-based routing protocol is proposed, where theoretical analysis guarantees packet delivery in a network where the ratio of communication range to sensing range is ≥ 4. In [79], Simplot-Ryl et al. proposed two kinds of backbones, namely neighbor and area dominating sets, where a node has coverage only if it is in the backbone or its neighbor is in the backbone. The retransmission of messages is therefore reduced, and energy waste is minimized.

In the above approaches, the main function of the protocol is to determine an appropriate set of backbone nodes to keep the network connected and operating with high capacity while enabling the remaining sensor nodes to sleep. While this approach is effective in saving large amounts of energy in dense networks, it is not effective in sparse networks due to lack of redundant routes. Therefore, in sparse networks all nodes must be activated to keep the network connected. However, even if all nodes must be considered as potential routers to connect the network, given the event-driven nature of some wireless sensor networks, oftentimes traffic is only coming from a small subset of the sensor nodes at any given time. Hence, the nodes that are required as routers can at times be set to sleep if information about the traffic in the network is known (i.e., any routers that are not currently being used to transfer data to the sink can be set to sleep). We will discuss such sleeping routing protocols in the following section.

4.2 Sleeping Routing

In a network where traffic patterns can be determined, it is possible to set all routing nodes that are currently not involved in data delivery to the sleep mode, saving energy and improving the network lifetime. Energy savings through in-network data processing can further benefit routing schemes. Sleeping routing can be achieved by incorporating power management techniques into the routing protocol.

Routing protocols such as [80–83] update routes periodically and save energy by placing routing nodes that are not involved in the data delivery to sleep. For example, the on-demand power management framework [81] configures the sensors to sleep by monitoring routing control messages and data transmissions such that only useful routers are kept active. In addition, connectivity is only maintained between pairs of senders and receivers and along the routes of data communication, thus further reducing the energy consumption. The Route-Oriented Sleep (ROS) approach [82] also utilizes the routing decisions so that sensor nodes do not wake up when they are not part of a routing path. Also, by switching a sensor's role such as sensing and forwarding, based on a critical "probability of percolation" [84], the ROS approach reduces communication interference and latency that occur during data transmission with minimal loss of sensing and communication coverage.

Another example is sleeping applied to the Directed Diffusion protocol [80], such that all nodes within an interest-query interval that are not reinforced by the sink as routers can go to sleep [85]. To achieve this, several timers are needed at the sensor nodes so that the nodes know when they must wake up to receive new interest queries from the sink or new path discovery messages. Sleeping Directed Diffusion ensures that the sleeping nodes do not negatively impact route discovery and maintenance. In addition, the sink node collects remaining energy information from the sensors to determine which nodes can be reinforced as routing nodes. This is done to ensure that the selected routing nodes have enough energy to remain active for the period when the other possible routing nodes are asleep.

In the Gossip-based sleep protocol (GSP) [86] all the receiving nodes go to sleep with a gossip sleep probability 'p'. GSP does not require neighborhood information or synchronization. During each round, every node chooses to sleep or stay awake with probability p. This probability is chosen with the goal of having the remaining awake nodes maintain network connectivity, considering a static network. Although this approach helps in energy conservation, due to the random method for selecting awake nodes, an optimal route cannot be achieved and long delays may be encountered.

An entirely different approach to sleeping routing is applied using radio controlled wake-up, STEM [29], where all nodes are in the sleep mode until they are woken up by their previous-hop neighbor using a wake-up radio. STEM is combined with topology control routing schemes such as SPAN and GAF to reduce energy consumption resulting in increased node lifetime. However, the STEM protocol is limited based on the wake-up range.

Sleeping routing protocols discussed in this section have the benefit of extending network lifetime by enabling more sensors to sleep (and hence be available for routing). However, they do not provide any guarantee about the quality of service (QoS). For example, when several nodes are sleeping, the routes may be longer than the optimal routes where all the nodes are awake and potential routers. Additionally, there may be reliability issues using the limited redundancy of the routes due to sensor sleeping. One approach to alleviate this issue and to trade off energy with reliability is to use multipath routing, where the data are sent over multiple paths to the sink. In the next section, we discuss choosing an optimal degree of redundancy and selecting the set of paths to achieve the best reliability-energy balance.

4.3 Sleeping Multipath Routing

Multipath routing [87–89] is used to increase the likelihood of reliable data delivery by sending multiple copies of data along different paths. To increase reliability, Forward Error Correction (FEC) codes can be combined with multipath routing protocols [87]. However, while multipath routing is effective in increasing reliability, it also requires more nodes to be involved in data delivery thus leading to higher energy consumption (less sleeping) and shorter network lifetime. Sleeping multipath routing approaches can therefore achieve minimal energy consumption while supporting network performance such as increased network throughput, and reliability.

A general Sleeping Multipath Routing approach to arbitrate this trade-off in reliability and energy consumption was proposed in [90]. This Sleeping Multipath Routing approach can be applied to any multipath routing protocol by discovering disjoint multiple routes between a source and the sink node. During this discovery, information about the reliability of each route is collected. The sink can then use this information, along with knowledge of the application's reliability requirement, to determine which set of disjoint routes meet the reliability requirement of the application. Among this set, the one that minimizes energy (e.g., uses the fewest sensor nodes) is selected, and the nodes on the remainder of the disjoint routes are put to sleep. When multiple subsets of disjoint routes exist, these subsets can serve the application one after the other. Thus, when the current disjoint routes deplete their energy, another set takes over, thereby prolonging the network lifetime multiple times depending on the available redundant subsets of disjoint routes.

This approach of sleeping multipath routing has been applied to a multipath version of the Directed Diffusion algorithm [90]. While Directed Diffusion finds multiple routes from a source node to the sink, there are no guarantees that these multiple routes are disjoint. In the multipath version of Directed Diffusion [90], some changes are made to the Directed Diffusion protocol to ensure disjoint routes are found and to ensure that during route discovery, information about the reliability of the routes is piggybacked onto the route discovery messages. This approach provides the sink with information about the available disjoint routes and about the reliability of each of these routes. With this information, the sink can determine which subsets of routes meet the application requirements and select the subset that uses the least amount of energy. Upon reinforcing the selected routes, the sink sends a negative reinforcement on the non-selected routes, such that all nodes on these non-selected routes can go to sleep until the next route discovery or route maintenance interval.

The key requirements for applying sleeping at the routing level are to maintain network connectivity and reliability while minimizing energy waste in the nodes not serving as routers (topology control) or those routers that are not currently routing data (sleeping routing and sleeping multipath routing). In the previous sections, we discussed the possibility for nodes to sleep at the MAC layer or the routing layer, and this discussion prompts some intriguing questions such as what happens if we employ sleeping at *both* the MAC and routing layers? Will we get more gain by sleeping at both layers, or will these different sleeping techniques interact in a performance degrading manner? How can we achieve even higher gains through coordinated sleeping at the MAC and routing layers? In the following section, we examine the effects of sleeping at multiple layers in an uncoordinated way as well as cross-layer sleeping for increased network lifetime.

5 Cross-Layer Sleeping Techniques

There are various steps involved in transmitting a message from a sender to a receiver, but the most important of all is coordination not just among the sensors but also among the protocols at each layer of the protocol stack. Unlike traditional networks, wireless sensor networks depend on the cooperation of individual nodes to provide the required accuracy, reliability and efficiency. Therefore, the distributed yet cooperative nature of the nodes can be exploited to the application's benefit. As discussed previously, sleeping at both the MAC and the Routing layers conserves energy; therefore, a cross-layered approach that exploits the cooperative nature of sensor networks promises even higher energy savings.

5.1 Effects of Uncoordinated Sleeping in the Stack Layers

The main factor that ensures improved performance for sleeping is coordination between multiple layers for any given network. In any application with uncoordinated sleeping between multiple layers, a node turned off by the routing layer can be woken up by the MAC layer based on its duty cycle, thus primarily following the MAC layer schedule and consuming more energy by idle listening [80]. On the other hand, cross-layer coordinated sleeping can significantly improve network throughput and data delivery, especially when the sleeping is tailored based on network conditions and application requirements that change over time. Therefore, a smart decision made by a power manager to switch between sleeping schemes with cross-layer coordination can outperform single layer sleeping schemes [80].

5.2 Cross Layer Sleeping

In this section, we examine approaches where the sleeping technique is extended as a cross layer approach to two or more layers of the protocol stack. For example, the network and MAC layers can be combined or interact to improve energy efficiency and network reliability. The primary concern is the coordination between different layers and the feasibility of such an approach in real-time environments. Cross-layered sleeping can be achieved by coordinating the sleeping in multiple stack layers.

Sleeping using scheduling algorithms combined with routing protocols is used to form a cross-layered approach in [91–93]. Zhu et al. [91], proposed a duty cycle based on the Connected K-Neighborhood (CKN) sleep scheduling algorithm [83] for a two-phase greedy forwarding (TPGF) geographical multipath routing algorithm, where the CKN algorithm aims at allowing a portion of the nodes to sleep while maintaining network connectivity among the rest of the nodes. Since the TPGF multipath routing focuses on exploring the maximum number of optimal node-disjoint routing paths to minimize path length and end-to-end transmission delay, using the CKN algorithm proves that waking up more sensor nodes does not always help to find more transmission paths, nor does it reduce the average length of the transmission paths found by TPGF.

Ha et al. proposed a cross-layer sleep scheduling technique based on an organization approach called Sense-Sleep Trees (SS-Trees) [92, 95], where an iterative algorithm is used to determine the feasible SS-tree structure. All the SS-trees originate from the sink, and the channel access is performed using a CSMA-based MAC with implicit acknowledgements. In the initialization stage, the sink gathers network connectivity information, computes SS-trees and disseminates sleep schedules to every node. In the operation stage, nodes switch between sleep and active modes, and during long inactive periods, i.e., when sensing is not needed, the entire network will enter a hibernation mode to conserve energy. In addition, the SS-trees must be constructed in such a way as to minimize the number of shared nodes to minimize the number of co-SS tree neighbors of each node, and the forwarding message cost between the data sink and each node. In [95], a greedy approach is proposed to compute the SS-trees. The optimal sleep schedule is applied on the computed SS-tree to maximize energy efficiency. This approach requires sensors to maintain local connectivity with their one-hop neighbors.

MAC and routing layer sleeping are combined in ORAS [93], where opportunistic routing and asynchronous sleeping are applied. Each node in ORAS independently maintains its own low duty cycle, with minimal inter-node coordination or synchronization clocks. ORAS employs an expected number of transmissions (ETX)-based forwarder election mechanism to have high throughput and low delay. Due to the opportunistic nature of ORAS, it exploits both the spatial and temporal diversity of traffic during transmission.

In [96], an energy efficient cross-layer MAC protocol called MAC-CROSS is proposed, where routing information at the network layer is used in the MAC layer

Table 1 ABBREVIATIONS

Abbreviation	Description
ACK	ACKnowledgement
ADV-MAC	ADVertisement-based MAC
AI-LMAC	Adaptive Information-centric and Lightweight MAC
ASCENT	Adaptive Self-Configuring sEnsor Networks Topology
ATMA	Advertisement-based TDMA
B-MAC	Berkeley MAC
BAW	Bulk Acoustic Wave
BMSI	Bio-Mechanical Signal Interpreter
CAP	Contention Access Period
CDMA	Code Division Multiple Access
CFP	Contention Free Period
CKN	Connected K-Neighborhood
CLEEP	Cross-Layer Energy-Efficient Protocol
CP	Communicating Parties
CSMA	Carrier Sense Multiple Access
CSMA/CA	Collision Sense Multiple Access with Collision Avoidance
CTS	Clear to Send
DCF	Distributed Coordination Function
DEE-MAC	Dynamic Energy Efficient TDMA-based MAC
EA-MAC	Energy Adaptive MAC
ETX	Expected number of transmissions (TX)
FEC	Forward Error Correction
FDMA	Frequency Division Multiple Access
FLAMA	Flow Aware Medium Access
GAF	Geography-informed energy conservation for Ad Hoc Routing
GSP	Gossip-based Sleep Protocol
ID	Identifier
ISTH	Incremental Shortest-path Tree Heuristic
LEACH	Low Energy Adaptive Clustering Hierarchy
LF	Low Frequency
LMAC	Lightweight MAC
LPL	Low Power Listening
MAC	Medium Access Control
MAC-CROSS	MAC CROSS-layer design Protocol
MCU	Microcontroller unit
MH-TRACE	MultiHop-Time Reservation using Adaptive Control for Energy efficiency

(continued)

Table 1 (continued)

Abbreviation	Description
ODMAC	On-Demand MAC
ORAS	Opportunistic Routing with Asynchronous Sleeping
PACT	Power Aware Clustered TDMA
PAMAS	Power Aware Multiple-Access protocol with Signaling
PDR	Packet Delivery Rate
PHY	Physical layer
PW-MAC	Predictive Wakeup MAC
RAM	Random Access Memory
RFID	Radio-Frequency IDentification
RI-MAC	Receiver-Initiated MAC
ROS	Route-Oriented Sleep
RTS	Request To Send
S-MAC	Sensor MAC
S-MAC/AL	Sensor MAC with Adaptive Listening
SCP-MAC	Scheduled Channel Polling –MAC
SS-TDMA	Self-Stabilizing TDMA
SS-Trees	Sense-Sleep Trees
STEM	Sparse Topology and Energy Management
SYNC	Synchronization
T-MAC	Timeout MAC
TA-MAC	Traffic Aware MAC
TDMA	Time Division Multiple Access
TP	Third Parties
PGF	Two-Phase Greedy Forwarding
TRAMA	Traffic Adaptive Medium Access
UHF	Ultra High Frequency
UP	Upcoming communicating Parties
WISP	Wireless Identification and Sensing Platform
WiseMAC	Wireless sensor MAC
Z-MAC	Zebra MAC

to maximize sleep duration in each node. In MAC-CROSS, nodes are categorized as communicating parties (CP), upcoming communicating parties (UP), and third parties (TP), based on any nodes that are currently participating in communication, nodes that are to be involved in upcoming data transmissions and nodes that are not included in any routing paths, respectively. The notification of role change from UP to TP is performed by the modified MAC protocol using RTS and CTS control frames, where new fields *Final_destination_Addr* and *UP_Addr* are added. This ensures that UP nodes are awake when TP nodes remain in their sleep modes. The process of adding information to the control frames also helps the intermediate or routing nodes to update their MAC about the neighboring node's role change. Thus, a cross-layer approach to extend the lifetime of the sensors is attained in MAC-CROSS.

CLEEP [97] combines the physical, MAC and network layer strategies to extend the network's lifetime. In the physical layer, the transmission power between two nodes is coordinated by maintaining a neighbor node information table and updating this periodically. The optimal routing path is formed by applying Incremental Shortest-path Tree Heuristic (ISTH) on the neighborhood information table formed by the PHY layer. In addition, ISTH requires different sensors to share forwarder nodes in order to reduce the number of active nodes in the network, thus conserving energy. Finally, the MAC layer uses duty cycle scheduling based on the routing information, and applies an RTS/CTS mechanism similar to the 802.11 MAC protocol to avoid collisions and overhearing.

Yang et al. proposed cross-layer coordination in [80] based on priority, using sleeping Directed Diffusion and S-MAC as the routing and MAC layer approaches, respectively. Higher priority is given to sleeping Directed Diffusion, where nodes that are not involved in data delivery are put to sleep during successive flooding that overrides the functionality of S-MAC. S-MAC effectively schedules a node only when sleeping Directed Diffusion needs to keep this node active, thus prioritizing sleeping Directed Diffusion over S-MAC sleeping.

Sleeping can be achieved using a layered approach or cross-layered coordination, where simple scheduling algorithms are used independently or in a coordinated manner to conserve energy and to attain a common goal of extending the lifetime of a sensor. Routing layer sleeping is more suited for networks with high redundancy or high contention, whereas MAC layer sleeping is more sensitive to contention [80]. Understanding such characteristics of the network can aid in making smart decisions to dynamically switch between different sleeping schemes. In the following section, a sleep manager technique is explored. This technique dynamically applies sleeping schemes to the protocol stack depending on the application and network requirements.

5.3 Sleep Manager

A sensor network environment can change over the course of the network lifetime. Hence, novel approaches are needed that can tolerate changes in the network conditions and application requirements, yet provide solutions to incorporate sleeping for energy savings without compromising performance or reliability.

The feasibility of applying an adaptive sleep manager that can dynamically switch sleeping schemes at the routing layer or the MAC layer or both layers with cross-layer coordination is proposed in [85]. This approach is demonstrated using a sleep manager that manages Sleeping Multipath Directed Diffusion at the routing layer and S-MAC at the MAC layer. The sleep manager estimates packet delay of S-MAC using the Markov model for duty-cycled MAC protocols proposed in [98]. This estimated delay is then used by the Sleeping Multipath Directed Diffusion protocol

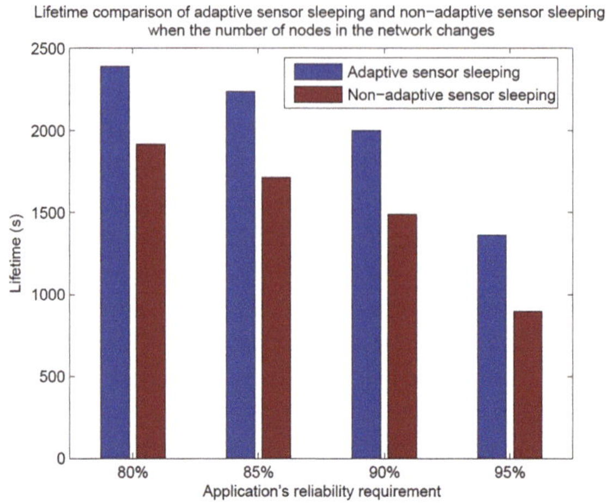

Fig. 9 Network lifetime comparison of adaptive and non-adaptive sensor sleeping as the number of sensors varies throughout the simulation time for different application reliability requirements [85]

to optimize the sleeping timers at the routing layer, so that sensors will neither be put to sleep at the routing layer when the MAC layer has not delivered the pending packets, nor be awake for an unnecessary time after the MAC layer finishes packet delivery. As the network conditions change over time, the estimation of S-MAC packet delay varies, and hence the sleeping timers at the routing layer are set accordingly. The sleep manager also (1) makes routing layer sleeping decisions overrule the MAC layer sleeping decisions, and (2) selects appropriate routing paths to support the application's reliability requirement, which may change over time. The simulation results in [85] show that the sleep manager can effectively extend the network lifetime. Moreover, the sleep manager significantly prolongs the network lifetime while meeting the reliability requirements under the scenarios of dynamic reliability requirements and varying network conditions. As shown in Fig. 9, the network lifetimes using adaptive Sleeping Multipath Directed Diffusion and S-MAC significantly improve as network size varies throughout the simulation when compared to the non-adaptive Sleeping Multipath Directed Diffusion and S-MAC, for various application reliability requirements. This shows that adaptive sensor sleeping can control key parameters in a dynamic manner to handle varying network conditions, thereby improving network lifetime.

6 Conclusions

Much energy is wasted in wireless sensor networks due to idle listening, relaying data and redundant data gathering by active sensors in the network. Sleeping is a key technique used to reduce this energy waste. In this chapter, the need for sleeping techniques is discussed, and different approaches to sleeping at the MAC and Routing layers as well as cross-layer sleeping are described and analyzed. Sleeping techniques prolong the network lifetime, thereby reducing network maintenance and cost, and thus they are crucial for future wireless sensor network deployments.

References

1. D. Dudek, C. Haas, A. Kuntz, M. Zitterbart, D. Kruger, P. Rothenpieler, D. Pfisterer, S. Fischer, A wireless sensor network for border surveillance, in *Proceedings of the 7th ACM Conference on Embedded Networked Sensor Systems (SenSys '09)*. (ACM, New York, 2009), pp. 303–304
2. P. Corke, T. Wark, R. Jurdak, D. Moore, P. Valencia, Environmental wireless sensor networks. Proc. IEEE **98**(11), 19031917 (2010)
3. N. Xu, S. Rangwala, K.K. Chintalapudi, D. Ganesan, A. Broad, R. Govindan, D. Estrin, A wireless sensor network for structural monitoring, in *Proceedings of the 2nd International Conference on Embedded Networked Sensor Systems (SenSys '04)*. (ACM, New York, 2004), pp. 13–24
4. U. Ramachandran, R. Kumar, M. Wolenetz, B. Cooper, B. Agarwalla, J. Shin, P. Hutto, A. Paul, Dynamic data fusion for future sensor networks. ACM Trans. Sen. Netw. **2**(3), 404–443 (2006)
5. R. Rajagopalan, P.K. Varshney, Data aggregation techniques in sensor networks: A survey. IEEE Commun. Surveys Tutorials, **8**(4) (4th Quarter) (2006)
6. M. Holland, T. Wang, B. Tavli, A. Seyedi, W. Heinzelman, Optimizing physical-layer parameters for wireless sensor networks. ACM Trans. Sen. Netw. **7**, 4, (Article 28 Feb. 2011), doi: 10.1145/1921621.1921622
7. Texas Instruments (2011) *Mixed Signal Microcontroller*. Revised Aug 2011: http://www.ti.com/lit/ds/symlink/msp430f2001.pdf
8. H. Ba, J. Parvin, L. Soto, I. Demirkol, W. Heinzelman, Passive RFID-based wake-up radios for wireless sensor network, in appears, in *Wirelessly Powered Sensor Networks and Computational RFID* (Springer, Berlin, 2011)
9. Y. Zhang, C.-H. Feng, I. Demirkol, W. Heinzelman, Energy-efficient duty cycle assignment for receiver-based convergecast in wireless sensor networks, in *IEEE GLOBECOM*. pp. 1–5 (2010)
10. M. Kuorilehto, M. Kohvakka, J. Suhonen, P. Hamalainen, M. Hannikainen, T.D. Hamalainen (2007) MAC protocols, in *Ultra-Low Energy Wireless Sensor Networks in Practice*. pp. 73–88. (Wiley, West Sussex, England, 2007)
11. M. Kohvakka, M. Kuorilehto, M. Hannikainen, T.D. Hamalainen, Performance analysis of IEEE 802.15.4 and zigbee for large-scale wireless sensor network applications, in *Proceedings of ACM International Workshop on Performance Evaluation of Wireless Ad hoc, Sensor, and Ubiquitous Networks*. Malaga, Spain. pp. 1–6 (2006)
12. M.R. Ahmad, E. Dutkiewicz, X. Huang A survey of low duty cycle MAC protocols in wireless sensor networks, in *Emerging Communications for Wireless Sensor Networks*, ed. by A. Foerster, A. Foerster, (2011) ISBN: 978-953-307-082-7
13. R. Zheng, J.C. Hou, L. Sha, Asynchronous wakeup for ad hoc networks, in *Proceedings of the ACM symposium on Mobile ad hoc networking and, computing (MobiHoc)*, pp. 35–45 (2003)
14. W. Ye, J. Heidemann, D. Estrin, An energy-efficient MAC protocol for wireless sensor networks, in *Proceedings of the 21st International Annual Joint Conference of the IEEE Computer and Communications Societies (INFOCOM)*, New York (2002)
15. Y. Sun, O. Gurewitz, D.B. Johnson, RI-MAC: A receiver-initiated asynchronous duty cycle MAC protocol for dynamic traffic loads in wireless sensor networks, in *The 6th ACM Conference on Embedded Networked Sensor Systems*, pp. 1–14 (2008)
16. S. Ray, I. Demirkol, W. Heinzelman, ADV-MAC: Analysis and optimization of energy efficiency through advertisements for wireless sensor networks. Elsevier Ad Hoc Netw. J. **9**(5), 876–892 (2011)
17. Y. Xue, N.H. Vaidya, A wakeup scheme for sensor networks: Achieving balance between energy saving and end-to-end delay, in *Proc. IEEE RTAS*, pp. 19–26 (2004)
18. L. Gu, J.A. Stankovic, Radio-triggered wake-up capability for sensor networks, in *Proceedings of the 10th IEEE Real-Time and Embedded Technology and Applications Symposium RTAS '04*. pp. 27–36 (2011) doi: 10.1109/RTTAS.2004.1317246

19. I. Demirkol, C. Ersoy, E. Onur, Wake-up receivers for wireless sensor networks: benefits and challenges. IEEE Wireless Commun. **16**(4), 88–96 (2009). doi. 10.1109/MWC.2009.5281260
20. R. Muraleedharan, L.A. Osadciw, Cognitive routing protocol for sensor based intelligent transportation system, in *Wireless Technologies in Intelligent Transportation Systems* ed. by M. Zhou, Y. Zhang, L.T. Yang (Nova Science Publishers, USA, 2009) ISBN 978-1-60741-588-6
21. B. Otis, Y.H. Chee, J. Rabaey, A 400μW-RX, 1.6mW-TX super-regenerative transceiv-er for wireless sensor networks. IEEE ISSCC **1**, 396–397 (2005). doi:10.1109/ISSCC.2005.1494036
22. N. Pletcher, S. Gambini, J. Rabaey, A 65μW, 1.9 GHz RF to digital baseband wakeup receiver for wireless sensor nodes. IEEE CICC, 539–542 (2007). doi: 10.1109/CICC.2007.4405789
23. S. Von der Mark, R. Kamp, M. Huber, G. Boeck, Three stage wakeup scheme for sensor networks. SBMO/IEEE MTT-S 205–208 (2005). doi. 10.1109/IMOC.2005.1579978
24. J. Ansari, D. Pankin, P. Mähönen, Radio-triggered wake-ups with addressing capabilities for extremely low power sensor network applications. PIMRC 1–5, (2008). doi:10.1109/PIMRC.2008.4699501
25. B. Van der Doorn, W. Kavelaars, K. Langendoen, A prototype low-cost wakeup radio for the 868 MHz band. IJSNET **5**, 22–32 (2009). doi:10.1504/IJSNET.2009.023313
26. Austria Microsystems (2010) AS3933 3-D Low Frequency RF Wake-up Receiver. http://www.austriamicrosystems.com/Wake-up-receiver/AS3933. Accessed 12 Dec 2010
27. A. Sanchez, J. Aguilar, S. Blanc, J.J. Serrano, RFID-based wake-up system for wireless sensor networks. Proc. SPIE 8067. 806708 (2011). doi. 10.1117/12.887039
28. G.U. Gamm, M. Kostic, M. Sippel, L.M. Reindl, Low power wireless sensor node for use in building automation, in *IEEE 12th Annual Wireless and Microwave Technology Conference (WAMICON)*, pp. 1–6 (2011). doi: 10.1109/WAMICON.2011.5872856
29. C. Schurgers, V. Tsiatsis, S. Ganeriwal, M. Srivastava, Optimizing sensor networks in the energy-latency-density design space. IEEE Trans. Mobile Comput. **1**(1), 70–80 (2002). doi:10.1109/TMC.2002.1011060
30. A.G. Ruzzelli, R. Jurdak, G.M.P. O'Hare, On the RFID wake-up impulse for multihop sensor network, in *Proceedings Convergence of RFID and Wireless Sensor Networks and Their Applications (SenseID) Workshop at ACM Int'l Conference Embedded Networked Sensor Systems (Sensys '07)*, Nov 2007
31. R. Jurdak, A.G. Ruzzelli, G.M.P. O'Hare, Multi-hop RFID wake-up radio: Design, evaluation and energy tradeoffs, in *ICCCN'08*, 1–8 (2008). doi. 10.1109/ICCCN.2008.ECP.124
32. H. Ba, I. Demirkol, W. Heinzelman, Feasibility and benefits of passive RFID wake-up radios for wireless sensor networks, in *Global Telecommunications Conference (GLOBECOM 2010)*, pp. 1–5 (2010). doi: 10.1109/GLOCOM.2010.5683585
33. H. Javaheri, G. Noubir, S. Noubir, RF control of biological systems: Applications to wireless sensor networks, in *Proceedings of Nano-Net*, Jan 2009
34. D.H. Goldberg, A.G. Andreou, P. Julian, P. Pouliquen, L. Riddle, R. Rosasco, VLSI implementation of an energy-aware wake-up detector for an acoustic surveillance sensor network. ACM Trans. Sensor Networks. **2**(4), 594–611 (2006)
35. M. Malinowski, M. Moskwa, M. Feldmeier, M. Laibowitz, J.A. Paradiso, CargoNet: a low-cost micropower sensor node exploiting quasi-passive wakeup for adaptive asychronous monitoring of exceptional events, in *Proceedings of the 5th International Conference on Embedded Networked Sensor Systems*, Sydney (2007). doi: 10.1145/1322263.1322278
36. S. Sudevalayam, P. Kulkarni, Energy harvesting sensor nodes: Survey and implications, Department of Computer Science and Engineering (CSE), Indian Institute of Technology Bombay (IITB), Tech. Rep. IITB/CSE/2008/December/19. TR-CSE-2008-19 (2008)
37. A. Mainwaring, D. Culler, J. Polastre, R. Szewczyk, J. Anderson, Wireless sensor networks for habitat monitoring, in *Proceedings of the 1st ACM International Workshop on Wireless Sensor Networks and Applications*, Sept28–28, Atlanta (2002). doi:10.1145/570738.570751
38. M. Perillo, W. Heinzelman, *Wireless Sensor Network Protocols* (Appears in Fundamental Algorithms and Protocols for Wireless and Mobile Networks). (CRC Hall, Boca Raton, 2005)

39. S.-C. Wang, A. Helmy, Performance limits and analysis of contention-based IEEE 802.11 MAC, in *Proceedings 31st IEEE Conference on Local, Computer Networks*, pp. 418–425 (2006)
40. S. Singh, C.S. Raghavendra, PAMAS–power aware multi-access protocol with signalling for ad hoc networks. ACM SIGCOMM Comput. Commun. Rev., **28**(3), 5–26 (1998). doi. 10.1145/293927.293928
41. H. Gong, J. Cao, M. Liu, L. Chen, L. Xie, A traffic aware, energy-efficient MAC protocol for wireless sensor networks. Intl J.Ad Hoc Ubiquitous Comput., **4**, 148–156 (2009). doi: 10.1504/IJAHUC.2009.024517
42. P. Lin, C. Qiao, X. Wang, Medium access control with a dynamic duty cycle for sensor networks, in *IEEE Wireless Communications and Networking Conference*. vol. 3, pp. 1534–1539 (2004)
43. T.V. Dam, K. Langendoen, An adaptive energy-efficient MAC protocol for wireless sensor networks, in *The First ACM Conference on Embedded Networked Sensor Systems (Sensys'03)*. ACM, New York, pp. 171–180 (2003). doi:10.1145/958491.958512
44. Y. Kwon, Y. Fang, H. Latchmun, Fast collision resolution (FCR) MAC algorithm for wireless local area networks. Proc. IEEE GLOBECOM. *3*, 2250–2254 (2002). doi: 10.1109/GLOCOM.2002.1189032
45. W. Ye, J. Heidemann, D. Estrin, Medium access control with coordinated, adaptive sleeping for wireless sensor networks. IEEE/ACM Trans. Networking, pp. 93–506 (2004). doi:10.1109/TNET.2004.828953
46. IEEE 802.15.4-(2006) IEEE Standard for Information Technology-Part 15.4: Wireless Medium Access Control (MAC) and Physical Layer (PHY) specifications for Low Rate Wireless Personal Area Networks (LR-WPANs)
47. J. Polastre, J. Hill, D. Culler, Versatile low power media access for wireless sensor networks, in *ACM SenSys'04*. ACM, New York, pp. 95–107 (2004). doi: 10.1145/1031495.1031508
48. M. Buettner, G.V. Yee, E. Anderson, R. Han, X-MAC: a short preamble MAC protocol for duty-cycled wireless sensor networks, in *ACM SenSys '06*, ACM, New York, pp. 307–320 (2006). doi:10.1145/1182807.1182838
49. W. Ye, F. Silva, J. Heidemann, Ultra-low duty cycle MAC with scheduled channel polling. *SenSys06*. ACM, New York, pp: 321–334 (2006). doi:10.1145/1182807.1182839
50. A. El-Hoiydi, J.-D. Decotignie, WiseMAC: An ultra low power MAC protocol for multi-hop wireless sensor networks, in *First Int. Workshop on Algorithmic Aspects of Wireless Sensor Networks (ALGOSENSORS)*. Lecture Notes in Computer Science. LNCS 3121 (2004)
51. Y. Sun, O. Gurewitz, D.B. Johnson, RI-MAC: A receiver-initiated asynchronous duty cycle MAC protocol for dynamic traffic loads in wireless sensor networks. *SenSys08*. ACM, New York, pp. 1–14 (2008). doi: 10.1145/1460412.1460414
52. L. Tang, Y. Sun, O. Gurewitz, D.B. Johnson, PW-MAC: An energy-efficient predictive-wakeup MAC protocol for wireless sensor networks, in *Proceedings of the 30th IEEE International Conference on Computer Communications (INFOCOM 2011)*, pp: 1305–1313 (2011)
53. K. Kunert, TDMA-based MAC protocols for wireless sensor networks, in *State of the Art and Important Research Issues*. http://www2.hh.se/staff/tola/ces_2005/papers/kristina_kunert_final.pdf (2005)
54. W.R. Heinzelman, A. Chandrakasan, H. Balakrishnan, Energy-efficient communication protocol for wireless microsensor networks, in *Proceedings of the HICSS '00. IEEE Computer Society*. Washington, DC (2000). doi: 10.1109/HICSS.2000.926982
55. G. Pei, C. Chien, Low power TDMA in large wireless sensor networks, in *Military Communications Conference, 2001. MILCOM 2001. Communications for Network-Centric Operations: Creating the Information Force. IEEE*, vol. 1, pp. 347–351 (2002). doi:10.1109/MILCOM.2001.985817
56. B. Tavli, W. Heinzelman, MH-TRACE: multihop time reservation using adaptive control for energy efficiency. Selected Areas in Communications. IEEE J. **22**(5), 942–953 (2004). doi:10.1109/JSAC.2004.826932
57. S.S. Kulkarni, M.U. Arumugam, TDMA service for sensor networks, in *Proceedings of the 24th International Conference on Distributed Computing Systems Workshops. IEEE Computer Society*. Washington, pp. 604–609 (2004)

58. S. Cho, K. Kanuri, J.-W. Cho, J.-Y. Lee, S.-D. June, *Dynamic Energy Efficient TDMA-based MAC Protocol for Wireless Sensor Networks* (Autonomic and Autonomous Systems and International Conference on Networking and Services). (Papeete, Tahiti, 2005), pp. 48–48
59. J. Mao, Z. Wu, X. Wu, A TDMA scheduling scheme for many-to-one communications in wireless sensor networks. Comput. Commun. **30**(4), 863–872 (2007). doi:10.1016/j.comcom.2006.10.006
60. V. Rajendran, J. Garcia-Luna-Aceves, K. Obraczka, Energy-efficient, application-aware medium access for sensor networks, in *2nd IEEE Conference on Mobile Ad-hoc and Sensor Systems (MASS 2005)* (2005). doi: 10.1109/MAHSS.2005.1542852
61. L. Campelli, A. Capone, M. Cesana, A receiver oriented MAC protocol for wireless sensor networks. In mobile adhoc and sensor systems. *IEEE International conference*, pp 1–10 (2007). doi: 10.1109/MOBHOC.2007.4428626
62. L.V.. Hoesel, P. Havinga, A lightweight medium access protocol (LMAC) for wireless sensor networks, in *1st Intl Workshop on Networked Sensing Systems (INSS) Tokyo, Japan*, pp 205–208 (2004)
63. S. Chatterjea, L.V. Hoesel, P. Havinga, AI-LMAC: An adaptive information-centric and lightweight MAC protocol for wireless sensor networks, in *2nd Int. Conf. on Intelligent Sensors, Sensor Networks and Information Processing*, pp. 381–388 (2004). doi: 10.1109/ISSNIP.2004.1417492
64. I. Demirkol, C. Ersoy, F. Alagoz, MAC protocols for wireless sensor networks: A survey. IEEE Commun. Magaz. **44**(4), 115–121 (2006). doi:10.1109/MCOM.2006.1632658
65. L. Deliang, P. Fei, *Energy-efficient MAC protocols for Wireless Sensor Networks* (Information and Communication Technologies). (Beihang University, Beijing, 2009)
66. I. Rhee, A. Warrier, M. Aia, J. Min, Z-MAC: A hybrid MAC for wireless sensor networks. Networking. IEEE/ACM Trans. **16**(3), 511–524 (2008)
67. S. Ray, I. Demirkol, W. Heinzelman, ATMA: Advertisement-based TDMA protocol for bursty traffic in wireless sensor networks, in *Global Telecommunications Conference (GLOBECOM 2010)* (2010). doi: 10.1109/GLOCOM.2010.5683930
68. S. Ray, I. Demirkol, W. Heinzelman, ADV-MAC: Advertisement-based MAC protocol for wireless sensor networks, *MobiQuitous '09*, pp. 1–2, July 2009
69. R. Jurdak, A.G. Ruzzelli, G. O'Hare, Radio sleep mode optimization in wireless sensor networks. IEEE Trans. Mobile Comput. **9**(7), 955–968 (2010)
70. A. Kansal, J. Hsu, S. Zahedi, M.B. Srivastava, Power management in energy harvesting sensor networks. Trans. Embedded Comput. Sys. **6**(4), 2007 (2007)
71. A. Eu, W.K. Seah, H.-P. Tan, A study of MAC schemes for wireless sensor networks powered by ambient energy harvesting, in *WICON '08 Proceedings of the 4th Annual International Conference on Wireless Internet*, pp. 1–9 (2008)
72. X. Fafoutis, N. Dragoni, ODMAC: An on-demand MAC protocol for energy harvesting - wireless sensor networks, in *Proceedings of the 8th ACM Symposium on Performance evaluation of wireless ad hoc, sensor, and ubiquitous networks. PE-WASUN '11*, pp 49–56 (2011)
73. J. Kim, J.-W. Lee, Energy adaptive MAC protocol for wireless sensor networks with RF energy transfer, in *Proceedings of the Third International Conference on Ubiquitous and Future Networks (ICUFN)*. Dalian, China, pp. 89–94 (2011)
74. P. Santi, Topology control in wireless ad hoc and sensor networks. ACM Comput. Survey (CSUR) **37**(2), 164–194 (2005). doi:10.1145/1089733.1089736
75. Y. Xu, J. Heidemann, D. Estrin, Geography-informed energy conservation for ad hoc routing, in *7th Annual International Conference on Mobile Computing and Networking*, pp. 70–84 (2001)
76. B. Chen, K. Jamieson, H. Balakrishnan, R. Morris, Span: An energy-efficient coordination algorithm for topology maintenance, in *Ad Hoc Wirleess Networks*, vol. 8, issue 5, pp. 481–494. Kluwer Academic Publishers, Dordrecht (2002)
77. A. Cerpa, D. Estrin, Ascent: Adaptive self-configuring sensor networks topologies. IEEE Trans. Mobile Comput. **3**, **3**, 272–285 (2002)
78. H. Frey, D. Gorgen, Geographical cluster-based routing in sensing-covered network. IEEE Trans. Parallel Distributed Syst. **17**(9), 899–911 (2006). doi:10.1109/TPDS.2006.124

79. D. Simplot-Ryl, I. Stojmenović, J. Wu, Energy-efficient backbone construction, broadcasting, and area coverage in sensor networks, in *Handbook of Sensor Networks: Algorithms and Architectures* ed. by I. Stojmenović. Wiley, Hoboken. doi: 10.1002/047174414X.ch11
80. O. Yang, W. Heinzelman, A better choice for sensor sleeping, in *6th European Conference on Wireless Sensor Networks (EWSN '09)* (2008)
81. R. Zheng, R. Kravets, On-demand power management for ad hoc networks, in *The 22nd Annual Joint Conference of the IEEE Computer and Communications Societies*, vol. 1, pp. 481–491 (2003)
82. H. Wang, W. Wang, D. Peng, H. Sharif, A route-oriented sleep approach in wireless sensor network, in *The 10th IEEE Singapore International Conference on Communication Systems*, pp. 1–5 (2006)
83. H. Cheng, X. Jia, An energy efficient routing algorithm for wireless sensor networks, in *International Conference on Wireless Communications. Networking and Mobile Computing*, vol. 2, pp. 905–910 (2005)
84. K.E. Haynes, R.R. Stough, R.G. Kulkarni, Towards a Percolation Model of Accessibility. http://www.geovista.psu.edu/sites/geocomp99/Gc99/034/abs99-034.htm (1999)
85. O. Yang, *Sleeping Strategies for Wireless Sensor Networks*, PhD Dissertation. University of Rochester, 2011
86. X. Hou, T. Tipper, J.F. Kabara, D. Yupho, GSP: Gossip-based sleep protocol for energy efficient routing in wireless sensor networks, in *The 16th International Conference on Wireless Communications*. Calgary, Alberta, Canada (2004)
87. S. Dulman, T. Nieberg, J. Wu, P. Havinga, Trade-Off between traffic overhead and reliability in multipath routing for wireless Ssensor networks, in *Wireless Communications and Networking Conference*, pp. 1918–1922 (2003)
88. B. Yahya, J. Ben-Othman, REER: Robust and energy efficient multipath routing protocol for wireless sensor networks, in *Global Telecommunications Conference*, pp. 1–7 (2009)
89. L. He, Efficient multi-path routing in wireless sensor networks, in *The 6th International Conference on Wireless Communications Networking and Mobile, Computing*, pp. 1–4 (2010)
90. O. Yang, W. Heinzelman, Sleeping multipath routing: A trade-off between reliability and lifetime in wireless sensor networks. *Global Communication Conference, Houston* (2011)
91. C. Zhu, L. Yang, L. Shu, L. Wang, T. Hara, Sleep Scheduling Towards Geographic Routing in Duty-Cycled Sensor Networks with A Mobile Sink, in *The 8th Annual IEEE Communications Society Conference on Sensor, Mesh and Ad Hoc Communications and Networks*. Salt Lake City (2011)
92. R.W. Ha, P.-H. Ho, X.S. Shen, Cross-layer organization for wireless sensor networks using sense-sleep trees. Proc. WirelessCom, **2**, 952–957 (2005)
93. S. Liu, M. Sha, L. Huang, ORAS: Opportunistic routing with asynchronous sleep, in *International Conference on Future Computer and Communication (ICFCC 2010)*. (IEEE Press, Wuhan, China, 2010)
94. S. Nath, P.B. Gibbons, Communicating via fireflies: geographic routing on duty-cycled sensors, in *Proceedings of the 6th International Conference on Information Processing in Sensor Networks. IPSN '07*, pp 440–449 (2007)
95. R. Ha, *A Sleep Scheduling Based Cross Layer Design Approach for Application Specific Wireless Sensor Network*, Dissertation. University of Waterloo, Canada (2006)
96. *International Workshops: XRA, IWSN, MEGA, and ICSE* (Springer, Harbin, 2006)
97. S. Liu, Y. Bai, M.o. Sha, Q. Deng, D. Qian, CLEEP: A novel cross-layer energy-efficient protocol for wireless sensor networks, in *4th International Conference on Wireless Communications. Networking and Mobile Computing. WiCOM '08*. pp. 1–4 (2008)
98. O. Yang, W. Heinzelman, Modeling and performance analysis for duty-cycled MAC protocols in wireless sensor networks, in *IEEE Transactions on Mobile Computing*, pp: 905–921 (2012)

Part III
Routing

Chapter 7
Energy-Aware Routing for Wireless Sensor Networks

Ahmed E. A. A. Abdulla, Hiroki Nishiyama, Nirwan Ansari and Nei Kato

Abstract Wireless sensor networks have attracted much attention due to their ability to collect data from areas of interest. The limited energy capacity along with the difficulty of charging batteries of deployed sensors render energy-aware routing essential for sustained operation of wireless sensor networks. In this chapter, we classify energy-aware routing algorithms into five categories according to their network architecture: flat multi-hop routing that finds paths to minimize energy consumption or increase sensor network lifetime, hierarchical routing that creates a hierarchy and applies *data-aggregation* to reduce energy consumption, hybrid multi-hop routing that is a combination of the former two and mitigates the energy hole problem, *data-centric* routing that performs in-network *data-aggregation* to eliminate wasteful transmissions, and location-based routing that uses location information to reduce the energy consumption of the wireless sensor network. Furthermore, we present a cross-cutting discussion which addresses *data-aggregation*, network lifetime definition, routing overhead, the energy hole phenomenon, and collisions/interferences.

A. E. A. A. Abdulla · H. Nishiyama · N. Kato
Graduate School of Information Sciences, Tohoku University, Sendai, Japan
e-mail: ahmed@it.ecei.tohoku.ac.jp

H. Nishiyama
e-mail: bigtree@it.ecei.tohoku.ac.jp

N. Kato
e-mail: kato@it.ecei.tohoku.ac.jp

N. Ansari (✉)
Advanced Networking Laboratory, ECE Department, New Jersey Institute of Technology, Newark, NJ, USA
e-mail: nirwan.ansari@njit.edu

H. M. Ammari (ed.), *The Art of Wireless Sensor Networks*,
Signals and Communication Technology, DOI: 10.1007/978-3-642-40009-4_7,

© Springer-Verlag Berlin Heidelberg 2014

1 Introduction

Advances in wireless communications and nanotechnology have facilitated the widespread use of wireless sensor networks [4, 6, 14, 48]. Wireless sensor networks rely heavily on battery power to drive their functionality. When the energy of a battery is depleted, the sensor loses its functionality. Replacing/charging the batteries of a large number of sensors is an insurmountable task in terms of time and cost; the task becomes infeasible in potentially dangerous terrain. Hence, severely limited energy capacities of wireless sensor networks render energy-efficient technologies indispensable for deploying wireless sensor networks. The energy consumption of a wireless sensor node can be attributed to the following major activities:

1. Information gathering: energy consumed by the sensors onboard the nodes for gathering information.
2. Computation: energy consumed for processing purposes, predominantly attributed to the basic system operation.
3. Communications: energy consumed to transmit data from sensors to their neighbors. This usually takes up the largest share of energy consumption of a wireless sensor network.

In this chapter, we focus our attention on the energy consumption associated with communications; in particular, we consider energy-aware routing for wireless sensor networks. This kind of routing algorithms has a very different objective from traditional routing algorithms; traditionally, routing has been designed to maximize throughput and/or scalability. Although the aforementioned objectives are important, communications is the major energy guzzler, and thus considering the energy consumption of routing is of significant importance. Therefore, this chapter addresses energy-aware routing for wireless sensor networks.

A wireless sensor network is usually deployed without the aid of infrastructure such that sensors cooperate to facilitate communications in the wireless sensor network. A wireless sensor network consists of two basic building blocks, namely, a sink and a number of sensors, all of which are capable of communicating with each other over a common wireless channel. The sink acts as the final point of collection, and from which data can be extracted for further processing and transmission. The sink assumes the role of a gateway because it is where all packets are routed, thus enabling connection to other networks such as the Internet. In a practical implementation, the sink has access to a virtually unlimited energy source. Although we have limited our discussion to a wireless sensor network with a single sink, the wireless sensor network can, in general, have more than one sink. With each sink responsible for collecting data from a sub-group of sensors, all the data collected from nodes of all sub-groups are gathered into a single node for processing. As a result, this mode of data gathering can be thought of as an integration of multiple wireless sensor networks, each with a single sink. The second component of the sensor network is a collection (hundreds or thousands) of sensors, which are responsible for collecting data from their surroundings; to enable communications in an infrastructureless network, they consume their limited energy reserves to relay data from other sensors, and thus decreasing the energy consumption of the sensors is the key objective of energy-aware routing for wireless sensor networks. Energy-aware routing algorithms can be classified into five categories according to their network architecture.

The first category is flat multi-hop routing, where routes from the source node to the destination node are selected with low energy consumption in mind. The second category is hierarchical multi-hop routing, where sensors take different roles and form hierarchies. Hierarchical multi-hop routing reduce energy consumption by decreasing the volume of data flowing within the wireless sensor network. The third is hybrid multi-hop routing, which is a combination of the first and second categories, and aims to mitigate the energy hole problem inherent to the many-to-one (convergecast) traffic patterns in wireless sensor networks. The forth category is *data-centric* routing that performs in-network *data-aggregation* in intermediate sensors to reduce the energy consumption inefficiencies in classical routing algorithms. The fifth is location-based routing, where location information is used to decrease the energy consumption of the wireless sensor network. In this chapter, we examine the landmark algorithms of each category that have largely shaped the roadmap for innovation in this area. Moreover, we examine recently proposed state-of-the-art routing schemes of each category.

The remainder of this chapter is organized as follows. Section 2 provides background materials of energy-aware routing in wireless sensor networks. Sections 3–7 examine various categories of energy-aware routing algorithms proposed for wireless sensor networks. We further discuss and compare these routing algorithms from different perspectives in Sect. 8, and finalize this chapter with a conclusion in Sect. 9.

2 Single-Hop Versus Multi-Hop Energy Consumption

Owing to the lack of infrastructure support in wireless sensor networks, sensors need to take the responsibility of transmitting the data they collected to the sink. In a relatively small-scale wireless sensor network deployment, it may be possible for all the nodes to transmit their collected data to the sink directly. For the majority of wireless sensor networks applications, where nodes are far away from the sink, the simple strategy of directly sending data to the sink does not work for a number of reasons. We mention the most relevant of them. Firstly, the sensors have a limited transmission range and cannot transmit data over this hardware-specific range. Secondly, long transmission distances are considered to be energy inefficient. Given a sending and receiving node, the following equations quantify the energy consumption of the sender and receiver [5, 34, 41],

$$e_s(i) = \varepsilon_1 d_{i,j}^{\phi} + \varepsilon_2 \tag{1}$$

$$e_r(j) = \varepsilon_3. \tag{2}$$

Here, $e_s(i)$ is the energy consumed for sending a unit of data by the sensor i to the sensor j. ϕ is the path loss exponent dependent on the wireless fading environment, and its value is usually from two to four, two for short distances and four for long distances. The term ε_1 is a constant specific to the specific wireless system. ε_2 is the electronics energy, characterized by factors such as digital coding, modulation, filtering, and spreading of the signal. $e_r(j)$ is the energy consumed by the receiving node, which is a constant, ε_3. Given that $\varepsilon_1 \gg \varepsilon_2$ and $\varepsilon_1 \gg \varepsilon_3$, Eq. (1) shows that the energy consumption has a growth rate of order $O(d_{i,j}^{\phi})$. In other words, the energy consumption efficiency degrades with the length of transmission distance, $d_{i,j}$.

Multi-hop transmission strategies are considered to be advantageous due to their energy-efficient transmission distances. In a multi-hop transmission strategy, rather than transmitting the data directly from the sending sensor to the receiving sensor, one long transmission is divided into multiple shorter transmissions with each having energy consumption according to Eq. (1). Evidently, transmitting at shorter distances is more energy efficient. Therefore for the above-mentioned reasons, multi-hop routing is suitable for wireless sensor networks, where sensors cooperate with each other to facilitate low-energy communications in wireless sensor networks.

3 Flat Multi-Hop Routing Algorithms

Flat multi-hop routing algorithms are based on concepts inherent from contemporary networks [25]. In traditional wired networks, if a set of nodes are directly connected together via a common medium, point-to-point communications between two neighboring nodes can be easily executed via a data-link layer algorithm. If the two nodes do not share a common wired link, the concept of routing, which is applying the point-to-point data-link algorithm iteratively, applies to a packet as it passes from one node to another till it reaches its destination. Since there are many possible paths, choosing the best possible path defined by a specific criterion is dictated by the routing algorithm.

The above-mentioned techniques are applicable to networks that are wireless and lack infrastructure support, i.e., wireless sensor networks. The set of nodes that are within the maximum transmission distance of each other are thought of as neighbors, and can directly be connected via the wireless medium. Since many paths exist between a source and destination pair, there must be criteria to select the most appropriate path. In traditional wired networks, an emphasis has been placed on choosing the path which maximizes the end-to-end throughput and minimizes the delay (by selecting the path with the minimum number of hops, or the path with the fastest links). These criteria are usually derived from the user requirements (users want to have a fast connection). In wireless sensor networks, although the end-to-end delay is important, the amount of energy consumed by the network is even more critical as exhausted nodes will greatly affect the lifetime of the network.

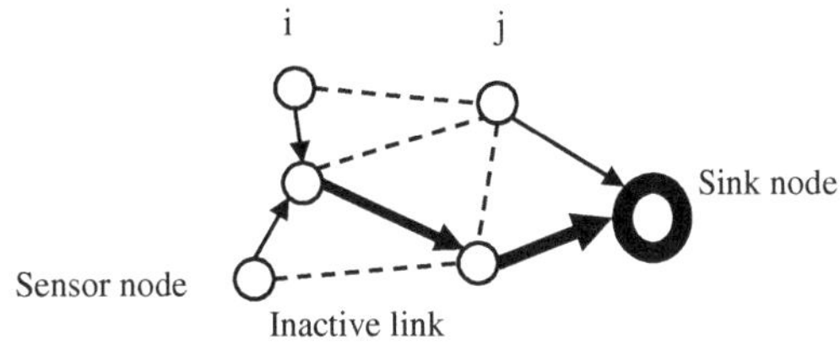

Fig. 1 An example of flat multi-hop routing. Each sensor can communicate with other sensors within its maximum transmission range. The *arrow's* width represents the amount of data that should flow through its associated link. Other links are not utilized

Specifically, the routing algorithm can evaluate a path from the viewpoint of energy consumption of a single link according to Eqs. (1) and (2).

3.1 Minimizing Energy Consumption

Toh [41] described a method to select paths that allow minimum energy consumption as shown in Fig. 1. In the illustration, the flat energy-aware routing algorithm utilizes the links indicated by arrows that minimize the energy consumed in the wireless sensor network, while the rest of the links are inactive. The energy calculation method is as follows. The energy-aware *link cost* is defined in terms of the amount of energy consumed by each wireless link. More precisely, the energy burden on the two end nodes, i.e., the sending and receiving nodes, can be quantified as

$$linkcost(i, j) = e_s(i) + e_r(j). \tag{3}$$

Thus, the total energy consumed by the wireless sensor network for using path l, P_l, can be quantified as

$$P_l = \sum linkcost(i, j) \quad \forall i, j \in l. \tag{4}$$

The desired route, which can minimize the energy consumption for sending data between any sensor, i, and the sink, $P^{min}_{(i,sink)}$, can be obtained from the following equation

$$P^{min}_{(i,sink)} = \min_{l \in L} P_l. \tag{5}$$

Here, L is the set of all possible paths from sensor i to the sink. Thus, by routing traffic through $P^{min}_{(i,sink)}$, the energy consumed by the wireless sensor network can be minimized, hence ultimately increasing the lifetime of the wireless sensor network.

3.2 *Maximizing Network Lifetime*

Chang and Tassiulas [10, 11] adopted concepts from *linear programming* to design a routing algorithm that maximizes the lifetime of the wireless sensor network. We first present their proposed model, followed by the theory and their proposed heuristic algorithm that spreads the data flow equally among sensors to increase the lifetime of the sensor network.

3.2.1 Linear Programming Model

Given a directed graph, $G(N, L)$, in which a set of sensors, N, exist and are connected together via a set of directed links (i, j), L. Here, i and $j \in N$, are the two sensors that communicate via this link. Each sensor i has a set of sensors, S_i, which can be reached within its maximum transmission range. A link (i, j) exists in L if $j \in S_i$. Denote E_i as the initial battery energy reserve of sensor i. The energy consumed for transmitting a message by sensor i to destination sensor j can be evaluated from Eq. (1) and is denoted as e_{ij}. Sensor i transmits to sensor j at the rate of q_{ij}. Data being transmitted from a source sensor to a destination sensor over a path is referred to as a flow. It has quantity and direction. If it is from the source to the destination, then it is referred to as a positive flow; otherwise, it is called a negative flow. Denote O as the set of origin sensors, from which data are originated, and D as the set of possible destination sensors.

We shall next present properties and the associated equations to model network behavior. Firstly, the conservation of flow (the summation of all incoming flows subtracting the sum of all outgoing flows in each node must be equal to the amount of data generated from the node itself), as illustrated in Fig. 2, can be expressed in form of a linear equation as

$$\sum_{j \in S_i} q_{ji} - \sum_{k \in S_i} q_{ik} = Q_i, \quad \forall i \in (N - D). \tag{6}$$

Here, Q_i is the information generation rate of sensor i. The time period till the energy of the sensor i is depleted, $T_i(\mathbf{q})$, is inversely proportional to the amount of

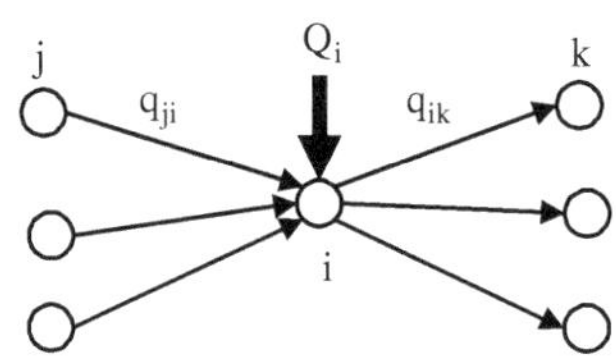

Fig. 2 Flow conservation at sensor i. The summation of all incoming flows to sensor i, $\sum_{j \in S_i} q_{ji}$, subtracting the summation of all outgoing flows from sensor i, $\sum_{k \in S_i} q_{ik}$, equals the information generated in sensor i, Q_i

data flowing through it, $\mathbf{q} = q_{ij} : \forall j \in S_i$, that is,

$$T_i(\mathbf{q}) = \frac{E_i}{\sum_{j \in S_i} e_{ij} q_{ij}}. \tag{7}$$

Network lifetime: The lifetime of the system is defined as the minimum lifetime of a sensor in a wireless sensor network, i.e.,

$$T_{WSN}(\mathbf{q}) = \min_{i \in N} T_i(\mathbf{q}). \tag{8}$$

A network designer would like to maximize the lifetime of the wireless sensor network, and thus the objective function can be formulated as follows,

$$\max_{\mathbf{q}} T_{WSN}(\mathbf{q}). \tag{9}$$

Furthermore, this can be expressed as

$$\max_{\mathbf{q}} \min_{i \in N} \frac{E_i}{\sum_{j \in S_i} e_{ij} q_{ij}}. \tag{10}$$

The above optimization problem also serves as a model for understanding how energy-aware routing algorithms can operate to maximize the lifetime of the wireless sensor network. The basic observation is that if a sensor would have to transmit more data than other sensors, it would live for a shorter time. Thus, the network lifetime would decrease according to Eq. (8). Ultimately, to achieve the maximum lifetime, a routing algorithm should equally spread the load among all sensors.

Theorem 1 (Necessary optimality condition [10]) *Given that paths are of positive flow to the destination. The minimum lifetime over all nodes is maximized* $\rightarrow$ *The lifetime of all paths from the source node to the destination node are equal.*

Proof We prove the above theorem by contradiction. Let the lifetime of a path be determined by the minimum lifetime over all the nodes in the path. Also, define a path with positive flow as one with flows originating from the source to the destination. Assume that the minimum lifetime over all nodes is maximized. Here, assume that the lifetime of all paths with positive flow to the destination are not equal (contrary to the conclusion of the above theorem). Then, there exists a path with positive flow that has a shorter lifetime as compared to all other paths. This path's lifetime, which is also the minimum lifetime over all nodes, can be increased by moving a small amount of positive flow from it to any of the other paths, thereby making its lifetime longer than the minimum lifetime over all nodes before moving the flow. Thus, this contradicts the first assumption that the minimum lifetime over all sensors is maximized. □

3.2.2 Algorithm

Chang and Tassiulas [10] proposed two algorithms that spread flows equally among all the paths. Both of them follow the same structure, as will be explained next; their differences will be described afterwards.

For each node $i \in (N - D)$ in the wireless sensor network,

1. Determine from which path to which path the flows should be redirected.
2. Determine the fraction of flows that should be diverted.
3. Redirect the fraction of flows as determined in Step 1 and Step 2.

The two algorithms differ in the way they implement Step 1. It may be implemented based on the lifetime, calculated by Eq. (7), of the sending node and the nodes along the path to the destination node. Additionally, it can be based on the residual energy of the sending node and the intermediate nodes along the path to the destination. The cost function is defined as

$$linkcost(i, j) = \frac{1}{E_i - e_{ij} n_{ij}}, \tag{11}$$

where E_i is the residual energy of the sensor i, and n_{ij} is the number of message units transmitted from node i to node j (i.e., the size of flow traversing the link).

3.3 Recent Innovations

The seminal work of Chang and Tassiulas [10, 11] has paved the way for many innovations for flat multi-hop routing algorithms. Indeed, many new routing algorithms have been proposed to tackle the short comings of the foundation laid out by them. We shall next highlight how these latest works advance the state of art in flat multi-hop routing for wireless sensor networks.

The linear programming model presented in [10, 11] models the lifetime of the wireless sensor network as the time when the first sensor dies. This definition of network lifetime is not an accurate one because the network can still be functional after the first sensor's death. Many researchers have tried to improve this definition. For example, Karkvandi et al. [23] proposed a novel network lifetime criterion based on Sensing Spatial Coverage (SSC), which refers to the ability of a sensor to monitor a phenomenon of interest in an area. By using the SSC-based lifetime definition, the network is able to improve the monitoring of the area of interest. Liu and Cao [27] pointed out that in a wireless sensor network with low density, the spacial temporal coverage requirements cannot be satisfied while satisfying the lifetime constraints. Therefore, they proposed to schedule sensors to sleep in order to increase their coverage while meeting network lifetime requirements. Furthermore, Naddafzadeh-Shirazi and Lampe [37] defined lifetime as the time till the network is unable to achieve given detection requirements (DRs), which are defined in terms of probabilities of detection and false alarm as dictated by application requirements.

Liu et al. [28] considered maximization of network lifetime by scheduling sensors to sleep during idle listening periods. They observed that the traffic is light most of the time in many sensor network applications and the idle sensors are wasting valuable energy during this period. Therefore, they proposed to include sleep cycle scheduling in the routing problem to eliminate the energy wasted in idle periods and thus improve the longevity of the wireless sensor network.

3.4 Summary

Flat multi-hop routing algorithms are based on classic concepts for traditional wired networks. The basic idea is to modify the link cost to reflect the energy consumption attributed to utilizing the wireless link between two sensors. After assigning link costs to each link, a shortest-path algorithm such as Dijkstra's algorithm [40] can be utilized to find the least energy consuming path among a set of available paths between a source node and a destination node. Generally, this category of routing algorithms fails to capitalize on the redundancy that is inherent in wireless sensor networks to reduce their energy consumption.

4 Hierarchical Routing Algorithms

Hierarchical routing assigns different roles to sensors. The hierarchy is formed when some nodes are chosen to act as gateways (an intermediary) for other nodes.

The concept of hierarchical routing has been readily applied in traditional wired networks [40]. The complexity of the routing process increases with the network size. The number of interactions among sensors increases owing to the increased interaction of the routing protocol initialization, thus leading to huge waste of computation resources. Based on a clever insight that there is no need for every node to know information about every other node, a hierarchy can be established. Following the divide-and-conquer concept, the network can be divided into smaller areas, sometimes referred to as *regions*, and then each region internally creates paths among individual nodes. At the *inter-region* level, each *region* establishes the routes to other *regions*, and individual nodes can communicate outside their *regions* through a special node, often referred to as a *gateway*. Consequently, the computation cost can be substantially decreased. The example presented here is a two-level hierarchy. Two levels are definitely not sufficient for huge networks. In general, the number of levels is dependent on the size of the network.

In the context of wireless sensor networks, *regions* are refereed to as clusters and gateways as cluster heads. Adopting hierarchical network architectures allows special operations to be assigned to the gateway of each cluster, viz, *data-aggregation*, where redundancy is capitalized as well as can be reduced, therefore reducing the volume of data flows in the wireless sensor network. This eliminates many unneeded network operations, and greatly reduces the energy consumption of the wireless sensor network.

4.1 LEACH

Low Energy Adaptive Clustering Hierarchy (LEACH) [18] is the most popular form of hierarchical routing for wireless sensor networks. LEACH as depicted in Fig. 3 is a two-level hierarchy, with the sink acting at the top of the hierarchy. Time is divided into time periods called rounds. In the beginning of each round, the sensors divide themselves into two groups, a group of Cluster Heads (CHs) and a group of Cluster Members (CMs).

The first group of sensors serve the role of CHs, and the remaining nodes assume the role of CMs. The number of CHs is generally less than the number of CMs. The wireless sensor network is divided into clusters. The distributed selection of cluster shapes results in their resemblance to Voronoi diagrams centered on each CH.

According to the location of each CM, it will choose to join its closest CH. Each CH, along with a number of CMs, forms a cluster. CMs act as normal sensors by collecting data from their surroundings. The CHs also function as normal sensors but, additionally, they act as gateways for their respective clusters. After each CM collects data from its surrounding, it transmits the data to its respective CH. After the CH collects data from its CMs, it aggregates them along with its own data, and sends them to the sink.

Subsequently, the volume of data flowing within the network is substantially reduced due to *data-aggregation*, thus significantly decreasing the energy consumption in the wireless sensor network.

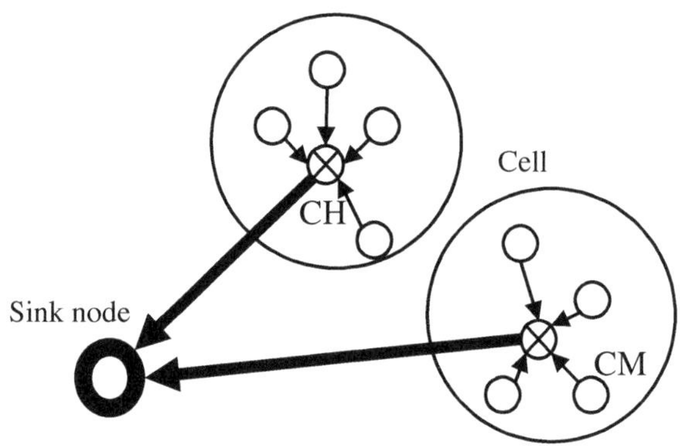

Fig. 3 An example of LEACH, where each CH collects data from its CMs to aggregate and send them to the sink

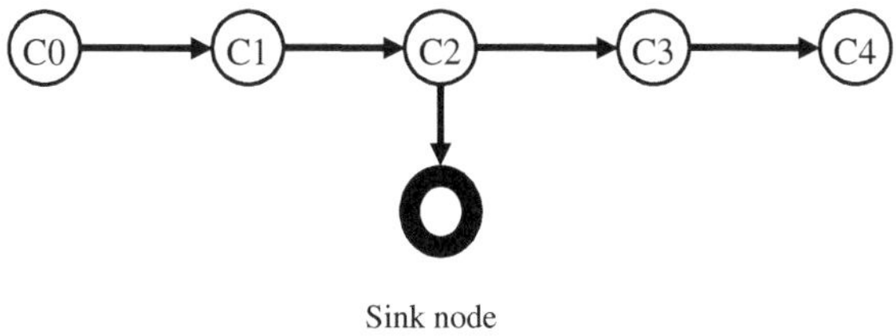

Fig. 4 An example of chain-structured clusters of PEGASIS, where nodes apply a greedy algorithm to find the closest node to them to form a cluster

4.2 PEGASIS

Power-Efficient Gathering in Sensor Information Systems (PEGASIS) [26] clusters nodes in a chain-based shape, differing from the cluster shapes adopted in LEACH. Fig. 4, redrawn from [26], illustrates PEGASIS.

The basic idea is that each sensor forwards its data to its neighbor; the neighbor adds its own data and aggregates both of them and sends a single packet to its neighbor. This process is repeated till the data is delivered to the leader (the sensor that is responsible for sending the data to the sink). The choice of neighbors follows a greedy method, in which each node finds the closest neighbor to itself. Each node in a chain takes turn to become a leader.

In PEGASIS, nodes only need to communicate with their closest neighbors, so that the transmission distance is short, thus decreasing the energy consumed for communication per unit of data. The advantage of using a chain-based design is to avoid cluster formation found in LEACH.

4.3 Recent Innovations

The cluster-based design of hierarchical routing pioneered in LEACH [18] is an effective solution to decrease the energy consumption of wireless sensor networks. LEACH, however, presents several drawbacks, and thus many researchers have proposed important improvements to LEACH. One major drawback is the distributed nature of cluster formation in LEACH that can result in uneven distribution of CHs, thus leading to variated transmission distances among CHs and their CMs. Consequently, energy consumption among CHs and CMs vary greatly, i.e., imbalanced energy consumption. Grid-based cluster design [53] mitigates this drawback. In a grid-based cluster wireless sensor network, the network is divided into grids of equal size. Sensors are aware of the grid they belong to by relating grid dimensions to their positions. Zhang et al. [53] proposed to optimize the grid size by using probabilistic distance models to achieve more efficient energy consumption.

Although clustering significantly reduces the energy consumption of individual sensors, it increases the communication burden on CHs. As illustrated in Fig. 5, once the CH has gathered information, it needs to transmit it to the sink either

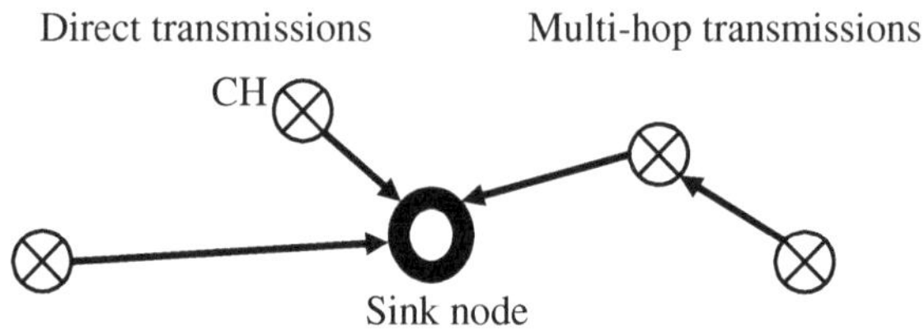

Fig. 5 An illustration of the two transmission schemes used for communication between CHs and the sink

via direct transmission or via multi-hop transmissions through intermediate nodes. Shu and Krunz [38] proposed to optimize the balance between the aforementioned CH transmission schemes to extend the lifetime of CHs.

CH selection has a great influence on the energy consumption of the wireless sensor network. Various CH selection schemes have been proposed [50], in which a sensor is elected to become a CH based on several criteria such as residual energy and node degree. Recently, Wei et al. [44] considered the case where sensors produce differing amounts of traffic load, and proposed to increase the probability of nodes with higher power and lower traffic generation rates to become a CH.

In most wireless sensor network scenarios, the correlation of sensor data makes collecting all the sensor data unnecessary. Also, collecting all sensor data is energy consuming. One technique to eliminate the transmission of unnecessary data is to employ data predictors. Data predictors use sensors' past inputs to estimate their future data. If the error bound (difference between the predicted value and actual value) is acceptable, the sensors need not transmit their sensed data. Thus, data predictors alleviate the traffic burden and subsequently reduce the energy consumption of the wireless sensor network. Jiang et al. [22] proposed to implement data predictors in CHs found in hierarchical routing, such as the one illustrated in Fig. 3. However, the energy consumption for training the data predictor (computation) is non-negligible, and therefore they have investigated which conditions render using data predictors in CHs energy efficient. They showed that energy efficiency is a function of both the correlation of sensors' collected data and the desired error bound.

4.4 Summary

Hierarchical multi-hop routing is a technique adapted from existing networks, where it has been employed for its superior scalability and low complexity. In wireless sensor networks, hierarchical multi-hop routing exhibits its merit in the form of *data-aggregation*, which reduces the volume of data transmissions, and in turn reduces the energy consumption of the wireless sensor network. The *data-aggregation* [7] scheme itself is dependent on the nature of data collected within the wireless sensor network. For examples, *data compression* is applicable when the data are correlated to a certain extent in environmental monitoring applications, *beamforming* when various signals are combined to produce a signal with a better signal-to-noise ratio in acoustic data, and *data fusion* when several messages contain the exact content in moving/migrating objects.

Some wireless sensor network environments only allow *data-aggregation* to reduce the volume of data by a small amount. For example, when data compression is employed and the correlation between the collected data is low, then the compression rate defined as

$$Compression\ rate = \frac{SIZE[Compressed\ Data]}{SIZE[Original\ Data]} \tag{12}$$

will be close to one, and hence the energy savings gained by transmitting a lower volume of data will be outweighed by the energy consumed by forming clusters. In summary, hierarchical multi-hop routing should only be employed in applications where the volume of data can be substantially reduced with *data-aggregation*.

5 Hybrid Routing Algorithms

The concept of hybrid routing for wireless sensor networks was first proposed in [33]. The motivation behind this strategy is to address the energy hole problem, which is also refereed to as the hotspot problem. This problem is inherent to the design of sink-based wireless sensor networks. Since all traffic originating from the sensors is destined to the sink, the nodes that are close to the sink consume more energy and exhaust their battery energy in a much more rapid manner than other sensors. If the sensors close to the sink die, the sink will be isolated. Thus, the wireless sensor network will loose its functionality, despite the fact that the rest of the wireless sensor network is left intact.

5.1 HYMN

HYbrid Multi-hop routiNg (HYMN) [1, 33], depicted in Fig. 6, is a hybrid of two categories of routing algorithms, namely, flat multi-hop routing, introduced in Sect. 3, and hierarchical routing, introduced in Sect. 4. A comparison among these three categories is shown in Table 1.

The area within the maximum transmission range of the sink is referred to as the Sink Connectivity Area (SCA). The sensors in this area allow the sink to connect to the sensors beyond its maximum transmission range. Generally, the number of sensors in the SCA is relatively much less than the remaining sensors in the wireless sensor network. Rationally, the largest part of energy consumption in the SCA is attributed to relaying traffic that originates from outside the SCA. On the other hand, the share of energy consumption attributed to transmitting data originating from the SCA itself is relatively much less.

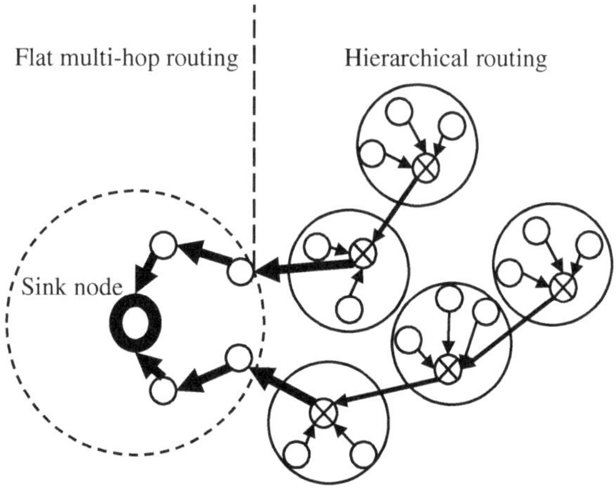

Fig. 6 An illustration of HYbrid Multi-hop routinNg (HYMN). HYMN combines two categories of routing algorithms

Table 1 A comparison among three types of energy-aware routing algorithms

Type	Data-aggregation	Transmission distance
Flat multi-hop routing	No	Short
Hierarchical routing	Yes	Long
HYMN	Yes	Short (in the SCA)

From the above discussion, to decrease the energy consumption of the SCA, the energy consumption per unit of data transmission must be decreased, and/or the volume of data flowing through the network must be limited. HYMN achieves the effect of both solutions. Outside the SCA, a hierarchical routing algorithm can be adopted to reduce the volume of influx going into the SCA, so that *data-aggregation* decreases the flow of data into the SCA, and flat multi-hop routing is used inside the SCA to achieve energy-efficient transmissions (short distances).

Consequently, by focusing on mitigating the energy hole problem, HYMN successfully decreases the energy consumption in the SCA, and increases the longevity of the wireless sensor network.

5.2 Summary

Hybrid multi-hop routing adopts two strategies of routing, namely, flat multi-hop routing and hierarchical routing. Although it has been shown that HYMN improves the longevity of wireless sensor networks, selecting the two respective routing algorithms still remains an open research issue.

6 Data-Centric Routing Algorithms

Owing to the large number of deployed sensors in a typical wireless sensor network, it is difficult to assign a global identification scheme such as an IP identifier. Additionally, although disseminating information from a source to a possible destination can be handled by applying the classical flooding method [19], this technique is energy inefficient. Thus, researchers [19, 20] have proposed a new addressing scheme, referred to as *data-centric* routing. In contrast with *address-centric*, in which each sensor independently transmits its data along a path towards a destination, *data-centric* routing algorithms scrutinize data-types, give each datum an identifier/name, and instead of identifying individual sensors, data are identified. Furthermore, these methods allow efficient energy consumption by eliminating redundant data transmissions. In the remainder of this section, we describe how basic schemes for information dissemination work, followed by prominent examples of *data-centric* routing algorithms.

6.1 Basic Schemes and Issues

A routing algorithm needs to find paths between a source node and a destination node; the intermediate sensors operate independently from other sensors with no prior knowledge to determine the path between the source sensor and the destination sensor. *Flooding* and *gossiping* [8, 36] are classical local techniques used for disseminating data throughout the network.

6.1.1 Flooding

Flooding [8] starts from the sensor that is the origin of the data; the origin broadcasts its message to its neighboring nodes. Each of these neighbors progresses by re-broadcasting the same message to all their neighbors. In effect, the message gets propagated throughout the entire network. *Flooding* clearly generates a large number of packets; furthermore, the algorithm can go on infinitely and ceases to stop unless a mechanism is used to halt it. The mechanism to halt the message from propagating forever can be provisioned by a *Time To Live (TTL)* mechanism. A TTL mechanism is a counter that is decremented every time a message is relayed; upon reaching zero, the message is no longer propagated and is discarded, thus resulting in the termination of the propagation. Generally, the TTL field should be approximately set equal to the number of hops, i.e., *hop-count*.

6.1.2 Gossiping

Gossiping [8, 36] is another dissemination algorithm that is based on local interactions. Generally, *gossiping* transmits a smaller number of packets as compared to *flooding*. In a gossip algorithm, each sensor, which has a message to share, periodically chooses one sensor from its neighbors as its peer. Then, the sensor transmits the message to its chosen neighbor. The receiving sensor re-transmits the message to one of its neighbors with probability p or drops the message with probability $1 - p$. Consequently, the message reaches its destination. The choice of which neighbor sensor to send to and p are design dependent parameters. The choice of which neighbor to send to can be random. p can be fixed or a function of network parameters such as the number of received duplicates, which can be determined by a unique ID for each message.

6.1.3 Energy Consumption Issues

The disadvantages from the viewpoint of energy consumption is the large amount of redundant transmissions that needlessly consume the energy of the wireless sensor network. Figures 7 and 8, redrawn from [19], illustrate the wasted energy in the wireless sensor network. Figure 7 shows the *implosion* problem. It is clear that only the transmissions on either of the right or left path are sufficient, and all other transmissions are extra transmissions that unnecessarily consume the energy of the wireless sensor network. Also, as illustrated in Fig. 8, the same data that was collected from the same area, i.e., area r, is delivered multiple times to the destination, i.e., sensor C, needlessly wasting the energy of the wireless sensor network. This phenomenon is referred to as *overlap*.

Energy-aware *data-centric* routing algorithms eliminate the energy consumed by the *implosion* problem by eliminating needless forwarding and the *overlap* problem by eliminating the transmission of duplicated data.

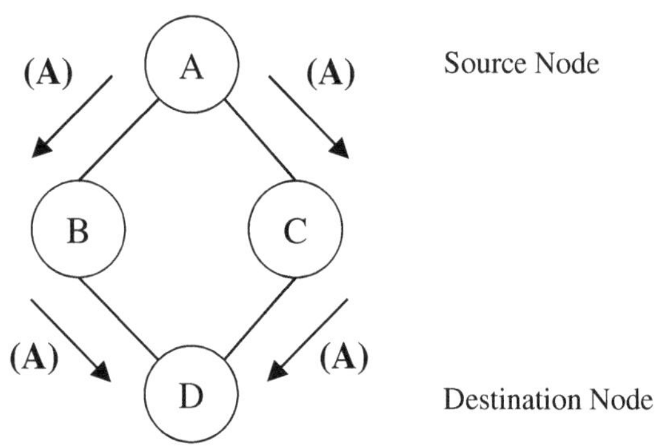

Fig. 7 The implosion problem in classical *flooding*. The destination node, D, gets the same data twice. The wireless sensor network wastes energy by the sending the same data twice

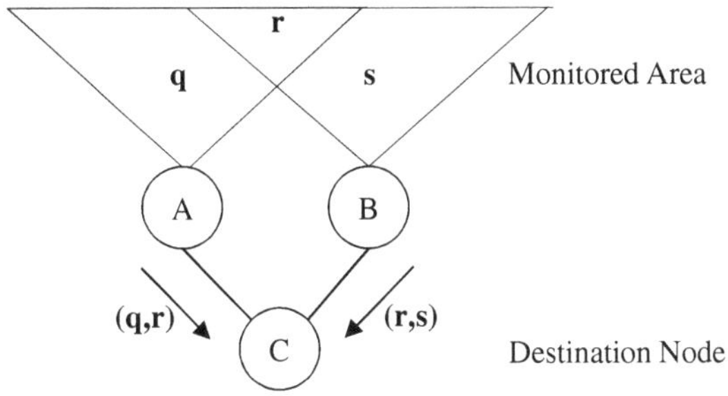

Fig. 8 The overlap problem in classical *flooding*. The destination node C, receives two copies of the data *r*

6.2 SPIN

The main idea behind Sensor Protocols for Information via Negotiation (SPIN) [19] is to give high-level data descriptors to identify each kind of data, referred to as *metadata*. Utilizing the *metadata*, the SPIN nodes negotiate with each other and insure that only required data are transferred, thus eliminating excess energy consumption caused by both the *overlap* and *implosion* phenomena. There is no standard definition for *metadata*, and they differ from applications to applications as well as vary from the types of data collected.

The negotiation process in the basic SPIN protocol, named SPIN-1, is conducted via a three hand-shake procedure, as illustrated in Fig. 9, which is redrawn from [19]. Each stage of the hand-shake has a defined message, as described below.

1. **ADV:** new data advertisement. This message begins the three-state handshake, and it is sent when a sensor has new information it would like to share. The sensor could have acquired the new information via monitoring its surroundings or from one of its neighbors. The ADV message contains *metadata*, which are sent to the sensor's one-hop neighbors.
2. **REQ:** request data. The second stage of the three-stage handshake is triggered when a node that has received an ADV message is interested in the data defined in the *metadata*. An interested node sends the REQ message to the ADV message sender.
3. **DATA:** data message. The third and final stage of the SPIN handshake. The DATA message contains the information defined by the *metadata*, and is sent by the ADV message sender.

After the DATA message is sent, the three-stage handshake is completed. Upon acquiring the DATA message, the receiver initiates the above-mentioned three-stage handshake; by iteratively applying the three-stage handshake mechanism, all the data are efficiently disseminated throughout the network. The above described mechanism avoids energy consumption attributed to unneeded transmissions, i.e., the *implosion* problem, since it eliminates redundant transmissions. Additionally, the *metadata*

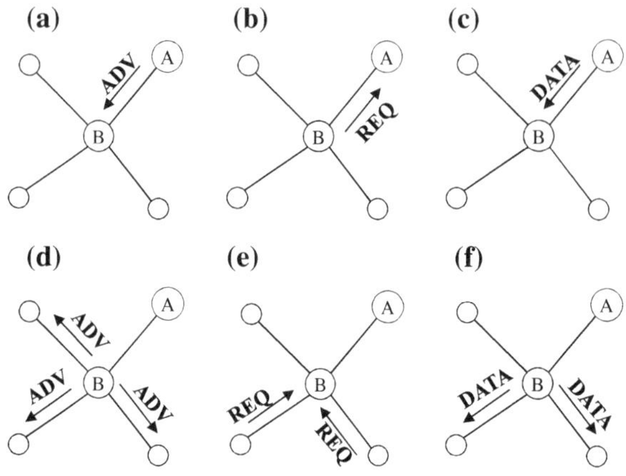

Fig. 9 The operation of the *data-centric* routing algorithm SPIN. **a** ADV stage: sensor A starts by sending an advertisement message indicating that it has new data to its neighbor, sensor B. **b** REQ stage: sensor B responds by requesting the data through a REQ message sent to sensor A. **c** DATA stage: after receiving the REQ message, sensor A sends the requested data to sensor B in a DATA message. **d** ADV stage: upon obtaining the data, sensor B advertises its new data through a REQ message sent to all its neighbors. **e** REQ stage: the process in **b** is repeated. **f** DATA stage: the process in **c** is repeated

enable a sensor to request only the data it requires and avoids wasting the energy of the wireless sensor network by receiving data that it already has, i.e., the *overlap* problem.

6.3 Directed Diffusion

Directed diffusion [20] is a great innovation over basic *data-centric* routing algorithms because it decreases the flow of data in the wireless sensor network by incorporating *data-aggregation*. In directed diffusion, the sink creates tree like routes throughout the wireless sensor network to eliminate the energy consumption associated with the *implosion* problem. Also, the *data-aggregation* scheme mitigates the excessive energy consumption associated with the *ovelap* problem.

The above mentioned tree structure is created by the sink when it advertises its *interests*. Upon receiving these *interests*, the sensors know what kind of information the sink requests. When the sensors reply, in-network *data-aggregation* is performed. In-network *data-aggregation* aggregates messages from different sources to decrease the amount of network operations. This form of aggregation utilizes knowledge of application requirements, and is conducted via local-interactions.

The algorithm, as illustrated in Fig. 10 redrawn from [20], is first trigged by the sink.

1. The sink broadcasts a message describing the information that it has interest in, and the message is intuitively referred to as an *interest* message. The sink's

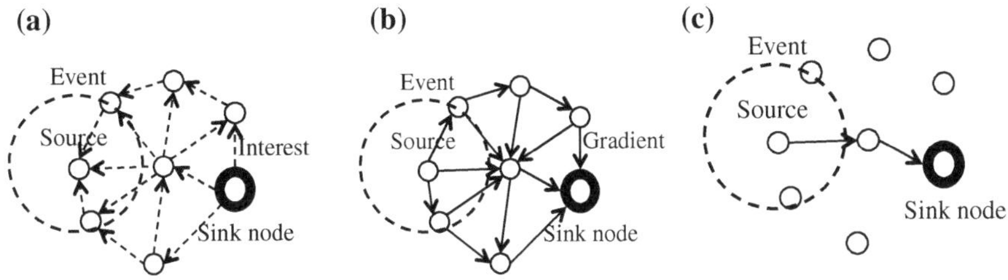

Fig. 10 The operation of the *data-centric* routing algorithm, directed diffusion. **a** Interest propagation from the sink throughout the network. **b** Gradient setup from the source to the sink. **c** Sending data to the sink and path reinforcement

interests are propagated through the network. An *interest* may contain the following information.
Type: The type of object to be monitored.
Interval: How often information should be reported back.
Duration: How long the sink is still interested in acquiring this information.
Location: The location of sensors where information is of interest.

2. Sensors within the one-hop range of the sink, i.e., within its maximum transmission range, receive the sink's interests directly. These one-hop neighbor sensors relay the *interests* to their neighboring sensors. Via relaying, the sink's *interests* get propagated throughout the wireless sensor network, and all sensors get to know about the *interests* of the sink. *Gradients* are created in each sensor, and indicate the source of the *interests*.
3. Data reporting is triggered when a sensor located within the field of interest receives an interest message. The sensor sends the data back to the neighboring sensor that is indicated in the gradient. An intermediate sensor which receives multiple reports corresponding to earlier interests it relayed can play an active role in decreasing the energy consumption of the wireless sensor network. Intermediate sensors are able to apply *data-aggregation* (e.g, looking into multiple reports and combining them, or forwarding reports with better confidence intervals).

As described above, the *gradients* are created after the *interests* propagate in the wireless sensor network. As there could be many paths from the source sensor to the sink, transmitting the messages from the source sensor through all the paths to the sink would lead to needless energy consumption associated to the *implosion* problem. Directed diffusion reinforces one path, thus eliminating the excessive energy consumption attributed to the *implosion* problem. Furthermore, as described above, the *interest* and *gradient* mechanisms allow intermediate sensors between the source sensor and the sink to apply *data-aggregation* to decrease the number of network operations needed to transmit messages in the wireless sensor network, thus reducing the energy consumption caused by the *overlap* problem. In summary, directed diffusion utilizes the *data-centric* communication paradigm and in-network *data-aggregation* to reduce energy consumption.

6.4 Recent Innovations

Since the groundbreaking work of directed diffusion [20], various advances within the realm of *data-centric* routing have emerged, and have made inroads in new applications. For example, Jiang et al. [21] have investigated the top-k problem, which aims to acquire the top most (or least) k-values from the data collected in a wireless sensor network (e.g., the top ten highest temperature readings). Since only the top k-values are needed (i.e., essential), collecting all data from the sensors is wasteful, and thus Jiang et al. [21] proposed to enable intermediate nodes along the path from the source node to the sink to filter/discard less significant data, viz. those having values less than the required top-k values. As a result, redundant transmissions of insignificant data that unnecessarily consume energy of the wireless sensor network are avoided.

Directed diffusion enforces a path from many available paths for data delivery. Yahya and Ben-Othman [47] pointed out that if the current drawn from a battery is decreased or halted, the battery can regain some of its energy back; this is called the relaxation phenomenon. RELAX [47] routes traffic through multiple paths so as to capitalize on the battery relaxation phenomenon to increase the lifetime of the wireless sensor network.

As illustrated in Fig. 10b, gradients in data-centric routing allow sensors to route their collected data to the sink via sink-bound paths. Since their formation is determined by the location of the phenomena under surveillance (e.g., object tracking or event monitoring) and the sink location, these gradients are far from optimal in terms of the energy consumption of the created path. Ren et al. [35] proposed to construct the gradients such that packets flow through the area with high residual energy density, i.e., an area with a large number of nodes and large residual energy. Furthermore, Wu et al. [45] proposed to construct the gradients so as to maximize the lifetime of the sensors. Lifetime is defined as the time until the first sensor has died. Chatzimilioudis et al. [12] investigated the energy loss associated with collisions. The occurrence of collisions causes more energy consumption for retransmission. They pointed out that the probability of collision increases with node degree, i.e., the number of links each node is connected. Also, as illustrated in Fig. 1, since all the data ends in the sink, the node degree of a sensor increases as the node's position gets closer to the sink. Therefore, they have proposed to construct gradients so as to minimize collisions by balancing the node degrees.

In directed diffusion, data can be opportunistically aggregated when they meet at any intermediate node. The formation of the aggregation tree is based on the chronological order of occurred events. However, the resulting tree structure produces non-optimal aggregation. Villas et al. [42, 43] proposed a method to increase the overlap between routes to enhance the quality of aggregation, thus leading to more energy savings.

6.5 Summary

Data-centric routing modeled after directed diffusion is one of the most popular routing algorithms with *data-aggregation* for wireless sensor networks. This class of routing algorithms are particularly suitable for query-based data collection. In contrast, LEACH-like routing algorithms are intended to be used for uniform reporting purposes.

Data-centric routing algorithms require data to be clearly defined by using *metadata*. By using the *metadata* field, sensors can do in-network *data-aggregation* to decrease the amount of network operations conducted in the network. There is no standard definition for the *metadata* field, and it is application specific. Thus, defining an efficient format for the *metadata* field to allow *data-aggregation* for complex schemes is a very important issue in *data-centric* routing algorithms.

7 Location-Based Routing Algorithms

Location information is essential to the functionality of most energy-aware routing algorithms for wireless sensor networks. It is used to calculate energy consumption of transmissions to be used to make path selection decisions as in flat multi-hop routing algorithms, discussed in Sect. 3. Location information can be obtained based on small low-power Global Positioning System (GPS)-enabled devices built into the sensors, from the relative signal strength of the received signals, and other methods. Location information can play a central role in the absence of IP-like addresses, and help reduce energy consumption. Location-based routing algorithms have been previously proposed for general ad hoc networks, but those that are energy-aware can be applied to wireless sensor networks.

7.1 GAF

Geographic Adaptive Fidelity (GAF) proposed by Y. Xu et al. [46] is a *location-based* routing algorithm implemented for general ad hoc networks, but is suitable for use in wireless sensor networks. GAF capitalizes on the spatial redundancy of sensors and reduces the number of unnecessary active sensors by setting some of them to sleep while insuring sufficient active sensors to achieve a constant level of routing *fidelity*. In general, deactivating redundant sensors substantially decreases the energy consumption of the wireless sensor network. This dividend is particularly distinctive in densely deployed networks, such as wireless sensor networks. This is attributed to the high correlation between node density and node redundancy. It is worth noting that GAF can integrate with other routing algorithms.

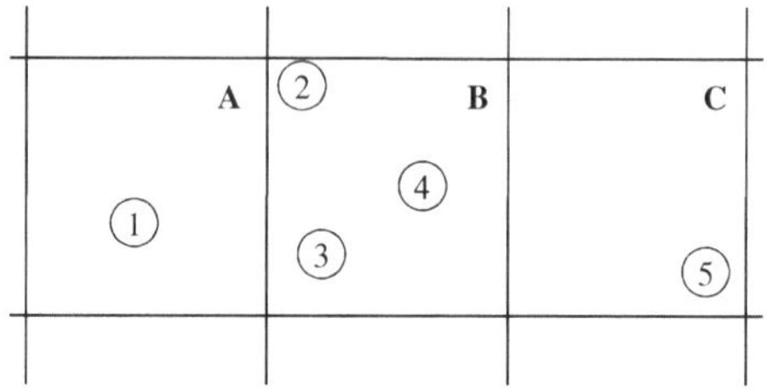

Fig. 11 An example of virtual grid in GAF

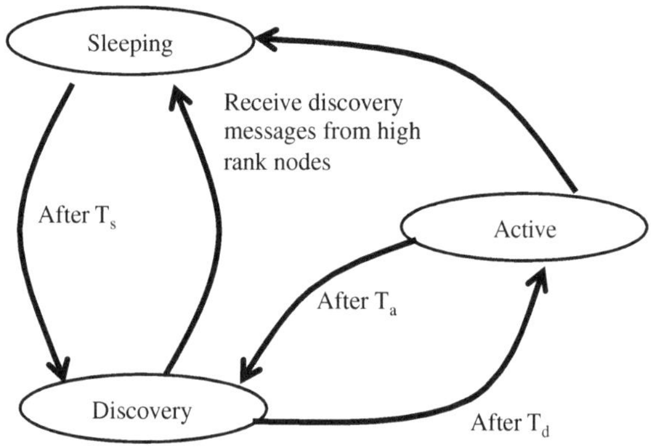

Fig. 12 The three states of GAF

GAF starts by dividing the wireless sensor network into virtual grids, as shown in Fig. 11. Each sensor in a virtual grid cell is able to directly communicate with all the sensors in the neighboring adjacent cells. As shown in Fig. 11, each cell contains several wireless sensors, and all sensors that are within the same cell are considered to be equivalent in terms of packet routing. It is worth noting that the maximum transmission distance of the sensors dictates the size of a block.

In the illustration, sensors 1 and 5 can relay data between each other by transmitting their packets to any of the sensors in the intermediate cell, i.e., sensors 2, 3, and 4. In other words, only one of these intermediate sensors is essential for inter-cell communications, and thus the remaining sensors can be put to sleep. This consequentially reduces the energy consumption of the wireless sensor network.

Sensors employing GAF enter a three-state process. As depicted in Fig. 12, the states of this process include discovery, active, and sleeping. The discovery state is when a sensor turns on its radio, waits for T_d seconds, and exchanges messages with other sensors to find out its neighbors within the virtual grid cell. Once a single active sensor is selected, this sensor becomes fully functional by participating in routing activities for a period of T_a seconds. The remaining nodes enter the sleep state, in which the sensors turn off their radio and save substantial energy for a period of T_s. The time spent in each state is application dependent and can be tuned by adjusting the values of T_a, T_d, and T_s. A node in the active or discovery states goes into the sleep state if it determines that some other high ranking node will take over the role of routing. A high ranking node is chosen by a ranking procedure, which is dependent on applications and is done via node negotiation. For example, ranking can be an arbitrary ordering of nodes or can be performed to optimize wireless sensor network lifetime.

Consequently by reducing the number of active sensors to only the essential number required to sustain routing fidelity, GAF is able to successfully reduce the energy consumption of the wireless sensor network.

7.2 GEAR

Geographic and Energy-Aware Routing (GEAR) was proposed in [16, 49], also independently in [39], as a wireless sensor network specific *location-based* routing algorithm. GEAR is based on the observation that usually queries include location information indicating the target area of the sensors. Intuitively, to be efficient, every query should only be propagated towards its targeted area, not to the entire network because doing so is aimless. This approach is in stark contrast with flooding in which data propagates throughout the entire network.

Each individual node maintains two values that quantify the energy consumption of each path. The first is a speculative cost, which is a function of the energy consumption of the sensor itself and the distance between the sensor and the destination. The second is an acquired cost that is the actual cost, and is learned from messages once they reach their destinations. The differences between the speculated cost and the acquired cost arise from holes in the topology. A hole in the topology is generated when a sensor does not have a next hop, which is closer to the destination, thereby forcing the sensor to divert the traffic around the hole.

The algorithm has two phases. The first phase delivers the packet to the target region, and the second distributes the packet within the region itself.

1. The first phase starts when a query is disseminated. Upon receiving a packet, the sensor reads its destination information and checks whether it has a neighboring node, which is closer to the destined region. If a sensor, which satisfies this criterion, exists, the packet is forwarded to that node. In the case where there is more than one sensor, the closest among them is chosen. On the other hand, if there are no neighboring nodes, then this implies the existence of a hole. When a hole exists, a sensor is chosen based on the speculated cost to detour the packet around the hole.
2. The second phase starts when the packet reaches its intended region. It can be diffused in the region by following one of two methods, restricted flooding or recursive geographic forwarding. Restricted flooding requires each sensor to broadcast once, and is not a wise choice when the sensor density is high. Recursive geographic forwarding, illustrated in Fig. 13, works by using a divide-and-conquer approach; first, the area to which the message is to be disseminated is divided into four regions, and a copy of this message is transmitted to one of the sub-regions. This procedure is repeated until each region has one sensor, and consequently the message is disseminated to all sensors in the targeted area.

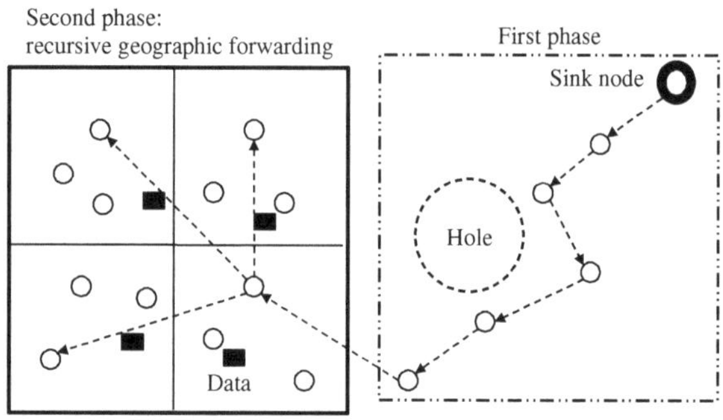

Fig. 13 An illustration of GEAR's two stages of routing. The first phase delivers the query message to the intended area. The second phase uses recursive geographic forwarding to distribute the query message to the intended area

In conclusion, GEAR efficiently reduces the amount of wasteful transmissions by limiting the query propagation towards its intended region only. Therefore, it decreases the wasteful energy consumption.

7.3 Recent Innovations

Geographic routing is a class of location-based routing algorithms that uses a greedy algorithm to forward its data to the sink through intermediate sensors closer to the sink. However, the existence of holes (dead ends) in topologies requires geographic routing to maintain extra non-local state information or employ other auxiliary techniques. Kermarrec and Tan [24] proposed to decompose a given network into Greedily Routable Components (GRC). GRC are paths where greedy routing is guaranteed to work. By routing packets through GRC, the overhead associated with non-local state information is removed, and energy consumption due to routing around holes is eliminated. Furthermore, Chang et al. [9] proposed an innovative scheme that to get around these holes. Sensors bordering around holes in their approach actively establish a forbidden region to enable packets to be guided around holes and move along a short path from the hole to the sink, and thus incurring less energy consumption. It has been shown that contemporary geometric algorithms that are designed to function in 2D environments perform poorly in practical 3D environments [52]. Zhou et al. [52] proposed a scheme that first forwards packets greedily as in [49] as long as it can find a node closer to the destination than itself. If a hole is reached and greedy forwarding fails, packets are routed deterministically using hull trees around the hole. A hull tree is a spanning tree where each node has an associated 2D convex hull that contains the positions from all its child nodes in the subtree rooted at the sink. A 2D convex hull is a geometric object that for any line drawn from two end points in it, the line will be in the 2D convex hull.

In an environment where there are multiple sinks that generate queries to nodes in an overlapped area, as shown in Fig. 14, the sensors in the overlapped area have to report the same data multiple times, thus incurring wasteful energy consumption.

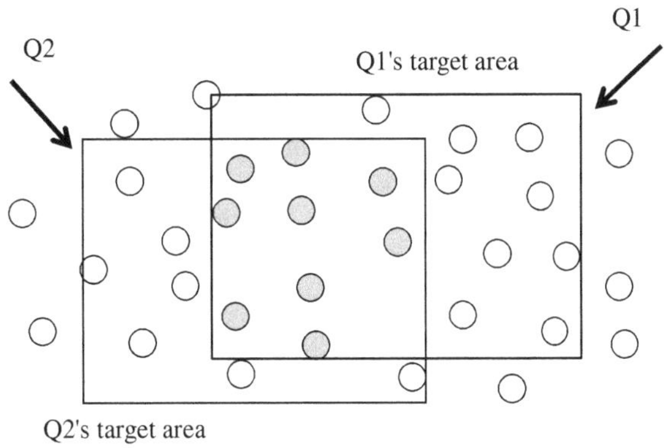

Fig. 14 An illustration of a wireless sensor network under two queries, Q1 and Q2. The sensors that are overlapping between the two queries are colored in *gray*

Zhang et al. [51] proposed to group nodes into zones according to their locations. In the event that there exists queries that overlap in an area, the sensors in the area only respond once, thereby eliminating the energy wasted from redundant transmissions.

7.4 Summary

Location-based routing algorithms are a class of routing algorithms that enhance energy consumption efficiency by capitalizing on location information. The GPS system might not function in some applications, such as ocean-bottom wireless sensor networks that place the sensors at the bottom of the sea, or applications where large obstacles hinder GPS's functionality. Thus, localization techniques to take the role of GPSs are of prime importance.

8 Discussion

Energy-aware routing is a challenging issue, which has attracted substantial research efforts. Research in this area has adopted many techniques from similar networks. Specifically, since some wireless sensor networks possess features similar to those of a wireless ad hoc network, many routing techniques employed in ad hoc networks can be adopted for wireless sensor networks.

The main concern in this research direction is low-energy communications. This is attributed to the large share of energy consumed for communications. Practically, sending a bit over 10 or 100 m can consume as much energy as millions of computational operations conducted in the processing unit of the sensor, referred to as the R4 signal energy drop-off [2].

Wireless sensor networks can capitalize on the application scope in their real world implementations to reduce energy consumption by taking into account of redundancy of their locations and collected data.

Radio operation schemes also play a major role in the energy consumed in a wireless sensor network. The amount of time a radio is on has a direct relationship with the energy it consumes. The longer it is on, the more energy it consumes. Generally, the radio can operate in always-on, synchronized radio [18, 28], or low-duty cycle [13, 17] operation modes. In always-on operation mode, the radio is always on. This consumes the maximum amount of energy. In the synchronized radio operation, the radio is on only when it is needed. This allows more efficient energy consumption as compared with always-on operation. In low-duty cycle radio operation, the radio is off most of the time and is only on for a relatively small amount of time. This operation mode is the least energy consuming.

Table 2 summaries the characteristics of the routing categories examined in this chapter, along with notable representative algorithms of each routing category. They are also characterized in terms of *data-aggregation*, location awareness, mobility, and network lifetime.

Location-interactions refers to the ability of a routing algorithm to function via local interactions executed in individual sensors without the need for global information about the wireless sensor network. Gathering global information about the entire wireless sensor network consumes a large amount of energy for information exchange. It is worth noting that although hybrid routing algorithms require global information for the flat multi-hop routing part to function, the area of the flat multi-hop routing part is relatively small in size.

Table 2 A general comparison among various classes of energy-aware routing for wireless sensor networks

Name	Class	Data-aggregation	Local-interactions	Mobility	Network lifetime
Toh [41]	Flat	No	No	Unsupported	ANL[a]
Chang et al. [11]	Flat	No	No	Unsupported	FNL[b]
LEACH [18]	Hierarchical	Yes	Yes	Unsupported	ANL
PEGASIS [26]	Hierarchical	Yes	Yes	Unsupported	ANL
HYMN [33]	Hybrid	Yes	No	Unsupported	SNI[c]
SPIN [19]	Data-centric	No	Yes	Limited	ANL
DD[d] [20]	Data-centric	Yes	Yes	Unsupported	ANL
GAF [46]	Location-based	No	No	Unsupported	ANL
GEAR [49]	Location-based	Yes	No	Unsupported	ANL

[a]Average node lifetime
[a]First node lifetime
[c]Sink node isolation
[d]Directed diffusion

Early research in wireless sensor networks mostly envisioned and considered inexpensive sensors with limited or no mobility. Consequently, as can be seen from Table 2, very limited support for mobility was considered in early wireless sensor network routing algorithms. Recent advances in this field have investigated many scenarios with mobility of sinks [3, 15, 29, 30]. In addition, wireless sensor and actuator networks [31, 32] have recently drawn much research attention, where the sensors are mobile and self-healing.

8.1 Data-Aggregation

The larger the volume of data to transmit, the larger the energy consumption of the network. Hence, *data-aggregation* is of paramount importance to achieve low-energy communications in wireless sensor networks. Four routing categories, namely, hierarchical, hybrid, *data-centric*, and location-aware routing algorithms, facilitate *data-aggregation*. Hybrid routing utilizes the *data-aggregation* function of hierarchical routing algorithms. Also, GEAR adopts the *data-aggregation* method of *data-centric* routing. Thus, it is worthwhile to elaborate on *data-aggregation* methods employed in both *data-centric* and hierarchical routing algorithms. As will be discussed later, the *data-aggregation* method methodology is rather application specific/dependent.

Data-aggregation in hierarchical routing algorithms is conducted in the CHs of each cluster. The reporting model is aimed at constant uniform reporting, in which sensors transmit data in each time interval; once the CH receives the sensed data from its CMs, it can utilize *data-aggregation*. As compared with the *data-aggregation* in *data-centric* routing, the *data-aggregation* method utilized in hierarchical routing can work in conditions where the sensors produce a low amount of data collected from overlapping areas. It can reduce the energy consumption in scenarios with high data correlation. This reporting model is particularly suited for applications such as environmental monitoring, where periodic information is required about the environment.

Data-aggregation in *data-centric* routing, viz, in-network *data-aggregation*, eliminates the overhead of cluster formation found in hierarchical routing algorithms. In in-network *data-aggregation*, sensors along the path to the destination do *data-aggregation* to reduce the flow of data in the wireless sensor network.

Generally, the in-network *data-aggregation* method considers overlapping data collected from different sensors and merges redundant reports to decrease the number of transmissions conducted in the wireless sensor network. The performance of this *data-aggregation* mechanism will degrade when the overlap between data collected from different sensors is small. The overlap between the data collected from different sensors decreases when the sensing area of sensors is small relative to their density (the ratio between the number of deployed sensors to the size of area they are deployed in). In such cases, the reduction of energy consumption gained by using the *data-aggregation* function of *data-centric* routing will become insignificant.

On the other hand, the query model adopted in *data-centric* routing is well suited for applications where need-based data reporting is conducted. For example, the sensor node observing a desired event only reports to the sink when the event occurs. It produces a low amount of transmissions and will consume a small amount of energy as compared to the uniform data reporting model.

8.2 Network Lifetime Definition

The objective of all the energy-aware routing algorithms for wireless sensor networks is to decrease the energy consumption, and therefore to prolong operation periods of the network. Furthermore, these routing algorithms can be evaluated under different metrics. Particularly, network lifetime is a widely accepted metric for evaluating the energy-aware routing algorithms. Network lifetime can have differing definitions, and some of these definitions can be misleading. It is important to understand how the wireless sensor network functions, and carefully define the network lifetime to accurately evaluate a given routing algorithm. Many researchers have defined network lifetime as the time that the first sensor dies, i.e., first node life [11]. However, in many scenarios, a wireless sensor network can still function even after the first sensor has died. Alternatively, defining network lifetime as the time when all sensors die does not give much insight on the functionality of the wireless sensor network since an isolated node collecting data and unable to transmit its collected data to the sink is of no use. Therefore, defining network lifetime as the time when the sink cannot collect data from the wireless sensor network, i.e., Sink Node Isolation (SNI), is more appropriate and accurate. Moreover, designing energy-aware routing algorithms to improve the average lifetime over all sensors is rather popular.

Table 2 shows various definitions of network lifetime that each routing algorithm has adopted. It can be seen that the most popular definition is average network lifetime, which does not necessarily result in longer lifetime. Note that only hybrid multi-hop routing is designed with the motivation to mitigate the energy hole problem, thus resulting in improved lifetime of the wireless sensor network.

8.3 Routing Overhead

Routing overhead is a major energy consumer in wireless sensor networks. Decreasing frequency of information updates necessary for routing can decrease the energy consumed by the routing overhead. However, decreasing their frequency leads to degradation of the energy-aware routing algorithm's performance due to inaccurate information or outdated information about the wireless sensor network.

In flat multi-hop routing algorithms, deciding which path to route traffic in order to achieve minimum energy consumption or maximum lifetime requires information about the energy consumed per unit in each link, which can be calculated from

Eq. (1), and the residual energy of each sensor. This information needs to be regularly updated to achieve minimum energy consumption when some nodes along a path die and the path no longer produces the minimum energy consumption and/or a sensor is overly energy exhausted and traffic must be directed from it to allow it live longer. The frequency of route information updates affects the accuracy of paths with the minimum energy consumption and the maximum lifetime. Obviously, requiring frequent updates is an energy intensive operation, and hence could pose a great drawback to these methods.

Hierarchical routing algorithms form clusters wherein a single sensor acts as a CH. To form a cluster, an election process needs to take pace where sensors present themselves as CHs, and then each CH manages a collection of CMs, and this process consumes energy of the wireless sensor network. Furthermore, since the role of CH is an energy consuming role with *data-aggregation* and inter-cluster communications, the sensors take turns in becoming a CH, thus reinitiating the energy consuming CH election process. Decreasing the frequency of CH election puts the elected CHs in risk of energy exhaustion (dying) and lost coverage before other sensors can take on the CH role. Alternatively, increasing the CH election process frequency would put a high energy burden on the wireless sensor network.

In *data-centric* routing, the sink sends queries to the wireless sensor network advertising its *interests*; such queries consume energy. Therefore, a relationship between the sink and sensors is created that can satisfy its *interests*, and afterwords data transfer occurs between sensors and the sink. Generally, this relationship has a predetermined time limit, and upon expiration a new relationship needs to be established. Thus, continuous relationship establishment is required, thereby consuming energy of the wireless sensor network. On the other hand, limiting relationship establishment results in failures of the wireless sensor network to fulfill its objective.

Location-aware routing algorithms are generally incorporated with other routing energy-aware routing algorithms, and thus inherit the energy consumption attributed to the routing overhead of the adopted energy-aware routing algorithm. Furthermore, this category employs additional schemes for energy savings such as allowing some sensors to sleep. These schemes require information exchange, and thus consume additional energy.

8.4 Energy Hole Phenomenon

The energy hole phenomenon is defined as the energy consumption imbalance among sensors. This inevitably leads to rapid energy exhaustion of sensors in the high-energy consuming areas, thus resulting in holes in these areas, and subsequently network partition. This phenomenon is attributed to the traffic patterns in wireless sensor networks, namely, the many-to-one (convergecast) traffic directed towards the sink.

In flat multi-hop routing, all nodes, except the sink, assume the same role and responsibility. If all the sensors transmit their data towards a central point, i.e., the sink, nodes closer to the sink will inevitably end up draining their energy faster.

Along with the lack of *data-aggregation* that decreases the volume of data flowing in the wireless sensor network, the sink is consequentially disconnected from the surviving sensors.

The application scope of hierarchical routing algorithms considers applications with uniform reporting directed to the sink that subsequently causes the energy hole phenomenon. Furthermore, CHs in hierarchical routing algorithms conduct inter-cluster communications, and their relatively smaller number leads to inefficient long-distance transmissions that in turn augments the severity of the energy hole phenomenon.

HYMN synergies two categories of wireless sensor network routing algorithms to mitigate the energy hole phenomenon by using energy efficient transmission distances and *data-aggregation*. Thus, HYMN surpasses the contemporary categories of energy-aware routing algorithms.

Data-centric routing algorithms adopt the query-based reporting model. In this model, the sink queries a specific area. As a direct result, the flow of traffic depends on the scenario under consideration. For example, if an application demands reporting of a certain object's movements, the areas where this object moves will incur higher energy consumption rate than other areas. This phenomenon is referred to as the query hotspot.

Location-based routing algorithms are typically coupled with other routing algorithms, and thereby inherit the energy hole phenomenon characteristics of the latter algorithm.

8.5 Collisions and Interferences

Wireless sensor networks can be categorized as a special case of ad hoc networks, and face the same issues of collisions and interference, which occur when two nodes within sufficiently close distance from each other try to communicate on the same channel; thus, energy is consumed for retransmitting the same message again. The higher the number of collisions, the larger the amount of energy is consumed in the wireless sensor network. Owing to the ad hoc nature of wireless sensor networks, adopting a centralized management schemes for Medium Access Control (MAC) is not feasible; it is practical to deploy a distributed MAC scheme. All the routing techniques, except hierarchical routing, introduced here employ such MAC schemes. In the case where distributed MAC schemes are implemented, a high amount of energy is consumed for MAC operations due to collisions. This is also applicable to HYMN as it is also partly composed of hierarchical routing.

In hierarchical routing, the CH takes a leading role by aggregating data and sending them to the sink. Furthermore, a CH is normally enabled with a centralized MAC scheme to manage the collision and interference issues. LEACH [18] adopts the Time Division Multiple Access (TDMA) MAC scheme for channel access. Upon cluster formation, the CH organizes a TDMA schedule and transmits this schedule to the CMs in its cluster. Applying TDMA ensures that there are no collisions when the CMs transmit their data to the CH, and thus avoids the energy consumed due to collisions. Moreover, the transmission circuitry of each CM can be turned off at most of the time except when it is its turn for transmission, thus reducing the energy consumed by the individual sensors. However, this scheme cannot avoid interferences or collisions caused by neighboring clusters.

9 Conclusion

In this chapter, we have addressed the crucial problem of energy-aware routing for wireless sensor networks. The limited energy capacity along with the difficulty of changing batteries of deployed sensors makes energy-efficient technologies essential for the longevity of wireless sensor networks. We classify energy-aware routing algorithms into five categories according to their network architecture; flat multi-hop routing that finds paths to minimize energy consumption or increase sensor network lifetime, hierarchical routing that creates a hierarchy and applies *data-aggregation* to reduce energy consumption, hybrid multi-hop routing that is a combination of the former two routing algorithms and mitigates the energy hole problem, *data-centric* routing where in-network *data-aggregation* is performed to eliminate wasteful transmissions, and location-based routing that uses location information to reduce the energy consumption of the wireless sensor network. Moreover, we have discussed how the various energy-aware routing algorithms perform from many different perspectives such as *data-aggregation*, network lifetime definition, routing overhead, the energy hole phenomenon, and collisions/interferences.

References

1. A. Abdulla, H. Nishiyama, N. Ansari, N. Kato, Hymn to improve the scalability of wireless sensor networks, in *IEEE Conference on Computer Communications Workshops (INFOCOM WKSHPS)*, pp. 519–524 (2011). doi:10.1109/INFCOMW.2011.5928868
2. J. Al-Karaki, A. Al-Karaki, Routing techniques in wireless sensor networks: a survey. IEEE Wireless Commun. **11**(6), 6–28 (2004). doi:10.1109/MWC.2004.1368893
3. H. Ammari, S. Das, Promoting heterogeneity, mobility, and energy-aware voronoi diagram in wireless sensor networks. IEEE Trans. Parallel Distrib. Syst. **19**(7), 995–1008 (2008). doi:10.1109/TPDS.2008.31
4. G. Anastasi, M. Conti, M.D. Francesco, A. Passarella, Energy conservation in wireless sensor networks: a survey. Ad Hoc Netw. **7**(3), 537–568 (2009). doi:10.1016/j.adhoc.2008.06.003
5. J. Aslam, Q. Li, D. Rus, Three power-aware routing algorithms for sensor networks. Wireless Commun. Mob. Comput. **3**, 187–208 (2002)
6. A. Bachir, M. Dohler, T. Watteyne, K. Leung, Mac essentials for wireless sensor networks. IEEE Commun. Surv. Tutor. **12**(2), 222–248 (2010). doi:10.1109/SURV.2010.020510.00058
7. M. Bhardwaj, A. Chandrakasan, Bounding the lifetime of sensor networks via optimal role assignments. In: INFOCOM 2002, in *Proceedings of Twenty-First Annual Joint Conference of the IEEE Computer and Communications Societies*, vol. 3, pp. 1587–1596 (2002). doi:10.1109/INFCOM.2002.1019410

8. B. Blywis, M. Gnes, F. Juraschek, O. Hahm, N. Schmittberger, A survey of flooding, gossip routing, and related schemes for wireless multi-hop networks. Tech. rep., Free University of Berlin (2011)
9. C.Y. Chang, C.T. Chang, Y.C. Chen, S.C. Lee, Active route-guiding protocols for resisting obstacles in wireless sensor networks. IEEE Trans. Veh. Technol. **59**(9), 4425–4442 (2010). doi:10.1109/TVT.2010.2068065
10. J.H. Chang, L. Tassiulas, Routing for maximum system lifetime in wireless ad-hoc networks, in *37th Annual Allerton Conference on Communication, Control, and, Computing* (1999)
11. J.H. Chang, L. Tassiulas, Maximum lifetime routing in wireless sensor networks. IEEE/ACM Trans. Netw. **12**, 609–619 (2004). doi:10.1109/TNET.2004.833122
12. G. Chatzimilioudis, D. Zeinalipour-Yazti, D. Gunopulos, Minimum-hot-spot query trees for wireless sensor networks, in *Proceedings of the Ninth ACM International Workshop on Data Engineering for Wireless and Mobile Access, MobiDE '10*, pp. 33–40. ACM, New York (2010). doi:10.1145/1850822.1850829
13. L. Chen, S. Guo, Y. Shu, F. Zhang, Y. Gu, J. Chen, T. He, Poster: Selective reference mechanism for neighbor discovery in low-duty-cycle wireless sensor networks, in *Proceedings of the 9th ACM Conference on Embedded Networked Sensor Systems, SenSys '11*, pp. 367–368. ACM, New York (2011). doi:10.1145/2070942.2070993
14. I. Dietrich, F. Dressler, On the lifetime of wireless sensor networks. ACM Trans. Sen. Netw. **5**(1), 5:1–5:39 (2009). doi:10.1145/1464420.1464425
15. S. Gandham, M. Dawande, R. Prakash, S. Venkatesan, Energy efficient schemes for wireless sensor networks with multiple mobile base stations, in *Global Telecommunications Conference, 2003. GLOBECOM '03*. IEEE, vol. 1, pp. 377–381 (2003). doi:10.1109/GLOCOM.2003.1258265
16. D. Ganesan, A. Cerpa, W. Ye, Y. Yu, J. Zhao, D. Estrin, Networking issues in wireless sensor networks. J. Parallel Distrib. Comput. **64**(7), 799–814 (2004). doi:10.1016/j.jpdc.2004.03.016
17. Y. Gu, T. He, Dynamic switching-based data forwarding for low-duty-cycle wireless sensor networks. IEEE Trans. Mob. Comput. **10**(12), 1741–1754 (2011). doi:10.1109/TMC.2010.266
18. W. Heinzelman, A. Chandrakasan, H. Balakrishnan, An application-specific protocol architecture for wireless microsensor networks. IEEE Trans. Wireless Commun. **1**(4), 660–670 (2002). doi:10.1109/TWC.2002.804190
19. W.R. Heinzelman, J. Kulik, H. Balakrishnan, Adaptive protocols for information dissemination in wireless sensor networks, in *Proceedings of the 5th Annual ACM/IEEE International Conference on Mobile Computing and Networking, MobiCom '99*, pp. 174–185. ACM, New York (1999). doi:10.1145/313451.313529
20. C. Intanagonwiwat, R. Govindan, D. Estrin, J. Heidemann, F. Silva, Directed diffusion for wireless sensor networking. IEEE/ACM Trans. Netw. **11**(1), 2–16 (2003). doi:10.1109/TNET.2002.808417
21. H. Jiang, J. Cheng, D. Wang, C. Wang, G. Tan, Continuous multi-dimensional top-k query processing in sensor networks, in *INFOCOM, 2011 Proceedings IEEE*, pp. 793–801 (2011). doi:10.1109/INFCOM.2011.5935301
22. H. Jiang, S. Jin, C. Wang, Prediction or not? An energy-efficient framework for clustering-based data collection in wireless sensor networks. IEEE Trans. Parallel Distrib. Syst. **22**(6), 1064–1071 (2011). doi:10.1109/TPDS.2010.174
23. H. Karkvandi, E. Pecht, O. Yadid-Pecht, Effective lifetime-aware routing in wireless sensor networks. IEEE Sens. J. **11**(12), 3359–3367 (2011). doi:10.1109/JSEN.2011.2159110
24. A.M. Kermarrec, G. Tan, Greedy geographic routing in large-scale sensor networks: a minimum network decomposition approach, in *Proceedings of the Eleventh ACM International Symposium on Mobile Ad Hoc Networking and Computing, MobiHoc '10*, pp. 161–170. ACM, New York (2010). doi:10.1145/1860093.1860116
25. J.F. Kurose, K.W. Ross, *Computer Networking: A Top-Down Approach Featuring the Internet*, 6th edn (Addison-Wesley, New York, 2012)
26. S. Lindsey, C. Raghavendra, Pegasis: power-efficient gathering in sensor information systems, in *Aerospace Conference Proceedings, 2002*. IEEE, vol. 3, pp. 3-1125–3-1130 (2002). doi:10.1109/AERO.2002.1035242

27. C. Liu, G. Cao, Spatial-temporal coverage optimization in wireless sensor networks. IEEE Trans. Mob. Comput. **10**(4), 465–478 (2011). doi:10.1109/TMC.2010.172
28. F. Liu, C.Y. Tsui, Y.J. Zhang, Joint routing and sleep scheduling for lifetime maximization of wireless sensor networks. IEEE Trans. Wireless Commun. **9**(7), 2258–2267 (2010). doi:10.1109/TWC.2010.07.090629
29. J. Luo, J.P. Hubaux, Joint mobility and routing for lifetime elongation in wireless sensor networks, in *INFOCOM 2005. 24th Annual Joint Conference of the IEEE Computer and Communications Societies. Proceedings IEEE*, vol. 3, pp. 1735–1746 (2005). doi:10.1109/INFCOM.2005.1498454
30. H. Nakayama, N. Ansari, A. Jamalipour, N. Kato, Fault-resilient sensing in wireless sensor networks. Comput. Commun. **30**(11–12), 2375–2384 (2007). doi:10.1016/j.comcom.2007.04.023. Special issue on security on wireless ad hoc and sensor networks
31. H. Nakayama, Z. Fadlullah, N. Ansari, N. Kato, A novel scheme for wsan sink mobility based on clustering and set packing techniques. IEEE Trans. Autom. Control **56**(10), 2381–2389 (2011). doi:10.1109/TAC.2011.2163872
32. A. Nayak, I. Stojmenovic, *Wireless Sensor and Actuator Networks: Algorithms and Protocols for Scalable Coordination and Data Communication*, 1st edn (Wiley-Interscience, Hoboken, 2010)
33. H. Nishiyama, A. Abdulla, N. Ansari, Y. Nemoto, N. Kato, Hymn to improve the longevity of wireless sensor networks, in *GLOBECOM 2010, 2010 IEEE Global Telecommunications Conference*, pp. 1–5 (2010). doi:10.1109/GLOCOM.2010.5683756
34. S. Olariu, I. Stojmenovic, Design guidelines for maximizing lifetime and avoiding energy holes in sensor networks with uniform distribution and uniform reporting, in *INFOCOM 2006. 25th IEEE International Conference on Computer Communications*, pp. 1–12 (2006). doi:10.1109/INFOCOM.2006.296
35. F. Ren, J. Zhang, T. He, C. Lin, S. Ren, Ebrp: Energy-balanced routing protocol for data gathering in wireless sensor networks. IEEE Trans. Parallel Distrib. Syst. **22**(12), 2108–2125 (2011). doi:10.1109/TPDS.2011.40
36. D. Shah, *Gossip Algorithms* (Now Publishers Inc., Norwell, 2009)
37. G. Shirazi, L. Lampe, Lifetime maximization in uwb sensor networks for event detection. IEEE Trans. Signal Processing **59**(9), 4411–4423 (2011). doi:10.1109/TSP.2011.2159212
38. T. Shu, M. Krunz, Coverage-time optimization for clustered wireless sensor networks: a power-balancing approach. IEEE/ACM Trans. Netw. **18**(1), 202–215 (2010). doi:10.1109/TNET.2009.2022936
39. I. Stojmenovic, X. Lin, Power-aware localized routing in wireless networks. IEEE Trans. Parallel Distrib. Syst. **12**(11), 1122–1133 (2001). doi:10.1109/71.969123
40. A.S. Tanenbaum, D.J. Wetherall, *Computer Networks*, 5th edn (Prentice Hall, New Jersey, 2010)
41. C.K. Toh, Maximum battery life routing to support ubiquitous mobile computing in wireless ad hoc networks. IEEE Commun. Mag. **39**(6), 138–147 (2001). doi:10.1109/35.925682
42. L. Villas, A. Boukerche, H. Ramos Filho, H. Oliveira, R. Araujo, A. Loureiro, Drina: a lightweight and reliable routing approach for in-network aggregation in wireless sensor networks. IEEE Trans. Comput. **99**, 1 (2012). doi:10.1109/TC.2012.31
43. L.A. Villas, D.L. Guidoni, R.B. Araújo, A. Boukerche, A.A. Loureiro, A scalable and dynamic data aggregation aware routing protocol for wireless sensor networks, in *Proceedings of the 13th ACM International Conference on Modeling, Analysis, and Simulation of Wireless and Mobile Systems, MSWIM '10*, pp. 110–117. ACM, New York (2010). doi:10.1145/1868521.1868540
44. D. Wei, P. Navaratnam, A. Gluhak, R. Tafazolli, Energy-efficient clustering for wireless sensor networks with unbalanced traffic load, in *Wireless Communications and Networking Conference (WCNC)*, 2010 IEEE, pp. 1–6 (2010). doi:10.1109/WCNC.2010.5506172
45. Y. Wu, Z. Mao, S. Fahmy, N.B. Shroff, Constructing maximum-lifetime data gathering forests in sensor networks. IEEE/ACM Trans. Netw. **18**(5), 1571–1584 (2010). doi:10.1109/TNET.2010.2045896

46. Y. Xu, J. Heidemann, D. Estrin, Geography-informed energy conservation for ad hoc routing, in *Proceedings of the 7th Annual International Conference on Mobile Computing and Networking, MobiCom '01*, pp. 70–84. ACM, New York (2001). doi:10.1145/381677.381685
47. B. Yahya, J. Ben-Othman, Relax: an energy efficient multipath routing protocol for wireless sensor networks, in *2010 IEEE International Conference on Communications (ICC)*, pp. 1–6 (2010). doi:10.1109/ICC.2010.5502156
48. J. Yick, B. Mukherjee, D. Ghosal, Wireless sensor network survey. Comput. Netw. **52**(12), 2292–2330 (2008). doi:10.1016/j.comnet.2008.04.002
49. Y. Yu, R. Govindan, D. Estrin, Geographical and energy aware routing: a recursive data dissemination protocol for wireless sensor networks. Tech. rep., University of California, Los Angeles (2001)
50. C. Zhang, E. Hou, N. Ansari, in *Chapter 6: Node Clustering in Wireless Sensor Networks*, ed. by J. Zheng, A. Jamalipour. Wireless Sensor Networks: A Networking Perspective (Wiley/IEEE Press, New York, 2009), pp. 173–214
51. Z. Zhang, A.D. Kshemkalyani, S.M. Shatz, Dynamic multiroot, multiquery processing based on data sharing in sensor networks. ACM Trans. Sens. Netw. **6**(3), 25:1–25 (2010). doi:10.1145/1754414.1754421
52. J. Zhou, Y. Chen, B. Leong, P.S. Sundaramoorthy, Practical 3d geographic routing for wireless sensor networks, in *Proceedings of the 8th ACM Conference on Embedded Networked Sensor Systems, SenSys '10*, pp. 337–350. ACM, New York (2010). doi:10.1145/1869983.1870016
53. Y. Zhuang, J. Pan, L. Cai, Minimizing energy consumption with probabilistic distance models in wireless sensor networks, in *INFOCOM, 2010 Proceedings IEEE*, pp. 1–9 (2010). doi:10.1109/INFCOM.2010.5462073

Chapter 8
Utility-Based Routing in Wireless Sensor Networks

X. Li and Jie Wu

Abstract *Wireless sensor networks* (WSNs) have been proposed for monitoring physical environments. The applications in WSNs have comprised a wide variety of scenarios. The design of routing protocols in WSNs becomes more complicated than the traditional network when we consider the energy cost, throughput, reliability, and delay as routing metrics. Selecting a particular routing protocol mainly depends on the capabilities of the nodes, and on the requirements of the application. In this chapter, we will briefly discuss the existing utility-based routing protocols for WSNs. We put them into several categories according to their utility properties, such as delay, cost, and packet delivery ratio. In addition, we will also cover the composition-based utility for wireless networks and its extensions in low duty-cycle WSNs.

1 Introduction

Over the last few years, *wireless sensor networks* (WSNs) have been used in many applications, such as military surveillance, infrastructure protection, and scientific exploration [1–4]. The major task of WSNs is to monitor environmental changes and report unexpected events to the destination.

The special features of WSNs bring out new challenges. One of the features is the lifetime of a sensor node, which is constrained by the battery. Thus, to reduce the energy cost, the consideration of energy efficiency is often preferred in a WSN design. Moreover, these problems are complicated by the lossy links and collisions during

X. Li · J. Wu (✉)
Department of Computer Information and Sciences, Temple University,
Philadelphia, PA 19122, USA
e-mail: jiewu@temple.edu

X. Li
e-mail: xiaoguang.li@temple.edu

H. M. Ammari (ed.), *The Art of Wireless Sensor Networks*,
Signals and Communication Technology, DOI: 10.1007/978-3-642-40009-4_8,
© Springer-Verlag Berlin Heidelberg 2014

communication among wireless sensor nodes. In practice, all of these utilities in WSNs are available in different forms with many individual peculiarities. Obvious trade-offs include accuracy, dependability, energy consumption, delay, reliability, and so on.

Unlike the prior works about WSNs, which mainly focus on the design of MAC protocols, we will briefly take an overview of algorithmic methods which are related to the routing protocols. The routing protocol is designed to obtain a route for data transmission from the source to the destination. The route is selected based on the routing metric for different application requirements. In multi-hop networks, when a source node wants to send its packets to a destination, the intermediate node has to decide which neighbor an incoming packet should be forwarded to, so that it eventually reaches the destination.

As the routing protocol plays an important role in determining the path, a good application is dependent on the routing efficiency. Challenges in routing protocol design are very critical, based on different characteristics in WSNs. This chapter presents a survey on the routing designs of WSNs, based on a selected utility. This chapter aims at providing the basic knowledge on utility-based routing designs in WSNs. The readers are expected to acquire the recent studies and techniques on developing routing protocols in WSNs. We will first introduce the delay-based, packet-delivery-ratio-based, and energy-based utilities for routing protocols. Then, we will offer a special type of routing protocol based on composite utility.

2 Background

In this section, we will offer the background of utility-based routing. In addition, we will also introduce the other related issues regarding the routing design in WSNs.

2.1 Utility-Based Routing

Intuitively, the utility-based routing is composed of routing and utility, as shown in Fig. 1. The routing designs are dominated by different forms of routing processes, such as unicast, multicast, and broadcast. The utilities are the routing metrics, such as cost, packet delivery ratio, and delay. Depending on the application of the sensor network, the utility-based routing can be continuous, event-driven, query-driven, or a hybrid. For the continuous model, the node sends data periodically. In the event-driven and query-driven models, the transmission is triggered when an event occurs or when a query is generated. However, the data delivery model mentioned above may coexist in the network. This is needed to accommodate different types of data delivery. There are two parts for utility-based routing: the routing protocols, and utilities, as shown in Fig. 1. Many routing protocols have been proposed, based on the

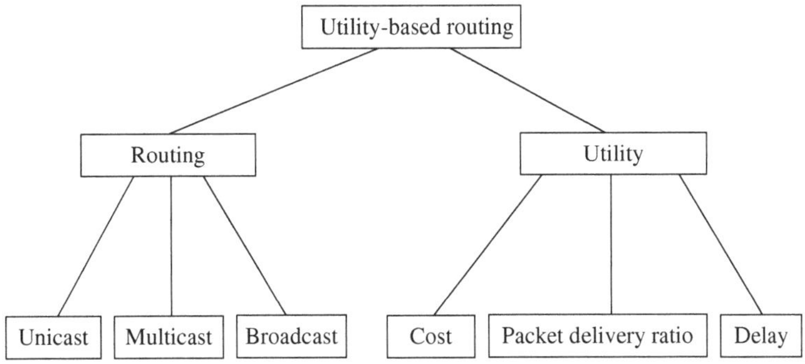

Fig. 1 Utility-based routing in WSNs

requirements of different applications and quality of service. In the real application, three transmission patterns are used for data delivery: unicast, multicast, and broadcast.

2.2 *Objectives of Different Utilities*

WSNs are widely used for environmental sensing and data processing, with extremely low energy and cost. The utilities of routing in WSNs are commonly discussed by delivery ratio, throughput, delay, saving cost (hop count, energy), and composition utility (combination of them). In this part, we will discuss the objectives of different metrics.

(1) Packet delivery ratio (PDR): The packet delivery ratio is estimated by sending a number of packets in a short period of time. The receiver will compute the percentage of received packets. Thus, the purpose of improving the delivery ratio is to reduce the delay and cost. Consequently, we can achieve both time and energy efficiency.

(2) Delay: During the transmission process, the packet is delayed at each node or during the data delivery. Especially in low duty-cycling, the sensor has to wait a certain time period to transmit the packet. Several issues need to be considered for the routing design, such as the expected end-to-end delay, packet delay, and sleep scheduling problems.

(3) Energy cost: The energy cost is also important in path-selection. As redundant packets may consume more energy, the metric can be designed by reducing any unexpected transmissions and real-time scheduling. In recent works, the expected transmission cost and real-time energy cost were proposed for selecting the path.

To represent the network topology, the weighted graph has been proposed. Given a weighted graph $G = (V, E, W)$, V is the set of vertices, E is the set of links, and W is the set of weights for the links. As shown in Fig. 2, the weight of a link could be energy cost, delay, reliability or other conditions.

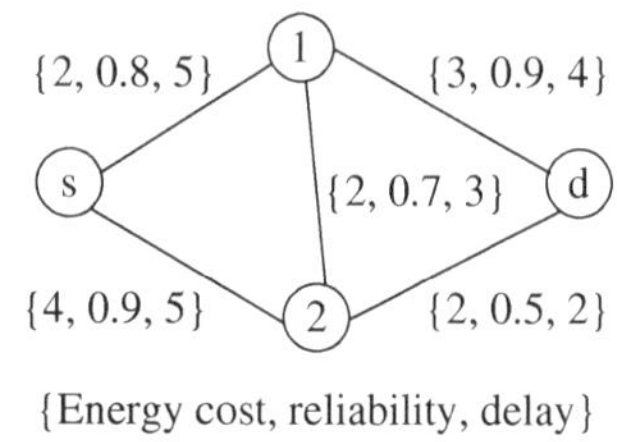

Fig. 2 An example of weighted graph

2.3 Reporting Model

Unlike wired networks, wireless channels are, by nature, error-prone [5]. Thus, neighboring nodes might not successfully receive messages. This means that transmission over a wireless connection is unreliable. Reliability is defined as the ability of the network to ensure reliable data transmission in a network structure that is continuously changing.

In low duty-cycle WSNs, many factors will affect the reliability of the link, such as *packet-error rate* (PER), buffer size, and duty-cycle. In wireless networks, packet errors are common, due to fading caused by the environment and interference from other wireless devices. Another problem is that if the nodes are randomly deployed, the inadequate locations and distances cause unreliability. Packet buffering in WSNs is limited, due to memory constraints. When a buffer is full, a packet must be dropped, which reduces reliability. To solve the reliability problem, several schemes have been proposed. Opportunistic and retransmission-based routing are two typical methods, as shown in Fig. 3.

If the transmission fails, a retransmission may be needed for the delivery. For the first case, the sender can select another forwarder to transmit the packet, as shown in Fig. 3a. This is called "opportunistic forwarding" [6]. In the other case, the node can simply retransmit the packet using the same link., as shown in Fig. 3b. The transmission count, delivery ratio, or other utilities are provided to measure the metric.

Apart from different kinds of utilities, many works have been conducted about forwarding methods. In WSNs, the simplest forwarding method is *flooding*. Once a node receives the packet, the node will forward the packet to all of its neighbors.

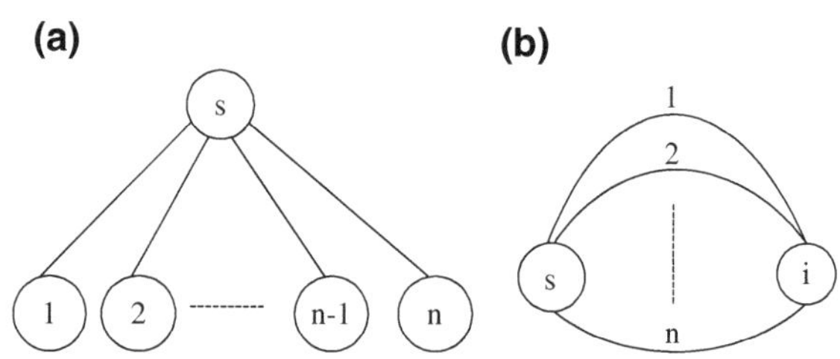

Fig. 3 Forwarding methods. **a** Opportunistic forwarding. **b** Retransmission

In this way, the packet is surely forwarded to the destination, as long as the network is connected. To avoid cyclic transmissions, the node should forward the packets to the neighbor it has not seen. To avoid useless propagation, packets usually have an expiration time, i.e, *time to live* (TTL) or maximum number of hops.

Probabilistic broadcast approaches, broadly called "gossip," offer a simpler alternative to deterministic routing. With gossiping, nodes in the network are required to forward packets with a pre-specified probability, $p_{gossip} \leq 1$. The main idea is that the proper p_{gossip} will make the entire network receive the broadcast message with q very high probability. Recent research [7] has mentioned that the correct value of p_{gossip} is closely associated with the topology of the network.

2.4 Additional Issues

In addition to the above utilities, other factors may also affect the network performance. In this subsection, we will offer other facts related to the routing design.

2.4.1 WSNs and Low Duty-Cycle WSNs

Many applications in WSNs need a long time energy conservation due to limited energy supply. The special feature of the applications in WSNs is that the sensors are equipped with limited energy. Thus, it is desirable to turn off the radios when the sensors do not need to participate in the data delivery.

To resolve the conflicts, it is necessary to reduce the communication cost and duty-cycles. B-MAC [8] is one of them. B-MAC supports dynamic reconguration and provides bidirectional interfaces for system services to optimize performance, whether it be for throughput, latency, or power conservation. These sensors construct the *low duty-cycle WSNs* [9]. There are two states for the sensors: active or sleep. Usually, when a sensor node is in active mode, it can listen to the channel to receive the packets or transmit the packets. Various MAC protocols have been proposed in low duty-cycle WSNs. The difference relies on the synchronization mechanism. Some MAC protocols require both of the nodes to be in active mode for data transmission, such as S-MAC [10]. The end-to-end delay is the least common multiple (LCM) of the two nodes. To lower the end-to-end delay latency, other MAC protocols were proposed. The sensors can still listen/overhear the packets by providing additional energy when they are in sleep mode. X-MAC [11] is one of them. With these MAC protocols, the end-to-end delay is the "wake up" period of the node. Figure 4 shows an example of two MAC protocols. Suppose nodes s and d are the neighboring nodes. The working schedules for nodes s and d are 2 and $3k$, respectively. The arrow lines show the schedule when two nodes can communicate with each other.

In this way, the energy can be consumed during: (1) Network setup: during this process, the sensors wake up, re-open network connections, and initialize the sensors. (2) Data processing: when the sensors are in active mode, they can transmit the

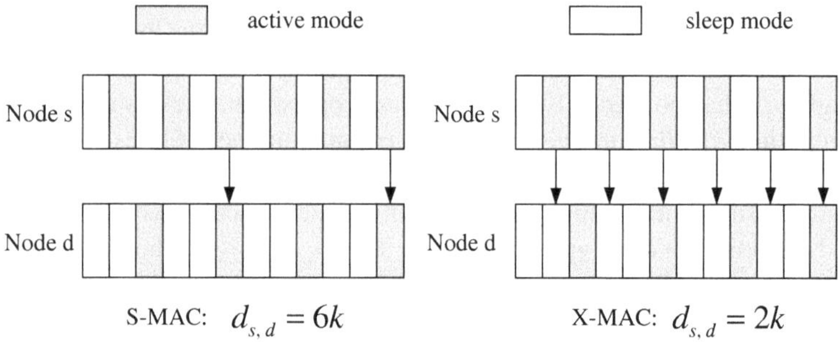

Fig. 4 Low duty-cycle WSNs

packets. (3) Tear down: the sensors close the network connection, reset, and go into sleep mode. (4) Maintenance: The sensor nodes are in sleep mode. Very little energy will be consumed.

2.4.2 Topology

The topological deployment of WSNs is important. This could either be determined or random. For determined deployment, the sensors can be deployed along the roadside, or at a metro station, etc. The Sand [12] sensor network for target tracking and the CitySense [1] network for urban monitoring are the instances where optimal patterns can be provided [13]. However, lots of routing protocols are designed for random deployment.

Considering different deployments, it is important that a path exist from the source to destination. In other words, it is necessary to ensure the connectivity of the network. Typically, there is an inverse relationship between scalability and reliability in WSNs. As the number of nodes in the network increases, it is more difficult to ensure reliability. More dynamics in the environment will increase the number of control packets during the routing process. Moreover, the network cannot afford the large amount of overhead caused by the dynamics, which will result in less reliability. The two basic problems in real applications are bad paths and links to the sink nodes.

As we have discussed, the routing design in WSNs mainly focuses on the *packet delivery ratio* (PDR), latency, and energy efficiency [14]. In the following, we will provide some recent work that relates to these metrics, plus some new ones through a composition metric. The notation list is provided in Table 1. To make it consistent, we use p, d, and c for the reliability, delay, and cost calculations, respectively. Table 1 shows the notation list used in this paper.

Table 1 Notation list

Parameter	Description
$p_{i,j}$	The reliability of link (i, j)
p_i	The packet reception probability at node i
t_i	The time slot at node i
ω_i	The working schedule of node i
$d_{i,j}$	The delay between nodes i and j
$c_{i,j}$	The cost of link (i, j)
$h_{i,j}$	The hop count between nodes i and j
τ	The time span for each time unit
$d(P)$	The delay for a path P
$\rho_{i,j}$	Tthe duty-cycle rate of link (i, j)

3 Single Utility-Based Routing

In this section, we will provide an overview of the recent works related with single utility-based routing design. The "single utility" means that the routing metric is designed for the purpose of improving the packet delivery ratio, lowering the end-to-end delay, or saving energy costs.

3.1 Packet Delivery Ratio

The packet delivery ratio has several cases: the *expected delivery ratio* (EDR), the *expected transmission count* (ETX), and the *quality of forwarding* (QoF). In the following, we will introduce several related approaches.

3.1.1 Expected Delivery Ratio

In opportunistic routing, each node s has n neighbors that construct the forwarding sequence, as shown in Fig. 3a. For a given forwarding sequence, suppose that P_i is the overall probability that a packet is successfully delivered by the ith forwarder. It can be represented as follows [14]:

$$P_i = (\prod_{j=1}^{i-1} (1 - p_j))p_i.$$

Therefore, the corresponding EDR for node s can be expressed as follows:

$$EDR = \sum_{i=1}^{n} P_i \cdot EDR_i. \quad (1)$$

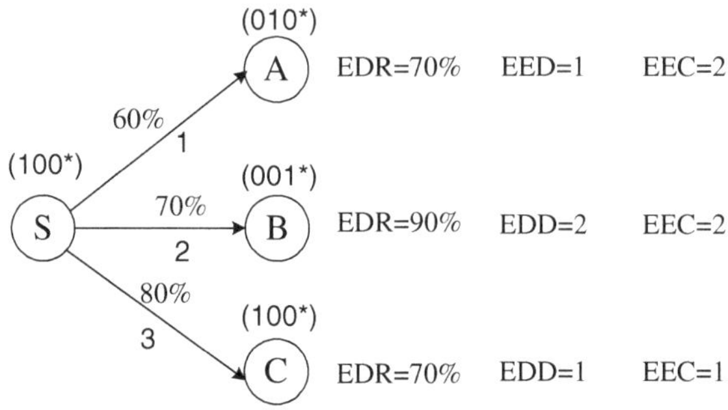

Fig. 5 An example of data forwarding in low duty-cycle WSNs

Figure 5 shows an example of computing EDR. According to Eq. 1, EDR for node s is $0.6 \cdot 0.7 + (1 - 0.6) \cdot 0.7 \cdot 0.9 + (1 - 0.6) \cdot (1 - 0.7) \cdot 0.8 \cdot 0.7 = 0.74$, where $P_A = 0.6$, $P_B = (1-0.6) \cdot 0.7 = 0.28$ and $P_C = (1-0.6) \cdot (1-0.7) \cdot 0.8 = 0.096$.

3.1.2 Link Correlation

The link correlation in low duty-cycle WSNs was first studied in [15]. In both indoor and outdoor experiments, they observed that if a packet is received by a sensor node with a low *packet reception ratio* (PRR), in most cases, this packet can also be received by the high PRR nodes. In order to quantify the relationship, conditional packet reception probability at node s was defined as $p_s(p_h|p_l)$, where p_h and p_l are for higher and lower link quality, respectively. p_h and p_l are the neighboring receivers of the sender s.

The work [15] was further extended to [16]. Traditionally, energy optimality in a designated flooding-tree is achieved by selecting parents with a smaller hop count and the best link quality. However, in this work, the authors studied the link correlation with which redundant transmission can be ignored. As an example, shown in Fig. 6, node E wants to select the senders from A and B. If we do not consider the link correlation, link AE should be selected for the transmission, since the quality of AE(85 %) is higher than BE(75 %). However, if link correlation is considered, the conclusion will not hold. Let p_1 and p_2 denote the link qualities for the two receivers, respectively. For k successful transmissions, the number of transmissions m needed for both of the receivers should satisfy the following equation:

$$\begin{aligned} p_r(m = k) &= p_r(m > k - 1) - p_r(m > k) \\ &= ((1 - p_1)^{k-1} + (1 - p_2)^{k-1} - p_{12}^{k-1}) \\ &\quad -((1 - p_1)^k + (1 - p_2)^k - p_{12}^k). \end{aligned}$$

Thus, the expected transmission $E(m)$ is:

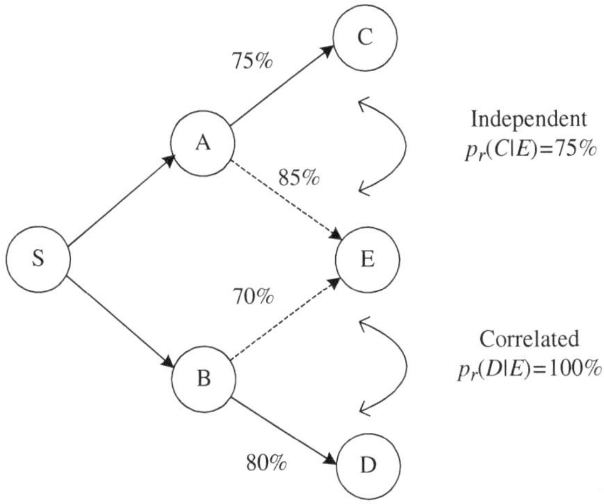

Fig. 6 An example of correlation

$$E(m) = \sum_{k=1}^{+\infty} k P_r(m = k) = \frac{1}{p_1} + \frac{1}{p_2} - \frac{1}{1 - p_{12}}$$

As shown in Fig. 6, nodes E and C are independent. if node E chooses A as the forwarder, $p_{12} = (1 - p_1)(1 - p_2)$. $E(m) = 1/0.75 + 1/0.85 - (1/(1 - (1 - 0.75) \cdot (1 - 0.85))) = 1.47$. In the other case, since nodes D and E are correlated, if node E selects B as the forwarder, $p_{12} = 1 - p_E - p_D + p_E \cdot p_r(D/E) = 1 - 0.7 - 0.8 + 0.7 \cdot 1 = 0.2$. Here, $p_r(D/E)$ is the probability that node D can receive the packet if the packet can be received by node E. Then, $E(m) = 1/0.7 + 1/0.8 - 1/(1 - 0.2) = 1.43$.

The proposed correlated flooding includes two parts. One part is to collect the link quality information and partition the receivers into different groups. The senders will send a hello message to their neighbors. The receivers will record the information in a bitmap format (1: success, 0: failure). The distance between two correlated bitmaps is called the Hamming distance. It is defined as the number of different positions between the bitmaps. For example, the distance of [0111] and [0111] is 0, while the distance of [0111] and [1000] is 4.

The other part is the sender selection process. Each receiver may belong to multiple groups. In the sender selection process, the receiver will select the sender with the highest priority. If there are more than one, it will choose the one with the best link quality. The advantage of this method is that it makes use of the link correlation, and reduces the number of transmissions.

3.1.3 Expected Transmission Count

Many routing metrics have been proposed to measure the link quality in wireless networks. The ETX [17] is one of the typical routing metrics. It can be represented as follows:

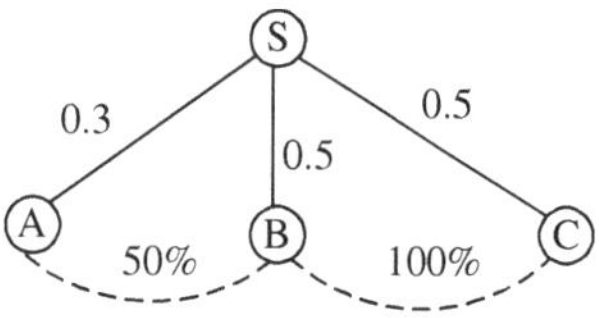

Fig. 7 An example of correlation

$$ETX = \frac{1}{p_f \cdot p_r},$$

where p_f is the probability that the packet can be successfully received. p_r is the reverse probability that the ACK can be successfully received. ETX selects paths with the minimum expected number of transmissions (including retransmissions) required to deliver a packet to its destination. For example, p_f and p_r equals to 0.7 and 0.8, respectively. Then, ETX of the link is $1/(0.7 \cdot 0.8) = 1.785$.

Basalamah et al. [18] proposed a new routing scheme which makes use of link correlation and opportunistic transmission scheme. The link correlation aware opportunistic routing was proposed to improve the performance exploiting the diversity gain. The motivation example can be explained as follows.

As shown in Fig. 7, the expected transmission count of the three links are $1/0.5 = 2$, $1/0.5 = 2$ and $1/0.3 = 3.33$ for nodes S to A, B, C, respectively. It is obvious that links (S, B) and (S, C) are better than link (S, C). Therefore, we have the expected transmission times:

$$ETX_{s,b,c} = \frac{1}{1 - \prod_i (1 - p_{s,i})}.$$

In this example, we have $ETX(s, b, c) = 1/(1 - 0.25) = 1.33$ for candidate set $\{B, C\}$. Likewise, we can also obtain a similar result by considering A and B as the candidate set. The result of $ETX_{s,a,b}$ is 1.176. However, this calculation is without link correlation. When we consider the link correlation, we could use the following equation:

$$ETX = \frac{1}{1 - Pr(\overline{E_{s,1}}, \overline{E_{s,2}}, \ldots, \overline{E_{s,n}})},$$

where $E_{s,i}$ is the event that the packet is successfully received by node i. Using this example, the links from node S to nodes B and C is 100 % correlated. Thus, $Pr(\overline{E_{s,B}}, \overline{E_{s,C}})$ is 0.5. The ETX reduces to 2. In this situation, the selection of A and B is the best choice.

3.1.4 Quality of Forwarding

In WSNs, some routing protocols are designed by allowing the retransmission strategy, as shown in Fig. 3b. The *quality of forwarding* (QoF) [19] was proposed to measure the path quality. The authors considered two kinds of PDR: one is for physical links and the other is for virtual links (inside the node). p is denoted as the probability that the packets successfully go through the link. r is denoted as the most retransmission times. Thus, the *packet delivery ratio* (PDR) over a link is

$$PDR = 1 - (1 - p)^{r+1}. \tag{2}$$

According to Eq. 2, the *expected transmission count* (ETC) was proposed using the following equation:

$$\begin{aligned} ETC &= (\sum_{k=1}^{r+1} kp(1-p)^{k-1}) + (r+1)(1-p)^{r+1} \\ &= \frac{1-(1-p)^{r+1}}{p}. \end{aligned} \tag{3}$$

Here, $\sum_{k=1}^{r+1} kp(1-p)^{k-1}$ represents the expected transmission times that the packet passes the link. $(r+1)(1-p)^{r+1}$ represents the expected transmission times where the packet will fail to to pass the link. Using Eqs. 2 and 3, the QoF is the ratio of the data delivery ratio to the actual transmission times:

$$QoF = \frac{PDR}{ETC}.$$

Thus, for a physical link, the $QoF = p$. Note that ETC is different from ETX in that it not only considers link quality, but also retransmission limit. When $r \to \infty$, ETX = ETC. To calculate the QoF of a path, the PDR of a node has also been integrated into the routing design. The path QoF considers both data delivery ratio and transmission cost. If the data delivery ratios of two paths are the same, QoF selects the path with lower transmission count. If the transmission count of two paths are the same, QoF favors the path with high data delivery ratio.

3.2 Delay

In this subsection, we will focus on the routing issues that relate to the *end-to-end* (E2E) delay. The E2E delay is one of the most fundamental issues for WSNs. Many applications in WSNs require an E2E delay guarantee for time sensitive data. For example, telemonitoring of human health status, vehicle anti-theft [20], and target tracking [4] are classified as the time-sensitive applications.

In low duty-cycle WSNs, due to the limitation of the energy budget, the sensors are scheduled to "sleep" or "active" states. When the sensors are in sleep mode, they cannot transmit the packets. The time spent on waiting for its neighbors to wake-up is called "sleeping latency." Thus, unlike the traditional wireless networks, the delay-based routing design also includes the sleeping latency, in addition to the transmission delay. The sleeping latency (in seconds), however, is much longer than the transmission delay and propagation delay (in milliseconds). Therefore, the E2E delays mainly dominate the sleeping latency.

In traditional wireless networks, the shortest path algorithm is used to find the optimal path in the weighted graph $G = (V, E)$. However, in the low duty-cycle WSNs, the graph changes over time, which is called a "time-dependant graph." $G = (V, E(t))$ [21] is used to represent the models, where V is a set of nodes, and $E(t)$ is a set of edges that appears at time t. Several works have been conducted using time-dependant graph models.

3.2.1 Sleep Scheduling

In [22], the authors provided delay-efficient sleep scheduling for WSNs. They consider the case of a single wake up schedule, where each sensor can choose exactly one of k slots to wake up. They also prove that minimizing the E2E is, in general, NP-hard. The interesting part is the time slot assignment for each node. The time slot assignment is to assign a slot to a certain node, and schedule the node to wake up. Let ω be a slot assignment function and $d_{i,j}$ be the delay in transmitting the data from i to j. The delay on a path P under the slot assignment function ω is defined in Eq. 4:

$$d(P) = \sum_{(i,j)\in P} d_{i,j}. \tag{4}$$

They use two models: *all-to-all communication* and *weighted communication*. In the all-to-all communication, the delay diameter is defined as $D_{i,j}$, which is the shortest delay path between nodes i and j under the slot assignment function ω. The problem is to find an assignment function ω that minimizes the delay diameter. This is called *delay-efficient sleep scheduling* (DESS). Figure 8 shows an example of the delay diameter. Among all of the pairs: (A, B), (A, C), (A, D), …, (B, A), (B, C), …, the delay diameter $D_{A,B} = d_{A,E} + d_{E,F} + d_{F,B}$ is 7. The property is non-symmetric, since $d_{A,B} \neq d_{B,A}$. Compared to the delay parameter, the hop diameter is symmetric. For example, $h_{A,B} = h_{B,A} = 3$.

In the weighted communication model, they defined the *average delay diameter*, which is $\sum_{i,j\in V} w_{i,j} \cdot d_{i,j}$. Like DESS, they also offer *average delay efficient sleep scheduling* (ADESS). ADESS is used to find the slot assignment function ω that

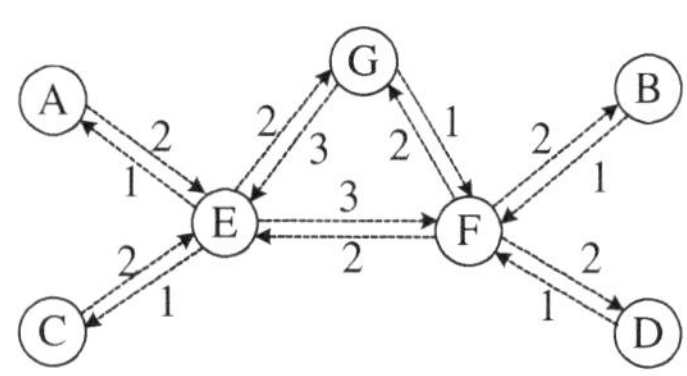

Fig. 8 An example of delay diameter

minimizes the average delay diameter. The main difference is that this method focuses on the fact that the communication between some pairs occurs more frequently than other pairs.

3.2.2 Pipeline Scheduling

In order to reduce unnecessary forwarding interruption, a state-of the-art solution has been provided by using the technique of pipeline scheduling. The most recent work is presented in [23]. Cao et al. proposed a *Robust Multi-pipeline Scheduling* (RMS) algorithm to coordinate multiple parallel pipelines. Pipeline scheduling-based routing is one of the multipath routing designs. Multi-path routing is the routing technique of using multiple alternative paths through a network. The path could be node disjoint, edge disjoint, or overlapped. The specialty of pipeline scheduling is to lower the switching delay by coordinating the transmission time.

The advantage of the pipeline scheme is the decrease in the sleep latency. The single pipeline scheme is always fragile, due to unreliable links. Therefore, multiple pipelines are provided to dynamically switch one forwarder to another forwarder.

Examples are provided in Fig. 9. We assume that the duty cycle is 100 s. In the original scheduling method, if the transmission from nodes A to B fails, the packet has to wait a longer time to be retransmitted, as shown in Fig. 9a. However, if we use multiple pipelines, we can see that the packet can dynamically switch to another forwarder E. Figure 9b, c show the process. Note that if we reschedule the transmission time using pipeline, such that if one pipeline has failed, the node can dynamically switch to another one. In this way, the latency can be reduced, while using the same energy cost. In this scheme, the route is decided dynamically by the timely results. There are three steps for the algorithm.

The first is selection of the virtual forwarding set. The virtual forwarding set is constructed using the link quality. Suppose that the link quality is $\{q_{s1}, q_{s2}, \ldots, q_{sn}\}$. The ϕ is defined as the threshold, which is the one-hop delivery ratio. As shown in Fig. 10, the first k attempts are $1-(1-q_{s1})(1-q_{s2})\cdots(1-q_{sk})$. Suppose that we have M links in the forwarding set, the following Eq. 16 needs to be satisfied:

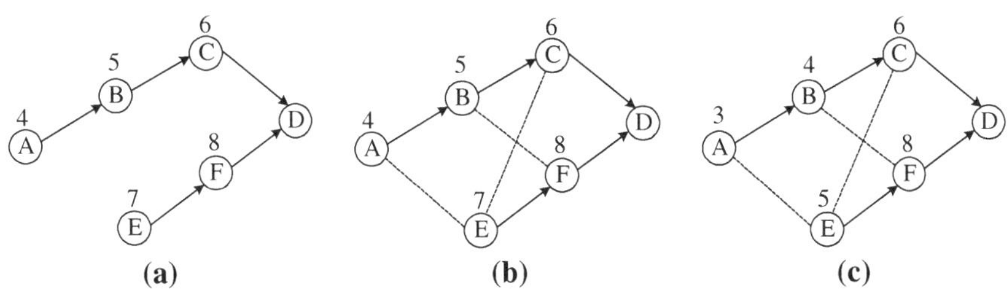

Fig. 9 An example of pipeline scheduling

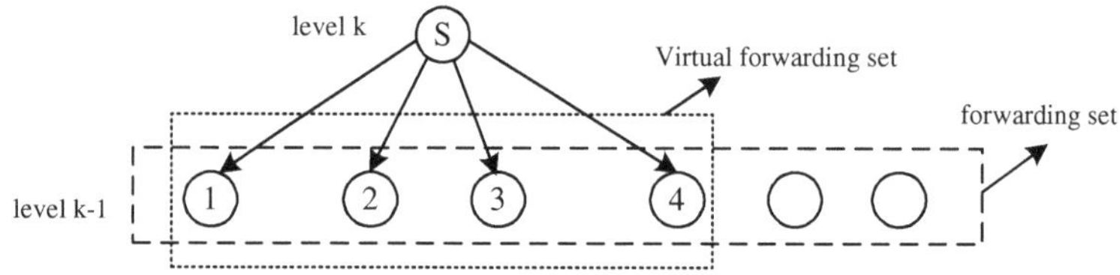

Fig. 10 An example of virtual forwarding set

Fig. 11 An example of simultaneous wake-up time

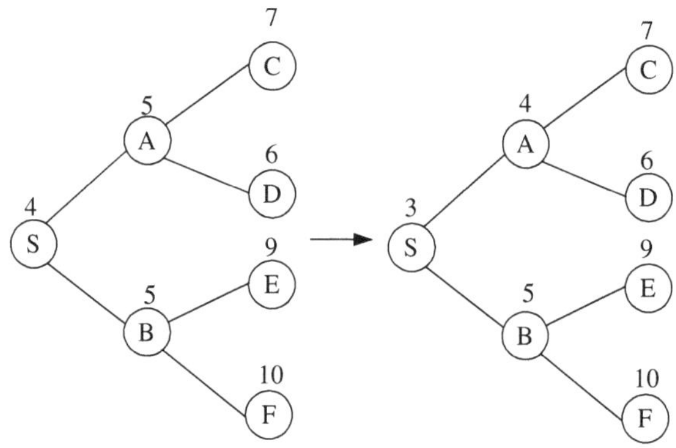

$$1 - \prod_{i=1}^{M}(1 - q_{si}) < \phi \tag{5}$$

The second step is propagative scheduling. The basic purpose is to let each node decide its wakeup schedule, and minimize the expected delay of two consecutive levels. For example, node A has two parents, whose wakeup time is $\{t_1\}$ and $\{t_2\}$. Then, the candidate time of node A is $\{t_1 - 1\}$ and $\{t_2 - 1\}$. Therefore, we have modular delay as defined in Eq. 6:

$$|t_1 - t_2|_T = \begin{cases} t_1 - t_2, \ t_1 \geq t_2 \\ t_1 + T - t_2, \ t_1 < t_2. \end{cases} \tag{6}$$

The third step is to avoid the simultaneous wakeup time. Suppose that node A has two parents who wakeup at the same time slot 4. Therefore, node A can either choose nodes B or C to forward the packet. To avoid this, we can shift the wakeup time of node B to time slot 3. Figure 11 shows an example of this process.

3.2.3 Collaborative Scheduling

The collaborative scheduling was studied in [24]. The authors provided a collaborative scheme based on the concept of error interference. They designed a sensing probability bound to control tolerable sensing errors. The proposed scheme aims at achieving low energy cost. The error interference is defined as the difference between the ground truth environmental data and corresponding values generated by the predictor of sensor nodes. There are four stages in the proposed algorithms.

The first step is to detect the neighbors. In this stage, each node recognizes its neighboring nodes, and assigns a table for each neighbor to build the weight graph. The neighborhood formation is a dynamic stage, which will be refreshed after a defined period.

The second step is to generate the node-pair weighted graphs. Specially, the following approach to calculate data correlation $C(i, j)$ between two observation vectors by node N_i and node N_j:

$$C(i, j) = \frac{m \sum o_k^i o_k^j - \sum o_k^i o_k^j}{\sqrt{m \sum (o_k^i)^2 - \sum (o_k^i)^2} \sqrt{m \sum (o_k^j)^2 - \sum (o_k^j)^2}}, \tag{7}$$

where $\{o_1^i, o_2^i, \ldots, o_k^i\}$ is the observation vector, obtained through discrete sampling.

The third one is to use the error bound to control the neighbors. The sensing node, can infer the prediction errors of correlated neighboring nodes by comparing its real-time sensing values with corresponding predicted values. Using the probability density function $\rho(x)$, each node i can evaluate the cumulative distribution function $PMF_i(e_i^m)$:

$$PMF_i(e_i^m) = \int_{-e_i^m}^{e_i^m} \rho(x)dx, \tag{8}$$

where e_i^m is observation error, based on the difference between actual sensing data and prediction values that are generated by our prediction model. Then, the inferred error between nodes i and j can be expressed as follows:

$$e_{ij} = PMF_j^{-1}(PMF_i(t[k])). \tag{9}$$

Here, $t[k]$ is the variable for each step:

$$t[k] = \begin{cases} 2 \cdot t[k-1] & PMF_j(t[k-1]) < PMF_i(e_i^m) \\ \dfrac{t[k-1] + t[k-2]}{2} & PMF_j(t[k-1]) > PMF_i(e_i^m), \end{cases} \tag{10}$$

where $PMF^1()$ is the inverse function of PMF, and $t[0] = 0$, $t[1] = e_i^m$. It is an iteration process. It will not stop until $PMF_j(t[k1]) = PMF_i(e_i^m)$. The basic process is that each time we will compare $PMF_j(t[k-1])$ and $PMF_i(e_i^m)$. Then, we can decide the next $t[k]$ until $PMF_j(t[k1]) = PMF_i(e_i^m)$.

The fourth step is to determine whether the sensors switch on/off. The node can remain turned off if the inferred error is smaller than the error tolerance. The error tolerance is a specified bound (e.g., the tolerance threshold of errors).

3.2.4 Expected End-to-End Delay

In [21], $\Gamma_i = (\omega_i, \tau)$ is used to represent the working schedule of node i. ω_i is an infinite binary string in which 1 denotes the active state, and 0 denotes the dormant state. τ denotes the time span for each bit. For example, the total time-span of the binary string (00101) with $\tau = 2$ s is 10 s, since there are totally five bits in the string. Thus, for a sender i and receiver j, if the working schedules are $\Gamma_i = ((10000)^*, \tau)$ and $\Gamma_j = ((00010)^*, \tau)$, respectively, the E2E should be $3\tau - 0\tau = 3\tau$. The main contribution is that they provide an E2E delay guarantee by adding extra active bits to nodes. For example, for the single link route $A \rightarrow B$, $\Gamma_A = ((010)^*, \tau)$ and $\Gamma_B = ((100)^*, \tau)$. By adding an extra active bit to node B, and by changing its working schedule from $(100)^*$ to $(101)^*$, the sleep latency can be reduced to τ.

The expected E2E was mentioned in [14]. It is formally defined as the expected data delivery delay from source node S to destination D over a multi-hop route. They proposed a time-expanded graph model to represent the low duty-cycle WSNs. In the ideal case, the E2E delay in this network is equal to $H \cdot \tau$, where H is the minimum number of hops between a source and destination. Figure 12 shows the process of the delivery from node A to node D. Since node B is in sleep mode, the packet can only be delivered at time 2. The transmission procedure is the same as for the following node. Thus, the E2E delay is 4, where node A sends the packet at time 1, and node D can receive it at time 4. The main concept of this paper is the forwarding sequence, which is maintained by each node. The forwarding sequences are constructed by the nodes' neighbors. During the transmission, the sink node will check the time associated with the first node in the sequence and forward the packet. If the transmission is successful, the forwarding is done. Otherwise, the node may check the second node in the sequence and forward the packet. If P_i' is the probability that the packet arrives at the ith forwarder, under this scenario, the routing metric of

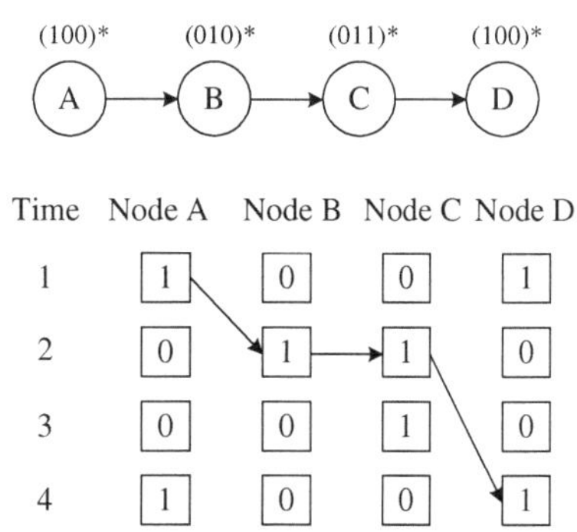

Fig. 12 Time-expanded network

expected E2E delay (EED) is

$$EED = \sum_{i}^{n} P_i'(EED_i + d_i), \tag{11}$$

where d_i is the waiting delay at node i. $P_i' = P_i \cdot EDR_i/EDR$ can be computed according to EDR and P_i, which is discussed in the subsection of packet delivery ratio. We use Fig. 5 as an example to compute EED. From Fig. 5, we have $d_A = 2$, $d_B = 4$, and $d_C = 6$. From the previous section, we can get $P_A' = 0.6 \cdot 0.7/0.74 = 0.57$, $P_B' = 0.28 \cdot 0.9/0.74 = 0.34$, and $P_C' = 0.096 \cdot 0.8/0.74 = 0.1$. According to Eq. 11, the EED of node S is $0.57 \cdot (2+1) + 0.34 \cdot (4+2) + 0.1 \cdot (6+1) = 4.45$.

3.2.5 End-to-End Delay in the Rrandom Walk Model

The E2E delay was also studied in random walk models. In the *i.i.d.* (independent and identically distributed) random duty-cycling model [25], ρ is the duty cycle rate of each node. Each node is in the active state with a probability ρ, while it is dominated with $1-\rho$. For the link (i, j), the probability of nodes i and j being in the wake mode is ρ^2. It is shown that the per-hop latency is $d_i = \frac{1}{1-(1-\rho^2)^{n_i}}$, where n_i is the number of neighbors of node i. Note that, in this case, the working schedule is previously unknown. In the pseudo-random duty-cycle model [26], the node has the knowledge of the working schedule. Therefore, the per-hop latency is $d_i' = \frac{1}{1-(1-\rho)^{n_i}}$.

In [26], the authors studied several aspects of latency in random walk models: (1) Hitting time: the expected time for source s to hit the destination d. (2) Commute time: the expected round-trip time between source s and destination d. (3) Cover time: the expected time from the source s to all of the other nodes in the network. Under the stochastic routing framework, the authors in [27] studied the routing problem using the Markov chain. They developed centralized and distributed implementations in low duty-cycle WSNs.

3.2.6 Communication Delay in Low Duty-Cycle Sensor Networks

In [28], the authors introduced a novel solution of communication delay in low duty-cycle sensor networks. They provide sink-to-one and sink-to-many solutions, and their distributed implementation.

The network topology is denoted as $G(t) = (V, E(t))$, where V is a set of nodes and $E(t)$ is a set of directed links at time t. Each edge $e_{ij}(t)$ belongs to $E(t)$ if node i and node j are in each other's communication ranges. Node j is in active mode, so as to receive the packets at time t. $d_{ij}(t)$ is defined as the delay when node i sends the packet at time t, when node j is in active mode to receive the packet.

Figure 13 shows an example of the proposed solution. In this example. the original E2E delay from node s to node b is 6. After argumentation of active mode at time

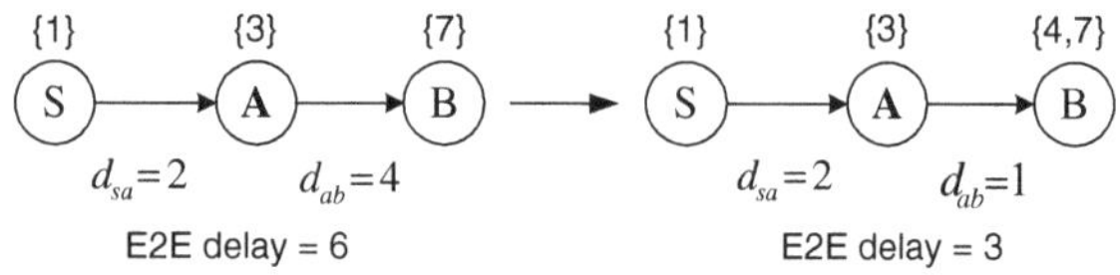

Fig. 13 Example of active argumentation

4 for node b, the E2E delay can be reduced to 3. In addition, the original schedule of node b is time 7. After argumentation, the schedule of node b is both time 4 and time 7. In the following, we will provide more details of the solution. The first is how to find the minimum delay for active instance augmentation. We define D_j^h as the minimal delay from the source to node j, with at most h active augmentation. The initial state is

$$D_j^h = \begin{cases} d_{sj}, \ h = 0 \\ 1, \ h = 1. \end{cases}$$

Based on this solution, we then offer the recursive solution. The main idea is that we could use intermediate node p to help the delivery. Then, we have

$$D_j^h = \min \begin{cases} D_j^h, \\ D_p^{h-1} + 1, \ h > 0 \\ D_p^h + d_{pj}(t). \end{cases}$$

An example can be presented in Fig. 14. In this example, the initial states are $D_a^0 = 5$, $D_a^1 = 1$, $D_b^0 = 2$, and $D_b^1 = 1$. Then, for node c, we have $D_a^0 + d_{ac}(6) = 5+8 = 13$. Using the above equation, we have:

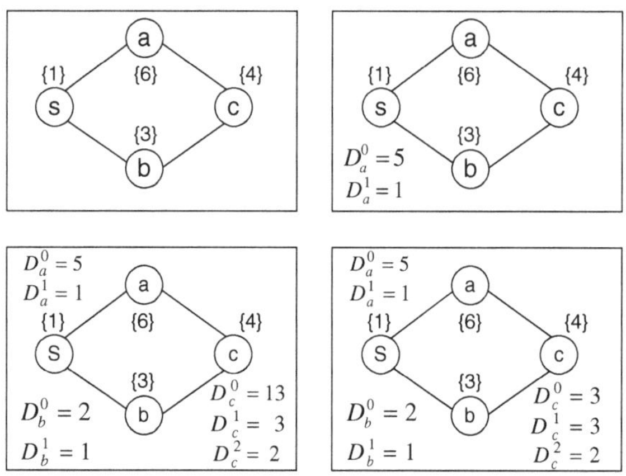

Fig. 14 Example of delay computation

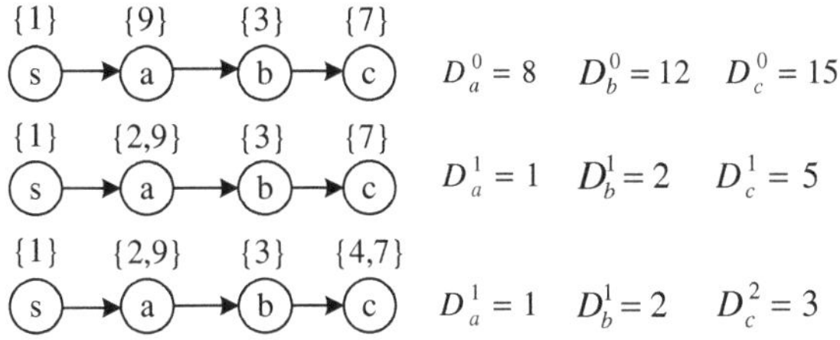

Fig. 15 Example of linear topology

$$D_c^1 = \min \begin{cases} D_a^0 + 1 = 5 + 1 = 6 \\ D_a^1 + d_{ac}(2) = 1 + 2 = 3 \end{cases} = 3.$$

Correspondingly, we have $D_c^2 = 1 + 1 = 2$. We also offer the example of linear topology shown in Fig. 15. In this example, the initial states are $D_a^0 = 8$, $D_b^0 = 12$, and $D_c^0 = 15$. After the first round of argumentation, we have $D_a^1 = 1$. Then, D_b^1 is 2. $D_c^1 = 5$. The algorithms stops at the second argumentation, where we set $D_c^2 = 3$.

3.2.7 Opportunistic Flooding and Expected Packet Delay

Flooding has been investigated extensively in wireless networks. However, there are several challenges when using low duty-cycle WSNs. Firstly, the nodes stay asleep most of time and wake up asynchronously. A broadcasting packet is rarely received by multiple nodes simultaneously. Secondly, the sender may have to wait for a certain period of time until its receiver becomes active. Finally, the wireless link is unreliable. The transmission may be repeated due to the low link quality.

Opportunistic flooding in low duty-cycle WSNs was proposed in [29]. The main objective is to reduce redundant transmissions, while achieving fast dissemination. As shown in Fig. 16a, the flooding structure of the network is a *directed acyclic graph* (DAG) of N vertices. Figure 16b offers the solution of the structure, which is called an "energy-optimal tree." The energy optimal tree is built based on a smaller hop count and the best link quality. The proposed opportunistic flooding consists of three

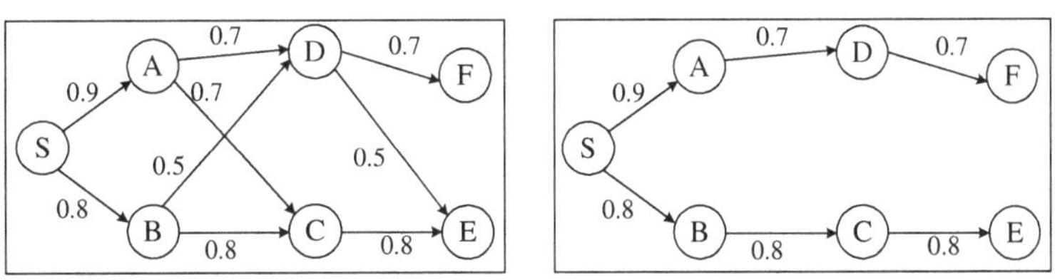

(a) Original flooding structure **(b)** Energy-optimal structure

Fig. 16 DAG-based flooding structure. **a** Original flooding structure. **b** Energy-optimal structure

parts: the *probability mass function* (pmf), decision making process, and decision conflict resolution.

The pmf is denoted as a set of tuples $\{(t_h(i), p_h(i))\}$, where $p_h(i)$ is the probability of receiving the packet at time $t_h(i)$. For the source node, it will always awaken. Therefore, the delay is 0 and the probability is 100 %. The pmf of the source is (0,100 %). Then, the probability that it receives the flooding packet at its jth active time is

$$p_{h+1}(j) = \sum_{i:t_h(i)<t_{h+1}(j)} p_h(i)p(1-p)^{n_{ij}}.$$

Figure 17 shows an example to compute pmf. The probability that node A receives the packet at time 10 is 0.9. At time 20, the probability is $0.9 \cdot (1-0.9) = 0.09$. For the node D, the probability is the sum of two cases: (i) node A receives the packet at time 10, or (ii) the probability that node A receives the packet at time 20. Then, pmf is $0.9 \cdot (1-0.7) \cdot 0.7 + 0.09 \cdot 0.7 = 0.252$. Similarly, all of the nodes in the network will compute their pmf.

In the decision-making process, the p-quantile delay (D_p) is a threshold delay. A node computes the *expected packet delay* (EPD) and makes a forwarding decision, based on the comparison between the EPD and D_p. The EPD will be introduced in the following section. If we have the transmission from A to B, the EPD can be computed by using the following equation:

$$EPD = \sum_{j:t_{h+1}(j)>t_l(i)} t_{h+1}(j)p(1-p)^{n_{ij}},$$

where p is the link quality. A node with hop count h is denoted as a level-h node. Then, n_{ij} is the level-$(h+1)$ node's active time units between $t_h(i)$ and $t_{h+1}(j)$. EPD is the sum of all the series. A simple way to obtain the EPD is to use the expected transmissions $\lceil \frac{1}{p} \rceil$ and get the time slot of the $\lceil \frac{1}{p} \rceil$th transmission. For example, as shown in Fig. 17, since $p = 0.5$, we have the expected transmission 2. Then, we have $EPD = 13$, as the 2rd try is at time 13. D_p can be computed by using the discrete quartile function:

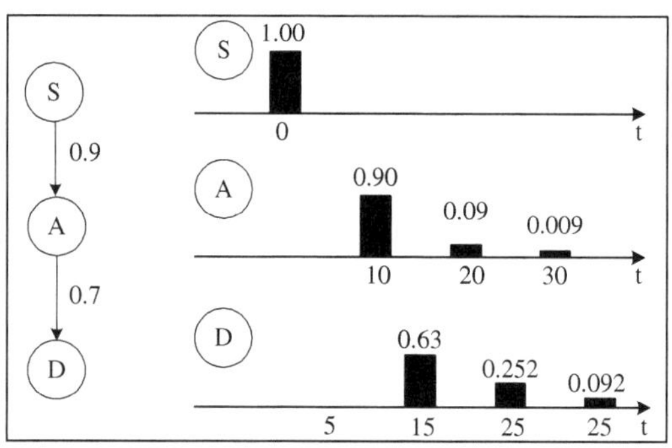

Fig. 17 The pmf computation

Fig. 18 An example of decision making

$$F^{-1}(p) = \min\{x \in R : P_r(t \leq x) \geq p\}.$$

Figure 18 shows an example of the decision making process. Thus, $D_p = F^{-1}(0.2 + 0.5) = 9$. Since EPD $= 13 > D_p = 9$, the second try is redundant.

The motivation for the decision conflict resolution is the hidden terminal problem. Since links are unreliable, the more nodes in the same set will make it possible for packets to be sent at the same time. It will be likely that a transmission is not sensed by all of the other nodes, leading to a collision. The link quality threshold h_{th} was proposed for the selection process. All of the links have a higher link quality than h_{th}. During the selection process, the selected candidate will follow the order of the link quality. To resolve the conflict in the sender set, the backoff function was designed to select the higher link quality function. This means that the duration of the backoff depends on the link quality between the sender and the receiver. When multiple nodes want to send the packet to the same node, the node with shortest backoff time $d_{backoff}$ can forward the packet first. Suppose that the bound of backoff is $D_{backoff}$ and the maximum size of sender set is n. Each node could compute its own backoff by using the following equation:

$$d_{backoff} = (\lfloor n(1-p) \rfloor)\frac{D_{backoff}}{n} + d_{rand},$$

where d_{rand} is a random time period, and $d_{backoff}$ is the backoff time for each node. The basic idea is that each node computes its own backoff time $d_{backoff}$ and transmit the packets after $d_{backoff}$.

3.3 Energy Cost

The energy cost is an important part of WSNs. While developing the routing protocols, it is crucial to ensure the power efficiency. The power consumption can be put into two categories: data transmission and data processing.

3.3.1 Real-Time Power-Aware Routing

The *real-time power-aware routing* (RPAR) scheme was proposed in [30]. This routing protocol is based on the tradeoff between transmission power and communication delay. This work focuses on the real-time applications in which meeting deadlines is more important than throughput. The goal is to increase the number of packets that meet deadlines, while minimizing the energy consumption.

The *delivery velocity* was proposed as the distance that a packet travels, divided by its packet delay. The slack is the time remaining until the deadline expires. It can be updated at each hop. With the source s and destination d, the required velocity of a packet is

$$v_{req}(s, d) = \frac{dis(s, d)}{slack_{rec} - (t_{head} - t_{rec})},$$

where $dis(s, d)$ is the Euclidean distance of s and d. t_{head} is the time when the packet becomes the head of the transmission queue. t_{rec} is the time to receive the packet. The proposed RPAR protocol uses the velocity assignment policy to map a packet's deadline. They also provide a delay estimator for different forwarding choices:

$$d_p = (d_{cont} + d_{tran}) \cdot ETX(p),$$

where $ETX(p)$ is the expected number of transmissions from node s to neighbor i at power p. d_{cont} and d_{tran} are the contention delay and transmission delay, respectively. Based on the velocity policy and delay estimator, RPAR computes the energy cost of all of the eligible choices. Then, it will forward the packet using the most energy-efficient choice:

$$C(s, d) = C(p) \cdot ETX(p) \cdot \frac{dis(s, d)}{dis(s, d) - dis(i, d)}.$$

3.3.2 Expected Energy Consumption

As shown in Fig. 3, we set $|S|$ as the forwarding sequence. The EEC [14] was defined as the energy consumption required to deliver a packet from node s to sink node d. For each node i, EEC_i is the expected energy cost, and c_i is the transmission cost. If P'_i is the probability that the packet arrives at the ith forwarder, then the EEC should be:

$$EEC = \sum_{i=1}^{n} P'_i \cdot (c_i + EEC_i), \tag{12}$$

where P'_i is the same as was discussed in the subsection about the packet delivery ratio. As an example, shown in Fig. 5, since $P'_A = 0.57$, $P'_B = 0.34$, and $P'_C = 0.1$, EEC of node s is $0.57 \cdot (1 + 2) + 0.34 \cdot (2 + 2) + 0.1 \cdot (3 + 1) = 3.47$.

So far, we have introduced the three routing metrics, proposed in [14]: expected delivery radio (EDR), expected E2E delay (EED), and expected energy consumption (EEC). Among all of the metrics, the basic idea is that we can select the *optimal* subsequence from the forwarding sequence, so as to achieve the optimal solution. The optimal sequence is in terms of the maximum EDR, minimum E2E delay, or minimum EEC.

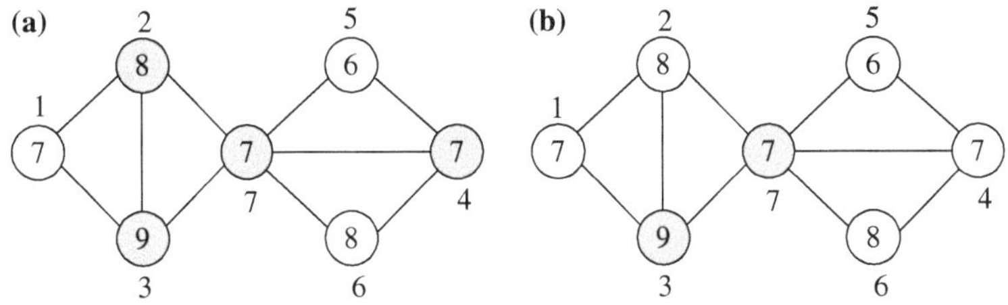

Fig. 19 An example for power-aware broadcasting

3.3.3 Power-Aware Broadcasting

As we mentioned in the previous section, a straightforward broadcasting scheme is flooding. However, this approach will result in redundant transmissions and more energy consumption. In [31], Rule 1 and Rule 2 are provided to select gateways based on the node priority $(id(v))$. Suppose that $N(v)$ represents the neighbor set of node v, and $N[v] = N(v) \cup \{v\}$ is a closed neighbor set of v; in Rule 1, if $N[v] \subseteq N[u]$ in G and $id(v) < id(u)$, then unmark v. In Rule 2, if $N(v) \subseteq N(u) \cup N(w)$ in G and $id(v) = \min\{id(v), id(u), id(w)\}$, then unmark v. To prolong the lifetime of the sensor nodes, the authors in [32] proposed the use of connected dominating sets (CDS) to reduce the number of transmissions, as well as conserve energy.

They propose saving energy by only allowing dominating nodes to retransmit the packets. In addition, they also provide the activity scheduling method to dynamically select the dominating nodes. The selected gateway nodes are based on marking process [33], where Rule 1 and Rule 2 are based on energy levels. Figure 19 shows an example of the procedure, where each node is assigned with a value inside each node representing energy level. Figure 19a shows the graph with marking process, and Fig. 19b presents the results after Rule 1 is based on energy level. That is, energy level is used as the primary priority, and node id is used as the secondary priority when energy levels are the same. The nodes in the pink color are the selected gateways. In this case, node 2 is removed as it is covered by node 3, which has a higher energy level. Nodes 4 and 7 have the same energy level, but node 4 is removed, for it has a smaller id.

3.3.4 Gossip

In WSNs, broadcast services should minimize energy consumption by reducing redundant transmissions. Flooding is considered the simplest method of broadcasting. However, this will lead to collisions and redundant packet receptions.

Smart gossip [7] was proposed to adapt transmission probabilities based on the underlying network topology. To obtain the p_{gossip}, the *average reception percentage* was proposed to measure the reliability, which is the reception percentage averaged over all nodes in the network. To evaluate the overhead, the *average forwarding*

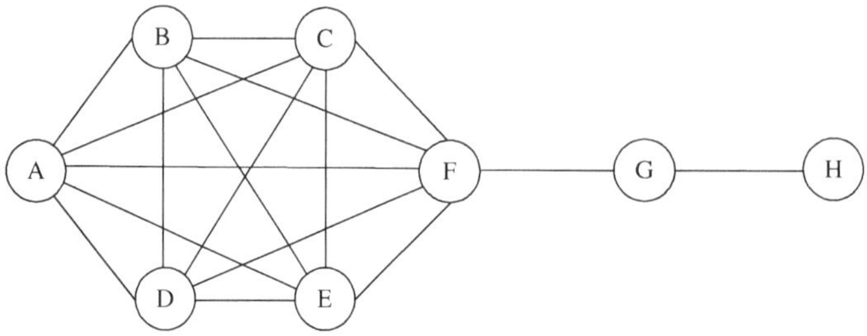

Fig. 20 An example representing the need for smart gossip

percentage was also proposed, which is the forwarding percentage averaged over all nodes in the network. It has been proven that the average forwarding percentage is also the measure of the average energy consumed at a node while transmitting gossip messages.

Figure 20 shows an example that represents the need for smart gossip. If node F can identify that node G depends only on F to receive the gossip, and nodes B, C, D, and E can identify that they are never required to forward the gossip, then we achieve the efficiency. If a node Y has k parents, then it suffices for each parent to use a gossip probability (p_{gossip}), which ensures the probability that at least one of them will transmit is greater than the per-hop reception probability (p_r). This idea can be presented as follows:

$$(1 - p_{gossip})^k < 1 - p_r,$$

where $(1 - p_{gossip})^k$ is the probability that all k parents choose not to transmit. This has to be less than $(1 - p_r)$ to meet the application reliability requirement.

4 Composite Utility-Based Metric

Although we have sorted the recent works into different categories, routing design may be involved in several factors. These factors may be related to each other. In this subsection, we offer additional discussions of them. We call the utility with multiple purposes a composite utility-based metric.

One of the composite utility-based routing schemes is presented in [34], which is also called "utility-based routing." The concept of utility in this work is different from our previous discussion. This utility-based routing scheme is a special routing approach that is based on the network topology, as well as the importance of the packet to be delivered [34, 35]. The composite utility model here is in terms of the expected benefit (of the routing source successfully forwarding a packet to the destination) minus the expected cost incurred by forwarding nodes. Unlike wired networks, wireless connections are unreliable, due to interference and coverage issues. With the utility-based routing metric, the more valuable packet will be delivered through a more reliable route at the expense of a higher transmission energy cost [34]; This is a common phenomenon in wireless communication. Utility-based routing is used to reflect the trade-off between a highly reliable route (which is usually more costly) and a less reliable route (which is usually less costly) based on the value of the packet. A simple analogy that relates to utility-based routing is the postal service: a high-value package (e.g., one that contains a passport for a visa application) usually uses registered mail for reliability at a higher premium cost. An ordinary package is usually mailed through a regular service.

There are two aspects to be considered in the routing design of wireless networks. Firstly, unlike the wired network, wireless connections are unreliable due to the interference or coverage issues. Another problem that cannot be neglected is the cost of transmission, due to limited energy supplies. In [34], the authors have the following assumptions. The reliability is the packet-error-rate associated with the link (i, j). The cost is the energy cost for transmitting a packet from sender i to receiver j. This consists of the transmission power, alone. In the following, we will offer the overview of several works related to utility-based routing. We will start with ad hoc networks. Then, we will focus on multi-hop networks and opportunistic routing. At last, we will discuss the utility-based approach in low duty-cycle WSNs.

4.1 Composite Utility-Based Routing in Ad hoc Networks

The utility-based routing in ad hoc networks was proposed in [36]. The basic idea of utility can be presented as follows. To illustrate the model, we use a single link route from s to d. The source node s wants to transmit a packet to the destination d. If this transmission is successful, we will offer a benefit v for the route. During this transmission, each link will incur a cost. The expected utility value can be derived from the benefit and expected cost. The idea can be presented in the following:

$$u_{s,d} = p_{s,d} \cdot (v - c_{s,d}) + (1 - p_{s,d}) \cdot (0 - c_{s,d}) = p_{s,d}v - c_{s,d}, \tag{13}$$

where $u_{s,d}$ is the utility for the transmission. $p_{s,d}$ and $c_{s,d}$ are the probability and cost of link (s, d), respectively. Thus, for a path $< 1, 2, \cdots, n >$, the expected utility U is:

$$U = \left(\prod_{i=1}^{n-1} p_{i,i+1}\right) \cdot v - \sum_{i=1}^{n-1} c_{i,i+1} \prod_{j=1}^{i-1} p_{j,j+1} = P_R \cdot v - C_R, \tag{14}$$

where P_R is the path reliability, and C_R is the path cost. When comparing Eq. 13 with Eq. 14, the expected utility has a similar form: stability times benefit minus cost. The stability P_R is the multiplication of reliability for each node on the path. The expected cost of each hop is dependant on the successful delivery of previous

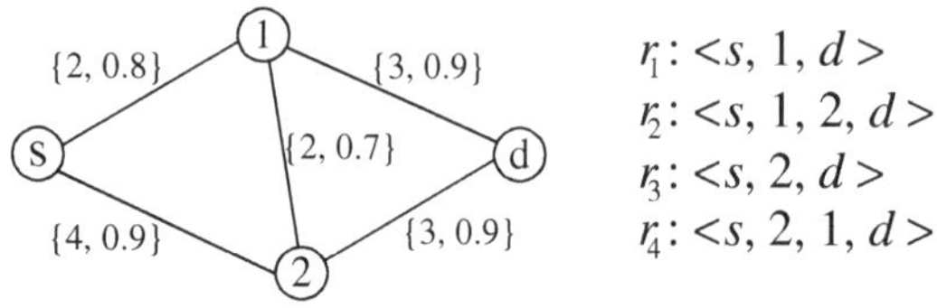

Fig. 21 An example of utility-based routing

Table 2 Utility in a simple wireless network

	u_d	u_1	u_2	U
r_1	20/30	15/24		10/17.2
r_2	20/30	8.5/14.8	15/24	4.8/9.8
r_3	20/30		15/24	9.5/17.6
r_4	20/30	15/24	8.5/14.8	3.7/9.3

Table 3 Utility in a simple wireless network

	u_d	u_1	u_2	u_3	u_4	u_s
u	20	10.6	17	11.3	14	9.54
opu	20	10.6	17	11.4	14	10.45

transmissions. For example, as shown in Fig. 21, there are four routes. For r_1, we have $U = 0.8 \cdot 0.9 \cdot 20 - 2 - 3 \cdot 0.8 = 10$. In a backward manner, we can view node 1 as the virtual source. By applying Eq. 13, we have $u_1 = 0.9 \cdot 20 - 3 = 15$. For the link $(s, 1)$, nodes s and 1 can be viewed as the source and destination. The utility U is $15 \cdot 0.8 - 2 = 10$. From the results, we know that the route with maximum utility is not the same when considering different benefits for the delivery. When comparing routes r_1 and r_3, r_1 is better than r_3 when we offer the benefit 20. However, route r_3 is the best if the benefit is 30. This means that the packets with more benefit will select a more reliable route, despite the cost being higher (Tables 2 and 3).

4.2 Composite Utility-Based Routing Using Opportunistic Routing

Utility-based routing in multi-hop networks mainly focuses on the *opportunistic routing* (OR). The opportunistic routing was proposed in [6]. The main features of OR routing is the rule of the selection of a relay set for each node, and the rule regarding relay prioritization. OR routing would achieve higher throughput, since each of the source's transmissions is likely to be received by at least one relay.

In [35], the authors explored the optimality of utility-based routing through OR without allowing retransmissions. By integrating the idea proposed in [36], OR routing with the utility-based model can be presented as follows:

$$opu_i = \sum_{j=i+1}^{i+k} (opu_j \cdot p_{i,j} \cdot \prod_{l=i+1}^{j-1} (1 - p_{i,l})) - c, \qquad (15)$$

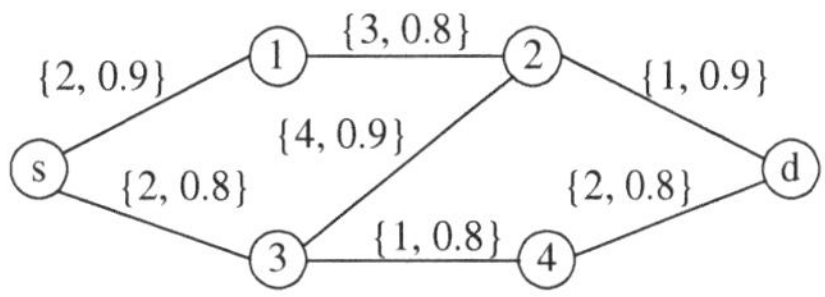

Fig. 22 An example of utility-based routing in multi-hop networks

where opu_i denotes the node i's *residual expected network utilities* (RENU). Note that the relays are prioritized in order from $i+1$ to $i+k$, with $i+1$ as the highest priority.

Figure 22 shows an example of multi-hop networks. The calculation starts from the node d. Suppose that the benefit is 20. For the node 2, the utility $u_2 = p_{2,d} \cdot v - c_{2,d} = 17$. The procedure is the same as for node 4. Thus, $u_4 = 14$. However, node 3 has two receivers. By using Eq. 21, we have $u_4 = 17 \cdot 0.9 + 14 \cdot 0.8 \cdot (1 - 0.9) - 4 = 11.4$. At last, we can get the utility $opu = 10.45$.

4.3 Composite Utility-Based Routing in Low Duty-Cycle WSNs

In low duty-cycle WSNs, the sensors are scheduled to be either in active or sleep mode to achieve low-energy consumption [11, 21]. When the sensors are in sleep mode, they cannot transmit the packets. The utility-based routing in low duty-cycle WSNs is more challenging when we consider the periodic delay of the active and sleep schedule on each node. In addition to the cost and reliability, the delay and deadline are also important for the routing design in low duty-cycle WSNs.

Although energy can be saved by putting the sensor into sleep mode, the cost still exists, due to the set-up and tear-down operations during the transition from active to sleep mode, and vice versa [37]. In fact, energy efficiency is highly related to time efficiency. It can be assumed that time and energy efficiency are equal during the normal operation mode. The time efficiency can be obtained by the ratio of the time spent on the data transmission over the total time for the ideal working schedule. In the randomized duty-cycle scheme, the Markov chain method was used for the analysis [37]. This provides the result of the expected time efficiency. The lower bound and upper bound on it are also provided. The intuitive idea is that the duty-cycle will cause additional delay, but the energy can be saved by putting the nodes into sleep mode.

From our discussion, we know that PDR is highly dependant on reliability. In the definition of PDR, it is measured by a given time period. Then, PDR and delay have an inverse relationship with each other. Higher PDR means a lower delay, and vise versa. As we have discussed in the previous section, the time efficiency and energy efficiency are equal. Thus, it can be derived that the energy cost is also lower. In other words, if the reliability is lower, additional delay and cost will be incurred. In this subsection, we will introduce the timed-based method. Since low duty-cycle WSNs are time-dependent networks, we use the concept of "journey" to represent a

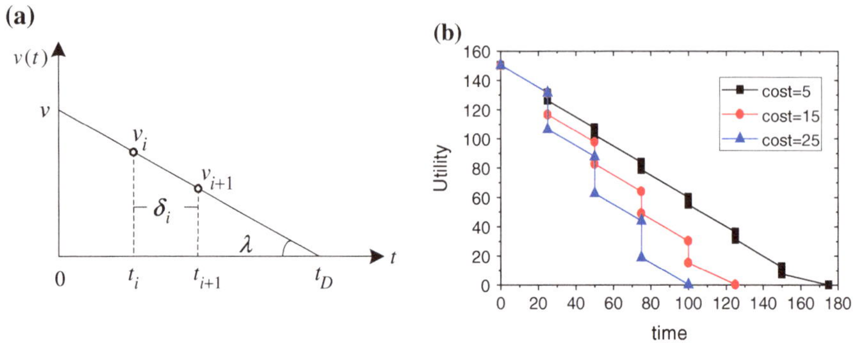

Fig. 23 An example of timed-utility. **a** Timed-benefit function. **b** Different costs

path. We set $J = (R, T)$ to be a journey where $R = \{0(= s), 1, \ldots, n(= d)\}$ and $T = \{t_0, \ldots, t_n\}$, such that $t_i \leq t_{i+1}$ $\forall i \in \{0, \ldots, n-1\}$. Here, t_i is the contact time at node i. In general, there are several possible journeys for a given path. Unlike single-utility, multiple factors can be jointly involved in the routing metric design.

We first provide the timed benefit function. This is used in the timed-utility model. In our timed-utility model, we measure the packet value, cost, and delay to make forwarding decisions. The timed-benefit function is to integrate the delay and deadline into the composite utility.

We use a single link route from s to d as an example. $v(t)$ is defined as the function of TB where the benefit linearly decreases over time, as shown in Fig. 23. t is denoted as the contact time slot for the transmission. The deadline t_D is set as the timeliness of a packet when the benefit is reduced to zero. Suppose that c is the cost for the transmission; the utility u is $u = v(t) - c$. We use $\lambda = v/t_D$ as the slope. Then, $u = v - \lambda \cdot t - c$. Correspondingly, for a journey J, the utility u_i for each node i could be presented as follows:

$$u_{i+1} = u_i - \lambda \cdot \delta_i - c_{i,i+1}, \tag{16}$$

where $c_{i,i+1}$ is the communication cost of link $(i, i+1)$. Figure 23b shows the results with different costs. There are two types of decay. The first one is related with $v(t)$, and the other one is for the communication cost. According to Eq. 14, for a journey J, we have the expected utility:

$$U = (\prod_{i=0}^{n-1} p_{i,i+1}) \cdot (v - \lambda \cdot (\sum_{i=0}^{n-1} \delta_i)) - \sum_{i=0}^{n-1} c_{i,i+1} \prod_{j=0}^{i-1} p_{j,j+1}, \tag{17}$$

where δ_i is the delay of link $(i, i+1)$. By applying the result, the forward solution is forwarded for each hop. It has been divided into two parts:

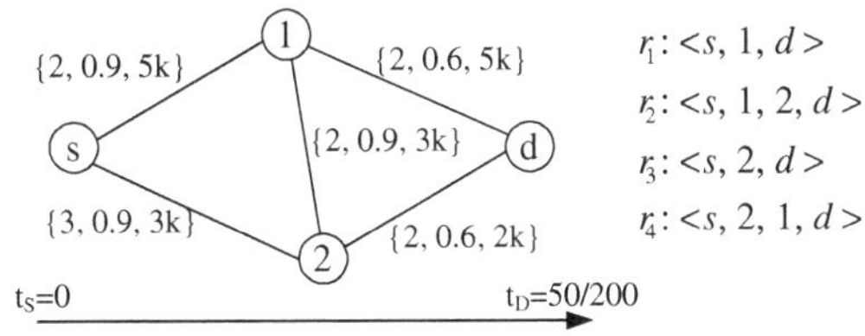

Fig. 24 An example for the TU model

$$v'_{i+1} = p_{i,i+1} \cdot (v'_i - \lambda \cdot (\prod_{j=0}^{i-1} p_{j,j+1}) \cdot \delta_i) \tag{18}$$

Further, we define a notation u_i as the expected utility. The expected utility value is where node i is treated as the virtual destination. Then, we have:

$$u_{i+1} = v'_{i+1} - \sum_{j=0}^{i}(\prod_{k=0}^{j-1} p_{k,k+1})c_{j,j+1}. \tag{19}$$

To explain our function, we use $r_1 :< s, 1, d >$ with $t_D = 50$ and $v = 50$. As shown in Fig. 24, we have $\delta_s = 5$, $\delta_1 = 5$ and $\lambda = 1$. According to Eq. 17, we obtain the utility $U_{r_1} = 0.9 \cdot 0.6 \cdot (50 - 10) - 2 - 2 \cdot 0.9 = 17.8$. Alternatively, we can also use the step-by-step formula, i.e., Eqs. 18 and 19, to obtain the utility. Since we have $t_D = 50$ and $u_s = 50$, then the expected remaining benefit of node 1 is $v'_1 = p_{s,1} \cdot (v - \lambda \cdot \delta_s) = 0.9 \cdot (50 - 1 \cdot 5) = 40.5$, and the expected utility is $u_1 = v'_1 - c_{s,1} = 40.5 - 2 = 38.5$. Further, the expected remaining benefit of node d is $v'_d = p_{1,d} \cdot (v'_1 - \lambda \cdot p_{s,1} \cdot \delta_1) = 0.6 \cdot (40.5 - 0.9 \cdot 5) = 21.6$. Then, we can get the expected utility of d, $u_d = v'_d - c_{s,1} - c_{1,d} \cdot p_{1,d} = 21.6 - 1.8 - 2 = 17.8$.

4.4 Composite Utility-Based Broadcast in Low Duty-Cycle WSNs

The duty-cycle-aware broadcast scheme was proposed in [38]. Given the node s to broadcast a message starting from time t_0, the forwarding schedule can be represented as

$$S = (1, t_1), (2, t_2), \ldots, (i, t_i), \ldots, (m, t_m),$$

where $(t_0 \le t_1 \le \cdots \le t_m)$. (i, t_i) denotes the ith forwarding, where node i forwards the packet at time t_i. We set S_i as the set of nodes that receives the broadcast message in the i-th forwarding. Then, $| \bigcup_{i=0}^{m} S_i | = n$ where n is the number of nodes. The function $f(|S|, t_m - t_0)$ is proposed to deal with the trade-off between the total message forwarding ($|S|$) and the total latency $(t_m - t_0)$. This paper focuses on a common linear combination, $f(|S|, t_m - t_0) = \alpha|S| + \beta(t_m - t_0)$. This function covers the demands for different applications. For example, if the broadcast message is about an emergency event and of small size, a small α with a large β will ensure that the message is quickly delivered to the whole network, though possibly with higher forwarding costs. On the other hand, if it is not an emergency message, a large α with a small β will work well to save forwarding costs.

Recent work in [39] deals with energy and delay constrained for WSNs. They provide a solution by integrating the reliability, cost, and delay constraints. The main idea is that they use the multi path scheme to solve the problem. The definitions are provided for the new scheme.

5 Additional Discussions

5.1 The Maximum Utility Model

According to the implementation scheme, the destination will return the routing information to the source by using the optimal path. We will only use the optimal path for the retransmission. In [34], the backward method was proposed for the utility-based routing. To simply our result, we also offer the backward method in order to compute the retransmission times. This process is similar with the discussion of preliminary. Then we offer the following equation:

$$u_i = p_i \cdot (u_{i+1} - \lambda \cdot (\prod_{j=i+1}^{n-1} p_j) \cdot \delta_i) - c_i. \tag{20}$$

Theorem 1 is provided for the proof of the equality.

Theorem 1 *Given a journey $J =< R, T >$, the utility with backward method and single journey are the same, $U = u_0$.*

Remember that the optimal route will be returned from the destination to the source. To simplify our analysis, we set $d_i = \lambda \cdot (\prod_{j=i+1}^{n-1} p_j) \cdot \delta_i$ There are two aspects for the retransmission: First, the reliability could be improved. If the retransmission time is r, the new reliability should be $1 - (1 - p_i)^r$. Second, it will produce the additional cost and delay, which is $r \cdot c_i$ and $r \cdot d_i$. Then, we have the utility function $u_i(r)$ with r retries:

$$u_i(r) = (1 - (1 - p_i)^r)(u_i - r \cdot d_i) - r \cdot c_i.$$

To get the optimal utility value, we need to investigate the difference between r and $r + 1$ retries:

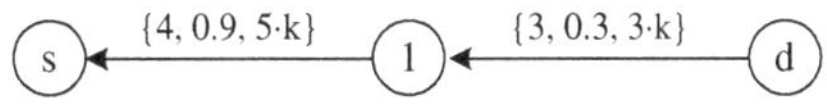

Fig. 25 An example for LDC-TU model

$$\begin{aligned} u_i(r) - u_i(r+1) &= (1-(1-p_i)^r)(u - rd_i) - rc_i \\ &\quad - (1-(1-p_i)^{r+1})(u_i - (r+1)d_i) - (r+1)c_i \\ &= d_i + c_i - (1-p_i)^r(p_i u_i - rp_i d_i + (1-p_i)d_i). \end{aligned} \tag{21}$$

For the analysis of Eq. 21, we know that when $u_i(r) > u_i(r+1)$, r is the amount of retries with maximum utility. The optimal r should exist, since when r increases, $u_i(r) - u_i(r+1)$ is also higher. Note that each time, p_i will be updated by $(1-(1-p_i)^r)$.

We use Fig. 25 as an example. We set $v = 30$ and $t_D = 30$. According to Eq. 1, when $r = 1$, we have $u_1(1) - u_1(2) = -1.14 < 0$ and $u_1(2) - u_1(3) = 0.443 > 0$. Thus, for the link $(1, d)$, the retransmission time is 2. However, for the link $(s, 1)$, we have $u_s(1) - u_s(2) > 0$. Thus, we have $r = 1$. For the implementation of the routing algorithm, when the destination obtains the path after the RREQ process, it will generate the *route response* (RREP) and send it back to the source. The number of retransmissions along the path can be computed during the RREP process.

Thus, we have provided the extension of maximum retransmission. In the following, we offer the extensions for other utility models.

5.2 The Other Utility Models

In this part, we offer the *expected retransmissions* (ExpR). From the aspect of reliability, we know that many applications need an end-to-end delivery guarantee. The following equation is used to obtain the expected retransmissions of link $(i, i+1)$:

$$r_i = 1/p_i.$$

Besides this, we also offer the *delay ratio* (DelayR)-based method, which is:

$$r_i = \delta_{i-1}/\delta_i,$$

where $1 \leq r \leq 1/p$. In DelayR, we allow partial retransmissions without extra delay. Then, we have provided three extensions: MaxU, ExpR, and DelayR.

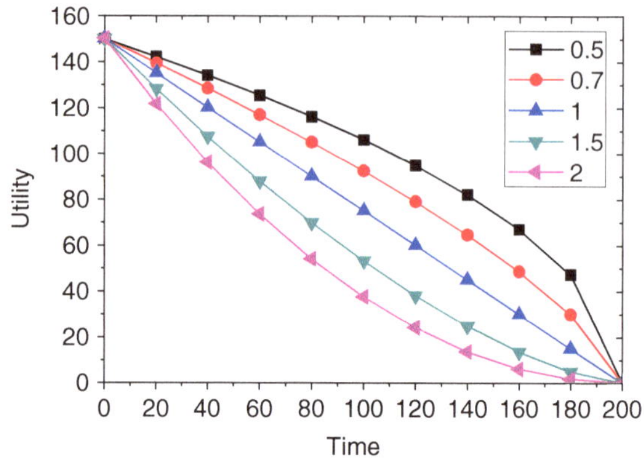

Fig. 26 The extension of timed-benefit function

5.3 The Timed-Benefit with Different Indices

In this part, we will offer the extension with different indices. The proposed benefit function will linearly decrease as the time increases. However, in other cases, the delay distribution may vary upon different cases. In the following, we will provide the extension of timed-benefit function. In the extension of timed-benefit function, the benefit is also zero when t arrives at deadline t_D. Thus, we have:

$$v(t) = v \cdot (1 - t/t_D)^k,$$

where k is the index. It can be varied according to different cases. Figure 26 shows an example of timed-benefit function with different k. Since $0 < 1 - t/t_D \leq 1$, it is a convex function when $k > 1$. It follows the concave function if $0 < k < 1$.

5.4 Comparisons

So far, we have introduced single utility-based routing and composite utility-based routing. The two kinds of utility-based routing are designed to achieve different objectives. Therefore, the comparisons among them become very interesting. Here, we settle them into several parts.

(1) Single utility vs Composite utility: A naive way is to analyze the joint performance (throughput, packet loss rate) or single performance (delay, cost, packet delivery ratio). For example, the paths selected from single utility, such as minimum E2E cost, minimum E2E delay, or maximum E2E reliability, can be compared with the paths chosen from composite utility.

(2) Utility under space view or time domain: As we have mentioned in previous chapters, node deployment serves an important role in the selection of a path. Several

nodes which are close to each other can construct a group, called a community. Different communities have their own utility value. The problem is how to build communities and compare them. Moreover, under different time domains, the variety of the communities is another challenge.

(3) Single copy vs Multiple copies: The utility models we have mentioned can be extended using multiple copies. The utility value is updated, since multiple nodes hold the packets. The E2E cost, E2E delay, and E2E reliability needed to be reconsidered. The composite utility model can be renewed, as well.

5.5 Future work

In this part, we will discuss future work related to routing in WSNs. Currently, lots of research is being conducted in the design of routing protocols. We provide Table 4 for the classification of recent works. As shown in Table 4, works are seldom designed for multicast transmission. Furthermore, the speciality of low duty-cycle WSNs requires a more feasible routing design. Thus, we provide the future works in several directions.

A promising direction lies in multicast routing. Unlike unicast and broadcast, multicast routing facilitates the packets to be forwarded to a group of destinations. In this situation, there are several paths for the different destinations. Reducing the redundant packets and selecting a feasible path becomes more challenging; especially in low duty-cycle WSNs, where many applications require the packets to be delivered before the deadline. This means that the end-to-end delay is very important for the selection of the route.

As we have mentioned before, the utility-based routing approach is designed for a single pair that consists of a source and a destination. In the unicast, we offer the reward if the packet arrives at the destination and satisfies certain requirements. This work can be extended by the consideration of several destinations, or all other nodes. A simple idea is to offer a reward when all of the destinations receive the packet. A more complicated way is that the reward could be divided by the destinations. Moreover, the optimal solution was proposed using a backward method. This method might not be suitable for multiple destinations. However, we still hope to find a feasible solution for this problem. The extensions could also be retransmission strategies. As shown in Fig. 3, there are two methods used upon the failure of a transmission. For the second method, the number of retransmissions can be investigated by achieving some goals: delay, cost, reliability, or a combination of them.

6 Conclusion

In this chapter, we studied the routing issues in WSNs. We covered several topics in the single utility design for different kinds of routing protocols. We discussed utility-based routing, which considers the value of packets. Composition-based methods

Table 4 Routing protocols

Routing	Latency	Energy cost	Reliability	Communication pattern	Types of WSNs
[21]	√			Unicast	Low duty-cycle WSNs
[29]			√	Broadcast	Low duty-cycle WSNs
[17]			√	Unicast	WSNs
[35]		√	√	Unicast	WSNs
[34]		√	√	Unicast	WSNs
[40]		√	√	Unicast	Low duty-cycle WSNs
[36]		√	√	Unicast	WSNs
[41]		√	√	Unicast	WSNs
[42]			√	Unicast	WSNs
[6]			√	Unicast	WSNs
[14]			√	Unicast	Low duty-cycle WSNs
[38]	√			Broadcast	Low duty-cycle WSNs
[22]	√			Unicast	WSNs
[5]			√	Broadcast	Low duty-cycle WSNs
[19]			√	Unicast	WSNs
[30]		√		Unicast	WSNs
[7]		√	√	Broadcast	WSNs
[16]	√	√	√	Broadcast	Low duty-cycle WSNs
[15]			√	Broadcast	WSNs
[43]	√		√	Broadcast	Low duty-cycle WSNs
[37]		√		Unicast	WSNs
[26]	√			Unicast	Low duty-cycle WSNs
[3]	√	√		Multicast	Low duty-cycle WSNs
[44]		√		Multicast	WSNs
[27]	√			Unicast	Low duty-cycle WSNs
[45]		√		Unicast	Low duty-cycle WSNs
[46]			√	Unicast	Low duty-cycle WSNs
[25]	√			Unicast	Low duty-cycle WSNs
[2]		√		Unicast	WSNs
[32]		√		Broadcast	WSNs
[31]		√		Broadcast	WSNs
[33]		√		Broadcast	WSNs
[18]			√	Unicast	WSNs
[23]	√			Unicast	Low duty-cycle WSNs
[39]	√	√		Unicast	WSNs
[24]	√			Unicast	WSNs
[28]	√			Unicast	Low duty-cycle WSNs

were also introduced. For example, delay and cost can jointly be considered in the routing design. In the future, we believe that new routing protocols can be provided for mobility control, as well as other factors involved in our discussion, for different kinds of applications.

Acknowledgments This research was supported in part by NSF grants ECCS 1231461, ECCS 1128209, CNS 1138963, CNS 1065444, and CCF 1028167.

References

1. R. Murty, A. Gosain, M. Tierney, A. Brody, A. Fahad, J. Bers, M. Welsh, Citysense: a vision for an urban-scale wireless networking testbed, in *Proceedings of the IEEE International Conference on Technologies for Homeland Security* (2008)
2. J. Redi, S. Kolek, K. Manning, C. Partridge, R. Rosales-Hain, R. Ramanathan, I. Castineyra, JAVeLEN-An ultra-low energy ad hoc wireless network. Ad Hoc Netw. **6**, 108–126 (2008)
3. A. Ruzzelli, G. O'Hare, R. Jurdak, MERLIN: cross-layer integration of MAC and routing for low duty-cycle sensor networks. Ad Hoc Netw. **6**, 1238–1257 (2008)
4. Z. Zhong, T. Zhu, D. Wang, T. He, Tracking with unreliable node sequences, in *Proceedings of IEEE Infocom* (2009)
5. F. Wang, J. Liu, On reliable broadcast in low duty-cycle wireless sensor networks. IEEE Trans. Mob. Comput. **11**, 767–779 (2011)
6. S. Biswas, R. Morris, ExOR: opportunistic routing in multi-hop wireless networks, in *Proceedings of ACM SIGCOMM* (2005)
7. P. Kyasanur, R. Choudhury, I. Gupta, Smart gossip: an adaptive gossip-based broadcasting service for sensor networks, in *Proceedings of IEEE MASS* (2006)
8. J. Polastre, J. Hill, D. Culler, Versatile low power media access for wireless sensor networks, in *Proceedings of the 2nd International Conference on Embedded Networked Sensor Systems* (Sensys) (2004)
9. M. R. Ahmad, E. Dutkiewicz, X. Huang, A survey of low duty cycle MAC protocols in wireless sensor networks, in *Book Chapter in, Wireless Sensor Network* (2009)
10. W. Ye, J. Heidemann, D. Estrin, An energy-efficient MAC protocol for wireless sensor network, in *Proceedings of IEEE Infocom* (2002)
11. M. Buettner, G. Yee, E. Anderson, R. Han, X-MAC: a short preamble MAC protocol for duty-cycled wireless sensor networks, in *Proceedings of ACM Sensys* (2006)
12. A. Arora, P. Dutta et al., A line in the sand: a wireless sensor network for target detection, classification, and tracking. Comput. Netw. **46**, 605–634 (2004)
13. X. Bai, C. Zhang, D. Xuan, J. Teng, W. Jia, Low-connectivity and full-coverage three dimensional networks, in *Proceedings of ACM Mobihoc* (2009)
14. Y. Gu, T. He, Data forwarding in extremely low duty-cycle sensor networks with unreliable communication links, in *Proceedings of ACM Sensys* (2007)
15. T. Zhu, Z. Zhong, T. He, Z. Zhang, Exploring link correlation for efficient flooding in wireless sensor networks, in *Proceedings of the 7th USENIX Conference on Networked Systems Design and Implementation* (2010)
16. S. Guo, S. Kim, T. Zhu, Y. Gu, T. He, Correlated flooding in low-duty-cycle wireless sensor networks, in *Proceedings of IEEE ICNP* (2011)
17. D. Aguayo, J. Morris, A high-throughput path metric for multi-hop wireless routing, in *Proceedings of ACM Mobicom* (2003)
18. A. Basalamah, S. Kim, S. Guo, T. He, A. Tobe, Y.Basalamah, S. Kim, S. Guo, T. He, Y. Tobe, Link correlation aware opportunistic routing, in *Proceedings of IEEE Infocom* (2012)
19. J. Wang, Y. Liu, M. Li, W. Dong, Y. He, Qof: Towards comprehensive path quality measurement in wireless sensor networks, in *Proceedings of IEEE Infocom* (2011)
20. H. Song, S. Zhu, G. Cao, Svats: a sensor-network-based vehicle anti-theft system, in *Proceedings of IEEE Infocom* (2008)
21. Y. Gu, T. He, M. Lin, J. Xu, Spatiotemporal delay control for low-duty-cycle sensor networks, in *Proceedings of IEEE Real-Time Systems Symposium (RTSS)* (2009)

22. G. Lu, N. Sadagopan, B. Krishnamachari, A. Goel, Delay efficient sleep scheduling in wireless sensor networks, in *Proceedings of IEEE Infocom* (2005)
23. Y. Cao, S. Guo, T. He, Robust multi-pipeline scheduling in low-duty-cycle wireless sensor networks, in *Proceedings of IEEE Infocom* (2012)
24. Q. Zhang, Y. Gu, L. Gu, Q. Cao, T. He, Collaborative scheduling in highly dynamic environments using error inference, in *IEEE Seventh International Conference on Mobile Ad-hoc and Sensor Networks (MSN)* (2011), pp. 105–114
25. P. Basu, C. Chau, Opportunistic forwarding in wireless networks with duty cycling, in *Proceedings of the 3rd ACM Workshop on Challenged Networks* (2008)
26. C. Chau, P. Basu, Analysis of latency of stateless opportunistic forwarding in intermittently connected networks. IEEE/ACM Trans. Netw. (TON) **19**, 1111–1124 (2011)
27. D. Kim, M. Liu, Optimal stochastic routing in low duty-cycled wireless sensor networks, in *Proceedings of the 4th Annual International Conference on Wireless Internet* (2008)
28. Y. Gu, T. He, Bounding communication delay in energy harvesting sensor networks, in *IEEE 30th International Conference on Distributed Computing Systems (ICDCS)* (2010), pp. 837–847
29. S. Guo, Y. Gu, B. Jiang, T. He, Opportunistic flooding in low-duty-cycle wireless sensor networks with unreliable links, in *Proceedings of ACM Mobicom* (2009)
30. O. Chipara, Z. He, G. Xing, Q. Chen, X. Wang, C. Lu, J. Stankovic, T. Abdelzaher, Real-time power-aware routing in sensor networks, in *Proceedings of IEEE IWQoS* (2006)
31. J. Wu, Extended dominating-set-based routing in ad hoc wireless networks with unidirectional links. IEEE Trans. Parallel Distrib. Syst. **13**, 866–881 (2002)
32. J. Wu, B. Wu, I. Stojmenovic, Power-aware broadcasting and activity scheduling in ad hoc wireless networks using connected dominating sets. Wirel. Commun. Mobile Comput. **3**, 425–438 (2003)
33. J. Wu, H. Li, On calculating connected dominating set for efficient routing in ad hoc wireless networks, in *Proceedings of the 3rd International Workshop on Discrete Algorithms and Methods for Mobile Computing and, Communications* (1999)
34. M. Lu, J. Wu, Social welfare based routing in ad hoc networks, in *Proceedings of IEEE ICPP* (2006)
35. J. Wu, M. Lu, F. Li, Utility-based opportunistic routing in multi-hop wireless networks, in *Proceedings of IEEE ICDCS* (2008)
36. M. Lu, F. Li, J. Wu, Efficient opportunistic routing in utility-based ad hoc networks. IEEE Trans. Reliab. **58**, 485–495 (2009)
37. G. Ghidini, S. Das, An energy-efficient markov chain-based randomized duty cycling scheme for wireless sensor networks, in *Proceedings of IEEE ICDCS* (2011)
38. F. Wang, J. Liu, Duty-cycle-aware broadcast in wireless sensor networks, in *Proceedings of IEEE Infocom* (2009)
39. S. Bai, W. Zhang, G. Xue, J. Tang, C. Wang, Dear: delay-bounded energy-constrained adaptive routing in wireless sensor networks, in *Proceedings of IEEE Infocom* (2012)
40. Y. Gu, T. He, Dynamic switching-based data forwarding for low-duty-cycle wireless sensor networks. IEEE Trans. Mob. Comput. **10**, 1741–1754 (2010)
41. M. Lu, J. Wu, Utility-based data-gathering in wireless sensor networks with unstable links. J. Distrib. Comput. Netw. 13–24 (2008)
42. Q. Cao, T. He, L. Fang, T. Abdelzaher, J. Stankovic, S. Son, Efficiency centric communication model for wireless sensor networks, in *Proceedings of IEEE Infocom* (2006)
43. S. Lai, B. Ravindran, On multihop broadcast over adaptively duty-cycled wireless sensor networks. Distrib. Comput. Sens. Syst. 158–171 (2010)
44. C. Feng, W. Heinzelman, Rbmulticast: Receiver based multicast for wireless sensor networks, in *Proceedings of IEEE WCNC* (2009)
45. W. Pak, K. Cho, S. Bahk, Energy efficient routing protocol for wireless sensor networks with ultra low duty cycle, in *Proceedings of IEEE 20th International Symposium on Personal, Indoor and Mobile Radio, Communications* (2009)
46. F. Yang, I. Augé-Blum, Delivery ratio-maximized wakeup scheduling for ultra-low duty-cycled WSNs under real-time constraints. Comput. Netw. **55**, 497–513 (2011)

Part IV
Topology and Mobility Management

Chapter 9
Topology Management Techniques for Tolerating Node Failure

Mohamed Younis, Sookyoung Lee, Izzet Fatih Senturk and Kemal Akkaya

Abstract In Wireless Sensor Networks (WSNs) sensor nodes often operate unattended in a collaborative manner to perform some tasks. In many applications, the network is deployed in harsh environments such as battlefield where the nodes are susceptible to damage. In addition, nodes may fail due to energy depletion and breakdown in the onboard electronics. The failure of nodes may have major consequences. First, some areas may be left uncovered. Second, the fidelity of the collected data gets degraded. And finally, the network may get partitioned into disjoint segments. In particular, losing network connectivity has a very negative effect on the applications since it prevents data exchange and hinders coordination among some nodes. Therefore, restoring the overall network connectivity with the least resource overhead and performance impact is very crucial. This chapter focuses on network topology management techniques for tolerating node failures. It analyzes the effects of node failure on network connectivity in WSNs, categorizes recently published recovery schemes, and outlines related open issues.

M. Younis (✉) · S. Lee
University of Maryland at Baltimore County, Baltimore, USA
e-mail: younis@cs.umbc.edu

S. Lee
e-mail: slee22@umbc.edu

I. F. Senturk · K. Akkaya
Southern Illinois University Carbondale, Carbondale, USA
e-mail: isenturk@cs.siu.edu

K. Akkaya
e-mail: kemal@cs.siu.edu

H. M. Ammari (ed.), *The Art of Wireless Sensor Networks*,
Signals and Communication Technology, DOI: 10.1007/978-3-642-40009-4_9,

© Springer-Verlag Berlin Heidelberg 2014

1 Introduction

The growing interest in the applications of Wireless Sensor Networks (WSNs) has motivated a lot of research work during recent years [1, 2]. For some of these applications, such as space exploration, coastal and border protection, combat field reconnaissance and search and rescue, it is envisioned that a set of mobile sensor nodes will be employed to collaboratively monitor an area of interest and track certain events or phenomena [3–9]. By getting these sensors to operate unattended in harsh environments, it would be possible to avoid the risk to human life and decrease the cost of the application.

Since a sensor node is typically constrained in its energy, computation and communication resources, a large set of sensors are involved to ensure area coverage and increase the fidelity of the collected data. Upon their deployment, nodes are expected to stay reachable to each other and form a network. Network connectivity enables nodes to coordinate their action while performing a task, and to forward their readings to in-situ users. In fact in many setups, such as a disaster management application, nodes need to collaborate with each other in order to effectively search for survivors, assess damage and identify safe escape paths. To enable such interactions, nodes need to stay reachable to each other. Therefore, the inter-sensor connectivity has a significant impact on the effectiveness of WSNs and should be sustained all the time.

However, a sudden failure of a node can cause a disruption to the network operation. A node may fail due to an external damage inflicted by the inhospitable surroundings or simply because of hardware malfunction. The loss of a node can break communication paths in the network and deem some of its neighbors unreachable. Moreover, WSNs operating in a harsh environment may suffer from large-scale damage which partitions the network into disjoint segments. For example in a battle field, parts of the deployment area may be attacked by explosives, and thus a set of sensor nodes in the vicinity would be destroyed and the surviving sensor nodes are split into disjoint segments. Restoring inter-segment connectivity would be crucial so that the WSN becomes operational again.

In this chapter we highlight the challenges that node failure introduces to the operation of WSNs and provide taxonomy of mitigation approaches. We categorize fault-tolerance schemes proposed in the literature according to the goals of the recovery process and the tolerance methodology. Since this process is in general a form of topology management (i.e., often leads to changes in the network topology parameters), we start with an overview of contemporary techniques and objective of topology management in wireless networks.

2 Topology Management in Wireless Networks

Networks require monitoring and maintenance whether they are wired or wireless. The service which provides these tasks is called network management. Network management includes five functional areas as identified by ISO: configuration management, fault management, security management, performance management and accounting management [6]. However, the unique requirements of wireless networks such as cellular networks, Mobile Ad Hoc Networks (MANETs), or WSNs have inspired a new functional area, namely *topology management*. This term is sometimes used interchangeably with topology control and refers to the management of parameters such as transmission power of the nodes, state of the nodes, role of the nodes, connectivity of the network, etc. By modifying these parameters, one can change the topology of the network and the state of the nodes. Note that this stage naturally follows the creation of topology. Before describing the topology management methodologies for WSNs, we would like to provide brief overview of the methodologies in Cellular networks and MANETs.

2.1 Topology Management in Cellular Networks and MANETs

In cellular networks, topology management is done in the form of location management. To establish the connection for an incoming call, cellular system must know the cell that the user currently resides in. Thus, the system must keep track of the users' locations to determine the cell to which the call is to be forwarded. The process of identifying the exact cell is called *paging* and the method to keep track of the user locations is called *location update*. Updating location whenever a user transitions to a new cell may save the paging delay; however it introduces extra overhead not only for the cellular system, in terms of processing and radio bandwidth, but also for the mobile device in terms of power consumption especially if the incoming call rate is low. On the other hand, leaving out the location updates completely necessitates a system wide paging which requires a broadcast to all cells. The incurred cost of location update and the paging can be balanced by location management which is a form of topology management in cellular systems. For this purpose, static and dynamic location management algorithms have been proposed [10, 11].

In Mobile Ad Hoc Networks (MANETs), wireless links can be subject to unpredictable node movements which lead to frequent link failures and arbitrary topology changes. Thus, maintaining the network connectivity can be challenging in MANETs. In addition, most of the time only a subset of the wireless links is indeed necessary for forming an efficient connected network. To avoid the excessive messaging overhead for frequent state update while efficiently providing network services, e.g., multicast, redundant and unnecessary topology information should be ignored. This implies the necessity of distributed self-configuring topology management techniques to track inter-node connectivity. To this end, topology management schemes in MANETs have been studied in two main categories: power control and hierarchical topology organization [12]. In power control schemes, nodes adaptively set their transmission ranges by adjusting the transmission power or turn off their radios periodically [13–15]. Hierarchical topology organization schemes structure the topology hierarchically by grouping the nodes in clusters. Selecting the optimal node to be a cluster head, setting the diameter of a cluster in terms of the number of hops from the cluster head, determining the optimal number of clusters to be formed, and maintaining the cluster membership in a highly dynamic MANET are deemed the main research issues [16–21]. Most of the work on topology management of MANET has focused on dealing with frequent loss of communication links due to mobility, with significantly less attention to network partitioning conditions. To interconnect disjoint network segments some nodes are equipped with long range or satellite transceivers in order to overcome the lack of line-of-sight links [22].

2.2 Topology Management Methodologies in WSNs

The primary objective of the topology management techniques in WSNs is achieving sustainable network coverage while maintaining network connectivity and conserving energy. Topology management in WSNs can be done through deterministic node placement or performed autonomously after random deployment given the limited human intervention [23]. Topology management techniques are employed to track the status of communication links among the nodes, to conserve energy by switching off some of the nodes without degrading network coverage and connectivity, to support a hierarchical task assignment for data aggregation, to balance the load on existing nodes and links, to provide scalability by minimizing medium access collision and limiting overhead. Topology management can also be considered together with configuration and fault management [24], as will be elaborated later in the chapter.

Existing topology management techniques/algorithms for WSNs can be classified into the following five categories:

- **Node Discovery**: Detecting the nodes and their locations is an essential function in a WSN not only after the initial deployment but also for integrating newly added nodes. The scope of node discovery is subject to certain trade-offs based on the application specific goals. For instance, for large networks resource savings in terms of energy and bandwidth can be achieved by not sharing some of the topology details that are deemed unnecessary for certain parts of the network [25].
- **Sleep Cycle Management**: By utilizing redundancy, some of the nodes in a WSN can be turned off as a means for saving energy. In addition, the number of transmitted messages will decline, causing less interference and fewer failed transmission attempts. Determining the sleep schedule while sustaining full area coverage and strong network connectivity is a popular topology management optimization that received quite an attention from the research community [26–29].
- **Clustering**: To achieve scalability and energy efficiency, nodes of a WSN may be grouped to form a hierarchical topology. In this way, nodes can send their reading to a cluster-head which in turn aggregates and forwards the data to the sink node after eliminating redundant data [21]. Although the failure of cluster-head often requires re-clustering, some approaches have provisioned the topology adjustment by associating a primary and backup cluster-heads for each sensor node [30–32].

- **Power Control**: The transmission range reflects the maximum distance at which a receiver can stay from the sender. The longer the range is, the higher the power consumption would be. Many of the advanced radios allow programmable transmission power so that a node can avoid consuming excessive energy in reaching nearby receivers. Low power transmission can also reduce interference and boost the network throughput. However, the use of low transmission power limits the network connectivity since nodes would have fewer directly reachable neighbors. Many power control optimization techniques have been proposed to exploit such trade-off to appropriately manage the WSN topology [33–35].
- **Movement Control**: Node mobility has been exploited as a means for optimizing the network performance. For example, in [36–39], the focus is on prolonging the network lifetime by reducing energy consumed by stationary sensors, whereas in [40] and [41] other metrics have been targeted. In addition, mobile relays with more capabilities than sensors are used as data forwarders in order to prolong the lifetime of a network of stationary sensors [42, 43] or to link disjoint batches of nodes [44–46].

3 Classification of Failure and Tolerance Techniques

3.1 Node Failure Models

A sensor failure implies a loss of functionality caused by damage, energy exhaustion or component breakdown. The impact of the failure reflects the negative consequences on the network operation and application. In the context of WSN, a sensor failure affects network connectivity and coverage. In WSNs, node failure can be classified into two categories; a single node failure and simultaneous multiple failures.

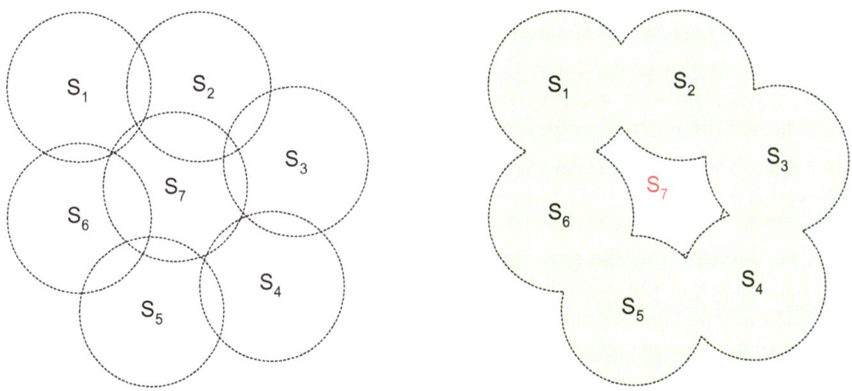

Fig. 1 Assuming a disc coverage model, the failure of S_7 causes a coverage gap in the network

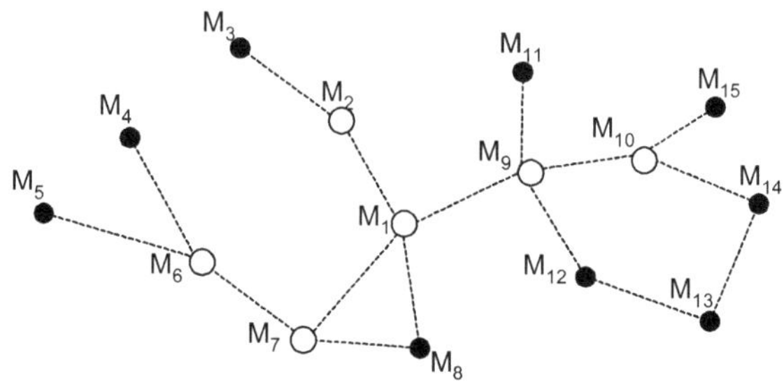

Fig. 2 Example of a single node failure in WSN; *white* nodes are cut vertices and the failure of those nodes thus causes a network to one or multiple network partitions, meanwhile *black* nodes are not essential to reaching other nodes and their failure does not affect network partition

A single node failure indicates that one node fails at a time. The type of failure can be simply detected using local heartbeat messages. Unless there is overlap in coverage, the failed node will leave out part of the area unmonitored as shown in Fig. 1. On the other hand, the node position within the network topology determines its criticality to connectivity. Considering the topology as a graph, a leaf node does not serve on the path between any two nodes and thus would not be critical to connectivity. Node M_{15} in Fig. 2 is an example leaf node. Some nodes like M_{13} are also not critical to network connectivity since its neighbors M_{12} and M_{14} have a path between them that does not include M_{13}. However, some nodes act as cut-vertices and when any of them fails the network gets partitioned into disjoint blocks. A cut-vertex in a graph is a vertex that if removed splits the graph into multiple connected sub-graphs. In other words, a cut-vertex node in the network plays the role of a gateway between two sub-networks. In Fig. 2, nodes M_1, M_2, M_6, M_7, M_9 and M_{10} are cut-vertices and are considered critical for connectivity. The failure of a critical node thus negatively affects the network operation and may deem the network useless. Schemes for mitigating the failure of a single node include provisioning bi-connectivity to avoid the presence of cut-vertices in the topology [47] and real-time connectivity restoration orchestrated by the neighbor of the failed node [48–51].

Classifying nodes based on criticality can be performed through depth-first-search algorithm, which is a centralized scheme that needs knowledge of the entire network topology. However, in highly autonomous networks and in networks involving mobile sensor nodes centralized management schemes are not favored due to the messaging overhead necessary for maintaining a network-wide state. This has motivated the use of probabilistic mechanisms for assessing the criticality of nodes. Probabilistic schemes generally trade off the accuracy of the assessment for the reduced messaging overhead [51–53]. The idea is to use *k*-hop neighbor information rather than the entire network. Obviously, the higher the value of *k* the better the accuracy will be. It has been shown that the probability of missing a cut-vertex is zero while a very high percentage of the picked nodes are really cut-vertices [51, 53]. Using 2-hop information, it was shown that the accuracy can reach 90 % [51].

The second type of node failure model is based on multiple simultaneous failures. WSNs operating in a harsh environment may be subject to damage that can be so significant in a part of the covered area that the network gets partitioned into

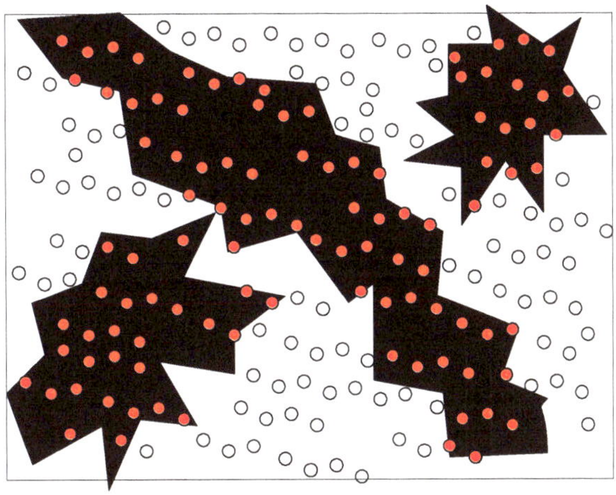

Fig. 3 Illustration of a segmented WSN due to damage; *solid circles* indicate failed sensors, *white ones* are operational nodes

disjoint segments. For example in a combat field, parts of the deployment area may be bombed, destroying the sensor nodes in the vicinity. Figure 3 shows an articulation, where the dark areas represent the extent of the damage, after which the surviving nodes are partitioned into disjoint segments due to the loss of connectivity. The simultaneous failure of multiple nodes is very challenging, not only in the recovery process but also in determining the scope of the failure. Repairing a network after a single node failure is neither able to handle simultaneous failure of multiple nodes, nor analyze the scope and recover from large-scale network damage. In addition, these schemes rely on maintaining 1 or 2-hop neighbors list in order for the recovery process to converge. Significant network damage may involve many hops, which mandates maintaining larger local state information that spans several hops, and imposes prohibitive overhead.

3.2 Taxonomy of Fault-Tolerance Techniques

The mechanisms for tolerating of node failure in WSNs can be categorized based on the following attributes:

- *Concerns a node failure may raise*: Generally coverage and connectivity are at stake when a node fails. The focus of the fault-tolerance differs based on whether coverage loss is to be mitigated or strong connectivity is to be maintained. Some techniques care for both coverage and connectivity, with one of them being as the primary concern and the second becoming a constraint, or a secondary concern while performing the recovery. Given the scope of this chapter, we shall focus more on connectivity-related fault-tolerance techniques, for which connectivity is a primary or a secondary concern.

- *The scope of the recovery*: Localized schemes limit the involvements of the nodes in the recovery and thus reduce the impact that a node failure and the associate recovery process have on the normal network operation. Obviously, in the presence of a resource-rich node that is aware of the state of the entire network, it can devise an optimized recovery procedure with the most suitable set of actions.
- *Recovery approach*: Unattended setups would favor autonomous recovery from a node failure that is orchestrated among the nodes in a distributed manner. Centralized recovery fits applications in which in-situ base-station or a remote command node are accessible and can determine the course of required actions.
- *Methodology*: Tolerance of a node failure can simply be reactive in nature, meaning that a recovery action is taken after a failure is detected. Another option is to provision the tolerance at time of network setup in order to prevent a failure from degrading connectivity and/or coverage. Hybrid strategies also exist.
- *Recovery Objective*: In addition to mitigating the effect of a failure, the recovery process may have additional objectives to be achieved. These objectives often shape up the recovery algorithm. Examples of these objectives include minimizing the overhead, limiting the involvement of healthy nodes, achieving certain level of quality of service (QoS), etc.

Figure 4 summarizes these classifications. The balance of this chapter summarizes the state of the art on tolerance of node failure and further highlights these attributes. The discussion will be structured based on the node failure model and the concern that triggers the recovery.

4 Tolerance of Coverage Loss Due to Node Failures

As pointed out earlier and illustrated by Fig. 1, the failure of a sensor node S_7 may result in coverage loss in the vicinity, i.e., a hole in the network coverage, unless some healthy sensors are located within the sensing radius of S_7. Since coverage is usually considered a QoS metric, the presence of holes in the network coverage needs to be avoided or at least their sizes should be minimized. In theory, failure of up to $k-1$ nodes can be tolerated if a region is within the sensing radius of at least k nodes [54, 55]. Even though the probability of coverage loss can be minimized through proactive methods such as redundant node deployment during the initial setup, we focus on reactive schemes that can be employed after the failure takes place. Deploying a substitute for the failed nodes is the most intuitive means for regaining lost coverage, where the failed node is simply replaced by a spare. However, in some setups augmenting the coverage by deploying additional nodes is not feasible and repositioning of healthy nodes is the only option. In the balance of this section, we discuss example techniques for minimizing the coverage loss.

Drougas and Kalogeraki [56] have proposed DÉCOR, a distributed DEpendable COverage Restoration algorithm, to minimize the number of nodes to be deployed to achieve/restore k-coverage in a region. Two cell models are considered. In the first,

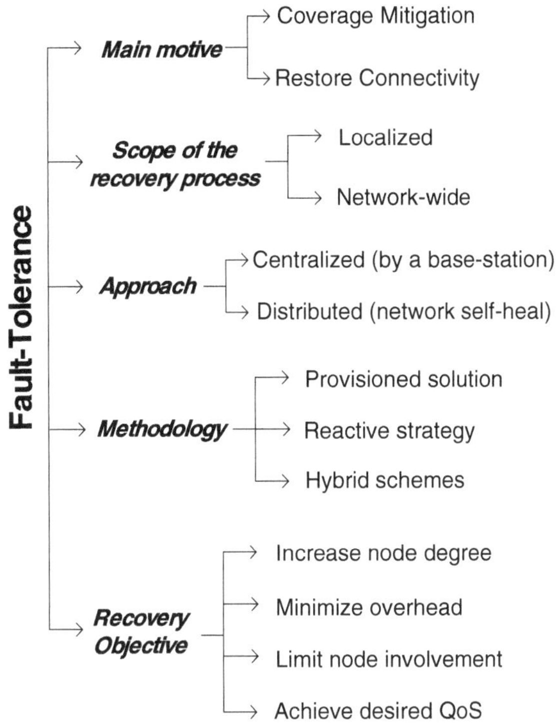

Fig. 4 Classification of the various fault-tolerance mechanisms for wireless sensor networks

the region is partitioned into square-shaped cells. A node is elected within each cell to be responsible for where coverage is lacking within its cell. The second model employs Voronoi tessellation to define cells based on the distance between existing sensors. The node of a Voronoi cell is responsible for identifying spots within its cell that are not k-covered. Assessing the coverage degree requires finding out the intersection of multiple circles, which can be very complex for large value of k. To reduce the coverage assessment complexity, discrepancy theory is employed and a set of points within a cell are used to find out where coverage is lacking. The two approaches are illustrated in Fig. 5a, b, respectively. A similar approach is proposed in [57] where the problem is formulated as a disc covering optimization and a heuristic solution is found with a bounded error.

As mentioned above, repositioning the healthy nodes is the most popular methodology for mitigating coverage loss. For networks with stationary sensors, the use of robots has been pursued in relocating sensors from area with redundant coverage to uncovered spots [58–61]. Nonetheless, the bulk of published schemes have exploited node mobility to restructure the network topology. For example, Ganeriwal et al. [62] deal with the potential of coverage loss before it happens. If the remaining energy of a node which exclusively monitors an area drops below the certain level, it sends a panic message to nodes in area that is known to have coverage overlap. A response to

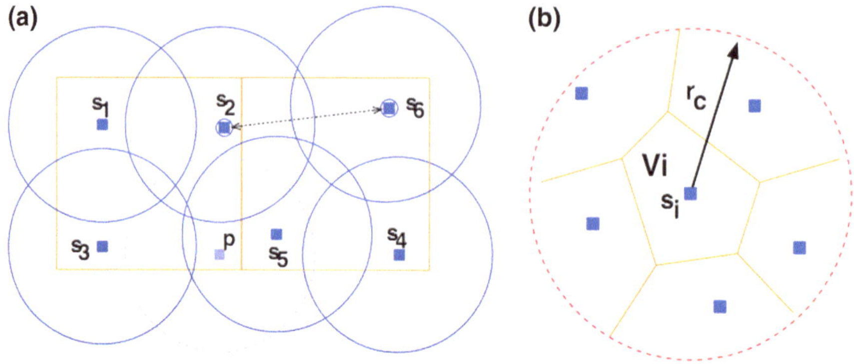

Fig. 5 Illustrating the two cell models used for assessing coverage; **a** a cell is a square where the nodes elect a leader, e.g., S_6, for assessing the coverage at the various points within the cell in collaboration with the leaders of neighboring cells, e.g., S_2; **b** a cell is defined based on Voronoi decomposition of the area based on the location of the individual nodes, e.g., S_i, with respect to neighbors, i.e., those with the commination range r_c. The figure is taken from [66]

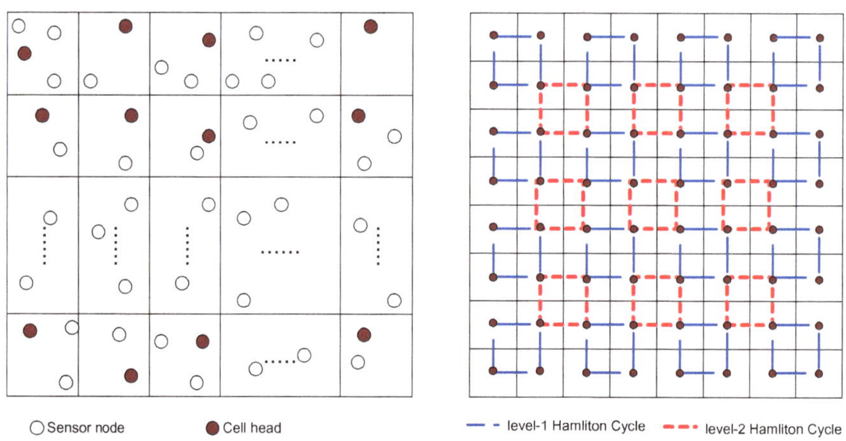

Fig. 6 Illustrating the virtual grid model, where the *dark* and *hallow circles* represent the cell heads and regular sensor nodes, respectively. Hamilton cycles are formed locally among the cell heads and aggregated in a hierarchical structure (only two levels are shown)

the panic is made by volunteers and the panicking node identifies the volunteer with the most energy resource and least motion overhead. Before volunteering, a node should ensure that its departure will not leave a hole in coverage.

To detect and mitigate node failure, Wu and Jiang pursue a hierarchical model [63]. The monitored area is modeled as a grid with cell sizes that ensure connectivity between nodes in adjacent cells. Each cell elects one node to be a head and cell-based Hamilton cycles are formed at different levels using cell heads as representatives. The lowest level involves 4 adjacent cells. One head in a lower cycle becomes a member of the next tier cycle and so on. Figure 6 illustrates the idea. When coverage is degraded in a cell, its head will negotiate with the fellow heads within the Hamilton cycle on getting a node to move in and fill the coverage gap. If not possible at a low cycle level, the request is elevated to the next level and so on. If any of the heads on a cycle dies, the Hamilton cycle becomes a Hamilton path and a spare node is moved to the vacant area along the path to restore the cycle and the coverage. Asim et al. [64] also model the network as a grid and designate a node per each cell for coverage maintenance by relocating nodes in case of a failure. A publish/subscribe abstraction is pursued for finding redundant nodes where the cells with redundant nodes are publishers and the cells that require more nodes are subscribers. Direct and cascaded movements are offered based on where the failure takes place and where the substitute node is located.

In [65], Sahoo et al. have presented a distributed coverage restoration algorithm which strives to reduce the distance that a node needs to travel. The scope of the motion is limited to 1-hop neighbors. The recovery problem is modeled using vector algebra. A node determines boundary gaps based on the intersection of its coverage range with that of its neighbors. A node then determines the movement direction using polygon laws of vector addition. The approach resembles the Grid-Quorum algorithm of Wang et al. [66], the self-spreading algorithm of Heo and Varshney [67], which model the node's motion using electrostatic interaction based on Coulomb's law between charges. Meanwhile, Nguyen et al. [68] assumes a hybrid network model where both stationary and mobile nodes exist. Mobile nodes are employed to fill coverage holes caused by the failure of stationary nodes. Like the work of Wang et al. [66] a virtual grid is assumed and the mobile nodes are relocated based on energy metric. However, the authors argue that minimizing the total energy is not effective and maximizing the remaining energy per node is more important to the network lifetime.

Rather than just re-acting the node failures, some approaches have exploited node mobility to improve the coverage after the network deployment. For instance, Wang et al. [69] utilize Voronoi diagrams of each sensor to decide where to move them so that they are dispersed as evenly as possible in the region. The goal is to maximize the coverage within the shortest time duration and with minimal overhead in terms of travel distances and message exchange. Another notable approach for the same problem followed the idea of two dimensional scanning of clustered networks [70]. The approach, which is called SMART, adopts a popular scheme for balancing load among nodes in parallel processing architectures by assigning equal number of tasks to each processor. This idea is used by creating a grid-based WSN where each cluster is represented by a square cell. SMART was shown to use the minimum number of moves when the network is converged.

Some work considered how a substitute for the failed node can be efficiently founded. For example, Li et al. [71] have focused on optimizing the discovery of a redundant node. They form an information mesh that facilitates the dissemination of a request by designating proxies that maintain a list of available spares in their proximity. Meanwhile, the ZONER approach [72] opts to overcome the presence of obstacles in the field that may hinder the discovery of redundant nodes and the repositioning of them to substitute the failed ones. The idea is to use limited flooding within a zone and employ face routing [73], which is a popular geographical routing scheme for MANET, in order to move the redundant node to the position of the faulty one. Again to avoid overburdening a single node, a shifting strategy is employed where the closest node S_1 on the route to the faulty node will move, node S_2 will then replace S_1, and so on until the picked redundant node S_n relocates to the position of S_{n-1}.

Some of the published work paid significant attention to the impact of the failure and the scope of the recovery. For example, the focus of [74] is on assessing the effect of the failed node and deciding on whether it is necessary to replace it. Three metrics are factored in, namely, the increased per-node energy for data dissemination, coverage loss at the network level, and the local level, i.e., in the vicinity of the failure. Meanwhile, the approach of Sekhar et al. [75] is to use the neighbor of the failed node in mitigating the coverage loss. Nodes determine how far they can move towards the failed node based on the coverage overlap in their vicinity. Four heuristics have been proposed for determining how much each neighbor of the failed node should contribute, i.e., how much distance it travels. These heuristics differ in the optimization metrics. A cascaded version is also proposed where the scope of the relocation includes nodes that are multi-hop away from the failed node.

The bulk of published work assumes that all nodes have the same sensing capabilities. Unlike that work, Kasinathan and Younis [76] address the mitigation of coverage loss in networks with heterogeneous sensor nodes. They have proposed a distributed algorithm which pursues a combination of proactive and reactive recovery strategies. Again, their Coverage loss Mitigation Algorithm after Node failure (CoMAN) avoids the deployment of redundant nodes and just re-organizes the available healthy nodes. Every node maintains a 2-hop network state, i.e., those 1 and 2-hop neighbors. The collected information about a neighbor includes its coverage capabilities, criticality to the network connectivity and proximity, i.e., how far it is. CoMAN performs some pre-failure planning. Based on the neighbors list, each node "A" identifies all nodes that share some of its sensing capabilities and form a list of Possible Replaceable Nodes (PRN). Each entry in the PRN list includes a set of nodes whose collective sensing capabilities match or surpass that of "A." Node "A" shares its PRN list with its direct neighbors in order to orchestrate a recovery if node "A" fails. In order to minimize the effect on the network performance, the picked set of backup nodes, i.e., selected PRN, is chosen such that repositioning these nodes causes the least coverage and connectivity impact on the network. In addition, the distance that these backups need to travel to replace "A" is also taken into consideration to limit the recovery overhead. Figure 7 illustrates how CoMAN performs the recovery through a detailed example.

A general classification for all approaches summarized in this section is that coverage is the only metric considered in the recovery process and connectivity is assumed not to be an issue. In other words, for a disc coverage model the communication range was assumed to be significantly longer than the coverage range. As we will discuss in the next sections, some work cared about connectivity in addition to coverage or some simply did not factor in coverage at all and focused on maintaining/restoring strong network connectivity.

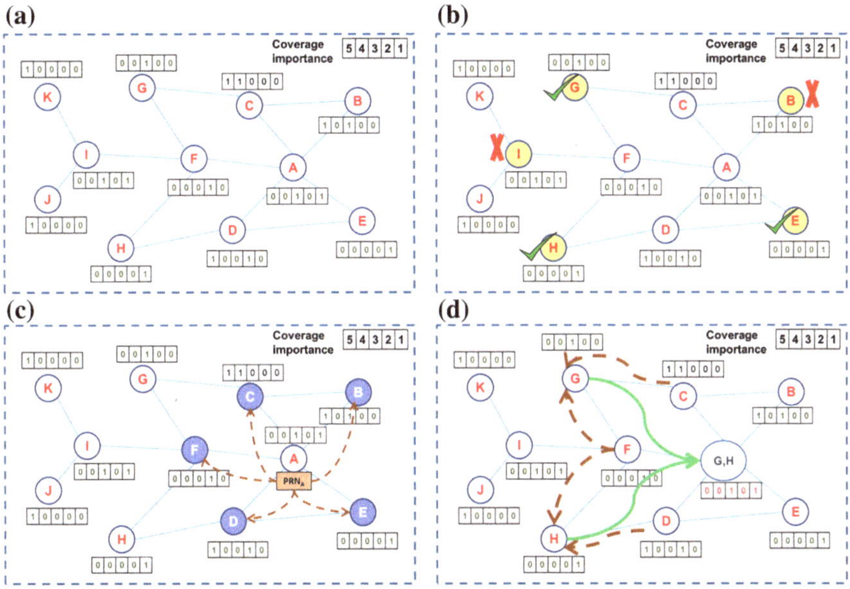

Fig. 7 Illustrating the operation of CoMAN [83]; **a** the sensing capabilities of the individual nodes are modeled using a binary string. Each node has a subset of the sensing capabilities—1's and 0's annotated next to each node represent the presence or absence of a particular capability, respectively. The importance of each sensing capability is inversely proportional to the capability number, i.e., capability #1 the most critical sensing capability, **b** each node determines its PRN list. For node *A*, node *I* is a cut vertex is thus disqualified as a backup given the negative impact of moving it. Meanwhile, node *B* has the capability 1 which is more important than all capabilities of *A*, and is therefore excluded as a viable recovery option. Hence, the PRNs of node *A* are {E, G}, and {G, H}; **c** node *A* exchanges its PRN list with its direct neighbors; **d** after the failure of *A*, the direct neighbors determine the best recovery option out of the entries in the PRN of *A*, which will be {E, G} since node *E* is closer to *A* than node *H*. The picked backups, i.e., nodes E and G, will relocate to the position of *A*

5 Connectivity-Centric Recovery from a Single Node Failure

In order to tolerate a node failure that causes network partitioning, two methodologies can be identified: (i) pre-cautionary and (ii) real-time restoration (repair). The pre-cautionary methodology strives to provision fault-tolerance in the network topology both at setup and during normal operation. The idea is to establish a *k*-connected topology such that every node can reach other nodes over at least *k* node independent paths. Such an arrangement will allow the network to seamlessly tolerate the failure of up to $k - 1$ nodes [47, 77]. For coverage, redundant nodes with overlapping range are deployed to achieve *k*-coverage property and withstand the failure of $k - 1$ nodes without suffering a hole in network coverage [54, 55]. However, when connectivity is also considered, the failure of even one among the *k* nodes can cause network partitioning regardless of the sustained coverage. In this and next sections, we direct our attention to tolerance techniques that have the network connectivity as a primary or secondary concern.

5.1 Provisioned Recovery Schemes

The popular proactive strategy for preserving the network connectivity in the presence of a faulty node is to carefully place redundant sensor nodes during the initial deployment of a WSN. The idea is to provide more than one routing path between any pair of sensors in the network. The route alternatives should also be node disjoint so that the failure of a single node will not break all viable routes. In this subsection, we summarize some of the published techniques.

In general, optimal node placement is a very challenging problem that has been proven to be NP-Hard for most of the formulations of sensor deployment [78–81]. To tackle such complexity, several heuristics have been proposed to find sub-optimal solutions [24, 78–87]. However, the context of these optimization strategies highly depends on how to assess the quality of candidate positions, which is based on a structural quality metric such as inter-node proximity, coverage, network connectivity, etc. Published work on sensor placement can be grouped into two categories. The first considers unconstrained setups and tries to just establish connectivity between end points [78–80]. In the second category either additional performance objectives are targeted [82, 83] or higher degrees of connectivity are to be achieved [24, 84, 85, 88]. Given the scope of this section, we shall focus on the second category.

Although provisioning k-connectivity enables the network to tolerate the failure of up to $k-1$ consecutive node failures without suffering partitioning, bi-connectivity has been the most popular goal given the complexity of the node placement problem and the increased node count in achieving high level of connectivity will involve. Overall, published proactive schemes strive to achieve the desired level of fault resilience, i.e., level of connectivity, using the least node count. A variant of the problem is considered in [24], in which source nodes are distinguished from relaying nodes on the route and k-connectivity is only applied to the inter-relay topology. Such variant is named *partial k-connectivity.* In some publications, k-connectivity is provided as byproduct of determining a connected dominating set to obtain a robust backbone [88]. In addition, some work assumed that the relays possess more capabilities than the sensor nodes [85].

The goal of Tang et al. [84] and Hao et al. [85] is to place the minimum number of relay nodes such that each sensor is connected to at least two relays and the inter-relay network is 2-connected. As a result, the failure of a relay node or a sensor node does not affect the overall network connectivity. Tang et al. divide the area into cells and find a position for a relay so that it becomes connected to all sensors in a cell and also to other relays in neighboring cells. The work is further extended in [85] by formulating the same placement problem as *2-Connected Relay Node Double Cover* (2CRNDC), which finds the fewest number of locations for placing relay nodes, so

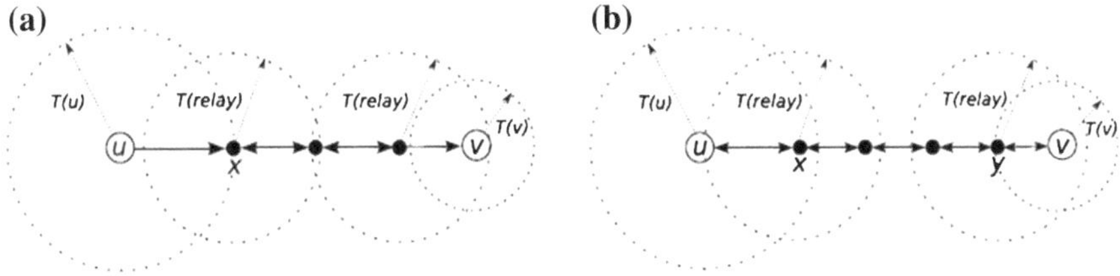

Fig. 8 Examples of a one-way (**a**) and two-way (**b**) steinerized path. Between two sensor nodes *u* and *v*, the least number of relay nodes (*black circles*) are placed to establish a one-way or two-way path. The figure is from [24]

that each sensor is covered by at least two relays and the group of relay nodes is bi-connected. This problem can be reduced to the well-known *Minimum Geometric Disc Cover problem* which is NP-complete and they thus present a polynomial time approximation algorithm. The algorithm computes a possible position *p* of a relay and *C*(*p*) which is a set of sensor nodes covered by a relay node locating at *p*. Then, the algorithm simply identifies positions that cover the maximum number of sensors, at which relays are virtually placed. By analyzing the inter-relay connectivity, the relays with most coverage are switched from virtual to real, in order to form a 2-connected graph.

Meanwhile, Han et al. [24] construct $k(k \geq 1)$ disjoint paths between any two sensor nodes. Unlike [84, 85], they consider a heterogeneous WSN where sensors have different transmission radii and opt to deploy the least number of relays. However, all relay nodes have the same communication range. The authors proposed an approximation algorithm by solving the minimum *k*-vertex connected spanning graph (MKCSG) problem [32, 89–91] and then placing the least number of relay nodes to establish $k(k \geq 1)$ vertex-disjoint paths between every pair of sensor nodes. Considering a directed or an undirected graph, the algorithm provides a one-way or a two-way steinerized path, i.e., populates nodes to establish connectivity among the two end node on the path, along with each edge of the found MKCSG as seen in Fig. 8. Since two sensor nodes *u* and *v* have different radio ranges, for building an asymmetric communication link between them more relays are required, as demonstrated in Fig. 8b.

5.2 Reactive Connectivity Restoration Schemes

Real-time connectivity restoration implements a recovery procedure when a node failure is detected. Such a reactive methodology better suits dynamic WSNs, since they are asynchronous in nature and it is difficult to predict the location and the scope of the failure. Therefore, adaptive schemes can best scope the recovery process depending on the effect of the failure on the network connectivity. Proposed reactive recovery schemes can be categorized as follows:

- *Determining the impact of the failure:* As indicated earlier, the loss of some nodes may not impact the network connectivity and thus would not warrant any recovery effort. However, the failure of a cut-vertex node causes the network to get partitioned and the connectivity would need to be restored. To qualify the impact of a node failure, a cut-vertex detection procedure should be applied. Recall that cut-vertices in a graph can be determined by forming a depth first search tree whose complexity grows exponentially with the depth of the graph. If the recovery process is not centralized, the complexity can grow significantly. In that case, each neighbor A_i of the failed actor A_f needs to independently assess whether A_f is a cut-vertex or not. To overcome such complexity, some of the distributed schemes in the literature pursue a probabilistic determination of cut-vertices using 2-hop neighbors information [51, 53].
- *Required network state*: As mentioned, identifying cut-vertices requires a network wide analysis. In addition, it is conceivable to generate an optimal recovery plan by considering the state of the entire network. Some of the published work strives to pursue a local analysis instead, in order to ensure scalability to large networks. Basically, the accuracy of determining a cut-vertex and the optimality of the recovery process are traded off for reduced overhead and rapid convergence. While the use of 2-hop information seems popular in the relevant literature [48, 50, 51], some allowed the quality of the solution to scale when further nodes, that are p-hop away, are known [50].
- *Scope of the recovery*: Another factor that differentiates among published schemes is how many nodes are involved in the recovery. Two main methodologies can be identified; block and sequential node movements. In block movement, a set of connected nodes travel together as unit. The idea is re-link disjoint partitions or to boost the network connectivity, e.g., eliminate cut-vertices, by moving one partition towards another [47]. The second methodology is to tolerate the degradation in connectivity, caused by the loss of the failed node, by relocating one or few nodes in a non-coordinated manner [48–50, 92]. Since the relocated nodes may get detached from their neighbors, a cascaded repositioning is pursued where neighbors follow through in order to sustain connectivity. The process is repeated recursively until reaching nodes whose movement would not violate the connectivity of the WSN.
- *Connectivity goal*: Published schemes also differ in the degree of network connectivity that ought to be sustained. Most efforts have been dedicated to repairing a partitioned network, i.e., to become 1-connected [48–51], and to restoring bi-connectivity [47, 50]. To the best of our knowledge, there is no published work that achieves general k-connectivity through controlled mobility of nodes. Note that there may be other goals in addition to connectivity. However, in this section we focus on the approaches which have connectivity as a primary goal.
- *Type of algorithm*: Some of the published work employs centralized algorithms where one of the nodes takes charge of generating the recovery plan and coordinating the relocation process [92]. These approaches rely on the availability of an alternate communication path to inform other nodes on what to do. For example in [47], the network was assumed to be bi-connected prior to the failure and thus the healthy nodes can still reach each other. On the other hand, distributed algorithms

have been the preferred choice for restoring connectivity of large networks and for repairing partitioned networks [48–51, 93–105]. In such a case, the nodes are assumed to have some pre-failure state (e.g., k-hop) information and utilize that information to detect and recover from network partitioning. The rationales are that localized and distributed algorithms scale well and that partitioned networks would not allow communication with a centralized coordinator for the recovery.

All published reactive approaches pursue node repositioning and fundamentally differ in the required network state and the performance objective of the recovery process, other than re-establishing strong network connectivity. Examples of the considered metrics include the number of nodes that get engaged in the recovery, the relocation overhead in terms of the travelled distance and messaging, the network coverage, etc. In the balance of this section we summarize sample of the published real-time connectivity restoration schemes.

Recovery Using Two-Hop State Information: DARA [48] is among the connectivity restoration approaches that require 2-hop information to assess the criticality of the node to the network connectivity and orchestrate recovery in a distributed manner. Two variants of the algorithm, namely DARA-1C and DARA-2C, are developed to address 1 and 2-connectivity requirements, respectively. The idea is to identify the least number of nodes that should be repositioned in order to re-establish a particular level of connectivity. DARA strives to localize the scope of the recovery process and minimizes the movement overhead imposed on the involved nodes. In other words, DARA pursues coordinated multi-node relocation in order to re-establish communication links among the impacted nodes. The main idea of DARA-1C is to replace the dead node by a suitable neighbor. The selection of the best candidate (neighbor) is based on the node degree and the physical proximity to the dead node. The relocation procedure is recursively applied to handle any nodes that get disconnected due to the movement of one of their neighbors (e.g., the best candidate that has replaced the faulty node).

When a critical node fails some of the nodes may be temporarily isolated until the network connectivity is restored and thus the network operation may be disrupted during the recovery. A way to avoid such short-lived disruption is to establish a 2-connected network at the time of node deployment. In a 2-connected network, there are at least two node-independent paths among each pair of nodes. This type of connectivity would ensure continual inter-node coordination even if a node fails. Such robust operation is particularly important in WSNs where real-time decisions are made. Since the network should sustain such robustness throughout its lifetime, 2-connectivity should be restored after a node fails. Similar to DARA-1C, DARA-2C identifies the nodes that are affected, i.e., lost their 2-connectivity property, due to the failed node [93]. Detecting such nodes and restoring their bi-connectivity is a very challenging problem. Nonetheless, through a careful analysis the solution space is proven to be limited to boundary nodes in the network. Such analysis has made DARA-2C a very efficient approach for restoring bi-connectivity. Basically, only a subset of the neighbors of the failed node is relocated in order to restore 2-connectivity.

Another node-relocation based solution for repairing a partitioned network is published in GLOBECOM'07 [94]. However, this approach is restricted to relinking only two partitions and does not handle multi-segments scenarios. The idea is to pick the closest node in the two partitions and move them towards each other until a communication link can be established. However, since such movement may disconnect the repositioned nodes from their partitions, additional nodes are then picked from each partition for performing cascaded movements in order to maintain the intra-partition connectivity. The collective effect is like stretching the topology of the participating sub-networks towards each other. In order to pick the appropriate node for the cascaded motion, a connected dominating set (CDS) of the individual partitions is considered. As long as the CDS is maintained for a partition, the effect of relocating a node will be limited in scope and would not risk the connectivity of the other nodes in the partition. Therefore, the approach opts to reposition non-CDS nodes. However, such repositioning is done in a cascaded manner by replacing multiple nodes in order to minimize the maximum distance a node has to move. The authors employ the approximate algorithm of Dia and Wu [106] for finding the CDS using 2-hop information. Figure 9 shows an illustrative example.

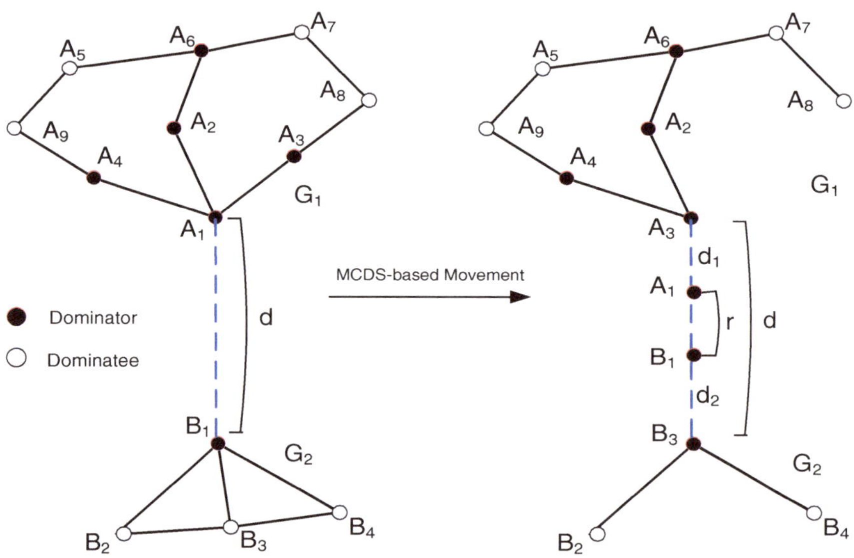

Fig. 9 Illustrating how to connect two disjoint partitions, G_1 and G_2, using the approach of [90]. The nodes in the two partitions that are closest to each other are A_1 and B_1. Let the distance between these two nodes be d, which is assumed to be greater than the transmission range of nodes, r. In order to establish connectivity some nodes from both G_1 and G_2 need to be positioned along the line A_1B_1 such that they remain connected to either G_1 or G_1 and have a distance less than r. Since B_1 is a dominator, when it moves, it has to be replaced by the closest dominatee if available, which is B_3 in the example. Moving the dominator A_1 will requires the relocation of the closest dominators, namely, A_3, since there is no dominatee connected to A_1. Because node A_3 is not connected to any other dominator, no further relocation is necessary

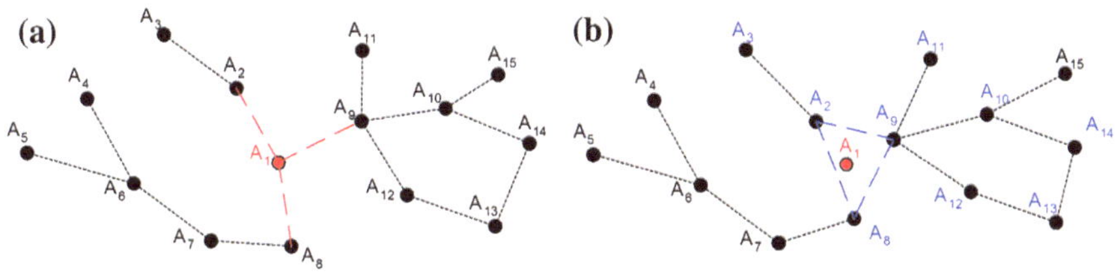

Fig. 10 Illustrating the basic idea for RIM; **a** When A_1 fails, its neighbors A_2, A_8, and A_9, applies RIM; **b** Nodes A_2, A_8, and A_9, move inward towards A_1 and. Such a motion, trigger relocation of A_3 to stay in contact with A_2, and the relocation of A_{10}, A_{11}, and A_{12}, to sustain their links with A_9

Recovery Without Explicit State Update: The approaches discussed above assume that every node knows its 2-hop neighbors and can assess the seriousness of the impact inflicted by the failure of one of the nodes. Unlike these approaches, the RIM algorithm, denoting Recovery by Inward Motion, requires just 1-hop information [49]. RIM is a localized scheme that limits the scope of the recovery process and operates in a distributed manner. Basically, RIM orchestrates a coordinated multi-node relocation in order to re-establish communication links in the neighborhood of the failed node. The main idea is to move the neighbors of a failed node A_1 inward towards the position of A_1 so that they would be able to reach each other. Figure 10 illustrates the idea. The rationale is that these neighbors are the ones directly impacted by the failure, and when they can reach each other again, the network connectivity would be restored to its pre-failure status. The relocation procedure is recursively applied to handle any node that gets disconnected due to the movement of one of their neighbors (e.g., those which moved towards the faulty node). The main advantages of RIM are its simplicity and effectiveness. RIM employs a simple procedure that recovers from both serious and non-serious breaks in connectivity without checking whether the failed node is a cut-vertex or not. The entire recovery process is distributed, enabling the network to self-heal without any external supervision. RIM has been mathematically analyzed and shown to correctly converge. The messaging overhead stays linear in the number of nodes. The maximum distance a node travels in RIM is $r/2$ where r is the node radio range. The simulation validation of RIM has shown that RIM does well in sparse networks and outperforms approaches such as DARA [48] that use 2-hop neighboring information. However, with high node density RIM tends to move many nodes and increases the total travel overhead on the network. Nonetheless, considering the overhead per node would still make RIM a favorite approach since it balances the load among nodes.

To limit the motion overhead imposed by RIM, a Volunteer-instigated Connectivity Restoration (VCR) algorithm has been proposed [95]. In VCR the neighbors of the failed node volunteer to restore connectivity by exploiting their partially utilized transmission range and by repositioning closer to the failed node. These neighbors volunteer by increasing their transmission power and moving towards the failed node A_f. In order to avoid increased medium access collision in the vicinity of A_f, VCR applies a diffusion force among volunteer nodes based on their transmission range so that they spread while staying connected. Another variant of RIM has been recently published [96]. The approach is called Least Distance Movement Recovery (LDMR) and operates in a distributed manner. The idea is for a set of direct neighbors of the failed node A_f to move toward A_f, very much like RIM [49], while their original positions are replaced with the nearest uncritical nodes. The recovery process starts with a search phase where each neighbor broadcasts a message containing the failed node ID, neighbor node ID and, Time-To-Live to limit the scope of the recovery. A candidate node responds to all requests. When a neighbor of A_f receives responses, it chooses the best candidate based on a certain criteria (e.g., distance). To avoid overbooking a candidate, a confirmation message is sent to ensure that no two neighbors of A_f will rely on the same candidate. If uncritical nodes are not available, the cascaded relocation of RIM is applied.

Another approach that does not require explicit state update is LeDiR [97]. LeDiR, denoting Least-Disruptive topology Repair, considers the connectivity restoration problem subject to path length constraints. Basically, in some applications such as combat robotic networks and search-and-rescue operation, timely coordination among the nodes is required and extending the shortest path between two nodes as a side effect of the recovery process would not be acceptable. For example interaction among nodes during a combat operation would require timeliness in order to accurately track and attack a fast moving target. LeDiR relies on the local view of a node about the network to relocate the least number of nodes and ensure that the shortest path between any pair of nodes is not extended relative to its pre-failure status. LeDiR is a localized and distributed algorithm that utilizes the typical accumulation of routing information, e.g., during path discovery, and avoids the explicit state update.

When a node A_f fails, its neighbors will individually consult their possibly-incomplete routing table to decide on the appropriate course of action and define their role in the recovery if any. If the failed node is a critical node, i.e., cut-vertex, the neighbor A_i that belongs to the smallest partition reacts. LeDiR limits the relocation to nodes in the smallest disjoint partition in order to reduce the recovery overhead. The smallest block is the one with the least number of nodes and would be identified by finding the reachable set of nodes for every 1-hop neighbor of A_f and then picking the set with the fewest nodes. Again, the routing table will be used for that. Intra-partition connectivity is sustained through cascaded relocation as in DARA [48] and RIM [49], where a node A_j that loses contact with a neighbor A_i travels toward the new position of A_i. A variant of LeDiR, called a Least-Movement Topology Repair (Le-MoToR) algorithm, has been recently published [98]. Compared to LeDiR, Le-MoToR relocates a node A_j in a partition on a different travel path by making moving parallel to the line $A_i A_f$. This modification reduces the number of moved nodes and sustains the coverage achieved by the node in the smallest partition by not shrinking the topology towards A_i after it replaces A_f.

Recovery Preplanning: Recovery from a node failure can be facilitated if some level of preplanning is performed. This is the methodology pursued by the Detection and Connectivity Restoration (DCR) algorithm [99]. DCR proactively identifies nodes that are critical to the network connectivity based on local topological information, and designates appropriate, preferably uncritical, backup nodes. Upon failure detection, the backup initiates a recovery process by replacing the failed node. The departure of the backup can trigger more relocation if it is also a critical node. Therefore, DCR prefers designating uncritical nodes as backups in order for the recovery process to terminate quickly and involve the fewest nodes. In other words, DCR strives to avoid procrastination, localize the scope of recovery and minimize the movement overhead. The performance advantage of the recovery preplanning of DCR has been captured by simulation. The simulation results have demonstrated that DCR reduces the travel distance overhead by 30–60 % compared to DARA [48] and RIM [49], depending on the network size. In addition, significantly fewer nodes are engaged and fewer messages are exchanged to coordinate the recovery process.

The approach of [94] is further extended in [51] to handle cases for which the network gets partitioned into multiple disjoint segments. The idea is do preplanning by designating nodes to lead the recovery when a failure takes place. The proposed Partition Detection and Recovery Algorithm (PADRA), not only uses 2-hop information in determining the impact of the failure on network connectivity, but also in identifying a CDS for the network. PADRA designates for each cut-vertex A_i a failure handler within the network that would start the recovery process when A_i fails. The ideal handler will be a dominatee neighbor of A_i that can simply replace A_i. If a dominatee is not available, the closest dominator is picked as the failure handler of A_i. Repositioning the failure handler at the position of A_i will then trigger cascaded relocation until a dominatee is encountered. There is no procedure followed for finding the closest dominatee to the failure handler as a means of minimizing the total traveled distance for all involved nodes. Instead, a greedy heuristic is pursued where the closest dominator is picked if a dominatee is not available. Figure 11 illustrates the operation of PADRA through an example. PADRA was extended in [100] by replacing the greedy heuristic with a dynamic programming-based approach. With such an extension, the total travel distance is significantly reduced as it considers all the alternatives at the expense of some extra messaging overhead.

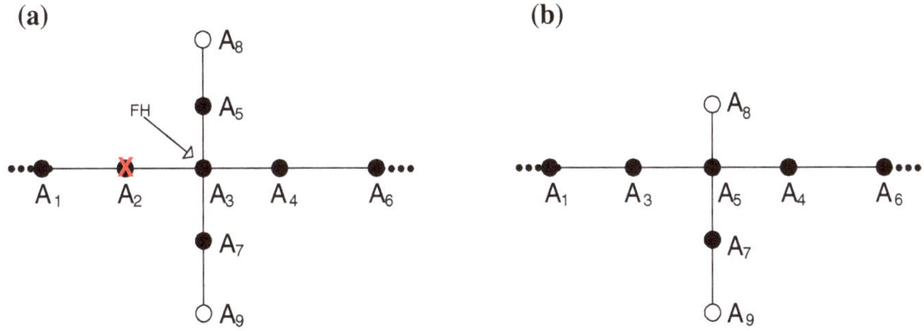

Fig. 11 A sample execution of PADRA [22]; *dark* nodes are dominators and *white* hallow are dominatees. **a** When A_2 fails, node A_3 becomes the failure handler and starts the recovery process. **b** A_3 replaces A_2, A_5 replaces A_3 and finally A_8, which is a dominatee, replaces A_5 and the connectivity is restored

Constrained Motion: Repositioning nodes to tolerate a failure may also be subject to application level constraints. For example, the node may be assigned an important task that is spatially tied to a certain spot. To handle such a scenario, the DARA approach is extended in [101] to factor in the importance of on-going tasks in the selection of a candidate node for replacing A_f. In addition, a hybrid Application-centric Connectivity Restoration (ACR) algorithm is proposed [102] that factors in application level interests besides efficient resource utilization while recovering from critical node failures. Like DCR [99], ACR determines critical nodes and designates for them backups as part of pre-failure planning. Each node discovers in a distributed manner whether it is critical or not based on local topology information. Each critical (primary) node picks a suitable backup that can satisfy application-level constraints. While choosing a backup, a primary node strives to find a nearby uncritical node in order to limit the scope of the recovery and reduce the overhead. Moreover, ACR strives to minimize the effect of node failure on coverage and connectivity by engaging strongly connected nodes with overlapped coverage.

With the exception of RIM and its variants, the other reactive approaches for restoring connectivity are based on a single underlying principle of replacing the failed node without considering the possible fact that the location of the failed node could have been the reason for its failure. These approaches also tend to trigger a cascaded relocation of many nodes resulting in increased overhead. For example, DARA [48] and RIM [49] often terminate after engaging the nodes at the network periphery. To counter such shortcoming, an algorithm for Connectivity Restoration through node Rearrangement (CRR) is proposed in [103]. CRR pursues rearrangement of nodes while limiting the scope of the recovery to the vicinity of the failed node. The main idea is to reposition the 1-hop neighbors of the failed node so that the topology becomes strongly connected, i.e., a path exists between every pair of nodes. The node rearrangement is modeled as a variant of the Steiner minimal tree formation problem to connect the 2-hop neighbors of A_f. The 1-hop neighbors of A_f are then placed at the identified Steiner points. If the 1-hop neighbors are fewer than the Steiner points, the 2-hop nodes are relocated.

Combined Coverage and Connectivity Schemes: A combined coverage and connectivity metric has also been considered in the recovery process. For example, the node recovery through active spare designation (NORAS) algorithm [104], factors in the coverage overlap in selecting a substitute for the failed node. Again the recovery is triggered by the failure of a critical node A_f that partitions the network, i.e., motivated by loss of connectivity. In NORAS, potential backups are qualified based on coverage redundancy and connectivity. A node that would cause the least coverage degradation is favored. In addition, low node degree would make a node an attractive backup since limited cascaded relocation may be required. NORAS opts to localize the scope of the recovery by picking backups within the 2-hop neighborhood of A_f. Upon detecting the failure of A_f, the designated spare travels to replace A_f or a series of cascaded relocation on the shortest route between A_f and backup will be triggered to split the load on multiple nodes.

Meanwhile, the coverage conscious connectivity restoration (C^3R) approach exploits both temporal and spatial domain [105]. The main idea is to exploit both temporal and spatial domain. Basically, the neighbors of A_f will collaborate in the recovery by taking turns. Each participating node will reposition to the vicinity of A_f, serve the network for some time and then go back to its original location. A heuristic solution has been proposed to identify the nodes that should be involved and a schedule is devised for them to serve the area covered by A_f. This leads to intermittent connectivity and monitoring of all the originally covered spots. An optimized version of C^3R, called ECR, is also proposed [105]. ECR is geared for energy efficiency at the expense of coverage and connectivity and is suitable for applications where network longevity is a prime objective. ECR devises a recovery schedule that minimizes the ratio of travel-imposed energy to that being consumed while a node is stationary. Such a scheduling problem has been formulated as a linear program.

To provide a performance bound for the schemes that tolerate a single node failure and handle network architectures in which the base-station can develop the recovery plan, the recovery problem has been formulated as an Integer Linear Program (ILP) in [92]. The objective of the ILP-based optimization model is to form a strongly connected topology while minimizing the distance that the individual nodes have to travel and minimizing the loss in coverage caused by the failure of some nodes. The proposed solution handles the failure of one or multiple nodes. The ILP formulation can be viewed as providing a lower bound on achievable total travel distance for node-relocation-based connectivity restoration.

6 Tolerating the Failure of Multiple Nodes

Due to the harsh surroundings, more than one sensor node may simultaneously fail. In addition, the network may suffer a large-scale damage that involves many nodes and would thus create multiple disjoint segments. Restoring connectivity is very challenging in this case and is an under-researched problem. Three recovery strategies have been pursued in the literature. In the first strategy, the network topology is restructured by repositioning nodes from the various segments in order to re-establish connectivity. This methodology would support network self-healing and enable distributed implementation. In the second strategy, additional relay nodes are deployed to interconnect the disjoint segments. Finally, the third strategy involves mobile data mules that tour the area and carry data from one segment to another. In the balance of this section we highlight the popular objectives of the recovery process in addition to restoring connectivity and summarize sample techniques for each recovery strategy.

6.1 Objective of the Recovery Process

In addition to restoring strong network connectivity that got lost due to multi-node failure, the recovery process often needs to meet some application-level objectives that the network should support. The followings are some of the most popular design factors that have been considered in the literature.

- *Restoration cost*: The cost of repairing a damaged WSN topology is an issue of concern. When relocation of existing nodes is pursued, the overhead will be limited to, pre-failure state update, recovery coordination messages and the travel distance [107]. On the other hand when deploying new nodes, it is desirable to minimize the number of the additional nodes due to logistics and budgetary concerns [108, 109].
- *Topology characteristics*: Fundamentally, establishing routes between the various segments is the main goal of the recovery. However, some work has considered achieving certain characteristics of the inter-segment topology. Having high level of connectivity is an example. Some work tried to form a bi-connected inter-segment topology to increase robustness against single node failure and to split the inter-segment traffic load on multiple data paths [110]. Meanwhile some just settled for achieving high node degree as a means for reducing the number of cut-vertices in the inter-segment topology and lowering the probability of having network partitioning when one of the deployed nodes fails [111, 112]. Obviously, the increased level of connectivity comes at the price of increased node count, which reflects the cost of the recovery, and would thus introduce some trade-off. When mobile data mules are employed, data delivery latency becomes the main concern and is often handled by careful scheduling of the tour [52].
- *Quality of service (QoS)*: QoS requirements are also important design factors which can affect the performance in WSNs. For example, there may be bandwidth requirements for the links between segments that happen to have voluminous data sources. In addition, the population of video and imaging sensors in the various segments would typically vary given the non-uniform distribution of nodes after the network is damaged. That will make the inter-segment links subject to latency and bandwidth constraints that have to be factored in the recovery process [113]. It is worth noting that robustness and load balancing are among the popular QoS goals and forming k-connected inter-segment topology can also be viewed as a means for achieving these QoS objectives.

6.2 Restoring Connectivity by Node Relocation

Restoring connectivity after the simultaneous failure of multiple nodes through the repositioning of some of the healthy nodes is very challenging. The main difficulty is determining the scope of the damage and coordinating the response of the individual nodes. Given the complexity of the recovery few attempts have been made to solve the problem in a distributed and reactive way [107, 111, 114]. However, the coordination of nodes in such a case is not possible. Therefore, nodes either have to move to the center of the region or utilize pre-failure information to apply an autonomous approach. Some approaches have solved it in a centralized manner [93, 100, 115] or considered a constrained form of the problem. For example, the DCR algorithm [99] discussed earlier has been extended in [116] to address one multi-node failure scenario in which no more than two of the failed nodes are adjacent. The developed algorithm, which is named RAM, identifies critical nodes and designates for them distinct backups. The key feature that enables RAM to handle the simultaneous failure of two adjacent nodes lies in the backup selection process. When a critical node chooses a backup, it prefers an uncritical node that is not serving another primary. Moreover, two adjacent critical nodes cannot serve each other as backup. This ensures that there will be some backup to recover in case adjacent nodes fail at the same time. If a critical node, i.e., primary, "A" picks an uncritical neighbor "B" as a backup, RAM requires "B" to also pick a backup "C" among its neighbors using the same criteria. The designated backups detect the failure of adjacent nodes and simultaneously execute the recovery process. A similar idea of backups (i.e., failure handlers) has been followed in [100] to handle multiple simultaneous node failures. Basically, the idea is to run the PADRA approach at multiple locations using the failure handlers as the initiators of the processes. However, when the nodes are relocated in a cascaded manner, two processes may need to relocate the same nodes for different failure recoveries and thus this creates race conditions. This problem is addressed via lock procedures meaning that the node does not respond to a request until it is done with another request.

The Distributed algorithm for Optimized Relay node placement using Minimum Steiner tree (DORMS) is another recovery approach that employs node repositioning to tolerate the failure of multiple nodes [111]. Since in autonomously operating network it is infeasible to perform a network-wide analysis to diagnose where segments are located, DORMS moves relay nodes from each segment toward the center of the deployment area. As soon as those relays become in range of each other, the partitioned network resume operation. The goal of DORMS is to design an efficient topology, in terms of the path length among segments, while minimizing the number of required additional nodes. Therefore, DORMS further models such initial inter-segment topology as a Steiner minimal tree in order to reduce the count of required relays. In order to find a topology which reduces the node count, DORMS employs k-LCA [117], which is the best known approximation algorithm for finding a minimum Steiner tree. The identified Steiner points are populated and the other initially-employed relays return to their respective segments to resume their pre-failure duties.

On the other hand, the main idea of the approach of autonomous repair (AuR) of damaged WSN topologies is to regroup the healthy nodes by moving the disjoint towards one another and towards the center of the deployment area [107]. The design principle of AuR is based on modeling connectivity between neighboring nodes as a modified electrostatic interaction based on Coulomb's law between charges. In AuR, the recovery is localized with nodes only interacting with their immediate neighbors.

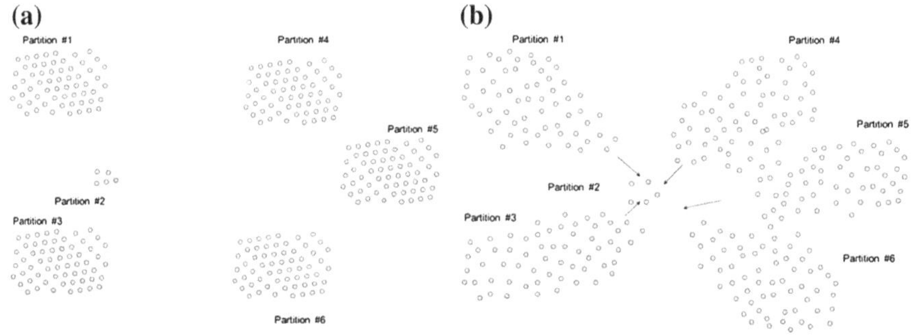

Fig. 12 AuR employs self-spreading and motion towards the center if the deployment to reconnect the disjoint segments of the network: **a** the damaged topology, **b** AuR in action

The neighbors of the failed nodes are to lead the recovery process by spreading out towards the lost nodes, causing the intra-segment topology to be stretched. If connectivity is not restored, the segment is then moved as a block towards the center of the deployment area. Moving all segments towards the center will increase the node density in the vicinity of the center point and ensure that the connectivity gets reestablished. Figure 12 illustrates the idea.

Another distributed approach to handle multiple failures is presented in [114]. This approach utilizes former route information in order to estimate the location of the damaged nodes. Before the recovery is initiated, a distributed partition detection algorithm is run and the nodes which cannot transmit their data are identified. Once the partitioning is detected, a leader node is elected among those nodes to start moving on the former paths until a live node or the sink is reached.

6.3 Recovery Through Deployment of Relays

In setups in which the nodes are not mobile or the number of healthy nodes is not sufficient to re-establish connectivity while meeting the application requirements, the deployment of additional nodes is inevitable. The deployed nodes would act as relays between the segments. The connectivity restoration problem then becomes determining the fewest number of relays and their locations so that data routes are formed between every pair of segments. Thus, the recovery problem is mapped to finding the Steiner Minimum Tree with Minimum number of Steiner Points (SMT-MSP) [79], which was shown to be an NP-Hard problem by Lin and Xue [118]. The SMT-MSP is a well-studied problem in the literature and a number of heuristics have been proposed. A summary and comparison of some of the published techniques can be found in [23]. To avoid repetition we summarize few of the recently-published solutions in the context failure recovery.

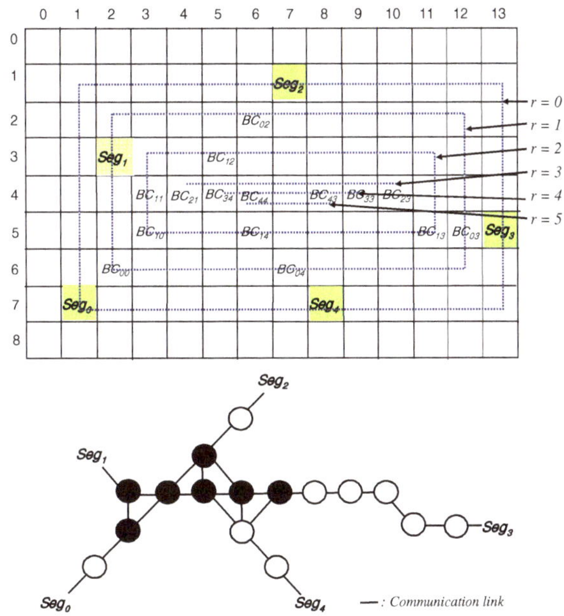

Fig. 13 Illustrating how CORP operates in rounds to form the topology shown on the *right*. The area is modeled as a grid and in each round a BC is populated for each segment. In the 5th round all segments become connected

To counter the complexity of the recovery problem, Lee and Younis [112] modeled the deployment area as a grid of equal-sized cells, and each network segment is assumed to be located in the middle of the cell. The optimization problem is then mapped to selecting the fewest number of cells for populating relays such that all segments are connected. A Cell-based Optimized Relay node Placement (CORP) algorithm is proposed. CORP is a polynomial-time algorithm that pursues greedy heuristics. It defines the best neighboring cell (BC) of a segment Seg_i as the one that lies on the shortest paths connecting Seg_i to the other segments. CORP operates in rounds. In each round, the best cells are selected and populated with relays based on where the most recently populated nodes are located. The overall placement process converges by populating relays inwards until all relays become reachable to one another. After all segments are connected, the algorithm prunes redundant relay nodes. Figure 13 illustrates the operation of CORP. A distributed implementation of CORP has also been proposed.

Another recent approach is FeSTA [108], which denotes Federating Network Segments via Triangular Steiner Tree Approximation. FeSTA deploys relay nodes and forms connected components of segments by finding local sub-optimal solutions for groups of three segments. A segment is represented by a terminal and all possible 3 distinct subsets of segments are listed. For every triangle, i.e., subset of 3 terminals, FeSTA decides either to form a new connected component or to incorporate the terminals of that triangle to an existing connected component based on the required number of relay nodes. Forming a new connected component, in essence, is equivalent to finding Steiner minimal tree of the subset of 3 terminals. The terminals (segments) join an existing connected component, if the cost for connecting these terminals individually is less than forming a distinct component. After all segments are covered, i.e., become part of a connected component, the scale of the problem is reduced to linking the individual connected components by steinerizing an edge between two nodes in distinct components.

Moreover, a two-step heuristic for forming a Connected Inter-Segment Topology (CIST) has been proposed in [109] in order to establish inter-segment connectivity by relay node placement. CIST also strives to minimize the required number of relays. Unlike other schemes, each segment is not represented by a single terminal, instead, all nodes located on the boundary of the individual segments are considered. CIST leverages the ideas promoted by FeSTA. First, CIST determines the minimal spanning tree (*mst*) edges between segments and estimates the corresponding number of relays required to establish connectivity on these edges. CIST then considers all combinations of three segments that include two segments with an *mst* edge. For each of these combinations, CIST determines the fewest relays needed to form SMT-MSP for a triangle whose vertices are located in distinct segments, and reports the reduction in the relay count relative to the *mst*-edge-based connection of the corresponding three segments. In the next step, CIST connects groups of three segments with positive gain, i.e., reduction in relay count, by steinerizing the corresponding triangle. Finally, the remaining segments are connected via steinerizing *mst* edges. Figure 14 illustrates the idea.

6.3.1 Recovery with QoS Objectives

In some setups it is desirable to interconnect multiple network segments to serve an emerging event. In addition, in some node failure scenarios the recovery process needs to re-establish connected inter-segment topologies with desired features, e.g., high node degree. In other words, unlike the work discussed above, support for some intra-network QoS requirements is desired. The employed relay nodes can be newly introduced or simply relocated from other parts of the network. The latter scenario will require some trade-off since some parts of the network may operate at some degraded levels of performance. Multiple heuristic solutions have been developed to tackle many variants of this problem.

The first approach, which is named Effective QoS-Aware Relay Node Placement Algorithm (EQAR), opts to form a connected topology using the least number of relays while meeting inter-segment capacity constraints [113]. The inter-segment QoS requirements may be just a byproduct of the damage since the segments may be of different sizes and in turn, the volume of the generated traffic may widely vary. In addition, each segment may have its own QoS requirement depending on the application and the number of video and imaging sensors that the segment has. The deployment area is modeled as a grid with equal-sized cells. The problem becomes identifying the cells that ought to be populated with relays so that the total number of deployed relays is reduced and the QoS goals are met. EQAR introduces a cell-based cost function based on the residual capacity of the relays which have been deployed

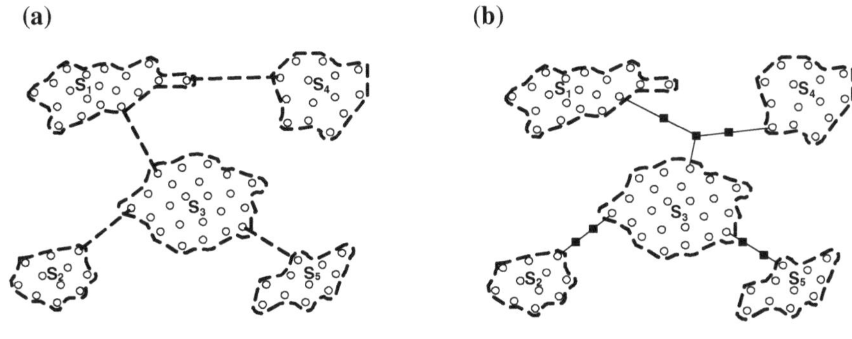

Fig. 14 An illustration of how CIST works. **a** The initial topologies of segments. *Dashed lines* represent *mst* edges. **b** CIST first processes the triangular subset $\{S_1, S_3, S_4\}$. The *dark rectangles* represent relays. CIST then processes the triangular subset $\{S_2, S_3, S_5\}$. Since the gain of the subset is turned to be zero, the *mst* edges that connect these segments are steinerized

in the cell. The optimization problem is then mapped to finding the cell-based least cost paths that collectively meet the QoS requirements. In other words, the objective of the optimization is to maximize the utilization of the residual relaying capacity. Increasing the utilization of relays also increases the connectivity and allows the resultant inter-segment topology to be more resilient to local damage.

An extended version of the CORP approach [112] discussed above is developed to support inter-segment capacity requirements. The new approach, which is called QoS-Aware Relay Node Placement (QRP) [119], pursues greedy heuristics to populate the least number of relays such that the disjoint segments are connected and the QoS requirements between every pair of two segments are met. Again, QRP models the area as a grid of equal-sized cells and defines the best neighboring cell of a segment Seg_i as the one that requires the least relaying capacity to connect Seg_i to the other segments with QoS values being met. Like CORP, QRP operates in rounds. In each round, the best cells are selected and populated with relays based on where the most recently populated relays are located. This process concludes when all segments are connected using the newly deployed relays.

Meanwhile the Spider Web approach opts to re-establish connectivity, i.e., 1-vertex connectivity, using the least number of relays while achieving high degree of connectivity in the formed topology [110]. Published schemes often form an *mst* among the isolated segments. An *mst*-based topology usually makes some nodes a hot spot in terms of the traffic load and limits the achievable network throughput, and may thus deem the inter-node collaboration insufficient for specific application tasks. Unlike these schemes, this approach establishes a topology that resembles a spider web, for which the segments are situated at the perimeter. Such a topology not only exhibits stronger connectivity than an *mst* but also achieves better sensor coverage and enables balanced distribution of traffic load among the employed relays. The simulation results have shown that these distinct features are provided with a comparable relay count to that an *mst*-based solution would involve. To further increase the robustness of the formed topology, the approach is extended so that the final topology is guaranteed to be 2-vertex connected. Figure 15 shows an example of the formed topology.

The problem considered by Al-Turjman et al. [120] is a bit different. A certain number of relays were assumed to be available. The goal is to re-establish connectivity among the segments and to achieve the most connected inter-relay topology using the allowed relay count. The area is modeled as a grid and relays are to be placed on the intersection points. An *mst* is formed using the edges of the grid and assuming all intersections have relays. Only the selected intersection points are populated with relays. In the second phase the unused relays, out of the allowed relay count, are populated so that the node degree of the inter-relay topology is maximized. To achieve that, the connectivity is modeled using the Laplacian matrix of a graph [121]. The connectivity is then measured by computing the second smallest eigenvalue λ_2, where λ_2 indicates the minimum number of links which if omitted the graph loses its strong connectivity, i.e., becomes partitioned.

6.3.2 Employing Mobile Data Mules

The final strategy for connecting the segments of a partitioned network is through the employment of mobile relays called mobile data mules (MDM) that tour the area and carry data from one segment to another. Mobility has been exploited for data delivery in sparse MANET, for data routing in delay tolerant networks, and for data collection in dispersed sensor networks. In those types of networks, an MDM plays one of three roles; a collector that tours the sensors and carries their data to a remote entity, a base-station that consumes the data, or a relay that forwards data from one node to another [23, 37, 43–46]. MDMs can be naturally available in the environment and move randomly in an uncoordinated manner. Shah et al. [122] took advantage of these MDMs to collect data from sensor nodes that happen to be on the MDM travel path and to carry these data to a base-station. Although, the data delivery is conducted in an opportunistic fashion, the authors argued that the high availability of mobile nodes that can serve as MDMs makes the approach viable for urban sparsely-populated WSNs. It is worth noting that MDMs can also be used for tolerating coverage loss, especially when the number of failed nodes is relatively small or when coverage of specific landmarks is of high importance. In that case, the MDM can provide intermittent coverage by periodically visiting certain spots [105].

Energy conservation has been a popular objective for employing MDMs. For example, in [123, 124], the authors have proposed energy efficient data collection protocols in single-hop WSNs by employing MDMs moving on a fixed path. In [125–127] the same problem has been tackled for multi-hop WSNs. In addition, MDMs are used as data forwarders for densely-populated nodes in order to prolong the lifetime of a network [43]. Meanwhile, Alsalih et al. [128] have proposed an MDM placement approach to provide energy efficiency in highly energy constrained sensor networks. Sensors are not assumed to spatially distant from each other so that their transmission intersects. Using the intersection points, the authors formulate the

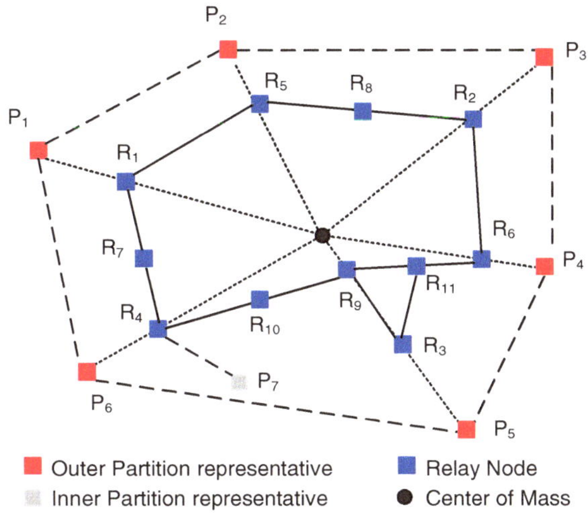

Fig. 15 An illustrative example of topology established by employing spider-web relay node deployment strategy

problem as a Mixed Integer Linear Program (MILP) to find an optimal solution. However, especially when the scope of the damage is so wide, the nodes that the MDM is going to visit may be so far apart that their transmission ranges do not intersect. Therefore, a MILP-based solution may not be feasible. On the other hand, Ekici at al. [129] have presented an algorithm for defining a tour that strives to minimize the data loss rates due to buffer overflow for a specific data generation frequency.

MDMs have also been used to link disjoint batches of nodes [44–46] which may be the result of a large-scale failure that damaged the WSN topology. The focus of [44, 46] is on studying the delay effect of using mobile relays. Analytical models were developed to form a stochastic distribution for data latency while the MDM serves as a data relay, data collector, and data sink. On the other hand, "Message Ferries" have been introduced in [45] to efficiently deliver data in sparse MANET using deterministic movement of MDMs. However, the travel route for an MDM is not derived to serve the network, instead the nodes adjust their wake-up and sleep schedule to connect with MDM when it comes to the vicinity. Overall, an approach that considers the availability of multiple MDMs and determines their optimal travel route and schedule is yet to be developed.

A variant of the MDM-based connectivity solution is considered in [22], where MDMs are placed to act as mobile access points in order to connect nodes in isolated networks through airborne units or satellites. The deployed nodes usually do not have expensive radios for long-haul communication and usually serve limited geographical areas. The limited communication range and the occasional failure of nodes may result in partitioning the network, leaving some nodes unreachable to some others. To overcome such structural weaknesses in the network, MDMs are employed to interconnect isolated sub-networks through an airborne relay, such as an unattended air vehicle (UAV) or satellite. As depicted in Fig. 16, which is redrawn from [22], a

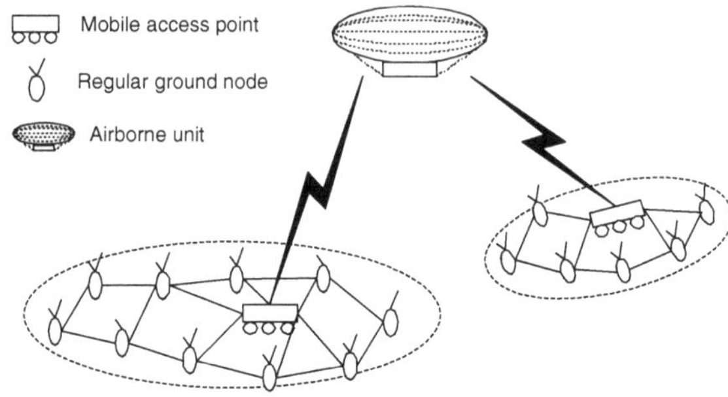

Fig. 16 MDMs are placed to connect isolated sub-networks through an airborne unit

MDM acts as an access point for the nodes in its neighborhood. A similar idea was proposed in [130].

7 Conclusion and Open Research Problems

This chapter has analyzed the impact of node failures on the operation of WSNs and categorized popular network topology management techniques for tolerating failures. The problem has been classified based on the scale and scope of the failure and popular objectives for the recovery process are enumerated. Mitigation schemes have been categorized into precautionary ones, that are performed before a failure happens, and reactive ones where the network deals with a failure only when it is detected. Sample recovery approaches for single and multi-node failures have been presented and compared. The followings are some open research problems that warrant additional investigation:

- As mentioned earlier, there is no work which can restore k-connectivity of a k-connected WSN in a distributed and efficient manner. Restoring k-connectivity through a generic algorithm that will work for any given k is certainly an interesting research direction.
- In underwater wireless sensor networks (UWSNs), the nodes are prone to failures more than terrestrial WSNs due to corrosion and fouling. Therefore, UWSN may get partitioned and some of the nodes may not be able to communicate with one another and with the surface station. Exploiting controlled mobility to restore the connectivity in such 3-D networks is very challenging.
- Combining hybrid proactive and reactive strategies for failure recovery is an under-researched area. For example, the network may start designating backups when nodes start to fail at a noticeable rate so that the recovery can be expedited.
- The robustness of the failure detection and tolerance can be enhanced when cross-layer techniques are leveraged. For example, distinguishing node and link failures is often difficult and may trigger many false alarms. A combined link and network layers methodology can significantly reduce the frequency of false positives.

Exploiting cross-layer techniques for fault tolerance is not a sufficiently explored area of research.

- While a number of published studies have employed MDMs for connecting disjoint network segments, many research questions are left unanswered, in particular, how to determine the optimal MDM count and how to coordinate their motion for optimal performance.
- The handling of security concerns when dealing with node failures is a tough and unexplored area of research. Security association, trust, and node vulnerability can constrain node placement/repositioning and thus complicate the recovery significantly.
- For most of the summarized approaches that pursue node repositioning, the movement is assumed to be obstacle free and no error is assumed in determining locations to move. Evaluation studies for assessing the effect of navigation and localization errors on the performance of recovery approaches are needed.

Acknowledgments This work was supported by the National Science Foundation (NSF) awards # CNS 1018171 and CNS 1018404.

References

1. I.F. Akyildiz et al., Wireless sensor networks: a survey. Comput. Netw. **38**, 393–422 (2002)
2. C.-Y. Chong, S.P. Kumar, Sensor networks: evolution, opportunities, and challenges. Proc. IEEE **91**(8), 1247–1256 (2003)
3. D. Estrin, L. Girod, G. Pottie, M. Srivastava, *Instrumenting the world with wireless sensor networks, Proceedings of the International Conference on Acoustics, Speech, and Signal Processing*, vol. 4 (Salt Lake City, UT, 2001), pp. 2033–2036
4. D. Ganesan et al., *Networking issues in wireless sensor networks* (J. Parallel Distrib, Comput, 2003). (Special Issue on Frontiers in Distributed Sensor Networks, Elsevier Publishers)
5. D. Estrin, *Next century challenges: scalable coordination in sensor networks, in Proceedings of the 5th Annual International Conference on Mobile Computing and Networks (MobiCOM 1999)* (Seattle, WA, August, 1999)
6. H. Karl, A. Willig, *Protocols and Architectures for Wireless Sensor Networks* (Wiley, New York, 2005)
7. R. Min et al., *Low power wireless sensor networks, in Proceedings of the International Conference on VLSI Design* (Bangalore, India, January, 2001)
8. K. Romer, F. Mattern, The design space of wireless sensor networks. IEEE Wirel. Commun. **11**(6), 54–61 (2004)
9. S. Tilak et al., A taxonomy of wireless micro-sensor network models. SIGMOBILE Mobile Comput. Commun. Rev. **6**(2), 28–36 (2002)
10. A. Bar-Noy, I. Kessler, M. Sidi, Mobile users: to update or not to update? ACM/Baltzer J. Wirel. Netw. **1**, 175–195 (1995)
11. P.G. Escalle, V.C. Giner, J.M. Oltra, Reducing location update and paging costs in a PCS network. IEEE Tran. Wirel. Commun. **1**(1), 200–209 (2002)
12. L. Bao, J.J. Garcia-Luna-Aceves, *Topology management in ad hoc networks, in Proceedings of the 4th ACM International Symposium on Mobile Ad Hoc Networking and Computing (MobiHoc 2003)* (Annapolis, MD, June, 2003)
13. A. Muqattash, M.M. Krunz, A distributed transmission power control protocol for mobile ad hoc networks. IEEE Trans. Mobile Comput. **3**(2), 113–128 (2004)

14. C. Sengul, R. Kravets, TITAN: on-demand topology management in ad hoc networks. ACM SIGMOBILE Mobile Comput. Commun. Rev. **9**(1), 77–82 (2005)
15. J. Wu, F. Dai, Mobility-sensitive topology control in mobile ad hoc networks. IEEE Trans. Parallel Distrib. Syst. **17**(6), 522–535 (2006)
16. A. Amis, R. Prakash, T. Vuong, D.T. Huynh, MaxMin D-cluster formation in wireless ad hoc networks, in Proceedings of the IEEE Conference on, Computer Communications (INFOCOM'99), March 1999.
17. S. Banerjee, S. Khuller, *A clustering scheme for hierarchical control in multi-hop wireless networks, in Proceedings of IEEE Conference on Computer Communications (INFOCOM'01)* (Anchorage, Alaska, April, 2001)
18. C.R. Lin, M. Gerla, Adaptive clustering for mobile wireless networks. IEEE J. Sel. Areas Commun. **15**(7), 1265–1275 (1997)
19. K. Guan, J. Hsu, R. Ghanadan, J. Gu, P. Khuu, G. Sadosuk, M.J. Weber, *Adaptive management of scalable mobile ad-hoc networks with non-homogeneous topology, in Proceedings of the IEEE Military Communications Conference (MILCOM 2007)* (Orlando, FL, October, 2007)
20. M. Chatterjee, S.K. Das, D. Turgut, WCA: a weighted clustering algorithm for mobile ad hoc networks. J. Cluster Comput. (Special Issue on Mobile Ad hoc Networks) **5**(2), 193–204 (2002)
21. A.A. Abbasi, M. Younis, A survey on clustering algorithms for wireless sensor networks. Comput. Commun. **30**(14–15), 2826–2841 (2007)
22. C.-C. Shen, O. Koc, C. Jaikaeo, Z. Huang, *Trajectory control of mobile access points in MANET, in Proceedings of the 48th IEEE Global Telecommunications Conference (GLOBECOM '05)* (St. Louis, MO, November, 2005)
23. M. Younis, K. Akkaya, Strategies and techniques for node placement in wireless sensor networks: a survey. J. Ad-Hoc Netw. **6**(4), 621–655 (2008)
24. X. Han, X. Cao, E.L. Lloyd, C.-C. Shen, *Fault-tolerant relay nodes placement in heterogeneous wireless sensor networks, in Proceedings of the 26th IEEE/ACM Joint Conference on Computers and Communications (INFOCOM'07)* (Anchorage, AK, May, 2007)
25. B. Deb, S. Bhatnagar, B. Nath, STREAM: sensor topology retrieval at multiple resolutions. J. Telecommun. (Special Issue on Wireless Sensor Networks) **26**(2–4), 285–320 (2004)
26. A. Cerpa, D. Estrin, *ASCENT: adaptive self-configuring sensor networks topologies, in Proceedings of the 21st International Annual Joint Conference of the IEEE Computer and Communications Societies (INFOCOM 2002)* (June, New York, NY, 2002)
27. P.B. Godfrey, D. Ratajczak, *Naps: scalable, robust topology management in wireless ad hoc networks, in Proceedings of the IEEE 3rd International Symposium on Information Processing in Sensor Networks (IPSN 2004)* (CA, April, Berkeley, 2004)
28. C. Schurgers, V. Tsiatsis, S. Ganeriwal, M. Srivastava, Optimizing sensor networks in the energy-latency-density design space. IEEE Trans. Mobile Comput. **1**(1), 70–80 (2002)
29. Y. Xu, J. Heidemann, D. Estrin, *Geography-informed energy conservation for ad hoc routing, in Proceedings of the 7th Annual International Conference on Mobile Computing and Networking (MobiCom'01)* (Rome, Italy, June, 2001)
30. Y. Lai, H. Chen, *Energy-efficient fault-tolerant mechanism for clustered wireless sensor networks, in Proceedings of the 16th International Conference on Computer Communications and Networks (ICCCN'07)* (Honolulu, Hawaii, August, 2007)
31. G. Gupta, M. Younis, *Fault-tolerant clustering of wireless sensor networks, in Proceedings of the Wireless Communications and Networking Conference (WCNC 2003)* (New Orleans, LA, March, 2003)
32. T. Bagheri, *DFMC: decentralized fault management mechanism for cluster based wireless sensor networks, in Proceedings of the 2nd International Conference on Digital Information and Communication Technology and its Applications (DICTAP)* (Bangkok, Thailand, May, 2012)
33. L.H.A. Correiaa, D.F. Macedoa, A.L. dos Santos, A.A.F. Loureiroa, J.M.S. Nogueiraa, Transmission power control techniques for wireless sensor networks. Comput. Netw. **51**(17), 4765–4779 (2007)

34. S. Lin, J. Zhang, G. Zhou, L. Gu, J.A. Stankovic, T. He, ATPC: adaptive transmission power control for wireless sensor networks, in Proceedings of the 4th ACM Conference on Embedded Networked Sensor Systems (SenSys 2006), Boulder, CO., October 2006
35. J. Jeong, D. Culler, J.H. Oh, *Empirical analysis of transmission power control algorithms for wireless sensor networks, in Proceedings of the 4th International Conference on Networked Sensing Systems (INSS '07)* (Braunschweig, Germany, June, 2007)
36. J. Luo, J.-P. Hubaux, *Joint mobility and routing for lifetime elongation in wireless sensor networks, in Proceedings of IEEE INFOCOM 2005* (Miami, FL, March, 2005)
37. I. Chatzigiannakis, A. Kinalis, S. Nikoletseas, *Sink mobility protocols for data collection in wireless sensor networks, in Proceedings of the 4th ACM International Workshop on Mobility Management and Wireless Access (MOBIWAC)* (Terromolinos, Spain, October, 2006)
38. Z.M. Wang, S. Basagni, E. Melachrinoudis, C. Petrioli, *Exploiting sink mobility for maximizing sensor networks lifetime, in Proceedings of the 38th Annual Hawaii International Conference on System Sciences (HICSS'05)* (Hawaii, January, 2005)
39. W. Alsalih, S. Akl, H. Hassanein, *Placement of multiple mobile base stations in wireless sensor networks, in Proceedings of the IEEE Symposium on Signal Processing and Information Technology (ISSPIT)* (Cairo, Egypt, December, 2007)
40. W. Youssef, M. Younis, K. Akkaya, *An intelligent safety-aware gateway relocation scheme for wireless sensor networks, in Proceedings of the IEEE International Conference on Communications (ICC 2006)* (Istanbul, Turkey, June, 2006)
41. K. Akkaya, M. Younis, Sink repositioning for enhanced performance in wireless sensor networks. Comput. Netw. **49**(4), 512–534 (2005)
42. W. Wang, V. Srinivasan, K.-C. Chua, *Using mobile relays to prolong the lifetime of wireless sensor networks, in Proceedings of the International Conference on Mobile Computing and Networking* (Cologne, Germany, 2005)
43. H. Jun et al., Trading latency for energy in densely deployed wireless ad hoc networks using message ferrying. J. Ad Hoc Netw. **5**(4), 444–461 (2007)
44. H. Almasaeid, A.E. Kamal, *Modeling mobility-assisted data collection in wireless sensor networks, in Proceedings of the IEEE Global Telecommunications Conference (GLOBECOM'08)* (New Orleans, LA, December, 2008)
45. W. Zhao, M. Ammar, E. Zegura, *A message ferrying approach for data delivery in sparse mobile ad hoc networks, in Proceedings of the 5th ACM international symposium on mobile ad hoc networking and computing (MobiHoc'04)* (Tokyo, Japan, May, 2004)
46. H. Almasaeid, A.E. Kamal, *Data delivery in fragmented wireless sensor netoworks using mobile agents, in Proceedings of the 10th ACM/IEEE International Symposium on Modeling, Analysis and Simulation of Wireless and Mobile Systems (MSWiM)* (Chania, Greece, October, 2007)
47. P. Basu, J. Redi, Movement control algorithms for realization of fault-tolerant ad hoc robot networks. IEEE Netw. **18**(4), 36–44 (2004)
48. A. Abbasi, K. Akkaya, M. Younis, *A distributed connectivity restoration algorithm in wireless sensor and actor networks, in Proceeding of the 32nd IEEE Conference on Local Computer Networks (LCN 2007)* (Dublin, Ireland, October, 2007)
49. M. Younis, S. Lee, A.A. Abbasi, A localized algorithm for restoring inter-node connectivity in networks of moveable sensors. IEEE Trans. Comput. **59**(12), 1669–1682 (2010)
50. S. Das, *Localized movement control for fault tolerance of mobile robot networks, in Proceedings of the 1st IFIP International Conference on Wireless Sensor and Actor Networks (WSAN 2007)* (Albacete, Spain, September, 2007)
51. K. Akkaya, A. Thimmapuram, F. Senel, S. Uludag, *Distributed recovery of actor failures in wireless sensor and actor networks, in Proceedings of the IEEE Wireless Communications and Networking Conference (WCNC'08)* (Las Vegas, NV, March, 2008)
52. S. Milenko Jorgić, M. Hauspie, D. Simplot-Ryl, I. Stojmenovic, *Localized algorithms for detection of critical nodes and links for connectivity in ad hoc networks, in Proceedings of the Third Annual IFIP Mediterranean Ad Hoc Networking Workshop* (Med-Hoc-Net, Bodrum, Turkey, June, 2004)

53. X. Liu, L. Xiao, A. Kreling, Y. Liu, *Optimizing overlay topology by reducing cut vertices, in Proceedings of ACM Workshop on Network and OS Support for Digital Audio and Video* (Newport, RI, May, 2006)
54. S. Kumar, T.H. Lai, J. Balogh, On k-coverage in a mostly sleeping sensor network. Wireless Netw. **14**(3), 277–294 (2008)
55. H.M. Ammari, Stochastic k-coverage in wireless sensor networks, in *Proceedings of the 4th International Conference on Wireless Algorithms, Systems, and Applications (WASA '09)*, ed. by B. Liu, A. Bestavros, D.-Z. Du, J. Wang (Springer, Berlin, 2009), pp. 125–134
56. Y. Drougas, V. Kalogeraki, *Distributed, reliable restoration techniques using wireless sensor devices, in Proceedings of the 21st IEEE International Parallel and Distributed Processing Symposium (IPDPS 2007)* (Long Beach, CA, March, 2007)
57. N. Kumar, D. Gunopulos, V. Kalogeraki, *Sensor network coverage restoration, in Proceedings of the International conference on Distributed Computing in Sensor Systems (DCOSS 2005)* (Marina del Ray, CA, June, 2005)
58. R. Falcon, X. Li, A. Nayak, Carrier-based focused coverage formation in wireless sensor and robot networks. IEEE Trans. Autom. Control **56**(10), 2406–2417 (2011)
59. G. Fletcher, X. Li, A. Nayak, I. Stojmenovic, *Randomized robot-assisted relocation of sensors for coverage repair in wireless sensor networks, in Proceedings of the 72nd IEEE Vehicular Technology Conference Fall (VTC 2010-Fall)* (Ottawa, Canada, September, 2010)
60. C.-Y. Chang, S.-W. Chang, M.-H. Li, Y.-C. Chen, *energy-efficient mechanisms for coverage recovery in WSNs, in Proceedings of the International Wireless Communications and Mobile Computing Conference (IWCMC 2008)* (Crete Island, August, 2008)
61. Y. Mei, C. Xian, S. Das, Y.C. Hu, Y.-H. Lu, Sensor replacement using mobile robots. Comput. Commun. **30**(13), 2615–2626 (2007)
62. S. Ganeriwal, A. Kansal, M.B. Srivastava, *Self-aware actuation for fault repair in sensor networks, in Proceedings of the IEEE International Conference on Robotics and Automation (ICRA 2004)* (New Orleans, LA, April, 2004)
63. J. Wu, Z. Jiang, *A hierarchical structure based coverage repair in wireless sensor networks, in Proceedings of the 19th IEEE International Symposium on Personal, Indoor and Mobile Radio Communications (PIMRC 2008)* (Cannes, France, September, 2008)
64. M. Asim, H. Mokhtar, M.Z. Khan, M. Merabti, *A sensor relocation scheme for wireless sensor networks, in Proceedings of the Workshop on Advanced Information Networking and Applications (WAINA2011)* (Singapore, March, 2011)
65. P.K. Sahoo, J.-Z. Tsai, H.-L. Ke, *Vector method based coverage hole recovery in wireless sensor networks, in Proceedings of the 2nd International Conference on Communication Systems and Networks (COMSNETS 2010)* (Bangalore, India, January, 2010)
66. G. Wang, G. Cao, T. La Porta, W. Zhang, *Sensor relocation in mobile sensor networks, in Proceedings of the 24th Annual IEEE Conference on Computer Communications (INFOCOM'05)* (Miami, FL, March, 2005)
67. N. Heo, P.K. Varshney, Energy-efficient deployment of intelligent mobile sensor networks. IEEE Trans. Syst. Man Cybern. A **35**(1), 78–92 (2005)
68. D.T. Nguyen, N.P. Nguyen, M.T. Thai, A. Helal, *An optimal algorithm for coverage hole healing in hybrid sensor networks, in Proceedings of the 7th International Wireless Communications and Mobile Computing Conference (IWCMC 2011)* (Istanbul, Turkey, July, 2011)
69. G. Wang, G. Cao, T. La Porta, *Movement-assisted sensor deployment, in Proceedings of 23rd International Annual Joint Conference of the IEEE Computer and Communications Societies (INFOCOM'04)* (Hong Kong, March, 2004)
70. J. Wu, S. Yang, *SMART: a scan-based movement assisted sensor deployment method in wireless sensor networks, in Proceedings of 24th International Annual Joint Conference of the IEEE Computer and Communications Societies (INFOCOM'05)* (Miami, FL, March, 2005)
71. X. Li, N. Santoro, I. Stojmenovic, *Mesh-based sensor relocation for coverage maintenance in mobile sensor networks, Proceedings of the 4th International Conferene on Ubiquitous Intelligence and Computing (UIC-07), Lecture Notes in Computer Science*, vol. 4611 (Hong Kong, July, 2007), pp. 696–708

72. X. Li, N. Santoro, *ZONER: a ZONE-based sensor relocation protocol for mobile sensor networks, in Proceedings of the 31st IEEE Conference on Local Computer Networks (LCN 2006)* (Tampa, FL, November, 2006)
73. I. Stojmenovic, Position based routing in ad hoc networks. IEEE Commun. Mag. **40**(7), 128–134 (2002)
74. S. Parikh, V. Vokkarane, L. Xing, D. Kasilingam, *Node-replacement policies to maintain threshold-coverage in wireless sensor networks, in Proceedings of 16th IEEE International Conference on Computer Communications and Networks (ICCCN 2007)* (Honolulu, HI, August, 2007)
75. A. Sekhar, B.S. Manoj, C.S.R. Murthy, *Dynamic coverage maintenance algorithms for sensor networks with limited mobility, in Proceedings of the 3rd IEEE Annual Conference on Pervasive Computing and Communications (PerCom 2005)* (Koloa, Kauai, HI, March, 2005)
76. K.R. Kasinathan, M. Younis, *Distributed approach for mitigating coverage loss in heterogeneous wireless sensor networks, in Proceedings of the 3rd IEEE International Workshop on Management of Emerging Networks and Services (MENS 2011)* (Houston, TX, December, 2011)
77. N. Li, J.C. Hou, *FLSS: a fault-tolerant topology control algorithm for wireless networks, in Proceedings of the 10th ACM Annual International Conference on Mobile Computing and Networking (MobiCom 2004)* (PA, September, Philadelphia, 2004)
78. X. Cheng, D.-z. Du, L. Wang, B. Xu, Relay sensor placement in wireless sensor networks. Wirel. Netw. 14(3), 347–355 (2008).
79. E.L. Lloyd, G. Xue, Relay node placement in wireless sensor networks. IEEE Trans. Comput. **56**(1), 134–138 (2007)
80. A. Efrat et al., *Improved approximation algorithms for relay placement, in Proceedings of the 16th European Symposium on Algorithms* (Karlsruhe, Germany, September, 2008)
81. S. Poduri, S. Pattem, B. Krishnamachari, G.S. Sukhatme, *Sensor network configuration and the curse of dimensionality, in Proceedings of the 3rd IEEE Workshop on Embedded Networked Sensors* (MA, May, Cambridge, 2006)
82. Y.T. Hou, Y. Shi, H.D. Sherali, On energy provisioning and relay node placement for wireless sensor networks. IEEE Trans. Wireless Commun. **4**(5), 2579–2590 (2005)
83. Z. Cheng, M. Perillo, W.B. Heinzelman, General network lifetime and cost models for evaluating sensor network deployment strategies. IEEE Trans. Mobile Comput. **7**(4), 484–497 (2008)
84. J. Tang, B. Hao, A. Sen, Relay node placement in large scale wireless sensor networks. Comput. Commun. (Special Issue on Wireless Sensor Networks) **29**, 490–501 (2006)
85. B. Hao, H. Tang, G. Xue, *Fault-tolerant relay node placement in wireless sensor networks: formulation and approximation, in Proceedings of the Workshop on High Performance Switching and Routing (Phoenix* (AZ, April, 2004)
86. J. Pan, L. Cai, Y.T. Hou, Y. Shi, S.X. Shen, Optimal base-station locations in two-tiered wireless sensor networks. IEEE Trans. Mobile Comput. **4**(5), 458–473 (2005)
87. Q. Wang, K. Xu, G. Takahara, H. Hassanein, *Locally optimal relay node placement in heterogeneous wireless sensor networks, in Proceedings of the 48th IEEE Global Telecommunications Conference (GLOBECOM 2005)* (St. Louis, Missouri, November, 2005)
88. F. Dai, J. Wu, On constructing k-connected k-dominating sets in wireless ad-hoc and sensor networks. J. Parallel Distrib. Comput. **66**(7), 947–958 (2006)
89. J.L. Bredin, E.D. Demaine, M. Hajiaghayi, D. Rus, *Deploying sensor networks with guaranteed capacity and fault tolerance, in Proceedings of the 5th ACM International Symposium on Mobile Ad Hoc Networking and Computing (MobiHoc 2005)* (Urbana-Champaign, IL, May, 2005)
90. G. Kortsarz1, Z. Nutov, Approximating node connectivity problems via set covers. Algorithmica 37, 75–92 (2003).
91. J. Cheriyan, S. Vempala, A. Vetta, Approximation algorithms for minimum-cost k-vertex connected subgraphs, in Proceedings of the 34th Annual ACM Symposium on the Theory of Computing, pp. 306–312, Montreal, Quebec, Canada, May 2002 Kindly update Refs. [98, 116] with volume number, page range and year of publication, if applicable.

92. A. Alfadhly, U. Baroudi, M. Younis, Optimal node repositioning for tolerating node failure in wireless sensor actor network, in Proceedings of the 25th Biennial Symposium on Communications, Kingston, Canada, May 2010 KIndly update Ref. [98] with volume number, page range and year of publication, if applicable.
93. A. Abbasi, M. Younis, K. Akkaya, Movement-assisted connectivity restoration in wireless sensor and actor networks. IEEE Trans. Parallel Distrib. Syst. 20(9), 1366–1379 (2009) Kindly update Refs. [98, 116] with volume number, page range and year of publication, if applicable.
94. F. Senel, K. Akkaya, M. Younis, *An efficient mechanism for establishing connectivity in wireless sensor and actor networks, in Proceedings of the 50th IEEE Global Telecommunications Conference (Globecom'07)* (Washington, DC, November, 2007)
95. M. Imran, M. Younis, A.M. Said, H. Hasbullah, *Volunteer-instigated connectivity restoration algorithm for wireless sensor and actor networks, in Proceedings of the IEEE International Conference on Wireless Communications, Networking and Information Security (WCNIS2010)* (Beijing, China, June, 2010)
96. A. Alfadhly, U. Baroudi, M. Younis, *Least distance movement recovery approach for large scale wireless sensor-actor networks, in Proceedings of the Workshop on Federated Wireless Sensor Networks (FedSenS 2011)* (Istanbul, Turkey, July, 2011)
97. A. Abbasi, M. Younis, U. Baroudi, *Restoring connectivity in wireless sensor-actor networks with minimal topology changes, in Proceedings of the IEEE International Conference on Communications (ICC 2010)* (Cape Town, South Africa, May, 2010)
98. A. Abbasi, M. Younis, U. Baroudi, A least-movement topology repair algorithm for partitioned wireless sensor-actor networks. Int. J. Sensor Netw. Inderscience, Switzerland, **11**(4), 250–262 (2012)
99. M. Imran, M. Younis, A.M. Said, H. Hasbullah, *Partitioning detection and connectivity restoration algorithm for wireless sensor actor networks, in Proceedings of the IEEE/IFIP International Conferences on Embedded and Ubiquitous Computing (EUC 2010)* (Hong Kong, China, December, 2010)
100. K. Akkaya et al., Distributed recovery from network partitioning in movable sensor/actor networks via controlled mobility. IEEE Trans. Comput. **59**(2), 258–271 (2010)
101. A. Abbasi, U. Baroudi, M. Younis, K. Akkaya, C^2*AM: an algorithm for application aware movement-assisted recovery in wireless sensor and actor networks, in Proceedings of the 5th International Wireless Communications and Mobile Computing Conference (IWCMC 2009)* (June, Leipzig, Germany, 2009)
102. M. Imran, M. Younis, A.M. Said, H. Hasbullah, *Application-centric connectivity restoration algorithm for wireless sensor actor networks, in Proceedings of the 6th International Conferences on Grid and Ubiquitous Pervasive Computing (GPC 2011)* (Oulu, Finland, May, 2011)
103. M. Younis, R. Waknis, *Connectivity restoration in wireless sensor networks using Steiner tree approximations, in Proceedings of the IEEE Global Telecommunications Conference (GLOBECOM'10)* (Miami, FL, December, 2010)
104. S. Vaidya, M. Younis, *Efficient failure recovery in wireless sensor networks through active spare designation, in Proceedings of the 1st International Workshop on Interconnections of Wireless Sensor Networks (IWSN'10)* (Santa Barbara, CA, June, 2010)
105. N. Tamboli, M. Younis, Coverage-aware connectivity restoration in mobile sensor networks. Elsevier J. Netw. Comput. Appl. **33**, 363–374 (2010)
106. F. Dai, J. Wu, An extended localized algorithms for connected dominating set formation in ad hoc wireless networks. IEEE Trans. Parallel Distrib. Syst. **15**(10), 1027–1040 (2004)
107. Y. Joshi, *Autonomous Recovery from Multi-Node Failures in Wireless Sensor Networks* (University of Maryland Baltimore Count, December, MS Thesis, Department of Computer Science and Electrical Engineering , 2011)
108. F. Senel, M. Younis, Relay node placement in structurally damaged wireless sensor networks via triangular Steiner tree approximation. Comput. Commun. **34**(16), 1932–1941 (2011)
109. F. Senel, M. Younis, *Optimized connectivity restoration in a partitioned wireless sensor networks, in IEEE International Conference on Global Telecommunications Conference (GLOBECOM 2011)* (Houston, Texas, December, 2011). (Submitted)

110. F. Senel, M. Younis, K. Akkaya, Bio-inspired relay node placement heuristics for repairing damaged wireless sensor networks. IEEE Trans. Veh. Technol. **60**(4), 1835–1848 (2011)
111. S. Lee, M. Younis, Recovery from multiple simultaneous failures in wireless sensor networks using minimum Steiner tree. J. Parallel Distrib. Syst. **70**, 525–536 (2010)
112. S. Lee, M. Younis, Optimized relay placement to federate segments in wireless sensor networks. IEEE J. Sel. Area Commun. (Special Issue on Mission Critical Networking) **28**(5), 742–752 (2010)
113. S. Lee, M. Younis, EQAR: effective Qos-aware relay node placement algorithm for connecting disjoint wireless sensor sub-networks. IEEE Trans. Comput. **60**(12), 1772–1778 (2011)
114. S. Vemulapalli, K. Akkaya, Mobility-based self route recovery from multiple node failures in mobile sensor networks, in Proceedings of 10th IEEE International Workshop on Wireless Local Networks (WLN 2010), Denver, CO., October 2010.
115. M. Sir, I. Senturk, E. Sisikoglu, K. Akkaya, *An optimization-based approach for connecting partitioned mobile sensor/actuator networks, in Proceedings of 3rd International Workshop on Wireless Sensor, Actuator and Robot Networks (WiSARN)* (Shanghai, China, April, 2011)
116. M. Imran, M. Younis, A.M. Said, H. Hasbullah, Localized motion-based connectivity restoration algorithms for wireless sensor actor networks. J. Netw. Comput. Appl. Elsevier Science, The Netherlands, **35**(2), 844–856 (2012)
117. G. Robins, A. Zelikovsky, Tighter bounds for graph Steiner tree approximation. SIAM J. Disc. Math. **19**(1), 122–134 (2005)
118. G. Lin, G. Xue, Steiner tree problem with minimum number of Steiner points and bounded edge-length. Inf. Process. Lett. **69**, 53–57 (1999)
119. S. Lee, M. Younis, *QoS-aware relay node placement for connecting disjoint segments in wireless sensor networks, in Proceedings of the First International Workshop on Interconnections of Wireless Sensor Networks (IWSN'10)* (Santa Barbara, CA, June, 2010)
120. F. Al-Turjman, H. Hassanein, M. Ibnkahla, *Optimized relay placement to federate wireless sensor networks in environmental applications, in Proceedings of the IEEE International Workshop on Federated Sensor Systems (FedSenS11)* (Istanbul, Turkey, July, 2011)
121. A. Ghosh, S. Boyd, *Growing well-connected graphs, in Proceedings of the 45th IEEE Conference on Decision and Control (CDC)* (San Diego, CA, December, 2006)
122. R.C. Shah, S. Roy, S. Jain, W. Brunette, Data MULEs: modeling and analysis of a three-tier architecture for sparse sensor networks. Elsevier Ad Hoc Netw. J. **1**(2–3), 215–233 (2003)
123. A. Chakrabarti, A. Sabharwal, B. Aazhang, Communication power optimization in a sensor network with a path-constrained mobile observer. ACM Trans. Sensor Netw. **2**(3), 297–324 (2006)
124. L. Song, D. Hatzinakos, Architecture of wireless sensor networks with mobile sinks: sparsely deployed sensors. IEEE Trans. Veh. Technol. **56**(4), 1826–1836 (2007)
125. A. Somasundara, A. Kansal, D. Jea, D. Estrin, M. Srivastava, Controllably mobile infrastructure for low energy embedded networks. IEEE Trans. Mobile Comput. **5**(8), 958–973 (2006)
126. S. Gao, H. Zhang, S.K. Das, Efficient data collection in wireless sensor networks with path-constrained mobile sinks. IEEE Trans. Mobile Comput. **10**(5), 592–608 (2011)
127. J. Li, S.M. Shatz, A.D. Kshemkalyani, Mobile sampling of sensor field data using controlled broadcast. IEEE Trans. Mobile Comput. **10**(6), 881–896 (2011)
128. W. Alsalih, H.S. Hassanein, S.G. Akl, Placement of multiple mobile data collectors in wireless sensor networks. J. Ad Hoc Netw. **8**(4), 378–390 (2010)
129. E. Ekici, Y. Gu, D. Bozdag, Mobility-based communication in wireless sensor networks. IEEE Commun. **44**(7), 56–62 (2006)
130. M. Ahmed et al., *Positioning range extension gateways in mobile ad hoc wireless networks to improve connectivity and throughput, in Proceedings of IEEE Military Communications Conference (MILCOM'01)* (Washington, DC, October, 2001)

Chapter 10
Mobility Management with Integrated Coverage and Connectivity

Yi Zou and Krishnendu Chakrabarty

Abstract Mobility management is a major challenge in mobile ad hoc networks (MANETs), due in part to the dynamically changing network topologies. For mobile wireless sensor networks (WSNs) that are deployed for surveillance applications, it is important to use a mobility management scheme that can empower nodes to make better decisions regarding their positions such that strategic tasks such as target tracking can benefit from node movement. In this chapter, we describe a purposeful and distributed mobility management scheme for mobile sensor networks. The proposed scheme considers node movement decisions as part of a distributed optimization problem, which integrates mobility-enhanced improvement in the quality of target tracking data with the associated negative consequences of increased energy consumption due to locomotion, potential loss of network connectivity, and loss of sensing coverage.

1 Introduction

The mobile wireless sensor network is a special type of wireless sensor networks where some or all of the nodes can move. Nodes may perform a one time move, such as in [53] to help deployment, or can move constantly, such as in scenarios discussed in [5]. In some cases, only a portion of the nodes can move [39], where in other cases, all nodes can move freely [43, 55]. With the advancement of new technologies bringing more energy efficient hardware with more powerful sensing, communication, and computation capabilities than ever before, mobile wireless sensor networks

Y. Zou
Intel Corporation, Portland, OR 97229, USA
e-mail: yi.zou@intel.com

K. Chakrabarty (✉)
Duke University, Durham, NC 27708, USA
e-mail: krish@ee.duke.edu

H. M. Ammari (ed.), *The Art of Wireless Sensor Networks*,
Signals and Communication Technology, DOI: 10.1007/978-3-642-40009-4_10,

© Springer-Verlag Berlin Heidelberg 2014

are pushed to a much wider range of applications for target monitoring, detecting, or tracking. However, since locomotion generally consumes more energy than sensing or communication, mobility in wireless network, in a way, is like a double-edged sword: when the holder is skillful enough to use it properly, it helps defeat more enemies; otherwise, it may actually hurt him- or herself. It is therefore important to understand the implications on wireless sensor networks when mobility is introduced.

The mobile wireless sensor network is a vast research topic; it is virtually impossible to cover all topics in a single chapter. While it is by no no means the goal to give a complete collection of all existing works in this area, this chapter is designed to build a foundation for readers to understand most important issues in mobility management for wireless sensor networks. It is to the authors' best hope that, after finish reading this chapter, readers would feel comfortable and encouraged to go beyond texts presented here to investigate further topics related to mobility management in wireless sensor networks.

1.1 Mobility in Sensor Networks

Mobile ad hoc networks have received considerable attention in the literature [11]. Most existing methods for mobility management focus on communication issues arising from dynamically changing topologies due to node mobility [11, 19, 22], especially in personal communication service (PCS) such as cellular phone networks [36]. Mobility management has long been recognized as a major challenge in mobile ad hoc networks (MANETs) [5, 31]. As discussed in [5], a MANET generally has the following characteristics: (1) new members can join and leave the network any time; (2) no base station is available to provide connectivity to backbone hosts or to other mobile hosts; (3) it is difficult to implement sophisticated scheme for handover and location management; (4) each node acts as a router, forwarding packets from others nodes; (5) communication connectivity is usually "weak" in the sense that it is easily broken due to node movement. As pointed out in [17], wireless sensor networks (WSNs) are auto-configurable networks of nodes like MANETs and are connected via wireless links.

Regardless of whether sensor nodes being static or mobile, the fundamental purpose of a wireless sensor network is to serve its designated application. Mobility may simply be a requirement even though the node itself does not move, such is the case in mobile cellular phones. Therefore, it is extremely important to understand the application first before adopting any wireless sensor network algorithms. To the best of our knowledge, one of the most popular and fundamental applications for wireless sensor networks is to monitor, detect, or track an object of interest. An example of such applications is described in the habitat monitoring project carried by the University of California, Berkeley [38]. In this chapter, target tracking is considered to be the designated type of applications for wireless sensor networks. Particularly, this chapter focuses on the case where every individual wireless sensor node deployed in the field can move freely as long as it has enough supporting energy. The implications from the mobility on the sensing, communication, and energy consumption are at center of discussion in this chapter.

1.2 Challenges in Mobility Management

Every action in wireless sensor network has a consequence of energy consumption. Locomotion consumes far more energy than sensing or communication or even combined. Therefore, it is crucial to know whenever a node makes a movement, it will help the application goal of target tracking that the wireless sensor network is deployed for. Obviously, one would start asking questions, such as the ones listed below:

1. How will an individual node move wisely?
2. What are the justifications for the node movement?
3. What is needed for an individual node to make the movement decision?
4. Can this be achieved in a distributed manner?

Mobility management in mobile wireless sensor networks is different from that in mobile ad hoc networks because the movement of sensor nodes here is not random; rather, the movement of sensor nodes is purposeful, e.g., to actively and better track an intruder. In such scenarios, it is important to have an efficient mobility management scheme to ensure that sensor node mobility is exploited in the best possible way, e.g., to improve the quality of target tracking. At the same time, the mobility management strategy should avoid inefficient usage of scarce resources, such as energy and network bandwidth. Furthermore, the mobility management scheme should also take into account the potential negative consequences of node movement, e.g., loss of area coverage, loss of connectivity, and degradation of network performance. In addition, node movement also involves locomotion energy and routing overhead, especially the need to reestablish routes. Therefore, a practical mobility management scheme should empower a node with the ability to determine whether it should move, and where it should move, such that the movement can enhance tracking quality without depleting scarce resources, or significantly compromising coverage and network connectivity.

For wireless sensor networks with scarce energy resources, it is not always favorable for nodes to move during field operation because the energy required for locomotion energy is often much higher than that for sensing and communication [29, 35]. However, as shown in [27, 42], when nodes can afford the energy cost associated with mobility, it is important to have a network management scheme that can make effective use of mobility to facilitate application objectives. For example, multiple mobile robots can be deployed in a battlefield for target tracking without human intervention [29]. These mobile robots can form an ad hoc sensor network for monitoring the region of interest. To ensure better tracking quality for a moving target, it is beneficial to dynamically move nodes to advantageous locations.

1.3 Chapter Layout

The rest of the chapter is organized as follows. Section 2 presents background material and existing work related to mobility management in wireless sensor networks. For readers looking for more extensive reading, Sect. 2 may be used as a shortcut to related topics. To prepare readers for a serious technical discussion and mathematical analysis to follow, Sect. 3 reviews the Bayesian estimation theory for target tracking [3, 7, 51]. Applying concepts and techniques taken from Bayesian estimation theory, Sect. 3 formulates the problem of mobility management in wireless sensor networks as an optimization problem based on predicated sensor measurements. The optimization criteria are further discussed in Sects. 4 and 5, where cost and negative consequences due to node movement are discussed. These analysis form the mathematical foundation for implementing the probabilistic mobility management scheme in a fully distributed manner. Some simulation results are presented in Sect. 6 as an illustration of metrics used in this study, where readers may use as a reference point for deriving their own criteria. Section 7 concludes this chapter and presents future research directions.

2 Literature Review

To prepare readers for technical technical discussions in later sections, Sect. 2 reviews related research work in the context of mobility management in wireless sensor networks. Rather than covering everything in current literatures, Sect. 2 section focuses on selected topics that authors think are potentially most relevant and useful to readers, namely in target tracking, routing protocols, and mobility management. Nevertheless, readers are encouraged to read further in other topics that this chapter is not able to cover due to limited space.

2.1 Target Tracking

One of the most popular and fundamental applications for wireless sensor networks is to monitor, detect, or even track an object of interest. Research on centralized target tracking has been carried on for many years, originating from early work on target tracking by radar during World War II [14]. Common tracking techniques include Kalman filtering, Bayesian estimation methods, and their variants [7, 14, 20]. The unique constraints of wireless sensor networks such as limited energy due to battery-based power supply, limited storage capacity for time-series data, scalability for network management, and distributed sensing and data processing, pose a number of new challenges for target tracking.

Recent research efforts on target tracking in wireless sensor networks [2] have focused on collaborative sensing [14, 28, 51], energy-efficient routing and management [8, 13, 21, 45, 48], and sensor node deployment [41, 42, 53]. Collaborative sensing and signal processing provide raw sensory data from the low-level sensing units on sensor nodes. In many cases, cheap sensors such as omnidirectional acoustic sensors [14, 51] are used since alternatives such as CCD cameras generally require more resources for power, memory, bandwidth and computation. Some earlier work [4] considers binary sensing model to take advantage of sensor geometric properties in target tracking. In many existing literatures, a wireless sensor network is viewed as a grid-based topology. A recent study [52] proposes to improve the energy efficiency based on distance distributions in grid-based wireless sensor networks. In Cao et al. [12], the authors derive closed-form formulas for sensor detection probability and delay to evaluate various sensing models from a probabilistic perspective. More recent work [44] proposes a distributed event detection mechanism with minimal energy cost. Sensor activities represented by these events are collaborated through the proposed method, which is an appealing solution for applications like site monitoring. Although the target information from a single node is generally limited, more useful information can be obtained through data exchange and aggregation between multiple nodes, based upon which higher-level strategic decisions can be made [23, 34]. In Jiang et al. [25], the authors combine the energy efficiency with the design of node sleep scheduling algorithm for target tracking. In this work, the tracking data is fed to the sensor network to select nodes to wake up and join the target tracking with more accurate sensing data.

2.2 Routing Protocols

Routing in ad hoc sensor networks has received a lot attention and is considered a great challenge for ad hoc sensor networks [2]. Many efforts have been made to achieve energy-efficient routing in data aggregation, especially for target tracking applications. For a brief review, several typical examples are discussed in the following. The LEACH protocol [21] forms a clustered hierarchy in sensor networks, where the cluster head will be responsible for transmitting sensor data for its cluster members. The energy saving is achieved because the data is consolidated through such clusterization. SPAN [13] is another energy-efficient routing protocol, where sensor nodes are selected to operate on off-duty and on-duty cycles for sensor nodes. By switching between off-duty and on-duty state, the energy level of all nodes are averaged, resulting in an extended sensor network lifetime. In Xu et al. [48], the authors introduced routing fidelity as measure to evaluate the routing cost in the sense of energy consumption. This is used to adaptively tune the routing, resulting in energy consumption reduction. This, however, may not perform well due to its dependency on geographic information of the network. Another effort for energy-efficient routing is the rumor routing protocol proposed in [8], where routing is based on reaction on events in the network. This frees the routing protocol from depending on any geographic information of sensor nodes when a coordinate system is not available. Backbone-based routing using connected-dominating-set (CDS) is proposed in [45]. After the backbone is established, routing and querying can be achieved via the backbone nodes, leaving none-backbone nodes in a energy-saving state. The energy conservation via node scheduling is also extended to more realistic scenarios of wireless sensor network applications, where both network connectivity and sensing coverage have to be satisfied. To achieve such goal, [46] investigates the geometrical relationships between sensing and communication and proposes a Coverage Configuration Protocol (CCP) integrated with the SPAN protocol. Similarly, [54] combines requirements for sensing coverage and communication connectivity into one set of input parameters of an optimization problems and proposes a distributed solution based on minimum dominating set heuristics.

2.3 Mobility Management

Relatively less attention has been devoted to the problem of mobility management for mobile sensor networks. Obviously, the network topology changes when nodes move, and this change in topology affects both sensing coverage and communication connectivity [16, 46, 54].

For static sensor networks, [37] provides an elegant analysis of the interrelated problems of sensing coverage and communication connectivity. In mobile sensor networks, however, these issues are complicated due to the changing topology, typically resulting in a node being disconnected, the network becoming partitioned, or loss of sensing coverage in some regions, making the mobility management a more interesting and more challenging topic in wireless sensor networks [1, 16]. In addition, the mechanical energy consumption due to node mobility is generally higher than the energy required for sensing, communication, and computation [29, 35].

In some earlier work, sensor motion is managed as an event-based mechanism [10], since it is natural to move sensors close to locations where these events of interest are happening, In Wang et al. [42], the authors used limited mobility to achieve a better topology that considers both sensing and communication. The mobility investment is traded-off with the improved topology, which subsequent operations can benefit from. The "dynamic enclose cell" routing protocol is introduced in [50] to reduce the overhead for routing due to increased complexity in mobile sensor networks. However, this work is focused on the adaptation of static network routing protocols, rather than protocol design from the perspective of a mobile sensor network. The idea of "virtual coordinates" is proposed in [32], where virtual coordinates are based on node connectivity; this forms an abstract layer that existing geometry-based routing protocols can use. In Jea et al. [24], the authors explored the use of the mobile element, i.e., data mule in the paper's context, to perform data collection. The control mobility is achieved by setting rules based on where the mobile element goes as well as how long it is expected to take. Recently in [35], the authors have shown that purposeful mobility can be used to achieve energy saving for routing in the sense of an amortized cost measure. It is also shown in [27] that mobility is helpful for maintaining sensing coverage for both static and mobile targets, particularly when the number of nodes is not sufficient for covering the complete area. This idea is similar to repositioning schemes for sensor node deployment as proposed in [42, 53]. Since mobile sensor nodes are normally more expensive than static nodes, hybrid wireless sensor networks also receive a good amount of attention. A good example is in [40], static sensors broadcast events to mobile sensors who bid based on Voronoi weights to be able to reply to the static nodes. Winning mobile nodes are then guided by the static nodes to perform the active sensing. Alternatively, probabilistic graphical model such as Bayesian network has also been proposed to provider a better inference of mobility management in wireless sensor networks. As illustrated in a recent study in [6]. However, the computation involved in this approach is quite significant and the results from [6] are also limited to a centralized implementation. In another recent study [30], the authors propose a distributed sensor fusion mechanism for mobile wireless sensor networks. The mobility management scheme combines two techniques, the weighted Voronoi diagrams-based broadcasting for node location updating, and the Kalman filtering-based node location prediction. The proposed scheme in [30] also integrates the mobility management with geographical routing for efficient data forwarding to achieve lower latency as well less energy overhead. A good summary of related mobility management algorithms in wireless sensor networks can be found in [43] with evaluation on pros and cons in many published mobility management algorithms. In [39, 47], authors looked into the assistance from collaborative sensing of static sensors to help mobile sensors to move to preferred locations. In a way, this approach balances the limited number of expensive mobile nodes with a large number of cheap static nodes. The mobile node in this work, acts as the collaborative sensing cluster head node from the perspective of the earlier IDSQ method [51] for collaborative sensing. Static nodes within a mobile node's communication range are like "local" sensors for the mobile node. This allows the distributed implementation of the proposed method.

Noticing that most existing work in this context is not considering a fully mobile wireless sensor network where every individual node can move freely at any time, authors in [55] proposed a generic distributed and probabilistic approach. In Zou and Chakrabarty [55], authors assume that mobility is a given existing capability for each node but each node makes its own local decision on where to move. Looking back at [39, 47], one would notice that [55] shares a lot of design challenges. Both authors believe the mobility of a node can be taken advantage of to better serve the application while paying the penalty for spending energy on locomotion. Both authors stress that sensing coverage, communication connectivity, and energy consumption are equally as important as each other. Both authors take all these factors into consideration in their design. Both authors attempt to solve the mobility management issue from a probabilistic approach that is more realistically matched in real-life scenarios. All existing work in this topic have shown that managed mobility leads to improved network topologies, which can in turn facilitate subsequent data collection and processing operations in sensor networks.

Without loss of generality, in the following sections, [55] is used as the basis for technical discussion for the following reasons. This chapter is designed for fully mobile wireless sensor network. This allows readers of this chapter to evaluate the challenges in mobility management from a much wider perspective. For example, based on the knowledge from this chapter, readers can easily apply restrictions to limit number of mobile nodes for their own applications, potentially reducing cost. Secondly, this chapter aims at the mobility management for the specific objective, i.e., target tracking, one of the most useful and popular applications for wireless sensor networks. Furthermore, this chapter presents a probabilistic analysis that evaluates the risks of losing connectivity and sensing coverage from the perspective of a mobile sensor network with an inherently dynamic topology. Eventually, this chapter presents a distributed mobility management scheme that unifies tracking quality, sensing coverage, network connectivity, and energy consumption. Readers are expected to find such general mobility management easily applicable, with modifications such as sensor characteristics or communication range or power constraints, to satisfy their own target tracking applications. For example, mobile robots deployed in a battlefield environment, mobility is affordable from cost considerations, but nodes should be carefully controlled to improve tracking quality.

3 Tracking Quality Improvement Due to Node Movement

To improve the quality of target tracking, a node can decide to move to another location at the next time instant. These locations are referred to as candidate locations. In the following discussion, First, Sect. 3 describes the tracking problem based on standard target estimation theory [3, 7]. This is then followed by formulating the problem of selecting the best candidate location for a node in a fully distributed manner.

3.1 Preliminaries

The target state is denoted by $\mathbf{x_t} \in \mathbb{R}^{d_x}$, where t is a discrete time sequence. For example, for target tracking in a 2D field, $\mathbf{x}_t$ can be defined as a column vector of $[x; y; x_a; y_a]$, where x, y are the target speeds and x_a, y_a are the corresponding accelerations in X and Y coordinates. The parameter $\mathbf{x}_t$ is described by the system model as:

$$\mathbf{x}_t = \mathbf{f}_t(\mathbf{x}_{t-1}, \mathbf{w}_{t-1}), \tag{1}$$

where $\mathbf{f}_t : \mathbb{R}^{d_x} \times \mathbb{R}^{d_w} \to \mathbb{R}^{d_x}$ and $\mathbf{w}_{t-1}$ represents i.i.d. process noise. The parameters d_x and d_w denote the dimensions of $\mathbf{x}_t$ and $\mathbf{w}_t$, respectively. Note that $\mathbf{x}_t$ is estimated recursively from sensor measurements given by the observation model as:

$$\mathbf{z}_t = \mathbf{h}_t(\mathbf{x}_t, \mathbf{v}_t), \tag{2}$$

where $\mathbf{h}_t : \mathbb{R}^{d_x} \times \mathbb{R}^{d_v} \to \mathbb{R}^{d_z}$ and $\mathbf{v}_t$ represents i.i.d. measurement noise. The parameters d_z and d_v denote the dimensions of $\mathbf{z}_t$ and $\mathbf{v}_t$, respectively. The statistics for both $\mathbf{w}_t$ and $\mathbf{v}_t$ are assumed to be known. In Bayesian estimation theory, $\mathbf{x}_t$ is estimated recursively by incorporating the new measurements to modify the prior, and thereby obtain the posterior [3, 7, 51]. Let us denote the estimated target state at time t as $\hat{\mathbf{x}}_t$. The Bayesian estimation is given by

$$p(\hat{\mathbf{x}}_t|\mathbf{z}_{1:t-1}) = \int p(\hat{\mathbf{x}}_t|\hat{\mathbf{x}}_{t-1})p(\hat{\mathbf{x}}_{t-1}|\mathbf{z}_{1:t-1})d\hat{\mathbf{x}}_{t-1}, \tag{3}$$

$$p(\hat{\mathbf{x}}_t|\mathbf{z}_{1:t}) = \frac{p(\mathbf{z}_t|\hat{\mathbf{x}}_t)p(\hat{\mathbf{x}}_t|\mathbf{z}_{1:t-1})}{p(\mathbf{z}_t|\mathbf{z}_{1:t-1})}, \tag{4}$$

$$p(\mathbf{z}_t|\mathbf{z}_{1:t-1}) = \int p(\mathbf{z}_t|\hat{\mathbf{x}}_t)p(\hat{\mathbf{x}}_t|\mathbf{z}_{1:t-1})d\hat{\mathbf{x}}_t, \tag{5}$$

where the likelihood is given by $p(\mathbf{z}_t|\hat{\mathbf{x}}_t)$. Equation (3) is the prediction step, Eq. (4) updates the prior using the new measurement $\mathbf{z}_t$ to obtain the posterior $p(\hat{\mathbf{x}}_t|\mathbf{z}_{1:t})$, and Eq. (5) is the normalizing factor. From these equations, readers can see that sensor measurements must be forwarded to a processing center for data integration, where the prior information is already available. In sensor networks, this can be implemented either in a centralized manner by designating one of the sensor nodes as the processing center [20, 23], or a cluster-based approach [51]. The first approach normally uses a node with more powerful computation capabilities for centralized processing; all other nodes forward their collected sensor data. The second approach depends on a dynamic clustering algorithm to select one of the nodes as the cluster head, i.e., the node that performs sensor fusion. When the cluster head is changed, usually in accordance with the estimated target track, it needs to pass the prior information to the next cluster head for continuous tracking [51].

The above scenario becomes more complicated when mobile sensor nodes need to make decisions locally about their movements to better track the target. Below, some challenges encountered are listed in the mobility management sensor nodes:

- Nodes must make decisions on where to move in a timely manner. Nodes may not be able to afford to wait for the posterior from the fully integrated sensor measurements on the processing center or the cluster head.
- In the prediction stage as shown in Eq. (3), $p(\hat{\mathbf{x}}_{t-1}|\mathbf{z}_{1:t-1})$ is assumed to be known to the processing center or the cluster head from previous estimation at $t-1$. This implies that if each mobile node needs to make a decision on its movement based on the result from the complete sensor integration, this information must also be available. However, because mobility management should be autonomous as well as distributed, it is difficult to decide to which nodes this information should be forwarded. Furthermore, continuously forwarding the prior based on all nodes' measurements will cause considerable burden on the communication bandwidth, and it will increase energy consumption.
- Node movements result in topology changes in the sensor network. This implies that the set of neighbor nodes for each node also changes.

To ensure that a node is able to make a local decision using only current local knowledge, it is assumed that every node has the capability to perform sensor integration locally. A node that has local sensor measurement exchanges sensor measurements within its own one-hop neighborhood. It also performs estimation using the Bayesian approach with possibly incomplete sensor measurements, i.e., $\mathbf{z}_t$ does not necessarily contain sensor measurements from all sensor nodes that have detected the target at time sequence t, but only nodes within the one-hop neighborhood. Note that sensor measurements can still be delivered to the designated processing server node for an estimation based on current complete sensor measurements.

Figure 1 illustrates how the mobility management works to help improve target tracking quality. Consider a mobile sensor node s_i, as shown in Fig. 1a. In order to improve the quality of target tracking data, node s_i moves to a location that leads to improved sensor measurement. In other words, s_i will expect a higher signal-to-noise S/N ratio at its location at time instant $t + 1$, compared to its sensor measurement at time instant t. Note that topology in a mobile sensor network is dynamic. Apart from the extra energy a node spends in movement, a node also faces risks of losing communication connectivity to its neighbors, as well as losing sensing coverage in certain areas. For the node s_i in this example, these are shown by Fig. 1b and Fig. 1c respectively.

The method described here is related to the Information Driven Sensor Query (IDSQ) method described in [51]. IDSQ is used to select the best sensor measurement from a set of fixed sensor nodes that have currently reported a target, i.e., sensor nodes that have new sensor measurements. The selection is based on the estimated information gain from the sensor measurement on the candidate sensor node, where the information gain is evaluated using well-defined rules.

In mobility management, the movement decision for a node is based on whether the new location will improve tracking quality. Since a node does not know *a priori* the quality of sensor measurements it will get at the new location, first a prediction can be made on all possible sensor measurements corresponding to all possible

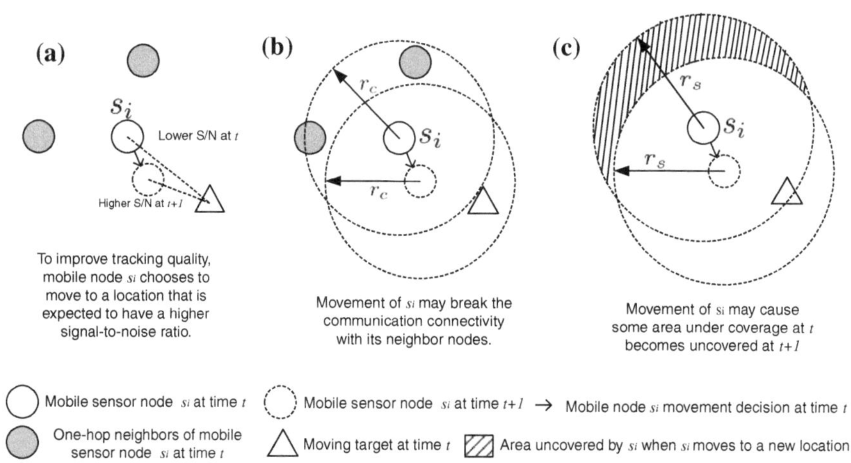

Fig. 1 An illustration of the mobility management method

candidate locations that the node might choose to move to. These predicted sensor measurements are then used as true measurements, as if they were from nodes currently located at these candidate locations. Thus, the problem of making decision on where to move is viewed as the problem of selecting one of the predicted measurement that is expected to best improve the quality of tracking data. In this sense, this problem is similar to the sensor selection strategy in [51]. However, in [51], the sensor measurements at time t are already available locally at those nodes to be selected, whereas in our case, the focus is on predicted measurements corresponding to candidate locations that a node has to decide to choose as its next location at time $t + 1$. Moreover, [51] does not consider mobile sensor nodes.

3.2 Assumptions

To simplify the discussion, this chapter makes the following assumptions for the sensor network:

1. It is assumed that both sensor nodes and the target are moving at constant speeds. This is justified since Bayesian estimation is not limited by this assumption.
2. It is assumed that the sampling interval of all sensor nodes is small enough such that there is no drastic change in sensor measurements of the target state.
3. All nodes have the same number of candidate locations where they can move. This is justified for a gridded sensing region.
4. Node s_i considers movement and carries out the evaluation process only if it detects a target.
5. A node uses the prior of its current location to predict the sensor measurements at its candidate locations.
6. A node uses the current sensor measurements from its current one-hop neighbor nodes.
7. When node s_i performs evaluation for movement decision at time instant t, it is assumed that the neighbor nodes of s_i at time instant $t + 1$ are the same as at t.
8. When node s_i performs evaluation for movement decision at time instant t, it is assumed that s_i has collected sensor measurements from its one-hop neighbors that have also detected the target.
9. When node s_i performs evaluation for movement decision at time instant t, it is assumed that s_i has complete knowledge about the candidate locations of its neighboring nodes.
10. When node s_i performs evaluation for movement decision at time instant t, for simplified discussion, this chapter only considers costs associated with locomotion, loss of communication connectivity, and loss of sensing coverage.

Note that it is possible that in certain time instances, the neighbor nodes can change. When this happens, errors are potentially introduced due to the fact that measurements from neighbor nodes are not consistent. However, one can always choose the value of Δt, i.e., the time difference two consecutive time instances, to be arbitrarily small such that the error introduced is negligible. Of course, in reality, making the Δt small imposes a much higher hardware requirement on the node's capabilities in sensing, computation, memory, and storage. Therefore, it is recommended to tune Δt to be just small enough to avoid the aforementioned side effect while still using most the affordable hardware.

Consider a node s_i located at $\mathbf{l}_t^i$ at time t. The prior from previous estimation, denoted by $p(\hat{\mathbf{x}}_{t-1}|\mathbf{z}_{1:t-1})$, is different for each sensor node since a node can start this process at any time when necessary and the sensor measurements are only from its one-hop neighborhood. This neighborhood might constantly change due to node movement. Let $\mathscr{N}_t^i$ be a function that maps from the discrete time t to a set of nodes that are one-hop neighbors of node s_i at t. Let us denote the sensor measurements available for s_i at time t as $\mathbf{z}_t^i$, i.e., $\mathbf{z}_t^i$ contains the measurement z_t^i from s_i and z_t^j from all nodes in $\mathscr{N}_t^i$. Let us use $\hat{\mathbf{x}}_t^i$ to denote the target estimate on node s_i at time t that is derived using $\mathbf{z}_t^i$. Figure 2 illustrates how a node uses the predicted measurement to decide its next movement.

To simplify the discussion, it is assumed that there are only a limited number of locations that a node can move to from its current position; these are referred to as candidate locations. Candidate locations can be determined from the speed of the node and the sampling frequency of the sensor on the node. At time t, let $\mathscr{L}_{t+1}^i$ be the set of candidate locations for node s_i at time $t+1$, i.e., $\mathbf{l}_{t+1}^i \in \mathscr{L}_{t+1}^i$. Our goal is then to find the location $\mathbf{l}_{t+1}^i$ such that the tracking quality is best among all candidate locations. Note that $\mathbf{l}_t^i \in \mathscr{L}_{t+1}^i$ has already included the case that a node may decide to stay in its current location. For a given grid point in the surveillance area, $\mathscr{L}_{t+1}^i$ can include locations that are just one step away from the current location, corresponding to due-east, north-east, due-north, north-west, due-west, south-west, due-south, south-east, and the current location, respectively. This is shown in Fig. 2 as the dotted squares where node s_i can move to.

Let us further denote the predicted sensor measurement for s_i at $\mathbf{l}_{t+1}^i$ as $\hat{z}_{t+1}^i(\mathbf{l}_{t+1}^i)$. Let us use $\hat{\mathbf{z}}_{t+1}^i(\mathbf{l}_{t+1}^i)$ to denote the vector containing all predicted sensor

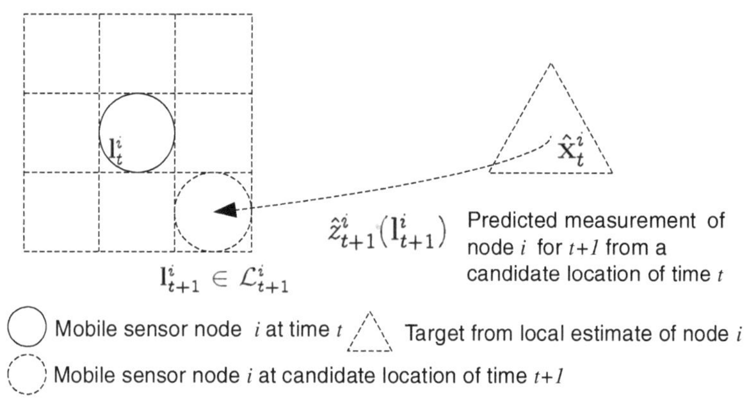

Fig. 2 Node s_i predicts its measurement at a candidate location based on its current target estimate

measurements within a one-hop neighborhood of s_i when s_i is located at $\mathbf{l}_{t+1}^i$ at time $t+1$. For example, assume that sensor nodes are equipped with acoustic sensors. The sensor measurement at time t on a node s_i located at $\mathbf{l}_t^i$ for a target located $\mathbf{x}_t$ is given by [14]

$$z_t^i = \frac{a}{\parallel \mathbf{x}_t - \mathbf{l}_t^i \parallel^{\frac{\alpha}{2}}} + v_t^i, \tag{6}$$

where a and α represent physical characteristics of the acoustic sensor, $\|\mathbf{x}_t - \mathbf{l}_t^i\|$ denotes the shortest distance between the target and the sensor at time t, and v_t^i is a measurement noise, assumed to be Gaussian. Based on Eq. (6), the predicted sensor measurement for node s_i at $\mathbf{l}_{t+1}^i$ can be obtained by

$$\hat{z}_{t+1}^i(\mathbf{l}_{t+1}^i) = \frac{a}{\parallel \hat{\mathbf{x}}_t^i - \mathbf{l}_{t+1}^i \parallel^{\frac{\alpha}{2}}} + v_{t+1}^i, \tag{7}$$

where $\hat{\mathbf{x}}_t^i$ represents the target estimate by node s_i using local knowledge, i.e., sensor measurements within the one-hop neighborhood of s_i. Let us use $\hat{\mathbf{x}}_t^i$ instead of $\hat{\mathbf{x}}_t$ due to the fact that s_i may not be able carry out target estimation to obtain $\hat{\mathbf{x}}_t$ based on all sensor measurements at time t; some nodes with the target data may not be within the one-hop neighborhood of s_i. Moreover, it is impossible to use the target estimate at time $t+1$, i.e., $\hat{\mathbf{x}}_{t+1}^i$, since sensor measurements at $t+1$ are not yet available. This implies that the error in the predicted sensor measurement is expected to be large if there is a drastic change in the target state, e.g., a sudden acceleration of the target.

Note that $\hat{\mathbf{z}}_{t+1}^i(\mathbf{l}_{t+1}^i)$ contains predicted sensor measurements from the neighbor nodes of s_i at $t+1$. However, for each candidate location $\mathbf{l}_{t+1}^i \in \mathscr{L}_{t+1}^i$, each neighbor node $s_j \in \mathscr{N}_t^i$ also has $|\mathscr{L}_{t+1}^j|$ candidate locations to choose from, which means that s_i has to calculate $|\mathscr{L}_{t+1}^j|$ predicted sensor measurements for each neighbor node $s_j \in \mathscr{N}_t^j$. This in turn implies that a complete computation on node s_i for all possible predicted sensor measurements requires a total of $|\mathscr{L}_{t+1}^i| \prod_{s_j \in \mathscr{N}_t^i} |\mathscr{L}_{t+1}^j|$ calculations. In this way, s_i is assumed to have complete knowledge of $\mathscr{L}_{t+1}^j$ for each $s_j \in \mathscr{N}_t^i$.

Furthermore, since neighbor nodes of s_i may also move, $\mathscr{N}_t^i$ and $\mathscr{N}_{t+1}^i$ are not necessarily the same. Since $\mathscr{N}_{t+1}^i$ is not yet available for s_i at t, It is assumed that $\mathscr{N}_{t+1}^i$ is the same as $\mathscr{N}_t^i$; the latter can be obtained by exchanging neighbor sensor measurements at current time instant t. The discussion is further simplified by assuming that s_i uses current sensor measurements from its current neighbor nodes in $\mathscr{N}_t^i$, i.e., for any two $\mathbf{l}_{t+1}^{i_1}, \mathbf{l}_{t+1}^{i_2} \in \mathscr{L}_{t+1}^i$, $\hat{\mathbf{z}}_{t+1}^i(\mathbf{l}_{t+1}^{i_1})$ and $\hat{\mathbf{z}}_{t+1}^i(\mathbf{l}_{t+1}^{i_2})$ only differ from each other in $\hat{z}_{t+1}^i(\mathbf{l}_{t+1}^{i_1})$ and $\hat{z}_{t+1}^i(\mathbf{l}_{t+1}^{i_2})$, and all other elements are the same as z_t^j, $\forall s_j \in \mathscr{N}_t^i$. These assumptions are valid because the interval between two consecutive discrete time instants is small; since the maximum displacement from node movement is also correspondingly small, it is expected that there is no drastic change in sensor measurements.

Table 1 List of notations used throughout the chapter

Notation	Description
n	Total number of sensor nodes
s_i	Sensor node i
r_c, r_s	Sensor node communication and sensing radius
X, Y	The x- and y-dimension of the 2D sensing grid
$g_k, \mathscr{G}$	Grid point with index k and the set of all grid points
$\mathbf{x}_t, d_x$	Target state vector at time t and its dimension
$\hat{\mathbf{x}}_t, d_x$	Estimated target state vector at time t and its dimension
$\mathbf{z}_t, d_z$	Sensor measurement vector at time t and its dimension
$\mathbf{w}_t, d_w$	Process noise vector at time t and its dimension
$\mathbf{v}_t, d_v$	Measurement noise vector at time t and its dimension
$z_t^i, \mathbf{z}_t^i$	Measurement from s_i and measurements available for s_i at t
m	Total number of candidate locations
$\mathbf{l}_t^i$	Candidate location vector for s_i at t
$\mathscr{L}_t^i$	The set of all candidate location vectors for s_i at t
$\mathscr{N}_t^i$	One-hop neighbors of node s_i at t
d_{t+1}^{ij}	The distance between node s_i and s_j
$\hat{z}_{t+1}^i(\mathbf{l}_{t+1}^i)$	The predicted measurement for s_i at $\mathbf{l}_{t+1}^i$
$\hat{\mathbf{z}}_{t+1}^i(\mathbf{l}_{t+1}^i)$	All predicted measurements for s_i at $\mathbf{l}_{t+1}^i$
ψ	Information utility function
δr_c	Communication radius threshold
c_k^i	Sensing coverage probability at g_k from s_i
$c_k^i(\mathbf{l}_t^i)$	Coverage probability of g_k from s_i at $\mathbf{l}_t^j$
S_k	The set of nodes that can detect g_k
c_k	The collective coverage probability of g_k from S_k
$\mathscr{S}_t^k$	The set of nodes that detect grid point g_k at time t
c_t^k	The collective coverage probability of g_k from $\mathscr{S}_t^k$
p_{th}	The required sensing coverage threshold
A_i	The sensing area of node s_i, $A_i \in \mathscr{G}$
$\mathscr{A}(\mathbf{l}_t^i)$	The sensing area of node s_i located at $\mathbf{l}_t^i$ at t

Table 1 presents a list of notation used throughout the chapter.

3.3 Probability of Node Movement to a New Location

Figure 3 illustrates how the predicted sensor measurement at candidate locations is used to obtain the predicted target estimate on the basis of Bayesian estimation, as described in Sect. 3.1. Next, the Bayesian estimation equations introduced earlier can be rewritten to calculate the target estimate based on the predicted sensor measurements. The node can then make a decision on where to move, i.e., select the best $\mathbf{l}_{t+1}^i \in \mathscr{L}_{t+1}^i$, by evaluating the estimated improvement in the target estimate for the next time instant $t+1$. This is shown as follows.

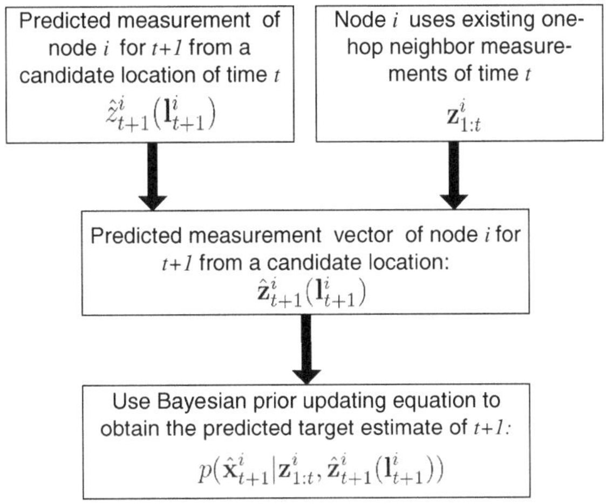

Fig. 3 An illustration of the use of the predicted measurement for updating the prior in Bayesian estimation

$$p(\hat{\mathbf{x}}^i_{t+1}|\mathbf{z}^i_{1:t}) = \int p(\hat{\mathbf{x}}^i_{t+1}|\hat{\mathbf{x}}^i_t)p(\hat{\mathbf{x}}^i_t|\mathbf{z}^i_{1:t})d\hat{\mathbf{x}}^i_t, \tag{8}$$

$$p(\hat{\mathbf{x}}^i_{t+1}|\mathbf{z}^i_{1:t}, \hat{\mathbf{z}}^i_{t+1}(\mathbf{l}^i_{t+1})) = \frac{p(\hat{\mathbf{z}}^i_{t+1}(\mathbf{l}^i_{t+1})|\hat{\mathbf{x}}^i_{t+1})p(\hat{\mathbf{x}}^i_{t+1}|\mathbf{z}^i_{1:t})}{p(\hat{\mathbf{z}}^i_{t+1}(\mathbf{l}^i_{t+1})|\mathbf{z}^i_{1:t})}, \tag{9}$$

$$p(\hat{\mathbf{z}}^i_{t+1}(\mathbf{l}^i_{t+1})|\mathbf{z}^i_{1:t}) = \int p(\hat{\mathbf{z}}^i_{t+1}(\mathbf{l}^i_{t+1})|\hat{\mathbf{x}}^i_{t+1})p(\hat{\mathbf{x}}^i_{t+1}|\mathbf{z}^i_{1:t})d\hat{\mathbf{x}}^i_{t+1}, \tag{10}$$

where $\hat{\mathbf{z}}^i_{t+1}$ is the vector containing predicted sensor measurements at time $t+1$ from node s_i, $\mathbf{z}^i_{1:t}$ is the vector containing previous measurements of node s_i and its one-hop neighbors, $\hat{\mathbf{x}}^i_t$ is the previous target estimate for node s_i, $\hat{\mathbf{x}}^i_{t+1}$ is the target estimate based on the predicted measurements $\hat{\mathbf{z}}^i_{t+1}$ at time $t+1$ from node s_i, and $p(\hat{\mathbf{z}}^i_{t+1}(\mathbf{l}^i_{t+1})|\hat{\mathbf{x}}^i_{t+1})$ is the likelihood based on the predicted measurements. Note that the above equations represent the Bayesian estimation method based on both predicted sensor measurements $\hat{\mathbf{z}}^i_{t+1}$ from node s_i and possibly incomplete previous target estimation $\hat{\mathbf{x}}^i_t$ from node s_i. However, if there exists a central processing node, it can still eventually send the posterior $p(\hat{\mathbf{x}}_t|\mathbf{z}_{1:t})$ to node s_i to replace $\hat{\mathbf{x}}^i_t$ for improving the local target estimate. In this way, even though s_i is able to make its movement decision based on local knowledge, future decisions can be improved when the target estimate based on complete sensor measurements is available to it.

After the target estimate based on predicted sensor measurements is obtained, i.e., $p(\hat{\mathbf{x}}^i_{t+1}|\mathbf{z}^i_{1:t}, \hat{\mathbf{z}}^i_{t+1}(\mathbf{l}^i_{t+1}))$, similar rules as proposed in [51] are used to select the best predicted sensor measurements $\hat{\mathbf{z}}^i_{t+1}(\mathbf{l}^i_{t+1})$, which subsequently gives us the best predicted sensor measurement $\hat{z}^i_{t+1}(\mathbf{l}^i_{t+1})$ from node s_i. Thus the corresponding candidate location $\mathbf{l}^i_{t+1}$ that is expected to best improve the tracking quality can be found from $\mathscr{L}^i_{t+1}$. This chapter uses the definition of the information utility function ψ described in [51]. The parameter ψ is defined as: $\psi : \mathscr{P}(\mathbb{R}^{d_l}) \rightarrow \mathbb{R}$, where d_l is the dimension of node location $\mathbf{l}^i_t$, and $\mathscr{P}(\mathbb{R}^{d_l})$ is a class of probability distributions.

In our case, $\mathscr{P}(\mathbb{R}^{d_l})$ corresponds to all posteriors calculated from predicted sensor measurements from all candidate nodes locations at $t+1$ in $\mathscr{L}_{t+1}^i$. Since the output of the utility function ψ is a real number, ψ maps the posterior based on the predicted measurement from the candidate location to a real number representing the improvement in tracking quality from this candidate location. In standard estimation theory, the trace or determinant of the estimation error covariance matrix is commonly used as a measure of the tracking quality [7, 51]. Let $\hat{\mathbf{e}}_{t+1}^i(\mathbf{l}_{t+1}^i)$ be the target estimate error from local estimation on s_i based on predicted measurements, and $\mathbf{R}_{t+1}^i(\mathbf{l}_{t+1}^i)$ be the corresponding covariance matrix of $\hat{\mathbf{e}}_{t+1}^i(\mathbf{l}_{t+1}^i)$, i.e., $\mathbf{R}_{t+1}^i(\mathbf{l}_{t+1}^i) = E\{(\hat{\mathbf{e}}_{t+1}^i)\cdot(\hat{\mathbf{e}}_{t+1}^i)'\}$, where $'$ denotes the transpose of a matrix. Suppose the posterior is Gaussian. One way of defining the utility function is given by:

$$\psi(p(\hat{\mathbf{x}}_{t+1}^i|\hat{\mathbf{z}}_{t+1}^i(\mathbf{l}_{t+1}^i)) \equiv -\mathbf{trace}(\mathbf{R}_{t+1}^i(\mathbf{l}_{t+1}^i)), \tag{11}$$

where **trace** is the trace of a matrix. Some alternative approaches can also be used to evaluate the improvement in tracking quality [7, 51]. In general, the selection of a candidate location should maximize the utility function, i.e.,

$$\bar{\mathbf{l}}_{t+1}^i = \max_{\mathbf{l}_{t+1}^i \in \mathscr{L}_{t+1}^i} \psi(p(\hat{\mathbf{x}}_{t+1}^i|\hat{\mathbf{z}}_{t+1}^i(\mathbf{l}_{t+1}^i)). \tag{12}$$

Since $\psi(p(\hat{\mathbf{x}}_{t+1}^i|\hat{\mathbf{z}}_{t+1}^i(\mathbf{l}_{t+1}^i))$ is associated with the posterior based on the predicted sensor measurements on candidate locations, Eq. (12) is essentiall a random variable. Thus, the probability of a node making a decision to move to $\mathbf{l}_{t+1}^i$ at $t+1$ is expressed as follows.

$$\begin{aligned} p(\mathbf{l}_{t+1}^i) &\equiv \Pr(s_i \text{ at } \mathbf{l}_{t+1}^i \text{ at } t+1) \\ &= \frac{\psi(p(\hat{\mathbf{x}}_{t+1}^i|\hat{\mathbf{z}}_{t+1}^i(\mathbf{l}_{t+1}^i))}{\sum\limits_{\mathbf{l}_{t+1}^i \in \mathscr{L}_{t+1}^i} \psi(p(\hat{\mathbf{x}}_{t+1}^i|\hat{\mathbf{z}}_{t+1}^i(\mathbf{l}_{t+1}^i))}, \end{aligned} \tag{13}$$

where $p(\mathbf{l}_{t+1}^i)$ represents the probability of node s_i being at $\mathbf{l}_{t+1}^i$ at $t+1$ to improve the quality of target tracking data. Note that in Eq. (13), it is assumed that by definition, $\psi(p(\hat{\mathbf{x}}_{t+1}^i|\hat{\mathbf{z}}_{t+1}^i(\mathbf{l}_{t+1}^i))$ yields a non negative real value. As shown in the next section, $p(\mathbf{l}_{t+1}^i)$ is used to evaluate the integrated cost when other factors are also considered in the discussed mobility management scheme. From Eq. (13), the computational complexity for obtaining $p(\mathbf{l}_{t+1}^i)$ depends on the complexity of the estimation algorithm as well as the size of set of candidate locations $\mathscr{L}_{t+1}^i$.

4 Estimation of Negative Consequences

As discussed in Sect. 1, it is required to consider the negative consequences of node movement, e.g., additional energy consumption, connectivity loss, and coverage loss. Additional risks include the need for reestablishing the route, the potentially higher rate of node failures due to node movement. This chapter limits the discussion to the three most important factors, namely energy, connectivity, and coverage issues. Impact from other factors may be considered as a topic for future investigation. In this section, probabilities associated with above-mentioned negative consequences, when node chooses to move to a candidate location, are derived. These probabilities are then used in an integrated cost evaluation described in the next section.

4.1 Energy Consumption

Obviously, nodes have to spend additional energy for movement. Even though sensor nodes on mobile platforms can carry more battery supplies, it is important to ensure that the available energy is properly used to best serve the purpose of surveillance tasks. It is assumed a simplified dynamics model for the sensor node movement; this model is similar to the one used in [35]. It is assumed that all nodes move at the same constant speed. Section 4.1 also ignores the energy consumption for acceleration when the node starts to move as well as for deceleration when the node stops to move. It is also assumed that the node always moves along a straight line, i.e., the distance that a node has moved during the interval between two consecutive time instants is the distance between the old location and the new location of the node.

Consider an arbitrarily chosen node s_i located at $\mathbf{l}_t^i$ at time instant t. Let $\mathscr{E}_{t+1}^i(\mathbf{l}_{t+1}^i)$ be a mapping from $\mathbf{l}_{t+1}^i$ to a real number representing the energy consumption on s_i when s_i decides to move to $\mathbf{l}_{t+1}^i$. The energy consumption is related to the distance that the node has moved from time t to $t+1$ as follows:

$$\mathscr{E}_{t+1}^i(\mathbf{l}_{t+1}^i) = \theta \|\mathbf{l}_t^i - \mathbf{l}_{t+1}^i\|, \tag{14}$$

where θ is a constant in unit of Joule per meter, and $\|\mathbf{l}_t^i - \mathbf{l}_{t+1}^i\|$ is the Euclidean distance between $\mathbf{l}_t^i$ and $\mathbf{l}_{t+1}^i$. Note that Eq. (14) indicates a linear relationship between the energy consumption and the distance moved because this chapter only considers a simplified energy consumption evaluation model. It is however not restricted to follow the energy consumption model in this chapter, based on nodes characteristics, readers can apply their energy consumption model in a similar way.

Since s_i may have multiple candidate locations to choose from as its next location, let us define $p_{t+1}^i(\mathrm{E}|\mathbf{l}_{t+1}^i)$ as the weighted probability for node s_i to move to $\mathbf{l}_{t+1}^i$ at $t+1$, where the weight indicates the energy consumption associated with this movement. The probability $p_{t+1}^i(\mathrm{E}|\mathbf{l}_{t+1}^i)$ is given by

$$p^i_{t+1}(\mathrm{E}|\mathbf{l}^i_{t+1}) = \frac{\bar{\mathscr{E}} - \mathscr{E}^i_{t+1}(\mathbf{l}^i_{t+1})}{\sum\limits_{\mathbf{l}^i_{t+1} \in \mathscr{L}^i_{t+1}} \left(\bar{\mathscr{E}} - \mathscr{E}^i_{t+1}(\mathbf{l}^i_{t+1})\right)}, \tag{15}$$

where $\bar{\mathscr{E}}(\bar{\mathscr{E}} \geq \mathscr{E}^i_{t+1}(\mathbf{l}^i_{t+1})$ is a known constant representing the maximum amount of energy that the node can afford for making the one-step movement. The parameter $\bar{\mathscr{E}}$ usually depends on the available battery and the operational lifetime requirement. Obviously, Eq. (15) yields the highest probability for a candidate location with minimum amount of energy consumption.

4.2 Probability of a Node Being Disconnected

Section 4.2 focuses on analyzing the risk that node becomes disconnected due to its possible movement for the next time instant $t + 1$. Once again, let us consider node s_i. and assume that the network is connected at current time instant t. The current work that analyzes connectivity in wireless sensor networks deals only with the relationship between the node density and the probability of the network being connected or disconnected [26, 49]. While these results are useful in random sensor deployment, they cannot be directly applied to mobile sensor networks where the topology is dynamically changing. This chapter simplifies the discussion by only considering the probability that node s_i is disconnected from all other nodes at the next time instant, i.e., the probability that $\mathscr{N}^i_{t+1} = \phi$, where $\mathscr{N}^i_{t+1}$ is defined earlier as the set of neighbor nodes for s_i at time $t + 1$. Let d^{ij}_{t+1} be the distance between node s_i and s_j at $t + 1$, and let r_c be the communication radius for a node. Then the probability that s_i is disconnected at $t + 1$ is given by

$$\Pr(\mathscr{N}^i_{t+1} = \phi) \equiv \Pr\left(d^{ij}_{t+1} > r_c, \forall s_j \in S \setminus \{s_i\}\right), \tag{16}$$

which requires the testing of connectivity from s_i to all other nodes in S since knowledge about $\mathscr{N}^i_{t+1}$ is generally not available until s_i has moved to the new location $\mathbf{l}^i_{t+1}$. To ensure that the discussed scheme is suitable for a distributed implementation, the calculation to the probability that s_i is disconnected is restricted to be from nodes only in the current neighbor set $\mathscr{N}^i_t$. Since it is possible that $\mathscr{N}^i_t \cap \mathscr{N}^i_{t+1} = \phi$ and $\mathscr{N}^i_{t+1} \setminus \mathscr{N}^i_t \neq \phi$, the above restriction is a conservation one in evaluating the probability of loss of connectivity. In this way, the calculation requires only current one-hop knowledge. Let $p^i_{t+1}(\mathrm{C}) \equiv \Pr(\mathscr{N}^i_{t+1} = \phi)$ be the probability that s_i is disconnected at time $t + 1$. Then $p^i_{t+1}(\mathrm{C})$ is givenby

$$p^i_{t+1}(\mathrm{C}) \equiv \sum_{\mathbf{l}^i_{t+1} \in \mathscr{L}^i_{t+1}} \Pr(\mathscr{N}^i_{t+1} = \phi|\mathbf{l}^i_{t+1})p(\mathbf{l}^i_{t+1})$$

$$= \sum_{\mathbf{l}_{t+1}^i \in \mathscr{L}_{t+1}^i} \Pr\left(d_{t+1}^{ij} > r_c, \forall s_j \in \mathscr{N}_t^i | \mathbf{l}_{t+1}^i\right) p(\mathbf{l}_{t+1}^i), \tag{17}$$

where $p(\mathbf{l}_{t+1}^j)$ is defined by Eq. (13) as the probability that neighbor node s_j is located at $\mathbf{l}_{t+1}^j$ at $t+1$. It can be argued whether $p(\mathbf{l}_{t+1}^j)$ is indeed available to s_i. The value of $p(\mathbf{l}_{t+1}^j)$ can be requested by s_i from its neighbors before it makes any movement, at the same time when it is receiving the sensor measurements. However, since $p(\mathbf{l}_{t+1}^j)$ is not available until the procedure described in Sect. 3 is completed, this implies that all nodes have to first wait for their neighbors to finish the evaluation of improvement in target tracking data. This also requires additional bandwidth. If retrieval of $p(\mathbf{l}_{t+1}^j)$ by s_i is not feasible, s_i does not have any *a priori* knowledge about how its neighbors are going to move. In this case, it can be simplified by letting $p(\mathbf{l}_{t+1}^j) = \frac{1}{|\mathscr{L}_{t+1}^j|}$, where $|\mathscr{L}_{t+1}^j|$ is a constant, and the same for each node s_j.

In Eq. (17), let $p_{t+1}^i(\mathrm{C}|\mathbf{l}_{t+1}^i) \equiv \Pr(d_{t+1}^{ij} > r_c, \forall s_j \in \mathscr{N}_t^i | s_i \text{ at } \mathbf{l}_{t+1}^i)$, which is the probability that s_i is disconnected at $t+1$, given that it moves to $\mathbf{l}_{t+1}^i$. Therefore, $p_{t+1}^i(\mathrm{C})$ is given by

$$\begin{aligned} p_{t+1}^i(\mathrm{C}) &= \sum_{\mathbf{l}_{t+1}^i \in \mathscr{L}_{t+1}^i} p_{t+1}^i(\mathrm{C}|\mathbf{l}_{t+1}^i) p(\mathbf{l}_{t+1}^i) \\ &= \sum_{\mathbf{l}_{t+1}^i \in \mathscr{L}_{t+1}^i} \left(\prod_{s_j \in \mathscr{N}_t^i} \Pr(d_{t+1}^{ij} > r_c | \mathbf{l}_{t+1}^i) \right) \cdot p(\mathbf{l}_{t+1}^i), \end{aligned} \tag{18}$$

where $\Pr(d_{t+1}^{ij} > r_c | \mathbf{l}_{t+1}^i)$ represents the probability that s_i is disconnected from its neighbor s_j, given that s_i moves to $\mathbf{l}_{t+1}^i$ at $t+1$. The probability $\Pr(d_{t+1}^{ij} > r_c | \mathbf{l}_{t+1}^i)$ can be obtained by

$$\Pr\left(d_{t+1}^{ij} > r_c | \mathbf{l}_{t+1}^i\right) = \sum_{\mathbf{l}_{t+1}^j \in \mathscr{L}_{t+1}^j} \Pr\left(d_{t+1}^{ij} > r_c | \mathbf{l}_{t+1}^j, \mathbf{l}_{t+1}^i\right) p(\mathbf{l}_{t+1}^j). \tag{19}$$

Now, $\Pr(d_{t+1}^{ij} > r_c | \mathbf{l}_{t+1}^j, \mathbf{l}_{t+1}^i)$, the probability that s_i is disconnected at time $t+1$ from a neighbor s_j, given that s_i is at $\mathbf{l}_{t+1}^i$ and s_j is at $\mathbf{l}_{t+1}^j$, is either 1 or 0 when $\mathbf{l}_{t+1}^i$ and $\mathbf{l}_{t+1}^j$ are given. However, to also include the differences in the distances between s_i and its neighbors in $\mathscr{N}_t^i$, the definition of this probability can be refined in several ways. Let $p_{t+1}^{ij}(\mathrm{C}) \equiv \Pr(d_{t+1}^{ij} > r_c | \mathbf{l}_{t+1}^i, \mathbf{l}_{t+1}^j)$. Several ways of determining $p_{t+1}^{ij}(\mathrm{C})$ are listed below.

1. Use predicted distance d_{t+1}^{ij} between s_i and s_j directly:

$$p_{t+1}^{ij}(\mathrm{C}) \equiv \Pr(d_{t+1}^{ij} > r_c | \mathbf{l}_{t+1}^i, \mathbf{l}_{t+1}^j) = \frac{d_{t+1}^{ij}}{\sum\limits_{\mathbf{l}_{t+1}^j \in \mathscr{L}_{t+1}^j} d_{t+1}^{ij}}. \tag{20}$$

Therefore, if s_i is at $\mathbf{l}_{t+1}^i$, it will have the highest probability of being disconnected from s_j when their mutual distance is the maximum among the candidate locations of s_j.

2. Alternatively, if you want to be more conservative, i.e., to inform the node of the possibility of being disconnected in advance of this event actually occurring at the next time instant, $p_{t+1}^{ij}(\mathrm{C})$ can be defined as

$$p_{t+1}^{ij}(\mathrm{C}) \equiv \Pr(d_{t+1}^{ij} > r_c | \mathbf{l}_{t+1}^i, \mathbf{l}_{t+1}^j) = \begin{cases} \dfrac{\frac{1}{\delta r_c - d_{t+1}^{ij}}}{\sum\limits_{\mathbf{l}_{t+1}^j \in \mathscr{L}_{t+1}^j} [\frac{1}{\delta r_c - d_{t+1}^{ij}}]} & \text{if } d_{t+1}^{ij} < \delta r_c \\ 1 & \text{if } d_{t+1}^{ij} \geq \delta r_c, \end{cases} \tag{21}$$

where δr_c is a fraction of r_c, representing an acceptable threshold on how far away the neighbor s_j can be to s_i, e.g., $\delta r_c = 0.9 r_c$. Equation (21) can be used in situations where connectivity is given high priority; a smaller value of δr_c imposes a more strict connectivity constraint on node movement.

Next, let us rewrite Eq. (18) for $p_{t+1}^i(\mathrm{C})$ as follows:

$$\begin{aligned} p_{t+1}^i(\mathrm{C}) &= \sum_{\mathbf{l}_{t+1}^i \in \mathscr{L}_{t+1}^i} p_{t+1}^i(\mathrm{C}|\mathbf{l}_{t+1}^i) p(\mathbf{l}_{t+1}^i) \\ &= \sum_{\mathbf{l}_{t+1}^i \in \mathscr{L}_{t+1}^i} \left(\prod_{s_j \in \mathscr{N}_t^i} \left(\sum_{\mathbf{l}_{t+1}^j \in \mathscr{L}_{t+1}^j} p_{t+1}^{ij}(\mathrm{C}) p(\mathbf{l}_{t+1}^j) \right) \right) p(\mathbf{l}_{t+1}^i). \end{aligned} \tag{22}$$

Note that Eq. (22) requires only local knowledge from one-hop neighborhood, which then can be implemented in a distributed manner. Also note that, from a general perspective of the connectivity in mobile sensor network, movement of nodes causes the partitioning of the network. As shown in the above discussion, the partitioning is avoided by evaluating the probability of the node being disconnected, which again is from the predicated target estimate. The connectivity problem in this case is integrated with the application goal for target tracking.

4.3 Potential Loss of Sensing Coverage

Another potential risk arising from node movement is the possible loss of sensing coverage in certain regions of the sensor field. When nodes move in the sensor field, the sensing area that is originally covered by these mobile nodes may not be covered by any other nodes. This implies that there may be some "holes" in the coverage over the sensor field. If these holes are not covered by any other nodes, a target that appears at the same time in this area will remain undetected. Therefore, the movement of nodes has to be managed in a way such that no such "hole" is formed.

The sensor field is represented as a 2D grid with dimension X by Y. Let there be a total of $n_g = XY$ grid points in the set $\mathcal{G}$, and let g_k be a grid point with index k. Let c_k^i be the probability that g_k is covered by node s_i, and $c_k \equiv p(S_k)$ be the mapping from the set of nodes S_k that detect the grid point g_k to the probability that g_k is covered by the set of nodes in S_k. When nodes in S_k move, c_k changes due to the fact that locations of nodes in S_k change as well; thus S_k may also change. Let $c_k^i(\mathbf{l}_t^i)$ map $\mathbf{l}_t^j$ to a probability representing the coverage probability of g_k due to s_i, when s_i is located at $\mathbf{l}_t^j$. Let $\mathcal{S}_t^k$ a mapping from time instant t to a set of nodes that detect grid point g_k at time t. Let us then define $c_t^k \equiv p(\mathcal{S}_t^k)$ be the mapping from a set of nodes $\mathcal{S}_t^k$ that detect g_k at t to the coverage probability at time t for g_k from nodes in $\mathcal{S}_t^k$. It is assumed that initially the sensing coverage requirement is achieved by the sensor deployment algorithm, i.e., after the sensor deployment, $\forall g_k \in \mathcal{G}, c_k \geq p_{th}$, where p_{th} is a given sensor deployment control parameter representing the required sensing coverage threshold. One way for calculating c_k and c_t^k is shown in [54] as:

$$c_k = 1 - \prod_{s_i \in S_k} (1 - c_k^i) \tag{23}$$

and

$$c_t^k = 1 - \prod_{s_j \in \mathcal{S}_k^t} (1 - c_k^j). \tag{24}$$

Consider an arbitrarily chosen node s_i at time t. Suppose that s_i moves from its current location $\mathbf{l}_t^i$ to $\mathbf{l}_{t+1}^i$. To simplify the analysis, it is assumed that the sensing radius r_s of a node remains constant when it moves. Therefore, the area that s_i is able to cover at any time instant is fixed. However, the global sensing coverage may be affected by node movement. The coverage in the sensing area centered at the current location of s_i, i.e., $\mathbf{l}_t^i$, may be reduced, e.g., there may not be enough nodes to provide coverage for the area centered at $\mathbf{l}_t^i$ after s_i has moved to $\mathbf{l}_{t+1}^i$. On the other hand, s_i may even improve the sensing coverage at its new location $\mathbf{l}_{t+1}^i$. A thorough evaluation of the loss (or gain) of global sensing coverage requires an exchange of global information for the current topology of the sensor network. Due to the need for a distributed implementation, only the knowledge of a limited number of hops is used for local sensing coverage evaluation, i.e., $\lceil \frac{2r_s}{r_c} \rceil$-hops neighbor information [54].

Let the sensing area of node s_i be $A_i \in \mathcal{G}$, i.e., A_i is the set of grid points in $\mathcal{G}$ covered by s_i. Let $\mathscr{A}(\mathbf{l}_t^i)$ be a mapping from $\mathbb{R}^{d_l}$ to the set of grid points, i.e., $\mathscr{A}(\mathbf{l}_t^i)$ is the set of grid points corresponding to the sensing area centered at $\mathbf{l}_t^i$. Let us denote the set of grid points that will not be covered by s_i after s_i moves to $\mathbf{l}_{t+1}^i$ as $\Delta\mathscr{A}(\mathbf{l}_{t+1}^i)$, which is given by

$$\Delta\mathscr{A}(\mathbf{l}_{t+1}^i) = \mathscr{A}(\mathbf{l}_{t+1}^i) \cup \mathscr{A}(\mathbf{l}_t^i) \setminus \mathscr{A}(\mathbf{l}_{t+1}^i). \tag{25}$$

To ensure that there is no hole in the sensing area originally covered by s_i at t, the following condition must be satisfied:

$$c_{t+1}^k \geq p_{th}, \ \forall g_k \in \Delta\mathscr{A}(\mathbf{l}_{t+1}^i). \tag{26}$$

The next task is to find the expected coverage for g_k at time $t+1$, i.e, $E\{c_{t+1}^k\}$. Obviously, $E\{c_{t+1}^k\}$ requires the knowledge of $\mathscr{S}_{t+1}^k$, which is not available to node s_i at t. However, the calculation of $E\{c_{t+1}^k\}$ can be restricted on nodes in $\mathscr{S}_t^k$ only, which contains information about the probability that any $s_j \in \mathscr{S}_t^k$ is not in $\mathscr{S}_{t+1}^k$. Since it is possible that $\mathscr{S}_{t+1}^k$ may include other nodes that are not in $\mathscr{S}_t^k$, the evaluation is more conservative. The value of $E\{c_{t+1}^k\}$ for grid point $g_k \in \Delta\mathscr{A}(\mathbf{l}_{t+1}^i)$ can be obtained as follows:

$$\begin{aligned}
E\{c_{t+1}^k\} &\equiv E\{c_{t+1}^k(\mathscr{S}_t^k)\} \\
&= E\left\{1 - \prod_{s_j \in \mathscr{S}_t^k} (1 - c_k^j(\mathbf{l}_{t+1}^j, \forall \mathbf{l}_{t+1}^j \in \mathscr{L}_{t+1}^j))\right\} \\
&= 1 - E\left\{\prod_{s_j \in \mathscr{S}_t^k} (1 - c_k^j(\mathbf{l}_{t+1}^j, \forall \mathbf{l}_{t+1}^j \in \mathscr{L}_{t+1}^j))\right\} \\
&= 1 - \prod_{s_j \in \mathscr{S}_t^k} \left(1 - E\left\{c_k^j(\mathbf{l}_{t+1}^j, \forall \mathbf{l}_{t+1}^j \in \mathscr{L}_{t+1}^j)\right\}\right) \\
&= 1 - \prod_{s_j \in \mathscr{S}_t^k} \left(1 - \sum_{\mathbf{l}_{t+1}^j \in \mathscr{L}_{t+1}^j} c_k^j(\mathbf{l}_{t+1}^j) p(\mathbf{l}_{t+1}^j)\right),
\end{aligned} \tag{27}$$

where $p(\mathbf{l}_{t+1}^j)$ is given by Eq. (13). Note that since nodes make movement decisions based on local knowledge only, it is assumed that the movement decision on all nodes are independent. Hence, the probability of the appearance of a hole in $\Delta\mathscr{A}(\mathbf{l}_{t+1}^i)$ can be obtained as:

$$\Pr(\Delta\mathscr{A}(\mathbf{l}_{t+1}^i) \text{ has a hole}) = \Pr(\exists g_k \in \Delta\mathscr{A}(\mathbf{l}_{t+1}^i) | c_{t+1}^k < p_{th})$$

$$\equiv \Pr(\exists g_k \in \Delta\mathscr{A}(\mathbf{l}^i_{t+1}) | E\{c^k_{t+1}\} < p_{th}). \qquad (28)$$

It is easy to see that Eq. (28) yields either 1 or 0 because $\Pr(\Delta\mathscr{A}(\mathbf{l}^i_{t+1})$ has a hole) denotes a binary outcome on the existence of a hole. Note that there may be more than one g_k in $\Delta\mathscr{A}(\mathbf{l}^i_{t+1})$ that satisfies $c^k_{t+1} < p_{th}$. Furthermore, different choice of $\mathbf{l}^i_{t+1}$ may yield different numbers of such grid points as g_k. To describe more accurately the loss of sensing coverage in $\Delta\mathscr{A}(\mathbf{l}^i_{t+1})$ due to the movement of node s_i, $p^i_{t+1}(\mathrm{S}|\mathbf{l}^i_{t+1})$ is introduced as the probability of the loss of sensing coverage for s_i, given that s_i is located at $\mathbf{l}^i_{t+1}$ at time $t+1$. The probability $p^i_{t+1}(\mathrm{S}|\mathbf{l}^i_{t+1})$ can be defined in several ways as listed below.

1. A straightforward method for obtaining $p^i_{t+1}(\mathrm{S}|\mathbf{l}^i_{t+1})$ is to count the number of grid points that fail to satisfy the coverage requirement. Therefore,

$$p^i_{t+1}(\mathrm{S}|\mathbf{l}^i_{t+1}) \equiv \frac{|\Delta\mathscr{A}_0(\mathbf{l}^i_{t+1})|}{|\Delta\mathscr{A}(\mathbf{l}^i_{t+1})|}, \qquad (29)$$

where $\Delta\mathscr{A}_0(\mathbf{l}^i_{t+1}) \subseteq \Delta\mathscr{A}(\mathbf{l}^i_{t+1})$, is defined as

$$\Delta\mathscr{A}_0(\mathbf{l}^i_{t+1}) \equiv \{g_k | g_k \in \Delta\mathscr{A}(\mathbf{l}^i_{t+1}), E\{c^k_{t+1}\} \leq p_{th}\}. \qquad (30)$$

Equation (30) gives the highest probability to the candidate location $\mathbf{l}^i_{t+1}$ that will have the highest number of grid points in the corresponding $\Delta\mathscr{A}(\mathbf{l}^i_{t+1})$ whose coverage are below the threshold p_{th}. Note that this approach does not consider the exact coverage on grid points.

2. To also include the drop of coverage on grid points that fail to meet the coverage threshold requirement, $p^i_{t+1}(\mathrm{S}|\mathbf{l}^i_{t+1})$ can be refined to express the coverage loss more precisely. Thus,

$$p^i_{t+1}(\mathrm{S}|\mathbf{l}^i_{t+1}) \equiv \Pr(\text{coverage drop of } g_k \in \Delta\mathscr{A}(\mathbf{l}^i_{t+1}) | E\{c^k_{t+1}\} \leq p_{th})$$

$$= \frac{\sum\limits_{\forall g_k \in \Delta\mathscr{A}(\mathbf{l}^i_{t+1}), E\{c^k_{t+1}\} < p_{th}} (p_{th} - E\{c^k_{t+1}\})}{\sum\limits_{\forall g_k \in \Delta\mathscr{A}(\mathbf{l}^i_{t+1})} E\{c^k_{t+1}\}}. \qquad (31)$$

3. Alternatively, since c^k_t is available to s_k at t, $p^i_{t+1}(\mathrm{S}|\mathbf{l}^i_{t+1})$ can then be defined to reflect the expected absolute loss of coverage per grid point. Let $\Delta c^k_{t+1} \equiv c^k_{t+1} - c^k_t$. Obviously, $\Delta c^k_{t+1} < 0$ represents the sensing coverage loss on grid point g_k. Since only $E\{c^k_{t+1}\}$ is available, the expectation of Δc^k_{t+1} is then used, i.e, $E\{\Delta c^k_{t+1}\}$. Therefore,

$$p^i_{t+1}(\mathrm{S}|\mathbf{l}^i_{t+1}) \equiv \Pr(\text{coverage loss of } g_k \in \Delta\mathscr{A}(\mathbf{l}^i_{t+1})|E\{\Delta c^k_{t+1}\} < 0)$$
$$= \frac{\sum\limits_{\forall g_k \in \Delta\mathscr{A}(\mathbf{l}^i_{t+1}), E\{\Delta c^k_{t+1}\}<0} |E\{\Delta c^k_{t+1}\}|}{\sum\limits_{\forall g_k \in \Delta\mathscr{A}(\mathbf{l}^i_{t+1})} |E\{\Delta c^k_{t+1}\}|}, \quad (32)$$

where $|E\{\Delta c^k_{t+1}\}|$ gives the absolute value of $E\{\Delta c^k_{t+1}\}$. Equation (32) shows that if a candidate location $\mathbf{l}^i_{t+1}$ gives the maximum total absolute coverage loss on all grid points, it has the highest probability of loss of sensing coverage $p^i_{t+1}(\mathrm{S}|\mathbf{l}^i_{t+1})$. Note that the denominator of Eq. (32) also includes the possible coverage gain on all grid points in $\Delta\mathscr{A}(\mathbf{l}^i_{t+1})$. For example, for two candidate locations with the same amount of coverage loss given by the numerator in Eq. (32), the one with larger coverage gain in the denominator has the lower probability of loss of sensing coverage.

Note that $\mathbf{l}^i_t \in \mathscr{L}^i_{t+1}$, which implies that $\Delta\mathscr{A}(\mathbf{l}^i_{t+1}) = \phi$ when $\mathbf{l}^i_{t+1} = \mathbf{l}^i_t$. So let us define $p^i_{t+1}(\mathrm{S}|\mathbf{l}^i_{t+1} = \mathbf{l}^i_t) = 0$. With $p^i_{t+1}(\mathrm{S}|\mathbf{l}^i_{t+1})$, the probability of loss of sensing coverage, denoted by $p^i_{t+1}(\mathrm{S})$, can be obtained as follows.

$$p^i_{t+1}(\mathrm{S}) = \sum_{\forall \mathbf{l}^i_{t+1} \in \mathscr{L}^i_{t+1}} p^i_{t+1}(\mathrm{S}|\mathbf{l}^i_{t+1})\, p(\mathbf{l}^i_{t+1}), \quad (33)$$

where $p^i_{t+1}(\mathrm{S}|\mathbf{l}^i_{t+1})$ can be obtained by one of the definitions described above.

5 Decision on Node Movement

In Sects. 3 and 4, we have derived the probabilities associated with tracking quality improvement, additional energy consumption, loss of connectivity, and loss of coverage. Next, we investigate the cost evaluation based using these probabilities.

5.1 Cost Evaluation

Recall that the optimal candidate location in the sense of tracking quality improvement is given by Eq. (12). However, this selection does not consider the negative consequences described in Sect. 4. Next, the selection rule is presented. The selection rule is based on the cost evaluation that takes into account of all negative consequences due to node movement. Let constants C_e, C_c, and C_s be the individual costs corresponding to energy consumption due to movement, loss of connectivity, and loss of coverage, respectively. To simplify the discussion, it is assumed that C_e, C_c, and C_s are already properly normalized. Note that C_e, C_c, and C_s have non negative values to indicate the costs associated with the energy consumption in movement and risks of losing connectivity and coverage. Let $\mathcal{C}^i_{t+1}(\mathbf{l}^i_{t+1})$ be the total cost for node s_i when s_i moves to $\mathbf{l}^i_{t+1}$ at $t+1$. Let $\mathcal{C}^i_{t+1}(\mathbf{l}^i_{t+1})$ be defined as

$$\mathcal{C}^i_{t+1}(\mathbf{l}^i_{t+1}) \equiv p^i_{t+1}(\mathrm{E}|\mathbf{l}^i_{t+1})\mathrm{C}_e w_e + p^i_{t+1}(\mathrm{C}|\mathbf{l}^i_{t+1})\mathrm{C}_c w_c + p^i_{t+1}(\mathrm{S}|\mathbf{l}^i_{t+1})\mathrm{C}_s w_s, \quad (34)$$

where w_e, w_c, and w_s are normalized weighting factors for energy consumption, connectivity and sensing coverage, respectively. In various types of application scenarios, w_e, w_c, and w_s can be used to reflect different priorities on these costs. Based on Eq. (34), the expected total cost for s_i when s_i moves to $\mathbf{l}^i_{t+1}$ can be found as:

$$E\{\mathcal{C}^i_{t+1}\} \equiv \sum_{\mathbf{l}^i_{t+1} \in \mathcal{L}^i_{t+1}} \mathcal{C}^i_{t+1}(\mathbf{l}^i_{t+1}) p(\mathbf{l}^i_{t+1}). \quad (35)$$

Equation (35) can be used as an extension to the discussed scheme in this chapter for group mobility management, where nodes can exchange their expected total cost and decide who should move. Similarly, if there is a need to impose a centralized control over node movement, the expected total cost given by Eq. (35) can be sent to the base station to guide the node movement. Hence the discussed scheme is flexible in various types of applications.

5.2 Decision on Movement

When the total cost is obtained for all candidate locations, the optimal selection of the candidate location for node s_i can be obtained by considering both positive and negative consequences. In practice, the decision on node movement depends on the actual requirement for particular operations in the sensor network. Considering both the positive and negative consequences, the selection rule for the candidate location is defined as a two-step selection process described below.

- **Selection Step 1**: We first find locations that are expected to improve the target tracking data. This is given by:

$$\bar{\mathcal{L}}^i_{t+1} = \{\mathbf{l}^i_{t+1} | \mathbf{l}^i_{t+1} \in \mathcal{L}^i_{t+1} \text{ and } \Psi(\mathbf{l}^i_{t+1}) \geq \delta_\psi \max_{\mathbf{l}^i_{t+1} \in \mathcal{L}^i_{t+1}} \Psi(\mathbf{l}^i_{t+1})\}, \quad (36)$$

  where $\Psi(\mathbf{l}^i_{t+1}) \equiv \psi(p(\hat{\mathbf{x}}^i_{t+1}|\hat{\mathbf{z}}^i_{t+1}(\mathbf{l}^i_{t+1}))$ is described in Sect. 3.3 and $\delta_\psi (0 \leq \delta_\psi \leq 1)$ is a given parameter, which represents how many candidate locations are considered good enough for the improvement of target tracking data. Obviously, $\bar{\mathcal{L}}^i_{t+1} \subseteq \mathcal{L}^i_{t+1}$. The next step then selects the candidate location in $\bar{\mathcal{L}}^i_{t+1}$ that has the minimum cost.

Input: node s_i

For $\forall \mathbf{l}^i_{t+1} \in \mathscr{L}^i_{t+1}$ **Do**

  Calculate the predicted measurement $\hat{z}^i_{t+1}(\mathbf{l}^{i_1}_{t+1})$ and construct $\hat{\mathbf{z}}^i_{t+1}(\mathbf{l}^i_{t+1})$

  Perform local target estimation using $\hat{\mathbf{z}}^i_{t+1}(\mathbf{l}^i_{t+1})$

  Evaluate the tracking improvement using Eq. (13)

  Calculate probability of movement $p(\mathbf{l}^i_{t+1})$

  Evaluate the $\mathscr{E}^i_{t+1}(\mathbf{l}^i_{t+1})$, $p^i_{t+1}(\mathrm{C}|\mathbf{l}^i_{t+1})$, and $p^i_{t+1}(\mathrm{S}|\mathbf{l}^i_{t+1})$

End

For $\forall \mathbf{l}^i_{t+1} \in \bar{\mathscr{L}}^i_{t+1}$ **Do**

  Evaluate the $p^i_{t+1}(\mathrm{C}|\mathbf{l}^i_{t+1})$ and the cost $\mathscr{C}^i_{t+1}(\mathbf{l}^i_{t+1})$

End

Evaluate the expected total cost $E\{\mathscr{C}^i_{t+1}\}$

Move to the determined location

Fig. 4 Pseudocode for distributed mobility management

- **Selection Step 2**: Secondly, the location that is found with the minimal cost among all candidate locations is selected as follows:

$$\bar{\mathbf{l}}^i_{t+1} = \min_{\mathbf{l}^i_{t+1} \in \bar{\mathscr{L}}^i_{t+1}} \mathscr{C}^i_{t+1}(\mathbf{l}^i_{t+1}), \tag{37}$$

where $\mathscr{C}^i_{t+1}(\mathbf{l}^i_{t+1})$ is given by Eq. (34).

Note that when $\delta_\psi = 1$, Eq. (36) is then reduced to be the following:

$$\bar{\mathscr{L}}^i_{t+1} = \{\bar{\mathbf{l}}^i_{t+1} | \bar{\mathbf{l}}^i_{t+1} = \arg_{\mathbf{l}^i_{t+1} \in \mathscr{L}^i_{t+1}} \max \Psi(\mathbf{l}^i_{t+1})\}. \tag{38}$$

Recall from Sect. 3, where Eq. (38) is the same as the selection algorithm given by Eq. (12) in Sect. 3.3. This corresponds to the case where user wants to highly prioritize the target tracking improvement regardless of the associated costs; the selection step 2 can then be omitted. On the other hand, when $\delta_\psi = 0$, it leads to $\bar{\mathscr{L}}^i_{t+1} = \mathscr{L}^i_{t+1}$, which implies that user are more concerned about the costs associated with all candidate locations. The selection step 1 can then be omitted. Also note that as discussed in Sects. 3 and 4, all evaluations require only local knowledge, therefore, the discussed mobility management scheme can be implemented in a distributed manner.

5.3 *Analysis of Time Complexity*

Section 5.3 analyzes the time complexity for the discussed mobility management algorithm. Figure 4 shows pseudocode for the distributed mobility management procedure.

The computational complexity for the discussed procedure illustrated in Fig. 4 depends on complexities of the individual computation for target estimation, movement probability, connectivity, coverage loss, and cost evaluation. Note that for any two nodes s_i and $s_j (i \neq j)$, $|\mathscr{L}^i_{t+1}| = |\mathscr{L}^j_{t+1}|$. Let $m = |\mathscr{L}^j_{t+1}|$. Assume that the target tracking algorithm has a complexity of $O(T)$, where T reflects the complexity of the estimation algorithm. For example, in Kalman filtering, $O(T) = O(2d_x^2 d_z) + O(2d_x d_z^2) + O(d_x^3) + O(d_z^3)$ [18]. As shown in Sect. 3.3 from Eqs. (12) and (13), for each candidate location l^i_{t+1}, node s_i has to perform a local target estimation using the predicted sensor measurements to obtain $p(\mathbf{l}^i_{t+1})$. Note that the node has to integrate all available sensor data within its one-hop neighborhood, which for the worst case may be as many as $n-1$. Therefore,the computational complexity for node movement probability is $O(mnT)$. From Eq. (15), evaluation of $\mathscr{E}^i_{t+1}(\mathbf{l}^i_{t+1})$ takes $O(1)$ time. Evaluation of $p^i_{t+1}(\mathrm{C}|\mathbf{l}^i_{t+1})$, i.e., the probability for s_i being disconnected, requires the calculation of distances between s_i and all of its neighbors at the current time instant for all their candidate locations. Assume that there are a total of n nodes, this procedure takes $O(mn)$ time. For coverage loss evaluation, the complexity depends on grid dimension X and Y representing the sensing area. It takes $O(mnXY)$ time for coverage loss evaluation. From Sect. 5, the cost evaluation and movement decision take $O(m)$ time. Therefore, the mobility management procedure takes $O(mnT) + O(mn) + O(mnXY) + O(m) = O(mnXY) + O(mnT)$ time.

Note that in the propose mobility management algorithm, there is no communication required at the time for a node to make its movement. This is especially important and useful because a node can react to targets in a timely manner. Though the node still needs the prior to obtain the posterior, as mentioned in Sect. 3.3, the target estimate based on complete sensor measurements can be forwarded to nodes at a later stage after node makes its decision on its movement to dynamically track the mobile target. In mobile sensor networks, where topology is constantly changing, a node depends on techniques such as periodical HELLO message for current neighborhood information. The HELLO message can be easily extended for exchange of sensor measurements and target estimate. The mobility management algorithm discussed in this chapter imposes virtually no communication overhead.

6 Simulation Studies

To give readers a more concrete view of the previous analysis, Sect. 6 presents simulation results based on the previously discussed mobility management scheme using MatLab. The setup of the simulation contains a wireless sensor network with 20 homogeneous nodes with a communication radius of 18m and a sensing radius of 9 m. Nodes are randomly deployed in a 20 m $\times$ 20 m sensor field represented by a 20 $\times$ 20 grid. A linear system model is considered in the simulation using Kalman filtering for target estimation. Recall the system and measurement model given by

Eqs. (1) and (2) from Sect. 3.1, the system model and measurement model used in our simulation are given as:

$$\mathbf{x}_t = \begin{bmatrix} 1 & 0 \\ 0 & 1 \end{bmatrix} \mathbf{x}_{t-1} + \begin{bmatrix} \Delta t & 0 \\ 0 & \Delta t \end{bmatrix} \mathbf{u} + \mathbf{w}_{t-1}, \tag{39}$$

$$\mathbf{z}_t = \begin{bmatrix} 1 & 0 \\ 0 & 1 \end{bmatrix} \mathbf{x}_t + \mathbf{v}_t, \tag{40}$$

where Δt is the sampling interval, $\mathbf{u} = [\mu_x; \mu_y]$ represents the constant speeds of the target in x and y direction, respectively. The state vector $\mathbf{x}_t$ contains the x and y coordinates of the target. It is assumed that $\mu_x = \mu_y$. It is assumed that initially $\hat{\mathbf{x}}_0 = [0; 0]$ for all nodes. The sensing model that is used for coverage evaluation is given by: $c_k^i = e^{-\alpha d_k^i}$, where c_k^i is the coverage probability, d_k^i is the distance between the grid point g_k and the node s_i, and α represents the physical characteristics of the sensor [54]. In the simulations, the parameter α is chosen such that $\alpha = -\frac{\ln p_{th}}{r_s}$.

Note that readers can follow steps given in Sect. 6 to feed their own system and measurement models as well as sensing models to the same mobility management framework using MatLab or other tools to perform a similar simulation run to fine-tune on parameters like α, grid point size, sensing range, communication range, etc., before deploying in the field.

6.1 Static Sensor Network Versus Mobile Network with Mobility Management

In Figs. 5 and 6, a comparison of the discussed mobility management approach is presented, where the baseline case is for the sensor network using only static nodes i.e., nodes remain at their original locations throughout the simulation. In addition to the localized approach, Sect. 6.1 also considers a centralized target estimation approach, which is based on all the sensor measurements. The target speed is set as $\mu_x = \mu_y = 1$ m/s and nodes have the same speed as the target. The sampling interval Δt is 1 sec. The target is initially placed at [2; 2]. The selection of candidate locations for target tracking data improvement is based on the trace of the error covariance matrix. Figure 5 shows the comparison of the trace of the error covariance matrix and position estimation error in a $\log_{10}$ scale. It clearly shows that the error for a mobile sensor network is less than that for the static network. Another good and well-accepted metric for evaluating the tracking quality is the norm of the position error [7, 20, 51], which is shown at the bottom of Fig. 5. The norm of the position error for the mobile network is roughly 72.5 % less during the time that target is moving through the sensor field.

Figure 6a shows the average of the expected probability that a mobile node is disconnected, i.e., $p_{t+1}^i(\mathrm{C}|\mathbf{l}_{t+1}^i)$, and the average distance moved by the mobile nodes. Note that node movement causes an increase in the expected probability of nodes

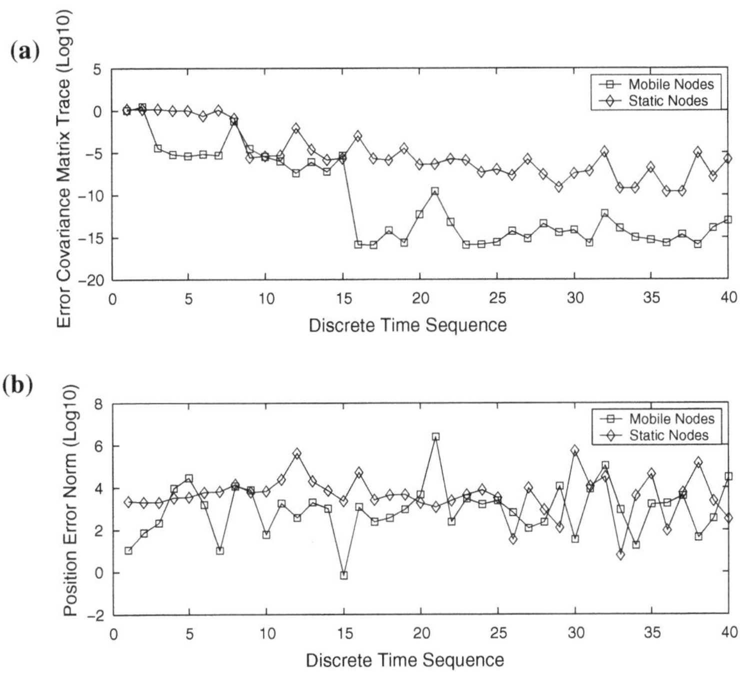

Fig. 5 **a** Trace of estimation error covariance matrix: mobility management versus static network. **b** Position error norm: mobility management versus static network

being disconnected, but since the sensor network is densely deployed and node movement is not drastic per time step, the probability of disconnection is still very small. Figure 6b illustrates the average of the expected probability of coverage loss, i.e., $p^i_{t+1}(\mathrm{S}|\mathbf{l}^i_{t+1})$ for all mobile nodes, and the average global coverage difference between the mobile network and the static network. The average global coverage is defined as the sum of individual grid points coverage on individual grid points over the total number of grid points on the example sensor field. Similar to Fig. 6a, $p^i_{t+1}(\mathrm{S}|\mathbf{l}^i_{t+1})$ in the top graph of Fig. 6b is very small. However, as shown by the bottom graph in Fig. 6b, the mobility management scheme improves the global coverage compared to the static network.

6.2 Random Mobile Sensor Network Versus Mobile Network with Mobility Management

Figure 7a, b illustrate the effects of mobility management for mobile sensor networks. In one case, the nodes in the mobile network uses the discussed mobility management algorithm to make movement decisions. In the baseline case, nodes randomly select a candidate location for the next move. Figure 7a shows that the

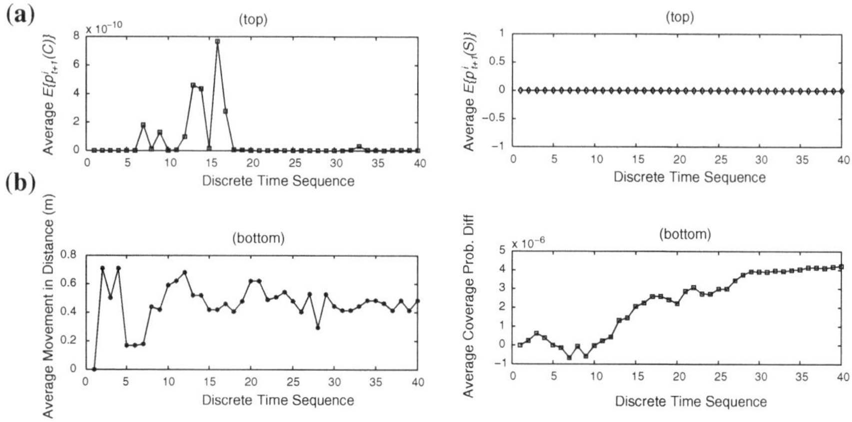

Fig. 6 **a** *Top*: Average expected probability of a node being disconnected. Bottom: Average distance moved by a node. **b** *Top*: Average expected probability of loss of sensing coverage. *Bottom*: Average global coverage difference between the example mobile sensor network and the static sensor

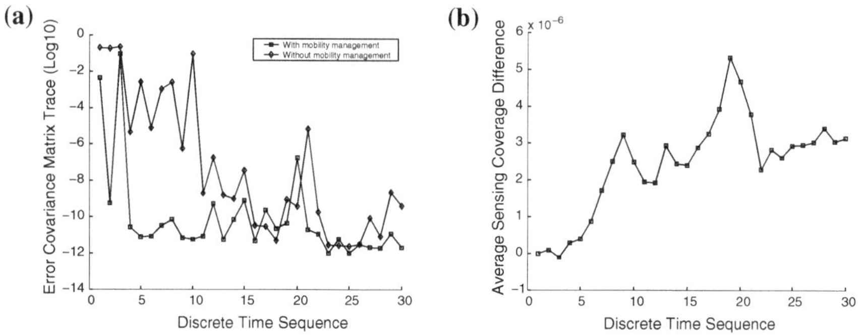

Fig. 7 **a** Trace of estimation error covariance matrix: with mobility. **b** Average global coverage difference between the mobile sensor network with mobility management and without mobility management.

mobility management algorithm achieves the selection of candidate locations that can provide better tracking quality. Figure 7b illustrates the effect of mobility management for improved sensing coverage, which is advantageous for surveillance. This benefit is not available for mobile sensor networks that use a random mobility management scheme.

6.3 Localized Versus Centralized Implementations

Next, Sect. 6.3 compares the localized implementation with a centralized implementation. For the centralized approach, it is assumed that here exists a base station that acts as the processing center. All nodes that have detected the target forward the data to the base station for processing. The base station uses the same evaluation rule as described in previous sections. However, the base station has global knowledge of the location of all the nodes and it can integrate sensor data from all the sensor data. It is expected that the centralized approach requires more bandwidth and energy for communication. Figure 8a, b show the simulation results. For simplicity, it is assumed that the base station is located at the center of the sensing region.

Fig. 8a shows the trace of estimation error covariance matrix and position error norm for the localized and centralized implementations. As shown in the top graph of Fig. 8a, the centralized approach outperforms localized implementation in tracking quality because in the localized algorithm nodes only have local knowledge. This is evident by the bottom of Fig. 8a, where the estimated error covariance matrix of localized implementation is always larger than that for the centralized case.

On the other hand, the energy consumption is considerably higher for the centralized method; see the bottom part of Fig. 8b. Equation (41) is used to calculate the energy consumption at time instant t for node s_i:

$$e(t,i) = e_{comm}(t,i) + e_{comp}(t,i) + e_{move}(t,i), \tag{41}$$

where $e_{comm}(t,i)$, $e_{comp}(t,i)$, and $e_{move}(t,i)$ account for energy consumption due to communication, computation, and movement, respectively. Note that sensing energy consumption is not considered here because it is not affected by whether the discussed mobility management scheme is implemented as a localized or a centralized algorithm. Let $\hat{\mathcal{N}}_t^i$ be the set of nodes that have detected the target and are also the one-hop neighbors of node s_i. Let $d_{ij}(t)$ be the distance between node s_i and s_j at time t. Let $\hat{d}_i(t) \equiv \max d_{ij}(t), \forall s_j \in \hat{\mathcal{N}}_t^i$, be the maximum distance between s_i and its neighbors. Let $\Delta d_i(t)$ be the distance that s_i moves at time t. The measures $e_{comm}(t,i)$, $e_{comp}(t,i)$, and $e_{move}(t,i)$ are evaluated as follows:

$$e_{comm}(t,i) = \hat{d}_i(t)^{\alpha} \times (\mathrm{E}_{send} + |\hat{\mathcal{N}}_t^i| \times \mathrm{E}_{recv}), \tag{42}$$

$$e_{comp}(t,i) = (|\hat{\mathcal{N}}_t^i| + 1) \times m \times \mathrm{E}_{comp}, \tag{43}$$

$$e_{move}(t,i) = \Delta d_i(t) \times \mathrm{E}_{move}, \tag{44}$$

where E_{send}, E_{recv}, E_{comp}, and E_{move} are power constants for communication, transmitting, computation, and locomotion, respectively. α is the attenuation constant. Example values chosen in the simulation run are as follows: $\alpha = 2.5$, $\mathrm{E}_{send} = \mathrm{E}_{recv} = 1$ mJ/m, and $\mathrm{E}_{move} = 50$ mJ/m, which are based on metrics presented in [35]. Note that E_{recv} and E_{send} are also related to the size of the data packet. In this simulation, it is assumed that a 100-byte packet is sufficient for a node to send its sensor response as well as its current location to its neighbors. It is assumed that $\mathrm{E}_{comp} = 0.1$ mJ. The number of candidate locations m, which is defined in Sect. 5.3, is set to 9 in this simulation.

As shown in top of Fig. 8b, since only local knowledge is used, the number of nodes involved in data integration for the localized algorithm is much lower than that

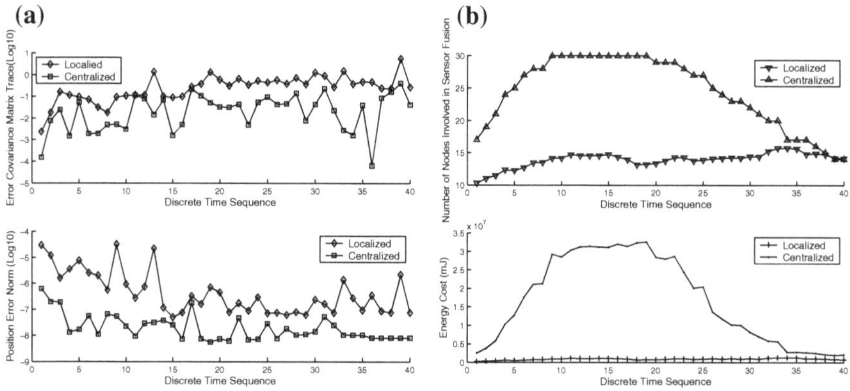

Fig. 8 Mobility management: localized versus centralized. **a** *Top*: Trace of estimation error covariance matrix. *Bottom*: Position error norm. **b** *Top*: Number of nodes involved in evaluation. *Bottom*: Energy cost evaluation

for the centralized case. As a result, the tracking quality is better for the centralized case, as shown in Fig. 8a. However, the centralized approach requires considerably more energy, as shown in the bottom of Fig. 8b. The results presented here can be used as a guideline to select an appropriate implementation strategy to trade-off energy consumption with tracking accuracy. Moreover, from the top of Fig. 8b, it is noticed that localized algorithm is scalable since the number of nodes involved in data integration does not increase with the total number of deployed nodes.

6.4 Discussion

The mobility management scheme discussed in this chapter determines the probability that a node moves to a certain location, i.e., $p(\mathbf{l}^i_{t+1})$. As shown in Sect. 3.3 $p(\mathbf{l}^i_{t+1})$ is obtained by making use of the predicted sensor measurement at the candidate location. Alternative techniques can be used to achieve the same purpose, for example, the distributed probability inference method as described in [33], where a distributed architecture is presented for message exchange among sensor nodes to solve problems such as probability inferencing. However, this architecture is not designed to handle mobility management for target tracking as described in this chapter. As discussed in Sect. 3, the predicted measurement from a node's candidate locations is used. Therefore, further improvement is expected for target tracking in mobile sensor networks by integrating the discussed mobility management scheme with the efficient probability inferencing technique based on the distributed architecture described in [33]. In view of the inherent complexity and the dynamic nature of mobile sensor networks, the algorithm is designed with only localized communication requirements. Every node can make its movement decision in a timely manner for dynamic

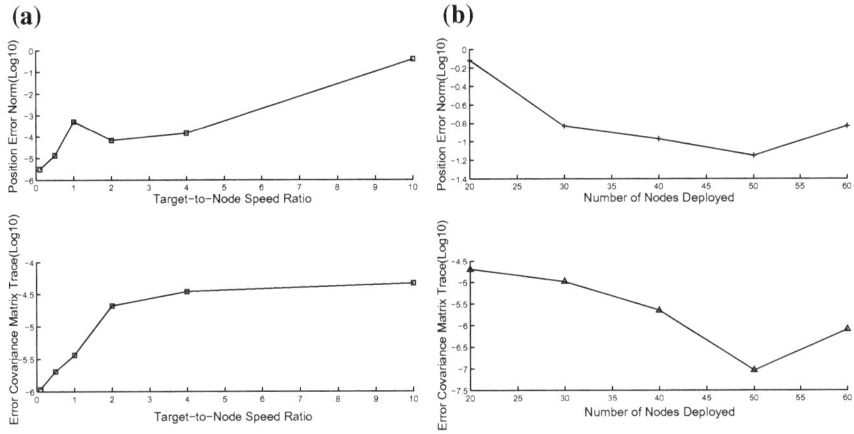

Fig. 9 **a** Mobility management for target at different speeds. Target-to-node speed ratio is 0.1, 0.5, 1, 2, 4, and 10. **b** Mobility management for different node densities. Random sensor deployment with 20, 30, 40, 50, and 60 nodes

target tracking without lengthy negotiation with neighbors for maintaining connectivity and sensing coverage. This ensures a flexible distributed implementation for mobility management in mobile sensor networks for target tracking.

Note that the performance of the discussed scheme is also related to factors such as the speed of the nodes, the target speed, and number of nodes deployed in the sensing region. Figure 9 shows the position error norm as well as error covariance matrix trace for target tracking using the discussed mobility management method for different target speeds and various number of nodes. Figure 9a illustrates the tracking performance when the target-to-node speed ratio is varied from 0.1 to 10. It is expected that when the target is moving slowly, nodes are able to track more accurately, as evident from the increase in the error covariance matrix trace. In Fig. 9b, when the number of nodes deployed increases, the target tracking quality is improved because more sensor data are available within one-hop neighborhood for the node to make movement decisions based on local knowledge.

7 Conclusions

The demand for mobile wireless sensor networks in various surveillance applications is growing rapidly with the advance of modern hardware and software in sensor networks. Mobility management, due to its inherent complexity, still presents itself as one of the most challenging topics in the wireless sensor network research community. This chapter bring to readers recent findings in mobility management for wireless sensor networks. It is important for reader to understand an optimal mobility management scheme must always consider at least three basic factors that are vital to the operation of a mobile wireless sensor network: effective sensing coverage, reliable communication connectivity, and efficient energy consumption. With every node in the field is potentially moving as well as is moving based on its local knowledge, the topology dynamics of such mobile sensor network makes the mobility management an formidable problem.

This chapter attempts to tackle the mobility management problem by mathematically formulating it as an optimization problem based on well-established Bayesian estimation theory for target tracking. Obviously, this approach imposes many limitations and simplifications, such as those described in Sect. 3.1. For example, consider the case where a target moves at a much higher speed than all sensor nodes, the difference between two consecutive measurements from a node is too large to be useful for the calculation of the predicted sensor measurements, as described in Sect. 3.3. In real world, this is mostly due to the lack of any *a priori* knowledge of the target. It is still possible to address this problem at the time when the sensor network is deployed, i.e., the sampling rate of the chosen sensor nodes should be fast enough to match the target speed. On the other hand, if the target is moving at a much lower speed, a single step of node movement will cause a drastic change in a node's two consecutive measurements. In this case, a node has to adjust its speed before it starts moving to avoid unnecessary locomotion energy consumption. One of the directions for future investigation is to add the robustness or adaptiveness to more complicated target dynamics model that is realistically non-linear in most practical applications. One direct approach is to linearize the target dynamics model to work directly with the scheme discussed in this chapter, where linearization error will have a significant impact on the effectiveness of this discussed scheme.

One rather strong assumption in this chapter is about the neighbor nodes. When trying to get predicted measurements from neighbor nodes, this chapter assumes neighbor nodes are not changing. In reality, this assumption may not be always valid. Imagine the node is moving at a very high speed that one movement may end up changing all neighbors completely. This potentially will cause huge errors in the predicted measurements that would eventually lead to erroneous movement decision. However, in theory, one can always choose the value of Δ, i.e., the time difference between two consecutive time instances, to be arbitrarily small such that the error introduced is virtually negligible. This work-around, unfortunately, is impractical for real-time embedded implementation since it imposes a higher expectation on the node's capabilities in sensing, computation, memory, and storage. This translates to more expensive hardware with more energy consumption and more bandwidth required. Alternatively, a more appropriate yet more complicated approach is to feed back global target tracking knowledge to individual nodes. One of the nodes, dedicated or not, has to take the role as the multiple sensor fusion node to perform target tracking based on all available sensor node measurement. The selection of such central sensor fusion node itself deserves a separate research effort where heuristically, the central sensor fusion node at any given time instant is located at the center of the sensor network such that it can communicate with all nodes with a minimum communication hops for better energy efficiency.

One other issue is the impact on existing routing protocol in wireless sensor networks. The topology of the sensor network is changing at any given time instant where mobility exists across all nodes. This chapter makes no effort on evaluating the impact on the routing protocols in the context of mobility management, e.g., quantifying the additional cost of reestablishing routes when previous routes become invalid due to node movement.

Last but not least, this chapter has focused on evaluating mobility management for target tracking for single target tracking scenarios. The discussions presented in this chapter is not limited to the case of multiple target tracking, it is yet to see how well the discussed mobility management scheme performs in multiple target tracking applications. Potentially, one new challenge in multiple target tracking is about partitioning the mobile wireless sensor networks to perform a single target tracking by a subset of mobile nodes. Another challenge is about target identification, particularly for homogeneous targets. However, without any *a priori* knowledge of target types, number of targets, the complexity of mobility management for multiple targets tracking is extremely difficult.

In summary, the constantly changing topology due to node movement makes mobility management difficult for mobile sensor networks. However, as shown in this chapter, when used properly, mobility can improve target tracking quality. This chapter presents a feasible design of such mobility management scheme for a distributed implementation with only one hop neighbor nodes needed. Each node makes its own decision using local knowledge based on optimization on the cost evaluation derived from Bayesian estimation theory. The cost evaluation technique allows the trade-off between target tracking quality improvement and the negative consequences of energy consumption, loss of connectivity and coverage. Even though certain compromises have to be made, such as those described in in Sects. 3.1 and 7, simulation results provide promising supports to the discussed mobility management scheme. It is beyond the scope of this chapter to solve all aforementioned problems, not to mention that none of these issues are trivial. Fortunately, this chapter has laid down the foundation and paved the way for the future study. It is the hope of the authors that this chapter brings more attention from the research community for more general and more graceful solutions to mobility management in wireless sensor networks.

Acknowledgments This research was supported by DARPA, and administered by the Army Research Office under Emergent Surveillance Plexus MURI Award No. DAAD19-01-1-0504. Any opinions, findings, and conclusions or recommendations expressed in this publication are those of the authors and do not necessarily reflect the views of the sponsoring agencies.

References

1. T. Abdelzaher, Y. Anokwa, P. Boda, J. Burke, D. Estrin, L. Guibas, A. Kansal, S. Madden, J. Reich, Mobiscopes for human spaces. IEEE Pervasive Comput. **6**(2), 20–27 (2007)
2. I.F. Akyildiz, W. Su, Y. Sankarasubramaniam, E. Cayirci, A survey on sensor networks. IEEE Commun. Mag. **40**(8), 102–114 (2002)

3. M.S. Arulampalam, S. Maskell, N. Gordon, T. Clapp, A tutorial on particle filters for online nonlinear/non-Gaussian Bayesian tracking. IEEE Trans. Sig. Process **50**, 174–188 (2002)
4. J. Aslam, Z. Butler, F. Constantin, V. Crespi, G. Cybenko, D. Rus, Tracking a moving object with a binary sensor network. in *Proceedings ACM Conference Embedded Networked Sensor Systems (SenSys)*, 2003, pp. 150–161
5. F. Baker, An outsider's view of MANET, IETF document ID: draft-baker-manet-review-01, IETF (2012), http://datatracker.ietf.org Accessed 18 March 2012
6. D. Ballari, M. Wachowicz, The design of a Bayesian network for mobility management in wireless sensor networks. in *Proceeding 6th International Conference Geographic Information Science (GIScience)*, 2010, Item ID 7424
7. Y. Bar-Shalom, W.D. Blair (eds.), *Multitarget-Multisensor Tracking: Applications and Advances-Volume III* (Artech House, MA, 2000)
8. D. Braginsky, D. Estrin, Rumor routing for sensor networks. in *Proceedings ACM International Workshop on Wireless Sensor Networks and Applications*, 2002, pp. 22–31
9. R.R. Brooks, C. Griffin, D. Friedlander, Distributed target classification and tracking in sensor networks, in *Proceedings IEEE*, 2003, pp. 1163–1171
10. Z. Butler, D. Rus, Event-based motion control for mobile-sensor networks. IEEE Pervasive Comput. **2**(4), 34–42 (2003)
11. T. Camp, J. Boleng, V. Davies, A survey of mobility models for ad hoc network research. J. Wireless Comm. Mob. Computing **2**, 483–502 (2002)
12. Q. Cao, T. Yan, J. Stankovic, T. Abdelzaher, Analysis of target detection performance for wireless sensor networks. in *Proceedings IEEE International Conference Distributed Computing in Sensor Systems (DCOSS)*, 2005, pp. 276–292
13. B. Chen, K. Jamieson, H. Balakrishnan, R. Morris, Span: an energy-efficient coordination algorithm for topology maintenance in ad hoc wireless networks. in *Proceedings ACM Annual International Conference Mobile Computing and Networking (MobiCom)*, 2001, pp. 85–96
14. J.C. Chen, K. Yao, R.E. Hudson, Source localization and beamforming. IEEE Signal Process. Mag. **19**, 30–39 (2002)
15. A.J. Coulson, A.G. Williamson, R.G. Vaughan, A statistical basis for lognormal shadowing effects in multipath fading channels. IEEE Trans. Comm. **46**, 494–502 (1998)
16. P. Dutta, D. Culler, Mobility changes everything in low-power wireless sensornets. in *Proceedings USENIX/IEEE Workshop on Hot Topics in Operating Systems (HotOS XII)*, Monte Verita, Switzerland, May 2009. http://static.usenix.org, Accessed 18 March 2012
17. J. Garcia-Macias, J. Gomez, MANET versus WSN, in *Sensor Networks and Configuration*, ed. by N.P. Mahalik (Springer, Berlin Heidelberg, 2007), pp. 369–388
18. M.J. Goris, D.A. Gray, I.M.Y. Mareels, Reducing the computational load of a kalman filter. IEE Electron. Lett. **33**, 1539–1541 (1997)
19. Z.J. Haas, B. Liang, Ad hoc mobility management with uniform quorum systems. IEEE/ACM Trans. Networking **7**, 228–240 (1999)
20. D.L. Hall, J. Llinas, *Handbook of Multisensor Data Fusion* (CRC Press, FL, 2001)
21. W. R. Heizelman, A. Chandrakasan, H. Balakrishnan, Energy efficient communication protocol for wireless micro sensor networks. in *Proceedings International Conference System Sciences*, 2000, pp. 1–10
22. X. Hong, M. Gerla, G. Pei, C. Chiang, A group mobility model for ad hoc wireless networks. in *Proceedings ACM/IEEE International Symposium Modeling, Analysis and Simulation of Wireless and Mobile Systems (MSWiM)*, 1999, pp. 53–60
23. S.S. Iyengar, R.R. Brooks (eds.), *Distributed Sensor Networks*, (Chapman & Hall/CRC Computer and Information Science Series, , CRC Press 2003)
24. D. Jea, A. A. Somasundara, M.B. Srivastava, Multiple controlled mobile elements (data mules) for data collection in sensor networks. in *Proceedings IEEE International Conference Distributed Computing in Sensor Systems (DCOSS)*, 2005, pp. 244–257
25. B. Jiang, K. Han, B. Ravindran, H. Cho, Energy efficient sleep scheduling based on moving directions in target tracking sensor network. in *Proceedings IEEE International Parallel & Distributed Processing Symposium (IPDPS)*, 2008, pp. 1–10

26. X. Y. Li, P. J. Wan, Y. Wang, C. W. Yi, Fault tolerant deployment and topology control in wireless networks. in *Proceedings ACM International Symposium Mobile Ad Hoc Networking and Computing (MobiHoc)*, 2003, pp. 117–128
27. B. Y. Liu, P. Brass, O. Dousse, P. Nain, D. Towsley, Mobility improves coverage of sensor network. in *Proceedings ACM International Symposium Mobile Ad Hoc Networking and Computing (MobiHoc)*, 2005, pp. 300–308
28. J. Liu, J. Liu, J. Reich, P. Cheung, F. Zhao, Distributed group management in sensor networks: algorithms and applications to localization and tracking. Telecommun. Syst. **26**(2–4), 235–251 (2004)
29. Y.G. Mei, Y.H. Lu, Y.C. Hu, C.S. George Lee, Deployment strategy for mobile robots with energy and timing constraints. in *Proceedings IEEE International Conference Intelligent Robots and Systems (ICRA)*, 2005
30. T. Melodia, D. Pompili, I.F. Akyldiz, Handling mobility in wireless sensor and actor networks. IEEE Tran. Mob. Comput. **9**, 160–173 (2010)
31. Mobile Ad-hoc Networks (MANET), IETF (2012), http://datatracker.ietf.org/wg/manet/charter. Accessed 18 March 2012
32. T. Moscibroda, R. O'Dell, M. Wattenhofer, R. Wattenhofer, Virtual coordinates for ad hoc and sensor networks. in *IEEE Foundations of Mobile Computing, Workshop*, 2004, pp. 8–16
33. M. Paskin, C. Guestrin, J. McFadden, A robust architecture for distributed inference in sensor networks. in *Proceedings ACM/IEEE International Conference Information Processing in Sensor Networks (IPSN)*, 2005, pp. 55–62
34. S. Phoha, T.F. La Porta, C. Griffin, *Sensor Network Operations* (Wiley, N.J, 2006)
35. R. Rao, G. Kesidis, Purposeful mobility for relaying and surveillance in mobile ad-hoc sensors networks. IEEE Trans. Mob. Comput. **3**, 225–231 (2004)
36. A. Roy, S.K. Das, A. Misra, Exploiting information theory for adaptive mobility and resource management in future cellular networks. IEEE Wirel. Comm. Mag. **11**, 59–65 (2004)
37. S. Shakkottai, R. Srikant, N.B. Shroff, Unreliable sensor grids: coverage, connectivity and diameter. in *Proceedings IEEE International Conference Computer Communications (INFOCOM)*, 2003, pp. 1073–1083
38. R. Szewczyk, E. Osterweil, J. Polastre, M. Hamilton, A. Mainwaring, D. Estrin, Habitat monitoring with sensor networks. Commun. ACM **47**, 33–40 (2004)
39. R. Tan, G. Xing, J. Wang, H.C. So, Exploiting reactive mobility for collaborative target detection in wireless sensor networks. IEEE Trans. Mob. Comput. **9**, 317–332 (2010)
40. A. Verma, H. Sawant, J. Tan, Selection and navigation of mobile sensor nodes using a sensor network. Pervasive Mob. Comput. **2**(1), 65–84 (2006)
41. X. R. Wang, G. L. Xing, Y. F. Zhang, C. Y. Lu, R. Pless, C. Gill, Integrated coverage and connectivity configuration in wireless sensor networks. in *Proceedings ACM Conference Embedded Networked Sensor Systems (SenSys)*, 2003, pp. 28–39
42. G. Wang, G. Cao, T. La Porta, Movement-assisted sensor deployment. IEEE Trans. Mob. Comput. **5**, 640–652 (2006)
43. Y.C. Wang, F.J. Wu, Y.C. Tseng, Mobility management algorithms and applications for mobile sensor networks. Wireless Communications and Mobile Computing **12**, 7–21 (2012)
44. G. Wittenburg, N. Dziengel, C. Wartenburger, J. Schiller, A system for distributed event detection in wireless sensor networks. in *Proceedings ACM/IEEE International Conference Information Processing in Sensor Networks (IPSN)*, 2010, pp. 94104
45. J. Wu, Extended dominating-set-based routing in ad hoc wireless networks with unidirectional links. IEEE Trans. Parallel Dist. Syst. **13**, 866–881 (2002)
46. G. Xing, X. Wang, Y. Zhang, C. Lu, R. Pless, C. Gill, Integrated coverage and connectivity configuration for energy conservation in sensor networks. ACM Trans. Sens. Netw. **1**, 36–72 (2005)
47. G. Xing, J. Wang, Z. Yuan, R. Tan, L. Sun, Q. Huang, X. Jia, H.C. So, Mobile scheduling for spatiotemporal detection in wireless sensor network. IEEE Trans. Parallel Dist. Syst. **21**, 1851–1866 (2010)

48. Y. Xu, J. Heidemann, D. Estrin, Geography-informed energy conservation for ad hoc routing. in *Proceedings ACM Annual InterntionL Confernece Mobile Computing and Networking (MobiCom)*, 2001, pp. 70–84
49. F. Xue, P.R. Kumar, The number of neighbors needed for connectivity of wireless networks. Wireless Netw. **10**(2), 169–181 (2004)
50. Y. Yang, D. Lee, M. Park, H. Peter, Dynamic enclose cell routing in mobile sensor networks. in *Proceedings IEEE Asia-Pacific Software Engineering Conference (APSEC)*, 2004, pp. 736–737
51. F. Zhao, J. Liu, J. Liu, L. Guibas, J. Reich, Collaborative signal and information processing: an information directed approach. Proc. IEEE **91**, 1199–1209 (2003)
52. Y. Zhuang, J. Pan, L. Cai, Minimizing energy consumption with probabilistic distance models in wireless sensor networks. in *Proceedings IEEE International Conference Computer Communications (INFOCOM)*, 2010, pp. 1–9
53. Y. Zou, K. Chakrabarty, Sensor deployment and target localization based on virtual forces. in *Proceedings IEEE International Conference Computer Communications (INFOCOM)*, 2003, pp. 1293–1303
54. Y. Zou, K. Chakrabarty, A distributed coverage- and connectivity-centric technique for selecting active nodes in wireless sensor networks. IEEE Trans. Comput. **54**, 978–991 (2005)
55. Y. Zou, K. Chakrabarty, Distributed mobility management for target tracking in mobile sensor networks. IEEE Trans. Mob. Comput. **8**, 872–887 (2007)

Part V
Localization and Task Management

Chapter 11
Range-Free Localization Techniques

Christian Poellabauer

Abstract In wireless sensor networks, sensors are often deployed without a priori knowledge of their locations or sensor node locations can change during the lifetime of a network. However, location information is essential for a variety of reasons. Sensors monitor phenomena in the physical world and given the location of the sensors, it is then possible to estimate the location of the observed phenomenon. For example, chemical and humidity sensors deployed on a farm can provide information about soil moisture, crop health, and animal movement if the sensor locations are known. Accurate location information is also needed for various sensor network management tasks such as routing based on geographic information, object tracking, and providing location-aware services. Frequently, sensor node localization is performed using ranging techniques, where the distances between a sensor device and several known reference points are determined to derive the position of a sensor. However, the cost and limitations of the hardware needed for range-based localization schemes often make them poor choices for WSNs. Therefore, a variety of localization protocols have been proposed that attempt to avoid the use of ranging techniques with the goal to provide more cost-effective and simpler alternatives. These range-free localization techniques estimate a node's position using either neighborhood information, hop counts from well-known anchor points, or information derived from the area a node is believed to reside in. This chapter introduces the basic concepts of range-free localization, surveys a variety of state-of-the-art localization techniques, compares qualitatively the characteristics of these protocols, and discusses current research directions in range-free localization.

C. Poellabauer (✉)
University of Notre Dame, Notre Dame, USA
e-mail: cpoellab@nd.edu

H. M. Ammari (ed.), *The Art of Wireless Sensor Networks*,
Signals and Communication Technology, DOI: 10.1007/978-3-642-40009-4_11,
© Springer-Verlag Berlin Heidelberg 2014

1 Overview

Wireless sensor networks are built and deployed to collect real-time information on the spatial-temporal characteristics of the physical environment. Therefore, the knowledge of the location of each sensor node in a network is critical to many WSN applications and services to appropriately interpret sensors readings. For example, environmental sensor networks measure air, water, and soil quality to detect the presence and sources of pollutants [16], but they require accurate spatial information to support precise modeling and simulation of the dispersion of the pollutants. Further, in battlefield scenarios, sensor locations are essential to accurately determining or predicting enemy movements [19] and in emergency response systems, sensor locations are needed to guide first responders towards survivors and away from harm [17]. Sensor node localization is also needed to support a variety of network management services, such as routing protocols (e.g., many routing protocols are based on the principle of geographic forwarding [34]), coverage area and topology control [33], energy management techniques (e.g., protocols that adjust in-network data management to preserve energy [13]), clustering [38], boundary detection [10], and various network security mechanisms and protocols [37].

As a consequence, localization has been the focus of a significant number of research efforts, ranging from manual configurations to distributed positioning algorithms. For example, Global Positioning System (GPS) has been a popular choice for localization in mobile devices, but it is often a poor choice for WSNs because of its high hardware cost (compared to the low cost of the miniaturized devices often used for sensing) and its inability to provide location services in indoor settings and other scenarios where no clear view of the sky is available. The type of location information provided by GPS can be expressed as *global* metric, i.e., a position within a general global reference frame (e.g., longitudes and latitudes). Another example of such a system is the Universal Transverse Mercator (UTM) coordinate system, which provides positions within zones and latitude bands. In contrast, *relative* metrics are based on arbitrary coordinate systems and reference frames, e.g., a sensor's location expressed as distances to other sensors without any relationship to global coordinates.

When systems such as GPS are unsuitable for a WSN, a network can also rely on a subset of nodes that know their global positions for localization. Other nodes in the network can then use these *anchor nodes* (or *reference nodes*) to estimate their own positions. Techniques that rely on such anchors are called *anchor-based localization* (as opposed to *anchor-free localization*). Many localization techniques, particularly in the category of anchor-based techniques, are based on range measurements, i.e., estimations of distances between several sensor nodes. These *range-based localization* techniques require sensors to monitor measurable characteristics such as received signal strengths of wireless communications or time difference of arrival of ultrasound pulses.

In contrast, *range-free localization* techniques do not rely on distance estimates, but instead rely on approaches such as estimating their relative positions to other nodes using connectivity-based algorithms. These range-free localization techniques are typically more cost-effective, because they do not require the nodes to have special hardware functionalities. On the downside, range-free localizations typically lead to coarser estimates. However, these estimates are sufficient for many types of WSN applications where limited localization accuracy is acceptable.

The remainder of this chapter discusses the fundamentals and state-of-the-art solutions for range-free localization techniques. When comparing localization techniques, two important qualities of localization information are the *accuracy* and *precision* of a position. For example, a GPS sensor indicating a position that is true within 10 m for 90 % of all measurements, the accuracy of the GPS reading is then 10 m (i.e., how close is the reading to the ground truth?) and the precision is 90 % (i.e., how consistent are the readings?). Apart from *physical* positions (such as the longitudes and latitudes provided by GPS), many applications (e.g., indoor tracking systems) may only require *symbolic* locations [7], such as "office building A14" or "mile marker 12 on Highway 50." Other metrics that determine the suitability of a localization technique for a specific WSN scenario include energy efficiency, support for Quality-of-Service, and localization overheads and costs.

2 Range-Free Versus Range-Based Localization

Most WSN localization techniques rely on variations of the same principle, i.e., they establish sensor node locations based on information exchange between neighboring nodes. When certain characteristics of the exchanged messages or signals are used to determine distances between nodes, a variety of localization techniques can be applied to determine a sensor's position, where the positioning accuracy depends on the quality of distance measurements. These types of localization techniques belong to the category of range-based localization. In contrast, if information exchange is only used to establish connectivity information (i.e., learn about the topology of a portion of the WSN), then a sensor node can learn its position in a network relative to other sensor nodes. While the accuracy of these range-free localization techniques is typically lower than the accuracy of range-based techniques, range-free approaches can provide more cost-effective solutions for large, low-cost wireless sensor networks. This section provides an overview of these two main categories of sensor node localization.

2.1 Overview of Range-Based Localization Techniques

In range-based localization, sensor nodes obtain distance estimates between themselves and other sensor nodes, e.g., using measurements of certain characteristics of radio signals, including signal propagation times, signal strengths, and angle of arrival. When the concept of *Time of Arrival (ToA)* is applied [3], the distance between the sender and receiver of a signal is determined by combining the signal propagation time with the known signal velocity. For example, an acoustic signal requires approximately 30 ms to travel a distance of 10 m (assuming a velocity of 343 m/s), while a radio signal requires only about 30 ns for the same distance (assuming a velocity of 300 km/s). Radio-based distance measurements require highly accurate clocks to measure such short propagation times, thereby adding to the cost and complexity of range-based localization techniques. Time of arrival methods can further be categorized as *one-way* methods and *two-way* methods. With one-way methods, the signal transmission time and reception time are measured at the sender and receiver, respectively, thereby requiring accurate time synchronization between sender and receiver. With the two-way method, the round-trip time of a signal is measured at the signal sender, therefore removing the need for time synchronization.

A second class of range-based localization techniques uses the *Time Difference of Arrival (TDoA)* approach [4], where two signals travel from the sender to the receiver with different velocities. For example, the first signal could be a radio signal, followed by an acoustic signal. The receiver is then able to determine its location similar to the ToA approach. TDoA-based approaches do not require the clocks of the sender and receiver to be synchronized and can obtain very accurate measurements. However, a disadvantage of the TDoA approach is the need for additional hardware, e.g., a microphone and speaker to transmit and receive acoustic signals.

Another popular technique is to determine the direction of signal propagation, typically using an antenna or microphone arrays. The *Angle of Arrival (AoA)* is then the angle between the propagation direction and some reference direction known as orientation [25]. For example, for acoustic measurements, several spatially separated microphones can be used to receive a single signal and the differences in arrival time, amplitude, or phase can then be used to determine an estimate of the arrival angle.

Finally, the *Received Signal Strength (RSS)* method determines distance based on signal attenuation. While RSS measurements may not lead to as accurate distance measurements as the ToA method (e.g., signal attenuation depends not only on distance, but also multi-path propagation effects, reflections, and noise), these measurements are often readily available in many wireless network cards.

Given a distance measurement between a sensor node and a reference node, the sensor node knows that its position must be along the circumference of a circle (in two-dimensional space) or sphere (in three-dimensional space) centered at the reference node, with the radius representing the distance between the reference node and the sensor node. To obtain a unique location in two dimensions, distance measurements from at least three non-collinear reference nodes are required (or distance measurements from at least four non-coplanar reference nodes in three dimensions). This process of determining a sensor node location is called *trilateration*. In contrast, a similar process called *triangulation* uses the geometric properties of triangles to estimate sensor locations. Triangulation requires at least two angles (or *bearings*) and the locations of the reference nodes or the distance between them to determine the sensor node location in two-dimensional space. The processes of trilateration and triangulation are illustrated in Fig. 1, where three reference nodes are used (identified by their known (x_i, y_i) coordinates).

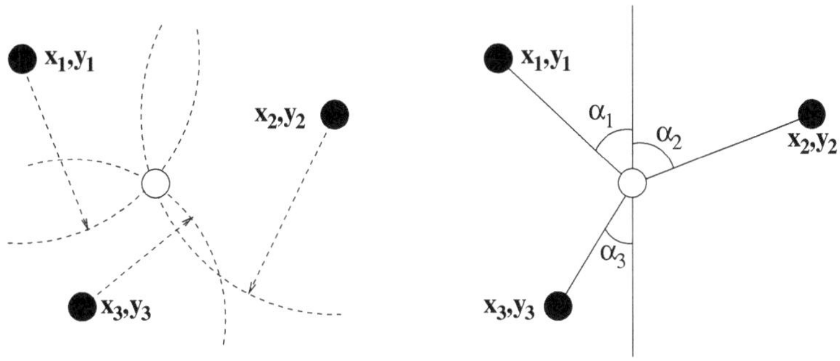

Fig. 1 Example of trilateration (shown on the *left*) and triangulation (shown on the *right*)

Triangulation and trilateration rely on the presence of at least three reference nodes. However, in many sensor networks, it is infeasible to assume that every sensor node can directly communicate with at least three reference nodes. Therefore, these techniques can further be extended, i.e., once a sensor has determined its own position (using either trilateration or triangulation with three or more reference nodes), this sensor can itself become a reference node for other sensors. This iterative process is called *iterative multilateration* [28]. A downside of this process is that every iteration contributes to the localization error.

A variation of this approach is called *collaborative multilateration*, where the goal is to construct a graph of *participating* nodes, i.e., nodes that are reference nodes or have at least three participating neighbors. A node can then estimate its position by solving the corresponding system of over-constrained quadratic equations relating the distances between the node and its neighbors.

Case Study: GPS-Based Localization

As an example of a trilateration system, consider the Global Positioning System (GPS) [8], which is the most widely publicized location-sensing system, providing an excellent lateration framework for determining geographic positions. GPS is the only fully operational global navigation satellite system (GNSS), consisting of at least 24 satellites orbiting the earth at altitudes of approximately 11000 miles. The satellites are uniformly distributed in a total of six orbits (i.e., there are four satellites per orbit) and they circle the earth twice a day at approximately 7000 miles per hour. Each satellite constantly broadcasts coded radio waves that contain information on the identity of the particular satellite, the location of the satellite, the satellite's status, and the date and time a signal has been sent. In addition to the satellites, GPS further relies on infrastructure on the ground to monitor satellite health, signal integrity, and orbital configuration, e.g., at least six monitor stations located around the world constantly receive the data sent by the satellites and forward the information to a master control station (MCS). This MCS uses the data from the monitor stations to compute corrections to the satellites' orbital and clock information, which are then sent back to the satellites via ground antennas.

Satellites and receivers use very accurate and synchronized clocks so that they generate the same code at exactly the same time. A GPS receiver compares its own generated code with the code received from the satellite, thereby determining the actual generation time of the code at the satellite and the time difference between the code generation time and the current time. This time difference expresses the travel time of the code from the satellite to the receiver. Since radio waves travel at the speed of light, given the time difference, the distance between receiver and satellite can be determined. Given this distance, the receiver knows that it must be positioned somewhere on the sphere centered on the satellite. With two more satellites, a receiver can determine two points where the three spheres intersect and typically one of these two intersections can be eliminated, e.g., because it would position the receiver far out in space. While three satellites appear to be sufficient for localization, a fourth satellite is required to obtain an accurate position.

While most GPS receivers available today are able to provide position measurements with accuracies of 10 m or less, advanced techniques to further increase the accuracy are also being used. For example, Differential GPS (DGPS) [20] uses land-based receivers with exactly known locations to receive GPS signals and compute correction factors, which are then broadcast to regular GPS receivers. Further, Real Time Kinematic (RTK) navigation is another technique used to provide centimeter-level accuracy by using a stationary GPS receiver together with one or more mobile units, where the stationary receiver re-broadcasts the signals it receives from the satellites as correction signals to the mobile units.

2.2 Overview of Range-Free Localization Techniques

The localization approaches discussed so far rely on ranging techniques such as RSS, Time of Arrival, Time Difference of Arrival, and Angle of Arrival. In contrast, *range-free localization techniques* do not require additional hardware and are therefore a cost effective alternative to range-based techniques. With range-free techniques, instead of estimating distances between sensor nodes, other approaches are used to determine a sensor node's location at a coarser granularity. These approaches can be grouped into techniques based on *area*, *hop count*, and *neighborhood* information [16].

In area-based localization, a network can be divided into areas or regions, and localization is then concerned with determining the region a node occupies. For example, a sensor node that is able to hear radio signals from two reference nodes can determine that it must reside in the overlapping region of the radio coverage areas of these reference nodes. The more reference nodes are available, the more accurately a node can determine its position (or the region it occupies becomes smaller).

Figure 2 shows two examples of area-based localization. On the left, a node (white node) determines that its position is in the region created by the overlapping radio coverage regions of its two neighboring nodes (black nodes). The figure on the right shows a similar situation with two nodes with directional antennas.

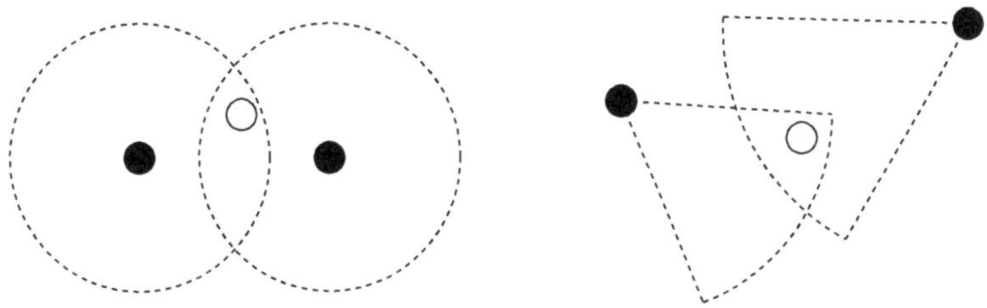

Fig. 2 Area-based localization using two reference nodes with omni-directional antennas (*left*) and directional antennas (*right*)

When directional antennas are used, the size of the overlapping region can be reduced, thereby increasing the positioning accuracy.

In localization techniques based on hop count, hop distances between two nodes A and B are used to estimate node distances and node positions. The hop count is the minimum number of hops h_{min} (i.e., the shortest route) that separates nodes A and B. The maximum distance between A and B is then $R * h_{min}$, where R is the radio coverage radius (assuming that all nodes have the same radio range).

Further, in neighborhood-based localization, reference nodes placed throughout the sensor network periodically issue beacons that include the emitting node's location. If a sensor node receives beacons from only one reference node, the sensor node simply assumes its own location to be identical to the reference node's location. However, if beacons from multiple reference nodes can be received, more sophisticated methods (e.g., the centroid technique) can be used to determine a node's position in the network.

Finally, recently a number of event-based localization techniques have emerged, where event emitters transmit signals that are being received by the sensors and these signals (or certain timing aspects of these signals) can be used to determine a sensor's location. In these techniques, localization is based on the presence of (often costly) event generator devices, therefore replacing the need for reference nodes. In general, in anchor-based techniques, the placement of reference nodes (anchors) needs to be carefully planned to ensure that sensors nodes can communicate with a sufficient number of reference nodes. In contrast, in event-based protocols (as an example for anchor-free techniques), it has to be ensured that all sensors can be in range (or line-of-sight) of one or more signal generator, which may limit the scale of the sensor network.

3 Case Studies of Range-Free Localization Techniques

In the remainder of this chapter, we discuss a variety of range-free localization solutions that utilize area, hop count, or neighborhood information to position sensors, but also more recent event-based techniques that rely on event detection by sensor nodes to determine sensors positions.

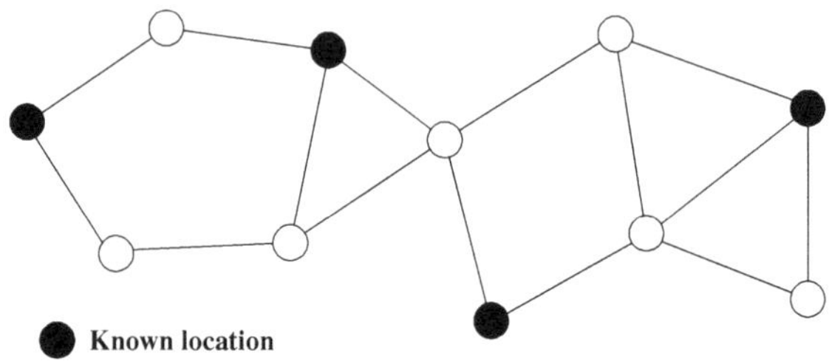

Fig. 3 Illustration of a sample network topology for the convex position estimation approach

3.1 Convex Position Estimation

One of the first area-based localization techniques determining positions exclusively on connectivity-induced constraints was introduced in [5]. In this approach, a network is represented as a graph with n nodes, with a subset of m nodes serving as reference nodes, i.e., their Cartesian positions are known (see Fig. 3, which shows reference nodes as black nodes, nodes with unknown positions as white nodes, and nodes that are able to communicate are connected via lines).

More formally expressed, the network is a graph with n nodes, where the positions of the first m nodes are known and expressed as $(x_1, y_1, \ldots, x_m, y_m)$. The challenge is then to find the unknown positions $(x_{m+1}, y_{m+1}, \ldots, x_n, y_n)$ of the remaining n-m nodes such that all proximity constraints are satisfied. In [5], a centralized solution to this challenge is proposed, where nodes communicate their connectivity information to a single computer that solves this optimization problem.

This approach is based on finding solutions to linear and semidefinite programs that can be used to generate feasible positions for the nodes in a network. The semidefinite problem (SDP) is a generalization of the linear program (LP) [21], with the objective function $c^T x$ and the following constraints:

$$\mathbf{F}(\mathbf{x}) = \mathbf{F}_0 + \mathbf{x}_1\mathbf{F}_1 + \cdots + \mathbf{x}_n\mathbf{F}_n < 0$$

$$\mathbf{Ax} < b$$

$$\mathbf{F}_i = \mathbf{F}_i^T.$$

The first constraint represents a matrix inequality on the cone of positive semidefinite matrices, i.e., the eigenvalues of F(x) must be nonpositive, which is known as linear matrix inequality (LMI). Each node has a position (x,y), allowing us to form a single vector containing all positions as

$$\mathbf{x} = [x_1 y_1 \ldots x_m y_m x_{m+1} y_{m+1} \ldots x_n y_n]^T.$$

Given this mathematical basis and assuming that the connectivity in a network can be represented as a set of convex position constraints (i.e., a convex set is one for which any two points in the set can be connected with a line entirely contained in the set), the work in [5] continues to propose solutions for convex constraint models for RF and optical communication systems. In the symmetric model where the communication range of each node can be represented as a circle, a connection between nodes can be expressed as a 2-norm constraint on the node positions. For example, the LMI can be determined as:

$$\|a - b\|_2 \leq R \Rightarrow \begin{bmatrix} I_2 R & a - b \\ (a - b)^T & R \end{bmatrix} \geq 0,$$

where R is the maximum range and node positions are expressed as a and b. Using Schur complements [2] transforms the quadratic inequality into an LMI with a 3×3 matrix, where I_2 represents the two-dimensional identity. Multiple LMIs can be stacked in diagonal blocks to form a single large SDP for the entire network.

This approach can further be extended, e.g., instead of maximum radio ranges R, smallest ranges r_{ab} are assigned to each constraint (i.e., replace R in the above formula with r_{ab}). Values for r_{ab} can be obtained during the initialization phase by varying the transmission power of the radios. Once a connection is first detected at a power P_0, the receiver node can calculate the maximum possible separation for reception at power P_0. This maximum separation r_{ab} can then be used to determine a tighter upper bound for each connection in the network. Another extension to this approach considers sensors with directional antennas or laser transmitters and receivers (optical communication), where antennas and detectors can be rotated until a signal is detected. This leads to cones representing the feasible sets and this can be used to reduce the size of the area where a sensor is believed to be, thereby reducing the localization error.

3.2 Ad Hoc Positioning System

As an example of a hop count based localization technique, the *ad hoc positioning system* (or *APS*) [22] provides a distributed connectivity-based localization approach that estimates node locations using a set of at least three reference nodes (the more reference nodes, the higher the accuracy obtainable in APS). An important aspect of APS is that it is based on the concept of *distance vector (DV)* exchange [18], where nodes periodically exchange their routing tables with their one-hop neighbors to obtain accurate network-wide connectivity (and routing) information.

APS supports various schemes, the simplest of them being the *DV-hop* approach. Here, each node maintains a table $\{X_i, Y_i, h_i\}$, where $\{X_i, Y_i\}$ represents the location of node i and h_i is the hop distance between this node and node i. Using the routing table updates exchanged in DV, reference nodes will learn about the presence (and location) of other reference nodes and can then calculate an average hop size, called

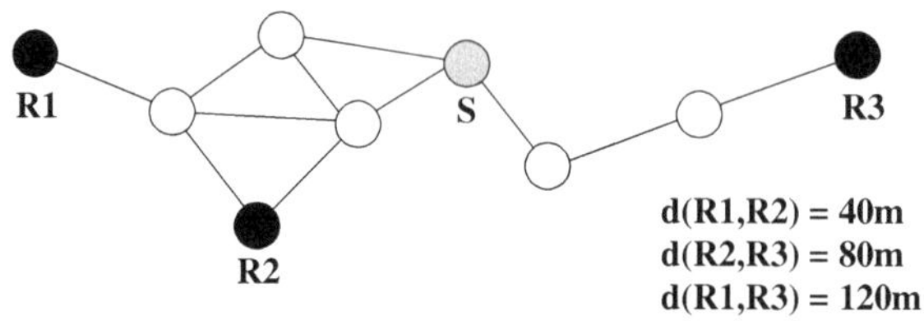

Fig. 4 Example of DV-hop localization

the *correction factor* c_i:

$$c_i = \frac{\sum \sqrt{(X_i - X_j)^2 + (Y_i - Y_j)^2}}{\sum h_i}$$

for all other reference nodes j ($i \neq j$). The computed correction factors are also propagated throughout the network and given the locations and correction factors of the reference nodes, a sensor node is then able to estimate its own location. Figure 4 illustrates an example of DV-hop localization using three reference nodes (R1, R2, R3) and six sensor nodes, where sensor node S is attempting to estimate its position.

Each reference node knows the Euclidean distance and the minimum hop count between itself and all other reference nodes. Given this information, reference node R1 is able to compute its correction factor as $(40 + 120)/(2 + 6) = 20$, which indicates the estimated hop distance in meters. Similarly, reference node R2 determines its correction factor as $(40 + 80)/(2 + 5) = 17.1$ and reference node R3 determines its correction factor as $(80 + 120)/(5 + 6) = 18.2$.

To ensure that each node will only use one correction factor (typically the one from the closest reference node), correction factors are propagated using controlled flooding, i.e., once a node has received a correction factor from one of its neighboring reference nodes, subsequent correction factors are ignored. For example, in Fig. 4, sensor node S may have received the correction factor from reference node R2 first (since it is closest in distance) and uses R2's correction factor to determine its distances to all reference nodes. That is, S computes is distance to R1 as 17.1*3, its distance to R2 as 17.1*2, and its distance to R3 as 17.1*3.

A variation of this approach is the *DV-distance* method, where distances between neighboring nodes are determined using radio signal strength measurements. As before, these distances are propagated throughout the network (but this times in meters instead of hops). While this approach provides finer granularity (not all hops are estimated to be the same size), it is also more sensitive to measurement errors. Another variation of this technique is to use true Euclidean distances. Here, a node must have at least two neighbors that have distance measurements to a reference node, where the distance between the two neighbors is known. Based on this information, simple trigonometric relationships can be used to determine the distance of a node to a reference node.

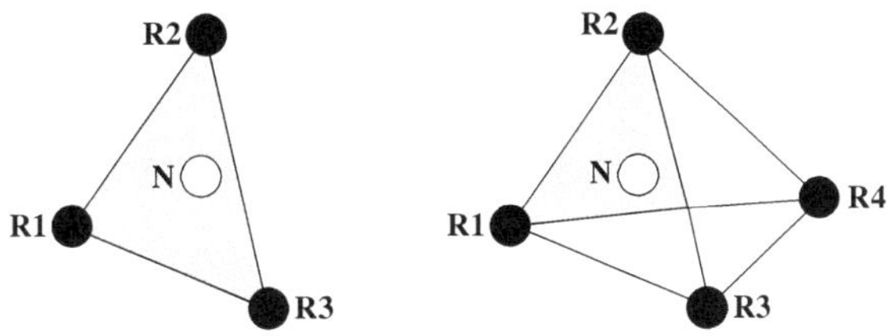

Fig. 5 Location estimation based on the intersection of the triangles formed by reference nodes

3.3 Approximate Point In Triangulation

A well-known example of an area-based range-free localization technique is *APIT* (or *Approximate Point In Triangulation*) [6], which, similar to APS, also requires the presence of reference nodes with well-known locations. The main idea behind APIT is to consider the triangles formed by different sets of three reference nodes and to determine whether a node resides within or outside of these triangles. By identifying the triangles a node resides in, it is possible to narrow down the node's potential locations.

The key procedure in this approach is the *Point In Triangulation* (*PIT*) test, which allows a node to determine these triangles. In this test, once a node has determined the locations of a set of reference nodes (via location updates similar to the APS protocol), it tests whether it resides within or outside of each triangle formed by each set of three reference nodes. Consider a set of three reference nodes *R1*, *R2*, and *R3*. A node *N* is then situated outside the triangle formed by these reference nodes, if there exists a direction such that a point adjacent to N is either further or closer to *R1*, *R2*, and *R3* simultaneously. If no such direction exists, node *N* is inside the triangle and the triangle formed by the three reference nodes can be added to the set of triangles in which *N* resides. This technique is illustrated in Fig. 5. In this example, node *N* resides within the triangle formed by reference nodes *R1*, *R2*, and *R3*, shown in the left graph. On the right of Fig. 5, a fourth node, *R4* has been added (i.e., *N* has learnt about the presence and location of the fourth reference node), which leads to additional triangles formed the sets *{R1, R2, R4}*, *{R1, R3, R4}*, and *{R2, R3, R4}*. Using this additional information, node N is able to reduce the size of the area it resides in.

However, this test as described above is not feasible in practice since it would require that nodes could be moved in any direction. Therefore, instead of this *perfect PIT* test, an *Approximate PIT* (*APIT*) test can be used as long as network density is sufficiently large. In APIT, node movement is emulated using neighbor information that is exchanged between nodes using beacon messages, e.g., using such beacon messages, a ranking among nodes in reach of a reference node can be established based on their signal strengths (and therefore distances). For example, if no neighbor of node *N* is further from or closer to all three reference nodes *R1*, *R2*, and *R3* simultaneously, node *N* assumes that it resides inside the triangle formed by the three reference nodes; otherwise node *N* assumes that it is outside the triangle.

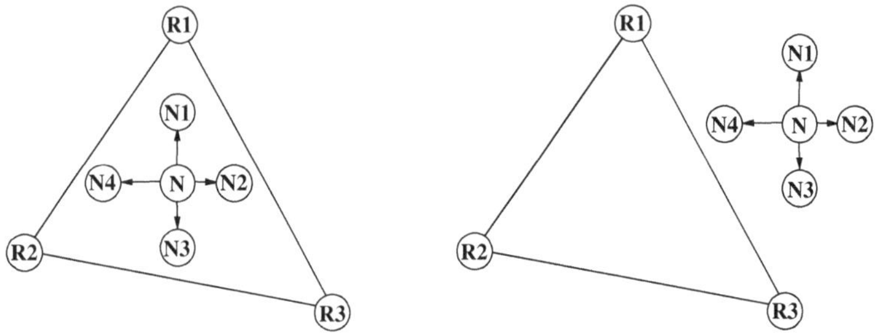

Fig. 6 Examples of APIT test scenarios

Figure 6 illustrates this approach, where the left graph shows an example where node *N* is within the triangle formed by the three reference nodes and the right graph shows an example where node *N* is outside. In the left example, node *N* has four neighbors *N[1..4]*, none of which is simultaneously closer to or further away from all three reference nodes. Therefore node *N* determines that it must reside within the triangle formed by the reference nodes. In contrast, the graph on the right shows an example where node *N* determines that it must reside outside the same triangle since node *N4* is closer to all three reference nodes than node *N*, while node *N2* is further away from all reference nodes, compared to node *N*.

In this approach, it is possible that a node's determination is wrong, simply because only a finite number of neighboring nodes (and therefore directions) can be used to evaluate whether the node is inside or outside the reference node triangle and small errors in signal strength measurements can falsify the result. For example, in the left graph of Fig. 6, if the signal strength measurements from node *N4* indicate that it must be further away from reference node *R2* than node *N*, node *N* would incorrectly conclude that it resides outside the triangle. Such measurement errors are very likely to occur due to the signal strength easily being affected by obstacles, multi-path propagation, etc.

Once the APIT test completes, a position estimate can be computed as the center of gravity of the intersection of all triangles in which M resides in. While the APIT approach provides a simple way of determining locations without the need for additional hardware on the sensor nodes, its main disadvantages are the effect of distance measurement errors on localization accuracy and the need for large network density (i.e., sensor nodes must have several reference nodes that can be reached and they must have several neighboring nodes to perform the APIT test).

3.4 Ring Overlapping Localization Techniques

Another area-based ranging technique that has been proposed recently [15, 39] is based on forming rings around reference nodes, where a sensor node estimates its position to be within the intersection of multiple such rings. The goal of these

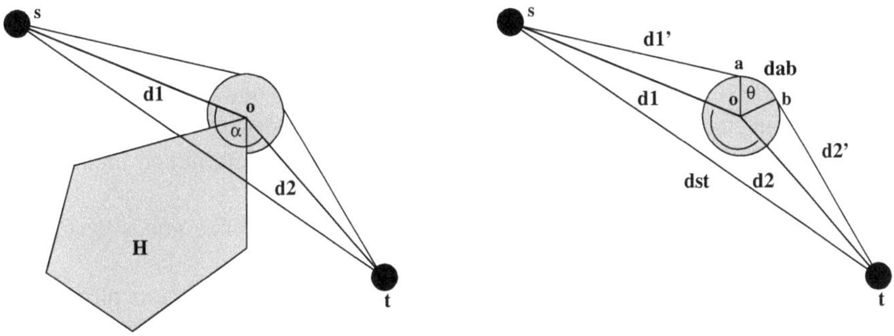

Fig. 7 Principle of the overlapping circles technique

techniques is to provide increased localization accuracy (e.g., compared to hop-based techniques), while keeping the cost of communication and computation low.

For example, in [15], the authors propose a technique called "Annulus Intersection and Grid Scan" or AIGS. In this technique, a reference node X is considered as the center of a circle, where the radius of the circle is the distance to another reference node Y. If the reference node density is sufficient, each reference can thereby form multiple such rings. In AIGS, a sensor node A with an unknown location tries to determine two specific rings around each neighboring reference node: (i) the circle with the shortest radius among all circles that contain node A and (ii) the circle with the largest radius among all circles that do not contain it. To select such rings, node A must be within the annulus of these circles and therefore, the annulus of the circles can be used to estimate the location of node A. In AIGS, the center of the intersections of the overlapping rings is taken as the estimated coordinate of the unknown node A.

This basic principle is shown in Fig. 7. This figure shows several reference nodes, with the circles indicating the radius of the closest and furthest neighbor beacon nodes. The shaded area indicates the intersection of the circles and the unknown node estimates its position to be in the center of this region. In detail, the algorithm performs the following steps. Note that the algorithm assumes that all nodes use omnidirectional antennas, all nodes are randomly deployed in 2-dimensional space, all nodes have the same transmit power, and there is no node mobility.

- *Step 1*: Localization is based on the transmission of two beacon messages by the reference nodes. The first one contains the coordinates of the reference node (x, y) and each receiver records the reference node's location and signal strength. In the second beacon, each reference node broadcasts its view of the network, i.e., the information previously received from other reference nodes. All sensor nodes receiving these beacons also record this information.
- *Step 2*: A sensor A with unknown location successively selects a reference node that is within its one-hop range and uses this node as the center of a number of circles, each with a radius that equals the distance between the reference node and one of its neighboring reference nodes. Then node A attempts to find the smallest annulus that exactly contains node A. If there is such an annulus, the algorithm continues

in Step 3, otherwise the process is repeated until all neighboring reference nodes have been investigated.

- *Step 3*: Node A now attempts to find the intersection area of the annulus; if it is the first annulus, then the intersection area will be itself, otherwise it will be the intersection of the newly discovered annulus the previously determined intersection area.
- *Step 4*: Node A repeats Step 2 until all neighboring reference nodes have been investigated.
- *Step 5*: If there is an intersection area, node A now computes the centroid area and takes the resulting location as its coordinate. Otherwise, node A computes its coordinates using a centroid localization algorithm.

In comparison to DV-Hop, AIGS performs well in both overhead and accuracy. The results presented in [15] indicate that in a network with 30 % reference nodes (out of 200 nodes in total), the localization error can be significantly reduced, particularly for large communication ranges. For example, while DV-Hop had an average localization error of 30 % (which stayed constant with varying communication ranges), the localization error for AIGS was less than 20 % for low communication ranges and less than 5 % for high communication ranges. The computational complexity of AIGS is similar to DV-Hop, while the communication overhead is lower (beacons are exchanged only within the one-hop neighborhood).

3.5 *Multidimensional Scaling*

A popular technique to achieve range-free localization (and an example for localization based on hop count) is multidimensional scaling (MDS) [1, 31], which has its roots in psychometric and psychophysics, and is a set of data analysis techniques that display the structure of distance-like data as a geometrical picture. MDS requires a powerful centralized device (e.g., a base station connecting a WSN to the rest of the world), which collects topological information from the network, determines the nodes' locations, and propagates this information back into the network.

With MDS, the network is represented as an undirected graph of n nodes, where the edges in the graph represent connectivity information. Further, a subset of m nodes represents reference nodes (i.e., nodes with well-known locations). The goal of MDS is to preserve the distance information, assuming that the distances between all pairs of nodes are known, such that the network can be recreated in the multidimensional space. The result is then an arbitrarily rotated and flipped version of the original network layout.

Classical MDS and MDS-MAP

One of the simplest versions of MDS, called classical MDS, has a closed form solution allowing for efficient implementations. In this version, the process is as follows. First, assume a matrix representing the squared distances between nodes:

$$D^2 = c1' + 1c' - 2SS',$$

where *1* is an *nx1* vector of ones, *S* is the similarity matrix for *n* points, where each row represents the coordinates of point *i* along *m* coordinates, *SS′* is called the scalar product matrix, and *c* is a vector consisting of the diagonal elements of the scalar product matrix. Next, by multiplying both sides of this equation by the centering matrix

$$T = I - \frac{11'}{n},$$

where *I* is the identity matrix and *1* is again a vector of ones, leads to

$$TD^2T = T\left(c1' + 1c' - 2SS'\right)T = Tc1'T + T1c'T - T\,(2B)\,T,$$

where $B = SS'$. Since centering a vector of ones yields a vector of zeros, this can be simplified to

$$TD2T = -T\,(2B)\,T.$$

Next, multiplying both sides with $-1/2$ results in

$$B = -\frac{1}{2}TD2T.$$

B is a symmetric matrix and can therefore be decomposed into

$$B = Q\Lambda Q' = Q'\Lambda^{\frac{1}{2}}\left(Q\Lambda^{\frac{1}{2}}\right)' = SS'.$$

This can then be decomposed into

$$B = Q\Lambda Q' = \left(Q'\Lambda^{\frac{1}{2}}\right)\left(Q\Lambda^{\frac{1}{2}}\right)' = SS'.$$

Finally, given *B*, the coordinates of *S* can now be determined by eigendecomposition:

$$S = Q\Lambda^{1/2}.$$

This approach is used in [31], where a localization method called MDS-MAP is proposed. In MDS-MAP, a distance matrix *D* is constructed using an all pairs shortest path algorithm (e.g., Dijkstra's), with d_{ij} representing the distance (i.e., the number of hops) between two nodes *i* and *j*. In the next step, the classical MDS approach as described in this section is used to arrive at an approximate value of the relative coordinate of each node. These relative coordinates can then be transformed into absolute coordinates by aligning the relative coordinates obtained for the reference nodes with their absolute coordinates. Using least-squares minimization, even more refined location estimates could be obtained. Another extension to this approach is to divide a WSN into overlapping regions, where localization is performed, as described above, separately for each of these regions. The resulting local *maps* can then be stitched together to arrive at a global network topology. This is achieved using nodes that appear in multiple maps, i.e., nodes that are shared between adjacent regions. The outcome is an improved performance for networks that are shaped irregularly, by avoiding the use of distance information of nodes that are far away from each other. Another modification of this approach is to implement it as a distributed solution (instead of the centralized solution used so far, which relies on collecting all nodes' information at a central location), which has been proposed in [29].

3.6 Rendered Path Localization

A challenge in range-free localization is to ensure that large numbers of reference nodes are uniformly distributed across the network. If this is not given, many solutions will fail or deliver poor results, particularly in *anisotropic* WSNs with possible holes. In anisotropic networks, the Euclidean distances between pairs of nodes may not closely correlate with the hop counts, because holes in the network force paths between the nodes to have curves. This is particularly common when nodes are deployed randomly, with denser areas of nodes alternating with sparser areas, leading to such holes.

As a consequence, the Rendered Path (REP) [14] protocol addresses this problem by path rendering and virtual hole construction operations in a distributed manner. The basic idea behind REP is to identify the boundaries of holes in a network and to label the boundary nodes of different holes with different colors. When the shortest path between two nodes passes the holes, it is rendered with the colors of the boundary nodes, i.e., a path can be rendered by multiple colors. By passing holes, the path is segmented according to the intermediate colorful boundary nodes. In addition, REP also generates virtual holes to augment and render the shortest path by calculating the Euclidean distance between two nodes based on the distance and the angle information along the rendered path.

Consider two nodes s and t, separated by a hole H (see Fig. 8), where REP renders a shortest path $P_G(s, t)$ between these two nodes.

Every point on the boundary of hole H is colored with the color of H and these points are said to be H-colored. When a hole between nodes s and t exists, the shortest path must intersect with the hole boundaries and REP knows (from the colored points and their colors) how many different holes a path has passed, i.e., the number of passed holes is equal to the number of different rendered colors.

Figure 8 illustrates a simple basic scenario for REP. The figure shows a convex hole H at point o, which is H-colored. The shortest path between s and t (P_{st}) is segmented into so and ot, where we assume that $|so| = d_1$ and $|ot| = d_2$. Based on the law of cosines, the triangle formed by s, t, and o has the following mathematical relationship:

Fig. 8 Basic scenario of REP

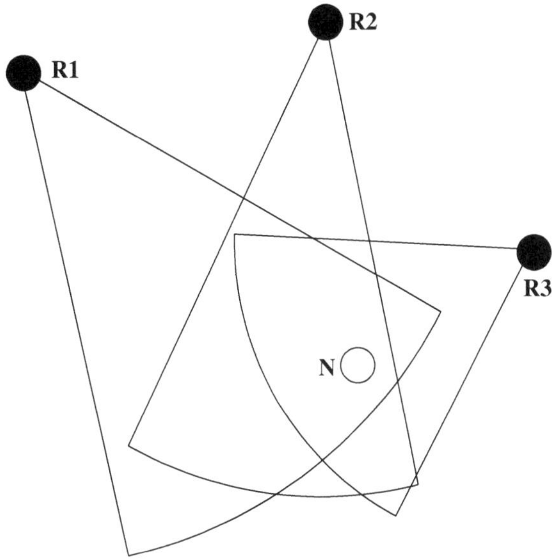

$$|st|^2 = |so|^2 + |ot|^2 - 2\,|so|\,.\,|ot|\,.cos\angle sot$$

and consequently:

$$d_{st} = \sqrt{d_1^2 + d_2^2 - 2d_1 d_2 cos\alpha}.$$

In order to obtain the angle between *so* and *ot*, REP creates a virtual hole (approximately round-shaped) around *o*, with radius *r*, such that the former shortest path *s-o -t* is blocked. The center *o* of this virtual hole is then called *focal point*. The color of the virtual hole is the color of *o* and the new shortest path between *s* and *t* is therefore augmented to bypass the enlarged hole. The right graph in Fig. 8 illustrates that the new shortest path P_{st}^* can be segmented into three pieces: an uncolored line *sa* (of length d_1'), an *o*-colored arc (of length d_{ab}), and another uncolored line *bt* (of length d_2'). The length d_{ab} reflects angle and can be determined as

$$\alpha = 2\pi - \frac{d_{ab}}{r} - arccos\frac{r}{d_1} - arccos\frac{r}{d_2}.$$

The distance d_{st} between nodes *s* and *t* can then be determined from the length information on the two rendered paths P_{st} and P_{st}^*.

The focus in this discussion and illustration has been on a simple scenario with a single hole separating two nodes. However, in real settings, it may often occur that paths interact with hole boundaries at more than one point, e.g., when a path intersects along a segment of a boundary with a convex hole or at several discrete points with a concave hole. However, solutions to these scenarios can be similarly derived as for the situation with a single boundary point.

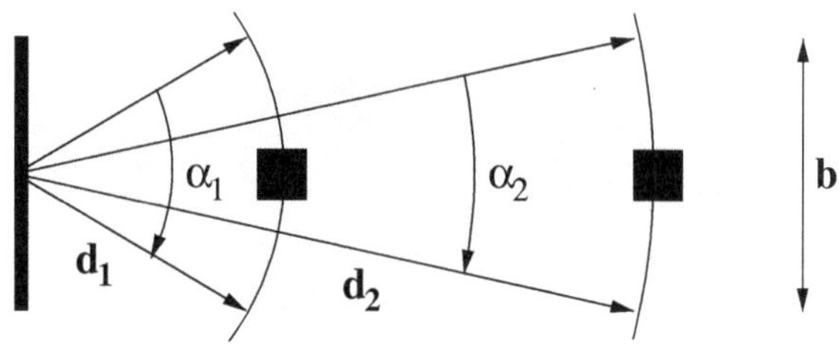

Fig. 9 Principle of SeRLoc localization

3.7 Secure Range-Independent Localization

The Secure Range-Independent Localization (or SeRLoc) protocol [12] is concerned with providing range-free localization in untrusted environments. To achieve secure localization, SeRLoc is primarily concerned with providing a decentralized solution that is resource-efficient, but also robust against security threats.

Toward this end, SeRLoc assumes that some nodes in the network know their location and orientation, e.g., the positions of these *locators* can be obtained using GPS receivers. Another assumption is that locators are equipped with sectored antennas with *M sectors* (while regular sensor nodes are equipped with omnidirectional antennas). A certain known directivity gain *G(M)* and an idealized angular reception are also assumed. Sensors then determine their positions using beacons that are transmitted by the locators; this principle is illustrated in Fig. 9. Each locator transmits different beacons at each antenna sector, where each beacon contains the coordinates of the locator and the angles of the antenna boundary lines. Sensor nodes collect these beacons and use them to determine whether they are within a specific antenna sector for the transmitting locator. Each sensor also knows the maximum transmission range for each locator, which can further be used to limit the size of the sector. Once beacons from all locators have been collected, a sensor node can determine the overlap between the sectors to identify the region within which they must reside.

Once this region has been determined, the sensor node then computes the center of gravity and assumes that this is the sensor's location.

As previously mentioned, a primary goal of SeRLoc is to provide localization in untrusted environments and therefore, SeRLoc includes a variety of security mechanisms. First, encryption is used to protect localization information, i.e., all beacon messages are encrypted using a shared global symmetric key (pre-loaded on sensor before deployment). Every sensor also shares a symmetric pairwise key with every locator.

Next, the use of a shared symmetric key does not identify the source of the messages a node receives. Therefore, a malicious node could impersonate multiple locators. To prevent malicious nodes from injecting false localization information into the network, sensors must authenticate the source of the beacons using collision-resistant hash functions. Each locator L_i has a unique password PW_i, which is blinded with the use of a collision-resistant hash function such as MD5 [26]. Due to the collision resistance property, it is computationally infeasible to find a value PW_j, such that $H(PW_i) = H(PW_j)$, with $PW_i PW_j$. The hash sequence is generated using the following approach:

$$H^0 = PW_i,\ H^i = H\left(H^{i-1}\right),\ i = 1, \ldots, n,$$

with n being a large number and H^0 never revealed to any sensor. Each sensor has a pre-loaded table containing the identifier of each locator and the corresponding hash value $H^n(PW_i)$. Now, assume that locator L_i wants to transmit its first beacon and sensors initially only know the hash value $H^n(PW_i)$. The beacon message contains $(H^{n-1}(PW_i), j)$ with the index $j = 1$ (i.e., the first hash value published). Every sensor receiving this beacon can now authenticate the locator only if $H(H^{n-1}(PW_i)) = H^n(PW_i)$. Once verified, the sensor replaces $H^n(PW_i)$ with $H^{n-1}(PW_i)$ in its table and increases the hash counter by one.

3.8 Event-Driven Localization

Recent work on localization has resulted in another type of range-free solutions, called *event-driven* localization. In these techniques, certain types of events, such as radio signals, light or laser beams, sound blasts, etc., are issued and nodes in the network perform event detection to determine their positions. This section provides an overview of solutions that belong to this category of localization algorithms.

The Lighthouse Approach

The idea behind event-based localization is to determine distances, angles, and positions using concrete events such as radio waves, beams of light, or acoustic signals transmitted by a reference node and received by a sensor node. In the Lighthouse location system [27], sensor nodes can determine their location with high accuracy without the need for additional infrastructure components besides a base station equipped with a light emitter.

The concept behind the Lighthouse approach is illustrated in Fig. 10. It uses an idealistic light source, which has the property that the emitted beam of light is parallel, i.e., the width b remains constant. The light source rotates and when the parallel beam passes by a sensor, it will see the flash of light for a certain period of time t_{beam}, which varies with the distance between the sensor and the light source (since the beam is parallel). The distance d between the sensor and the light source can then be expressed as

$$d = \frac{b}{2sin\frac{\alpha}{2}},$$

where expresses the angle under which the sensor sees the beam of light as follows:

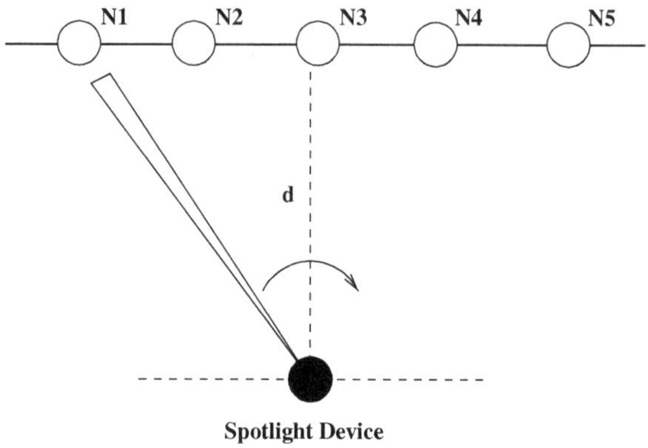

Fig. 10 The lighthouse localization approach (*top view*)

$$\alpha = 2\pi \frac{t_{beam}}{t_{turn}}.$$

Here, t_{turn} is the time the light source takes to perform a complete rotation. The beam width b is known and a constant, therefore, a sensor can then calculate the value of t_{beam} as

$$t_{beam} = t_2 - t_1$$

and

$$t_{turn} = t_3 - t_1,$$

where t_1 is the time the sensor sees the light for the first time, t_2 is the time the sensor no longer sees the light, and t_3 is the time when the sensor sees the light again.

So far, we assumed that the beam width b is constant, independent from the distance between light source and receiver node. However, perfectly parallel light beams are difficult to realize in practice and even very small beam spreads can result in significant localization errors. For example, a beam width of 10 cm and a beam spread of leads to a beam width of 18.7 cm at a distance of 5 m. An additional requirement is that the beam width should be as large as possible to keep inaccuracies small. To achieve this, two laser beams can be used to create the outline of a "virtual" parallel beam since the sensor nodes are only interested in detecting the edges of the virtual beam represented by the two laser beams.

The Spotlight Approach

Similar to the Lighthouse approach, another example for a range-free localization method that does not require reference nodes and uses events for localizations is Spotlight [35, 36]. The key concept behind this approach is to generate events (e.g., light beams), where the time when an event is detected, together with certain spatio-temporal properties of the generated events, can be used to derive sensor node locations. Key assumptions for this approach are that a Spotlight device can

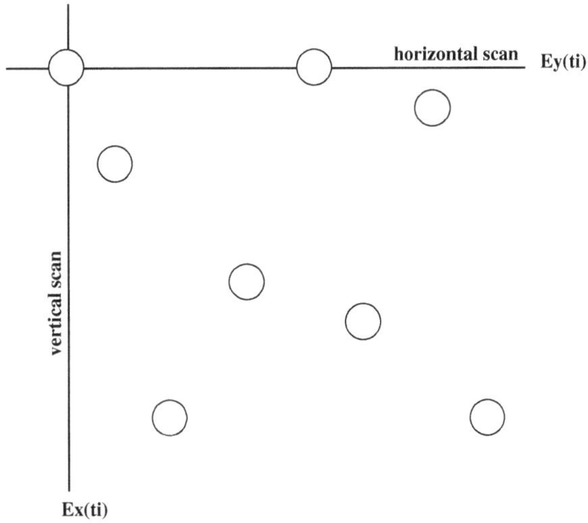

Fig. 11 Point scan event distribution function

generate events that can be detected by sensor nodes, the Spotlight device knows its location and the time when the event was created, the sensors in the network are time-synchronized, and the sensors have line-of-sight (LOS) to the Spotlight device.

The Spotlight protocol performs the following steps:

1. A Spotlight device distributes events *e(t)* in the space *A* over a certain period of time.
2. Sensor nodes record the time sequence $T_i = \{t_{i1}, t_{i2}, \ldots, t_{in}\}$ at which events are detected.
3. After the event distribution, these collected times are transmitted to the Spotlight device.
4. The Spotlight device determines the locations of the sensor nodes using the collected time sequences and the known Event Distribution Function *E(t)*.

The Event Distribution Function *E(t)* is the core technique and an essential component of the Spotlight approach and multiple solutions for *E(t)* have been proposed:

- In the **Point Scan** function, it is assumed that the sensors nodes are placed along a straight line, as shown in Fig. 11. The Spotlight device generates events, such as light spots, along this straight line with a constant speed. This leads to a series of event detections at the sensor nodes, where the set of timestamps when such events were detected by node *i* is described as $T_i = \{t_{i1}\}$. The event distribution function is described as:

$$E(t) = \{p \mid p \in A,\ p = t * s\},$$

where *A* describes the straight line going from coordinates *(0,0)* to *(0,l)*, i.e., $A = [0, l]$ and The localization function can then be derived as

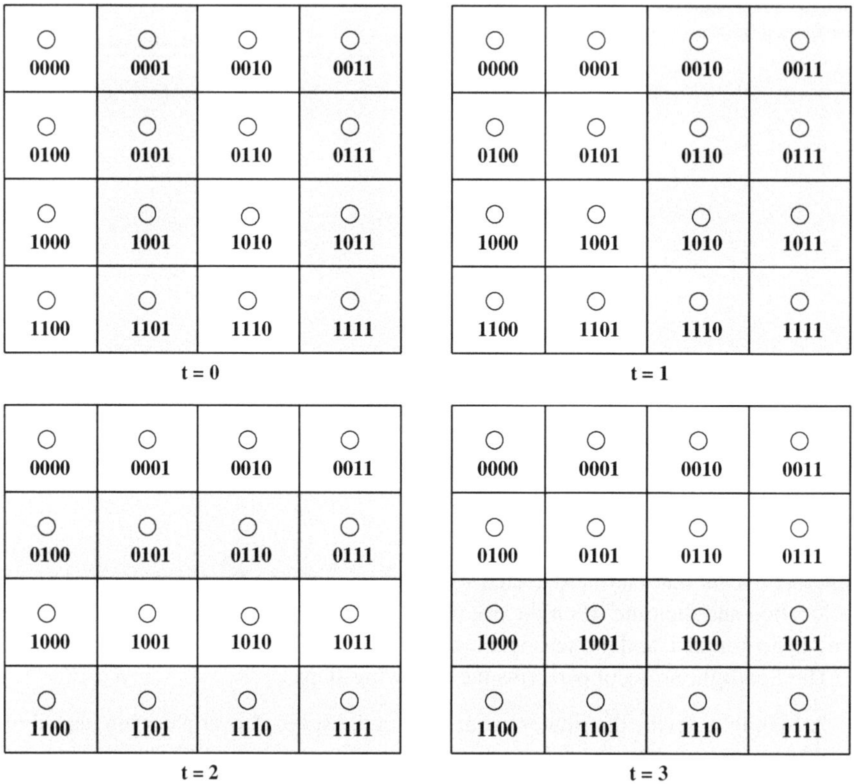

Fig. 12 Line scan event distribution function

$$L(T_i) = E(t_{i1}) = \{t_{i1} * s\}.$$

- In the **Line Scan** function (Fig. 12), nodes are distributed in a two dimensional plane ($A = [l \times l] \subset R^2$) and it is assumed that the Spotlight device can generate an entire line of events simultaneously (e.g., such as a laser).
  Given a scanning speed of s and a set of event detection timestamps (for node i), the line scan function is defined as

$$E_x(t) = \{p_k \mid k \in [0, l], \ p_k = (t * s, k)\} \ for \ t \in 0, l/s \text{ and}$$

$$E_y(t) = \{p_k \mid k \in [0, l], \ p_k = (k, \ t * s - l)\} \ for \ t \in l/s, 2l/s.$$

  The localization function can then be derived from the intersection of the vertical and horizontal scans, i.e.,

$$L(T_i) = E(t_{i1}) \cup E(t_{i2}).$$

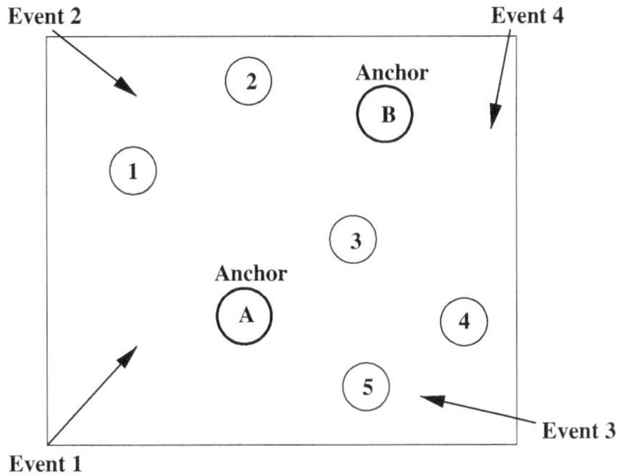

Fig. 13 The area cover implementation

- In the **Area Cover** function, Spotlight devices that can cover an entire area are being used. With this approach, a sensor field A is partitioned into multiple sections, each assigned a unique binary identifier, called *code* (Fig. 13). Each section S_k within area A has then a unique code k and events are generated according to the encoded bit in the bitstring. The area cover function is then:

$$BIT(k, j) = \begin{cases} true \ if \ jth \ bit \ of \ k \ is \ 1 \\ false \ if \ jth \ bit \ of \ k \ is \ 0 \end{cases} \text{ and}$$

$$E(t) = \{p \mid p \in S_k, BIT(k,t) = true\}.$$

The localization function can then be described as

$$L(T_i) = \{p \mid p = COG(S_k), BIT(k,t) = true \ if \ t \in T_i, \ BIT(k,t) = false \ if \ t \in T - T_i\}.$$

Multi-Sequence Positioning

Finally, the Multi-Sequence Positioning (MSP) approach is another example of event-based localization [39] and works by extracting relative location information from multiple simple one-dimensional orderings of sensor nodes.

For illustration of the concept behind MSP, consider Fig. 13, which shows a small sensor network with five nodes with unknown locations and two reference nodes with known locations. There are multiple event generators placed around the network and each generator produces an event at different points in time. For example, such events can be ultrasound signals or laser scans with different angles. The nodes in the sensor network detect these events at different times, depending on their distances to the event generators. For each event, we can then establish a node sequence, i.e., a node ordering (which includes both the sensor and the

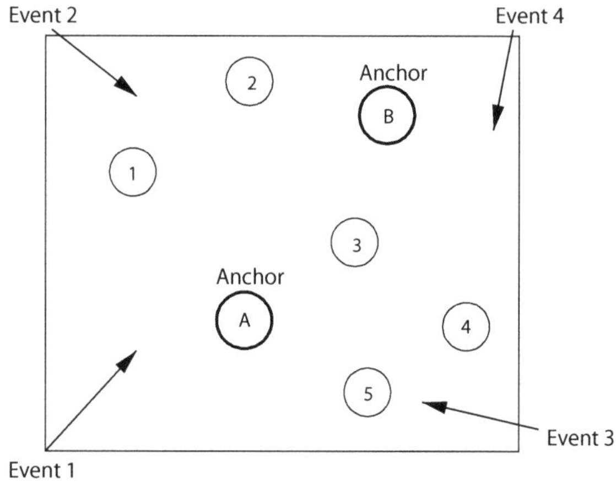

Fig. 14 Basic concept of MSP

Table 1 Node sequence for the four events generated in Fig. 14

Event 1	A – 1 – 5 – 3 – 4 – 2 – B
Event 2	1 – 2 – B – A – 3 – 5 – 4
Event 3	4 – 5 – B – 3 – A – 2 – 1
Event 4	B – 2 – 3 – 1 – 4 – A – 5

anchor nodes) based on the sequential detection of the event. An example of such ordering for the situation shown in Fig. 13 is presented in Table 1. Then, a multi-sequence processing algorithm narrows the potential locations for each node to a small area and finally, a distribution-based estimation method estimates the exact locations.

The main concept behind the MSP algorithm is to split a sensor network area into small pieces by processing node sequences. For example, in the example in Fig. 14, performing a straight-line scan from top to bottom results in a node sequence 2-B-1-3-A-4-5. The basic MSP algorithm uses two straight lines to scan an area from different directions, treating each scan as an event. A left-to-right scan leads to a node sequence 1-A-2-3-5-B-4.

Since the reference nodes have well-known locations, the two reference nodes in Fig. 14 split the area into nine parts. This process can be extended to cut the area into smaller pieces by increasing the number of reference nodes and scans (from different angles). The basic MSP algorithm processes each node sequence to determine the boundaries of a node (by searching for the predecessor and successor reference nodes for the node) and shrinks the location area of this node according to the newly obtained boundary information. Finally, a centroid estimation algorithm sets the center of gravity of the resulting polygon as the estimated location of the target node.

4 Discussion and Comparison

While the primary goal of range-free techniques is to produce localization solutions that are inexpensive (and ideally also with low computational and communication overheads), while typically providing less accurate results compared to range-based techniques. This section qualitatively compares the different localization schemes described in this chapter, discusses their underlying techniques, including the advantages and disadvantages of these techniques.

Centralized Versus Distributed Localization

In centralized localization schemes, node information is collected by a centralized server, which then executes the localization algorithm. Then, the server can use the computed locations for a variety of sensor network management tasks (e.g., to establish routes) or it can send the locations to the individual sensor nodes. Since the server can learn the entire network topology, the outcome can be optimized. On the other hand, the computational requirements of the server are typically very high. Additional downsides of a centralized approach are the large communication overheads (i.e., topology information must be collected centralized and location information may have to be disseminated to all nodes) and that location changes may take a long time to be recognized (i.e., these techniques work best in stationary networks).

In a distributed localization scheme, each sensor node independently determines its location, typically by exchanging information with neighboring sensor nodes, communicating with neighboring reference nodes, or by observing events (such as light beams). In contrast to centralized solutions, the overheads (communication and computation) are typically lower (making distributed solutions also more scalable), but since every node only knows localized information, the localization error tends to be larger.

Localization protocols such as MDS, convex position estimation, and Spotlight are centralized solutions due to their reliance on network-wide information for positioning. Other protocols such as APIT, APS, REP, and SeRLoc are distributed by nature. Finally, some protocols, e.g., hop-based positioning schemes, can be implemented both in distributed or centralized fashion.

Table 2 summarizes some of the key advantages and disadvantages of typical state-of-the-art centralized and distributed localization schemes. Generally, the increased communication overhead in centralized localization schemes limits the scalability of these schemes, but can lead to significantly more accurate positions. On the other hand, with distributed schemes, localization is performed locally only, leading to typically scalable solutions that produce results with larger errors.

Communication and Computational Complexity

The localization schemes discussed in this chapter vary widely in their communication and computational overheads (and consequently also their energy requirements).

Table 2 Comparison of centralized and distributed localization schemes

	Centralized schemes	Distributed schemes
Sensor cost	Low	Low-Medium
Number of anchors	Small	High
Accuracy	Medium	Low
Scalability	Medium	High
Computation overhead	Sensor: Low Server: High	Low
Communication overhead	High	Low
Robustness to anisotropic topologies	Low	Medium

For example, centralized solutions typically have large communication overheads, where network-wide information is collected centrally and positioning results may have to be reported back to each individual sensor node in the network. Algorithms based on graph theory (e.g., convex position estimation) or data analysis techniques (e.g., MDS) also have high computational requirements on the centralized server. For example, MDS has a computational cost of [24, 30], although variants of MDS with improved scalability have also been proposed [32], including a distributed version of MDS [9]. Due to the high communication overheads, centralized solutions typically scale poorly and react very slowly to topology changes. However, compared to the requirements for the server, the computational requirements for sensor nodes are usually very limited, which can limit the energy costs.

In contrast, in decentralized solutions, each sensor nodes communicates with neighboring nodes to determine its position. The overall communication overheads in the network can therefore be lower, but each node must execute its own version of the localization scheme and the overheads introduced depend on the complexity of the localization algorithm. For example, APIT is a rather computationally intensive distributed algorithm that performs best when the number of reference nodes within the vicinity of a sensor node is large (i.e., greater than 20). However, with 20 reference nodes, the intersections of 1140 areas need to be analyzed. On the positive side, since localization is performed locally, distributed schemes can detect changes in topology rather quickly (e.g., depending on the frequency of information exchange between reference nodes and sensor nodes) and therefore the latencies in positioning can be kept low.

Localization Accuracy

The accuracy of localization schemes and the error in the computed positions depend on a variety of parameters, including the number of reference nodes. For example, as mentioned previously, APIT performs best when the number of reference nodes "visible" to a sensor node is rather large, which requires a highly dense network of reference nodes. Further, schemes such as APIT perform best when the reference nodes are randomly distributed throughout the network, which is not always feasible in a real deployment. Compared to distributed solutions, centralized schemes have the advantage that they collect network-wide information, thereby allowing them to optimize their positioning algorithms. As a consequence, centralized localization algorithms typically provide positions with higher accuracy than distributed solutions that determine sensor location using only limited (local) information.

The achievable accuracy also depends significantly on the network topology, e.g., REP applies a path rendering technique to remove the impact of holes in the network. Simulations have shown that the accuracy of REP is significantly better than both APS and APIT in isotropic networks [6, 23]. However, REP assumes (i) that the network has high node density to allow it to achieve accurate boundary recognition, (ii) that the shortest path between two nodes is close to a straight line (if there are no holes), and (iii) that the length of an arc can be estimated from node connectivity along the circle boundary.

Schemes based on concepts in graph theory, such as MDS, can provide accurate results if the network is not too sparse. For example, simulation results [30] have shown that connectivity-based MDS methods achieve an average localization error of 0.31R (where R is the radio range in a randomly deployed network with mean 1-hop connectivity of 12.1).

In localization schemes that rely on localization events, a typical goal is to utilize an asymmetric system design to keep the overheads at the sensors low, i.e., sensor nodes can use simple techniques to detect the events. In the Lighthouse approach, an accuracy of 10 cm in a square space of 5 m*5 m has been demonstrated when the event emitter is carefully calibrated. The event distribution needs to be very accurately timed and typically line-of-sight is required for event detection. However, uneven terrain and the reliance on mechanical parts (e.g., rotating emitter) make careful calibration difficult. In contrast, MSP has the advantage that events are emitted by multiple event sources in different locations of the network (however, a sensor node must be able to detect events from at least two different emitter sources) and that precise control of event generation is not required, i.e., events can be generated anytime (and only event observation is necessary for localization).

Tables 3 and 4 summarize key advantages and disadvantages of the localization protocols discussed in this chapter. The goal is to emphasize the key functional and performance differences between the different techniques, to facilitate the selection of an appropriate localization solution given a specific network topology or application and to help identify potential research directions.

5 Research Directions

The last decade has seen the emergence of a variety of localization techniques and protocols, most of which belong to the class of area- or hop-based schemes. Event-based systems belong to a newer category that has received a significant amount of attention recently. In the area of event-based systems, more works is needed to address challenges such as accurate calibration, positioning accuracy, and how to maintain line-of-sight. When line-of-sight is not given, a secondary localization technique can

Table 3 Summary of localization techniques (part 1)

Localization protocol	Type	Centralized or distributed	Reference nodes	Security features
Convex Point Estimation	Area	Centralized	Medium-high number required	No
APS	Hop	Distributed	Medium-high number required	No
APIT	Area	Distributed	Many required	No
MDS	Hop	Centralized	None	No
REP	Hop	Distributed	Few required	No
Ring Overlapping	Area	Distributed	Medium number required	No
SeRLoc	Area	Distributed	Sectored antennas	Yes
Lighthouse	Event	Distributed	Light emitter (1+)	No
Spotlight	Event	Centralized	Event generator	No
MSP	Event	Centralized	Event generator (2+)	No

Table 4 Summary of localization techniques (part 2)

Localization protocol	Overhead (comput.)	Overhead (commun.)	Accuracy	Scalability
Convex point estimation	High (server) Low (sensor)	Medium	Good	Medium
APS	Low	Low	Medium-Good	Good
APIT	High	Low	Medium	Good
MDS	High	Medium	Medium	Medium
REP	High	Medium	High (anisotropic) Medium (regular)	Good
Ring Overlapping	Medium	Low	Medium-Good	Good
SeRLoc	Medium	Low	Medium-Good	Good
Lighthouse	Low	Low	Good	Medium
Spotlight	Medium (server) Low (sensor)	Low	High	Medium
MSP	Medium (server) Low (sensor)	Medium	Good-High	Good

be employed, possibly with reduced accuracy. The concept of *multimodal localization* (i.e., use of multiple localization techniques simultaneously) is not only attractive because of its robustness, but also its ability to further reduce localization error and to provide verifiability. Such integration of multiple techniques could also be used to combine range-based with range-free technologies in order to take advantage of the strengths of each approach. For example, range-based techniques can provide high accuracy, but often require additional ranging hardware, extensive environmental profiling, and careful system tuning and calibration. On the other hand, range-free techniques can be more economic at the resource-constrained sensors. More research is required to investigate new protocols that combine cost-effectiveness and accuracy.

Among the protocols discussed in this chapter, only SeRLoc addresses the need for security in wireless sensor networks. Localization is critical to most sensor networks and attacks can render them ineffective. However, very little work has been done to provide robust and secure localization schemes. SeRLoc has the ability to protect against wormholes, Sybil attacks, and attacks intended to compromise sensor nodes. While it provides some security, more work will be needed, especially for sensor network application in emergency response or military settings. In [11], the authors attempt to address some of the shortcomings of SeRLoc, specifically focusing on providing robust localization and verification of the location claim of a sensor node. The work in [11] does not require centralized management and is also not vulnerable to jamming.

More work is also needed to limit the impact of localization techniques on the communication and computation overheads, hardware costs, the need for centralized computation, and the need for dedicated event generators. Most sensor networks are composed of very low-cost technology, where the accuracy of localization can be affected by the hardware limitations of the sensors or by environmental noise affecting signal or event detection. By applying techniques such as uncertainty-based averaging, error propagation control, or data fusion, these impacts could be mitigated. Finally, there remain many challenges to be addressed in network topology and density, e.g., REP is an effort to address holes and irregular shapes in a network that could be problematic for connectivity-based localization techniques. These are complex challenges that remain open problems, especially for large networks and networks with varying densities.

Finally, most localization techniques focus on position estimation in two-dimensional space and very little prior work has focused on three-dimensional space. Techniques based on graph theory (e.g., convex position estimation) can be extended to three-dimensional space when the positions are extended to include a third component (x, y, z), however leading to significantly increased computational complexity. Hop-based techniques can also be extended rather easily if all three coordinates of the reference nodes are known. Area-based techniques will typically require more significant changes to compute the intersections of spheres and other three-dimensional objects instead of simple circles and triangles. Finally, while event-based techniques such as MSP may rather easily be adapted to three-dimensional scenarios, other techniques that are based on light events (e.g., Lighthouse) may be difficult to implement or require significant amounts of additional hardware (e.g., multiple simultaneous light beams in different directions in the Lighthouse scheme).

6 Summary

Localization in sensor networks is essential for many sensing applications and network management activities. This chapter provided a survey of range-free localization techniques in wireless sensor networks, including techniques based on hop count, area information, neighborhood information, and event detection. For many of these techniques, it is required that there are sufficient reference nodes and that those nodes are evenly distributed throughout the network. While the accuracy obtained by range-free localization techniques is typically lower than the accuracy of range-based techniques, a main advantage of range-free localization is that typically no additional hardware is needed and localization can therefore be performed at a lower cost. This chapter compared a variety of centralized and distributed localization techniques, including techniques belonging to the event-based positioning category. These techniques vary significantly in their overheads, performance, scalability, and the number of reference nodes. This chapter concluded with a comparison of these techniques, including a discussion of open research questions.

References

1. I. Borg, P. Groenen, in *Modern Multidimensional Scaling: Theory and Applications* (Springer, New York, 1997)
2. S. Boyd, L. El Ghaoui, E. Feron, V. Balakrishnan, Linear matrix inequalities in system and control theory. in *Society for Industrial and Applied Mathematics (SIAM)* (Philadelphia, PA, 1994)
3. Y.-T. Chan, W.-Y. Tsui, H.-C. So, P.-C. Ching, Time-of-arrival based localization under NLOS conditions. IEEE Trans. Veh. Technol. **55**(1), 17–24 (2006)
4. C.R. Comsa, A.M. Haimovich, S.C. Schwartz, Y.H. Dobyns, J.A. Dabin, Time difference of arrival based source localization within a sparse representation framework. in *Proceedings of the 45th Annual Conference on Information Sciences and Systems (CISS)* (Baltimore, MD, 2011)
5. L. Doherty, K.S.J. Pister, L. El Ghaoui, Convex position estimation in wireless sensor networks. in *Proceedings of the 20th Annual Joint Conference of the IEEE Computer and Communications Societies* (Anchorage, AK, 2001)
6. T. He, C. Huang, B.M. Blum, J.A. Stankovic, T. Abdelzaher, Range-free localization schemes for large scale sensor networks. in *Proceedings of the 9th Annual International Conference on Mobile Computing and Networking (MobiCom)* (San Diego, CA, 2003)
7. J. Hightower, G. Borriello, Location system for ubiquitous computing. Comput. J. **34**(8), 57–66 (2001)
8. B. Hofmann-Wellenhof, H. Lichtenegger, J. Collins, in *Global Positioning System: Theory and Practice*, 5th edn., (Springer, 2008)
9. X. Ji, H. Zha, Sensor positioning in wireless ad-hoc sensor networks using multidimensional scaling. in *Proceedings of the 23rd Annual Joint Conference of the IEEE Computer and Communications Societies (INFOCOM)* (Hong Kong, China, 2004)
10. I. Khan, H. Mokhtar, M. Merabti, A new self-detection scheme for sensor network boundary recognition. in *Proceedings of the 34th Conference on Local Computer Networks* (Zurich, Switzerland, 2009)
11. L. Lazos, S. Capkun, R. Poovendran, Robust position estimation in wireless sensor networks. in *Proceedings of the International Workshop on Information Processing in Sensor Networks (IPSN)* (2005)
12. L. Lazos, S. Capkun, R. Poovendran, SeRLoc: Secure range-independent localization for wireless sensor networks. in *Proceedings of the ACM Workshop on Wireless Security (WiSe)* (2004)
13. H. Lee, A. Klappenecker, K. Lee, L. Lin, Energy efficient data management for wireless sensor networks with data sink failure. in *Proceedings of the 2nd IEEE International Conference on Mobile Ad-hoc and Sensor Systems (MASS)* (Washington, DC, 2005)

14. M. Li, Y. Liu, Rendered path: range-free localization in Anisotropic sensor networks with holes. in *Proceedings of the Thirteenth Annual International Conference on Mobile Computing and Networking (MobiCom)* (Montreal, Canada, 2007)
15. C. Liu, K. Wu, T. He, Sensor localization with ring overlapping based on comparison of received signal strength indicator. in *Proceedings of the IEEE International Mobile Ad-Hoc and Sensor Systems (MASS) Conference* (Fort Lauderdale, FL, 2004)
16. Y. Liu, Z. Yang, X. Wang, L. Jian, Location, localization, and localizability. J. Comput. Sci. Technol. (IJST) **25**(2), 274–296 (2010)
17. K. Lorincz, D. Malan, T.R.F. Fulford-Jones, A. Nawoj, A. Clavel, V. Shnayder, G. Mainland, S. Moulton, M. Welsh, Sensor networks for emergency response: challenges and opportunities. IEEE Pervasive Comput. **3**(4), 16–23 (2004)
18. Y. Lu, W. Wang, Y. Zhong, B. Bhargava, Study of distance vector routing protocols for mobile ad hoc networks. in *Proceedings of the First IEEE International Conference on Pervasive Computing and Communications (PerCom)* (Dallas - Fort Worth, TX, 2003)
19. V. Mhatre, C. Rosenberg, D. Kofman, R. Mazumdar, N. Shroff, Design of surveillance sensor grids with a lifetime constraint. in *Proceedings of the First European Workshop on Wireless Sensor Networks (EWSN)*, (2004), pp. 263–275
20. L.S. Monteiro, T. Moore, C. Hill, What is the accuracy of DGPS? J. Navig. **58**(2), 207–225 (2005)
21. Y. Nesterov, Interior point polynomial algorithms in convex programming. in *Society for Industrial and Applied Mathematics (SIAM)* (Philadelphia, PA, 1994)
22. D. Niculescu, B. Nath, Ad hoc positioning system (APS). in *Proceedings of the IEEE Global Telecommunications Conference (GLOBECOM)* (San Antonio, TX, 2001)
23. D. Niculescu, B. Nath, DV based positioning in ad hoc networks. Telecommun. Syst. **22**(1–4), 267–280 (2003)
24. N. Patwari, A.O. Hero, J.A. Costa, Learning sensor location from signal strength. in *Secure Localization and Time Synchronization for Wireless Sensor and Ad Hoc Networks, Advances in Information Security Series*, vol. 30 (Springer, 2006)
25. R. Peng, M.L. Sichitiu, Angle of arrival localization for wireless sensor networks. in *Proceedings of the 3rd Annual IEEE Communications Society Conference on Sensor and Ad Hoc Communications and Networks* (2006)
26. R. Rivest, The MD5 message-digest algorithm. RFC 1321, April 1992
27. K. Roemer, The lighthouse location system for smart dust. in *Proceedings of the 1st International Conference On Mobile Systems, Applications And Services* (2003), pp. 15–30
28. A. Savvides, C.-C. Han, M.B. Strivastava, Dynamic fine-grained localization in ad-hoc networks of sensors. in *Proceedings of the 7th Annual International Conference on Mobile Computing and Networking* (Rome, Italy, 2001)
29. Y. Shang, W. Ruml, Improved MDS-based localization. in *Proceedings of the 23rd Annual Joint Conference of the IEEE Computer and Communications Societies (INFOCOM)* (2004)
30. Y. Shang, W. Ruml, Y. Zhang, M.P. Fromherz, Localization from mere connectivity. in *Proceedings of the 4th ACM International Symposium on Mobile Ad Hoc Networking and Computing (MobiHoc)* (Annapolis, Maryland, 2003)
31. Y. Shang, W. Ruml, Y. Zhang, M. Fromherz, Localization from connectivity in sensor networks. IEEE Trans. Parallel Distrib. Syst. **15**(11), 961–974 (2004)
32. Y. Shang, W. Ruml, M.P. Fromherz, Positioning using local maps. Ad Hoc Netw. **4**(2), 240–253 (2006)
33. I.G. Siqueira, L.B. Ruiz, A.A.F. Loureiro, J.M. Nogueira, Coverage area management for wireless sensor networks. Int. J. Netw. Manag. **17**(1), 17–31 (2007)
34. I. Stojmenovic, Position based routing in ad hoc networks. IEEE Commun. Mag. **40**(7), 128–134 (2002)
35. R. Stoleru, T. He, J.A. Stankovic, Range-free localization, Chapter in secure localization and time synchronization for wireless sensor and Ad Hoc networks. in *Advances in Information Security Series*, ed. by R. Poovendran, C. Wang, S. Roy, vol. 30 (Springer, 2007)

36. R. Stoleru, T. He, J.A. Stankovic, D. Luebke, A high-accuracy low-cost localization system for wireless sensor networks. in *Proceedings of the ACM Conference on Embedded Networked Sensor Systems (SenSys)* (San Diego, CA, 2005)
37. J. Undercoffer, S. Avancha, A. Joshi, J. Pinkston, Security for Sensor Networks. in *Proceedings of the CADIP Research Symposium* (2002)
38. O. Younis, S. Fahmy, Distributed clustering in ad-hoc sensor networks: a hybrid, energy-efficient approach. in *Proceedings of the 23rd Annual Joint Conference of the IEEE Computer and Communications Societies (Infocom)* (Hong Kong, 2004)
39. Z. Zhong, T. He, MSP: multi-sequence positioning of wireless sensor nodes. in *Proceedings of the 5th International Conference On Embedded Networked Sensor Systems* (2007), pp. 15–28

Chapter 12
Energy-Efficient Task Management

Hady S. AbdelSalam and Stephan Olariu

Abstract In numerous applications of wireless sensor networks, the reliability of the data collected by sensors is cast as specific QoS requirements expressed in terms of the minimum number of sensors needed to perform various tasks. Designing a long-lived sensor network with reliable performance has always been challenging due to the modest non-renewable energy budget of individual sensors. In order to promote network longevity, this chapter looks at two energy-aware task management protocols: the first protocol is centralized, while the second one is fully distributed. Both protocols assign sensors to tasks based on their remaining energy so that energy expenditure among neighboring sensors is as even as possible. We compare the network longevity, i.e., the functional lifetime of the sensor network, achieved by assigning tasks to sensors using the proposed protocols against an optimal task assignment and also against energy-oblivious protocols. Extensive simulation results have revealed that the performance of the proposed protocols is very close to that of the optimal task assignment. Furthermore, our simulations have shown that the proposed protocols can increase the functional longevity of the network by about 16 %.

1 Introduction

An important research direction in wireless sensor networks (WSNs) is to develop techniques that increase the useful lifetime (a.k.a. functional longevity) of the network, while maintaining reliable performance satisfying application-specific QoS

H. S. AbdelSalam (✉)
Microsoft Corporation, One Microsoft Way, Redmond, WA 98052, USA
e-mail: hasalam@microsoft.com

S. Olariu
Computer Science Department, Old Dominion University, Norfolk, VA 23529, USA
e-mail: olariu@cs.odu.edu

H. M. Ammari (ed.), *The Art of Wireless Sensor Networks*,
Signals and Communication Technology, DOI: 10.1007/978-3-642-40009-4_12,
© Springer-Verlag Berlin Heidelberg 2014

requirements. The major challenge in the design of efficient protocols that satisfy these requirements is the modest, non-renewable energy budget of individual sensors. A great deal of research has been conducted to optimize sensor energy expenditure at different levels of the network stack: MAC [19, 20], routing [25], avoiding energy holes [16, 24], data aggregation [8] among others. In spite of all this, the problem of designing energy efficient protocols that promote network longevity and that satisfy the stipulated QoS requirements of the application is still very much open. As it turns out, in such a context, energy-oblivious task management protocols may lead to uneven expenditure of sensor energy by assigning uneven workloads to sensors. This, in turn, translates into reduced sensor density around those heavily loaded sensors and may, eventually, lead to the creation of energy holes that partition the network into disconnected islands. We refer the reader to relevant chapters in this handbook where many interesting facets of the energy expenditure problem are discussed in some detail. We note, in passing, an unmistakable trend in recent network research: namely, persistent and sustained efforts directed at making the sensors energetically self-sufficient by harvesting energy from the ambient environment. Notwithstanding these efforts, that are still very much in their infancy, the problem of designing energy-efficient protocols for sensor networks will continue to be relevant for the foreseeable future. This is acutely perceived in contemporary research in nano-sensor technologies where rather minute quanta of energy are being harvested from the environment; moreover, the rate at which energy is harvested is low, making the design of energy-efficient protocols for nano-scale sensors imperative.

In this chapter we explore a different angle of energy-related problems in sensor networks that, to the best of our knowledge, has been overlooked in the sensor networks literature. Specifically, we look at *task assignment* protocols, namely strategies for assigning sensors to tasks. We show that improper task assignment can result in severely skewed energy consumption of sensor energy due to tasking a group of sensors more than others. Figure 1 illustrates a simple example of the problems that could well arise due to improper task assignment. Assume that sensors were deployed as shown in Fig. 1a; two tasks T_1 and T_2 are issued at two different locations. Without

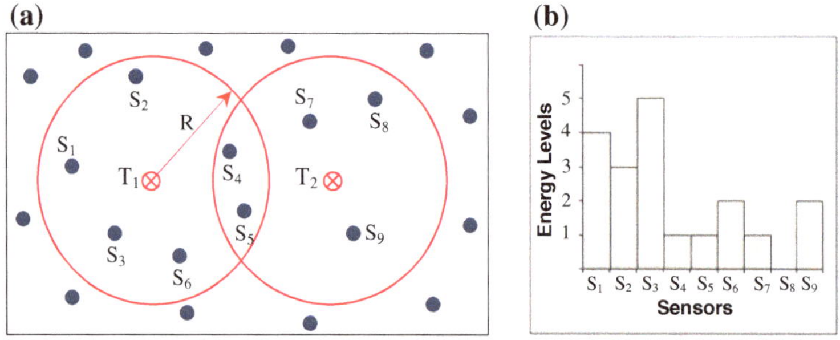

Fig. 1 An example of improper task assignment

loss of generality, we assume that sensor energy comes in discrete units and that each task consumes one unit of energy of each participating sensor and requires the cooperation of at least three sensors to collect accurate results (QoS requirements for reliable performance). Furthermore, we assume that the sensing range of sensors is R, so sensors within range R from the task location are allowed to participate in the task, other sensors cannot participate because the monitored phenomenon is outside their sensing area. In Fig. 1b, the vertical axis, labeled "Energy Levels," keeps track of the remaining energy units of sensors S_1 through S_9. For illustration purposes, we assume that sensors S_1 through S_9 are awake during the recruiting process for tasks T_1 and T_2.

Now, assume that the recruiting protocol for task T_1 selected sensors S_4, S_5, and S_6. Although the recruited sensors still have enough energy to complete task T_1, the energy of sensors S_4 and S_5 will be totally depleted after the task completion, hence they will not be able to participate in any subsequent task. Moreover, there is no way to recruit three sensors for task T_2. On the other hand, if the recruiting protocol for task T_1 had selected sensors S_1, S_2, and S_3, then S_4 and S_5 would have saved their energy to participate in task T_2. In fact, a more careful protocol would not recruit sensors S_4 and S_5 together for task T_2; instead it would recruit sensors S_7, S_9 and either S_4 or S_5. This latter choice guarantees that if there was a subsequent task T_3 at the same location as task T_2, there would be sensors with sufficient energy to execute the new task.

In spite of its simplicity, the previous example shows that task assignment can be very tricky and that performing it improperly may cause many problems, ranging from reducing network density in different areas of the network to creating energy holes that might segment the network into disconnected sub-networks. In turn, this might have a serious impact on network reliability and on longevity by increasing the task failure rate. For these reasons, when it comes to selecting the required workforce for a given task, task assignment protocols must take into consideration the difference in remaining energy at the individual sensor level.

In many WSN applications, the sensors will provide information in response to queries or tasks assigned to them by *Task-Issuing Entities* (TIEs) placed, for example, in the helmet of fire-fighters, first responders, paramedics and other authorized personnel. For each task, a workforce consisting of some of the sensors in the area of interest is to be recruited according to the QoS requirements specific to the task at hand. In this chapter we assume that QoS requirements are expressed in terms of the minimum number of sensors needed to perform a task. In this scenario, the naive approach that allows all the sensors in the same neighborhood to participate in all the tasks regardless of QoS requirements seems to be completely inappropriate, especially when the remaining sensor energy is very limited and network longevity is an important consideration. Hence, efficient task management, in conjunction with an energy-aware workforce selection, play a vital role in the success of this kind of networks.

The remainder of this chapter is organized as follows, a summary of related work is presented in Sects. 2, 3 we state our assumptions about sensor capabilities and the underlying network model. In Sect. 4 we use renewal theory to reason about the time-independent awake probability of sensors. Further, Sect. 5 discusses theoretical results involving task management policies that maximize network longevity. In Sect. 5.1 ideal protocols that guarantee an optimal task assignment are discussed. These theoretical results motivate our efforts, detailed in subsequent sections of the chapter, to keep the energy span of sensors as low as possible. In line with this goal, the next two section present our two energy-efficient task assignment protocols. Specifically, in Sect. 6, we discuss the details of our centralized workforce selection protocol. A distributed version of workforce selection protocol is presented in Sect. 7. Our extensive performance evaluation and simulation results are discussed in Sects. 8, 9 offers concluding remarks and discusses areas for future investigations.

2 Related Work

Due to the modest non-renewable energy budget at the individual sensor level, there has been a consensus in the literature that if the functional longevity of the WSN is to be promoted, sensor energy must be used wisely. This state of affairs has motivated extensive research devoted to optimizing sensor energy expenditure and to promoting network longevity. This flurry of research activity has produced a good number of energy-efficient or energy-aware protocols for WSN. Somewhat surprisingly, less attention has been devoted to developing efficient protocols for task management and workforce selection in WSN. To the best of our knowledge, we are the first researchers to consider the workforce selection problem in WSN [6, 7]. Consequently, we could not find other protocols against which to compare the performance of our proposed protocols. For this reason, we had to compare the performance of our protocols against the performance achieved when using ideal assignment based on sensor remaining energy.

In previously-published work [7], we proposed a centralized task management and workforce selection protocol for special purpose single-hop sensor-based mission-critical networks. The proposed protocol was used to recruit sensors that reside within a disc centered at the current location of the TIE. A major drawback of the previous protocol, it can only recruit sensors that lie within a single-hop from each other and also from the TIE.

In later work published in [6], we tried to remove the single-hop restriction by using the concept of network training [21]. The main idea was to divide the deployment area around TIEs using a set of concentric coronas and wedges. The cells which are formed from the intersections of coronas and wedges are used as the basic tasking units. The sizes of the coronas and wedges are chosen such that sensors within each tasking unit are assumed to be within the same transmission range. After selecting an appropriate coordinator, the workforce protocol proposed in [7] can be used within each cell independently. Although, the protocol looks simple and straightforward, it has a major scalability issue which is inherent in network training. In particular, as we move from the TIE outward, the size of the tasking cells increases from one corona to the next. After a certain point, some sensors within the same cell will not be within the communication range of each other, hence the workforce selection protocol described in [7] cannot be used.

In this chapter, we attempt to overcome the shortcomings mentioned in the previous protocols. First, we develop a tasking model that is appropriate for the alternating wakeup/sleep duty cycles inherent to sensor design. On top of this model, we propose two workforce selection protocols for sensor networks, the first protocol is centralized while the other is distributed. The new protocols do not have the single-hop restriction we had in the previous protocols. Hence, they can be used to recruit sensors anywhere in the network. Furthermore, the distributed protocol has the advantage of reducing the overhead imposed on any central nodes selected to be responsible for the coordination of the workforce selection process. We also provide a simple probabilistic analysis to estimate protocol necessary parameters. Finally, through simulation we compare the performance of the proposed protocols against ideal workforce selection and energy-oblivious protocols in order to verify and measure the increase in the functional longevity of the network.

3 The Network Model

In our network model, we assume the following:

- The sensors are powered by a non-renewable on-board energy source. When its energy supply is exhausted, the sensor becomes in-operational; hence the sensors sleep and wake up alternatively to save their energy budget. Sleep and awake cycles for different sensors occur asynchronously. For example, the sensors may sleep for a random amount of time distributed uniformly in the range $[T_s, T_S]$ and may stay awake for another random amount of time uniformly distributed in the range $[T_l, T_L]$;
- The sensors are deployed uniformly in the deployment area, assumed to be a planar region;
- The sensors may or may not have IDs. In any case, our protocols do not use such IDs;
- As already mentioned, in addition to the sensors, the WSN contains a small number of sinks, hereafter referred to as TIEs, responsible for assigning sensors to tasks and for getting the aggregated result back. The TIEs have a steady energy supply and are awake all the time. In addition, the TIEs have powerful transceivers with application-dependent ranges that can be used to issue tasks to sensors within different areas of the network;
- The sensors have a maximum transmission range denoted by t_x, and a maximum sensing range denoted by R. The communication and sensing ranges of sensors are governed by the inequality ($t_x \geq 2R$). This inequality guarantees that whenever two sensors are within the same sensing range (i.e., they can cooperate in the same task), they will be able to communicate with each other during the workforce selection process;

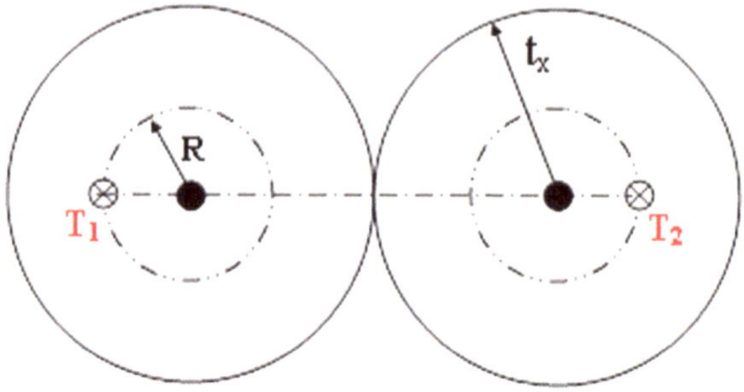

Fig. 2 Minimum distance between two concurrent tasks

- At deployment time, all sensors pack the same amount of energy. It is helpful to imagine that, initially, each sensor has L units of energy and that each time the sensor participates in a task it expends one unit of energy. That is, each sensor can participate in L tasks before its energy drops to zero, at which moment the sensor expires;
- Each sensor is aware of its remaining energy level;
- Each task is associated with a certain location, and if the distance between the locations of two tasks is less than $2(t_x + R)$, then they are not allowed to run concurrently to avoid interference (see Fig. 2). This constraint can be partially lifted if the TIEs can communicate between themselves through a separate channel. If this communication channel is available and if the distances between the centers of the tasks to be performed concurrently are relatively small, then these tasks can be combined into one single task. The QoS requirements of the combined task is chosen to be the maximum of the QoS requirements of the individual tasks. Only one TIE should be responsible for running the combined task. Once the aggregated result arrives at the TIE, it can be forwarded directly to other TIEs;
- Each task requires the cooperation of a workforce of sensors and consumes one unit of energy of each participating sensor. The size of the workforce is reflecting the QoS requirements specific to the task (see below);
- The network uses any of the well known localization protocols to localize sensors [2, 4, 5, 9–11, 14, 21], so each sensor has a rough estimate of its current position;
- We assume that sensors are able to distinguish between background noise and the collisions [22] that occur when there is an overlap between the transmissions of two or more sensors;
- Finally, we assume that QoS requirements of sensing tasks is expressed in terms of the minimum number of sensors that must participate in each task.

4 The Time-Invariant Awake Probability

In deployments populated by energy-constrained sensors it is of paramount importance to design energy-aware protocols that promote the functional longevity of the underlying network. This typically means that the sensors spend their lifetime alternating between two modes: in *sleep mode* the sensor turns off its radio interface

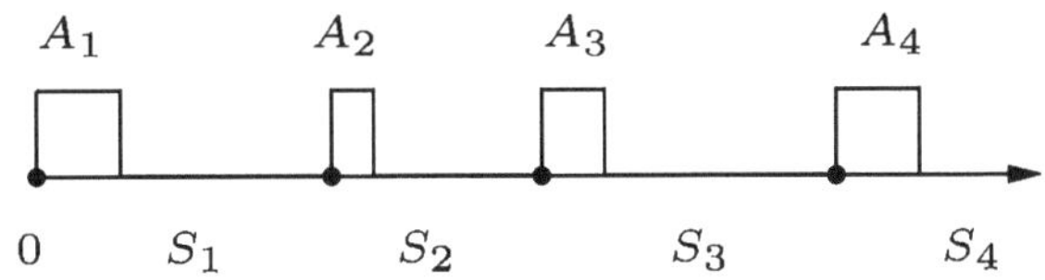

Fig. 3 Illustrating the renewal process of wake-up times

and clocks down its processor; in *awake* mode the sensor is fully functional, with its processor running at top speed and its radio interface turned on.

Let the sequence of random variables $A_1, A_2, \cdots, A_n, \cdots$ denote the duration of consecutive *awake* times of a given sensor. It is quite natural to assume that these awake times are independent identical distributed (i.i.d.) random variables with a common distribution function F_A and that A has *finite* expectation. Similarly, let the sequence of random variables $S_1, S_2, \cdots, S_n, \cdots$ denote the duration of consecutive *sleep* times of a given sensor. We assume that the sleep times are i.i.d. with a common distribution function F_S. As before, we assume that S has *finite* expectation.

To make a choice, we assume that the sensor wakes up for the 0-th time at $t = 0$. As it turns out, this assumption is adopted for convenience only; as far as the limiting probability of the sensor to be awake at some arbitrary time t this assumption is irrelevant. Referring to Fig. 3, we find it useful to model the lifetime of the given sensor as a renewal process $\{N(t),\ t \geq 0\}$, where the renewal points are the moments when the sensor wakes up (other than the 0-th wake-up at time $t = 0$). Let

$$X_0 = 0,\ X_1 = A_1 + S_1,\ X_2 = A_2 + S_2,\ \cdots,\ X_n = A_n + S_n,\ \cdots$$

be the inter-arrival times of the renewal process where for $n \geq 1$, X_n is the time interval between the $(n-1)$-th and the n-th renewals where, recall, $t = 0$ is taken as the 0-th renewal. It is clear that the random variables $X_1, X_2, \cdots, X_n, \cdots$ have a common distribution F_X, which is the convolution of F_A and F_S. Since, by assumption, both A and S have finite expectation, so does X. In fact, we can write

$$\mu = E[X] = E[A + S] = E[A] + E[S] < \infty. \tag{1}$$

We are interested in the limiting probability $h(t)$ that our sensor is awake at time t. To derive the integral equation satisfied by $h(t)$, we condition on the time of the first renewal, X_1. Indeed, we can write

$$\begin{aligned} h(t) &= \int_0^\infty \Pr[\{\text{awake at time } t\} \mid \{X_1 = s\}]\, \mathrm{d}F_{X_1}(s) \\ &= \int_0^t \Pr[\{\text{awake at time } t\} \mid \{X_1 = s\}]\, \mathrm{d}F_{X_1}(s) \\ &\quad + \int_t^\infty \Pr[\{\text{awake at time } t\} \mid \{X_1 = s\}]\, \mathrm{d}F_{X_1}(s). \end{aligned} \tag{2}$$

Observe that for $0 \leq s \leq t$, we can write

$$\Pr[\{\text{awake at time}t\} \mid \{X_1 = s\}] = h(t - s). \tag{3}$$

On the other hand, for $t > s$ we have

$$\begin{aligned}
&\int_t^{\infty} \Pr[\{\text{awake at time } t\} \mid \{X_1 = s\}]\, \mathrm{d}F_{X_1}(s) \\
&= \Pr[\{A_1 > t\} \cap \{X_1 > t\}] \\
&= \Pr[\{A_1 > t\}] \;\; [\text{since } \{A_1 > t\} \subseteq \{X_1 > t\}] \\
&= 1 - F_A(t).
\end{aligned} \tag{4}$$

By virtue of (3) and (4) combined, (2) becomes

$$h(t) = [1 - F_A(t)] + \int_0^t h(t - s)\, \mathrm{d}F_X(s). \tag{5}$$

Writing

$$Q(t) = 1 - F_A(t),$$

it is easy to see that $h(t)$ satisfies the renewal integral equation

$$h(t) = Q(t) + \int_0^t h(t - s)\, \mathrm{d}F_X(s) \tag{6}$$

and that $Q(t)$ has the following properties:

- $Q(t) \geq 0$;
- $Q(t)$ is non-increasing for all $t \geq 0$;
- $\int_0^{\infty} Q(t)\, \mathrm{d}t = \int_0^{\infty} [1 - F_A(t)]\, \mathrm{d}t = E[A] < \infty$.

Assuming that X is non-lattice,[1] the Key Renewal Theorem [18] guarantees that

$$\begin{aligned}
\lim_{t\to\infty} h(t) &= \lim_{t\to\infty} [1 - F_A(t)] + \lim_{t\to\infty} \frac{1}{\mu} \int_0^{\infty} Q(s)\, \mathrm{d}s \\
&= \lim_{t\to\infty} \frac{1}{\mu} \int_0^{\infty} Q(s)\, \mathrm{d}s \\
&= \frac{1}{\mu} E[A] \\
&= \frac{E[A]}{E[A] + E[S]} \; [\text{by (12.1)}].
\end{aligned} \tag{7}$$

[1] A discrete random variable is said to be *lattice* if all the values it can assume with positive probability are of the form nh for some $h > 0$ and integer n.

To summarize our findings, we have proved the following result.

Theorem 1 *Assuming that A has finite expectation and that* $T = A+S$ *is non-lattice with finite expectation, the limiting probability, p, that a given sensor is awake at time t equals*

$$p = \frac{E[A]}{E[A] + E[S]}.$$

We note that the conclusion of Theorem 1 is intuitively satisfying and that the result itself has been a part of the sensor network folklore where it had been used without proof. It is also worth mentioning that Theorem 1 has an unmistakable PASTA [23] "flavor": for, consider an observer external to the network that is watching and noting the behavior of our sensor. The observer will note that the long-term probability of the sensor to be awake is $\frac{E[A]}{E[A]+E[S]}$. On the other hand, Theorem 1 tells us that sufficiently far away from the origin (i.e., $t = 0$), the probability that the sensor is awake at an arbitrary time t is also $\frac{E[A]}{E[A]+E[S]}$;

Finally, we note that Theorem 1 is a very convenient tool since it allows one to use the time-invariant probability of a sensor being awake in lieu of the instantaneous awake probability. As a result, in practice it is often natural to assume that a sensor sleeps for a random amount of time uniformly distributed in the range $[T_s, T_S]$ and stays awake for a random amount of time uniformly distributed in the range $[T_l, T_L]$. Theorem 1 guarantees that we can evaluate p by writing

$$p = \frac{\frac{m(T_l+T_L)}{2}}{\frac{m(T_l+T_L+T_s+T_S)}{2}} = \frac{T_l + T_L}{T_l + T_L + T_s + T_S}. \tag{8}$$

5 Network Longevity

Consider a deployment area populated by N sensors, each having the same initial on-board energy. We assume that energy is expended only by participating in various tasks and that communication, housekeeping and governance activities do not involve any energy expenditure. While this is unrealistic, we feel the assumption is justified on two grounds: first, that sensing devices are mechanical devices (e.g., a thermometer) that, as a rule, require a substantial amount of energy to collect data; and, second, because task management establishes a lower bound on energy expenditure.

Assume a possibly infinite sequence of tasks $T_1, T_2, \ldots, T_k, \ldots$ issued sequentially, without the benefit of lookahead or speculation. Moreover, even though in practice the tasks may be correlated and, as a result of executing a task, a known sequence of other tasks is likely to be executed, we assume that the tasks are *stochastically independent*. This assumption is justified by the fact that any correlation between tasks can only help with scheduling and task management. This is, indeed, the case as correlation between tasks can lead to sharing sensor data or even aggregated data, reducing the need to expend sensor energy for these tasks. Likewise, if lookahead is permitted, task management is easier and can lead to a more efficient management of sensors' energy.

In the literature, *network longevity* is defined in various ways [1], including:

Definition 1: The longevity of the network is the number m of tasks $T_1, T_2, \cdots, T_m$ that are executed until the energy of some sensor is totally depleted;

Definition 2: The longevity of the network is defined as the number m such that tasks $T_1, T_2, \cdots, T_m$ are executed but task T_{m+1} cannot be executed;

Definition 3: For α, $(0 < \alpha \leq 1)$, the α-longevity (or α-lifetime) of the network is the number of tasks the network can perform until the sensor density drops below $\alpha \times \rho$ where ρ is the initial density.

It is plain that Definition 1 is rather pessimistic since, in general, the fact that one sensor has depleted its energy budget does not mean that no further tasks can be performed. On the other hand, a sensor whose energy has been depleted while other sensors still have a lot of remaining energy, reflects poorly on the effectiveness of the task management strategy. Next, Definition 2 is useful in proving theoretical results about the intrinsic properties of any task management strategy. The most flexible definition, Definition 3, will be adopted in this chapter and all our simulation results will be expressed in terms of α-longevity for various values of α.

5.1 Optimal Task Management Strategies

Before we start discussing task management strategies, it is appropriate to establish notation and terminology. Assume that, initially, all sensors have L units of energy and that each sensor expends one unit of energy for each task in which it participates.

Consider a task management strategy S and assume for some positive integer n, $(n \geq 0)$, tasks $T_1, T_2, \cdots, T_n$ have been performed under S:

- let $\max(S, n)$ be the largest remaining energy of any sensor. Recall that $\max(S, 0) = L$;
- let $\min(S, n)$ be the smallest remaining energy of any sensor; Recall that $\min(S, 0) = L$;
- let $\sigma(S, n) = \max(S, n) - \min(S, n)$ be the *energy spectrum* of S after having performed n tasks;
- let $r(S, n)$ denote the residual number of energy units available, collectively, to all the sensors;
- let $a(S, n)$ be the number of sensors with nonzero remaining energy.

For a task T, the number, $w(T)$, of sensors that need to perform the task is referred to as the size of the *workforce* of T. We say that a task T_{n+1} is *feasible* if $w(T_{n+1}) \leq r(S, n)$. In other words, a task is feasible if the size of its workforce does not exceed the residual energy units available in the network.

We now define formally the concept of an optimal task management strategy.

Definition 1 A task management strategy S is optimal if a feasible task can always be performed.

We are now ready to give a necessary and sufficient condition for the optimality of a task management strategy.

Theorem 2 *A task management strategy S is optimal if, and only if, for all $n \geq 0, \sigma(S, n) \leq 1$.*

Proof First, let S be a task management strategy such that for every sequence of tasks and for all $n \geq 0$, $\sigma(S, n) \leq 1$. Let m be the first subscript for which the tasks $T_1, T_2, \cdots, T_m$ have been performed but T_{m+1} cannot be performed. We need to show that T_{m+1} cannot be feasible. Indeed, the only way in which T_{m+1}, while feasible, cannot be executed is that

$$a(S, n) < w(T_{m+1}) \leq r(S, m).$$

By definition, this implies that $\max(S, m) \geq 2$ and $\min(S, n) = 0$.[2] However, this is a contradiction since now $\sigma(S, m) = \max(S, m) - \min(S, m) \geq 2$. Thus, the strategy S must be optimal.

Conversely, let S be a task management strategy and let j be a subscript for which $\sigma(S, j) > 1$. We propose to show that S cannot be optimal by exhibiting a feasible task that cannot be performed. Let N be arbitrary. We distinguish the following cases:

Case 1: $r(S, j) \leq N$

Since $\sigma(S, j) > 1$, it follows that $a(S, j) < r(S, j) \leq N$. Now consider a task T_{j+1} with $w(T_{j+1}) = a(S, j) + 1 \leq r(S, j)$. While, clearly T_{j+1} is feasible it cannot be performed by the system since the number of sensors with nonzero energy is smaller than the required size of its workforce.

Case 2: $r(S, j) > N$

If $a(S, j) < N$ then we proceed as in Case 1 above. Assume, therefore, that $a(S, j) \geq N$. Consider the tasks $T_{j+1}, T_{j+2}, \cdots, T_k$ with

$$w(T_{j+1}) = w(T_{j+2}) = \cdots = w(T_k) = N$$

in such a way that $a(S, k - 1) \geq N$ and $a(S, k) < N$. Since it must be the case that $a(S, k) < r(S, k)$, consider the task T_{k+1} with $w(T_{k+1}) = r(s, k)$. Clearly, T_{k+1} is feasible, yet is cannot be performed since there are too few sensors with nonzero energy.

We just proved that S cannot be optimal and the proof of the theorem is complete.

Theorem 2 suggests the following simple optimal task management strategy:

Optimal Task Management Strategy: when selecting sensors for task T_n with $n \geq 0$, select as many sensors as possible at energy level $\max(S, n)$. If $w(T_n)$ exceeds

[2] To see why $\min(S, n) = 0$, observe that if $\min(S, n) > 0$ then all the sensors have a nonzero energy budget and so the feasible task T_{m+1} can definitely be performed.

the number of sensors at energy level $\max(S, n)$, select the balance from the sensors with energy level $\max(S, n) - 1$.

Notice that, the Optimal Task Management Strategy is a nice theoretical tool that, unfortunately, is not directly implementable. This is because, in order to save energy, the sensors with the largest remaining energy may not be awake when we need them. However, motivated by Theorem 2, our task management strategies will strive to keep the energy spectrum as small as possible. While this is not guaranteed to achieve optimality unless the energy spectrum is always one, the intuition is that, in most cases, it will come close.

In the next two sections we shall discuss two very different task management strategies that were, originally, presented in [1] and [3]. In Sect. 6 we present a centralized task management strategy, while in Sect. 7 we discuss the details of a distributed such strategy.

6 Centralized Task Management

In our centralized task management protocol each task is issued by one of the TIEs, usually the TIE closest to the center of the Area-of-Interest (AoI) considered to be a disk of a radius compatible with the sensing range R of sensors [15]. We refer to the task management strategy detailed in this section as *centralized* because the the selection of the set of sensors (a.k.a. the workforce) to participate in a given task occurs in a centralized fashion, coordinated by one of the sensors within the vicinity of the center of the AoI. As discussed in [15], the TIE need not be co-located with the AoI: this is because the TIE has a sufficiently large transmission range that can target (by broadcasting) the sensors around the AoI. Of course, the sensors are assumed to have at least a coarse-grain awareness of the location [4, 17]. We refer to the sensor that coordinates the task as *task leader* or *task coordinator*. The task leader is selected using any of the well-known leader election protocols [12, 13]. We begin by describing our assumptions about the tasking model.

Each sensing task is associated with a certain location chosen to be at the center of the AoI. Based on the location of the AoI, the location of the sensor, and the sensing range R, each sensor can determine whether it can participate in a given task or not. We assume that each task is associated with predetermined QoS requirements that have to be satisfied and that determine the size, w, of the workforce that will perform the task.

Under the assumptions described above, the TIEs issue tasks and later receive the aggregated results for decision making. Figure 4 illustrates the different stages of running a task under this model.

A task starts when one of the TIEs sends a sequence of Call-To-Work (CTW) messages to get the attention of a sufficiently large number of sensors that we will refer to as "candidate sensors." After that, the candidate sensors run among themselves a leader election algorithm to select the task coordinator or the task

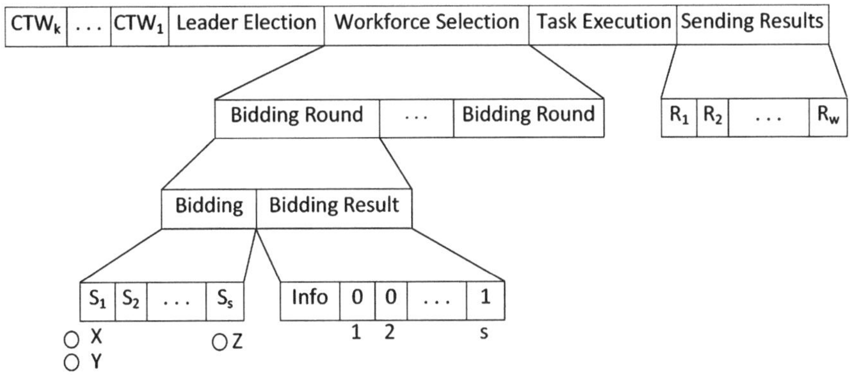

Fig. 4 Tasking model in the centralized protocol

leader which will be responsible for coordinating the workforce selection process. After selecting the leader, a contention-based workforce selection mechanism is used to recruit the required workforce based on sensors' remaining energy. The workforce selection process involves one or more bidding rounds coordinated by the task leader. After collecting the required workforce, task execution begins immediately by the recruited sensors. Task execution time might vary based on the type of the sensing task and the sensor capabilities, especially in a heterogeneous deployment. When the sensors complete execution, they start sending their sensory data back to the task leader in the order in which they joined the workforce. In the next subsections, we provide the technical details of each of these steps.

6.1 CTW Messages

Recall that in our tasking model, we assume that the TIE sends a sequence of k CTW messages to attract the attention of a sufficient number of sensors for the next task. An important parameter that the TIE has to evaluate is the parameter k.

An interesting problem is to determine the probability distribution of the number of candidate sensors "collected" at the end of k CTW messages. Assume a network density of ρ $sensors/m^2$, that the sensing area of any task is πR^2, and that the probability that a sensor is awake to receive a CTW message is p, independent of other sensors.

Recall that, as mentioned earlier, we assume that in order to save energy, a generic sensor alternates between sleep and awake periods: the sensor sleeps for a random amount of time uniformly distributed in the range $[T_s, T_S]$. Upon waking up, the sensor stays awake for a random amount of time uniformly distributed in the range $[T_l, T_L]$, after which is goes to sleep and the whole process is repeated. Observe that the TIE can use (8) to estimate the limiting probability, p, that a (generic) sensor is awake.

Theorem 3 *Let $X_1, X_2, \cdots, X_m$ be the number of candidate sensors collected, out of a population of N sensors, by the first m CTW messages. Then, $X_1+X_2+\cdots+X_m$ is binomially distributed with parameters N and $1-(1-p)^m$.*

Proof The proof is by induction on m. To settle the basis, we observe that

$$\begin{aligned}
\Pr[\{X_1+X_2=j\}] &= \sum_{i=0}^{j} \Pr[\{X_1+X_2=j\}|\{X_1=i\}]\Pr[\{X_1=i\}] \\
&= \sum_{i=0}^{j} \Pr[\{X_2=j-i\}]\Pr[\{X_1=i\}] \\
&= \sum_{i=0}^{j} \binom{N-i}{j-i} p^{j-i}(1-p)^{N-j} \binom{n}{i} p^i (1-p)^{N-i} \\
&= \sum_{i=0}^{j} \binom{N}{j}\binom{j}{i} p^j (1-p)^{2N-j} \left(\frac{1}{1-p}\right)^i \\
&= \binom{N}{j} p^j (1-p)^{2N-j} \sum_{i=0}^{j} \binom{j}{i} \left(\frac{1}{1-p}\right)^i \\
&= \binom{N}{j} p^j (1-p)^{2N-j} \frac{(2-p)^j}{(1-p)^j} \\
&= \binom{N}{j} p^j (2-p)^j (1-p)^{2(N-j)} \\
&= \binom{N}{j} \left[1-(1-p)^2\right]^j \left[(1-p)^2\right]^{N-j}.
\end{aligned}$$

Thus, X_1+X_2 is binomially distributed with parameters N and $1-(1-p)^2$.

Next, let m, $(m \geq 2)$, be arbitrary and assume that the convolution $X_1+X_2+\cdots+X_{m-1}$ is binomially distributed with parameters N and $1-(1-p))^{m-1}$. As before, we write

$$\begin{aligned}
&\Pr[\{X_1+X_2+\cdots+X_m=r\}] \\
&\quad= \sum_{k=0}^{r} \Pr[\{X_1+X_2+\cdots+X_m=r\}|\{X_1+X_2+\cdots+X_{m-1}=k\}] \\
&\Pr[\{X_1+X_2+\cdots+X_{m-1}=k\}] \\
&\quad= \sum_{k=0}^{r} Pr[\{X_m=r-k\}]\Pr[\{X_1+X_2+\cdots+X_{m-1}=k\}]
\end{aligned}$$

$$
\begin{aligned}
&= \sum_{k=0}^{r} \binom{N-k}{r-k} p^{r-k} (1-p)^{N-k} \binom{N}{k} \left[1-(1-p)^{m-1}\right]^{k} \left[(1-p)^{m-1}\right]^{N-k} \\
&= \sum_{k=0}^{r} \binom{N}{r}\binom{r}{k} (1-p)^{N-r} p^{r-k} \left[1-(1-p)^{m-1}\right]^{k} \left[(1-p)^{m-1}\right]^{N} \left[(1-p)^{m-1}\right]^{-k} \\
&= \binom{N}{r} (1-p)^{N-r} \left[(1-p)^{m-1}\right]^{N} \sum_{k=0}^{r} \left[\frac{1-(1-p)^{m-1}}{(1-p)^{m-1}}\right]^{k} p^{r-k} \\
&= \binom{N}{r} (1-p)^{N-r} \left[(1-p)^{m-1}\right]^{N} \left[p + \frac{1-(1-p)^{m-1}}{(1-p)^{m-1}}\right]^{r} \\
&= \binom{N}{r} \left[1-(1-p)^{m}\right]^{r} \left[(1-p)^{m}\right]^{N-r}.
\end{aligned}
$$

We just proved that $X_1 + X_2 + \cdots + X_m$ is binomially distributed with parameters N and $1 - (1-p)^m$, completing the proof of the theorem.

In addition to being an interesting, and somewhat surprising, result in its own right, Theorem 3 has the following useful consequence.

Corollary 1 *The expected number of sensors collected by k CTW messages is*

$$
\rho \pi R^2 \left[1 - (1-p)^k\right]. \tag{9}
$$

The TIE can use equation (9) to estimate k, the *least* number of CTW messages needed to attract the attention of at least c candidate sensors as follows:

$$
\begin{aligned}
c &\le \rho \pi R^2 \left[1 - (1-p)^k\right] \\
(1-p)^k &\le 1 - \frac{c}{\rho \pi R^2} \\
k &\ge \frac{\ln(1 - \frac{c}{\rho \pi R^2})}{\ln(1-p)} \\
k &= \left\lceil \frac{\ln(1 - \frac{c}{\rho \pi R^2})}{\ln(1-p)} \right\rceil.
\end{aligned} \tag{10}
$$

At this point it is important to observe that c does not represent the number of sensors required for the task; instead, c represents the number of candidate sensors from which the required workforce w is to be selected. Hence, if the TIE wants to collect a workforce of w sensors for the next task, it substitutes c in Eq. (10) by $c = f(w) \ge w$ (the derivation of the function $f(w)$ is presented later). This way the protocol probabilistically attracts the attention of more than w candidate sensors. From these sensors, only w sensors will be selected based on the difference in sensors' remaining energy.

6.2 Leader Election

Immediately after the last CTW message, candidate sensors willing to participate in the next task go through a simple leader election procedure to elect the leader sensor that will be responsible for coordinating the workforce selection process. The leader election technique proposed here is a simplified version of a more general technique that was previously proposed in [13].

The leader election process starts when each candidate sensor initializes an internal countdown timer to an appropriate random value that is inversely proportional to its remaining energy. Several formulas can be used for this purpose, for instance: $(\frac{A}{E_s}+r)$, or $A(E_{est}-E_s)+r)$, where A is a constant, E_s is the remaining energy of the sensor, E_{est} is the current estimate of the maximum energy in the task neighborhood, and r is a random value.

It is clear that the previous formulas probabilistically give preference to sensors with relatively high energy over other sensors; at the same time they produce random values which can help reduce transmission collisions between sensors. The value of the parameter A should be appropriately selected based on the maximum energy of sensors and the range of the random variable r. In our simulation we used $A = 10$ while r was randomly selected in the range $[1, 10]$.

When the timer of some sensor expires, the sensor transmits a message to all its neighbors announcing itself as a task leader. When other sensors receive this message, they realize that the leader has already been selected and they stop their internal timers. When the timers of two or more sensors expire at the same time and they start sending their messages concurrently, collisions occur. In such a situation, collisions can be easily resolved by allowing colliding sensors only to go through another round of counting down starting from randomly selected values. This process continues until a leader is elected.

It is important to note that the leader election protocol described above can be extended, in the obvious way, to allow the election of more than one leader forming what we refer to as a *coordination committee*. The members of the coordination committee share the same responsibilities as the task leader. For instance, when the coordination committee has m members, leader sensor i should be responsible only for coordinating bidding rounds $i, i+m, i+2m$ and so on. However, in the single task leader approach it would be responsible for coordinating all the bidding rounds for this task. Using a coordination committee of three or four members instead of a single task leader can distribute the task coordination load over different sensors which can help balance energy consumption among the sensors. Moreover, the coordination committee members still have the option to join the workforce if their energy allows them to do so. This flexibility can help reduce the running time of the workforce selection protocol.

6.3 Workforce Selection

After the leader election stage, the workforce selection proceeds through one or more *bidding rounds*. A bidding round is a contention-based mechanism used to select a subset of sensors from a larger set based on a certain criterion. Each bidding round has a number of bidding slots, explicitly specified in the CTW messages or the bidding result messages of the previous rounds. Candidate sensors willing to participate in a task show their interest by bidding randomly in one of the bidding slots. The leader sensor of any round is responsible for coordinating the bidding process (i.e., it announces the winning bidders at the end of each round, determines whether there is a need for another bidding round and announces the number of bidding slots in the next bidding round). Only single-bidder slots are considered winning slots. Once the leader sensor announces the winning slots in a round, the winning bidders (sensors which bid on any of the winning slots) immediately join the workforce. If the required workforce is not fully recruited, bidding continues for another round. In the new bidding round, not only previous round's losers are eligible to bid but also additional sensors who received the bidding results message but were asleep during the CTW stage can place their bids. If the bidding extends beyond the maximum number of bidding rounds allowed, there are two options available: (1) cancel the task, (2) execute the task with the currently recruited workforce. Either way the TIE must be informed that the required level of QoS may not be satisfied.

Immediately after a task leader has been elected (for the first bidding round) or after the bidding result message (for subsequent bidding rounds), the timeline is divided into a number of bidding slots; each bidder selects one of these slots at random and transmits a short frame that contains the sensor current energy level and the slot number (to avoid any confusion due to the lack of accurate synchronization between sensors).

As a result of the bidding process, each bidding slot can have zero, one, or multiple bids. Slots with no bids are useless, while those with multiple bidders result in garbled messages and are ignored. Only messages in single-bidder slots can be received correctly. At the end of each bidding round, the task leader records the number, G, of single-bidder slots. If G is greater than the required workforce w, then the task coordinator sensor selects the workforce among bidder sensors with the highest energy level. If G is less than w, then it selects all the winning sensors and announces another bidding round to collect the remaining workforce.

Each bidding result message contains a vector v that has s elements corresponding to the s bidding slots in the preceding bidding round, as illustrated in Fig. 6. For bidding slot i, the corresponding element $v[i]$ is set to 0 to indicate that the bidding sensor was not selected at this round, otherwise $v[i]$ is set to a non zero temporary ID to indicate that the bidding sensor was selected by the task coordinator to join the workforce. A sensor starts task execution immediately after it joins the workforce. When the sensors finish the execution of the required task and based on their temporary IDs they sequentially send their results to the task coordinator for local data

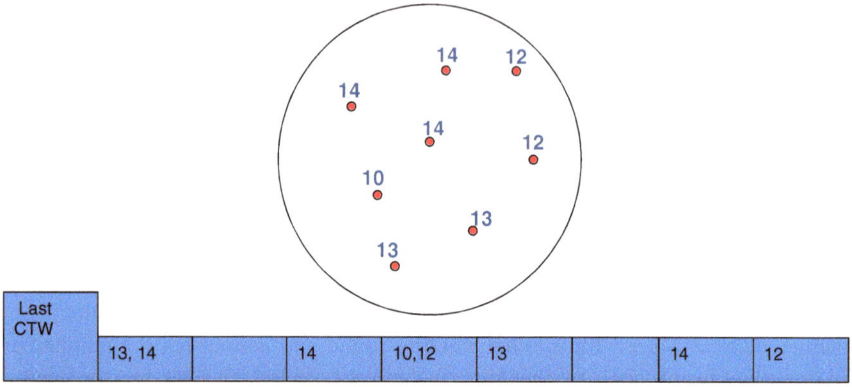

Fig. 5 Illustrating the bidding process

aggregation. Finally, the task coordinator (the leader sensor) sends the aggregated results to the TIE for further processing if necessary.

Candidate sensors participate in bidding with probabilities that are proportional to the difference between their current energy and the maximum energy among candidate sensors E_{max}. The TIE piggybacks on CTW messages its estimate of the value of E_{max} in the neighborhood of the required task. In the first bidding round of the first task within a certain area, E_{max} is unknown to the TIE, so candidate sensors participate in bidding with probabilities $\frac{1}{2}$ (we will justify this choice later). In subsequent rounds, the leader sensor can estimate E_{max} from the bids received so far and transmits the estimated value of E_{max} within the bidding results message. This value is also transmitted with the aggregated result to the TIE which uses the received values to update an internal two dimensional matrix that keeps track of the current estimate of E_{max} in different regions of the network.

6.3.1 An Example

We feel it is appropriate, at this time, to give the reader an idea of how the bidding process work. Referring to Fig. 5, assume that eight candidate sensors were identified in the AoI; these sensors are denoted in Fig. 5 by their remaining energy levels. Suppose further that eight slots were announced for bidding after the last CTW message. In the first slot two sensors (one at level 14 and the other at level 13) have transmitted; in the second slot no sensor has transmitted; in the third slot one sensor has transmitted, and so on. It is important to note that since each sensor is transmitting in one slot, the total number of bidders is eight. Now, slots with two transmissions result in a collision at the receiver (i.e., at the task leader), while a single-bidder slot result in a clear reception.

As illustrated in Fig. 6, after the last bidding slot, the task leader announces the results of the bidding process. Specifically, the task leader announces that the sensors that have bid in the first slot were unsuccessful, that the sensors that has bid in the

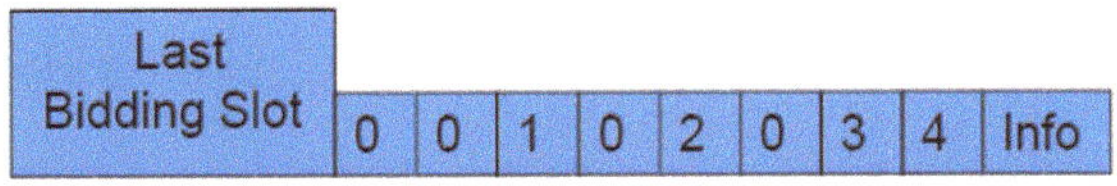

Fig. 6 Announcing the results of the bidding round

third slot has been successful and is assigned and ID of 1 in the workforce associated with the task. Similarly, the sensors that have bid in slots five, seven and eight have been successful and were assigned IDs 2, 3 and 4 in the workforce. Assuming that the workforce associated with the task must contain more than four sensors, the bidding process is continued, as above, until the workforce has been selected. It is important to observe that in addition to selecting a workforce for the task the bidding process also allows us to assign temporary IDs to the participating sensors. By using these IDs the task leader can address each of the sensors individually and also it allows the sensors to report their results to the task leader without collision.

Notice that as a result of the bidding process the task leader can estimate the largest remaining energy budget of the sensors in the AoI. Since the highest "bid" in the various slots was 14, the task leader knows that there are sensors with a remaining energy level of 14. Since some of them have collided with other sensors (see the first slot), not all the sensors at level 14 will participate in the current task. When the time comes to select sensors for a subsequent task, the task leader will remember that $E_{max} = 14$, which will be broadcast to all the sensors in the AoI as part of the CTW messages.

6.4 Estimating the Number of Bidding Slots

Next, we show how the number of bidding slots in any bidding round is estimated. Assuming that for a general bidding round we have n bidders and s slots, we calculate the expected number of single-bidder slots in this round. We assume that a bidder can bid on any slot with equal probability, more precisely $\frac{1}{s}$, and since we have n bidders that bid independently of each other, the probability that a specific slot has only one bidder equals $n(\frac{1}{s})(1-\frac{1}{s})^{n-1}$. Since we have s slots, the expected number of single-bidder slots, G, can be expressed as

$$E[G] = n\left(1-\frac{1}{s}\right)^{n-1}. \tag{11}$$

Recall that in our workforce selection only single-bidder slots matter. Hence, maximizing the number of these slots will definitely reduce the number of bidding rounds needed. Since n is a discrete variable, we define the continuous variable x such that $x = n$ for all the values of n. Now, we can differentiate $E[G]$ with respect to x

$$E[G] = x\left(1-\frac{1}{s}\right)^{x-1}$$

$$\frac{\mathrm{d}E[G]}{\mathrm{d}x} = \left(1 - \frac{1}{s}\right)^{x-1} \left[1 + x \cdot \ln\left(1 - \frac{1}{s}\right)\right]. \tag{12}$$

The maximum value of $E[G]$ occurs when its first derivative equals 0.

$$\begin{aligned}
\frac{\mathrm{d}E[G]}{\mathrm{d}x} &= \left(1 - \frac{1}{s}\right)^{x-1} \left[1 + x \cdot \ln\left(1 - \frac{1}{s}\right)\right] = 0 \\
1 &= -x \cdot \ln\left(1 - \frac{1}{s}\right) \\
1 &= \ln\left(1 - \frac{1}{s}\right)^{-x} \\
e &= \left[\left(1 - \frac{1}{s}\right)^{s}\right]^{-\frac{x}{s}} \\
e &\approx (e)^{\frac{x}{s}} \Rightarrow x \approx s \approx n.
\end{aligned} \tag{13}$$

Obviously, the approximation in Eq. (13) is valid for large values of s. Moreover, Eq. (13) shows that the maximum value of $E[G]$ occurs when the number of bidders n equals the number of bidding slots s and that the maximum number of single-bidder slots expected is given by

$$E[G]_{max} = s \cdot (1 - \frac{1}{s})^{s-1}.$$

For a task that requires a workforce of size w, the required number of single-bidder slots is also w. Based on this, we can determine the number of slots to use as follows

$$\begin{aligned}
G = w &= s\left(1 - \frac{1}{s}\right)^{s-1} \\
w &\approx \frac{s \cdot e^{-1}}{\left(1 - \frac{1}{s}\right)} \\
e \cdot w &\approx \frac{s^2}{s - 1} \\
s^2 - e \cdot w \cdot s + e \cdot w &\approx 0 \\
s &\approx \frac{e \cdot w\left(1 + \sqrt{1 - \frac{4}{e \cdot w}}\right)}{2}.
\end{aligned} \tag{14}$$

For $w > 3$, $(1 + \sqrt{1 - \frac{4}{e \cdot w}}) \approx 2$, hence Eq. (14) can be simplified to

$$s \approx e \cdot w \tag{15}$$

6.5 Determining a Suitable Number of Candidates

Recall that the bidding sensors are a subset of the candidate sensors, selected according to some criteria. In this subsection we turn our attention to the relation between n, the number of bidders and c, the number of candidate sensors. Previously, we mentioned that each candidate sensor participates in bidding with probability proportional to the difference between its energy (i.e., E_s) and the maximum energy among all candidate sensors (i.e., E_{max}). In particular, a candidate sensor with energy level E_s should bid with probability $\frac{1}{1+E_{max}-E_s}$. Assuming that the energy of candidate sensors is uniformly distributed across σ consecutive levels, the expected number of bidders can be related to the number of candidate sensors as follows,

$$\begin{aligned} E[n] &= \sum_{i=0}^{\sigma-1} p_i \cdot c_i \\ &= \sum_{i=0}^{\sigma-1} \frac{1}{1+i} \cdot \frac{c}{\sigma} \\ &= \frac{c}{\sigma}\left[1+\frac{1}{2}+\frac{1}{3}+\cdots+\frac{1}{\sigma}\right] \\ &= \frac{c \cdot H_\sigma}{\sigma} \\ &\approx \frac{c \cdot [\ln \sigma + \gamma]}{\sigma}. \end{aligned}$$

Solving for c, yields

$$c = \frac{\sigma \cdot E[n]}{\ln \sigma + \gamma} \approx \frac{n \cdot \sigma}{\ln \sigma + \gamma}, \tag{16}$$

where $\gamma \approx 0.57721$ is Euler's constant. The TIE uses Eq. (16) to estimate the required number of candidate sensors it has to collect through CTW messages in order to select the workforce for the next task. The parameter σ in Eq. (16) reflects the average width of the spectrum of sensors energy at different points in network lifetime. Since the task management protocol is designed to balance energy expenditure among sensors by minimizing variations in their energy, it is expected that the value of σ will remain small most of the time. Simulation results verified this expectation and showed that σ did not exceed 4 levels in all our experiments. More details will be presented in Sect. 8.

Recall that in Sect. 6.3, in the presence of an unknown value for E_{max}, the participation probability of sensors was taken to be $\frac{1}{2}$. Equation (16) can be used to justify our choice of this value as follows. Equation (16) expresses the ratio between the number of bidders and the number of candidate sensors as $\frac{\ln \sigma + \gamma}{\sigma}$. The average value of this ratio for small values of σ (i.e ≤ 6) is $0.515 \approx 0.5$. Hence, at the very early stage of the network lifetime when the differences between sensor energy levels are

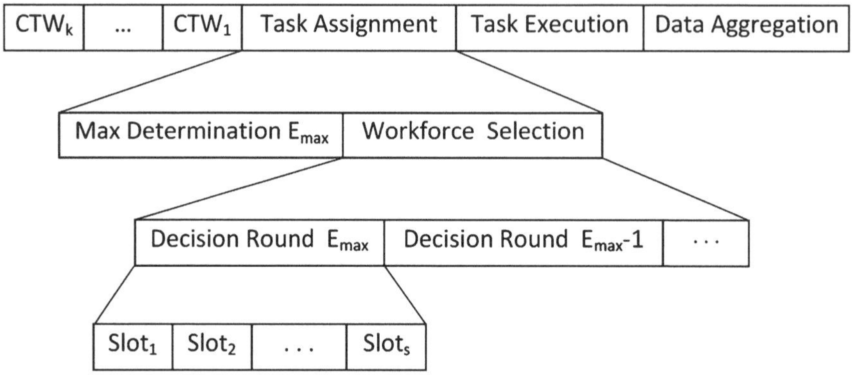

Fig. 7 Illustrating the distributed task management protocol

minor (i.e., σ is small) and when E_{max} is unknown, if each of the candidate sensors participates in bidding with probability $\frac{1}{2}$, then the expected number of bidders will be close to n, justifying our choice.

7 Distributed Task Management

One of the shortcomings of the centralized task management strategy is the excessive load it imposes on leader sensors for coordinating workforce selection and data aggregation. To overcome this problem, there are two possible approaches. As already mentioned, one natural approach is to elect a task coordination committee (instead of a single task leader) and to spread the workload across the members of the committee. A second approach, that will be pursued in this section, is to design a distributed version of the workforce selection protocol. In this version, we adopt the tasking model depicted in Fig. 7.

The distributed task management model is similar to the centralized one described in Sect. 6. However, it differs from it in the task assignment stage and in the way data is aggregated. In the distributed model, task assignment involves two phases. In the first phase, candidate sensors collected in the CTW stage run a distributed protocol to determine the maximum energy among themselves (see Sect. 7.1). In the second phase, each sensor decides whether or not to participate in the current task based on:

- the difference between its current energy and the maximum energy determined in the first phase;
- its distance to the position of the AoI (see Sect. 7.2).

Our previous assumptions about the capabilities of sensors still hold. In the following subsections, we present the details of the different phases of the task assignment protocol.

7.1 Phase 1: Estimating the Maximum Energy

The main goal of this phase is to run a fully distributed protocol in order to determine E_{max}, the maximum energy among the candidate sensors. Assume that sensor energy E_s can be quantized into 2^n levels (i.e., E_s can be encoded in a string of n bits). The idea is to let candidate sensors transmit the strings that represent their energy levels bit by bit in a sequence of very short n packets (n iterations). The timeline is divided into n slots, and sensors start the encoding process immediately after the last CTW message.[3] Sensors start their transmission from the most significant bit to the least significant bit as follows: a value of 0 is not transmitted while a value of 1 is transmitted. Sensors that pick up the values transmitted use the following disambiguation scheme:

- No packets received: 0 is recorded;
- A single packet received: 1 is recorded;
- Two or more packets received or a collision detected: 1 is recorded.

A sensor drops out if the binary representation of its energy has a 0 in its kth most significant bit, and it detected a collision or received one or more packets in the kth iteration (time slot). It the end, each sensor stores the maximum energy among all candidate sensors in the sensing area. Note also that there is no loss of information in the process of estimating the maximum.

At this point, one might argue that we cannot trust the synchronization achieved using the last CTW message because packets received by different sensors suffer from different propagation and processing delays. Although this seems to be true, we still can argue that the achieved level of synchronization is more than sufficient for our purpose especially when there is no actual payload (data) in the packets transmitted. In this interpretation, detecting a collision is equivalent to receiving a packet. Hence, if the iteration slot length is T, and the transmission time to send any of these small packets is T_t, the only way in which this protocol fails is when a sensor is delayed for a period longer than $T - T_t$. In this case its string is transmitted and interpreted shifted by one or more bits. However, since the slot time T can be chosen arbitrarily, we can choose T such that the probability that a sensor will be delayed longer than $T - T_t$ is very small, especially when the sensors are within the same sensing area.

Next, we illustrate by an example how this protocol works. Referring to Fig. 8, we assume a scenario where there are five sensors (i.e., S_1, S_2, S_3, S_4 and S_5) that lie within the sensing area of a specific task. We assume that the respective energy levels of these sensors are 11101, 10111, 11011, 01111 and 11100. To determine the maximum, we need five iterations. In the first iteration, only sensors S_1, S_2, S_3 and S_5 transmit. A collision is recorded and all the sensors set the most significant bit of the perceived maximum to 1. Moreover, sensor S_4 realizes it should drop out since its most significant bit is a 0. In the second iteration, only sensors S_1, S_3, and S_5 transmit. Again a collision is recorded, and all the sensors set the second most

[3] We assume that CTW messages include information about the number of remaining CTW messages.

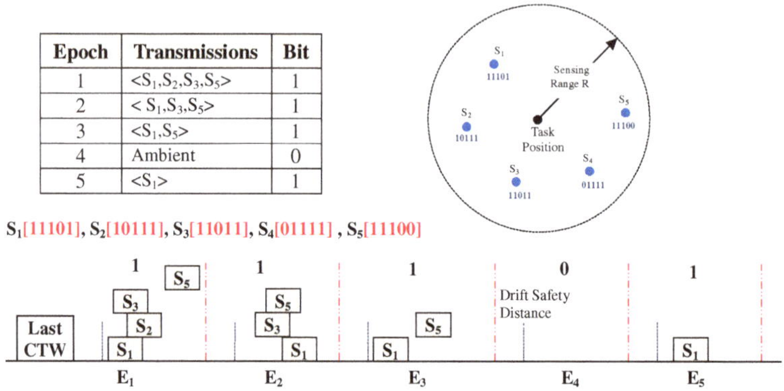

Fig. 8 Estimating the maximum energy among candidate sensors

significant bit to 1. Sensor S_2 drops out. In the third iteration, only sensors S_1 and S_5 transmit. All the sensors set the third most significant bit to 0. Sensor S_3 remains inactive for the remaining iterations. In the fourth iteration, there is no transmissions, so 0 is recorded. Finally, in the fifth iteration, only sensor S_1 transmits. Sensors set the least significant bit of the perceived maximum to 1 and reach the consensus that the maximum energy is 11101.

It is worthwhile to mention that when the sensors know the maximum energy within each sensing area, they can benefit from this in many different ways. For example, this knowledge allows each sensor to adjust its duty cycle (the ratio between its sleep and awake time) based on the difference between its remaining energy E_s and the maximum energy in its sensing area E_{max}. As a result, sensors with relatively low energy ($E_i < E_{max}$) can sleep for longer periods than sensors with relatively high-energy.

The previous protocol suggests an easy way to select a group of sensors to be responsible for sending the aggregated sensory data to the next hop in its way to the TIE. Obviously, many sensors may have energy equal to E_{max}, so an appropriate mechanism should be used to allow one sensor only to transmit. An easy way to do this is by letting each of these sensors initialize a countdown timer with a random value. The sensor whose timer expires first transmits a a special short message to announce itself. Sensors with non-expired timers turn off their timers immediately when they receive such message. If two or more timers expired simultaneously, a collision occurs and the colliding sensors go into another round till the packet is transmitted successfully.

7.2 Phase 2: To Participate or Not to Participate

After having determined the maximum energy in the sensing area, each sensor has to decide whether or not it is going to participate in the task at hand. Each sensor

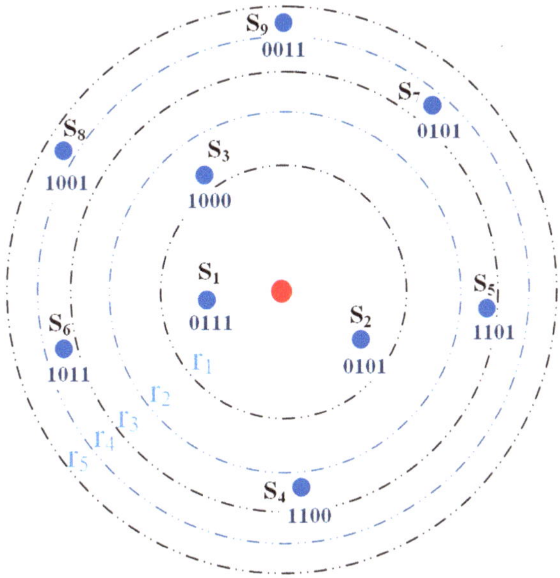

Fig. 9 Partitioning the AoI into k disjoint regions

makes this decision based on the difference between its energy E_s and the maximum energy E_{max} determined in the previous phase.

Since our protocol is fully distributed and there is no central node to coordinate sensor participation, the sensor will have to make decisions independently of each other. Unfortunately, under these conditions we cannot guarantee that the number of participating sensors matches the required workforce. The best we can do while keeping our protocol distributed is to keep the actual number of recruited sensors as close as possible to the required workforce.

To achieve this goal we divide the sensing range into k disjoint regions of equal size using concentric circles of radii $r_1 < r_2 < \cdots < r_k = R$ as illustrated in Fig. 9. The radii that divide the sensing area into regions are determined in such a way that the number of sensors in these regions are as even as possible. The best way to do this under uniform distribution is to choose these radii such that the areas of the regions are equal. Mathematically, this can be expressed as follows:

$$\pi \cdot (r_i^2 - r_{i-1}^2) = \pi \cdot (r_{i-1}^2 - r_{i-2}^2)$$
$$r_i^2 = 2r_{i-1}^2 - r_{i-2}^2 \qquad (17)$$

substituting in (17) for $r_{i-1}, r_{i-2} \ldots$

$$r_i^2 = 3r_{i-2}^2 - 2r_{i-3}^2$$
$$r_i^2 = 4r_{i-3}^2 - 3r_{i-4}^2$$
$$\vdots$$

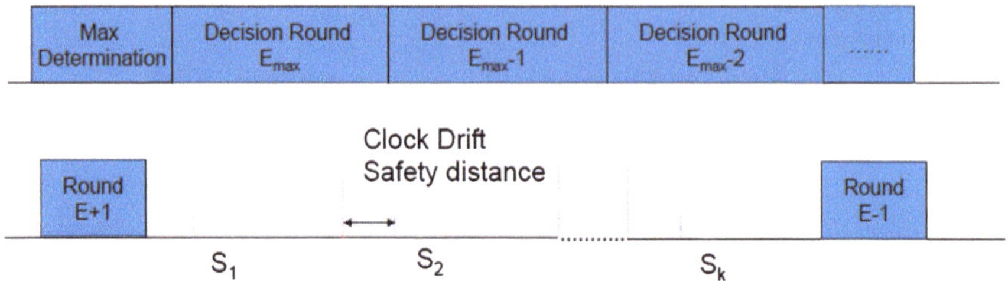

Fig. 10 Illustrating the various decision rounds

A simple inductive argument shows that

$$
\begin{aligned}
r_i^2 &= i \cdot r_1^2 - (i-1)r_0^2 \\
r_i &= \sqrt{i} \cdot r_1.
\end{aligned} \tag{18}
$$

If the entire sensing area is divided into k regions of equal area (i.e., $\frac{\pi R^2}{k}$), then

$$
\begin{aligned}
R = r_k &= \sqrt{k} \cdot r_1 \\
r_1 &= \frac{R}{\sqrt{k}}.
\end{aligned} \tag{19}
$$

Finally, substituting r_1 in (18) yields

$$
r_i = R\sqrt{\frac{i}{k}}. \tag{20}
$$

Based on the AoI position and its own position, each sensor can determine its region. Participation decisions are made in decision rounds that run immediately after estimating the maximum energy in the sensing area. Each decision round is associated with an energy level that determines the set of sensors that can join the selected workforce.

Referring to Fig. 10, in the first decision round, only sensors with energy equals to E_{max} are allowed to transmit; in the second decision round, only sensors with energy equals to $E_{max} - 1$ are allowed to transmit, and so on. The packets transmitted by sensors are very short and contain no payload, the payload is implicitly encoded in the transmission itself. Each decision round has k time slots corresponding to the k regions described above. In slot i in decision round $E_{max} - j + 1$, only sensors that are in region i with energy level equals $E_{max} - j + 1$ are allowed to transmit.

Each sensor initializes an internal counter to 0 and waits for the decision round corresponding to its energy and the time slot corresponding to its region. It is possible that the protocol collects the required workforce and terminates before the decision round and slot of a sensor comes. The idea of the protocol is to recruit sensors one by one based on their energy and using regions to reduce the number of sensors that are being added in each step (only one sensor at each step if possible). The sensors that pick up the packets transmitted use the following disambiguation scheme:

- No packets received: do nothing;
- A clear packet received: increment the counter by 1;
- A collision recorded: increment the counter by 2.

The protocol terminates when the internal counter of a sensor is greater than or equals the required workforce. Only sensors that transmitted join the workforce. To guarantee that the protocol terminates in a finite number of decision rounds, there should be a maximum number of decision rounds that the protocol allows. If the last decision round is reached, then all the sensors that are in the sensing area and have not already transmitted, transmit in their corresponding slot irrespective of their energy level.

7.3 Average Size of the Over-recruited Workforce

Next, we estimate the average number of over-recruited workforce by evaluating the probabilistic distribution of the number of sensors in each region and show how it changes with the number of regions k. Assuming a uniform distribution of sensors in the sensing area, we can map the problem to the classical "balls and bins" problem in which n balls are distributed uniformly at random into k bins. In our scenario we have n sensors that are deployed randomly in k regions. The reader should note that here n refers to the number of sensors in the sensing area with energy level that matches the energy level determined by the decision round. Since sensors in the sensing area may have different energy levels, n only represents a fraction of the total number of nodes in the sensing area.

The event that a given sensor will be deployed in a particular region is a Bernoulli trial with probability of success equal to the ratio between the region area to the whole sensing area (i.e., $\frac{1}{k}$ since the regions areas are equal). Thus, the random variable X_i which represents the number of sensors in region i follows a binomial distribution $B(n, \frac{1}{k})$.

$$Pr[\{X_i = j\}] = \binom{n}{j}\left(\frac{1}{k}\right)^j\left(\frac{k-1}{k}\right)^{n-j}. \tag{21}$$

To understand the source of over-recruiting notice that, as illustrated in Fig. 11, it is quite possible that more than two sensors in the same region transmit in the same slot, as happened in slot S_4 where three sensors have transmitted. While all three of these sensors will be part of the workforce, only two are counted and the protocol proceeds to recruit more sensors even though it may have reached the desired size of the workforce.

With this in mind, we are interested in evaluating the probability that more than two sensors fall in the same region because this may result in recruiting more sensors than the required size of the workforce. The sought probability is

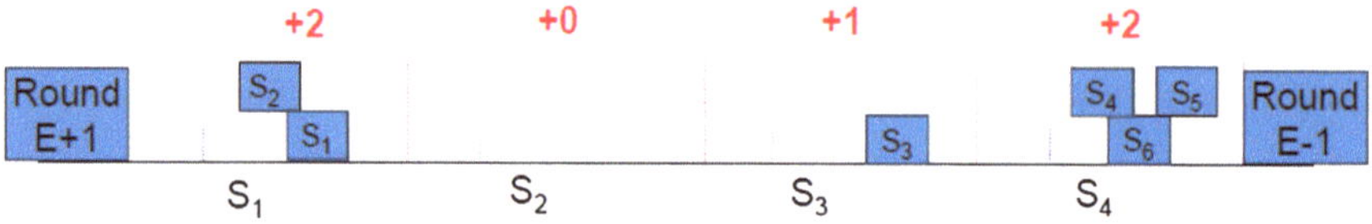

Fig. 11 Illustrating the source of over-recruiting

$$
\begin{aligned}
Pr[\{X_i > 2\}] &= 1 - Pr[\{X_i \le 2\}] \\
&= 1 - \sum_{j=0}^{2} \binom{n}{j} \left(\frac{1}{k}\right)^{j} \left(\frac{k-1}{k}\right)^{n-j}.
\end{aligned} \tag{22}
$$

To simplify our notation we define $\alpha_i(n, k)$ as follows:

$$
\begin{aligned}
\alpha_i(n,k) &= \sum_{j=0}^{i} \binom{n}{j} \left(\frac{1}{k}\right)^{j} \left(\frac{k-1}{k}\right)^{n-j} \\
&= \frac{1}{k^n} \sum_{j=0}^{i} \binom{n}{j} (k-1)^{n-j}
\end{aligned}
$$

$$
\alpha_1(n,k) = \frac{1}{k^n} \left[(k-1)^n + n(k-1)^{n-1} \right] \tag{23}
$$

$$
\alpha_2(n,k) = \frac{(k-1)^{n-2}}{k^n} \left((k-1)^2 + n(k-1) + \frac{n(n-1)}{2} \right). \tag{24}
$$

By substituting (24) into (22) we obtain

$$
Pr[\{X_i > 2\}] = 1 - \alpha_2(n,k). \tag{25}
$$

We are now ready to evaluate the expected size of the extra workforce recruited for a given task. For this purpose, we define the random variable Y_i that represents the number of extra workforce in region i and also define $Y = \sum_{i=1}^{k} Y_i$, the size of extra workforce in the whole sensing area. Easy manipulations show that

$$
\begin{aligned}
E[Y_i] &= \sum_{j=0}^{n-2} j \cdot Pr[\{Y_i = j\}] = \sum_{j=0}^{n-2} j \cdot Pr[\{X_i = j+2\}] \\
&= \sum_{j=1}^{n-2} j \cdot \binom{n}{j+2} \left(\frac{1}{k}\right)^{j+2} \left(\frac{k-1}{k}\right)^{n-(j+2)} \\
&= \sum_{l=3}^{n} (l-2) \cdot \binom{n}{l} \left(\frac{1}{k}\right)^{l} \left(\frac{k-1}{k}\right)^{n-l}
\end{aligned}
$$

Fig. 12 $E[Y] vs. k/n$

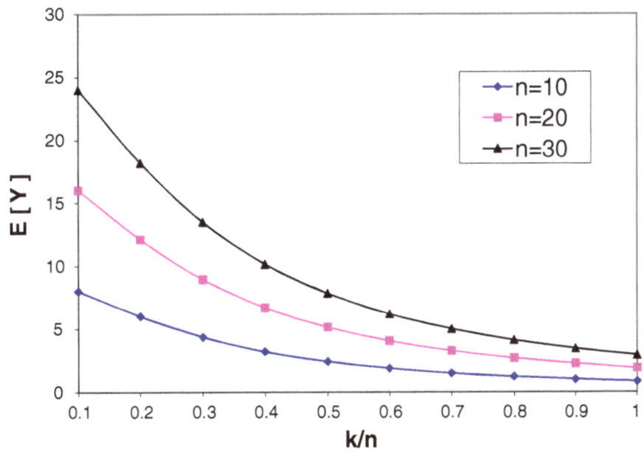

$$
\begin{aligned}
&= \frac{n}{k} \sum_{l-1=2}^{n-1} \binom{n-1}{l-1} \left(\frac{1}{k}\right)^{l-1} \left(\frac{k-1}{k}\right)^{(n-1)-(l-1)} \\
&\quad - 2 \sum_{l=3}^{n} \binom{n}{l} \left(\frac{1}{k}\right)^{l} \left(\frac{k-1}{k}\right)^{n-l} \\
&= \frac{n}{k} \left[1 - \alpha_1(n-1,k)\right] - 2\left[1 - \alpha_2(n,k)\right]. \qquad (26)
\end{aligned}
$$

By linearity of expectation

$$
\begin{aligned}
E[Y] &= \sum_{i=1}^{k} E[Y_i] = k \cdot E[Y_i] \\
&= n\left[1 - \alpha_1(n-1,k)\right] - 2k\left[1 - \alpha_2(n,k)\right]. \qquad (27)
\end{aligned}
$$

Figure 12 gives more insight into how $E[Y]$ changes with changes in k and n. Obviously, as the number of regions k increases the expected number of extra workforce recruited decreases to match the required workforce. It is also important to note that $E[Y]$ represents the expected number of extra workforce recruited assuming that the protocol continues to run in all the slots of the decision round. However this is not the case for many tasks in which the protocol terminates in the first few slots of the first decision round. So, in fact, $E[Y]$ is more like an upper bound on the extra workforce recruited.

8 Performance Evaluation

Using C++, we have built a wireless sensor networks simulator that implements different workforce selection protocols. As we mentioned earlier, to the best of our knowledge we are the first to address workforce selection in WSN. Hence, we could not compare our protocols against other protocols. Hence, we compared the performance of our proposed protocols to the performance achieved using ideal workforce selection in which the workforce is selected from sensors with the highest energy in the sensing range of the task being executed. In addition, we compare our protocols to energy-oblivious protocol which uses the same CTW and bidding rounds mechanisms described in our centralized protocol. The only difference is that the bidding decisions of candidate sensors are made irrespective of their remaining energy. This is different from our approach in which the estimate of the maximum energy among candidate sensors, E_{max}, is used to control the probability by which candidate sensors participate in bidding for the next task.

Our performance evaluation begins by specifying, in Sect. 8.1, the simulation model along with the system parameters used in our simulations. Later, in Sect. 8.2 we present a detailed performance evaluation of our two task management protocols obtained using the system model in Sect. 8.1.

8.1 Simulation Model

An important parameter which we need to consider when dealing with energy-neutral protocols is the "Participation Factor" (PF). PF defines the probability by which candidate sensors participate in bidding. In our simulation, we run several experiments assuming PF takes the values 0.5, 0.75, and 0.90.

Before presenting our results, we have to define network longevity. In a typical sensor network, sensors are deployed with a predetermined density ρ which is usually chosen in a way that satisfies QoS requirements expressed in terms of number of sensors participating in different sensing tasks. An appropriately chosen value of ρ can provide a reliable network performance by guaranteeing that a sufficient number of sensors will be available to perform upcoming tasks at the required level of QoS. Unfortunately, sensors have limited and non-renewable energy budget, once the energy of a sensor is entirely depleted, it dies and, eventually, the network density decreases. When sensor density drops below a certain threshold, the number of sensors available may not suffice to satisfy the QoS requirements of upcoming tasks and at this point the sensing results cannot be reliable anymore. Based on this, we define the α-reliable lifetime of a network as the average number of tasks the network can perform until the sensor density goes below α of its initial value. For instance, the 0.1-reliable lifetime of a network with density 0.5 $sensors/m^2$, is the average number of tasks that can be performed on this network before the sensor density become less than 0.05 $sensors/m^2$.

We conducted several experiments to evaluate the performance of our proposed protocols. In our simulation, sensors were deployed uniformly at random in a square with side length 200m. We used different sensor densities ranging from 0.3 to 1.5 sensors/m^2. The network has a single TIE placed at the origin (0, 0) and which is responsible for tasking sensors across the deployment area. QoS requirements of generated tasks were expressed in terms of the minimum number of sensors needed

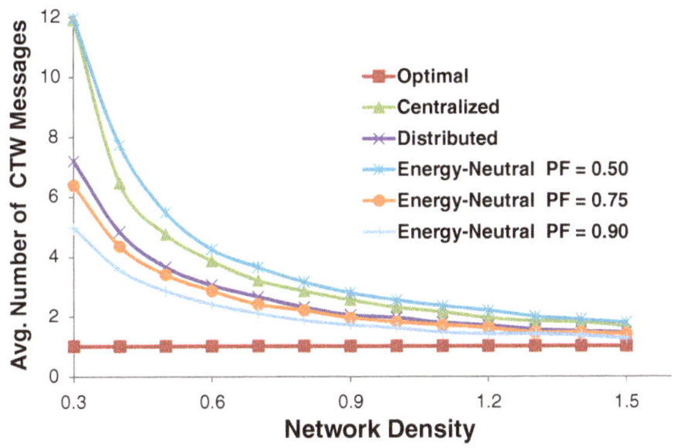

Fig. 13 Average number of CTW messages

to participate in each task. The required workforce for different tasks was selected randomly from the range [1, 20]. We tried to balance tasking load on different spots of the network by selecting the positions of the AoIs (task centers) uniformly at random across the whole deployment area. The sensors were assumed to have a fixed sensing range of 10.0 meters – beyond this range sensor readings may not be reliable. The sensors sleep and wake up alternatively and asynchronously in a way that makes them active for only 10 % of their lifetime. Initially, each sensor has exactly 30 units of energy (i.e., each sensor can at most participate in 30 tasks).

8.2 Simulation Results

Using the deployment and network parameters mentioned in Sect. 8.1 we have evaluated a number of performance metrics of our protocols.

8.2.1 Average Number of CTW Messages

We begin by investigating the number of CTW messages needed to attract the attention of a suitable number of candidates. Figure 13 shows, for different network densities, the average number of CTW messages needed to get the attention of sufficient number of sensors in order to execute the next upcoming task. The optimal protocol assumes that a single CTW message is sufficient to collect a suitable number of candidates. However, this is not the case for other protocols which substitute its specific estimate of the number of candidate sensors into Eq. (10) to evaluate the required number of CTW messages. Each of the shown protocols has its own way to estimate, c, the number of candidate sensors. The centralized protocol depends on Eqs. (15) and (16) to estimate c. However, in the distributed protocol, c is evaluated by multiplying the workforce size, w, by some constant factor f_1 (in our simulation experiments, we assume $f_1 = 4$).

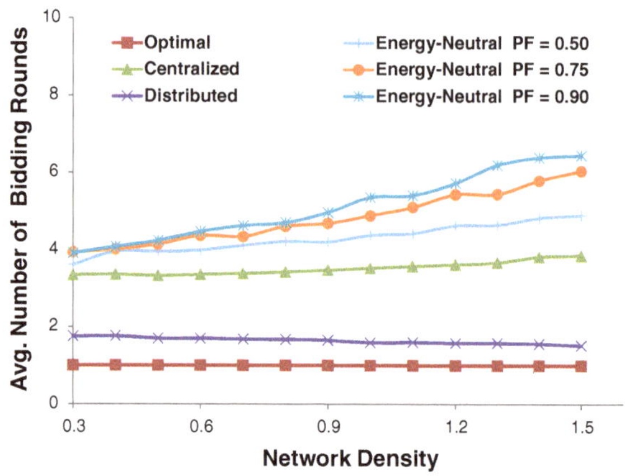

Fig. 14 Average number of bidding rounds

The energy-oblivious protocol uses Eq. (15) to estimate the number of bidders, n, from the workforce size. After that, it uses the participation factor to relate the number of bidders to the number of candidate sensors ($n = PF \cdot c$). From Fig. 13, we can see that the average number of CTW messages needed decreases as the network density increases with very minor differences between different protocols.

8.2.2 Average Number of Bidding Rounds

Figures 14 and 15 show, respectively, the expected number of bidding rounds along with the expected number of bidding slots within each round. In the optimal scenario, the workforce selection protocol ends using a single bidding round which has a number of bidding slots equal to the size of the required workforce. However, for other protocols, typically more than one bidding round is needed to compensate for any empty or garbled slots that might arise due to bidding collisions (i.e., when two or more sensors bid in the same slot). As the value of the participation factor of energy-neutral protocols increases, the expected number of bidding rounds increases. In turn, this increases the number of sensors willing to bid within the same round. Hence, the resulting number of single-bidder slots decreases and more rounds are needed to select the remaining workforce.

Figure 15 confirms that the average number of bidding slots used in the centralized protocol bidding rounds is very close to its counterpart in the energy-oblivious protocol with very minor changes due to using different participation factors. It is also important to understand that the curve associated with the distributed protocol in Fig. 14 shows the number of decision rounds used in workforce selection since in this protocol there is no bidding. And for the same reason, the number of bidding slots for this protocol is always zero.

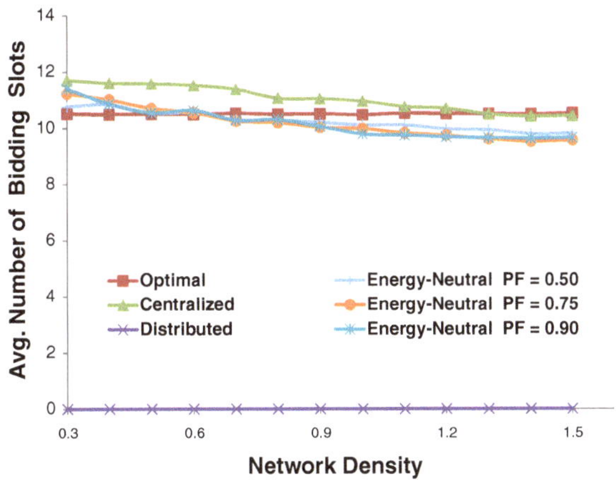

Fig. 15 Average number of bidding slots in bidding rounds

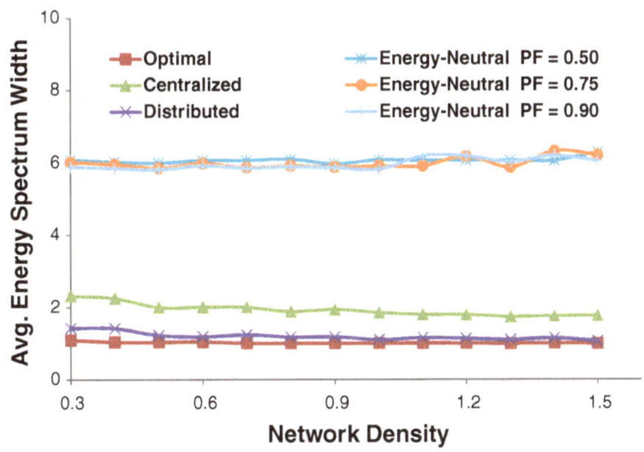

Fig. 16 Average width of sensor energy spectrum

8.2.3 Average Width of the Energy Spectrum

Figure 16 shows how our centralized and distributed protocols can preserve network density for longer periods by balancing the rate at which sensor energy is consumed. In particular, the figure compares the average width of sensor energy spectrum throughout the network 0.2-reliable lifetime for different network densities. To estimate the average width of sensor energy spectrum, we evaluated the width of the spectrum after the execution of every task using Eq. (28).

$$w_t = \begin{cases} \frac{1}{n_1} \sum_{s=1}^{n_1} \left(E_s - \overline{E}\right) & E_s \geq \overline{E} \\ \frac{1}{n_2} \sum_{s=1}^{n_2} \left(\overline{E} - E_s\right) & E_s < \overline{E}, \end{cases} \tag{28}$$

where $\overline{E}$ is the average energy of sensors immediately after the execution of task t. Here, n_1 is the number of sensors whose remaining energy is larger than or equal

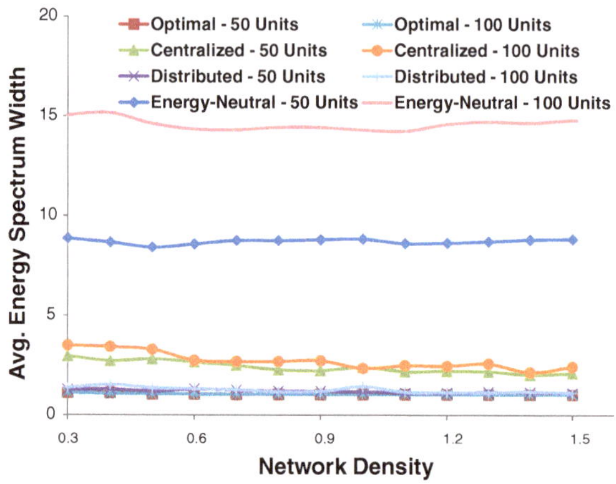

Fig. 17 Average width of energy spectrum for different initial energy levels

to $\overline{E}$. Similarly, n_2 is the number of sensors whose remaining energy is less than $\overline{E}$. Finally, the average spectrum width among all executed tasks was estimated as $\frac{1}{n}\sum_{t=1}^{n} w_t$, where n is the total number of tasks executed during the network lifetime.

Figure 16 tells us that the average width of sensor energy spectrum of the distributed protocol is much narrower than the width obtained when using any of the other protocols. Moreover, the average spectrum width is very close to the optimal value (i.e., 1). The superior performance of the distributed task management protocol over other protocols can be attributed to the following factors:

- An accurate estimation of the maximum energy among candidate sensors;
- Selecting the workforce using decisions rounds that give priority to sensors with higher energy levels over sensors with lower energy. Although, the average spectrum width of the centralized protocol is slightly larger than that of its counterpart, in the distributed protocol, it is much narrower than the spectrum of energy-oblivious protocols. It is worthwhile to mention that in both the centralized and the distributed protocols, the average spectrum width hardly changes with sensor initial energy. This is not the case for energy-oblivious protocols, where the average spectrum width increases when sensor initial energy increases as confirmed by Fig. 17.

The significant differences between the energy of sensors when using energy-oblivious protocols makes one expect that many of the heavily loaded sensors would die at an early stage of the network lifetime. Typically, when a large number of sensors which reside at some spot die, the network density at this spot decreases. Figures. 18, 19 and 20 capture this phenomenon when using different workforce selection protocols with initial deployment densities of 0.3, 0.7, and 1.0 respectively.

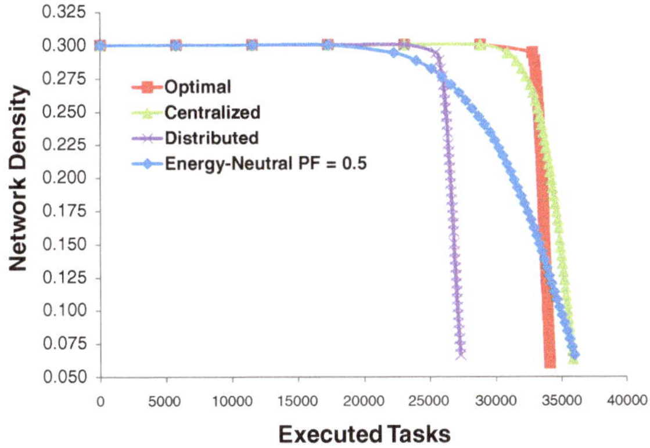

Fig. 18 Average density degradation throughout network 0.2-reliable-lifetime using different protocols for initial network density $\rho = 0.3$

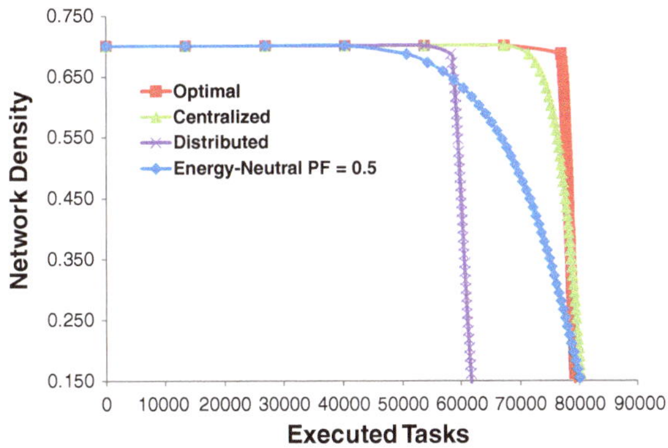

Fig. 19 Average density degradation throughout network 0.2-reliable-lifetime using different protocols for initial network density $\rho = 0.7$

8.2.4 Average Density Degradation for Various Initial Deployment Densities

As shown in Figs. 18, 19, and 20, the degradation in network density for energy-oblivious protocols starts at a relatively earlier stage compared to other protocols. Table 1 shows the percentage of the network lifetime before the network density starts to degrade.

The continuous degradation in network density can eventually create energy holes. We conducted a set of experiments to capture the impact of using different workforce selection protocols on the rate at which holes grow in the network. Figures 21, 22 and 23 offer a comparison of the growth rate of energy holes using the four protocols under different deployment densities. The initial deployment densities of Figs. 21,22 and 23 are respectively 0.3, 0.7 and 1.0. The steep slopes of the curves in Figs. 21,22 and 23 show that our centralized and distributed protocols can reduce the rate at which energy holes grow in the network specially under small and medium network densities. Under dense deployments, our protocols tends to be less effective in reducing the growth rate of energy holes. In order to explain the reason behind this, we recall that energy holes grow only when all the sensors within the hole are dead. Although the

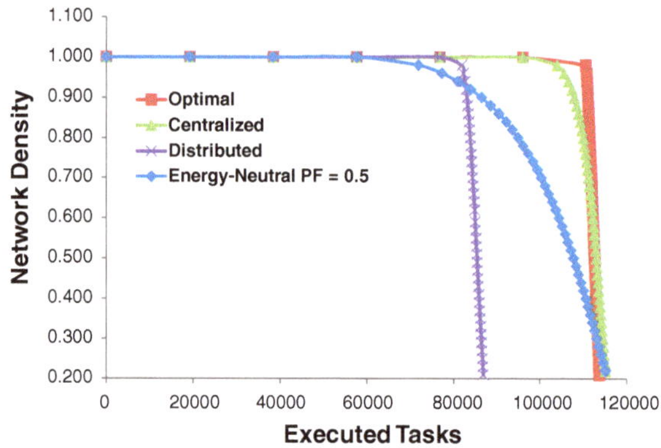

Fig. 20 Average density degradation throughout network 0.2-reliable-lifetime using different protocols for initial network density $\rho = 1.0$

Table 1 Percentage of network lifetime before density degrades

Density	Optimal	Centralized	Distributed	Energy-oblivious
0.3	96.1	86.0	93.2	61.7
0.7	97.0	88.8	94.5	63.4
1.0	97.2	83.3	94.3	62.3
1.5	97.3	90.9	94.3	61.9

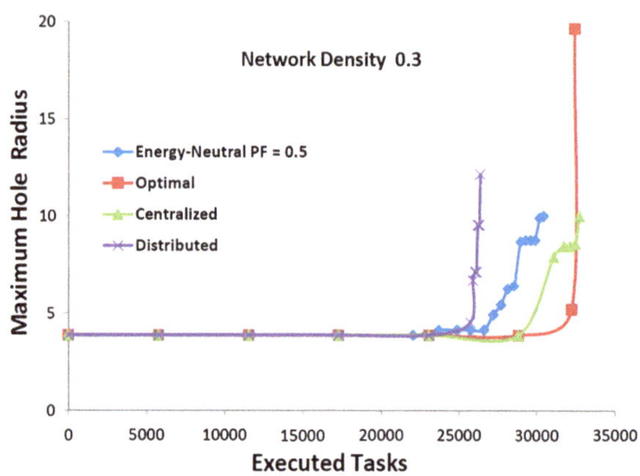

Fig. 21 A comparison between the growth rate of energy holes throughout 0.2-reliable-lifetime of the network using different protocols for initial network density $\rho = 0.3$

degradation in network density when using energy-oblivious protocols starts at an earlier stage, there is always a nonzero probability to find at least one alive sensor that can restrict the growth of the energy holes. This probability increases under dense deployments which in turn delays the appearance and growth of energy holes making our techniques less effective.

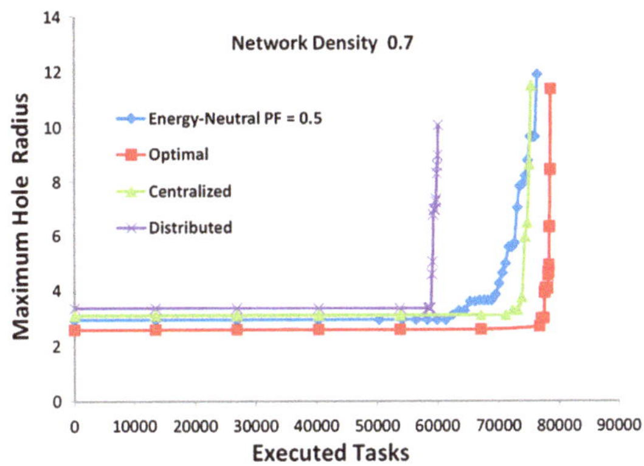

Fig. 22 A comparison between the growth rate of energy holes throughout 0.2-reliable-lifetime of the network using different protocols for initial network density $\rho = 0.7$

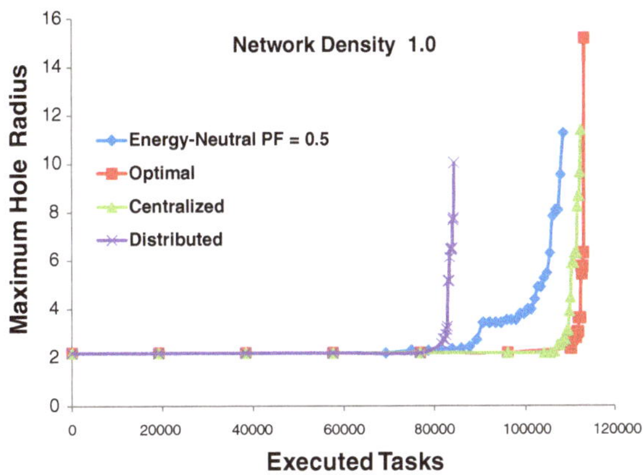

Fig. 23 A comparison between the growth rate of energy holes throughout 0.2-reliable-lifetime of the network using different protocols for initial network density $\rho = 1.0$

8.2.5 Growth Rate of Energy Holes

One of the interesting metrics of protocol performance is its capacity to delay the formation of energy holes in the network [16]. We have evaluated, by simulation, this parameter for various deployment densities. The results are presented in Figs. 21, 22 and 23.

8.2.6 The Total Number of Tasks Executed

Recall that the adopted definition of network longevity was the total number of tasks executed until the network density drops below a given value.

Figure 24 compares the maximum number of tasks the network can execute throughout its 0.2-reliable-lifetime using the four different protocols. It is interesting to note that some protocols are able to execute more tasks than the optimal

Fig. 24 Total number of executed tasks using different workforce sizes

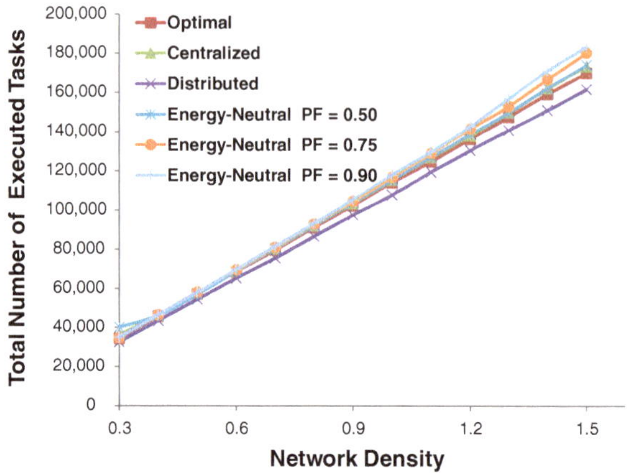

protocol. To understand how this could happen, we recall our tasking model in which a sequence of CTW messages are transmitted by the TIE to attract the attention of sensors willing to participate in the execution of the next task. In some cases, especially at later stages of network lifetime, the number of collected sensors is less than the required size of the workforce as determined by QoS. Hence, those under-recruited tasks are executed using whatever was collected even if the collected number of sensors was less than what was specified in the CTW messages. For example, if at a late stage of the network lifetime all sensors within a certain spot died except for a single sensor which has E units of energy remaining, then the network can assume falsely it can execute up to E additional tasks at this spot irrespective of the workforce size required by these tasks. Fortunately, the same scenario cannot happen using any of the optimal, centralized, or distributed approaches because of the minor differences between sensor energy. By the time the first sensor within a certain spot dies, the remaining sensors within the same spot will be about to die as well. Hence under-recruiting occurs for a small number of tasks.

Our explanation is confirmed by the results in Fig. 25 in which we show the number of tasks executed using the exact workforce size. From the figure, it is obvious that the optimal protocol has superior performance over other protocols. Although, the centralized approach is the closest protocol to optimal in performance, the distributed approach seems to perform poorly and even worse than energy-oblivious protocols. The reason for this state of affairs can be traced back to under-recruiting. Because of the distributed nature of the protocol, sensors join the workforce independently and, as a result, the number of recruited sensors may be larger than the required workforce size. Simulation results showed that for 38.3 % of the total executed tasks, the size of the workforce recruited by the distributed protocol is larger than the required workforce size by around 17.4 %. We confirmed this by adding another curve to Fig. 25 that shows the total number of tasks executed using a workforce size that is at least as large as the required workforce size. Surprisingly, after adding under-recruited tasks, the performance of the distributed protocol

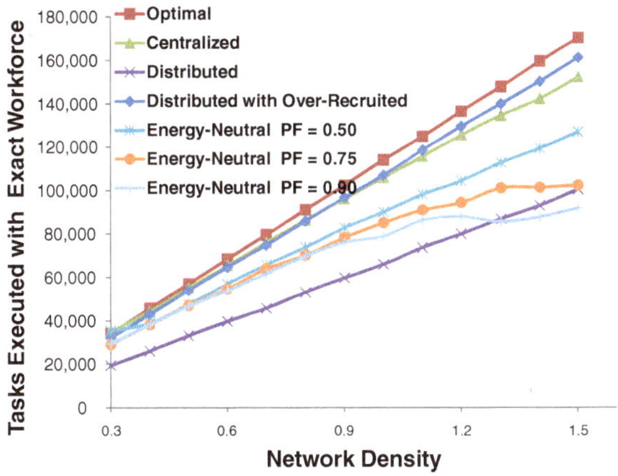

Fig. 25 Number of tasks executed using exact workforce size

has increased tremendously to the extent it became slightly better than the centralized protocol. On the average our centralized approach can increase the network 0.2-reliable lifetime by around 16.5 % while our distributed approach by around 17.2 %.

9 Concluding Remarks and Open Problems

In this chapter we showed how tasking sensors improperly can affect the reliability and the durability of the network by reducing network density, creating energy holes, and partitioning the network into isolated islands. We also discussed a centralized and a distributed workforce selection protocols for maximizing network lifetime by balancing task load among sensors within the same sensing area.

The centralized approach depends on running a contention-based bidding rounds to select required workforce for any task based on sensor remaining energy. On the other hand, the distributed approach works in two phases. In the first phase, the sensors within the task sensing range run a distributed protocol to estimate the maximum energy among them. After that, and in the second phase sensors join the workforce in decision rounds based on their distance to the AoI and the difference between their current energy and the maximum energy determined in the first phase. Simulation results demonstrated that the proposed protocols can increase the longevity of the network by evenly expending sensor energy and by reducing energy differences between sensors within the same sensing area.

In spite of these encouraging results, a lot remains to be done. At the moment, we are trying to extend this work by adjusting sensor sleeping time based on the relative difference between sensor remaining energy and the energy of other surrounding sensors. In particular, we try to prolong the sleeping time of sensors with low energy, and compensate for their absence by shortening the sleeping time of sensors with relatively high energy. The main advantage of this approach is to balance energy consumption among sensors without changing the effective density of the network.

Acknowledgments The work presented in this chapter was funded, in part, by NSF grants CNS-0721563 and CNS-1116238.

References

1. H.S. Abdelsalam, A virtual infrastructure for mitigating typical challenges in sensors networks, Old Dominion University, December 2010
2. H. S. AbdelSalam, S. Olariu, Hexagon-based localization technique for wireless sensor networks. in *PERCOMW '09: Proceedings of the Seventh IEEE International Conference on Pervasive Computing and Communications Workshops*, Galveston, Texas, IEEE Computer Society, March 2009
3. H.S. AbdelSalam, S. Olariu, Toward efficient task management in wireless sensor networks. IEEE Trans. Comput. **60**, 1638–1651 (2011)
4. H.S. AbdelSalam, S. Olariu, Bees: Bioinspired backbone selection in wireless sensor networks. IEEE Trans. Parallel Distrib. Syst. **23**, 44–51 (2012)
5. H.S. AbdelSalam, S. Olariu, S.R. Rizvi, in *Tiling-based localization scheme for sensor networks using a single beacon*, IEEE Globecom, Dec, 2008
6. H.S. Abdelsalam, S.R. Rizvi, Energy efficient workforce selection in special-purpose wireless sensor networks. in *INFOCOM Student Workshop 2008*, IEEE, 1–4, April 2008
7. H.S. AbdelSalam, S.R. Rizvi, S. Ainsworth, S. Olariu, A durable sensor enabled lifeline support for firefighters. in *INFOCOM Workshops (MCN)*, IEEE, 1–6, April 2008
8. H.M. Ammari, S.K. Das, Scheduling protocols for homogeneous and heterogeneous k-covered wireless sensor networks. Pervasive Mob. Comput. **7**(1), 79–97 (2011)
9. F. Barsi, A. Bertossi, C. Lavault, A. Navarra, S. Olariu, M.C. Pinotti, V. Ravelomanana, Efficient location training protocols for heterogeneous sensor and actor networks. IEEE Trans. Mob. Comput. **10**(3), 377–391 (2011)
10. T. He, C. Huang, B. Blum, J. Stankovic, T. Abdelzaher, Range-free localization schemes in large scale sensor networks, 2003
11. R. Nagpal, H.E. Shrobe, J. Bachrach, Organizing a global coordinate system from local information on an ad hoc sensor network. In IPSN, pp. 333–348, (2003)
12. K. Nakano, S. Olariu, Randomized o (log log n)-round leader election protocols in packet radio networks. in *ISAAC '98: Proceedings of the 9th International Symposium on Algorithms and Computation*, pages 209–218, London, springer, 1998
13. K. Nakano, S. Olariu, Leader election protocols for radio networks. Handb wirel networks mob. comput. pp. 219–242, 2002
14. D. Niculescu, B. Nath. Ad hoc positioning system (aps) using aoa. in *INFOCOM 2003. Twenty-Second Annual Joint Conference of the IEEE Computer and Communications Societies*. IEEE, 3:1734–1743 vol. 3, 30 March-3 April 2003
15. S. Olariu, M. Eltoweissy, M. Younis, ANSWER: AutoNomouS netWorked sEnsoR system. J. Parallel Distrib. Comput. **67**(1), 111–124 (2007)
16. S. Olariu, I. Stojmenovic, Design guidelines for maximizing lifetime and avoiding energy holes in sensor networks with uniform distribution and uniform reporting. in *INFOCOM 2006. 25th IEEE International Conference on Computer Communications. Proceedings*, pp. 1–12, April 2006
17. S. Olariu, A. Wada, L. Wilson, M. Eltoweissy, Wireless sensor networks: leveraging the virtual infrastructure. Network, IEEE, 18(4):51–56, July-Aug, 2004
18. E. Parzen, Stochastic Processes, pp. 182–183, Holden-Day, Inc., 1962

19. V. Rajendran, K. Obraczka, J.J. Garcia-Luna-Aceves, Energy-efficient collision-free medium access control for wireless sensor networks. in I.F. Akyildiz, D. Estrin, D.E. Culler, M.B. Srivastava, eds. SenSys, ACM, pp. 181–192, 2003
20. T. van Dam, K. Langendoen, An adaptive energy-efficient mac protocol for wireless sensor networks. in *SenSys '03: Proceedings of the 1st international conference on Embedded networked sensor systems*, pp. 171–180, New York, USA, ACM, 2003
21. A. Wadaa, S. Olariu, L. Wilson, M. Eltoweissy, K. Jones, Training a wireless sensor network. Mob. Networks Appl. **10**, 151–168 (2005)
22. K. Whitehouse, A. Woo, F. Jiang, J. Polastre, D. Culler, Exploiting the capture effect for collision detection and recovery. in *EmNets '05: Proceedings of the 2nd IEEE workshop on Embedded Networked Sensors*, IEEE Comput. Soc. pp. 45–52, Washington, DC, USA, 2005
23. R.W. Wolff, Poisson arrivals see time averages. Oper. Res. **30**(2), 223–231 (1982)
24. X. Wu, G. Chen, S.K. Das, On the energy hole problem of nonuniform node distribution in wireless sensor networks. IEEE Int. Conf. Mob. Adhoc Sen. Syst. (MASS), 2006 , pp 180–187, Oct. 2006
25. L. Yuan, W. Cheng, W. Cheng, X. Du, An energy-efficient real-time routing protocol for sensor networks. Comput. Commun. **30**(10), 2274–2283 (2007)

Part VI
Data Management

Chapter 13
Quality-Aware Sensor Data Management

Zhijing Qin, Qi Han, Sharad Mehrotra and Nalini Venkatasubramanian

Abstract In this chapter, we provide a data management perspective on large-scale sensor environments applications posing non-functional requirements to meet the underlying timeliness, reliability and accuracy needs in addition to the functional needs of data collection. Due to the large-scale regional spread, we need methods that will allow scaling of today's systems to large-scale deployments. Our data management techniques have solved a fundamental challenge in such situations, that is the ability to handle the explosion of sensor data in sensor networks, either due to scaling of the network or due to increased data generation by highly capable and "media-rich" nodes.

1 Motivation

Continuing advances in computational power, radio components, and reduction in the cost of high-performance processing and memory elements has led to the proliferation of portable devices (e.g., intelligent sensors, actuators, and sensory prosthetics) with substantial processing capabilities. By providing the ability to monitor phenomena in close proximity with multihop wireless communication (enabled by on-board radios) and collaborative in-network computation (enabled by on-board processing), embedded networked sensors are able to achieve accuracy, latency, and coverage in

Z. Qin (✉) · Q. Han · S. Mehrotra · N. Venkatasubramanian
Colorado School of Mines, University of California at Irvine, Irvine, CA, USA
e-mail: zhijing.qin@gmail.com

Q. Han
e-mail: qhan@mines.edu

S. Mehrotra
e-mail: sharad@ics.uci.edu

N. Venkatasubramanian
e-mail: nalini@ics.uci.edu

H. M. Ammari (ed.), *The Art of Wireless Sensor Networks*,
Signals and Communication Technology, DOI: 10.1007/978-3-642-40009-4_13,

© Springer-Verlag Berlin Heidelberg 2014

monitoring real-world cluttered environments that a small number of complex sensors cannot. Such devices are rapidly permeating a variety of applications domains such as avionics [1], environmental [2], structural sensing [3], telemedicine [4], space exploration [5], and command and control [6]. Popularly used wireless sensor devices include Mica motes from Crossbow, Tmote Sky from Moteiv, the MKII nodes from UCLA, and SunSpot from Sun.

In this chapter, we provide a data management perspective on large-scale sensor environments applications posing non-functional requirements to meet the underlying timeliness, reliability and accuracy needs in addition to the functional needs of data collection. Consider the following use-case of sensor networks in the domain of public health and safety where one needs to monitor chemical and biological contaminants in soil, ground water, streams, etc. The application scenarios range from long-term monitoring for slowly evolving disasters (e.g., leakage of industrial contaminants into ground water) to detection of sudden disasters (e.g., bio-chemical terrorist attack). Since the phenomena that we seek to monitor and respond to may involve large-scale regional spread (tens to hundreds of kilometers), we need methods that will allow scaling of today's systems to large-scale deployments. A fundamental challenge in such situations is the ability to handle the explosion of sensor data in such networks. This explosion in data occurs either due to scaling of the network or due to increased data generation by highly capable and 'media-rich' nodes. Since data movement costs precious network resources (e.g., energy, storage, bandwidth), data management must be part of the overall system and software architecture.

In circumstances where the presence of a human in the loop is either too expensive or too slow, embedded sensing systems must also respond autonomously and flexibly to unanticipated combinations of events at run-time. These systems are networked to form long-lived "systems of systems" that must run unobtrusively and autonomously, shielding operators from unnecessary details, while simultaneously communicating and responding to application-critical information at heretofore infeasible rates. An important design challenge for such complex distributed computing systems is to satisfy performance and reliability constraints while ensuring efficient exploration through a very large space of device and network operational choices. This process, unless addressed by the software development and execution architecture, is likely going to be manually driven thus limiting the application potential due to prohibitive application development and interoperability across the sensor networks.

There are several reasons why a data management perspective on sensor networks is increasingly important. Given the proliferation of sensing capabilities, we are observing a transition from a device centric view of sensor systems to an information centric view where sensors generate large amounts of data that is stored and processed to generate higher level information for applications. The notion of what is a sensor is changing (above and beyond mote-like devices) to accommodate multimodal sensors, human sensors, smartphones, all of which are capable of capturing, storing, processing, and communicating the sensed data. Managing heterogeneous sensor data and extracting information from it is key to many real world applications—e.g., cyberphysical systems [7].

Networks composed of large numbers of wireless sensors present significant challenges to the designers of data management systems which aim to incorporate data produced by such networks. We envision future generations of sensor networks that would take input from many remote sensors, and provide geographically-dispersed operators with the ability to interact with the collected information and to control remote actuators. As shown in Fig. 1, possible sensor sources may be smart phones, lap top, usb sensor, etc. Sensor data generated by those sources is collected via heterogeneous network (e.g., wifi, cellular, Ethernet, etc). A bunch of context aware applications use those data for various kinds of purposes. Ideally, user applications will interact with sensor-generated data in a high-level language appropriate to their application domain. This will then be translated to data management primitives, e.g., queries in an SQL-like language which will then be evaluated by the data management system, by some appropriate processing strategy. Such strategies should take into account (a) the quality requirements of applications, e.g., the level of precision requested for an average temperature value, (b) the underlying observed physical phenomena, whose properties may suggest a processing strategy, and (c) the characteristics and current state of the sensor network, e.g., its scale, degree of heterogeneity, processing/memory/energy capabilities of sensors.

In this chapter, we will first illustrate the different kinds of applications likely to use sensor data, and in particular, real-time monitoring, e.g., pollutant tracking using chemical sensors, data archival for future use, e.g., the recording of natural phenomena for future analysis by qualified scientists, and forecasting which extrapolates into the future, as in e.g., the prediction of the likely trajectory of a hurricane. We will characterize applications' non-functional needs in terms of quality of service and quality of data, describe how to specify these application needs and how to translate high level application needs to requirements for sensor data. We will further develop a model of a sensor system and sensor data. We will evaluate alternative architectures which can achieve application goals, spanning the spectrum from traditional fully-centralized approaches whose simplicity is counterbalanced by their inability to leverage sensor capabilities to fully decentralized ones which lack global scope

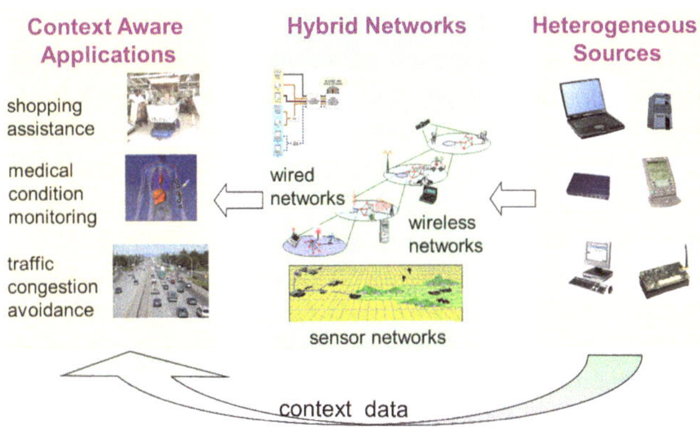

Fig. 1 Emerging applications in pervasive sensing environments

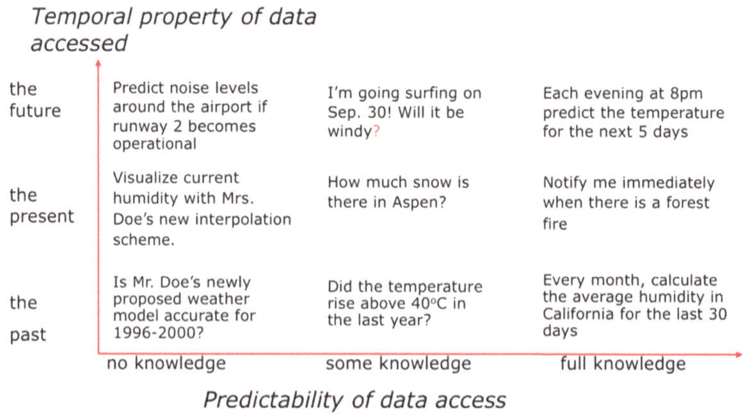

Fig. 2 Sensor application landscape

but may be resilient to failures. Finally, we will examine how alternative evaluation strategies, given an architectural choice, affect optimization goals such as timeliness, energy efficiency and resilience to faults.

2 A Landscape of Distributed Sensor Applications

Often, sensor-based systems are built with narrow application goals in mind. Consider for instance a simple application using chemical sensors to track and report pollutants in water streams periodically. We anticipate that as sensor network infrastructures become more sophisticated, they will have to accommodate several concurrent applications, some of which may have conflicting requirements in terms of timeliness, reliability and data accuracy. It is important for future sensor systems to accommodate alternative application types, and ensure that their conflicting requirements mesh with each other gracefully.

In order to cover a wide spectrum of sensor applications, we first distinguish between different application types based on the temporal aspect of sensor data (Fig. 2): archival applications focus on historical (past) data, e.g., in order to detect patterns over time and build time-varying models. Real-time data collection is not critical here, but high quality and reliable archival of sensor-generated data is; monitoring applications are interested in current sensor values such as intrusion detection systems; and forecasting applications are interested in predicting future sensor values, where human operators involved in decision making processes can avail of information about trends in sensor values. An example of this is route selection in intelligent transportation applications, where techniques to predict traffic conditions (estimated from traffic monitoring devices) among various route candidates can help in reducing travel times and latencies.

We can also classify applications based on the pattern of access to sensor data (Fig. 2). In some cases, e.g., pollutant tracking, the target application is known *before* sensor data is collected. Thus, one can set up the data collection framework in a manner that is optimal for the particular application. In particular, if multiple such applications co-exist, as in e.g., multiple continuous queries then one could exploit the overlap in applications' data needs to optimize the amount of data communication. In other cases, the application type (e.g., requests for aggregate pollutant levels over spatial grids) is known, but not the query instances, which are unknown and arrive in an *ad hoc* manner. Finally, there is the case where the specific application type may not be known beforehand. For example, one might instrument a network of traffic monitoring sensors for the purpose of monitoring speed levels in different segments of the road. However, sensors in such an infrastructure may potentially be used in the future for very different applications: e.g., to control traffic signals, or disseminate optimal routes in real-time to drivers.

The basic functionality of the sensor network is to sense, capture, and communicate data to answer application queries. The ability to use sensors in a wide range of applications has expanded the way sensing (and sensors) are used today. Traditionally, sensor data is viewed as being small—it was designed to encompass parameters such as temperature and smoke density; more recently, it has expanded to incorporate diverse and richer context data from cyberphysical systems, e.g., smartspace video surveillance data and power grid voltmeter. Further, sensor data can also be data captured by smart phones either from user input or from built in sensors such as accelerometers. The sensor application landscape discussed above is valid regardless of data sources. Increasingly, it has become more important to enable the functionality of the sensor network while addressing the tradeoffs and non-functional needs of various sensor applications. By non-functional needs, we mean applications' timeliness, reliability, and data accuracy needs. This can be achieved by taking advantage of a few key intuitions and building upon the wealth of techniques developed over the years in real-time systems, fault tolerant distributed computing and dynamic data management.

Consider the following example of sensing for emergency response at a crisis site. (1) Medical sensing: medical sensors can be deployed on a patient's body to monitor health related parameters. These data is collected via wireless personal area network for the doctors to monitor the patient's health status in real time. (2) Deployed environmental sensors at crisis site: smoke sensor can detect fire in a building, and it can also work with camera sensors to help determine a rescue route in a timely manner. The above example illustrates the utility of sensing in emergency response applications. The example also illustrates the need for timeliness, prioritization and fault tolerance in sensing. It illustrates that the networked sensor environment in practice is highly heterogeneous. It illustrates the need for careful and judicious utilization of scarce communication resources. In fact, it also illustrates that in the immediate aftermath of a disaster, energy efficiency of the sensors monitoring patients and environmental phenomena is of less importance than the timely and accurate communication of sensed information. It illustrates that these needs shift over time where enhancing sensor lifetimes through careful energy management at the sensor node is critical to monitor remnant or new phenomena in the days and weeks after the disaster. It illustrates that the sensing data has changing needs which will bring about different desired tradeoffs. For example, before a fire starts, smoke sensors only need to report when the smoke density exceeds a specified threshold. While during the fire, smoke sensors need to periodically report the smoke density with a high frequency. The need of the sensing changes from reliability to timeliness.

The first step towards the goal of supporting sensor applications with different quality needs is to fully understand the diverse needs of monitoring, archiving or forecasting sensor applications. This requires a careful analysis of a wide range of performance requirements, which define the extent to which performance specifications such as timeliness, reliability, and accuracy may be violated. We explore applications' quality requirements from several dimensions.

- *QoS-timeliness* may be specified in the format of periodicity, deadline, or a certain relative order of different tasks. For instance, in the subsurface contaminant tracking scenario, conductivity readings will only be collected after the temperature readings suggest the existence of a potential plume. Current sensor systems support a basic form of time constraints—data collection frequency as in TAG [8] and Cougar [9]. It is worthwhile to investigate other timing notions used in temporal, real-time databases [10–12] and active databases [13, 14], and then develop a variety of timing semantics appropriate for distributed sensor environments.
- *QoS-reliability* is most commonly defined as the percentage of nodes participating in the collection among all the nodes in the sensor network [15, 16], or as a set of nodes that cover the entire sensor network [17]. In addition to supporting these reliability specifications, we have developed more informative reliability metrics with well-defined semantics. For instance, the recall metric used in information retrieval may be used to indicate the desired completeness of the answer set, if the application is gathering all the readings that meet certain conditions [18]; reliability can also be specified as tolerable thresholds on "false-alarm" or "missed-event" probabilities (i.e., bounds on detection or estimation accuracy) [19].
- *QoD* (Quality of Data) desired from the sensing substrate may be imposed on individual sensor values, or on an answer computed over readings from a set of sensor reports. QoD requirements may be specified as desired data freshness, absolute or relative accuracy bounds. For instance, an application may be satisfied with a report that is off the true value by ± 5 [20] (i.e., absolute accuracy) or by 10 % [21] (i.e., relative accuracy).
- *Cost* is simply defined as energy consumption since energy is the most stringent resource constraint in sensor networks.

A sensor application may have requirements for one of the multidimensional parameters (QoS, QoD, Cost), or a combination of them with different preferences towards different constraints. These preferences will be used to guide future monitoring, sensing and collection plans. In practice, it may be very difficult (if not impossible) to satisfy all the specified requirements simultaneously, to reach the multi-constraint optimal point in reality given the dynamic network conditions and severe resource constraints in the sensor network. Therefore, we allow applications to specify their preference towards different constraints.

One of the important sensor application programming methodologies that have emerged in recent research is a separation between application logic and the sensor data acquisition that drives the application. Examples of such efforts include the Cougar [9] and Berkeley TinyDB projects [22] that have promoted the view of sensors as data producers and support declarative database languages such as SQL, suitably extended and modified, as a way for applications to specify their sensor data needs (e.g., ACQL [23] developed as part of TinyDB supports event-based queries, lifetime queries, etc.) which are of specific interest in sensor environments. An important research direction is extensions of both syntax and semantics of such languages to enable specification of not just their functional but also non-functional needs of applications. A step in that direction is TiNA [24] that extended ACQL to support applications with data accuracy requirements. However, constructs to specify other aspects including timeliness and reliability constraints are missing. Developing such a language requires a careful analysis of diverse needs of sensor applications ranging from monitoring, archival, prediction, actuation, etc.

To ensure end-to-end support for applications with multiple non-functional needs, it is essential to provide a channel for applications to specify what they need and in what manner. First, we will need a high level declarative language, using which applications can specify both functional and non-functional needs. Second, we will also need a compiler that takes in an application program and generates executable code. The specification of high level application needs will be either translated to data needs or tradeoffs to guide sensing and collection planning. We next discuss why existing work in the literature cannot be directly used and suggest possible ideas to address this issue.

Specification of application needs: Sensor applications are often interested in data of certain type (e.g., temperature, or humidity data), and they are not interested in where and how the data is obtained. TinyDB uses a modified version of traditional SQL, focusing on issues related to when and how often data are acquired [23]. It supports several new features that are unique to sensor applications, such as event-based queries, lifetime-based queries and actuation queries, etc. Worthy of mention is the "LIFETIME" clause, which provides an intuitive way for users to reason about power consumption. This language has been furthered extended for continuous aggregate queries by TAG [8] and Cougar [9] where sampling period is specified in a clause beginning "EPOCH DURATION" or "EVERY". Building on top of these, TiNA [24] introduces the "TOLERANCE" clause in the query specification, using which users can specify the temporal coherency tolerance for the query. For example, if the user specifies the tolerance to be 10 %, the sensor network will only report sensor readings that differ from previous reported readings by more than 10 %.

It is necessary to further extend the query language and provide other clauses for users to specify other non-functional needs in addition to accuracy and energy efficiency: timeliness, reliability.

- Specifying accuracy: The concept of relative accuracy introduced by TiNA allows a uniform and easy to understand definition of user tolerance on heterogeneous data sources, where the domain of sensed values is different from one sensor to another. However, there do exist other applications that prefer to specify their absolute tolerance (the maximum deviation of sensor reports from actual measurements). Therefore, we should add the support for absolute accuracy tolerance.
- Specifying timeliness: Specification of time constraints has been used extensively in real-time databases and active databases, where time constraints of transactions take the form of periodic, non-periodic, deadline based or non-deadline based; they can also be specified as relative relationships among different transactions. For example, transaction A must be finished 10 s before transaction B. We should leverage those well-studied methodologies and build time constraints into sensor query language. In addition, we should provide an approximate means for specifying time constraints. When there is uncertainty in the exact timing of event occurrences [25], each event occurrence can have a timestamp given by a time interval. We believe this will provide more flexibility for applications.
- Specifying reliability: Faults in sensor networks create ambiguity in answers to queries. For example, it is impossible for applications to know whether the answer is based on partial reports from sensors, whether the unreported readings are missing or just because those sensors do not satisfy the predicates. Reliability requirements can take the form of event detection probability, or percentage of readings from a certain region, etc. The "recall" metric used in information retrieval can be used to measure the completeness of answer to selection queries. That is, applications can specify in the language their desired recall requirements.

Our discussion so far about language support assumes deterministic guarantees to these non-functional needs. In fact, these assumptions can be relaxed. Some applications would be satisfied with probabilistic guarantees [26]; in addition, applications may have various preferences towards these non-functional needs. Some type of probabilistic reasoning will be necessitated by the fact that sensors record measurements at points in space and time, whereas applications are interested in the underlying phenomena which are continuous in nature, and can be approximated, e.g., by interpolation [27]. Therefore, the language should include clauses to applications to specify their preferences, which can later be used to guide sensing and collection.

Translation of Application Needs: In fact, an application's non-functional needs manifest themselves in the several layers of the system; by adapting and translating non-functional requirements between different layers in the system, we are able to satisfy the application requirements in a manner that is sensitive to the underlying resource availability. We next use the accuracy requirement as an example to explain the implication of accuracy needs at application layer, query service layer and data collection layer. We then illustrate how an application's accuracy needs can be mapped to data accuracy needs.

Application accuracy need is related to the real-life goals of applications. For example, a particular application might be interested in detecting a certain event in a region. Accuracy here can be defined as event detection accuracy. Data collection system can adjust accuracy settings at the various service layers with the ultimate goal of satisfying the application accuracy requirement while minimizing system resource assumption. Conversions between accuracy representations at query service and data collection layers depend on the particular query types.

Query accuracy requirements are specified by each query and each answer has certain accuracy guarantees. The querying service uses the data produced by the data collection service to produce answers to queries posed by users. There are several types of queries considered in our system: (a) Queries on an individual sensor may ask for its value at a given time at some accuracy level. When the query can be satisfied using the collected data, these queries can be processed and answered by utilizing only data obtained from the data collection service. Otherwise, the data collection service may begin sampling at a higher rate in order to provide the desired level of quality. (b) A set-based query asks for the set of sensors that possess certain property. The accuracy on the answer set can be usually captured by using "precision and recall" [28]. If E is the set of sensors which possess the property, e.g., "with its sensing value greater than 100" and A is the set of sensors returned as an answer, then "precision" measures the fraction of A which should have been returned, i.e., $\frac{A \cap E}{A}$, which measures the purity of the answer, whereas recall measures the fraction of E that was returned, i.e., $\frac{A \cap E}{E}$, which quantifies the completeness of the answer. (c) Aggregate queries are also quite common, which computer count, sum, average, min or max over a set of data. All the queries discussed above can be either one-time queries or continuous queries. Dealing with continuous queries is more challenging due to the fact that storing and updating data demand more system resource. Adapting data accuracy to meet query accuracy requirements is one of the main focuses in this research thrust.

Data collection accuracy is upper bounded by sensing technology, as it is impossible to obtain a better estimate of observed phenomena than that which is technologically feasible using the best sensing technology. However, invoking the sensing, processing the sensed data, and transmitting it, all require significant amount of resources, including energy, CPU cycles and disk space. This can become an excessive burden for resource-limited sensor devices or damaged/overloaded sensing infrastructures. Thus, the data collection should be flexible by: (i) switching between different sensors if more than one are available; (ii) approximating data time series in order to minimize transmission; finally,(iii) adapting the rate at which samples are obtained, saving on the cost of using the sensors, but producing a coarser approximation of sensed phenomena. The sensing accuracy is governed mostly by the deployed sensors. The accuracy of sensing corresponds to the "measurement error" in a traditional scientific experiment.

The complexity of mapping an application's accuracy requirement to accuracy requirements over sensor measurements depends on both the application as well as the nature of the underlying sensor network. For example, when we use acoustic sensors to track a moving object, the measurement at the sensor cannot directly translate to the displacement of the object. We should allow the application writer to adapt the application logic to deal with imprecise data without worrying about the underlying translation of data to measurement accuracy. We have demonstrated how such a translation can be achieved using a mathematical framework for a target tracking application willing to tolerate a bounded inaccuracy (e.g., tolerance to within 10 m from the target trajectory) [29]. Similar with accuracy need mapping, we can translate timeliness and reliability requirements as well. For queries with time constraints, we can avoid probing those nodes whose responses are slow [30]; for queries with reliability requirements, we can decide whether those missing reports need to be re-transmitted.

3 Sensor Data Models and Representation

A data model is used to store and represent sensor data at different levels of the architecture. Data representation plays an important role in answering user queries—a good model not only reduces the amount of efforts needed to answer queries, but also facilitates the satisfaction of non-functional needs, such as accuracy and reliability of the sensor data. There are several questions to be answered in developing a data representation model. (a) Shall we choose a simple or a complex model? A complex model might be more accurate, but typically requires more parameters that will need to be exchanged between sensors and servers, incurring extra overhead. It is preferred to use a simple model if it can suffice. (b) How is a model generated? A data model can be static and chosen from a set of fixed models; a model can also be dynamically learned from the changing sensor readings. (c) When should a model or the model related parameters be changed? If it is changed immediately upon a model violation, that would be too aggressive since the violation may be temporary a phenomena; if it is never changed, then it would be too conservative and sensor does not always comply with a single model. (d) Who updates the model? A server may have global knowledge and can maintain history, so a long-haul model is possible; however it does not have the most recent changes in sensor readings. Keeping the data at the servers and sensors consistent requires additional communication overhead. If a sensor updates the model, the advantage is that the sensor has better knowledge of recent history. However, a long-haul model may not be feasible if the sensor does not have enough memory to store the history.

Traditional sensor applications such as temperature and humidity monitoring require the sensor to report the exact data to the server, and the update may be periodically or event-triggered. Such kind of data enables the server to have an accurate view of the sensor data. However, given the limited computational, communication, and storage resources at the sensors, it is expensive to collect and communicate the exact data to the server, especially when the sensor data becomes voluminous. Moreover, there are inherent errors existing in sensors and a single exact data may be incorrect. For example, in an application such as target tracking in a sensor network, error in sensor intensity readings may result in error in localizing the object. Similarly, the result of a query for average temperature in a given region may be imprecise due to data error. Fortunately, one key observation is that a large number of

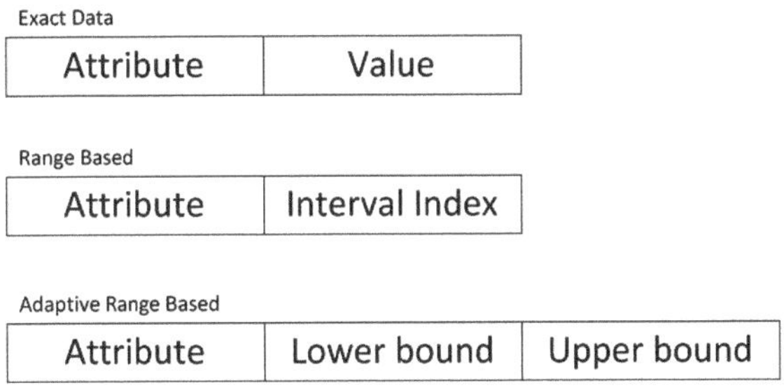

Fig. 3 Sensor data representations

sensor applications can tolerate a certain degree of error in data. The communication overhead between the data sensors and the server can be alleviated by exploiting the applications' error tolerance. This leads the possibility of using a range to represent the collected data. A larger range will cause less messaging overhead and energy consumption, but will lower the data quality. In contrast, small range will provide higher data quality while incurring bigger overhead and energy consumption.

Related research efforts [20, 31] explores the natural tradeoff between application quality and energy consumption at the sensors. With the dynamicity on both the application requirement and the data source, an adaptive range representation model is proposed [20, 31].The range of the collected data depends on both the application requirement and the cost. An optimized range can be chosen by satisfying the application requirement while minimize the cost (Fig. 3).

SensorML: More recently, increasingly ubiquitous sensors brought us sensor data with various content, quality and formats from heterogeneous data source [32], e.g., video surveillance data from pervasive computing space and speech data from firefighters' interphones. To integrate and utilize these various kinds of data, one single sensor data model is not sufficient. More generic and standardized sensor data models should be proposed. Senor Modeling Language (SensorML)[33] is an XML-based modeling language designed for such a purpose. SensorML provides the mechanism for describing the whole range of sensing from simple sensors to arbitrary complex sensor systems and its corresponding platforms. Each sensor is modeled as an operator that is an integral part of a system. Operators consist of input and output behavior, i.e., describe the stimulus received by the sensor and its subsequent action, additional parameters and the input-output transformation function. SensorML meta-data can be used to provide answers to the questions, (1) what is measured (phenomenon); (2) how is it measured (calibration, quality); (3) where is it measured (geometry, spatial response, and sampling); (4) when is it measured (temporal sampling, impulse response); and (5) why is it measured (target application, future processing).

However, the limitation of this standard is that it assumes that users and application writers are able to specify their needs by describing the sensors and operators required for the task at hand. It does not address middleware level challenges such as the need to represent high level concepts, such as entities and activities, which are much more natural for application designers to reason about. Furthermore, the specification does not address adaptivity challenges, hence assuming that there are no resource constraints.

The Virtual Sensor Abstraction: To capture sensing needs at a higher levels of abstraction, SATware [34] abstracts specific sensor formats, data types and the representation of sensor readings so that developers can build applications at the logical level without specific details about the acquisition process or the data formats of the underlying sensor(s). SATware suggests a conceptual level abstracting the raw sensor streams from application writers. More specifically, SATware encapsulates a query into a single operator: ***a virtual sensor***. This is the similar idea to encapsulation in object-oriented programming and aims to hide complexity of operators and simplify the design of applications. It also speeds up application development in SATware and reduces probability of having faulty applications.

Applications can also be modularized and tested independently. Also, reusability is increased since (1) operator topologies (and not only operators) can be now reused, and (2) virtual sensors can be replaced without changing the rest of the application. Virtual sensors provide a controlled avenue toward producing more semantic views closer to the problem domain of the users of sensor data steam processing application.

An example of a virtual sensor is

$$\text{WhoLeftCoffeeBurning} = 010(\text{CoffeeBurning}, \text{Person\_Id})$$

where O10 is an operator that detects if a person has let the coffee burn more than three times, given a person ID and the burning event. Additionally, the notion of a virtual sensor enables optimizations at other system levels to deal with non-functional constraints. For example the middleware services can be utilized to adapt the data collection process, perform sensor actuations and enable re-calibration of sensors to perturbations and errors.

Semantics-Based Data Models: While virtual sensors provides a higher level abstraction of the sensor data, there is still a significant semantic gap between the information of interest for users (e.g., "where is the evacuation warden?") and the data produced by sensors ("RFID reader 57 and read tag 0815"). A key reason for this gap is that traditional database systems, particularly those focused on capturing and managing data from the real world, are not good at dealing with the noise, loss, and uncertainty in data inherent in information that is obtained from sensors. We argue that a semantics-based data model should be used to represent the uncertainty that is inherent in information obtained from sensors, as a way of bridging this gap. Recently, there has been renewed interest in modeling the uncertainty of a sensor data value using a Bayesian framework. For example, Graphical models [35] have been used to managing and querying large-scale uncertain databases; these models capture not only tuple-level and attribute-level uncertainties, but can also represent arbitrary correlations that may be present among the data. Efficient strategies for query evaluation over such probabilistic databases are also been studied. For example, Deshpande et al. [36] have proposed a suite of techniques based on probabilistic models that are designed to allow database to tolerate noise and loss, exploit correlations to predict missing values and identify outliers [37]. Such correlations also provide a way to give approximate answers to users at a significantly lower cost and enable a range of new types of queries over the correlation structure. SATWare extends the semantic ideas in SensorML to capture concepts relevant to sensor-based system. It employs several ontologies and concepts to represent the physical world, the sensor, the data and their correlations.

Figure 4 illustrates the key elements of a semantics-based sensor data model that include (a) the sensor; (b) the environment model in which the sensor is embedded; (c) the phenomena (observed by a sensor) and the associated phenomena semantics; (d) the observability of a phenomena, an observation generated by a sensor and the corresponding observation extraction process. We describe these concepts in detail below.

***Phenomena and Phenomena Semantics*:** Phenomena represent any kind of feature property whose value is amenable to observation or estimation, including physical properties, existence and occurrence assessments, etc. Phenomenon has a type, for example, a person walked in certain space, meeting has started in a certain room, number of people in a region. The phenomena is a measureable value instance that happens in the pervasive space and changes over time. Almost any phenomena can be captured and represented by discretizing the value range. For example, the level of coffee in a coffee pot can be categorized to one of the following values: "full," "half-full" and "empty." The coffee level changes due to brewing of fresh coffee and consumption of coffee and an accurate representation models these state transitions as a function of time. *Phenomena semantics* represent the way that the phenomena evolves as a function of time and space. The phenomena semantics capture the state transition nature of the phenomena as a function of time and space. Consider the coffee level detection task described above, a full coffee level can transition to half-full, empty and overflowing states. The likelihood of each of the possible future states based on the current and past states gives us a meaningful context and understanding of the monitored system, in this case the coffee machine. The semantics of a

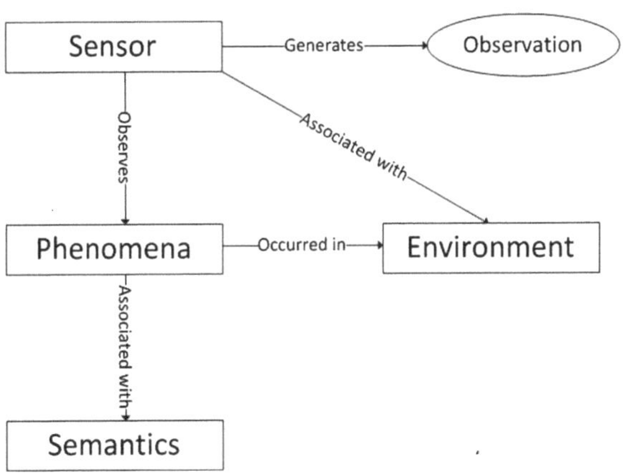

Fig. 4 A generalized semantics-based sensor data model

phenomena give the application a "window" to the future states of the phenomena in the expected sense. This, in turn, is used to guide the middleware in tasks such as data collection—if it knows that an entity is more likely to appear in a certain camera we can allocate resources accordingly.

Environment Model: creates an abstract representation of the monitored environment. The abstract environment representation allows the modeling of the phenomena state independent of the sensing infrastructure. The environment model creates unique space identifiers for different regions in the monitored space—r1,…,rk. For example, in a building setting the environment model is used to refer to different regions such as "ṙ1 = south west hallway, second floor" and "r2 = second floor kitchen".

Sensors: Sensors represent physical sources (e.g., devices) that provide input to observe a phenomena. The sensor is associated with metadata information that is used by the different processes that extract observations. Metadata information includes the

- Sensor type. Sensor types include video, motion, temperature, etc. Different sensor types are capable of observing different types of phenomena, for instance, measuring the temperature can be done using a thermometer and not a video camera.
- Environment Cover—the regions of the space that are covered by this sensor. For example, in the case of a camera sensor the regions covered include the parts of the environment in the field of view of the camera.
- Sensor parameters—the state of the sensor (e.g., in the case of camera sensor the state can be zoomed in or zoomed out, titled at 65 %, etc.).

Virtual Sensors and the Observation Extraction Process: Virtual sensors abstract the physical source of data and capture the extraction process that is used to observe a phenomena. Observations are generated following an extraction process—D that is controlled by extraction parameters. The extraction process is responsible of generating a digital representation of the high level event that was captured by the sensor. Generating the observation might include multiple operators and processes, however, for our problem we abstract that process and consider that the observation process contains an extraction process as well. For example, detecting if there is a face in the field of view requires processing by a face detection operator. We assume that the observation extraction process also executes all the relevant operators to generate the observation of interest to the application.

Observability of a Phenomena: A sensor has a finite number of possible states that control the way the sensor observes the monitored phenomena. Given a phenomena at a given location and time, a sensor would be able to observe it and hence will be in the observability set of the phenomena if the following four conditions hold:

(1) The sensor type can observe the phenomena of interest.
(2) The coverage of the sensor includes the location of the phenomena as specified by the environment model.
(3) The state of the sensor is such that the phenomena type is observable in that state by that sensor type.

(4) The observation parameters are calibrated to extract the observation correctly.

For example, collecting a high resolution frontal face image of an entity in MeerkatU, can only be done if (1) the sensor is of video camera type, and (2) the face of the entity appears in a region that is part of the field of view of the camera, as specified by our environment model. (3) The camera is zoomed into the region in which the face is present. (4) The face detection software is calibrated to detect that there is face in the field of view of the camera. The observability of a phenomena is related to observability in control systems [38] in which the internals of a system need to be observed using external measurements.

To exploit the above data model concepts effectively in gathering and processing data from large multisensory systems. one must understand the architecture of the distributed sensor system in more detail—this is the topic of the following section.

4 Architectures for Executing Sensor Applications

Given heterogeneous sensor platform distributed over a space and applications data needs expressed as queries, multiple execution architectures are possible. A traditional client-server approach would collect data at a (logically) centralized repository and evaluate queries over such a repository. In such a centralized sensing architecture, sensors are passive units that send data to the central server. Applications subsequently execute over this server, operating over the data stored there. This approach simply uses the minimum functionality expected of sensors, namely their ability to communicate their captured data to the world at large. Such an approach does not exploit the computation and storage capabilities available within the sensor network at sensor nodes.

In recent years, the availability of cheap wireless sensors has led many researchers to re-evaluate this architecture. Several observations led to this: first, current sensor designs include processing and memory components which are not leveraged if the sensor is considered as a passive beacon of data; second, bandwidth and energy limitations of battery-powered sensors may make the centralized approach very inefficient, as it entails the transmission of every value from its source, via multiple hops, to the server; third, servers themselves would not be able to scale gracefully to the large number of data sources anticipated for future sensor networks due to the low cost of these devices; fourth, time delays for transmitting data to a server before processing it may be prohibitive, and this may be especially critical in applications where real-time response, e.g., actuation, is driven by the sensors themselves.

As a result, many researchers have adopted approaches in which processing is pushed to the sensors themselves, either by eliminating the need for a server altogether, or by introducing some intelligence in the data flow structure, e.g., routing tree, used to transmit data from the sensors to the central server. In the traditional centralized approach, sensors sample periodically and transmit them to the server who then does all the processing. In a completely decentralized approach, data is transmitted between sensors and processed by the sensors themselves to determine whether there is an event of interest [29]. The third and more popular approach is to introduce some intelligence in the routing tree leading to the server; for example, a sensor routes values of other sensors towards the server, and it might avoid doing so if it determines that a sensor's value places it well outside the application's interest. This splits the processing between the server and the sensors, and would result at lower data transmission costs, as fewer values need to be transmitted.

Such "in-network" computation can effectively exploit resources at the sensors to trade computation for reduced communication. By computing directly in the sensor network, the data routing and computing can be co-optimized, resulting in higher scalability. There are, however, limitations of such an approach including limitations on types of data access that can be supported in-network, complexity of optimally splitting computation between sensors and servers, and the lack of a direct way of exploiting applications tolerance to errors and faults. Yet another approach is a hierarchical view in which the actual placement of the distributed server functionality is a function of the node capabilities in a hierarchical setting, the architecture allowing for dynamic migration of functionality based on the well-developed client server model. The exploration of such architectures has, in the literature, been motivated from a narrow perspective of suitability for one (or more) sensor application scenarios (e.g., monitoring, archiving), a comprehensive understanding of the suitability and feasibility of diverse architectures under different situations and blend of application loads is missing. Additionally, the implications of supporting non-functional requirements over these architectures is not well understood in many cases. Therefore, there is a need for a thorough comprehensive analysis of possible architectures with the focus on supporting non-functional application requirements.

There is an intuitive simplicity in the centralized approach, since any application can be conceivably built on top of it. By contrast, approaches which push processing to the sensors entail the need to create distributed versions of every task posed to the network, each of which must (i) work correctly, i.e., produce a result equivalent to that produced by a centralized approach, and (ii) work efficiently and robustly, especially in the sense of minimizing energy drain and being resilient to failures. It is not immediately clear which architecture is the most appropriate for what type of applications. A methodology for systematic evaluation of different architectural approaches must be based on the following factors:

- ***Application needs***: Consider an application monitoring certain phenomenon within a geographical region for real-time response. Imagine that a routing tree is used to achieve this goal, with each node in the tree aggregating data received from its children. There are many ways to build such a tree with different node placement and tree construction strategies. To minimize makespan, i.e., the completion time of collecting all data, it is essential to enable and exploit parallel transmissions as much as possible. It seems intuitive that a routing tree with large node fanouts may help increase parallel transmissions and reduce the depth of the tree, thus reduce the notification delay for changes in the observed phenomenon. However, when interference in wireless communication is considered, it is not obvious that large fanouts can always decrease makespan. Given a monitoring application, it is challenging to design a sensor placement and tree construction algorithm under constraints such as network coverage and wireless interference, to achieve the ultimate goal of timeliness. Furthermore, large fanouts may impose a great burden on individual parent sensors which must listen for and aggregate over several values. Therefore, an archival application would prefer a different architecture which prolongs the longevity of the sensor network by optimizing for energy drain alone.

- ***Node density of sensor networks***: Imagine two sensor networks that differ in the number of nodes which they contain. This point is salient in view of the fact that many current sensor deployments and many research efforts are evaluated over networks of small size, of up to at most a few hundred nodes. Much of the initial excitement about wireless sensors centered around the prospect that networks involving thousands and millions of sensors would be built. What is the best architecture for networks of different size? A centralized approach might impose tremendous processing costs on the central server over a large network. On the other hand, a fully distributed solution may be completely impractical for networks of this size, e.g., because of the long paths of lateral (sensor-to-sensor) communication. It appears that network clustering, which groups sensor nodes into clusters, is the most feasible approach. However, identifying the criteria that determine the sizes and even hierarchies of clusters is a challenging issue. Dynamically maintaining and changing the clusters is also an immediate challenge to accommodate node failures and balance energy drain across sensors.
- ***Heterogeneity of sensor networks***: Current approaches often assume the existence of a network monitoring a single environmental attribute, e.g., temperature. If multiple sensors exist, e.g., for temperature and atmospheric pressure, then integration of their data might need to happen de facto in a central server, depending on the ability of sensor nodes for temperature/pressure to communicate with each other. Some efforts [39–41] address issues emerging from the co-existence of multiple sensing modalities within a single sensor node. Yet, there exist many different possible gradations between a situation where temperature and pressure can only be correlated at a central server and one in which they are measured on the same node. Consider the example of an application for detecting anomalous behavior, e.g., intrusions, in sensitive installations. Light intensity, sound, smoke, video, motion detection and numerous other sensor types may co-exist in such a setting, and data from them ought to be integrated either in-network or at a central site for the purpose of e.g., raising an alarm or tracking an intruder. Bearing certain precedence constraints, e.g., acoustic sensors are much cheaper to operate than video sensors which allows us to check sound intensity first to detect an intrusion event with minimum energy consumption, data may be routed in a specific way, e.g., from acoustic sensors to video sensors, to enable in-network integration for the purpose of e.g., raising an alarm or tracking an intruder. Decisions on the best architecture for these applications have to be made considering all potential cooperation between sensor nodes.

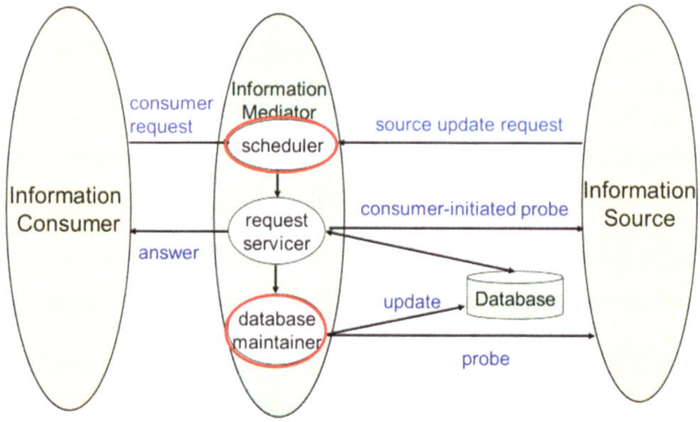

Fig. 5 A mediation-based architecture for sensor systems

A Mediation-Based Architecture for Sensor Data Management: Concurrently supporting multiple applications is complicated; it is desirable to have a middleware component bridging the applications and the network layer. There are plenty of works on WSN middleware. From the paradigm perspective, we can classify these middlewares into several categories: event-based [42], application driven [43], component based [44] and mediation based. In this chapter, we use a mediator to shield applications from underlying complexity; it is the module where the context collection process is executed. Figure 5 depicts the architectural components of a mediator-based framework for context collection. The efficiency of the system depends on specific algorithms applied in each component of the framework. We describe the functionality of each component.

The Information Sources correspond to different components in the distributed sensor system, such as the server, link, mobile or stationary host, and sensors. The context information about sources includes network parameters (such as residual link bandwidth, end-to-end delay on links, link load, link packet drop rate.), server parameters (such as CPU utilization, butter capacity, disk bandwidth.), stationary host parameters (such as client capacity, connectivity), mobile host parameters (such as mobile host location, connectivity, power level), and any data that sensors can capture (such as temperature, humidity, habitat activities). These sources can be programmed to send out information updates periodically, to respond to value requests, or to send out notifications when their values are beyond a certain pre-specified threshold.

The Information Consumers are application and system level tasks that use the data collected from the information sources. For instance, a traffic monitoring application is an application level task that obtains data from highway sensors periodically to assist in traffic planning and routing. Consumer requests are read-only requests that obtain current context, and they are often associated with QoS and/or QoD requirements. The arrival time of one-shot consumer requests are unpredictable, while continuous consumer requests are often in the format of interest registration, i.e., they register desired context data with specified triggering conditions.

The *Data Base* is also called *Context Repository* consisting of distributed databases which hold context information from information sources. Approximate data representation in the repository can be used to lower the cost of maintaining the repository. For example, each data item can be represented using an interval bounded by an upper and a lower value. The information in the context repository is updated based on current network conditions and user requirements using different update policies.

The *Information Mediator* serves as the decision point for the information collection process. When a number of users request dynamic data at varying requirements under constantly changing system/network conditions, mediation functionality is crucial to deliver the right data to the right user at the right time. Therefore, a key component of context collection architecture is an information mediator, which connects information sources and consumers and serves as a crux of the information collection process where collection decisions are instrumented.

Information Flow of a Context Collection Process: A typical context collection process works as follows. Information sources communicate changes in source values to the mediator, and the mediator uses the most recent value to update the context repository representation if deemed necessary. Information consumers forward their requests to the mediator which in turn retrieves the requested data from the repository. If the repository values meet user specified quality, the mediator returns the answer; otherwise, the mediator probes the sources for current values. In next section, we will show how we collect sensor data on such a mediation-based system.

Distributing the Mediator Functionality: Although logically a mediator resides between applications and sensors, in reality, it can be at a very powerful server, or at a resource sufficient base station, or at resource constrained sensors. Since sensor applications may involve a large number of sensors (especially when the observed phenomena moves spatially), it is desirable to use ***distributed cooperating mediators*** to ensure system scalability. Given a large number of consumers (query points) and sensor networks, the choice of how to design the mediation-based architecture has multiple challenges—it involves determining an optimal total number of mediators and their locations should be identified that minimizes the overhead involved in satisfying the functional and non-functional needs.

Given multiple mediators, techniques are required to decide where to place data from each sensor and how (e.g., at what accuracy level) the data should be maintained. In addition, we need to maintain a consistent view among all the replicas; with multiple mediators, we need to select for each user request where to retrieve the requested data in order to ensure applications' non-functional needs while balancing the load among distributed mediators. As sensor network infrastructures get more sophisticated, they must provide seamless access to data dispersed across a hierarchy of sensors, mediators, servers and archives—from sensor devices where data originates to large databases where data is stored and/or analyzed. Sensors are more than just passive beacons, but can perform useful work, so we need to consider where to store data and how to push part of the computation to sensors. Archival applications require large amounts of sensor data to be stored for future analysis. If this information is stored at sensors, then we are restricted by sensors limited storage, network

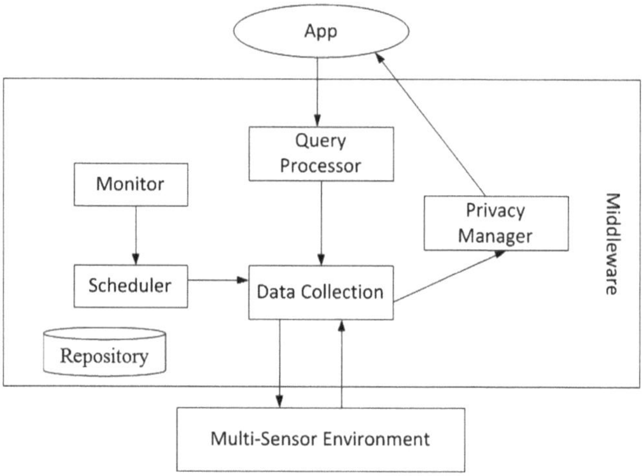

Fig. 6 SATware mediation middleware

and computation resources. If this information is stored at servers, the architecture must be able to cope with high data production rates, and prevent data staleness and/or wasted resources. Several research efforts have been devoted to this area. Dimensions [45] uses in-network wavelet-based progressive aging of summaries in support of long-term querying in storage. DCS [46] stores data at a node determined by the name associated with the sensed data. In SDCT [47], finding locations of the nodes for caching data to minimize communication cost corresponds to finding the nodes of a weighted Minimum Steiner tree whose edge weights depend on the edge's Euclidean length and its data refresh rate. Based on this, a dynamic distributed energy-conserving application-layer service for data caching and asynchronous multicast is presented. TSAR [48] is a storage architecture designed for multi-tier sensor network where an application comprises tens of tethered proxies, each managing tens to hundreds of untethered sensors. TSAR separates data from metadata by employing local archiving at the sensors and distributed indexing at the proxies. At the proxy tier, TSAR employs a novel multi-resolution ordered distributed index structure, the Interval Skip Graph, for efficiently supporting spatio-temporal and value queries. At the sensor tier, TSAR supports energy-aware adaptive summarization that can trade off the cost of transmitting metadata to the proxies against the overhead of false hits resulting from querying a coarse-grain index. The data placement problem should be addressed together with non-functional needs. Specifically, the goal should be to identify the most appropriate data placement strategies to achieve timeliness, accuracy, reliability needs while ensuring energy efficiency.

SATware [34] is a distributed multilevel mediation-based middleware for multi-sensor environment. If can efficiently capture, represent, process, and store information from the various data producers (e.g., cameras, motes, mesh routers) at desired levels of accuracy and granularity to meet the information quality and dependability needs of consumers (e.g., video data for surveillance or link congestion levels for routing) given storage and communication constraints. Shown in Fig. 6, the data collection module can adjust the collection policy to achieve this. The privacy issues (e.g., in video data) are solved by privacy manager. At the meantime, given the application requirement and the low level sensor environment, optimal sensing schedule can be made to achieve efficient data collection. In general the architecture of SAT-ware mediation middleware that translates application needs (expressed via queries) to a corresponding data collection plan to be executed on the multisensor environment. Such data collection techniques to deal with the multiple needs of sensor applications is the focus of the following section.

5 Sensor Data Collection

Given the importance and potential of the impact of sensor technologies, over the past decade, significant progress has been made on techniques to architect and program large-scale sensor systems. Important developments include design of light-weight operating systems for sensor devices, powerful programming frameworks that isolate application logic from the complexities of optimizing the computation over sensor networks, techniques for in-network processing that exploit computational resources at the sensors to reduce communication and preserve energy.

While substantial progress has been made, current research has primarily considered *functional* aspects of distributed sensor systems focusing on techniques to sense, capture, communicate, and compute over sensor networks. As sensor applications become more complex and diverse, *non-functional* application needs (such as timeliness, reliability, accuracy and privacy) become important. As an illustrative example, consider a network of sensors monitoring ground movement to detect presence/arrival of enemy forces in a given region in a command and control application. Timeliness and reliability of sensing (in presence of failures) might be of essence here if the countering maneuver requires immediate detection. Such timeliness and reliability requirements, however, come at certain costs, namely additional communication overheads, energy costs, etc. Furthermore, different applications over a given sensor infrastructure may have differing non-functional requirements. For instance, an online monitoring and actuation application might have real-time requirements, an analysis application over the same sensor system might only require that data be collected in a repository (eventually) at a given level of accuracy or spatial and temporal frequency. At the meantime, the more accurate the less privacy the system can provide. For example, an accurate location sensing would expose people's privacy to externals. Such differing application requirements may pose competing requirements on the underlying sensor data collection, coordination, and storage mechanisms. For instance, from the perspective of the archival application, it might be both feasible and desirable that the data be collected, temporarily stored, compressed and then transmitted to the repository. A real-time monitoring/actuation application, however, may demand low latency.

5.1 *Quality-Aware Sensor Data Collection*

Wireless sensor networks have typically been built with a high degree of dependency between applications and the underlying communication protocols. Such dependency is justified as necessary to achieve energy efficiency. However, it generates rigid systems with sensor networks specifically designed to suit a particular application. While providing a platform that accommodates all types of sensor applications is very difficult, one idea is to build a middleware architecture that can support a representative class of sensor applications—those with multiple performance requirements (in particular QoS, QoD, Cost). We observe that there exists a fundamental tradeoff between the overhead introduced in supporting the application and the QoS/QoD achieved. We refer to this characteristic as the QoS-QoD-Cost tradeoff. If we consider Cost as one dimension and composite performance as another dimension, the application fixes the position of one dimension, and the system is expected to maximize the position along the other dimension.

The Quality-aware Sensing Architecture (QUASAR) [49] is a framework that aims to provide end-to-end support of sensor data collection with data quality specifications [50] and varying QoS and QoD needs (Fig. 7). We envision two complementary techniques for quality aware sensor data collection.

The first category includes applications that aim to maximize the QoS/QoD without exceeding the energy budget: This applies when the lifetime of a sensor network is known and the application would like to get as high-quality data as possible. Providing desired timeliness, reliability, accuracy to applications while conserving energy continues to be an important goal. For instance, in the immediate aftermath of a toxic chemical leakage, timely and accurate communication of collected data is much more important than energy efficiency, hence the application would like to maximize QoS-timeliness and QoD subject to the constraint of remaining energy. With the finite remaining energy level on each sensor node, we are faced with a joint optimization problem when the objective of an application is to maximize more than one metric (i.e., two or three among reliability, timeliness, accuracy).

Existing work has addressed limited subsets of the problem space. Using decoupled strategies that optimize each performance goal in separate phases can unfortunately lead to very expensive data collection plans, since these performance goals are often interdependent. The decision on achieving one objective affects the decision on achieving the other. For instance, improving reliability might entail retransmission of packets, which may lead to increased latency. One strategy is to design a new composite evaluation metric. Most of the existing work focuses on one single performance metric, such as data freshness (i.e., the time elapsed from data generation time to the data collection time) [51], data fidelity (i.e., ratio of nodes participating in the collection to all the nodes in the area) [51], data accuracy measured as how much the obtained data deviates from the real value [20], or as combined spatial and time distortion [35]. We believe it is a better approach if we take into account pre-specified application preferences towards different performance goals and define a composite system performance metric to facilitate the identification of an "optimal" point

Fig. 7 QUASAR architecture

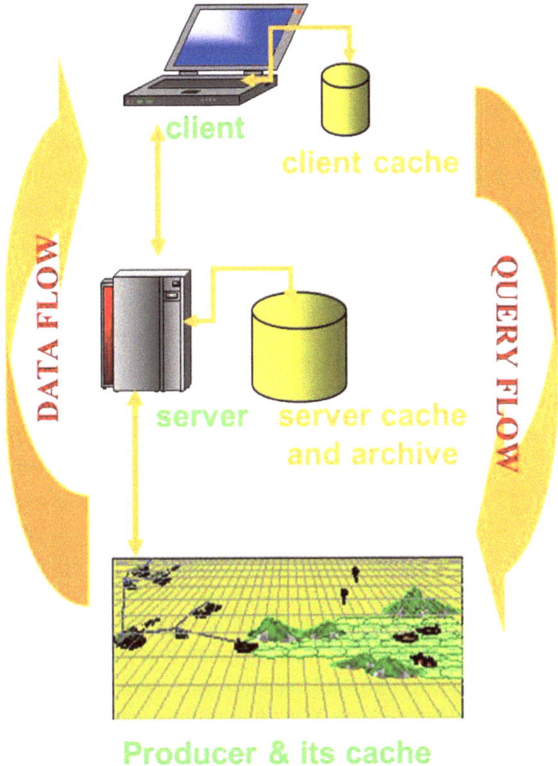

that would maximize the multiple QoS or QoD requirements. Alternatively, we can investigate the tradeoff between these performance goals and develop a collection plan that optimizes all the goals. More specifically, we might be able to derive theoretical upper bounds on QoS-timeliness, QoS-reliability and QoD, given the energy constraint. We can further design online algorithms that attempt to either maximize the composite performance metric as described above, or jointly optimize multiple goals simultaneously.

An alternate formulation for quality aware data collection can be developed where the goal is to minimize energy consumption while achieving a minimum acceptable level of QoS/QoD: This applies when a sensor network needs to consume as little energy as possible in order to last longer while ensuring the satisfaction of application needs. For instance, careful energy management of sensor nodes is critical to monitor remnant or new plumes in the days and weeks after the plume disaster; hence, the application would like to minimize energy consumption as long as QoD reaches a certain level. The multiple constraints can be any combination of requirements for QoS-timeliness, QoS-reliability and QoD, e.g., an application might want to minimize energy consumption while ensuring the minimum data accuracy and maximum tolerable latency. Since the granularity at which the sensor data is maintained at the server directly affects the amount of communication, and a sensor consumes energy even when it is idling, the total energy consumption is a function of data granularity and sensor idling time at different power saving states. Therefore, this is a multi-variable optimization problem subject to multiple constraints. An arbitrary combination of algorithms for satisfying timeliness requirements and those for meeting data accuracy needs will result in undesirable system performance, since these constraints are not orthogonal and varying one will typically affect the others. Hence, it is crucial to understand the interplay between different application needs.

In the following, we describe how individual non-functional needs have been supported in the literature.

Dealing with Data Accuracy Requirements. Since sensors are resource constrained, sensor data is often collected into more powerful servers. A natural tradeoff exists between the sensor resources (bandwidth, energy) consumed and the data accuracy collected at the server. Blindly transmitting sensor updates at a fixed periodicity to the server results in a suboptimal solution due to the differences in stability of sensor values and due to the varying application needs that impose different accuracy requirements across sensors. The idea of approximate caching in traditional databases has been applied to resource-constrained sensor networks [19, 49]. The novelty of that work is the application-aware integration of data accuracy satisfaction and power management of sensors [20, 52]. They have developed an optimal data collection protocol that depicts how the server and the sensors collaborate to maintain an optimal representation of sensor data. In concert with this, an optimal algorithm was designed for managing sensor states, which determines the length of sensor idling or sleeping.

Dealing with Reliability Requirements. With the increasingly popular use of WSNs, data is available as never before in many fields of study; practitioners are now burdened with the challenge of doing data-rich research rather than being data-starved. However, in situ sensors can be prone to errors, links between nodes are often unreliable, and nodes may become unresponsive in harsh environments, leaving to researchers the onerous task of deciphering often anomalous data. To this end, the REDFLAG fault detection service is developed that is a Run-time, Distributed, Flexible, detector of faults, that is also Lightweight And Generic [53]. REDFLAG addresses the two most worrisome issues in data-driven wireless sensor applications: abnormal data and missing data. REDFLAG exposes faults as they occur by using distributed algorithms in order to conserve energy. There are also approaches to guaranteeing an application's reliability requirement [15, 54]. Further, the problem of evaluating continuous selection queries over sensor data has been considered in the presence of faults [18, 55]. Reports produced by small sensors may not reach the querying node, resulting in an incomplete and ambiguous answer, as any of the non-reporting sensors may have produced a tuple which was lost. Fault Tolerant Evaluation of Continuous Selection Queries (FATE-CSQ) is a protocol that guarantees a user-requested level of quality in an efficient manner. FATE-CSQ is designed to be resilient to different kinds of failures.

Dealing with Timeliness Requirements. The nature of many sensor applications as well as continuously changing sensor data often imposes real-time requirements on WSN protocols. Due to numerous design constraints, such as limited bandwidth, memory and energy of sensor platforms, and packet collisions that can potentially lead to an unbounded number of retransmissions, timeliness techniques designed for real-time systems and real-time databases cannot be applied directly to WSNs. In order to design a protocol for sensor applications that require periodic collection of raw data reports from the entire network in a timely manner, previous work has formulated the problem as a graph coloring problem and then developed TIGRA (Timely Sensor Data Collection using Distributed Graph Coloring)—a distributed heuristic for graph coloring that takes into account application semantics and special characteristics of sensor networks [56, 57]. TIGRA ensures that no interference occurs and spatial channel reuse is maximized by assigning a specific time slot for each node. Although the end-to-end delay incurred by sensor data collection largely depends on a specific topology, platform, and application, TIGRA provides a transmission schedule that guarantees a deterministic delay on sensor data collection.

5.2 Dealing with Composite Requirements

A more challenging issue is to ensure that the multiple non-functional needs posed by application queries: timeliness, data accuracy, reliability are met in a cost-effective manner. However application requirements may conflict with each other. Consider the following cases.

- *Accuracy versus timeliness*: Frequent updates from sensors will undoubtedly improve accuracy, however, it might also leave the server not enough time to process user queries, hence violating many time constraints. A straightforward application of existing algorithms for dynamic data management that address the cost and accuracy tradeoffs attempt to keep the database "reasonably" accurate. However, it ignores the arrival order and time constraints of user requests. i.e., an urgent request might have to wait for a long time before it can be processed; even though the request can be processed by merely retrieving values from the repository, the long waiting time may lead to the violation of time constraints. There are other subtle implications that arise when timing needs are ignored. For instance, handling update requests from sensors prior to retrieval requests might either improve or worsen timing and cost depending on how many update requests remain to be processed. Furthermore, the cost involved in sensor data collection is not only introduced by communication, but also sensor idling at different power saving states.
- *Timeliness versus reliability*: In order to ensure reliability in the presence of faults, there may be a need to re-transmit data; however, this recovery can be time-consuming, hence leading to degradation in timeliness satisfaction. Directly applying existing fault tolerant techniques developed for delivering sensor data may provide certain reliability guarantees. However, those techniques typically are ignorant of timing needs of applications. For instance, in order to ensure

reliable data delivery, hop-by-hop recovery is often applied; however, the final arrival of the data might lag behind significantly and become useless since the timing constraints are violated.
- *Reliability versus accuracy* queries may derive some missing sensor data based on historical or neighboring reports, and this might be sufficient for reliability needs; however, this implies that the derived data may not be accurate.

Existing and ongoing research has addressed the tradeoffs between two dynamic factors in different context. For instance, the tradeoffs between transaction timeliness and data freshness have been studied via different real-time scheduling algorithms in real-time and temporal databases [36, 58, 59]; the accuracy and cost tradeoffs are explored by developing various cache management algorithms in dynamic data management [31, 49, 50, 60–67], the tradeoffs between reliability and energy efficiency in sensor networks have been investigated [8, 16, 37, 68–76]. However, the problem of supporting multi-dimension (possibly conflicting) non-functional needs—timeliness, accuracy, reliability and cost-effectiveness—simultaneously in the context of heterogeneous sensor networks remains a challenge. While real-time scheduling algorithms determine the order of request processing with the objective of increasing the number of requests with their time constraints met, they are incognizant of communication and data processing overheads involved in the data collection process in highly unreliable distributed sensor environments. The overheads come from of sampling sensors, retrieving data from repositories, updating repositories, etc. To address the cost constraint, these overheads should be kept to a minimum while addressing the timeliness/accuracy/reliability needs. In addition, the current approaches do not address the issues of how the scheduled request should be processed. For instance, should it obtain information from the repository and sacrifice accuracy for cost or should it probe the sensor and sacrifice cost for current and possibly future accuracy?

Determining an appropriate composition of the services for reliable accuracy-aware scheduling and cost-aware database maintenance is challenging due to the following reasons. Firstly, achieving optimality (minimized overall cost while meeting timelines s needs and accuracy constraints) in sensor data collection is impossible. Addressing the timeliness/accuracy/reliability/cost tradeoffs simultaneously is in fact NP-hard, since the problem is a superset of the classical scheduling problem with mutual exclusion constraints (which has been proven to be NP-hard [77, 78]). Secondly, how the composition should work needs to be determined. Should the services operate independently with no interaction (resulting in poor performance) or should each service be extended to work in concert with the others? Each service becomes more complicated when extended and composed and there are no straightforward rules for how to adjust parameters for each service in order to achieve the overall best performance. It is also unclear exactly what information is needed to enable the composition to make the parameter adjustments. Finally, there are a number of dynamic factors affecting the data collection process: the network latency and link quality are unpredictable and variable, the observed phenomena are highly dynamic, and sensors are failure prone. Hence, adapting the system while balancing the composite tradeoffs of timeliness, accuracy, reliability and cost is not straightforward. Lots of current research has primarily considered functional aspects of distributed sensor systems focusing on techniques to sense, capture, communicate, and compute over sensor networks. To support different non-functional (i.e., quality) needs of sensor data collection, most schemes are implemented at different layers such as MAC layer, routing layer, or data management layer. In fact, these non-functional needs are cross cutting issues that are better addressed by using cross-layer approaches. Motivated by the observation that various biological systems have developed mechanisms to meet conflicting requirements simultaneously, a biologically-inspired solution has been proposed to balance the tradeoffs among conflicting requirements (reliability, timeliness, energy efficiency) and govern mobile agent behavior in sensornets [79, 80] .

To ensure judicious composition, several general strategies can be exploited.

- **Error-Aware prediction:** Here, the individual sensor and mediator dynamically agree upon a model that predicts sensor measurements over the immediate future. If the sensor reading adheres to the predicted model within an error bound, the communication of the sensor readings to the mediator is avoided conserving energy. Prediction-based mechanisms can exploit local compute capabilities at the sensor to further optimize communication (the efficacy of the optimization process is dependent on the accuracy of the prediction model). While prediction is useful, there are many issues that need to be resolved in incorporating prediction models for dynamic data collection. For instance: Who determines the prediction models, the sensor or the mediator? What is the protocol by which a sensor and mediator agree on a model M? How does one do dynamic model switching to improve accuracy of the prediction process? What is the energy/performance overhead due to model maintenance and synchronization?
- **Spatiotemporal Sensor Correlation:** As sensor networks scale in size and density, there will be increasing redundancies (correlations) between different sensors [81]. This can be exploited in a principled manner to help recover from failures or provide a quick estimate of the sensor value when probing the sensor may result in violation of timeliness needs. We will develop a probabilistic approach where we model the correlations between different sensors using a probabilistic network (Bayesian net or Markov random field). Using a compact summary of the (joint) distribution of all the sensors, we will attempt to provide reasonably accurate answers to queries such as: Given that sensor X has value X, what is the most likely value of sensor Y? (to fill in missing values, for instance). As a by-product, such an approach can also detect failures. Consider the case where the readings of sensor X have changed significantly, but X is failing to communicate this for some reason. The shift in its data distribution will be reflected in other correlated sensors as well. If these sensors are still functioning, then the changes they signal, used in conjunction with the probabilistic model, will clearly indicate that communication is to be expected from X. The absence of such communication is a sign of the possible failure of X.

- **Application Aware Caching:** Caching part of the sensor data can be performed at different levels: at the sensor node, at the intermediate mediator, or at a centralized server. Cached data may be accurate enough to answer some queries, this may enable better timely results to queries. Cached data can be used to infer current sensor readings as well, which improves timeliness while avoiding the overhead of re-transmitting data from sensors. Questions that need to be answered while developing these techniques include where to cache, what to cache and how to cache (i.e., at what level of resolution).

6 Querying and Query Optimization in Sensor Networks

User tasks submitted in a high-level language appropriate to the application domain, will be mapped to appropriate data management primitives by the application software which will then be posed to the sensor database management system in a declarative (e.g., SQL-like) language. In this section, we will describe the various types of sensor queries and potential optimizations for query processing in sensor networks.

A query from a sensor application may ask for a data value produced by a sensor device at any time instant including the future, the present or the historical past. While queries about the future cannot be answered precisely at the present time by either the producer or the archiver, queries about the present can be answered by the data producer or the device in our case. For historical queries, the only choice is to answer them *approximately* at the archive since the producer discards the time series once it has sent a compressed representation to the archive. For queries seeking future data values, we seek to build evaluation strategies that allow tremendous scaling to large-scale data networks. Thus, we use an approximate representation of the actual data at the sources *within bounded quality guarantees* using a *quality-aware* client-server architecture to achieve scalable, energy and bandwidth efficient query processing. The data collection architecture as described above allows us to capture current and past data with a given quality. Applications need to be able to run on this imprecise data producing approximate results. While there are many types of queries are possible, we consider three types of queries over the imprecise data representation of particular interest to sensor applications: continuous monitoring queries, general purpose aggregate and SQL queries and monitoring applications.

Continuous queries for monitoring applications: Continuous queries are those in which a query runs periodically over a time period (e.g., summary of the traffic information on a freeway every 5 min). Such queries may be continuous aggregate queries with different precision bounds on their answers. Different queries will involve different sets of source/sensor data with possible overlaps. If the communication cost to collect a single data on any specific source is known, it is an interesting research problem to look for the optimal precisions on each data source so that the total communication cost is minimized and all query quality requirements are satisfied as well. Again the issue is given the error threshold, how to push the computation to the sensors and exploit inter-sensor communication to minimize the net communication overhead between the sensors and the server.

An example of continuous queries is the tracking of mobile objects over a period of time. The basic approach to tracking using sensors works as follows: at any instance, a set of sensors whose sensing range the object to be tracked is located in are activated. The activated sensors communicate their readings to the tracking module periodically which fuses the sensor readings to determine the location of a mobile object at any given time. The more frequent the communication, better is the track, where quality is defined as the reciprocal of the deviation from the actual trajectory. However, it incurs higher cost. Depending upon the type of sensor used, in its simplest form, the location can be triangulated from three sensor values.

In contrast to the above, an alternative tracking framework is to explore a systematic mechanism using which the energy consumption of the tracking module can be controlled based on the application's quality/accuracy requirement. Specifically, an application may specify its willingness to tolerate a bounded inaccuracy—e.g., tolerance to within 10 meters from the target trajectory. The application tolerance can be exploited to reduce communication between sensors and the tracking module. A key aspect of this approach is the translation of application quality to measurement quality at the sensor. At any stage during tracking, each enabled sensor and the tracking module dynamically agree upon a model that predicts the object's mobility in the immediate future. Such a model can either be communicated by the tracking module to the sensor when the sensor is activated or alternatively determined by the sensor using model fitting techniques based on a set of readings. If the object adheres to the predicted model within the error bound, the communication of the sensor readings to the tracking module is avoided thereby conserving energy. Note that the correctness of the approach is independent of the efficacy of the prediction model in predicting the target motion, though the effectiveness of the strategy (in terms of reducing communication) is directly related to the model. Previous work [29] has established the feasibility of the above envisioned adaptive tracking framework on top of adaptive precision data collection mechanisms. Using even the very conservative policies to adapt the sensor precision based on application tolerance, we can get many-fold improvement in power conservation.

Answering Value-based Queries with Bounded Quality: A query may ask for a data value (or aggregation of a set of data values) produced by a device at any time instant including the future, the present or the historical past. While queries about the future cannot be answered precisely at time n (which is the present time) by either the producer or the archiver, queries about the present can be answered by the data producer or the device in our case. For historical queries, the only choice is to answer them approximately at the archive since the producer discards the time series once it has sent a compressed representation to the archive. If a query asks for a data value at a time instant, it may be answered using any of the three following evaluation strategies:

Probe: The server issues a direct request to the device for the data item at the time instant. However, this requires that the producer, or the device, maintain the samples it

has not yet sent to the archive and listen continuously on the communication channel, thus making it in appropriate for the energy-constrained wireless sensor nodes
Wait: The server may wait for the data item to arrive. However if the quality requirements of the query is more stringent than ε(precision at the sensor), then the answer would be incorrect.
Predict: A predictive model (M, θ) is stored at the archive. Thus the archive or the server predicts the future value of the device and answers the query. The use of prediction to answer queries is motivated by the communication latency between the producer and the archiver of the time series. It does away with either doing a probe or waiting for a value to e sent by a sensor. Also, the archive can answer to interested applications, within ε_{pred} estimation of time series values before these values arrive at the archive.

Answering Set-based Queries with Bounded Quality: For queries that return a set of answers, (e.g., sample points in space-time where temperature exceeds 100 degrees), one of the key issues is how to measure the quality of results. Many metrics to measure differences between two sets have been proposed in the literature (e.g., Earth movers distance, match-and compare, etc.). While such measures could be used *diagnostically* to quantify the closeness of results to the true answer (i.e., quality of the results) when the true answers are known, there is no straightforward mechanism to estimate the quality of result at the server using these measures in a *prescriptive* setting such as ours when the true answers are not known a priori. An ideal measure, for our setting is such that (1) it is easy to compute, (2) does not require advance knowledge of exact answers, and (3) can be used to support progressive improvements of results. We have recently shown that *precision* (that measures the purity of results) and *recall* (that measures the completeness of results) suitably adapted to measure the set discrepancy metrics satisfy the above requirements [82]. The key aspect of these measures that makes them suitable for our purpose is that even though precision and recall cannot be accurately computed without knowledge of the true answer set, a range that provides a lower and upper bound on them can be determined without explicit knowledge of the true answers. Thus, at any stage of query processing, a tight bound on accuracy of the returned results can be determined. If the bound is not sufficient to meet application needs, results can be suitably refined. Previous work has explored such a mechanism for a simple selection queries that select certain sensor readings based on whether or not they satisfy a specified predicate. One direction of future research is to generalize this approach to a larger class of SQL queries including join queries.
Energy-aware query processing: Since the most critical resources in WSNs are energy and communication capabilities, research efforts in query optimization have aimed to minimize energy consumption and communication loads. *In-network data aggregation* is one such technique; it relies on the observation that a sensor node consumes much more energy to communicate than process data. Here sensor nodes operate in multiple phases to process the received sensor data and then send out the aggregated information. During the first phase, a hierarchical tree structure rooted at an access point is discovered via techniques such as network clustering [83, 84]. For

example, in TAG [8], a routing tree is formed during the query dissemination phase. A node always aggregates incoming data before sending it to its parent. Approximation techniques [21, 85] have also been explored to achieve energy-efficient data aggregation by eliminating unnecessary messages. X. Yu. S. Mehrotra et al. [86] tackles issues that arise due to shared media access by exploring sensor state scheduling for data collection and aggregation within sensor networks. They first identify potential collisions relevant to aggregate monitoring in the deployed wireless environment, and then propose TDMA scheduling algorithms based on greedy heuristics to determine sensor node operation states to achieve energy-efficient and collision-free communication for data collection and control message dissemination with short makespans. Other work [87–89] organizes the queries into groups according to their attributes so that similar queries results can be collected using the same path.

Distributed query processing: is an alternative approach to processing triggered queries. Although tree-based query processing has been shown to work well for monitoring queries in general, utilizing a fixed network topology for triggered queries is not always the best solution. SURCH (SURfing and searCHing) [90] is a fully decentralized peer-based algorithm for processing triggered queries in wireless sensor networks. SURCH combines query dissemination and processing so that a query can be partially processed on the fly in a sensor network. Partial results are delivered to the query initiator or a designated proxy for final processing. SURCH avoids the overheads of network topology construction or interaction with a server by exploiting local communication. The novelty of SURCH is fourfold. First, it is capable of processing ad hoc in-network generated queries more efficiently than existing tree-based techniques in addition to continuously monitoring queries. Second, by implementing prioritization, when only a small number of sensor nodes contribute to the query result, SURCH demands very little communication. Third, sensor workloads can be balanced to maximize the overall life-time of sensor networks through carefully designed propagation policies. Finally, SURCH has an inherent resilience to sensor failures, as it does not depend on any particular node, but works "around" failed nodes, discovering the ones that are still operational.

7 Conclusions

In this chapter, we discussed techniques and solutions for effective data management in distributed sensor networks. In particular, novel approaches to data representation, distributed sensor architectures, and query processing will trigger the development of alternative protocols/services for adaptive optimization engine that can be used over a wide range of sensor networks and applications. Sensor networks are an example of a complex technical system with multiple operational constraints and performance goals dictated by the deployed application at hand. In general, the dynamic nature of observed phenomena under varying system and network conditions implies that policies instrumented in the system should be dynamic and customizable.

Multiple system and application activities can interfere with each other when they occur concurrently in distributed sensor systems. Fundamental problems arise in the concurrent execution of multiple distributed services and protocols that manage multiple needs; further complications arise in the dynamic customization of these services and protocols. Arbitrary composition of those techniques developed for addressing each individual non-functional need will result in unsatisfactory system performance, when there are multiple needs. More research is needed to address the complexity of supporting distributed sensor data collection application under multiple conflicting requirements: timeliness, accuracy, reliability and cost-effectiveness.

A widescale and effective deployment of sensing infrastructures has the potential to impact the scientific community (disaster management, ecosystem monitoring, command and control) at large; further research in this area will help generate guidelines for the development of sensor architectures for specific disciplines. Novel challenges for the new decade include sustainability of large sensing infrastructures that provide intelligence to applications while preserving the privacy of individuals in sensed spaces.

References

1. J. Jude, et al., Sensorflock: an airborne wireless sensor network of micro-air vehicles. in *Proceedings of the 5th international conference on Embedded networked sensor systems, ACM* (2007)
2. J. Polastre, et al., Analysis of wireless sensor networks for habitat monitoring. Wireless Sensor Netw. 399–423 (2004)
3. K. Chintalapudi et al., Monitoring civil structures with a wireless sensor network. Internet Comput. IEEE **10**(2), 26–34 (2006)
4. C. Otto, et al., System architecture of a wireless body area sensor network for ubiquitous health monitoring. J. Mobile Multim. **1.4**, 307–326 (2006)
5. J. Henaut, D. Dragomirescu, R. Plana, Fpga based high date rate radio interfaces for aerospace wireless sensor systems. *Fourth International Conference on Systems, ICONS'09, IEEE,*(2009)
6. W.-A. Geoffrey et al., Deploying a wireless sensor network on an active volcano. Internet Comput. IEEE **10**(2), 18–25 (2006)
7. E.A. Lee, Cyber-physical systems-are computing foundations adequate, vol. 2. Position Paper for NSF Workshop on Cyber-Physical Systems: Research Motivation, Techniques and Roadmap (2006)
8. S. Madden, M.J. Franklin, J.M. Hellerstein, W. Hong, Tag: a tiny aggregation service for adhoc sensor networks. in *Proceedings of USENIX Symposium on Operating Systems Design and Implementation (OSDI)* (2002)
9. A. Demers, J. Gehrke, R. Rajaraman, N. Trigoni, Y. Yao, The cougar project: a work-in-progress report. in ACM SIGMOD Record **32**(4), (2003)
10. G. Ozsoyoglu, R.T. Snodgrass, Temporal and real-time databases: a survey. in *Proceeding of IEEE Transactions on Knowledge and Data Engineering (TKDE)***7**(4) (1995)
11. K. Ramamritham. Real-time databases. Int. J. Distribut. Parallel Databases **1**(2) (1996)
12. A. Tansel, J. Cliord, S. Gadia, S. Jajodia, A. Segev, R.T. Snodgrass, *Temporal Databases: Theory* (Design and Implementation, Benjamin/Cummings, 1994)
13. S. Chakravarthy, V. Krishnaprasad, E. Anwar, S.K. Kim, Composite events for active databases: semantics, contexts and detection. in *Proceedings of the International Conference on Very Large Data Bases (VLDB)* (1994)

14. A.P. Sistla, O. Wolfson. Temporal conditions and integrity constraints in active database systems. in *Proceedings of ACM International Conference on Management of Data (SIGMOD)* (1995)
15. Q. Han, I. Lazaridis, S. Mehrotra, N. Venkatasubramanian, Sensor data collection with expected reliability guarantees. in *Proceedings of IEEE International Workshop on Sensor Networks and Systems for Pervasive Computing (PerSeNS)* (2005), pp. 374–378
16. Y. Sankarasubramaniam, O.B. Akan, I.F. Akyildiz. Esrt: event-to-sink reliable transport in wireless sensor networks. in *Proceedings of ACM International Symposium on Mobile Ad Hoc Networking and Computing (MobiHoc)* (2003)
17. S.J. Park, Ra. Vedantham, R. Sivakumar, I.F. Akyildiz, A scalable approach for reliable downstream data delivery in wireless sensor networks. in *Proceedings of ACM International Symposium on Mobile Ad Hoc Networking and Computing (MobiHoc)* (2004)
18. I. Lazaridis, Q. Han, S. Mehrotra, N. Venkatasubramanian. Fault-tolerant queries over sensor data. in *Proceedings of International Conference on Management of Data (COMAD)* (2006), pp. 104–116
19. I. Hwang, Q. Han, A. Misra. MASTAQ: a middleware architecture for sensor applications with statistical quality constraints. in *Proceedings of IEEE International Workshop on Sensor Networks and Systems for Pervasive Computing (PerSeNS)*, (2005), pp. 390–395
20. Q. Han, S. Mehrotra, N. Venkatasubramanian. Energy efficient data collection in distributed sensor environments. in *Proceedings of IEEE International Conference on Distributed Computing Systems (ICDCS)* (2004), pp. 590–597
21. C. Olston, J. Jiang, J. Widom, Adaptive filters for continuous queries over distributed data streams. in *SIGMOD* (2003)
22. S. Madden, M.J. Franklin, J.M. Hellerstein, W. Hong, Tinydb: an acqusitional query processing system for sensor networks. ACM Trans. Database Syst. **30**(1), (2005)
23. S.R. Madden, M.J. Franklin, J.M. Hellerstein, .W. Hong, The design of an acquisitional query processor for sensor networks. in *Proceedings of ACM International Conference on Management of Data (SIGMOD)* (2003)
24. M. Sharaf, J. Beaver, A. Labrinidis, P.K. Chrysanthis, Balancing energy efficiency and quality of aggregate data in sensor networks. VLDB J. **13**(4), 384–403 (2004)
25. C.G. Lee, A.K. Mok, P. Kanana. Monitoring of timing constraints with confidence thresholds. in *Proceedings of IEEE Real-Time Systems Symposium (RTSS)* (2003)
26. R. Cheng, D.V. Kalashnikov, *S* (Evaluating probalistic queries over imprecise data. in ACM SIGMOD, Prabhakar, 2003)
27. S. Grumbach, P. Rigaux, L. Segoufin, Manipulating interpolated data is easier than you thought. VLDB J. 156–165 (2000)
28. R. Baeza-Yates, B. Ribeiro-Neto, *Modern Information Retrieval* (Addison Wesley, New York, 1999)
29. X. Yu, K. Niyogi, S. Mehrotra, N. Venkatasubramanian, Adaptive target tracking in sensor networks. in *CNDS* (2004)
30. Q. Han, M.B. Nguyen, S. Irani, N. Venkatasubramanian. Time-sensitive computation of aggregate functions over distributed imprecise data. in Proceedings of IEEE International Workshop on Parallel and Distributed Real-Time Systems (WPDRTS) (2004)
31. C. Olston, B.T. Loo, J. Widom. Adaptive precision setting for cached approximate values. in *Proceedings of ACM International Conference on Management of Data (SIGMOD)* (2001)
32. E. Bouillet, M. Feblowitz, Z. Liu, A. Ranganathan, A. Riabov, F. Ye, A semantics-based middleware for utilizing Heterogeneous sensor networks. DCOSS, 174–188 (2007)
33. M. Botts, OGC Implementation Specification 07–000: OpenGIS Sensor Model Language, SensorML (2012)
34. http://www.ics.uci.edu/projects/SATware/ (2012)
35. R. Cristescu, M. Vetterli, On the optimal density for real-time data gathering of spatiotemporal processes in sensor networks. in *Proceedings of IEEE International Conference on Information Processing in Sensor Networks (IPSN)* (2005)

36. A. Datta, I. Viguier, Providing real-time response, state recency and temporal consistency in databases for rapidly changing environments. Inform. Syst. **22**(4) (1997)
37. J. Considine, F. Li, G. Kollios, J. Brers, Approximate aggregation techniques for sensor databases. in *Proceedings of the IEEE International Conference on Data Engineering (ICDE)* (2004)
38. http://en.wikipedia.org/wiki/Control_system (2012)
39. B. Hore, et al. SATware: middleware for sentient spaces. in *Multimodal Surveillance: Sensors, Algorithms and Systems* (2007)
40. M. Kim, et al., A semantic framework for reconfiguration of instrumented cyber physical spaces. in *Workshop on Event-based Semantics, CPS Week* (2008)
41. V. Ronen, *Towards Adaptation in Sentient Spaces*. Dissertation (University of California, 2012)
42. E. Brewer, M. Demmer, B. Du, M. Ho, M. Kam, S. Nedevschi, J. Pal, R. Patra, S. Surana, K. Fall, The case for technology in developing regions. in IEEE Computer (2005)
43. W. Heinzelman, A. Murphy, H. Carvalho, M. Perillo, Middleware to support sensor network applications. IEEE Netw. **1**(18), 6–114 (2004)
44. G. Heineman, W. Councill, *Component-Based Software Engineering: Putting the Pieces Together* (Addison-Wesley, Reading, 2001)
45. D. Ganesan, B. Greenstein, D. Perelyubskiy, D. Estrin, *J* (An evaluation of multi-resolution search and storage in resource-constrained sensor networks. in SenSys, Heidemann, 2003)
46. S. Ratnasamy, B. Karp, S. Shenker, D. Estrin, R. Govindan, L. Yin, F. Yu, Data-centric storage in sensornets with ght, a geographic hash table. Mobile Netw. Appl. (2003)
47. K.S. Prabh, T.F. Abdelzaher, Energy-conserving data cache placement in sensor networks. ACM Trans. Sensor Netw. **1**(2) (2005)
48. P. Desnoyers, D. Ganesan, P. Shenoy, Tsar: a two tier storage architecture using interval skip graphs. in *ACM Sensys* (2005)
49. I. Lazaridis, Q. Han, X. Yu, S. Mehrotra, N. Venkatasubramanian, D. Kalashnikov, W. Yang, Quasar: Quality aware sensing architecture. ACM SIGMOD Rec. **33**(1), 26–31 (2004)
50. D. Carney, U. Cetintemel, M. Cherniack, C. Convey, S. Lee, G. Seidman, M. Stonebraker, N. Tatbul, S. Zdonik, Monitoring streams: a new class of data management applications. in *Proceedings of the 28th International Conference on Very Large Data Bases (VLDB)* (2002)
51. C. Lu, G. Xing, O. Chipara, C.L. Fok, S. Bhattacharya, A spatiotemporal query service for mobile users in sensor networks. in *Proceedings of International Conference on Distributed Computing Systems (ICDCS)* (2005)
52. Q. Han, S. Mehrotra, N. Venkatasubramanian, Application-aware integration of data collection and power management in wireless sensor networks. J. Parallel Distribut. Comput. **67**(9), 992–1006 (September 2007)
53. I. Urteaga, K. Barnhart, Q. Han, REDFLAG: a run-time, distributed, flexible, lightweight, and generic fault detection service for data-driven wireless sensor applications. Pervas. Mobile Comput. **5**(5), 432–446 (2009)
54. L. Paradis, Q. Han, A survey of fault management in wireless sensor networks. J. Netw. Syst. Manage. **15**(2), 171–190 (2007)
55. I. Lazaridis, Q. Han, S. Mehrotra, N. Venkatasubramanian, Fault-tolerant evaluation of continuous queries over sensor data. Int. J. Distribut. Sensor Netw/ **5**(4), 338–360 (2009)
56. L. Paradis, Q. Han, Tigra: timely sensor data collection using distributed graph coloring. in *Proceedings of IEEE International Conference on Pervasive Computing and Communication (PerCom)*, pp. 264–268 (2008)
57. L. Paradis, Q. Han, A data collection protocol for real-time sensor applications. Pervas. Mobile Comput. **5**(1), 369–384 (2009)
58. B. Adelberg, H. Garcia-Molina, B. Kao, Applying update streams in a soft real-time database system. in *Proceedings of ACM International Conference on Management of Data (SIGMOD)* (1995)
59. R. Zhang, C. Lu, T.F. Abdelzaher, J.A. Stankovic, Controlware: a middleware architecture for feedback control of software performance. in *Proceedings of International Conference on Distributed Computing Systems (ICDCS)*, (2002)

60. Y. Huang, R. Sloan, O. Wolfson. Divergence caching in client-server architectures.in *Proceedings of the IEEE Third International Conference on Parallel and Distributed Information Systems (PDIS)* (1994)
61. Q. Han, N. Venkatasubramanian, Autosec: an integrated middleware framework for dynamic service brokering. IEEE Distribut. Syst. Online **2**(7) (2001)
62. C. Olston, J. Widom, Best-effort cache synchronization with source cooperation. in *Proceedings of ACM International Conference on Management of Data (SIGMOD)* (2002)
63. L. Liu, C. Pu, W. Tang, Continual queries for internet scale event-driven information delivery. Proc. IEEE Trans. Knowl. Data Eng. **11**(4), 610–628 (1999)
64. J. Chen, D.J. DeWitt, F. Tian, Y. Wang. NiagaraCQ: a scalable continuous query system for internet databases. in *Proceedings of ACM International Conference on Management of Data (SIGMOD)* (2000)
65. S.R. Madden, M.J. Franklin. Fjording the stream: an architecture for queries over streaming sensor data. in *Proceedings of the IEEE International Conference on Data Engineering (ICDE)* (2002)
66. B. Babcock, S. Babu, M. Datar, R. Motwani, J. Widom, Models and issues in data stream systems. in *Proceedings of ACM Symposium on Principles of Database Systems (PODS)* (2002)
67. P. Bonnet, J.E. Gehrke, P. Seshadri, Towards sensor database systems. in *Proceedings of IEEE International Conference on Mobile Data Management (MDM)* (2001)
68. N. Tatbul, U. Cetintemel, S. Zdonik, M. Cherniack, M. Stonebraker. Load shedding in a data stream manager. in *VLDB Conference* (2003)
69. B. Babcock, M. Datar, *R* (Load shedding for aggregation queries over data streams. in Proceedings of ICDE Conference, Motwani, 2004)
70. C.Y. Wan, A.T. Campbell, L. Krishnamurthy. Psfq: a reliable transport protocol for wireless sensor networks. in *Proceedings of ACM International Workshop on Wireless Sensor Networks and Applications (WSNA)* (2002)
71. S.J. Park, Ra. Vedantham, R. Sivakumar, I.F. Akyildiz, A scalable approach for reliable downstream data delivery in wireless sensor networks. in *Proceedings of ACM International Symposium on Mobile Ad Hoc Networking and Computing (MobiHoc)* (2004)
72. S. Tilak, N.N. Abu-Ghazaleh, W. Heinzelman. Infrastructure tradeos for sensor networks. in *Proceedings of ACM International Workshop on Wireless Sensor Networks and Applications (WSNA)* (2002)
73. C.Y. Wan, S.B. Eisenman, A.T. Campbell. Coda: congestion detection and avoidance in sensor networks. in *Proceedings of the ACM Conference on Embedded Networked Sensor Systems (SenSys)*, (2003)
74. D. Ganesan, R. Govindan, S. Shenker, D. Estrin, Highly-resilient, energy-efficient multipath routing in wireless sensor networks. in *ACM Mobile Computing and Communications, Review*, vol. 1(2) (2002)
75. F. Ye, G. Zhong, S. Lu, L. Zhang. Gradient broadcast: a robust data delivery protocol for large scale sensor networks. ACM Baltzer J. Wireless Netw. **11**(2) (2003)
76. A. Deshpande, C. Guestrin, S. Madden, J. Hellerstein, W. Hong. Model Driven Approximate Query in Sensor Networks. VLDB J. 14(4) (2005).
77. A.K. Mok. Fundamental design problems of distributed systems for the hard-real-time environment. Ph.D. thesis (MIT, 1983)
78. J.A. Stankovic, M. Spuri, M.D. Natale, G.C. Buttazzo, Implications of classical scheduling results for real-time systems. IEEE Comput. **28**(6), 16–25 (1995)
79. P. Boonma, Q. Han, J. Suzuki, Leveraging biologically-inspired mobile agents supporting composite needs of reliability and timeliness in sensor applications. in Proceedings of IEEE International Conference on Frontiers in the Convergence of Bioscience and Information Technologies (FBIT) (2007), pp. 851–860
80. Q. Han, D. Hakkarinen, P. Boonma, J. Suzuki, Quality-aware sensor data collection. Int. J. Sensor Netw. **7**(3), 127–140 (2010). Special Issue on Data Quality Management in Wireless Sensor Networks

81. G. Hartl, B. Li, Infer: a bayseian inference approach towards energy efficient data collection in dense sensor networks. in *Proceedings of International Conference on Distributed Computing Systems (ICDCS)* (2005)
82. I. Lazaridis, Sharad mehrotra: capturing sensor-generated time series with quality guarantees. ICDE 429–440 (2003)
83. Y. Chen, A. Liestman, Approximating minimum size weakly-connected dominating sets for clustering mobile ad hoc networks. in *The 3rd ACM International Symposium on Mobile Ad Hoc Networking and, Computing* (2002)
84. K. Kalpakis, K. Dasgupta, P. Namjoshi. Maximum lifetime data gathering and aggregation in wireless sensor networks. in *The IEEE International Conference on Networking* (2002)
85. A. Deligiannakis, Y. Kotidis, *N* (Hierarchical in-network data aggregation with quality guarantees. EDBT, Roussopoulos, 2004)
86. X. Yu. S. Mehrotra, N. Venkatasubramanian, Sensor scheduling for aggregate monitoring in wireless sensor networks. SSDBM (2007), p. 24
87. B. Bonfils, P. Bonnet, Adaptive and decentralized operator placement for in-network query processing. in *Proceedings of IPSN International conference on information processing in sensor networks* (2003), pp. 47–62
88. S. Pattem, B. Krishnamachari, R. Govindan, The impact of spatial correlation on routing with compression in wireless sensor networks. in *Proceedings of IPSN (International conference on information processing in sensor networks)* (2004), pp. 28–35
89. M.A. Sharaf, J. Beaver, A. Labrinidis, P.K. Chrysanthis, Balancing energy efficiency and quality of aggregate data in sensor networks. VLDB J. **13**(4), 384–403 (2004)
90. X. Yu, S. Mehrotra, N. Venkatasubramanian, SURCH: distributed aggregation over wireless sensor networks. IDEAS 158–165 (2006)
91. C. Bisdikian, L.M. Kaplan, M.B. Srivastava, D.J. Thornley, D. Verma, R.I. Young, Building principles for a quality of information specification for sensor information. in *Proceedings of the 12th International Conference on, Information Fusion* (2009)
92. A. Deshpande, L. Getoor, P. Sen Book Chapter. in Graphical Models for Uncertain Data; Managing and Mining Uncertain Data, ed. by C. Aggarwal (Springer, New York, 2009)
93. A. Deshpande, C. Guestrin, *S* (Using probabilistic models for data management in acquisitional environments. in *The Proceedings of CIDR* (Madden, 2005)
94. A. Deshpande, C. Guestrin, S. Madden, W. Hong, in Exploiting correlated attributes in acquisitional query processing, ICDE (2005)

Chapter 14
Geometric Methods of Information Storage and Retrieval in Sensor Networks

Rik Sarkar

Abstract Sensor networks collect data from their environment. Locations of the sensors are an important attribute of that information and provide a context useful to understand, and use sensor data. In this chapter, we will discuss geometric ideas to organize sensor data by using their locations. We will consider distributed methods for managing queries about isolated events, queries about mobile objects, and queries for general signal fields. Location-based methods often require suitable simple geometric domains to operate, and we will discuss how they can be adapted to networks with complex shapes.

1 Introduction

The data collected by a sensor has to be available to others, since the sensor that produces a piece of datum is not always the one that uses it. The consumer may be far from the source of data and have no idea of how to find the one relevant source in a large network. A similar problem arises when the consumer asks for aggregate information, for example: sum or average or maximum, which need contribution a large group of sensors—at a large communication cost.

Methods of data pre-processing distribute hints about sensor information across the network. When done properly, pre-processing can make it easier to answer consumer queries—search and aggregation—and avoid the large costs.

Locations are a useful tool in pre-processing and an important aspect of sensor data. Physical events are naturally associated with coordinates instead of ids: the location of a fire alarm is critical, the sensor identity is not. Locations let us associate events with the physical world—they give us an *index* of the environment. As a result, locations are essential to *Cyber Physical Information.* Interpreting and processing

R. Sarkar (✉)
Department of Informatics, The University of Edinburgh, Edinburgh, UK
e-mail: rsarkar@inf.ed.ac.uk

H. M. Ammari (ed.), *The Art of Wireless Sensor Networks*,
Signals and Communication Technology, DOI: 10.1007/978-3-642-40009-4_14,
© Springer-Verlag Berlin Heidelberg 2014

data by locations gives a geometric structure to the otherwise amorphous sensor readings. As an added bonus, basic network operations of routing, communication, scheduling and many others can benefit from use of locations. Since wireless communication works only between nearby nodes, their locations have a close relation to the overall structure of a network.

In this chapter, we will examine how geometric ideas can leverage the locations of nodes for better utilization of sensors. We will see that geometry plays a role in processing the sensor data as well as in answering queries.

These methods can be divided into two categories by their applicability. In Sect. 2, we discuss ways of managing data in a static scenario where sensors measure general physical quantities or events such as temperature, pressure or occurrence of fire. In Sect. 3 we will consider methods for processing data about mobile objects. With the increasing popularity of mobile devices, this is an important sensing category, and carries interesting relations to the general case. In each of Sects. 2 and 3, we will consider two different styles of data organization— hierarchic data structures, and data distribution in a flat structure. In Sect. 4, we will discuss how these methods are adapted to complex networks by construction of virtual coordinates and segmentation of networks.

Our goal is to keep the methods as general as possible, so that they are applicable to the widest variety of scenarios. Therefore we will treat our network as general communication-capable sensors distributed in a plane, without any assumption on their specific sensing capabilities or other features. Nonetheless, we will also mention relevant example applications in each case to illustrate the methods and their uses.

In the rest of this section we discuss the general model and scenario used to describe the different methods. Locations and distributions of nodes are important to storage schemes, as is routing. In the following let us briefly review the aspects of these topics that will be relevant to our main discussion. Readers generally familiar with location-based algorithms and routings methods may wish to skip ahead to the next section.

1.1 Distribution and Location of Nodes

Finding locations of nodes is a challenge on its own and much research has been devoted to it. The methods and protocols vary by the requirements and the infrastructure available, and the theoretical questions related to localizing nodes in reasonable models are often intractable problems [4, 10]. For more related works on this topic see a recent survey [15]. From a practical point of view, GPS is becoming more affordable, and localizing sensors by collaborating with nearby and passing GPS enabled devices is often a possibility. Wireless and cellular signal-based localizations are also becoming and common and fairly reliable. Very accurate localization will not be important to our discussions, we therefore leave the question of localization here and assume that some form of approximate locations are available.

We do need some idea of how sensors are distributed in the domain. For our discussion, let us assume that the nodes are distributed over large area and with bounded density—the number of neighbors of any node is bounded by some constant number. This is a reasonable assumption, since in a small region we would like to have only a limited number of sensors. Too many devices in close proximity increase costs and reduce communication efficiency [16], but cannot provide corresponding sensing benefits.

For simplicity, we can discuss the performance of algorithms with respect to a specific sensor configuration. Let us suppose n sensors are distributed in a large enough square and with uniform density as above. If communication ranges of sensors are bounded above and below by some constants, this would mean that the sides of this square are of length $\Theta(\sqrt{n})$, and the diameter of the network is of the same order. The expected distance between any two random nodes in the network is also $\Theta(\sqrt{n})$.[1] We will discuss more general types of networks in Sect. 4.

1.2 Communication and Routing

To make best use of sensor networks, a sensor needs to communicate with other nodes. It needs support from the network to forward its messages. Multi-hop routing in wireless and sensor networks is a widely studied subject that we will not attempt to survey here, but will mention briefly a few concepts important to our later discussion.

Flooding. This is perhaps the simplest communication technique, where the message is sent to all neighbors of the source, and a receiver always sends it to all its own neighbors. As a result, the message reaches all nodes in the connected network, including the intended destination. Once the first message is delivered, one of the paths along which it traveled can be used for further communication between source and the destination. This is the basis of classic ad hoc routing protocols such as AODV [35] and DSR [18]. The cost of such a protocol is $\Theta(n)$ per communication pair, since the first message goes to all nodes.

Geographic routing. To make routing more efficient in sensor networks, several methods have been proposed making use of node locations. These schemes preprocess the network to compute a planar graph whose edges consist of communication links in the network (see [2]). The routing itself follows a two phase method. Suppose node s currently has a message for location t, then s uses one of the following tactics:

1. *Greedy routing*: Node s checks all its neighbors and finds neighbor w that is nearest to t. If $|wt| < |st|$, that is, w is nearer to t, then s sends the message to w.

[1] A function $f(n)$ is said to be $\Theta(g(n))$, if there are constants a, b, N such that for all $n > N$, $a \cdot g(n) \le f(n) \le b \cdot g(n)$. That is, for large enough n, $f(n)$ behaves like $g(n)$ to within constant factors. More details can be found in books on algorithms. See for example [7].

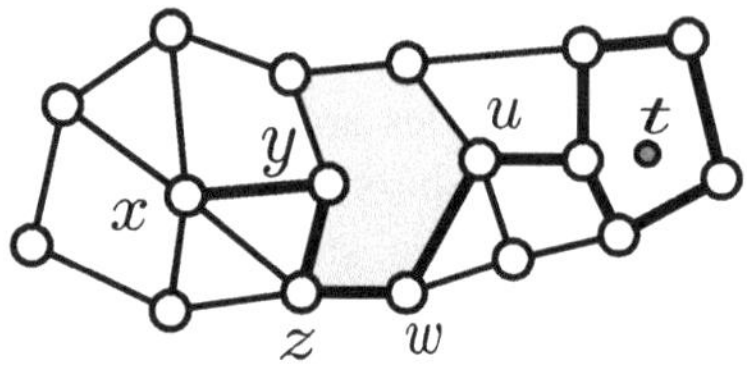

Fig. 1 Greedy routing and face routing. Message path is shown in *bold*, and goes through the following steps. A message for t starts from x and reaches y by a greedy step. Greedy step fails at y, so perimeter mode is initiated for the shaded face. Message reaches w while traversing the face. Greedy mode resumes at w since $|wt| < |yt|$. The message traverses the face containing location t looking for the node nearest to t

2. *Perimeter mode routing*: If no such w is available, the routing enters perimeter mode, where the message moves along the face of the planar graph containing s, until it finds w with $|wt| < |st|$.

See Fig. 1 for an example.

There are several methods [2, 19, 22, 25, 26], that are variations of this essential strategy. The most popular among these is GPSR [19], which we will use as our standard reference for this class of greedy plus perimeter mode routing. We will assume for simplicity that the sensors are distributed with sufficient density that there are no large "holes" in the square network. As such, we can say that if two nodes are distance d away in the plane, the GPSR path between them has a length $O(d)$—which is the communication cost between these two sensors. If there is no node at the destination location t, then GPSR traverses the face containing t, and arrives at the node closest to t.

Virtual coordinates and handling network shapes. A real sensor network will typically not be so simple. It will have unexpected shape, and will likely have coverage holes where there are no sensors. Is it still possible to apply the routing methods and the geometric data storage methods to these networks?

Protocols have been designed to compute *virtual coordinates*— assignment of a logical or virtual location to nodes in an abstract plane [5, 6, 34, 36, 39, 40, 45]. These methods morph the virtual network into more standard shapes, making it easier to use routing and geometric data storage methods. A different approach is proposed in [46, 47]—to decompose a complex shape into pieces, each of which is relatively simple, and easy to apply geometric methods.

We will discuss these methods of handling complex shapes in Sect. 4.

2 Information Brokerage and Range Queries

An important sensor network question is searching for particular pieces of information. Sensors detect and store significant *events*, and sometimes we need to find a particular type of event—for example, a tourist on safari may wish to find the elephants.[2] We have a sensor network monitoring the park, and some sensor P near a group of elephants has useful information—it has stored an "elephant detection" event. But sensor P is not aware where an interested tourist may be. Symmetrically, node C is in communication with the tourist, but has no idea where P is.

This general problem is called *Information Brokerage*: information *producer P* has some data, and information *consumer C* would like to learn this data. We need a general method by which producer and consumer can find each-other easily.

The conceptually simplest protocol to handle this problem is for the consumer to check every node in the network by flooding a query. This approach was taken in Directed Diffusion [17] and TinyDB [31]. It works well when data is updated often, but queries are occasional—a sensor stores updated information, and responds to the queries it receives. However, all other nodes in the network are also required to receive and forward the query, whether they have relevant data or not. When searches are frequent, this becomes an unnecessary load on the nodes, most of whom have nothing to do with the query.

We wish to have an information brokerage scheme that makes better use of network resources—uses less communications, and balances the load across the network. We would like to avoid consuming energy at nodes that can not help us in our search. All nodes reporting their data to a single server achieves this in a way—it avoids flooding the network, but also overloads the server and the nodes close to it. All the updates and queries have to be forwarded by the few nodes leading to the server, which will quickly run out of battery. Information should be spread out to balance the storage and communication loads in the network.

2.1 Hashing Data to Points and Curves

One method to coordinate producers and consumers is by using consistent hashing over the entire network. The idea is used in [37]. All nodes know a hash function that can be applied to the data query or key, and it returns a location. Let us denote this function as h.

The sensor P performs the hash h (*"Elephant"*), and obtains a location — a pair of coordinates: (x, y). Node P sends a message to the sensor at (x, y) that "Elephant" or relevant data is available at P's location. The consumer C performs the same hash when searching and obtains the same (x, y); thus C knows which sensor must have the information.

Location (x, y) need not be a sensor location at all, a random hash h will almost certainly give us an empty location. We can resolve this by storing the data at the sensor nearest to the location (x, y)—see Fig. 2. Finding this nearest sensor, called

---

[2] This commonly used example is from [37].

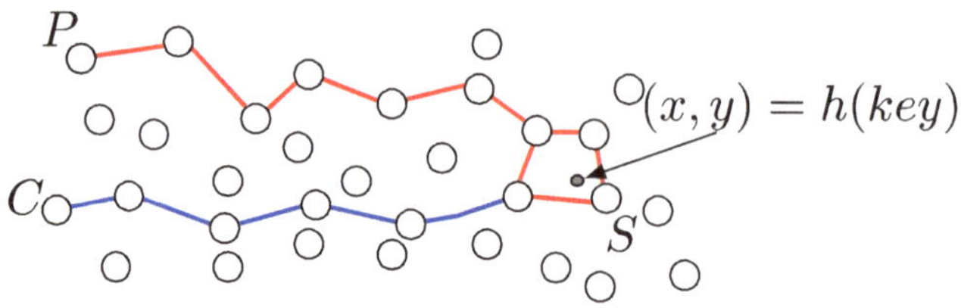

Fig. 2 Geographic hash tables. The data producer P sends a message to location $(x, y) = h(key)$, using GPSR. The trajectory is shown as the path in *red*. The data may be stored at the node S nearest to the location, or on the entire perimeter around the hash location. The consumer also performs GPSR with (x, y) as target and arrives at the data

the home node—we saw in Sect. 1.2 how this can be done using GPSR [19]. The consumer can also send a message via GPSR routing to (x, y) to arrive at this same sensor, and a return message retrieves the data. Since hash locations are completely random, the expected cost for a node—either a producer or a consumer—to send a message to the hash location is $\Theta(\sqrt{n})$. Therefore, the cost of a producer storing a piece of information at the hash location, or that of a consumer retrieving it are both asymptotically $\Theta(\sqrt{n})$ in expectation.

Storing a piece of data at only the home node is fragile—a failure of the node can destroy all the stored information. For better fault tolerance, GHT utilizes the GPSR's perimeter routing mechanism. While searching for the home node, GPSR traverses the perimeter of the face that contains the hash location, and GHT stores the data at all nodes on this home perimeter at no extra communication cost. Periodically, a probe is sent around the home perimeter, checking the health of the perimeter nodes. If some nodes have failed and the perimeter changed, then the new perimeter nodes are given a copy of the data.

The perimeter mode storage of data has an additional advantage that the consumer can get the data as soon as the search message hits some part of the perimeter curve. If we had gone a step ahead and stored the data at all the nodes on the path from the producer up to the home perimeter, and sent the consumer's search message in a path likely to touch the storage path, it could have collected the data without visiting the home region.

Can we stretch this idea further? Maybe we can deliberately send the producer's message along a path that the consumer can find. If done suitably, this method may have several additional benefits. In Fig. 2, the producer and consumer are fairly close, yet the search message has to travel to the hash location and back; which means slower response to the query, and additional load on the home nodes and those on the path. If the search found a nearby node on the producer's storage path, these unnecessary penalties would be avoided.

Our next brokerage technique, called Double Rulings [41, 43] expands this concept—data is stored along a path instead of at a node. The challenge is to design the trajectories such that the search paths find the storage paths easily.

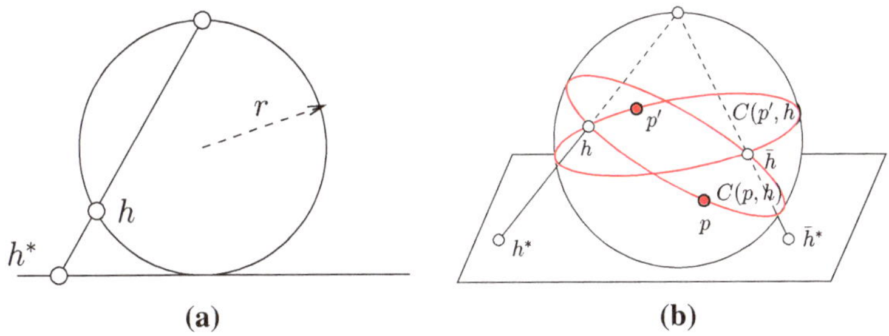

Fig. 3 **a** Stereographic projection in side-view. **b** Producer storage curves (*red*) for p and p'. They are great *circles* passing through hash location h

2.1.1 Double Rulings

Double Rulings is the general idea of storing data on a path such that the consumer's search finds it easily. Simply taking the GPSR path to the hash location does not suffice—the goal is to have the intersection of search and storage paths close to the consumer. The close intersection means less communication and faster response to the query. It will be best to have the intersections at different points on the storage path and have the load of responding to queries more evenly distributed. The specific scheme we will discuss is from [41, 43].

The intuition is to design the paths as abstract curves on a sphere. A type of curves we use are called *great circles*. These are circles that lie on the sphere and have the largest possible diameter. On the earth for example, the equator and the longitude circles are all great circles. The great circles on a sphere are analogous to straight lines on a plane—shortest distances are measured along them. Just as with the plane, given two points on a sphere, there is a unique great circle through them.[3]

The sensor network itself lies on a plane. To make use of curves on a sphere, we first need a correspondence between the sphere and the plane. This is done through a mapping called *stereographic projection*. Imagine the sphere is placed on the plane. For any point h^* on the plane, we imagine the straight line from h^* to the north pole of the sphere—its top most point. This line intersects the sphere at another unique point h, which is the image of h^* under the stereographic projection. This idea is shown in Fig. 3a.

With this map between the plane and the sphere, we can now define the storage and search curves in terms of curves on the sphere. An example of storage curves is shown in Fig. 3b where two producers are mapped to points p and p' on the sphere. In double ruling, the hash function gives a location h on the sphere. The data is stored by a producer along the great circle curve on the sphere passing through itself and h. To be precise, data is stored along nodes on the curve in the plane that is the stereographic map of the storage curve on the sphere. By properties of stereographic projection,

[3] The exception to this is the special case when the two points in question are antipodes to each-other on the sphere, and there are an infinite number of great circles passing through them.

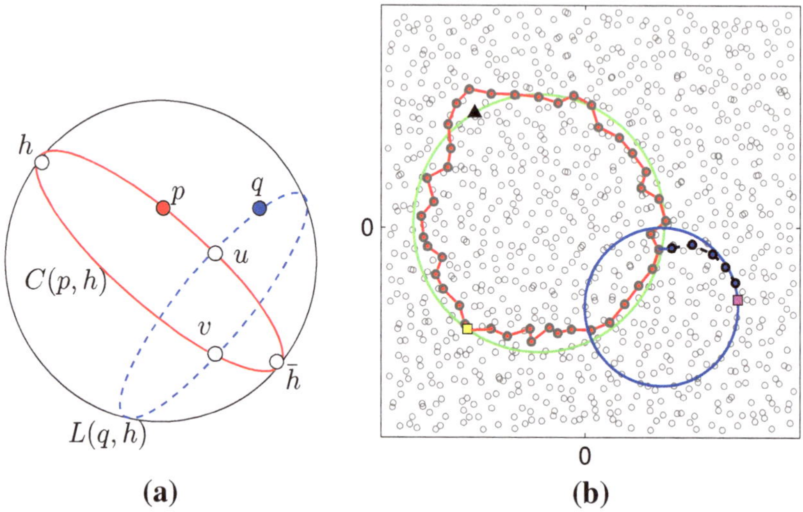

Fig. 4 **a** Double Rulings p and q are stereographic maps of producer and consumer; h is the hash, $\bar{h}$ is its antipode. $C(p, h)$ is storage curve (a great *circle*) and $L(q, h)$ is the retrieval *curve*, intersecting at u and v. **b** Actual Network storage path (*red*) and search and retrieval path (*blue*) in the plane. The hash location is shown as *black triangle*

the circular curves on the sphere map to circles in the plane. Thus the producer's cost for storing data along any circle in the network field is at most $O(\sqrt{n})$.[4]

Using curves on a sphere makes data search and retrieval easy—a great circle intersects any other great circle. Thus, the consumer doesn't even have to consider the hash h—it can simply send a message on a great circle and find any data in the network.

The knowledge of the hash function can be used to create smarter paths that have provably small search cost. We make use of *latitude curves* for this efficient search and retrieval in place of great circles. A latitude curve is defined to be one that maintains constant distance to the hash location on the sphere—see Fig. 4a.[5] The advantage of such curves is that they intersect any producer curve within a distance $O(d)$ from the consumer, where d is the distance between the consumer and the producer on the sphere. In Fig. 4a, this means that either u or v—must be close to the consumer q. The location and radius of the sphere can be chosen such that distance between any two points on the plane is at most a constant times the distance between corresponding points on the sphere, and the length of a path on the sphere is within a constant factor of the length of the corresponding path in the plane produced by the stereographic projection. Thus, the cost of search and retrieval for the consumer is

---

[4] A function $f(n)$ is said to be $O(g(n))$, if there are constants b, N such that for all $n > N$, $f(n) \leq b \cdot g(n)$. That is, for large enough n, $f(n)$ is less than $g(n)$ to within constant factors. See [7] for details. The largest possible circular arc has length $O(\sqrt{n})$ in a square of length $O(\sqrt{n})$.

[5] The name derives from latitudes on the earth, that maintain a constant distance to the poles.

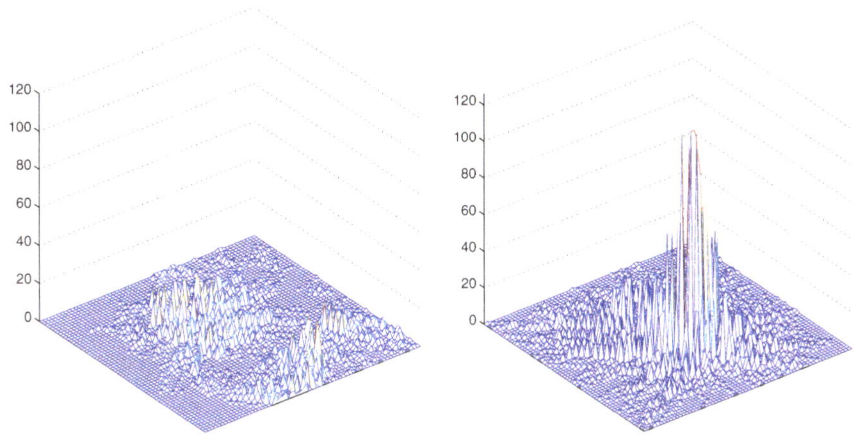

(a) (b)

Fig. 5 Communication load at different nodes. Information generated by one producer, and queries issued by 500 consumers. **a** Load for double rulings is distributed. **b** Load when using GHT is clustered around the hash location, and therefore poorly balanced

$O(d^*)$, where d^* is its distance to the producer in the plane. This property is called *distance sensitive retrieval.*

The storage and search paths in the plane are shown Fig. 4b. In a discrete network the paths cannot be the smooth curves we construct by projection. The network path following the curves as shown in the figure are obtained using a general strategy of "Routing along a curve" described in [33].

Double Rulings includes GHT as a subcase—every storage curve passes through the hash location, thus the consumer can always retrieve data from there. This is a useful feature when a consumer wants all the data of a particular type instead of just one—this can be done by visiting the hash location. On a sphere, its antipode $\bar{h}$ serves as an additional proxy hash location—any great circle passing through h also passes through $\bar{h}$, and the GHT style retrieval can be performed by visiting its image in the plane.

The great circle curves are useful for certain types of searches. For example, when the consumer asks for multiple types of data at once, a great circle retrieval can find all of them at once, since the a great circle will intersect all other great circle producer curves.

One of our goals in using in-network storage is to balance the data handling load among the nodes. Double ruling has the advantage that different consumers retrieve data from different parts of the producer curve, so that the tasks of responding to query are better distributed. This effect can be seen in Fig. 5. GHT creates high load near the hash location. The load from double rulings is more balanced. In fact, the overall load from double rulings is less, since it requires less communication per query.

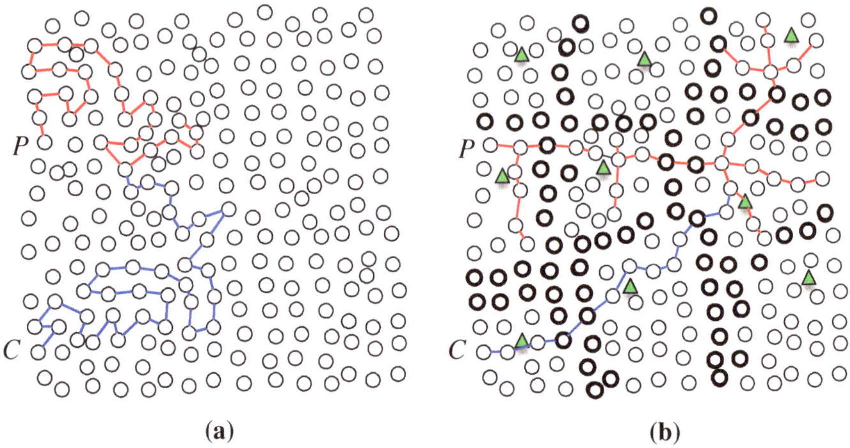

Fig. 6 **a** Rumor routing: Producer follows a random walk. Consumer also follows a random walk until it hits the producer. **b** Landmark-based double ruling. Landmark nodes are shown as green triangle. Hash to a tile instead of location, then follow GLIDER routing to reach tile. The nodes in bold are on boundary of one or more tiles

Location-free double ruling schemes. Intersection of paths to achieve information brokerage has been used based on other structures than the stereographic projection [41, 43]. We describe these methods very briefly, making use of Fig. 6. The first method, shown in Fig. 6a is called Rumor Routing, described in [3]. Here, the storage path is a random walk from the producer, and of some maximum length determined beforehand. The retrieval path is also a random walk, starting at the consumer. When the retrieval path meets the storage path, it finds the data, which is returned to the consumer.

The second method (Fig. 6b) relies on landmarks to decompose the network, and is described in [12]. A few nodes in the network are selected as *landmarks*. All other nodes determine the landmark nearest to them, and are grouped into "tiles" accordingly. This is easy to do by flooding messages from the landmarks, and using that to measure how far each node is from the different landmarks. The hash in this case is a complete tile. The producer sends the data using a related landmark-based routing scheme described in [11]. In each tile the path passes through, it shoots off additional branches so that a consumer path finds an easy intersection. The consumer uses the same hash and routing scheme, and intersects a storage node in the hash tile or an earlier one.

These methods have the advantage that they do not need locations to operate. Thus, they can be used when GPS or similar infrastructure are not present. However, when locations are available, they are substantially more expensive than the methods we discussed making use of locations. In the next subsection we return to our main discussion of location-based schemes—how they can be divided recursively for better query response.

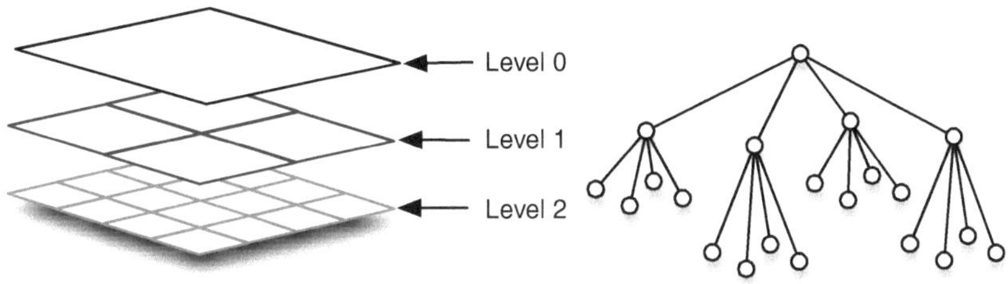

Fig. 7 A 3 level quadtree. On the left is the recursive partitioning—each level consists of one or more square areas. At the next level each square is split into 4 congruent *squares*. On the *right* is the tree—each square becomes a node, with edges to its children, and the parent

2.2 Hierarchical Partitions

Recursively partitioning a space is a common technique in data storage and search mechanisms. The reader may be familiar with the binary partitioning used in binary search on an array—where at each step the array is divided into 2 parts. By creating an abstract node for each such part, we get a corresponding abstract structure called a binary search tree, that represents the recursive partitioning of the array. To apply similar techniques to sensor nodes in a plane we need a two dimensional version of this idea.

Figure 7 shows a *quadtree partitioning*. We start with a square space at the top level, and recursively partition it into smaller congruent squares at each level. Recursive partitioning gives rise to an abstract tree structure shown on the right. Each square at each level corresponds to a node in the tree, thus each node other than the leaves has 4 children. We can interchangeably refer to a square or corresponding quadtree node as convenient. The partitioning in Fig. 7 has two levels in addition to the root, in general we can have any number of levels. It is reasonable to assume that the final level is one where the squares are unit sized. In our constant density square network model, unit sized leaves imply a quadtree with $\Theta(\log n)$ levels.

This general partitioning scheme has been used in different ways for sensor data handling. Here we briefly discuss a few of these.

2.2.1 Structured Replication in GHT

GHT [37] has a variant called structured replication (Fig. 8) designed to handle the hotspot problem of many producers trying to transmit updates to a single hash location. In this method, each node considers a quadtree partitioning of the square. On each level of the partitioning, it is possible to perform the hash on each square at that level, giving us 4^i hash location at each level i.

The producer P stores the data at the single nearest hash location among all the locations. In a k level tree storage cost is reduced to $O(\sqrt{n}/2^k)$, but the search

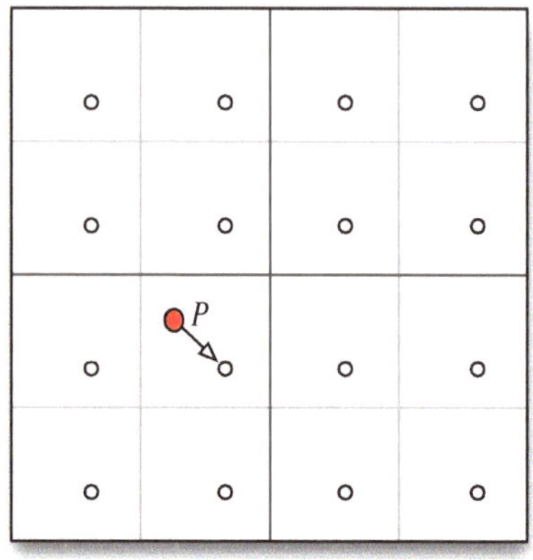

Fig. 8 Structured replication in GHT has a hash location in each square of a quadtree partitioning. A producer (shown *shaded*) stores the data at the nearest hash location among all these

cost increases—now the consumer has to search multiple locations. The protocol dictates that the consumer searches the hash at level 0, which automatically forwards the message to the level 1 hashes, each of which forwards the message to level 2 hashes and so on. Thus the search cost is $O(2^k\sqrt{n})$. This method therefore decreases storage or update cost, but increases the search cost. The authors point out that if the levels are such that at the lowest level squares are constant sized, then this costs $O(n)$—asymptotically the same flooding to retrieve data from the source node.

2.2.2 Fractional Cascading and Aggregate Information

Let us consider now a different question—answering aggregate queries using a hierarchical structure. Imagine all our sensors are monitoring a signal that has an attribute value, such as temperature. We can thus ask a question "Which sensors in region R have temperatures above T?" or, "How many sensors in region R have temperatures above T?" To answer these, we need a different type of data storage and retrieval than the information brokerage schemes.

The fractional cascading method [14] suggests storing at each sensor, aggregate data about exponentially growing regions: a constant number of values for each quadtree square that contains it. A sensor has detailed information about the local neighborhood, and progressively coarser information about larger regions.

The method starts with partitioning the square with a quadtree. Remember that a node u in the quadtree corresponds to a square in the partitioning. Its parent $p(u)$ is the larger square at the previous level that contains the square u. Fractional cascading then proceeds by storing some values at each sensor in a square. At each sensor in u, it stores the maximum in the square u, and the maxima for each sibling of u in the quadtree. That is, each node saves 4 values for each level in the tree—the quadtree nodes on the path from the leaf to the root, and their siblings (See Fig. 9). This means

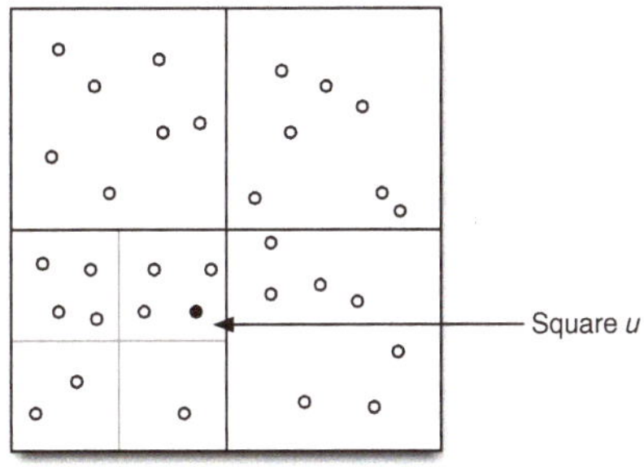

Fig. 9 The network from the point of view of the *shaded node* in quadtree square u. It stores one aggregate (for example maximum) for each square in this figure

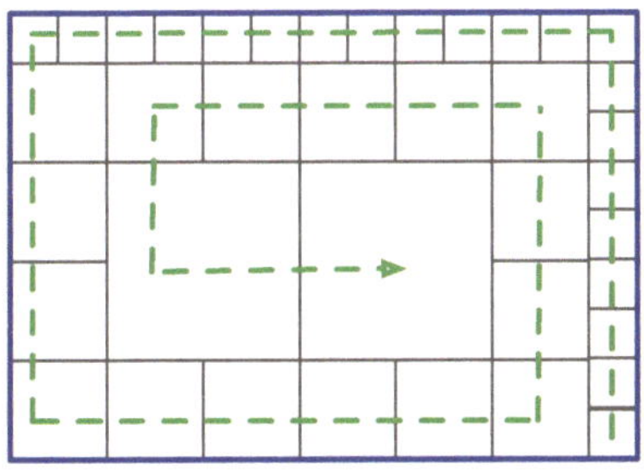

Fig. 10 Query response in fractional cascading. The query is to find an aggregate (e.g., sum of values) of all sensors in the outer rectangle R shown in bold. Each square in the figure is canonical piece—its parent square is not in R. The method needs to visit each canonical square once, this is done by following the spiral path shown as *dashed segments*, that has length $O(P \log P)$

that on a typical tree in our scenario, a node needs to store $O(\log n)$ values, and the communication cost of storing all the data in this format is $O(n \log n)$.

The query response is done in terms of *Canonical Pieces*. Given a region R, a canonical piece is a square in the partitioning that fits completely in R, but its parent does not fit—see Fig. 10. To find the true answer, we check each canonical piece. The cost of this traversal can be shown to be $O(D + \sqrt{Ak} + P \log P)$. Here A and P are the area and perimeter of R, while k is the number of sensors with temperature above T. The parameter D is the distance from the query source to the query range—it is the unavoidable cost of communicating with the query region.

Sometimes we may not be interested in a such a detailed report. We may just ask "What is the maximum temperature in R?" to find out if there is a fire in the region. On such a query, it is sufficient to check just one node from each canonical piece, since every node stores the maximum value of each square it belongs to. The traversal can be done nicely by following a spiral path in R visiting the smaller pieces at the edges first, and traversing progressively larger squares inwards. The cost of visiting all the canonical pieces will be simply $O(D + P \log P)$, taking into account the distance of R from the query point. Similarly, it is possible to compute sums, where each node stores the sum of values in its square and their siblings at all levels.

This is useful in answering a question of type "How many animals are there in region R?" and can be answered at the same cost.

Fractional cascading is a fundamental concept in computer algorithms [8], and have been used in different fields in different ways. It will make further appearances in the next section when we discuss tracking mobile devices.

2.2.3 DIM: Locality Preserving Storage of Multidimensional Data

Sensors in a network are likely to have many different sensing capabilities. And queries may be with respect to multiple parameters. Whether it is a data center or a wildlife preserve, we need to keep track of many different parameters that will help us understand animal behaviors, hardware failures and other events in the network.

For such data, it is useful to be able to make range queries: which hardware failures in the data center typically happen at high temperature and load conditions? At which locations has the bird been spotted when temperature was between 30 and 35 °C and humidity in range 80–100 %?

To answer the queries efficiently, it helps if similar data are stored close together. For example, if events at similar humidity and similar temperature are stored nearby, the cost of answering the queries above will be low.

Based on this idea, DIM [28] suggests a locality preserving hashing scheme for events. Here "locality preserving" means that it tries to place together events that are similar in some parameters.

Let us suppose for the moment that we have k different binary parameters. We can ask: "Was the temperature low or high?", "Was the humidity above or below 50 %?" and similar questions about every parameter. Thus each event is accompanied by a bit vector b of length k. To each possible bit vector, we will assign a zone—a region of the network—where the corresponding events will be stored.

This method is inspired by a different type of space partition called *kd–trees* [8]. We start with our square network region R, and partition it recursively, diving each region into two according to the next bit $b[i]$ of b. If i is even, we split R with a vertical line, and depending on $b[i]$ being 0 or 1, we choose left or right. Similarly, if i is odd, we split R with a horizontal line and depending on if $b[i]$ being 0 or 1, we choose bottom or top. Figure 11 shows an example how we can map any bit vector to a unique region of the network.

Thus, given an event whose binary properties are represented by b, we have a way to map it to a region, and events with same properties will be mapped to and stored in the same region. Observe that parameters whose values are earlier in the sequence have greater weight in determining the neighborhood of storage, creating an imbalance in the significance of different parameters.

The case where parameter values are not binary can be handled by considering the binary representation of the values. For simplicity, let us say each value is an integer in the range [0 to 2^v-1], represented by v bits. The first bit is the most significant—it determines if the value is in the range [0 to $2^{v-1}-1$] or [2^{v-1} to 2^v]. Given the first bit determining if the value is in the left or right half of the range, the second bit

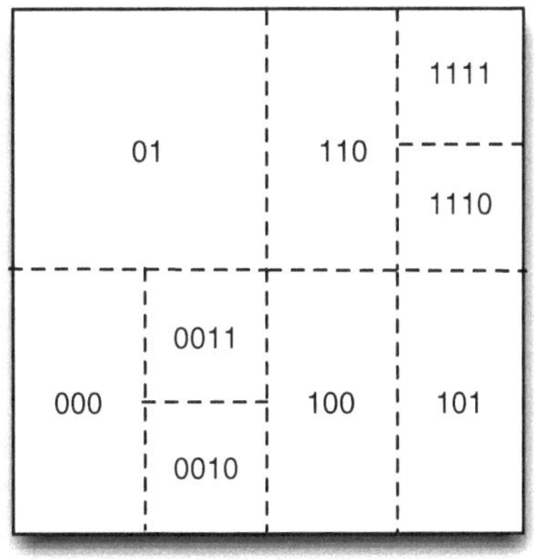

Fig. 11 Any bit vector can be assigned to a unique region in the network

determines if the value is in the lower or upper half of the reduced range, and so on. To map k such values to the network, we utilize this bit representation concept. The first k bits of b are the most significant bits if the k values, the next k bits store the second most significant bits of the values and so on.

3 Mobility Management and Tracking

Let us revisit a question we had considered in the previous section. A tourist asks "Where can I find an elephant?" We discussed some methods of brokerage that helps the tourist to find the animals of interest. These brokerage methods work well as long as the animals stay in their place, or move very rarely. What happens when the animals are active and move continuously? In such cases, methods such as GHT and Double Rulings have to continuously update the storage data—by sending messages on long paths or curves.

This is identical to a common question in mobile networks. "Where is user x?" It is important, when placing a call to user x in cellular networks. Cellular networks handle mobility by assigning to each phone a "home" server and having the phone update this server suitably. In a sensor network the corresponding strategy will be updating a hash location in GHT or sending a message along a double ruling path every time x moves, which is impractical for frequently moving targets.

The general problem of tracking and finding mobile objects is therefore a challenging topic in sensor networks. It is particularly important and difficult in the case when the tracked object is a frequently moving device such as phone, or a GPS in a fast moving car. The question of detecting a nearby target and detecting its movements and location are themselves subjects of extensive research. However, our topic of discussion in this chapter is sensor data, we will therefore focus on managing the tracking information obtained by the sensors. For simplicity, we can assume for example, that the mobile devices are GPS enabled and are willing to cooperate by communicating their true locations.

3.1 Hierarchic Tracking Data

Hierarchic data in quadtree format is relevant to tracking mobile objects as it is to tracking isolated data. The methods using hierarchic information fundamentally use the fractional cascading concept of storing more detailed information about the local neighborhood, and lower resolution data about regions farther away.

GLS: Grid location service. This method was originally described for mobile ad hoc networks in [27], but is based on the same essential ideas that we are using in sensor networks.

GLS assumes a global total ordering of n node ids in a cyclic directed list: $L = 0, 1, 2, \ldots, n-1 (\mathrm{mod}\ n)$. Suppose the node with id x belongs to a square s_i^x at level i. GLS stores x's location at the node whose id is the first after x in the sequence L among nodes in s_i^x. Then it repeats the procedure for the siblings of s_i^x and similarly stores the id of x at the successor of x in each of these squares.

You may have already noticed the similarity to fractional cascading that we saw in Sect. 2.2.2. The information in a square s_i at level i is replicated in each of its siblings. The difference is that in Sect. 2.2.2 we dealt with only the aggregate value, and every node in a s_i stored the same level i value. In GLS, there is no aggregation. The information about node x is stored at exactly one node in square s_i^x, and at one node in each sibling square.

When node y searches for x, it sends a search message to the node first in L after x for which y has location information. This node performs the same operation again, looking for x. It can be shown that this is guaranteed to reach a location server of x, which will be able to forward the message directly to x. The initial registration of x's location and updates on moving can be executed using the same basic operation. When x wants to select and update a location server in a square s, it sends an update message to s. The message starts as a search for x inside s and will find the node that should be x's location server. Thus, using the same elegant primitive, GLS handles both the fundamental operations of searches and updates.

The difficulty in GLS is that the search cost can be disproportionately large compared to the distance between x and y. When these two nodes are close but lie in different squares of the quadtree partitioning, the search may have to take a long path. The same problem can arise when x moves. A small move of x can produce a costly update. We will discuss next a method that solves this problem.

LLS: The locality aware location service. This hierarchic method [1] uses location servers at different levels of the quadtree. For a mobile node x, there is a hash location $h(x)$ at the root level of the quadtree that stores its data. Similarly, there is a hash location $h_s(x)$ for any square s at any other level of the partitioning. These locations act as location servers for x. Any other node that looks for x can get the information by communicating with a few of these servers.

At any time, if x is in square s_i^x at level i, then the location server at $h_{s_i^x}(x)$ stores location of x. In fact, it only notes a pointer to the square at level $i + 1$ that contains x. There are $O(\log n)$ different location servers—one at each level—that

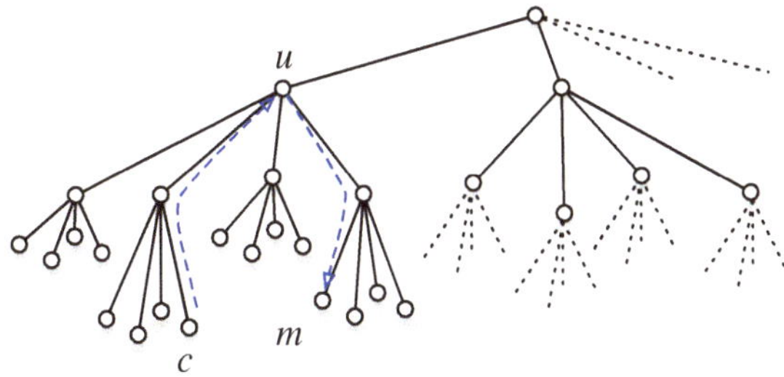

Fig. 12 Basic search in hierarchic mobility tracking. Only relevant part of quadtree is drawn. Suppose m is the lowest level square (leaf node in the quadtree) that contains the mobile user. The quadtree nodes on the path from m to the root have information about the mobile user. The consumer at node c searches ancestors of c until it hits p—the common ancestor with m. The search then proceeds down the tree to find m and the precise location of the user

carry information about x's location at any time. This is different from structured replication in GHT. In structured replication, there is a similar hierarchy of hash locations, but the data is stored at only one of them, whereas in LLS, information is stored at one server for each level. And in contrast to GLS, the server $h_{s_i^x}(x)$ at level i only records the level $i+1$ square s_{i+1}^x containing x, not its exact address. This has the advantage that when x moves anywhere inside that square, $h_{s_i^x}(x)$ does not need to be updated.

When a different node y wants to find the location of x, or to communicate with it, an inductive search is performed. The lowest level square s_j^y containing y is checked first. To be precise, the location server at $h_{s_j^y}(x)$ is asked for x's location. If this fails, the higher level server in s_{j-1}^y is checked and so on, until a server with information about x is found at some level—which may be the top level $h(x)$ in the worst case. Next, the query has to follow pointers down the levels to the leaves. An example is shown in (Fig. 12).

This basic search idea has a disadvantage that in certain cases, the search cost may be much higher than the distance between x and y —this search is not distance sensitive. See for example the case in Fig. 13: x and y are geographically close, but y has to communicate with the distant location server "1" to find x. This happens because the two nodes are separated at a high level of the partitioning, although they are quite close. The same issue can cause the update cost of x to be high when x crosses this boundary.

To make the search distance sensitive, LLS replicates x's location data at the eight level i squares neighboring s_i^x. This guarantees that the search incurs a communication cost $O(d)$ where the distance between x and y is d. This also makes it possible to keep the update costs distance sensitive. Updates are performed in a lazy manner. When x moves to one of the neighboring square, no updates are performed. When x moves out of this neighborhood, an update is performed to remove the outdated information and to enter the new information in the new neighborhood. Therefore x drags with it a sliding window of servers at

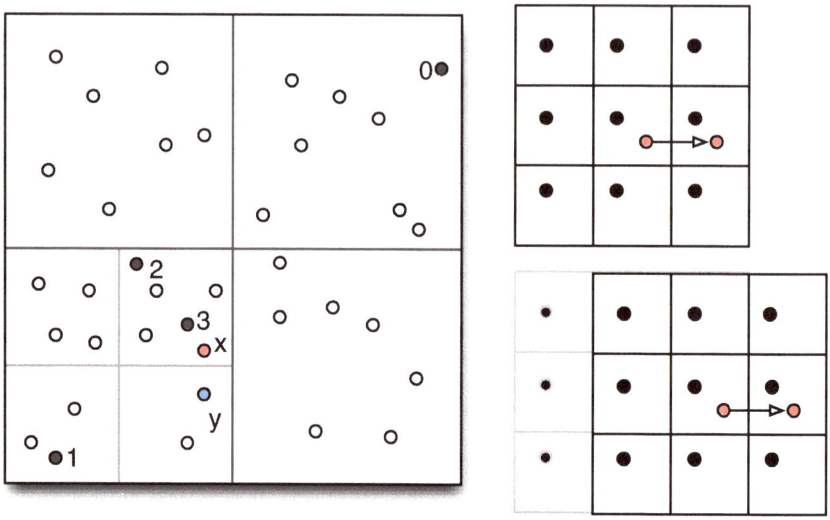

(a) (b)

Fig. 13 **a** The black shaded nodes are x's location servers. The numbers next to them are the level of the quadtree that they correspond to. y searches for x, and finds a trace of x at level 1. **b** At any level, x's location is stored at eight adjacent squares in addition to the host, When x moves to one of these squares, no updates take place. However, on the next move an update is triggered and all twelve squares are updated

each level—see Fig. 13b. The idea is to delay updates to avoid unnecessary communication. On average, if a node moves a distance d, then this scheme can be shown to have update costs of $O(d \log d)$. The cost is amortized. That means, the average cost over many moves is guaranteed to be low, but the update cost at a particular step can be arbitrarily high compared to the movement at that step.

During a search, it is possible that due to the lazy update scheme, a server claiming to have the target is in fact in error. However in such a case, the target is guaranteed to be in one of the neighboring squares. It can be shown that this does not increase the asymptotic search cost.

In the next section we will consider a different problem of aggregate queries: answering a question of type: "How many device are in an arbitrary region R of the network?" LLS, based on its information of node locations can answer the same way that range queries are answered in the fractional cascading method of Sect. 2.2.2—using the spiral traversal of canonical pieces shown in Fig. 10. While this works, the query process is not efficient since the preprocessing carried out by LLS was not designed for such questions. The algorithm in [29] is more efficient in answering such queries. There the preprocessing is based on computing a harmonic function that makes it easy to answer such aggregate questions. The computation of a harmonic function and its subsequent updates can be expensive; we will instead look at a different method for answering counting queries in the next subsection.

In summary, the utility of LLS and GLS is in search and communication with mobile nodes. The hierarchic partitioning scheme provides the benefit that the search process operates in a small neighborhood of the query origin, and goes to broader regions only when that is necessary. The updates are performed lazily—delayed as much as possible to minimize costs. Thus the location servers provide an essential service of point to point communication—they find a particular user based on the device identifier.

3.2 Differential Tracking Forms: Aggregate Tracking

Beyond the question of tracking movements of individual users, we can ask questions about aggregates of users. This is analogous to the range queries we had considered in the previous section. Now we wish to answer questions of type: "How many users are there in a region *R*?" This is useful when we wish to quickly find the traffic density in arbitrary regions of the network, or the number of people in the neighborhood of a sporting venue.

It is possible to answer the question using the location server hierarchy of LLS. We can use the spiral traversal method shown in Fig. 10, and answer the question the same way. However, there are a few aspects of LLS that we wish to improve upon:

1. **Search Costs**: While the canonical traversal is nice, we do not know where the different location servers in a canonical square may be. We would need to search all nodes in each square, and therefore all nodes in *R* to find the answer.
2. **Update Costs**: The costs of updating location servers may be high, even when a node moves a small distance. This is particularly significant when updates are very frequent, for example when tracking moving vehicles—with high speeds and large numbers.
3. **Privacy**: LLS requires tracking each device at every moment. The users may wish not to be tracked with such precision. We would like to keep the counting information, without following the precise movement of individual devices.

For this problem, we find it beneficial to leave the hierarchic fractionally cascaded data format and return to a flat data model once more. The model is based on the concept of a *differential form* in mathematics. This fundamental concept can be adapted to sensor networks by interpreting it as weights on edges of a planar graph [38]. We will address a more general problem of counting the total *weight* of targets in an arbitrary region. Counting targets is a special case of this question with each target having weight 1.

We first compute a planar graph in the network the way we discussed for routing (Sect. 1.2). This graph needs to be one such that when a target crosses an edge of this graph, one or more sensor detects this fact—for example, by an explicit update from the target itself, or by a localization carried pout by the sensors. We assign a weight

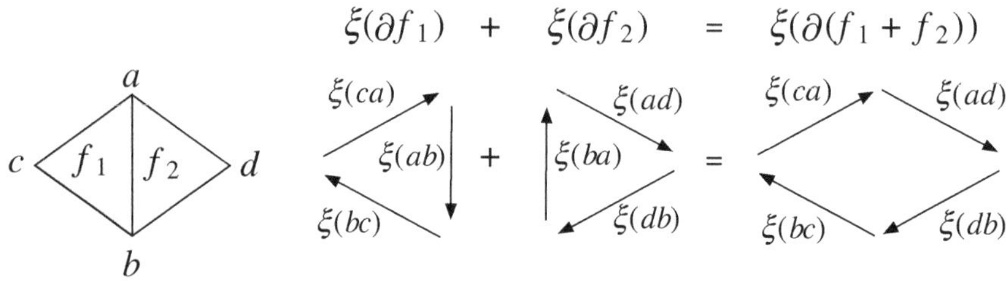

Fig. 14 The weight inside a collection of faces $f_1, f_2, \ldots$ can be obtained by simply adding the weights on the boundary of the collection. For every edge ab in the interior, weights $\xi(ab)$ and $\xi(ba)$ cancel, leaving only the outer perimeter edges

to each edge. This weight function ξ is a special one; it is constructed to have two important local properties:

1. The weight is directed: $\xi(ab) = -\xi(ba)$. That is, if a message moving from a to b sees a weight w, then a message moving from b to a sees a weight $-w$.
2. If the weight of targets in face f is w, then a message traveling along the boundary edges of f in a clockwise direction sees a total edge weight of w. A message traveling along the boundary of a face without targets sees a total weight zero.

It turns out that having these two simple properties allow very sophisticated tracking: we can find the weight of targets inside any region R just from the weights of edges on the boundary of R. For a face f, let us refer to the boundary of f traversed clockwise as ∂f. We can treat the region R to be a collection of faces, and denote its boundary by ∂R.

To obtain the total weight on faces inside R, it is sufficient to add the weights of edges on ∂R. Thus, we find the total weight inside R simply by making a clockwise tour of its boundary, without ever entering R at all! Fig. 14 shows the intuition behind this idea. For a formal proof, see [38].

When a target moves, we need to update these weights. Suppose a target of weight w moves from face f_1 to face f_2 above, crossing the edge ab, the function ξ is updated by $\xi(ab) \leftarrow \xi(ab) - w$ (or equivalently, $\xi(ba) \leftarrow \xi(ba) + w$). It is easy to check that this update reduces the weight on the boundary of f_1 by w, and increases the weight around f_2 by the same amount. The function ξ is called a *differential tracking form.*

Initializing the function ξ requires setting its value on the edges suitably. To do this, we can assume that the target arrived at its current position through some arbitrary path from the exterior, and update the edges it would have crossed on such a path (see Fig. 15). This is done in a more consistent and efficient way by constructing a dual of the planar graph and using a spanning tree of this dual to find paths for all targets. This method works at $O(n)$ communication complexity.

The differential tracking form has several nice properties. It is tolerant to coverage holes: even if a target enters a hole and is currently not in the range of any sensor, its information is stored at the boundary of the hole, and for any region that

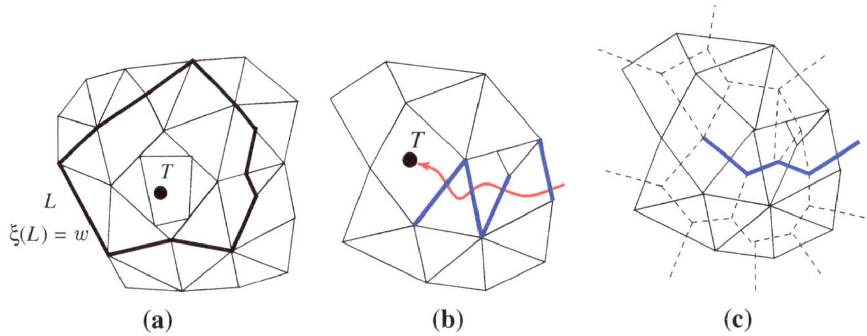

Fig. 15 **a** Suppose the single target T has weight w. Then for a loop L that contains T, $\xi(L) = w$. Such a loop is shown in *bold*. **b** A target T enters the network along the *red curve* updates a sequence of edges shown in *bold blue*. If the T is already at the position, we can imagine that T entered through some such path. **c** The trail corresponds to a path in the dual graph. The edges to be updated are precisely the duals of the edges on this path

contains the hole, this method still returns the correct answer. For the same reasons, it is also tolerant to sensor failures. The method can be made locally adaptive to insertion of new sensors and movements of the sensors themselves. The tracking is also anonymous—we update someone crossing an edge, but not the identity of the user or device. In fact, the entire protocol works without any need for identification.

The most important property, from a performance point of view, is that updates are very inexpensive. The weights on the edges can be maintained locally by the sensors that are the end points of the edge. When a target crosses the edge, this is detected locally, and this weight is modified. That is all that is needed. When a target moves a distance d, the update cost is $O(d)$—the number of edges crossed by the target. Efficient updates are critical to tracking mobile objects, since movement of devices or vehicles are generally much more frequent that queries. The completely local update has the additional advantage that it does not require participation of sensors far from the region of activity, which can happen, for example in LLS. It is therefore possible for sensors to stay in sleep mode while there is no activity in the neighborhood.

In general, it is useful to maintain different tracking forms for different categories of objects which we may want to search or count. This is particularly useful for example for the tourist looking for elephant type of query we discussed earlier, or for a traveller searching for a cab. It is shown in [38] that by repeated application of the counting query, such questions can be answered at a distance sensitive cost of $O(d)$.

Other than moving targets, tracking forms are also useful in aggregating general signals monitored by sensors. We simply need to treat the signal value at a sensor as the target weight at that location. For example, suppose we wish to find the average temperature. We need two tracking forms: one for temperature, one for node count. By finding the sum of temperatures and dividing by the number of sensors in any region, we easily find the average.

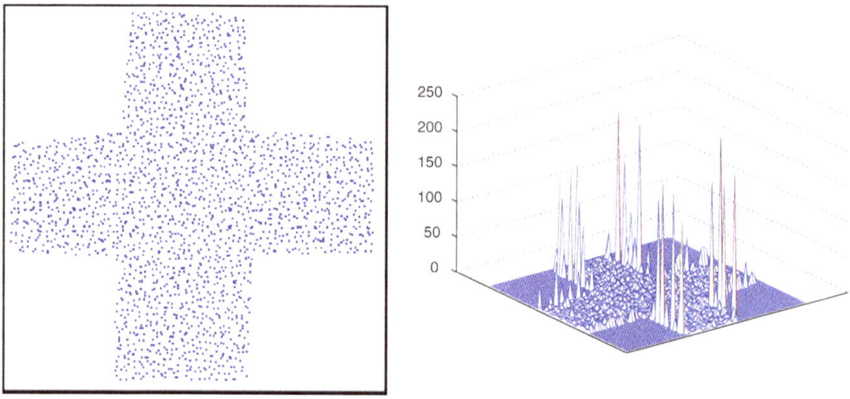

(a) (b)

Fig. 16 **a** A network whose boundary does not match the square. **b** Load distribution of DIM for a set of random data. The boundaries are overloaded while interior nodes have much lower workload. The variation in load between different boundary points comes from the design of the algorithm

On the whole, this method is very robust and flexible. The ability to compute sums of values can potentially be extended to compute other types of functions of sensor values. Exactly how to achieve such extensions and exactly what can be achieved with this method remains to be investigated.

4 Networks with Complex Shape: Segmentation and Virtual Coordinates

We had started off assuming that our sensors are in a nice square field, and we applied plane geometry without regard for obstacles. Reality may not be so accommodating. Obstacles like buildings may create *holes* in the sensor network or the perimeter may be irregular instead of a square. What happens to the hierarchic and flat structure algorithms when they encounter the boundary of such a gap in the network?

Empty regions in a sensor network are often governed by the nearest sensor. Recall what DIM [28] does: it stores data at zones determined by its attributes $(attr_1, attr_2, \ldots)$. In irregular networks the data aimed at zones in large empty regions are recorded by sensors at their boundary, and are also retrieved at the boundary. While this works, it is not the most efficient.

See for example Fig. 16. The network in Fig.16a has an irregular shape, and we apply DIM to the bounding box shown as the black square. All data hashed to the empty white regions are eventually stored at nodes nearest to these regions—at the network boundaries. These boundary nodes store a higher fraction of data than others as shown in Fig. 16b.

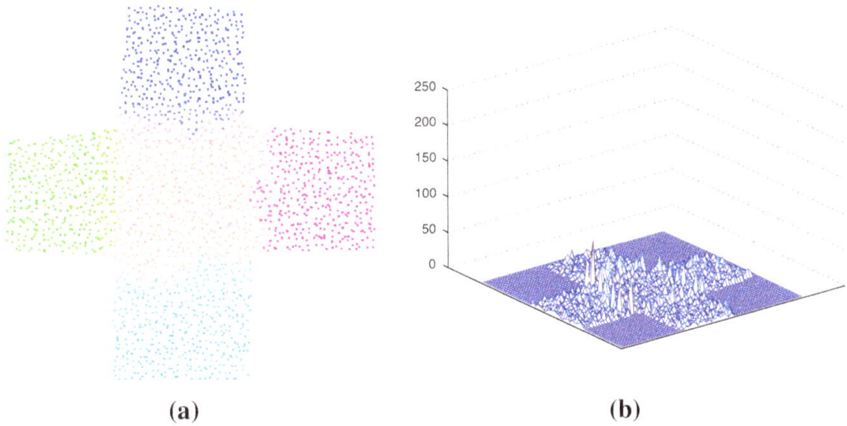

Fig. 17 **a** The network decomposed into segments forming simple shapes. Each segment is shown in a different color, along with its own bounding box. **b** Load distribution when DIM is applied to segment-wise bounding boxes instead of the global range. Bounding boxes may have partial overlaps, and those nodes have to operate for both the squares and therefore take double the load as seen above. But load balance is still better than Fig. 16

The problem is not specific to DIM. GHT and all the hierarchic data handling methods rely on "nearby" sensors in some way or the other, and therefore show similar imbalance against boundary nodes. Double ruling and other path-based methods are also not immune. A path that encounters a hole will typically move along the boundary to find a way to the destination. Routing methods like GPSR frequently move along hole boundaries in perimeter mode, and create heavy loads there. Without any special care, workload distribution will generally be heavily against boundary nodes, resulting in hotspots, delays, low efficiency, and possibly loss of packets.

The difficulties appear largely from our misplaced assumption of a simple square network. If we take the precise network shape and adjust our algorithms not to stray outside, they may operate in a balanced way. However, there are several practical and theoretical hurdles to this ideal approach. Efficiently describing a convoluted boundary and storing it at each sensor is difficult in general, and impossible for a sufficiently complicated boundary. Even if we can somehow figure out the boundaries and their descriptions, it is not clear how to adjust our algorithms. Hashes and space partitions that are natural in a square, are hard to adapt to arbitrary shapes. Instead we will discuss two other methods of handling complex shapes. The first is to decompose a network into simpler pieces and apply the algorithms independently in each piece. The second is to create virtual coordinates with simple shapes thus eliminating the problems of complicated boundaries.

4.1 Network Decomposition

The network of Fig. 16a has a natural structure made of five approximately square shapes. Fig. 17a shows a decomposition of this type. Nodes belonging to the same segment after decomposition are shown in the same color. The segmented network can now be covered by five bounding boxes— one for each segment. Instead of dividing event ranges into two parts and allocating to two parts of the network, we can now divide them into 5 parts of suitable sizes, and allocate to different segments proportional to sizes. In each segment, we follow the usual DIM method of binary space partitions to allocate events to zones in the bounding box. As a result, the allocation of zones is tight with respect to the actual sensor distribution and the imbalance at boundaries disappears—see Fig. 17b. Other data storage schemes will have similar improvements.

The network segmentation idea and the results above are from [46, 47]. The decomposition was obtained by taking the distance of nodes from the boundaries, and constructing a discrete representation of the gradient vector field of this distance. The different segments correspond to partitioning the vector field according to its different *sinks* or end points, and computed distributedly in-network. This method simply tries to divide the network into its constituent large pieces, and not specifically into squares. Smaller segments generally have tighter bounding boxes, and as a result better balanced performance.

Managing the network in suitable partitions has been considered in different scenarios. Decomposing the network in convex regions can help in routing, since simple greedy routing works well in convex shapes. This has led to the effort of decomposing the network into Greedily Routable Regions [21]. Other convex decompositions have helped in localization methods [30].

As we saw above, once the network is segmented, the data storage algorithm takes into account this split at the top level of operation and divides data into the segments. This method needs the storage algorithm to be aware of the segmentation and may be impractical and inefficient when many segments are connected in complicated ways. Our next method tries to avoid these issues by creating virtual coordinates.

4.2 Virtual Coordinates

Virtual coordinates are a simple concept. Instead of using the real locations, each sensor is assigned a different logical coordinate to simplify data processing. In our case, we want coordinates that give a simple virtual shape to a network in place of the original complex one—hopefully one that balances the load for our storage algorithms.

Figure 18b shows a type of virtual coordinates where each hole boundary of Fig. 18a is converted into a circle, and the outer boundary of the network becomes the outer circular boundary. This configuration is easier to describe than the original

Fig. 18 a Triangulated network. b Virtual coordinates with circular holes. The shape of the network is transformed by *Ricci Flow* to obtain coordinates with circular holes

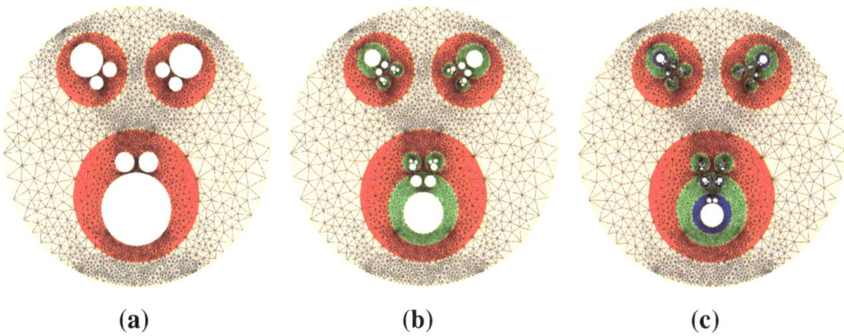

Fig. 19 The virtual coordinates can be reflected in a circular boundary. a At each reflection, a boundary goes to a *circle*. b & c The reflections can be repeated in the new circular boundaries until holes are negligibly small

one. A circle is represented by its center and radius, thus the entire network is now described by just three numbers per boundary.

The transformation of Fig. 18 is obtained by *conformal deformations* of the original network. A conformal deformation is equivalent to local scaling—either contraction or expansion—at each point in the network. *Ricci Flow* is a particular technique of repeated applications of such deformations that achieves the transformation of arbitrary boundaries to circle. Application of this method requires a triangulated planar graph. Thus the Ricci Flow algorithm described in [39] starts with a distributed method of triangulating networks. This triangulated state is visible in Fig. 18—each face of the graph, except the holes, is a triangle.

The network still has large holes and we need a method to solve the load imbalance of storage schemes. This is done by a technique called *circular reflections*. Similar to reflecting figures about a line in the plane, it is possible to perform reflections about a circle—points outside the circle are sent inside and vice versa. For each hole boundary, we perform such a reflection that creates a copy of the network inside it, see Fig. 19a. There are two nice effects of this reflection. First, it fills up much of the holes, reducing empty space, and second, the circular boundaries map to circular hole boundaries inside. This second property means that we can repeat the entire process in these new circular boundaries and reduce empty space indefinitely: Fig. 19b, c.

Finally, we can now have a network with only small and insignificant amount of empty spaces inside. For any point that was in a large empty space previously, now it is possible to find a reflected image of a node close to it. A message intended for this point goes to this node instead of the boundary, thus giving us a more balanced storage. Routing to a point inside a "hole" is simple—reflection preserves continuity at the boundary, and thus it is possible to reach the interior point by simple greedy routing. Either the storage or the routing do not require us to compute the reflected images of the nodes beforehand—they can be decided on the fly as needed. And we also do not need to consider an infinite number of reflections, a few levels of reflections suffice in most cases. See the discussions in [40] for more details of this method.

Several other methods of computing virtual coordinates exist, with different properties and applications. The reader is may find interesting ideas in hyperbolic virtual coordinates [24, 45]. However, these and most other virtual coordinates [11, 13, 32, 36] are designed for routing, and not for data storage.

5 Discussion

We hope that the reader got an impression of the variety of techniques and ideas that can be brought into play with locations as the primary index. Beyond the natural localization of sensor data by coordinates, many interesting queries are location oriented. Whether we look for the nearest cab or ask for the average temperature in a part of the city; we are interested not in the whole of network data, but only in a geometric local part of it, and our methods should be designed accordingly.

The impact of locations and geometry is almost always useful when looking for simple yet efficient methods to handle sensor data. See for example [9], which uses a gossip algorithm on sensor networks to compute averages of sensor values. Instead of exchanging information with neighbors as in a traditional gossip algorithm, here the nodes perform gossip with random locations in the network. As a result, the algorithm converges to the desired result, such as the average temperature, much faster in this method. Fractionally cascaded hierarchic information can also be computed, using a different gossip technique called spatial gossip [42, 44]. This protocol computes and stores $\log n$ aggregates at each node. For any level i of the hierarchy, a node p receives the aggregate of all nodes within a distance 2^i of p. This is analogous to the quadtree-based methods we saw earlier, except that unlike a quadtree, here the average is of a disk centered exactly at each node p. This technique in fact relies on an idea from social networks: a model for creating small world graphs [20, 23].

The concepts in processing information in sensor networks are more general than the applications to sensor networks themselves. With low power nodes and the need to process large quantities of data at minimal communication, sensor networks are a more restrictive and challenging platform than most others. A method that is suitable for sensor networks is also likely to be efficient in other networks with suitable adaptation. The approach of treating all nodes as minimal and equivalent aids this generality. Since we do not assume specific network configurations or special abilities on part of some nodes, these methods are likely to be easily adaptable. Such flexibility is important in the ever changing world of modern wireless and sensor networks.

Here we mentioned only a few works in this domain, and that too only superficially. Our goal was to introduce some elementary yet important ideas in the topic. The reader is encouraged to look into the original articles and the many other works for more details, subtleties and elegant ideas in this fast developing area.

As sensor networks and similar systems become common in the real world, our views on them will surely change. We will learn how they are deployed and used, and face new challenges. We will need to adapt our existing algorithms, and develop new models, applications and algorithms to adjust to the new networks.

With the popularity of location enabled portable devices, the tracking, storing and managing of location data are becoming more important. Since much of our data, including photos, notes and messages are now location tagged, we can develop new types of applications taking advantage of these features. At the same time, we can rethink our existing ideas and protocols for a location-aware world.

References

1. I. Abraham, D. Dolev, D. Malkhi, LLS: a locality aware location service for mobile ad hoc networks, in *DIALM-POMC '04: Proceedings of the 2004 Joint Workshop on Foundations of Mobile Computing*, (2004), pp. 75–84
2. P. Bose, P. Morin, I. Stojmenovic, J. Urrutia, Routing with guaranteed delivery in ad hoc wireless networks. Wireless Netw. **7**(6), 609–616 (2001)
3. D. Braginsky, D. Estrin, Rumor routing algorithm for sensor networks, in *Proceedings of the 1st ACM International Workshop on Wireless Sensor Networks and Applications (WSNA)*, Sept 2002, pp. 22–31
4. J. Bruck, J. Gao, A. Jiang. Localization and routing in sensor networks by local angle information, in *Proceedings of the 6th ACM International Symposium on Mobile ad hoc Networking and Computing (MobiHoc'05)*, May 2005, pp. 181–192
5. J. Bruck, J. Gao, A. Jiang, MAP: Medial axis based geometric routing in sensor networks. Wireless Netw. **13**(6), 835–853 (2007)
6. M. Caesar, M. Castro, E.B. Nightingale, G. O'Shea, A. Rowstron, Virtual ring routing: network routing inspired by DHTs, in *SIGCOMM'06*, (2006)
7. T.H. Cormen, C.E. Leiserson, R.L. Rivest, C. Stein, *Introduction to Algorithms* (MIT Press/McGraw-Hill, Massachusetts, 2001)
8. M. de Berg, M. van Kreveld, M. Overmars, O. Schwarzkopf, *Computational Geometry: Algorithms and Applications* (Springer-Verlag, Berlin, 1997)
9. A.G. Dimakis, A.D. Sarwate, M.J. Wainwright, Geographic gossip: efficient aggregation for sensor networks, in *IPSN '06: Proceedings of the 5th International Conference on Information Processing in Sensor networks*, (2006), pp. 69–76

10. T. Eren, D. Goldenberg, W. Whitley, Y. Yang, S. Morse, B. Anderson, P. Belhumeur, Rigidity, computation, and randomization of network localization, in *Proceedings of the 23rd Annual Joint Conference of the IEEE Computer and Communications Societies (INFOCOM'04)*, vol. 4, March 2004, pp. 2673–2684
11. Q. Fang, J. Gao, L. Guibas, V. de Silva, L. Zhang, GLIDER: Gradient landmark-based distributed routing for sensor networks, in *Proceedings of the 24th Conference of the IEEE Communication Society (INFOCOM)*, vol. 1, March 2005, pp. 339–350
12. Q. Fang, J. Gao, L.J. Guibas, Landmark-based information storage and retrieval in sensor network, in *The 25th Conference of the IEEE Communication Society (INFOCOM'06)*, vol. 1, April 2006, pp. 339–350
13. R. Fonesca, S. Ratnasamy, J. Zhao, C.T. Ee, D. Culler, S. Shenker, I. Stoica, Beacon vector routing: scalable point-to-point routing in wireless sensornets, in *Proceedings of the 2nd Symposium on Networked Systems Design and Implementation (NSDI)*, May 2005, pp. 329–342
14. J. Gao, L. Guibas, J. Hershberger, L. Zhang, Fractionally cascaded information in a sensor network, in *Proceedings of the 3rd International Symposium on Information Processing in Sensor Networks (IPSN'04)*, April 2004, pp. 311–319
15. J. Gao, L.J. Guibas, Geometric algorithms for sensor networks. Philos. Trans. R. Soc. A **370**, (1958), (2012), pp. 27–51
16. P. Gupta, P.R. Kumar, The capacity of wireless networks. IEEE Trans. Inf. Theory **46**(2), 388–404 (2000)
17. C. Intanagonwiwat, R. Govindan, D. Estrin, Directed diffusion: a scalable and robust communication paradigm for sensor networks, in *ACM Conference on Mobile Computing and Networking (MobiCom)*, (2000), pp. 56–67
18. D.B. Johnson, D.A. Maltz, in *Mobile Computing*, ed. by T. Imielinski, H. Korth. Dynamic Source Routing in Ad Hoc Wwireless Networks, vol. 353 (Kluwer Academic Publishers, Dordrecht, 1996)
19. B. Karp, H. Kung, GPSR: Greedy perimeter stateless routing for wireless networks, in *Proceedings of the ACM/IEEE International Conference on Mobile Computing and Networking (MobiCom)*, (2000), pp. 243–254
20. D. Kempe, J. Kleinberg, A. Demers, Spatial gossip and resource location protocols, in *STOC '01: Proceedings of the 33rd Annual ACM Symposium on Theory of Computing* (ACM Press, New York, NY, USA, 2001), pp. 163–172
21. A.-M. Kermarrec, G. Tan, Greedy geographic routing in large-scale sensor networks: a minimum network decomposition approach, in *Proceedings of the 11th ACM International Symposium on Mobile ad hoc Networking and Computing, MobiHoc '10* (ACM, New York, NY, USA, 2010), pp. 161–170
22. Y.-J. Kim, R. Govindan, B. Karp, S. Shenker, Geographic routing made practical, in *Proceedings of the 2nd USENIX/ACM Symposium on Networked System Design and Implementation (NSDI 2005)*, May 2005
23. J. Kleinberg, The small-world phenomenon: an algorithm perspective, in *STOC '00: Proceedings of the 32nd Annual ACM Symposium on Theory of Computing*, 2000, pp. 163–170
24. R. Kleinberg, Geographic routing using hyperbolic space, in *Proceedings of the 26th Conference of the IEEE Communications Society (INFOCOM'07)*, (2007), pp. 1902–1909
25. F. Kuhn, R. Wattenhofer, Y. Zhang, A. Zollinger, Geometric ad-hoc routing: of theory and practice, in *Proceedings of 22nd ACM International Symposium on the Principles of Distributed Computing (PODC)*, (2003), pp. 63–72
26. F. Kuhn, R. Wattenhofer, Y. Zhang, A. Zollinger, Geometric ad-hoc routing: of theory and practice, in *PODC '03: Proceedings of the 22nd Annual Symposium on Principles of Distributed Computing*, (2003), pp. 63–72
27. J. Li, J. Jannotti, D.S. J. De Couto, D.R. Karger, R. Morris, A scalable location service for geographic ad hoc routing, in *Proceedings of the 6th Annual International Conference on Mobile Computing and Networking, MobiCom '00*, (2000)
28. X. Li, Y.J. Kim, R. Govindan, W. Hong, Multi-dimensional range queries in sensor networks, in *Proceedings of the 1st International Conference on Embedded Networked Sensor Systems*, (2003), pp. 63–75

29. H. Lin, M. Lu, N. Milosavljević, J. Gao, L. Guibas, Composable information gradients in wireless sensor networks, in *Proceedings of the International Conference on Information Processing in Sensor Networks (IPSN'08)*, April 2008, pp. 121–132
30. W. Liu, D. Wang, H. Jiang, W. Liu, C. Wang, Approximate convex decomposition based localization in wireless sensor networks, in *INFOCOM*, (2012), pp. 1853–1861
31. S. Madden, M.J. Franklin, J.M. Hellerstein, W. Hong, TAG: a tiny aggregation service for ad-hoc sensor networks. SIGOPS Oper. Syst. Rev. **36**(SI), (2002), 131–146
32. T. Moscibroda, R. O'Dell, M. Wattenhofer, R. Wattenhofer, Virtual coordinates for ad hoc and sensor networks, in *Proceedings of the 2004 Joint Workshop on Foundations of Mobile Computing, DIALM-POMC '04* (ACM, New York, NY, USA, 2004), pp. 8–16
33. B. Nath, D. Niculescu, Routing on a curve. SIGCOMM Comput. Commun. Rev. **33**(1), (2003), 155–160
34. J. Newsome, D. Song, GEM: graph embedding for routing and data-centric storage in sensor networks without geographic information, in *SenSys '03: Proceedings of the 1st International Conference on Embedded Networked Sensor Systems*, (2003), pp. 76–88
35. C.E. Perkins, E.M. Royer, Ad hoc on-demand distance vector routing, in *Proceedings of the 2nd IEEE Workshop on Mobile Computing Systems and Applications*, (1999), pp. 90–100
36. A. Rao, C. Papadimitriou, S. Shenker, I. Stoica, Geographic routing without location information, in *Proceedings of the 9th Annual International Conference on Mobile Computing and Networking*, (2003), pp. 96–108
37. S. Ratnasamy, B. Karp, L. Yin, F. Yu, D. Estrin, R. Govindan, S. Shenker, GHT: A geographic hash table for data-centric storage in sensornets, in *Proceedings of 1st ACM Workshop on Wireless Sensor Networks ands Applications*, (2002), pp. 78–87
38. R. Sarkar, J. Gao, Differential forms for target tracking and aggregate queries in distributed networks, in *Proceedings of the 16th Annual International Conference on Mobile Computing and Networking (MOBICOM)*, (2010)
39. R. Sarkar, X. Yin, J. Gao, F. Luo, X. D. Gu, Greedy routing with guaranteed delivery using ricci flows, in *Proceedings of the 8th International Symposium on Information Processing in Sensor, Networks (IPSN'09)*, April 2009
40. R. Sarkar, W. Zeng, J. Gao, X.D. Gu, Covering space for in-network sensor data storage, in *Proceedings of the 9th International Symposium on Information Processing in Sensor, Networks (IPSN'10)*, April 2010
41. R. Sarkar, X. Zhu, J. Gao, Double rulings for information brokerage in sensor networks, in *Proceedings of the ACM/IEEE International Conference on Mobile Computing and Networking (MobiCom)*, Sept 2006, pp. 286–297
42. R. Sarkar, X. Zhu, J. Gao, Hierarchical spatial gossip for multi-resolution representations in sensor networks, in *Proceedings of the International Conference on Information Processing in Sensor, Networks (IPSN'07)*, April 2007, pp. 420–429
43. R. Sarkar, X. Zhu, J. Gao, Double rulings for information brokerage in sensor networks. IEEE/ACM Trans. Netw. **17**(6), 1902–1915 (2009)
44. R. Sarkar, X. Zhu, J. Gao, Hierarchical spatial gossip for multiresolution representations in sensor networks. ACM Trans. Sen. Netw. **8**(1), 4:1–4:24 (2011)
45. W. Zeng, R. Sarkar, F. Luo, X.D. Gu, J. Gao, Resilient routing for sensor networks using hyperbolic embedding of universal covering space, in *Proceedings of the 29th Annual IEEE Conference on Computer Communications (INFOCOM'10)*, April 2010
46. X. Zhu, R. Sarkar, J. Gao, Shape segmentation and applications in sensor networks. in *Proceedings of the 26th Conference of the IEEE Communications Society (INFOCOM'07)*, May 2007, pp. 1838–1846
47. X. Zhu, R. Sarkar, J. Gao, Segmenting a sensor field: algorithms and applications in network design. ACM Trans. Sen. Netw. **5**(2), 12:1–12:32 (2009)

Part VII
Data Gathering

Chapter 15
Data Gathering, Storage, and Post-Processing

Marcus Chang and Andreas Terzis

Abstract In this chapter we give an overview of the different components needed in an end-to-end wireless sensor network monitoring system. Using two case studies with vastly different data throughput we show selected examples of components commonly found in data collection networks and use actual deployments as motivating examples. Specifically, the Life Under Your Feet soil monitoring project focus on extreme duty-cycling and low data rate communications while the data center monitoring network RACNet emphasizes high throughput and efficient channel utilization.

1 Introduction

For monitoring applications (e.g., environmental and industrial/process monitoring) wireless sensor networks (WSNs) promise inexpensive, hands-free, low-cost, and low-impact data collection—an attractive alternative to manual data logging—in addition to providing considerably richer data.

Obviously, not all monitoring applications have the same objectives and each data collection system should be designed with the users' requirements in mind. For example, in a healthcare body sensor network (BSN) application [18] data privacy is a legal requirement which must be built into all steps of the data gathering process from the moment the sensor is sampled and until it reaches an authorized user, while such a requirement is rarely seen in environmental monitoring since these are often deployed in public areas anyway.

M. Chang · A. Terzis (✉)
Johns Hopkins University, Baltimore, MD, USA
e-mail: terzis@cs.jhu.edu

M. Chang
e-mail: mchang@cs.jhu.edu

H. M. Ammari (ed.), *The Art of Wireless Sensor Networks*,
Signals and Communication Technology, DOI: 10.1007/978-3-642-40009-4_15,

© Springer-Verlag Berlin Heidelberg 2014

We focus on data collection for industrial and scientific purposes since these particular WSNs tend to have more stringent data quality requirements than those found in more consumer oriented WSNs, meaning solutions for the former will often be usable in the latter but not vice versa.

In fact, when asking domain scientists about their requirements to the collected data the common response is that *all* raw measurements should be collected and persistently stored [3]. However, given the lossy nature of wireless links, even a reliable network protocol cannot prevent data losses since the nodes' resources, such as buffers and energy reserves, have a finite capacity. Therefore, in order to minimize the risk of data loss the data gathering protocols and algorithms must be chosen to suit the observed phenomena and the actual network deployment. For example, for battery driven nodes deployed in yearlong environmental monitoring projects the network protocol have to be both fast enough to sustain the data generated from the network (with minimal risk for buffer overflows) and power efficient enough to last the entire duration of the project.

In this chapter we will mainly focus on gathering data from sensor networks in the two opposite ends of the data generation spectrum, and orthogonal to this, we will focus on the large and dense networks from environmental and industrial monitoring (as opposed to smaller networks such as BSN), as described in Sect. 2.

Furthermore, the raw measurements need to be precisely timed and calibrated, to give scientists and plant managers high confidence that measured variations reflect actual changes in the underlying processes rather than random noise, systematic errors, or drift. Hence, in Sect. 3 we will explore timestamping methods specifically tailored for monitoring applications where data is analyzed offline, independent of when and where the data was actually collected, as opposed to sensor network applications that do in-network processing and actuation.

Last, for environmental monitoring, given the communal nature of field measurement locations, other scientists might use the data in ways unforeseen when the original measurements were taken. Generally speaking, techniques that distill measurements for a specific purpose potentially discard data that are important for future studies. Hence the need for persistence storage, even after the data has been collected from the sensor network, which we present an example of in Sect. 4.

To illustrate the effect different system requirements have on the design and implementation we will use two vastly different scientific and industrial monitoring applications as case studies for the rest of this chapter, namely the soil monitoring application *Life Under Your Feet* and the data center monitoring application RACNet , both presented below.

1.1 Soil Monitoring

Figure 1 depicts the overall architecture of the system that we developed to monitor soil abiotic factors. This system has operated since the Fall of 2005 in multiple deployments with durations ranging from a few weeks to 2–3 years at seven different

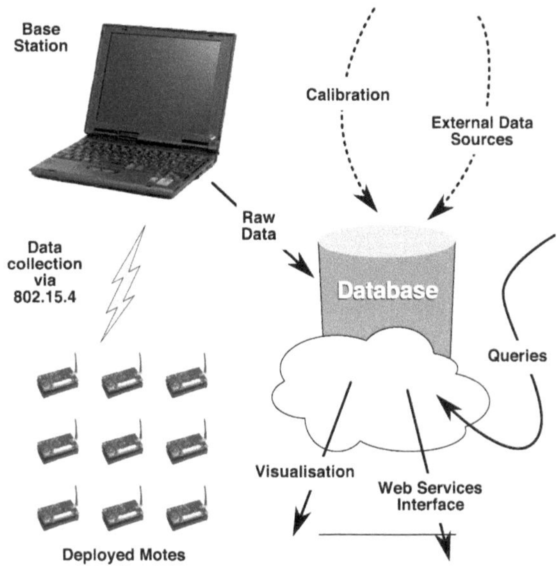

Fig. 1 Architecture of the end-to-end Life Under Your Feet data collection system

locations, across two continents and biospheres ranging from an urban forest at sea level, to a high-altitude desert, and moist South-American rain forests.

Using TelosB motes running TinyOS, we equipped the nodes with EC-5 Decagon soil moisture sensors, whose resistance varies with soil moisture, and soil thermistors whose resistance varies with temperature. We chose the Decagon soil moisture sensor because of its high accuracy and reliability.

Soil conditions are measured by each of the nodes deployed over the covered area. The collected measurements are stored on the nodes' local flash memory and are periodically retrieved by a base station over single- or multi-hop wireless links. Once the base station successfully retrieves the raw measurements, it inserts them into a SQL database. At this point, raw measurements are calibrated using sensor-specific calibration tables and are cross-correlated with data from external data sources (i.e., data from the weather service). The database acts not only as a repository for collected data, but also drives visualization tools and provides access to the data through SQL-query and Web Services interfaces.

1.2 Data Center Monitoring

RACNet [27] is a large-scale sensor network for high-fidelity data center environmental monitoring. RACNet uses custom-made *Genomote* sensor nodes (similar to TelosB) that employ a combination of wired and wireless communications to scale. A wireless master node and several wired sensors form a wired daisy chain to cover one side of a rack, collecting data at different heights. This design increases sensing

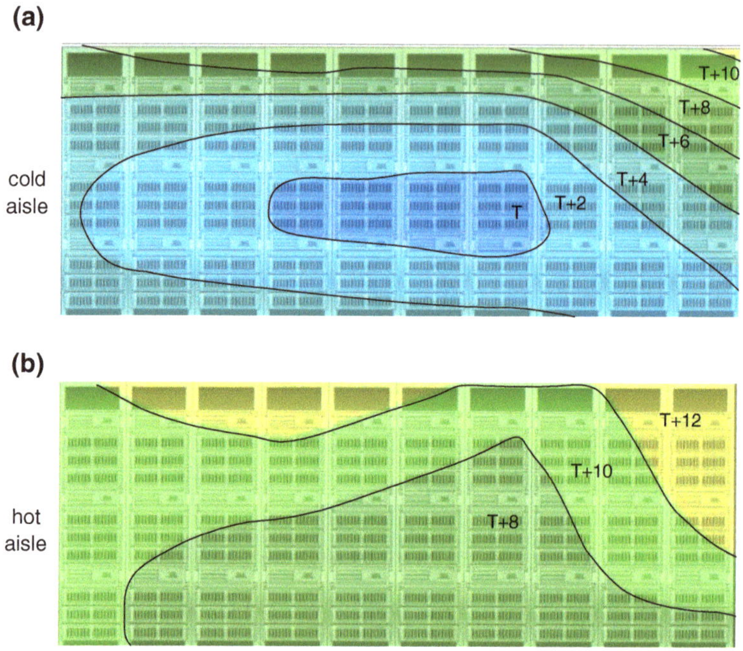

Fig. 2 Temperature distribution over the front (cold aisle) and back (hot aisle) of a row of 10 racks in RACNet . Significant spatial variation demands dense networks of temperature sensors

coverage and reduces the number of contending radios in the same space, without sacrificing deployment flexibility.

RACNet has been deployed in several data centers, including a production deployment at a 12,000 sq-ft. facility comprising 696 Genomotes, whereas 174 of these were wireless Genomotes . The system has been running since mid-2008, collecting more than 2.5 million measurement records per day. Each Genomote chain collects the following measurements every 30 s: three temperature readings collected at different rack heights, one humidity measurement, and a measurement indicating the availability of the USB power for network monitoring purposes. Figure 2 presents heat maps generated from 24 sensors located at the front and back of a row of 10 server racks.

2 Collection

Although WSN protocols have matured to a point where data collection protocols, such as the Collection Tree Protocol (CTP) [13] described below, and even the more general purpose IETF 6LoWPAN/RPL protocol can be used out-of-the-box and still perform well in terms of low duty-cycle and high data delivery rate for a wide range of networks, more challenging applications can benefit from more specialized network protocols.

For example, for extremely duty-cycled networks where data is only collected once per week or month due to lack of infrastructure, which often happens in environmental monitoring, the control packets in CTP would dominate the network traffic between collection rounds. Similarly, for dense networks with high data throughput the lack of channel diversity in CTP will increase interference.

In this section we will explore two diametrically opposite data collection protocols, namely Koala, specifically designed for infrequent burst transfers in deployments where data is collected sparingly and where network connectivity between collection rounds is unnecessary, and the Wireless Reliable Acquisition Protocol (WRAP), designed for high data rate and dense networks where throughput has higher priority than power consumption.

Related Low-Power Collection Protocols

Data collection has been thoroughly studied in the sensor network literature. A large portion of the work in this area focuses on the power aspect of the problem, aiming to minimize energy consumption through data aggregation (e.g., [30]), ultra-low duty-cycles (e.g., [2]), or optimal sensor placement (e.g., [12]).

For long-term data gathering applications, such as environmental monitoring, routing protocols need to support low duty-cycles ($<1\,\%$), reliably deliver collected measurements, operate unattended for long periods of time, and support tens to hundreds of nodes per gateway, deployed in sparse topologies. In response to these requirements, some WSN networking stacks employ techniques to coordinate the nodes' sleep schedules and maintain states at each of the network's nodes (e.g., routing entries, link quality information, etc.). These protocols synchronize the sleep schedules of neighboring nodes, collect and disseminate link quality information, and calculate routes through distributed min-cost routing algorithms.

Generally speaking, routing can be divided into the *data* and *control planes*. The first includes all the components necessary to forward packets along a multi-hop path. To do so, it relies on a forwarding table whose entries list the next-hop(s) on the path towards a particular destination. This forwarding table is maintained by the control plane which includes routing protocols that discover and select network paths.

The following list data collection protocols are some of the most prominent in the WSN literature:

Dozer [2] is the first data gathering protocol that achieves permille ($\sim 0.1\,\%$) duty-cycles by synchronizing sleep schedules and require nodes to persistently maintain routing trees. This synchronization also allows Dozer to continuously inform the gateway about the network's health and deliver the measurements with low latency.

Werner-Allen et al. proposed a request-reply collection protocol called Fetch in the context of their volcano monitoring project [47]. The base station first floods the network with the request, which triggers the target node to return the data. Since data collection occurs infrequently, Fetch does not maintain a dissemination topology.

Flush is a reliable, single-flow transport protocol for bulk downloads in sensor networks [17]. Flush supports flow control but lacks mechanisms for network-wide wake ups. Lance is a data driven collection protocol that schedules downloads based on the value of the data and the cost of delivery (e.g., energy) [46]. Meliou et al. introduced the concept of data gathering tours, whereby a network's gateway gathers data from a subset of the network's nodes [32]. To do so the gateway calculates a source route that visits all the nodes in the tour.

CTP is a best-effort data collection protocol that relies on hop-by-hop retransmissions to reduce packet loss [13]. It has been deployed in large WSNs with hundreds of nodes [28]. While CTP does not support low-power operation by itself, it can be combined with Low Power Listening (LPL) [1, 37] which power cycles the nodes' radios to achieve low duty-cycles.

In *Life Under Your Feet* we developed the Koala data collection protocol [35], which decouples the control and data planes at the node level, by implementing a *network-wide routing control plane*. The majority of the functionality of this network-wide plane is implemented at a centralized location (e.g., a gateway) using information collected by the network's nodes. This information is then used to calculate and disseminate the end-to-end paths that nodes will use.

Since experience has shown that implementing complex networking protocols on nodes can lead to unexpected failures in the field [19, 43, 45] decoupling the control and data plane should increase robustness. Compared to routing protocols implemented at the node level, this approach provides multiple benefits in addition to reduced complexity. First, nodes do not incur the overhead of persistently maintaining routing state. Moreover, the network-wide view provides the ability to perform other optimizations. For example, a node can establish two disjoint paths to the same destination to improve reliability and/or load balancing.

Related High-Throughput Collection Protocols

Unlike outdoor environmental monitoring deployments, in which sensors are sparsely deployed and power is the primary concern, industrial monitoring has distinctly different trade-offs. First, power consumption is no longer the determining factor. Instead, performance issues such as data yield and latency become critical. Second, to monitor large industrial installations at fine spatial granularities, large and dense sensor deployments are necessary. In turn, this dramatic increase in scale leads to solutions that are qualitatively different from sparse outdoor deployments.

Several companies have started to offer wireless sensor networks for industrial monitoring. Among them, Federspiel Controls [10] uses OEM sensors from Dust Networks, which incorporate a frequency-hopping protocol called Time Synchronized Mesh Protocol (TSMP) [7]. A TSMP network can support up to 255 nodes with a fixed TDMA schedule. Unfortunately, no results on the performance of TSMP are publicly available. SynapSense [42] provides the LiveImaging solution for monitoring data center environment conditions. Little information is known on the networking details of LiveImaging. To the best of our knowledge, LiveImaging supports only 5 min sampling intervals and does not support multiple frequency channels. Both solutions use battery powered sensors, which limit their sampling rate and system lifetime.

A number of multi-channel protocols have been proposed to address the challenges associated with high densities in sensor networks. First, several general multi-channel MAC protocols [21, 52] assign nearby nodes to different channels to improve spatial reuse. The frequent channel switching required in such node-based channel assignment protocols can generate large overhead.

Work from Le et al. [20] and Wu et al. [49], uses channel assignment strategies and results from control theory to achieve load balancing among different trees. However, neither of the two approaches offer reliable data delivery.

For RACNet we developed the *Wireless Reliable Acquisition Protocol* (WRAP). Like many data collection protocols, WRAP has a network layer that controls the topology and a transport layer for data retrieval. Nevertheless, WRAP is unique in the way it combines centralized and distributed decision making to achieve scalability and responsiveness. Specifically, the network layer constructs collection trees across multiple channels in a distributed way. On the other hand, the transport layer relies on a centralized token passing mechanism to prevent network congestion and reliably retrieve data from each of the network's nodes. Note that WRAP takes advantage of the energy supply from server USB ports, and does not exercise duty-cycling.

Next, we will present Koala and WRAP in detail.

2.1 Koala: Low-Power Data Collection

Koala [34] is a system for reliably downloading bulk data that targets data gathering applications with no real-time requirements. Koala uses the *Flexible Control Protocol (FCP)*, a signaling protocol we developed to install routing paths on the network's nodes. Koala uses FCP to create the multi-hop paths over which a WSN gateway downloads data from the nodes. FCP supports ephemeral paths that transmit a single datagram and persistent paths that persist until explicitly torn down. Both paths can offer reliable transfers. Koala uses Low Power Probing (LPP), an efficient technique to wake up the network's nodes before a download occurs and leverages the availability of multiple channels in IEEE 802.15.4 radios to perform data downloads over different channels, thereby minimizing overhearing costs.

Figure 3 represents the position of the Flexible Control Protocol (FCP) relative to other protocols in the TinyOS networking stack. Specifically, FCP sits directly above the Active Message layer which offers the ability to send unicast and broadcast messages to nodes within the same broadcast domain. In turn, FCP provides upper layers the ability to send one or more messages across multi-hop paths with or without end-to-end reliability.

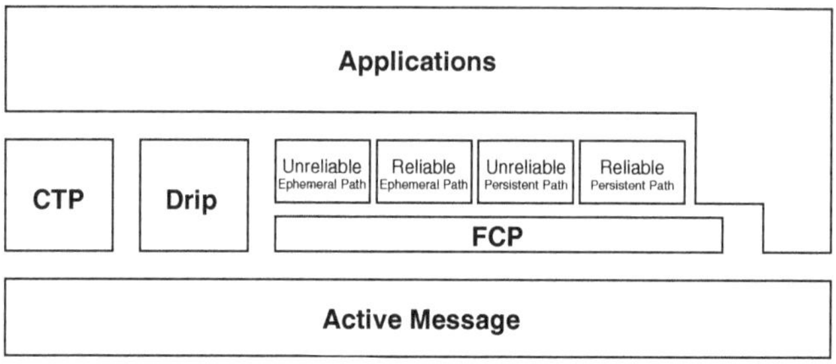

Fig. 3 The Flexible Control Protocol (FCP) used in Koala and its relations to other protocols in the TinyOS network protocol architecture

Because nodes do not keep any persistent routing state, a network path[1] must be established before it can be used to carry traffic. For this reason, FCP includes a path establishment phase that installs entries on the path tables of each of the nodes on the path. Nodes subsequently use these entries to forward packets, until they are explicitly or implicitly removed, thus relinquishing allocated resources.

Depending on whether paths are used to transmit one or multiple data packets, FCP provides *ephemeral* and *persistent* network paths.

Ephemeral Network Path. This service is equivalent to a source route since the data packet carries the network path it should follow in addition to the application's data. Intermediate nodes do not establish any state but rather forward the packet based on the path encoded in it. This service is useful when a node wants to send a short message, such as a command, to another node in the network. It has the advantage of not incurring the delay and the overhead associated with establishing the path. On the other hand, the maximum amount of application data that an ephemeral path can carry is limited to a single radio packet, minus the space necessary to store the path information.

Persistent Network Path. Unlike ephemeral paths, persistent network paths must be established before they can be used. This establishment phase requires intermediate nodes to allocate entries on their path tables. These entries are used to forward subsequent packets that do not carry source routes. At the end of a successful establishment phase the nodes at both ends of the path receive a path identifier that they use to forward traffic in both directions.

Both path types can offer end-to-end reliability, meaning that the destination will generate acknowledgments for each of the packets it receives. Moreover, intermediate FCP nodes will attempt to deliver the packets up to a maximum number of times and notify the sender if the path is no longer available.

[1] FCP network paths are analogous to virtual circuits or MPLS label-switched paths (LSPs) but with no QoS attributes associated to them.

2.1.1 Network Wide Wake-up

Because current node radios consume as much energy in idle listening mode as when they transmit or receive [44], nodes must maximize the time they keep their radios turned off. This means that a mechanism is necessary to wake up the network prior to a download operation. One potential solution would be for nodes to keep synchronized sleep schedules as in [2, 50]. Doing so would however require nodes to maintain persistent network state (i.e., their neighbors' sleep schedules) which contradicts our philosophy of simplifying node-level networking code.

Low Power Listening (LPL), in which nodes periodically sample the channel for signs of activity and transmitters send long preambles to generate such activity, offers an appealing alternative for waking up non-synchronized nodes. While initially presented in the context of bit-stream radios [37], LPL has been adopted to packet-based radios, in which case the preambles consist of a continuous stream of packets [1].

LPL however was designed for waking up individual nodes rather than the whole network, as Koala requires. While using LPL in broadcast mode is possible, doing so requires transmission of maximum length preambles, leading to packet storms that impede the collection of neighborhood data (described next). The underlying reason is that, unlike the unicast case, the sender does not know when all intended receivers have woken up. For this reason, it cannot terminate the preamble's transmission early. False negatives, situations in which a node fails to correctly detect a preamble, represent an even bigger threat. While not a significant issue in the unicast case—a false negative can be detected due to the lack of an acknowledgment, thus scheduling a retransmission at the sender—missed detections can cause nodes to completely miss the opportunity to wake up and participate in a download.

Low Power Probing (LPP), a technique we developed, addresses these problems by replacing passive channel probing at the receiver with active probing. Specifically, nodes periodically broadcast short packets requesting acknowledgments. If such an acknowledgment is received, the node wakes up and starts acknowledging other nodes' probes, otherwise it goes back to sleep. Figure 4 provides a graphical representation of LPL and LPP. LPP replaces Clear Channel Assessment (CCA) samples at the receiver with transmissions of short packets. In turn, this obviates the need for long preambles thus reducing the level of contention on the radio channel.

Algorithm 1 presents LPP in pseudo-code. Enabling and disabling of acknowledgments is necessary to avoid false positives when the probes of two or more nodes cause them to wake each other up by mistake. When the SLEEP() procedure returns, the node keeps its radio on until the next time the procedure is called. This procedure executes only at the network's nodes. The wake up operation is initiated by a gateway which enables its radio's acknowledgments and starts listening for probes from the network's nodes.

Algorithm 1 Lower Power Probing

```
procedure SLEEP(interval)
  loop
    TURNRADIOOFF()
    DEEPSLEEP(interval)
    TURNRADIOONWITHACKDISABLED()
    r ← SENDPROBE()
    if WASACKED(r) then
      ENABLERADIOACKS()
      return
```

2.1.2 Neighborhood Discovery

While the gateway selects the routes in Koala, its decisions are driven by information that the network's nodes collect. Specifically, once awake, each node collects its neighborhood by recording the identities of its neighbors as well as the quality of its links from these neighbors, defined as the Received Signal Strength Indicator (RSSI) of the received packets. These RSSI values are collected from the wake up probes (and the acknowledgments to these probes) received by the node's neighbors. Furthermore, to accelerate the neighborhood collection process, nodes send periodic beacons which are also acknowledged, generating bi-directional link information.[2]

We require two properties from the beaconing scheme: to generate a bounded amount of traffic overhead, independent of node density, and to be fair. To achieve these properties, nodes select their beaconing intervals from an exponential distribution and suppress their transmission if they receive a beacon before their timer expires. The memoryless property of the exponential distribution ensures fairness, while suppression limits the total number of beacons. Generating an exponential distribution from the uniform distribution requires computing $\log(x)$ with $x \in [0, 1]$. In

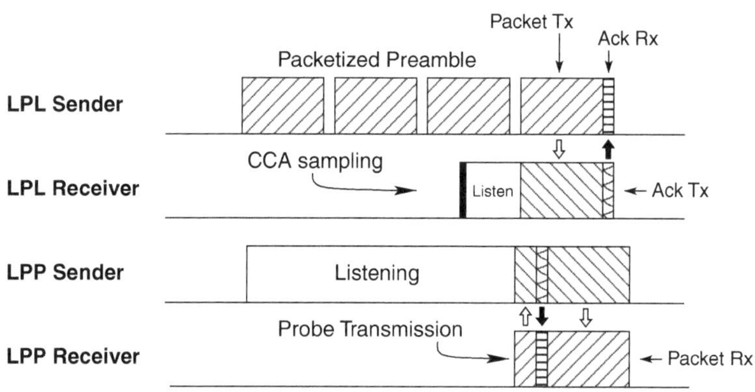

Fig. 4 A simplified representation of LPL for packet-based radios and LPP. Preamble and packet durations are not drawn to scale

[2] A node stops transmitting beacons once it participates in a download operation.

Algorithm 2 Neighborhood Collection

```
procedure NEIGHBORHOODCOLLECTION(bs)
  add ← INITQUEUE(bs)
  while QUEUEEMPTY(add) = False do
    node ←POPQUEUE(add)
    path ←BUILDPATH(node, parent, bs)
    r ←SENDNEIGHBORHOODREQ(node, path)
    if r ≠ Empty then
      for each (n, rssi) in R do
        UPDATENEIGHBORHOOD(n, rssi)
        if INQUEUE(n, add) = False then
          parent[n] ← node
          r ←APPENDQUEUE(n)

procedure BUILDPATH(n, p, s)
  r ←INITLIST(s)
  while n ≠ s do
    r ←APPENDLIST(p[n])
    n ← p[n]
  return r
```

practice, we found that approximating $\log(x)$ with the first term of its Taylor series

$$\log(x) = (x-1) - \frac{(x-1)^2}{2} + \frac{(x-1)^3}{3} - \frac{(x-1)^4}{4} \dots \quad (1)$$

produced satisfactory results.

The gateway uses unreliable persistent FCP paths to collect the nodes' neighborhood information every time it wakes up the network. It does so by following the procedure outlined in Algorithm 2. In summary, the gateway uses the neighbor information it collects directly, to download the neighborhoods of its immediate neighbors. Using this information, the gateway extends its network knowledge by another hop. Then, for each two-hop neighbor x, the gateway selects the link between x and its existing one-hop neighbors which has the highest RSSI value (say y). The path to x then is built by extending the path to y. The advantage of this approach is that new paths are always constructed by extending existing high-quality paths. The algorithm terminates after the gateway retrieves neighborhood information from all the nodes.

2.1.3 Routing Path Selection

Routing path selection is a two-step process that starts once the gateway retrieves neighborhood information from all the network's nodes. The gateway first computes the depth of each of the network's nodes through a breadth-first search (BFS) of the collected network topology, in which all links are initially considered equivalent. During the second step, we compute a path from each of the network's nodes back to the gateway. We do this by starting from the selected node and randomly following a good link towards a neighbor that is closer to the gateway (i.e., at a higher level of the BFS tree). In this context, good links are those with RSSI values higher than −70 dBm. We use this threshold to ensure that only stable links with low packet loss are used for data downloads [26]. We randomly choose among good links rather than selecting the best link to exercise multiple links. This way, the load of retrieving data is distributed more evenly among the network's nodes. If a path fails, indicated by an FCP error, the gateway selects an alternate path to download data from the current node.

2.1.4 Channel Switching

Once the network is active and the gateway has selected the paths it will use, it starts to download data sequentially from each of the network's nodes. However, downloading large blocks of data (∼100–200 KB) over multi-hop paths can take from tens of seconds to minutes depending on link conditions. Nodes that do not participate in the download waste energy during this time. Therefore, it is desirable to put these nodes to sleep. However, due to the way LPP works, even if such nodes go to sleep, they will be awaken because active nodes will acknowledge their probes.

To avoid these spurious wake ups the gateway instructs all nodes on the current download path to switch to a different frequency channel[3] before the download starts. Once the download completes, the gateway instructs the nodes to return to the common command channel. Both operations use reliable ephemeral FCP network paths. To switch the path $N_1, N_2, \ldots N_k$, the gateway initiates k sequential channel switch requests starting from the node farthest from it (i.e., N_k) and ending with N_1. If all channel switch operations are successful, the gateway initiates the download operations. Because the same path is re-used to download data from all k nodes, as soon as a download completes the gateway asks the source node to return to the command channel and go to sleep.

Algorithm 3 provides a formal description of both channel switch operations. While a number of failures can occur during a channel switching operation, Koala will eventually recover from all of them because nodes return to the command channel if no FCP activity is detected within a certain amount of time.

2.1.5 Data Download

The gateway uses reliable persistent FCP paths to download the data from the network's nodes. The only remaining challenge is to select the appropriate inter-packet interval with which the source should inject packets to the network, to avoid collisions with copies of its packets forwarded upstream. It is easy to show that in a download path in which each node interferes only with its predecessor and successor, the source should inject one packet for every three time slots (i.e., time necessary to transmit a single packet over a single hop) to avoid collisions. However the correct

[3] IEEE 802.15.4 radios provide 16 non-overlapping frequency channels.

Algorithm 3 Channel Switching

```
procedure PATHSWITCHING(path)
   s ← INITSTACK()
   t ← INITSTACK(path)                          ▷ Last element is the top
   c ← RANDOMCHANNEL()
   while STACKEMPTY(t) = False do
      node ← POPSTACK(t)
      r ← SENDCHSW(c, s)
      if r = Failed then
         break
      else
         PUSHSTACK(s, node)
   while STACKEMPTY(s) = False do
      node ← POPSTACK(t)
      if NEEDSDOWNLOAD(node) then
         DOWNLOAD(s)
      NODEDESWITCHING(s)

procedure NODEDESWITCHING(p)
   for node in p do
      if NEEDSDOWNLOAD(node) then
         SENDCHSWWITHSLEEP(CmdChannel, p)
         return
   SENDCHSWITCH(CmdChannel, p)
```

inter-packet delay is not known in the general case, because the interference graph is not known.

One could derive the optimum delay $OptDelay(m)$ for paths of length m for the worst case scenario in which all nodes interfere with each other. In practice however deployments are sparse and therefore this approach will produce sub-optimal results. One way we can reduce this delay bound is by using the collected neighborhood information. To do so, for each node n on the download path p, we compute $PathNeighbors(n, p)$ which is the number of neighbors that n has in p. The inter-packet delay can then be set to the largest $PathNeighbors()$ value since it represents the maximum interference at any single hop on the path.

However, both approaches do not consider the impact of hop-by-hop retransmissions in the face of packet loss. Such retransmissions increase the time required for a packet to "clear" an upstream hop and thus require larger and, more importantly, dynamically adjustable inter-packet delays at the source. For this reason, we opted for an alternative approach in which the gateway uses the acknowledgments that FCP generates to keep a running estimate of the path's round trip time (RTT). The source then injects a new packet every RTT/2 s. The rational is that after RTT/2 s the last packet most likely has exited the path, and it is safe to send the next packet. Algorithm 4 presents the full gateway's logic. We acknowledge that this approach is sub-optimal, transmitting slower than the optimal sending rate especially over paths with long RTTs. Alternatively, one could use more sophisticated rate control algorithms such as the ones in [17, 36] to overcome this limitation.

Once the gateway finishes downloading data from all the intended nodes, it leaves the network or goes to sleep. The rest of the network should also go to sleep when this happens. We achieve this behavior using the Drip dissemination protocol [23].

Algorithm 4 Data Download

```
procedure DOWNLOAD(path)
    start ← GETTIME()
    r ← SENDNEIGBORHOODREQ(p)
    rtt ← GETTIME()−start
    if r = Failed then
        return
    SENDDOWNLOADREQ(p, rtt/2)
    repeat
        r ← RECEIVEDATADOWNLOAD()
    until r = Failed ⋁ LASTPACKET(r)
```

Specifically, the gateway periodically (once every 5 s) disseminates monotonically increasing values for key K. Each node that receives an updated K value resets its internal timer (set to 15 s). If on the other hand, the node does not receive a new value before its timer expires, it goes to sleep.

2.1.6 Koala Summary

Koala provides the flexibility of running the network at different duty-cycles by adjusting the amount of data retrieved during each download operation. Intuitively, downloading larger blocks of data in a single operation is more efficient than using multiple smaller operations because the constant overhead of waking up the network is incurred only once. On the other hand, one needs to wait until the motes have collected the appropriate amount of data before a download operation can occur. Therefore, the data retrieval latency is a function of the data acquisition rate and the download size.

Figure 5, originally from [35], illustrates these trade-offs by presenting the overall duty-cycles achieved by different download sizes as a function of the amount of data nodes collect per day. The computed duty-cycles include all the system costs, including the cost of (repeatedly) waking up the network, collecting neighborhood information and installing routes, and the cost of downloading data. This is the reason why all duty-cycles start at 0.1 %, because this is the cost of running LPP alone. Interested readers can find a description of this experiment's detailed setup as well as thorough analysis of Koala in [35].

2.2 WRAP: High-Throughput Data Collection

For RACNet we developed the Wireless Reliable Acquisition Protocol (*WRAP*) [24] for scalable data collection given the unique radio characteristics found inside data centers. To tackle the challenge of self-interference caused by contention in dense wireless networks, WRAP uses multiple IEEE 802.15.4 frequency channels simultaneously and adaptively balances the number of nodes on each channel based on traffic load. WRAP also implements coordinated data collection through a *token*

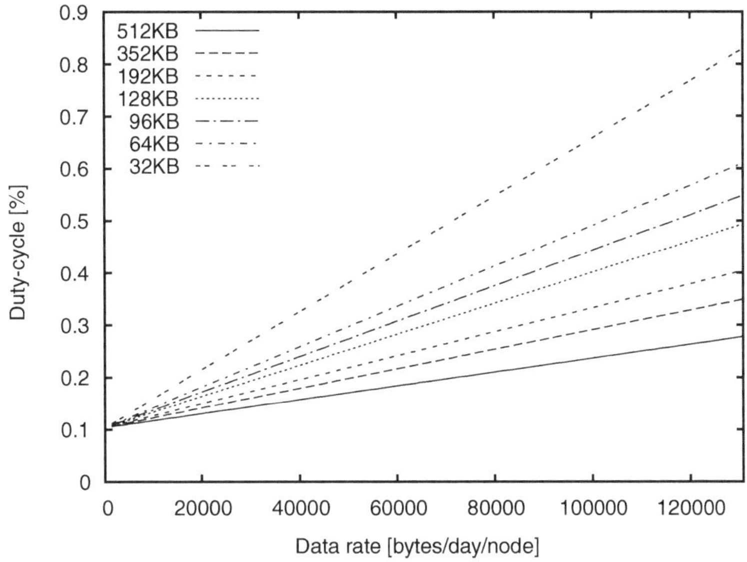

Fig. 5 Overall duty-cycle of Koala as a function of the per-node data acquisition rate. Each *line* represents a different download block size

passing protocol that provides an implicit network arbitration mechanism, allowing only one active packet flow per frequency channel.

From an architectural perspective, WRAP's design is at the center of the spectrum between distributed and centralized data collection. At one end of this spectrum, the nodes participating in a distributed data collection protocol collaborate to construct a common routing tree and independently forward data as soon as possible [13, 48].

At the other end of the design space lies the centralized approach, in which the gateway controls the operation of the entire network leveraging its ample computational resources and complete knowledge of the network topology [35, 41].

WRAP follows a hybrid approach whereby nodes determine the routing topology in a distributed way while the gateway coordinates data transport using a centralized token passing mechanism. Specifically, the gateway periodically generates a token that traverses the derived routing tree in a depth-first manner. Only the tree node that holds the token can transmit one or more packets before passing the token to the next node. By allowing only a single node to transmit at any point in time, token passing bypasses the inter-flow contention that can lead to congestion and packet loss. This is especially important close to the root of a dense network whereby concurrent flows are very likely to interfere with each other. In this respect WRAP is a congestion avoidance mechanism, unlike existing centralized [36] or distributed [38] congestion control protocols. Moreover, by eliminating congestion as a possible cause of packet loss, WRAP removes the ambiguity that complicates the response of congestion control protocols to missing data.

This division of responsibilities ensures timely adaptation to link quality variations and at the same time gives every network node a fair share of the network's resources without contention. RCRT is another example of a hybrid protocol that adds centralized coordination—the rate control information—to an otherwise distributed protocol [36]. However, imposing a single transmission rate for the entire network is inevitably biased towards the weakest node (i.e., the one behind the most lossy link) and artificially degrades the overall network throughput.

2.2.1 BiTree: Bi-directional Data Collection Tree

The network layer maintains robust data collection trees rooted at the network's gateways. The mechanism's distributed nature allows nodes to independently react to topology changes including degraded link qualities and node failures. Since WRAP also uses the tree to deliver downstream traffic such as requests for lost data packets, we focus on building bi-directional trees (BiTrees) with high quality bi-directional links.

Gateways initiate BiTree construction by broadcasting HEARTBEAT messages. HEARTBEATs include fields that represent the node's status, including its hop distance from the root, its parent node ID, the number of children, and the path quality metric to the root.

Upon receiving a HEARTBEAT message, a node takes the following steps: first, the incoming HEARTBEAT message needs to have a RSSI above a threshold to avoid links with high loss rates. Next, the node checks whether the potential parent has already reached its maximum number of children. If not, the next step is to evaluate the *path* quality to the gateway via this potential parent by computing the *path expected transmission count* (PETX) as follows:

$$PETX_j = \sum_{l \in P} \frac{1}{PRR_l} = PETX_i + \frac{1}{PRR_{i,j}}, \tag{2}$$

where j is the current node, i is its potential parent, and P is the path from j to the gateway via i.

To compute $PETX_j$ recursively, the $PETX_i$ is included in the HEARTBEAT messages. However, estimating $PRR_{i,j}$ directly from HEARTBEATs would require multiple message rounds. Instead, we take advantage of the Link Quality Indicator (LQI) available from radio chips such as the TI CC2420 [44], to reduce control message overhead. Specifically, we use the piece-wise linear approximation shown in Fig. 6 to estimate a link's PRR based on its LQI. We note that while the curve shown in Fig. 6 was derived from site survey data, it resembles the approximation used in [5].

The node selects the upstream neighbor with the smallest PETX as its *potential* parent and initiates a TREE_JOIN request. The parent also estimates the link quality from this potential child in the upstream direction before replying with a GRANT message. Otherwise, the TREE_JOIN operation times out. This two-way handshake has two benefits. First, it serves as an explicit agreement between the parent and the child node that both have the resources to relay messages for each other. Second,

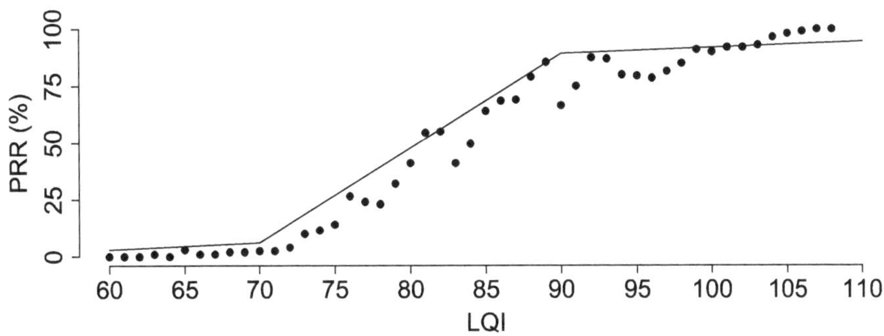

Fig. 6 Piece-wise linear approximation of PRR from LQI in RACNet. The *dots* are the average PRR for each LQI obtained from a site survey

since we require a BiTree for data downloading, it is important to ensure the link quality in both directions, as wireless links can be asymmetric [51].

As described above, nodes broadcast HEARTBEAT messages to construct and maintain BiTrees. It is therefore desirable to transmit multiple HEARTBEATs in a short amount of time, to accelerate the tree construction process. However, in large and dense networks, this can lead to broadcast storms and severe collisions, eventually affecting the quality and stability of the resulting tree.

A simple and low-maintenance solution would be to adopt a contention-based approach, in which nodes contend for the radio medium. However, this approach is ill-suited for dense networks because the large number of HEARTBEATs is likely to cause collisions and large delays. At the same time, a TDMA-based protocol that assigns exclusive time slots to each node within the same interference range is cumbersome as it requires tight time synchronization and additional control traffic to set up the schedule.

Instead, WRAP uses a reduced contention mechanism to regulate the broadcast of HEARTBEAT messages. Specifically, WRAP defines a time slot of length T that starts immediately after a node P broadcasts its HEARTBEAT message. The time slot is further divided into two uneven sections according to the number of children that P already has and the number of additional children that P can support. The first section is reserved for the HEARTBEAT messages sent by P's children, while the second is used by nodes that are not part of the tree to initiate the handshake process with P. Nodes that receive P's HEARTBEAT randomly select a time within the appropriate section to transmit their message.

While this mechanism reduces contention, it does not guarantee a collision-free network. Specifically, we do not coordinate among nodes within the same broadcast domain that connect to different parents. Instead, we let them contend for the medium.

2.2.2 Channel Diversity

A RACNet system may consist of many hundreds of nodes within one data center. One way to increase data throughput and reduce data latency is by using multiple gateways. To do so, we take advantage of channel diversity to build multiple BiTrees rooted at different gateways, each on a different channel frequency. Previous work has shown that simultaneous communications over two-channel-apart IEEE 802.15.4 channels do not interfere with each other [49].

In WRAP, every gateway has a fixed channel assigned by the operator. Non-gateway nodes start by scanning channels sequentially and looking for a tree to join. Since gateways continuously perform data collection, a node can first overhear the network traffic and decide whether the channel potentially has a tree that it can join. In addition, a node can bound its wait time on each channel to (little over) one HEARTBEAT time interval because gateways periodically initiate new rounds of HEARTBEAT beaconing. A node joins the first tree using the two-way handshake mechanism described above. However, the node joins any subsequent trees only if the estimated quality of the new path is better than the one on the current tree.

WRAP follows a transaction model when constructing BiTrees across different channels. It is possible that a node (temporarily) joins multiple trees as it actively scans all available channels. However, nodes in this state do not broadcast HEARTBEAT messages to recruit children. This is to limit further disturbance in the candidate trees that the node later decides not to join. When the scanning phase ends, the node switches to and stays in the last tree it joined. Finally, a node's parent takes its HEARTBEAT transmissions as an indication of its commitment to the tree. Other candidate parents eventually time out and remove the node from their children lists.

Nodes can significantly reduce their channel scanning time with the gateways' help. Specifically, gateways maintain the list of all channels they collectively occupy and include this information in their HEARTBEAT messages. Therefore, after receiving one HEARTBEAT message, nodes immediately know all available channels.

2.2.3 Load Balancing Multiple BiTrees

As nodes join and leave the network or link qualities change, the sizes of different BiTrees can become unbalanced. We quantify the size of a tree by its sum of hops Δ, or the total path length from each node in the tree to the root. This Δ largely determines the overall time necessary to finish a data collection round and for this reason WRAP uses Δ to balance the load among all the network's trees.

WRAP implements a distributed algorithm for balancing BiTrees. WRAP periodically checks the Δ's of different trees, and it initiates the channel-balancing process by sending a START_BAL message that propagates through the tree with the largest Δ. WRAP utilizes two mechanisms to avoid network instability: (i) it restricts the channel-balancing process to the gateway with the largest Δ, and (ii) it tolerates certain amount of imbalance in Δ. Let Δ_{avg} be the average among all trees. A gateway b^* starts the channel-balancing process only under the following conditions:

$$\Delta_{b^*} - \Delta_{avg} > \delta, \text{ and}$$
$$b^* = argmax_{b \in B}(\Delta_b), \tag{3}$$

where B is the set of all gateways and δ is a threshold parameter that controls the amount of tolerable imbalance.

The START_BAL message contains the probabilities for switching to each of the other channels. Switching probabilities are defined to be higher for more under-utilized channels.

Specifically, a node connected to the tree rooted at b^* will decide to switch out with probability $P_{out} = \frac{\Delta_{b^*} - \Delta_{avg}}{\Delta_{b^*}}$. Once the node decides to leave b^*'s tree, the probability that it switches to another tree $B_i \neq b^*$ is set asfollows:

$$P_i = 0, \text{ if } \Delta_i \geq \Delta_{avg}$$
$$P_i = \frac{\Delta_{avg} - \Delta_i}{\sum\limits_{b \in B \text{ and } \Delta_b < \Delta_{avg}} (\Delta_{avg} - \Delta_b)}, \text{ if } \Delta_i < \Delta_{avg}. \tag{4}$$

Intuitively, we attempt to migrate the extra nodes from gateway b^* to under-loaded gateways, based on their degrees of under-utilization. In other words, more nodes will attempt to join the tree with fewer nodes. Finally, if the node cannot find a parent in the target channel, it returns to its original channel.

2.2.4 Token-Based Data Download

The transport layer reliably collects data to RACNet gateways along the network's BiTrees. Rather than having nodes initiate data uploads asynchronously, WRAP coordinates the network traffic to reduce radio contention. At the same time, pull-based approaches in which gateways initiate data collection by sending requests to individual network nodes can incur significant overhead including the cost of one downstream message per node and the round-trip delay for transmitting each node's measurements. WRAP addresses these two sources of overhead by adopting a token passing approach.

The token passing mechanism does not require the gateways to have a priori knowledge of the network topology. Rather, it relies on the network to determine the next node that should hold the token. Since gateways continuously retrieve data from the network, this property also removes the overhead of having a separate phase for collecting neighborhood information from all nodes in the network. The basic protocol works as follows.

Gateways initiate a data collection round by passing the token to the first node on their list of children. Each token contains an unique 32-bit token ID so that nodes know when a new round of data collection has started. WRAP tokens traverse the tree in a depth-first order. After receiving the token, the node passes it sequentially to

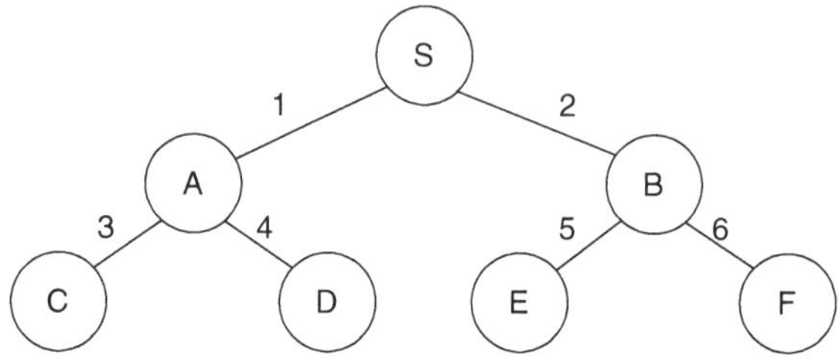

Fig. 7 Two-level binary tree. Passing the token in depth-first-search will yield the sequence 1, 3, 3, 4, 4, 1, 2, 5, 5, 6, 6, and 2 (12 edges in total)

all of its children in the tree. Once all of its children have finished transmitting their measurements the node streams the measurements it has accumulated since the last data collection round to the gateway. To minimize the number of packet transmissions, nodes aggregate as many measurements as possible in one packet. In practice, up to five such measurements fit in one maximum-size, 128-byte IEEE 802.15.4 frame. This ability to aggregate multiple measurements to a single packet is a side benefit of the architectural decision to decouple data collection from data generation.

When the gateway eventually receives the token back from the network, it scans the measurements received and recovers lost packets by requesting any missing sequence numbers.

Passing the token in a depth-first fashion ensures that the token travels each network edge exactly twice. For example, in the two level binary tree shown in Fig. 7, the edge visiting order is 1, 3, 3, 4, 4, 1, 2, 5, 5, 6, 6, and 2 (12 edges in total). If breadth-first traversal was used instead, the token would travel each edge at least twice, because the token has to travel back to the gateway. In the case of Fig. 7 this would lead to 16 edge traversals compared to 12. WRAP further reduces the token passing overhead through inference. First, since a parent forwards all the measurements from its children, it has the opportunity to inspect the MORE_DATA field in their packets and determine when the current child has sent its last packet. When this happens, the parent assumes that the child has released the token. Second, since children can overhear packets sent by the parent, the parent piggy-backs the next child node ID when it is ready to pass the token.

Although WRAP aggressively performs link-level retransmissions, the network can still lose the token for various reasons such as node failure. WRAP puts the burden of token recovery on the gateway. Since nodes stream data only when they hold the token, if the gateway's idle timer expires while waiting for incoming data, it assumes that the token has been lost and regenerates a token with the same ID. Since each token carries an unique ID, nodes that have held a token with the same ID will immediately release it.

2.2.5 End-to-End Reliability

WRAP implements a NACK-based, end-to-end data recovery scheme, whereby gateways request end-to-end retransmissions for missing sequence numbers. To amortize the round trip time incurred in the data retransmission process, WRAP accumulates multiple data retransmission requests destined to the same node.

WRAP encapsulates downstream data requests inside source-routed packets. Doing so, requires gateways to have knowledge of their tree topology. To do so, nodes in the tree piggy-back their parent node ID to the end of the data stream that they send to their gateway. Based on this process, a gateway can rebuild the complete tree topology at the end of a data collection round.

To stream data efficiently, the source node must determine the inter-packet transmission interval that minimizes self-interference and end-to-end delay. WRAP adapts a technique similar to the one proposed in [17] whereby a node estimates the inter-packet delay by measuring the time between transmitting the last packet of a batch and the last time it overhears the same packet forwarded by nodes upstream. To take into account the whole path, parents propagate the local estimates downstream via the HEARTBEAT message. Then, each child node updates its local inter-packet value to the maximum of the previous local value and the one in the HEARTBEAT message.

2.2.6 WRAP Evaluation

We compare WRAP to the Collection Tree Protocol (CTP) [13] and Rate-Controlled Reliable Transport (RCRT) [36]. CTP is a best-effort data collection protocol that does not implement end-to-end retransmissions but rather relies on hop-by-hop retransmissions to reduce packet loss. RCRT implements end-to-end reliability as well as congestion control by controlling the senders' transmission rates. We implemented the same application that periodically samples a set of sensors on top of all three protocols. All of our code is written in TinyOS 2.1 [22]. We use the default CTP version included with the TinyOS 2.1 distribution. We also ported RCRT to TinyOS 2.1 for fair comparison. In all cases, we use a single frequency channel (26) because CTP and RCRT do not have a channel balancing capability.

Sampling Intervals. We first stress the three protocols by increasing the application's sampling rate. Figure 8 presents the three protocols' behavior as we vary the

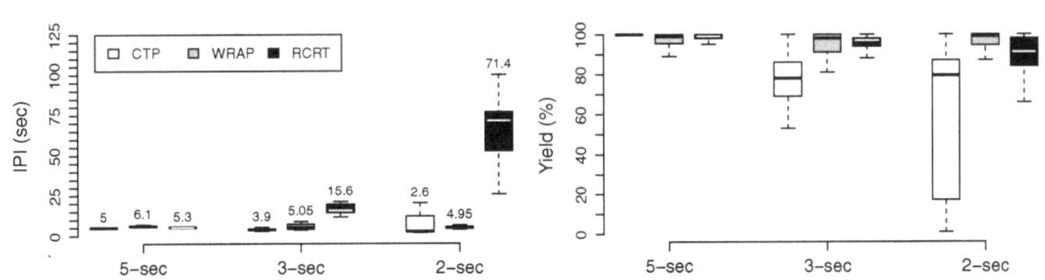

Fig. 8 Boxplots of the inter-packet interval (IPI) and data yields under various sampling intervals on the 62-node lab testbed, comparing WRAP to CTP and RCRT. The number on top of each IPI plot shows the average

application's sampling interval on a 62-node lab testbed. In each case, the experiment ran for at least 2 h.

While CTP achieves low inter-packet intervals at low network loads, packet losses increase drastically as the network becomes congested. RCRT reacts to this congestion by lowering nodes' transmission rate. We noticed that the RCRT gateway instructed the nodes to reduce their rate to the minimum configured rate (i.e., one packet per 60 s) when a 2-s sampling interval was used. This dramatic reaction was due to the fact that the gateway experienced multiple timeouts while waiting for nodes to acknowledge its requests to lower their rates. Because RCRT treats such timeouts as further signs of congestion, the gateway reacts by lowering the nodes' rates even further. However, even this drastic reaction did not achieve perfect yield in the case of 2-s sampling.

While CTP ignores congestion, leading to lower yields, and RCRT reactively lowers the nodes' transmission rate, leading to higher inter-packet intervals, WRAP prevents congestion from occurring in the first place by allowing only one network flow at any point in time. As the results from Fig. 8 suggest this strategy achieves high data yields and low inter-packet intervals across all sampling rates.

Network Density. Next, we stress the protocols by increasing the network's density. We do so by uniformly arranging an increasing number of nodes, following a grid pattern, over the same physical area in the data center testbed. Having more nodes in the same space not only increases the amount of traffic that the network must deliver, but also increases contention when node communications are not coordinated, as in the case of CTP and RCRT. To evaluate the effects of network size on performance we fix the application sampling rate to one packet every 10 s and increase the number of nodes from 50 to 150. We did not perform the RCRT experiment with 150 nodes because the performance of the protocol (inter-packet interval) degraded appreciably even with a network of 100 nodes. In each case, the experiment ran for at least 2 h.

As Fig. 9 shows, the inter-packet interval increases slightly with the network size. Interestingly, this increase was due to packet loss in the case of CTP and to the longer time necessary to service the whole network in the case of WRAP which achieved 100 % yields for all network sizes. As in the previous set of experiments, RCRT reacted to the increased levels of contention by reducing the nodes' transmission rate. Specifically, for the 100-node network the gateway set the nodes' rate to minimum, or one packet every 60 s. In practice however the inter-packet interval was even higher

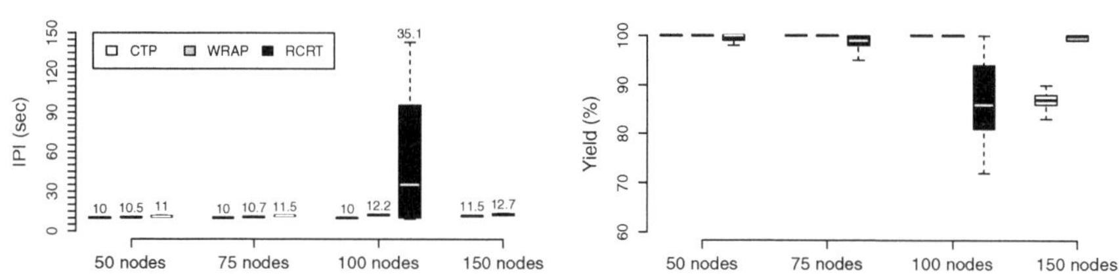

Fig. 9 Boxplots of the inter-packet interval (IPI) and data yields under various network sizes on the data center testbed, comparing WRAP to CTP and RCRT. Each node generates one packet every 10 s. The number on top of each IPI plot shows the average

because even this decreased rate was not able to prevent network losses and decreases data yields.

2.2.7 WRAP Summary

Decoupling the topology control from the data collection and using a token passing mechanism to provide network-wide arbitration allows WRAP to avoid congestion more efficiently than previous data collection protocols.

When compared with two leading data collection protocols, one using uncoordinated transmissions (CTP [13]) and the other using rate-based congestion control (RCRT [36]), WRAP is able to maintain high yield at various load and network sizes. As Fig. 8 shows, with a fixed inter-packet-interval of 10 s, increasing the network size has little effect on WRAP's yield while both RCRT and CTP starts to degrade when the network size reaches 100 and 150 nodes respectively. The interested reader can find details about this experiment and a detailed evaluation of WRAP in [24].

3 Time Reconstruction

In WSNs, each node's clock (referred to as local clock henceforth) is typically implemented as a free running counter that monotonically increases and resets to zero upon reboot. Since nodes do not reboot simultaneously nor count at the same speed, measurements timestamped with one local clock cannot be correlated with measurements from another.

In order to accomplish this correlation, nodes can either synchronize their clocks to a global reference clock before the sensors are sampled (so that measurements can be directly timestamped with this global clock) or during a post-processing step where the global time is reconstructed from each node's local clock.

In environmental monitoring networks, such as *L*ife Under Your Feet, the desired level of accuracy in this context is in the order of milliseconds to seconds. In order to reduce complexity of the code running on the mote and minimize the network overhead, it is more efficient to record sensor measurements using the mote's local time frame and perform a postmortem reconstruction to translate them to global time.

Postmortem time reconstruction schemes collect ⟨*local, global*⟩ pairs during a node's lifetime, using a global clock source (typically, an NTP-synchronized PC). These pairs (referred to as "anchor points") are then used to translate the collected measurements to the global time frame by estimating the nodes' clock skew and offset.

In the absence of reboots, this naive time reconstruction strategy performs well. However, in practice, nodes reboot due to low battery power, high moisture, and software defects. Even worse, when nodes experience these problems, they may remain completely inactive for non-deterministic periods of time. Measurements collected during periods which lack ⟨*local, global*⟩ anchors (due to rapid reboots and/or basestation absence) are difficult or impossible to accurately reconstruct. Unfortunately, such situations are not uncommon based on our deployment experiences and those reported by others [47].

To remedy this problem we developed the *Phoenix* [15] offline algorithm for reconstructing global timestamps that is robust to frequent node reboots and does not require a persistent global time source.

Nodes in Phoenix exchange their time-related state with their neighbors only, thereby establishing a chain of transitive temporal relationships to one or more nodes with references to the global time. These relationships allow Phoenix to reconstruct the measurement timeline for each node. The offline nature of Phoenix has two advantages: (a) it reduces the complexity of the software running on the node, and (b) it avoids the overhead associated with executing a continuous synchronization protocol.

Related Time Synchronization Protocols

Assignment of timestamps in sensor networks falls under two broad categories. Strict clock synchronization aims at ensuring that all the mote clocks are synchronized to the same clock source. Flooding Time Synchronization Protocol (FTSP, [31]), Reference Broadcast Synchronization (RBS, [9]), and the Timing-sync Protocol for Sensor Networks [11] are examples of this approach. These systems are typically used in applications such as target tracking and alarm detection which require strong real-time guarantees of reporting events. The second category is known as postmortem time reconstruction and it is mostly used due to its simplicity. While strict synchronization is appropriate for applications where there are specific events of interest that need to be reported, postmortem reconstruction is well-suited for applications where there is a continuous data stream and every measurement requires an accurate timestamp.

Phoenix falls under the second class of methods. The idea of using linear regression to translate local timestamps to global timestamps was first introduced by Werner-Allen et al. in a deployment that was aimed at studying active volcanoes [47]. This work, however, does not consider the impact caused by rebooting motes and basestation failures from a time reconstruction perspective. More recently, researchers have proposed data-driven methods for recovering temporal integrity [16, 29]. Lukac et al. use a model for microseism propagation to time-correct the data collected by their seismic sensors. Although data-driven methods have proved useful for recovering temporal integrity, they are not a solution for accurate timestamping.

Routing integrated time synchronization protocol (RITS, [39]) spans these categories. Each mote along the path (to the basestation) transforms the time of the reported event from the preceding mote's time frame, ending with an accurate global timestamp at the basestation. RITS does not consider the problem of mote reboots, and is designed for target tracking applications. The problem of mote reboots have been reported by a number of research groups. Chang et al. report that nodes rebooted every other day due to an unstable power source [4], whereas Dutta et al. employed the watchdog timer to reboot nodes due to software faults [8]. Allen et al. report an average node uptime of 69 % [47]. More recently, Chen et al. advocate *Neutron*, a solution that detects system violations and recovers from them without having to reboot the mote [6]. They advocate the notion of preserving "precious" states such as the time synchronization state. Nevertheless, Neutron cannot prevent all mote reboots and therefore Phoenix is still necessary.

3.1 Phoenix Postmortem Timestamp Reconstruction

The relationship between a mote's local clock, *LTS*, and the global clock, *GTS*, can be modeled with a simple linear relation: $GTS = \alpha \times LTS + \beta$, where α represents the mote's skew and β represents the intercept (global time when the mote reset its clock) [39]. This conversion from the local clock to global clock holds as long as the mote's local clock monotonically increases at a constant rate. We refer to this monotonically increasing period as a *segment*. When a mote reboots and starts a new segment, one needs to re-estimate the fit parameters. If a mote reboots multiple times while it is out of contact with the global clock source, estimating β for these segments is difficult. While data-driven treatments have proven useful for recovering temporal integrity, they cannot replace accurate timestamping solutions [16, 29]. Instead, time reconstruction techniques need to be robust to mote reboots and not require a persistent global time source.

The Phoenix [15] postmortem time reconstruction algorithm consists of two stages: an online collection stage and an offline reconstruction stage.

Online Anchor Collection

Each node operates solely with respect to its own local clock. A new *segment* (uniquely identified by ⟨*nodeid, reboot count*⟩) begins whenever a node reboots: each segment starts at a different time and may run at a different rate. Our architecture assumes that there is at least one node in the network that can periodically obtain references from an accurate global time source. This source is used by Phoenix to establish the global reference points needed and may be absent for long periods of time. The global time source can be any reliable source (a node equipped with a GPS receiver, NTP-synced basestation, etc.) and does not need to be identical to the basestation. Without loss of generality, we assume that the global time source is a GPS receiver connected to a node different from the basestation.

All nodes (including the GPS-connected node) broadcast their local clock and reboot-count values every T_{beacon} seconds. Each receiving node stores this information (along with its own local clock and reboot counter) in flash to form anchor records. The format of these records is $\langle nodeid_r, rc_r, lc_r, nodeid_s, rc_s, lc_s\rangle$; where rc, lc, r, and s refer to the reboot counter, local clock, receiver and sender, respectively. Periodically, nodes turn on their radios and listen for broadcasts in order to anchor their time frame to those of their neighbors. Each node tries to collect this information from its neighbors after every reboot and after every T_{wakeup} seconds ($\gg T_{beacon}$). The intuition behind selecting this strategy is as follows. The reboot time determines the offset (i.e., the value of the global clock when the node (re)boots). The earliest opportunity to extract this information is immediately after a reboot. To get a good estimate of the skew, one would like to collect multiple anchors that are well distributed in time. Thus, T_{wakeup} is a parameter that governs how far to spread out anchor collections. In the case of a GPS node, the $nodeid_r$, rc_r and $nodeid_s$, rc_s are identical, and lc_r, lc_s represent the local and global time respectively.

The basestation periodically downloads these anchors along with the sensor measurements. This information is then used to assign global timestamps to the collected measurements using Algorithm 5. If the rate of reboots is known, the anchor collection frequency can be fixed conservatively to collect enough anchors between reboots. One could also employ an adaptive strategy by collecting more anchors when the segment is small and reverting to a larger T_{wakeup} when an adequate number of anchors have been collected. It is advantageous for a node to attempt to collect anchors from a small set of neighbors (to minimize storage), but this requires a node to have some way of identifying the most useful segments for anchoring.

Offline Timestamp Reconstruction

The Phoenix algorithm is intuitively simple. We will outline it in text and draw attention to a few important details. For a more complete treatment, please refer to the pseudocode in Algorithm 5. Phoenix accepts as input the collection of all anchor points *AP* (both ⟨*local, neighbor*⟩ and ⟨*local, global*⟩). It then employs a least-square linear regression to extract the relationships between the local clocks of the segments that have anchored to each other (*LF*, for Local Fit). In addition to $LF_a(i, j)$ (slope), $LF_b(i, j)$ (intercept), Phoenix also obtains a goodness-of-fit (GOF) metric, $LF_\chi(i, j)$ (unbiased estimate of the variance of the residuals) and LF_{df} (degrees of freedom). For segments which have global references, Phoenix stores this as *GF* (for Global Fit).

The algorithm then initializes a queue with all of the segments which have direct anchors to the global clock. It dequeues the first element q and examines each segment c that has anchored to it. Phoenix uses the transitive relationship between $GF(q)$ and $LF(q, c)$ to produce a global fit $T(c)$ which associates segment c to the global clock through segment q. If $T_\chi(c)$ is lower than the previous value for $GF_\chi(c)$ (and using q would not create a cycle in the path used to reach the global clock), the algorithm replaces $GF(c)$ with $T(c)$, and places c in the queue. When the queue is empty, no segments have "routes" to the global clock which have a better goodness-of-fit than the ones which have been previously established. At this point, the algorithm terminates.

The selection of paths from an arbitrary segment to a segment with global time references can be thought of as a shortest-path problem (each segment represents a vertex and the fit between the two segments is an edge). The *GOF* metric represents

Algorithm 5 Phoenix

```
Ensure:
  a, b: alpha and beta for local-local fits;
  P: parent segment; Π: Ancestor segments

procedure PHOENIX(AP)
  for each (i, j) in KEYS(AP) do                        ▷ All unique segment pairs in AP
    LF_{a,b,χ,df}(i, j) ← LLSE(AP(i, j))                ▷ Compute the local-local fits
  for each s ∈ S do                                     ▷ Set of all unique segments
    GF_{α,β,P,Π,χ,df}(s) ← (∅, ∅, ∅, s, χ_MAX, ∅)       ▷ Initialize global fits
  for each g ∈ G do                                     ▷ All segments anchored to GTS
    INITGTSNODES(g, LF, GF)
    ENQUEUE(Q, g)                                       ▷ Add all the GTS nodes to the queue
  while NOTEMPTY(Q) do
    q ← DEQUEUE(Q)
    C ← NEIGHBORANCHORS(q)
    for each c ∈ C do
      T_{α,β,P,Π,χ,df}(c) ←GLOBALFIT(c, q, GF, LF)
      if (UPDATEFIT(c, T, GF)) then                     ▷ Check for a better fit
        ENQUEUE(C)
  return GF

procedure INITGTSNODES(g, LF, GF)
  GF(g) ← (LF_a(g, g'), LF_b(g, g'), ∅, g, LF_χ(g, g'), LF_df(g, g'))   ▷ g' is GTS, g is LTS

procedure GLOBALFIT(c, q, GF, LF)
  if q > c then                                         ▷ Smaller segment is the independent variable
    α_new ← GF_α(q) * LF_a(q, c)
    β_new ← GF_α(q) * LF_b(q, c) + GF_β(q)
  else
    α_new ← GF_α(q)/LF_a(q, c)
    β_new ← GF_α(q) − α_new * LF_b(q, c)
  χ ← (GF_df(q)*GF_χ(q)+LF_df(q,c)*LF_χ(q,c)) / (GF_df(q)+LF_df(q,c))   ▷ Compute the weighted GOF metric.
  df ← GF_df(q) + LF_df(q, c)
  return (α_new, β_new, q, {c ∪ GF_Π(q)}, χ, df)        ▷ Update parent/ancestors

procedure UPDATEFIT(c, T, GF)
  if c ∈ T_Π(c) then                                    ▷ Check for cycles
    return false
  if T_χ(c) < GF_χ(c) then
    GF_{α,β,P,Π,χ,df}(c) ← T_{α,β,P,Π,χ,df}(c)
    return true
  else
    return false
```

the edge weight. The running time complexity of the implementation of Phoenix was validated experimentally by varying the deployment lifetime (thereby varying number of segments). The runtime was found to increase slower than the square of the number of segments.

3.1.1 Phoenix Evaluation

We evaluate the effect of Phoenix using an actual deployment (running Koala) and compare it against the Robust Global Timestamp Reconstruction (RGTR) [16] algorithm, a time reconstruction algorithm we used before developing Phoenix.

We deployed a network (referred to as the "Olin" network) of 19 motes arranged in a grid topology in an urban forest near the Johns Hopkins University campus in Baltimore, MD. Anchors were collected for the entire period of 21 days using the methodology described above. The basestation collected data from these motes once every 4 h and the NTP-corrected clock of the basestation was used as a reliable global clock source. The motes rebooted every 5.7 days on average, resulting in a total of 62 segments. The maximum segment length was 19 days and the minimum was 2 h.

In RGTR, when a mote is contacted for a download, the basestation records an anchor point, i.e., it records the mote's current local clock and the basestation's global clock. Motes that are poorly connected to the basestation can remain out of contact with the basestation for several download rounds before they can transfer their outstanding data. When motes reboot at a rate faster than the frequency with which they contact the basestation, there exist periods which lack enough information to accurately recover the timestamps. Upon acquiring the anchor points, the measurements are converted from their local clock to the global clock.

We note that, in order to estimate the fit parameters (α, β) for the segments, RGTR requires at least 2 anchor points. If we assume that α is stable per mote for small segments. Using this assumption, at least one anchor point is needed to estimate the β for any given segment, provided that α has been estimated accurately for the mote.

Since it is very difficult to establish absolute ground truth in field experiments, we establish a synthetic ground truth by reconstructing timestamps using all the global anchors obtained from the basestation.[4] We record the α and β values for each segment and use these values as ground truth. Because we downloaded data every 4 h we obtained enough global anchors from the motes to be confident with the derived ground truth estimates.

In order to emulate a GPS mote, we selected a single mote (referred to as G-mote) that was one hop away from the basestation. We used the G-mote's global anchors obtained from the basestation as though they were taken using a GPS device. We ignored all other global anchors obtained from other motes. Furthermore, to emulate the absence of the basestation for N days, we discarded all the anchors taken by the G-mote during that N-day long period. We tested for values of N from one to eighteen.

After simulating the basestation failure, we reconstruct the timestamps by applying Phoenix using only the $\langle$*local, neighbor*$\rangle$ anchors, and global anchors available from the G-mote. This provides us with another set of α and β estimates for each of the segments. We compare these estimates with the ground truth estimates (pair-wise comparison). In order to provide a deeper insight, we decompose the average PPM error metric into its constituent components—α and β errors. Furthermore, we report the median and standard deviation of these α and β errors. Table 1 reports the results of these experiments. We found that the median α error stayed as low as 5.9 ppm, while the median β error stayed as low as 6.4 s for $N = 18$. In general, α_{med}, β_{med} and β_{std} increased as N increased and α_{std} stayed relatively consistent for different values of N. The stability of the α estimates using Phoenix with $N = 0$ and $N = 18$

[4] Note that every time a mote contacts the basestation, we obtain a global anchor for that mote.

Table 1 Phoenix accuracy using the Olin dataset as a function of the number of days that the basestation was unavailable

Error\Days	2	4	6	8	10	12	14	16	18
α_{med} (ppm)	1.73	1.73	1.85	1.70	1.96	2.20	4.36	5.47	5.93
α_{std} (ppm)	3.41	3.40	3.40	3.39	3.30	3.26	3.17	3.00	3.00
β_{med} (s)	0.88	0.88	0.91	0.94	1.16	1.55	4.52	6.02	6.44
β_{std} (s)	0.58	0.57	0.58	0.57	0.65	0.91	2.43	3.11	3.45

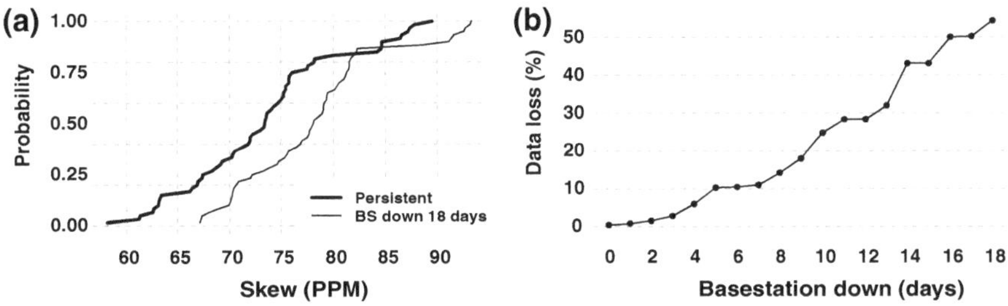

Fig. 10 The stability of the α estimates using Phoenix and the data loss using RGTR in comparison to Phoenix. **a** The CDF of α estimates on the Olin deployment. **b** Data loss using RGTR. Data loss from Phoenix was <0.06 %

is shown in Fig. 10a. The CDF shows that median skew was found to be around 75 ppm and the two curves track each other closely.

The data loss using Phoenix was found to be as low as 0.055 % when N was 18 days. In comparison, we found that there was significant data loss when the timestamps were reconstructed using RGTR. Figure 10b shows the data losses for different values of N. The figure does not report the Phoenix data loss as we found it to be 0.055 % irrespective of N. This demonstrates that Phoenix is able to reconstruct more than 99 % of the data even when motes reboot frequently and the basestation is unavailable for days. We note that in comparison to Phoenix, RGTR does not incur any additional storage and duty-cycle overheads as anchors are recorded at the basestation directly as part of the data downloads.

3.1.2 Phoenix Summary

Phoenix is an offline time reconstruction algorithm that assigns timestamps to measurements collected using each mote's local clock. One or more motes have references to a global time source. All motes broadcast their time-related state and periodically record the broadcasts of their neighbors. If a few mote segments are able to map their local measurements to the global time frame, this information can then be used to assign global timestamps to the measurements collected by their neighbors and so on. This epidemic-like spread of global information makes Phoenix robust to random mote reboots and basestation failures. We found that in practice there are more than enough possible ways to obtain good fits for the vast majority of data segments.

4 Storage and Access

After the measurements have been sampled and collected they need to be stored in an efficient and scalable way that also allows flexible post-processing, analysis, and visualization.

4.1 Data Storage

The database design, visualized in Fig. 11, follows naturally from the experimental design and the WSN. Each entry in the `Site` table is a geographic region.

Each site is partitioned into `Patches` which in turn contain `Nodes`. A particular Node has an array of `Sensors` that report environmental measurements. Each patch is a coherent deployment area, defined through its GPS coordinates. Sensor locations are relative to the reference coordinates of a patch.

The Node and Sensor types (metadata) are described in corresponding `Type` tables in Fig. 11. Each node has a record in the `Nodes` table describing its model, deployment, and other metadata. Each `Sensor` table entry describes its type, position, calibration information, and error characteristics. The `Event` table records state changes of the experiment such as battery changes, maintenance, site visits,

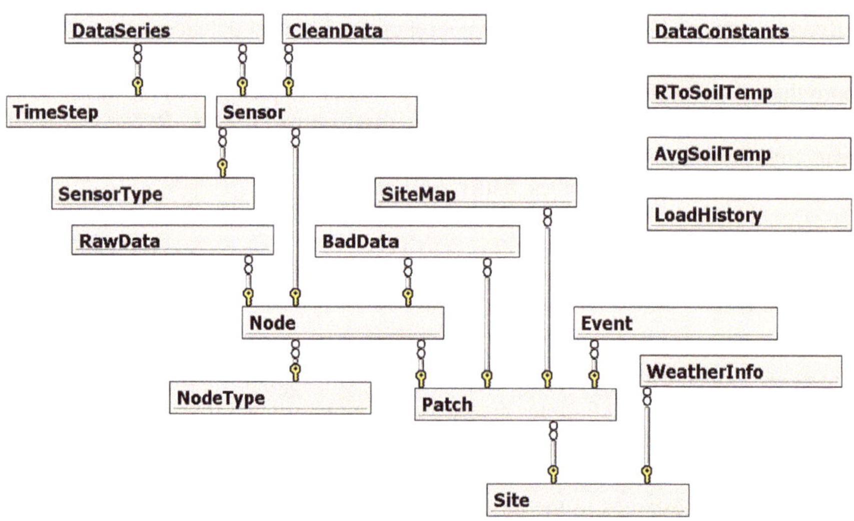

Fig. 11 Sensor Network Database Schema used in *Life Under Your Feet*. The different *boxes* correspond to database tables, while the *arrows* correspond to relations between tables

replacement of a sensor, sensor failure, etc. Global events are represented by pointing to the NULL patch or NULL node. The site configuration tables (`Site, Patch, SiteMap`) and hardware configuration tables (`Node, Sensor, NodeType, SensorType`) are loaded prior to the data collection. The `DataConstants` and `RToSoilTemp` contain constants that are used in the calibration process mentioned earlier and are also loaded before measurements are added to the database. As new nodes or sensors are added, new records are added to those tables. When new types of nodes or sensors are added, those types are added to the database type tables.

Raw measurements arrive at the database as comma-separated-list ASCII files. They are then loaded into the database using a two-step process common to data warehouse applications. (1) The data are first loaded into a quality-control (QC) table (`RawData`) in which duplicate records and other erroneous data are removed. (2) Next, the quality-controlled data are copied into the `CleanData` table, while faulty data (e.g., duplicates) are inserted into the `BadData` table. The contents of the `CleanData` table are then inserted into the `DataSeries` table after converting the timestamps of the collected measurements from "sensor time" (i.e., the node's local clock) to GMT using Phoenix. Finally the contents of the `DataSeries` table are calibrated using stored procedures.

The database, implemented in Microsoft SQL Server 2005, benefits from the http://www.skyserver.sdss.org database that was built for Astronomy applications [40]. It inherited a self-documenting framework that uses embedded markup tags in the comments of the DDL scripts to characterize the metadata (units, descriptions, enumerations, for the database objects, tables, views, stored procedures, and columns). The DDL is parsed a second time, and the metadata information is extracted and inserted into the database itself. A set of stored procedures generate an HTML rendering of the hyper-linked documentation (see *Schema Browser* on our website [25]).

4.2 Data Access

Since sensor-based experiments have inherent spatial dimensions map-based interfaces (such as Microsoft Bing Maps and Google Earth) offer an intuitive interface for exploring such experiments. We implemented such an web-interface to our database called Grazor, where users can search for specific locations and sensors with specific capabilities or just browse through the map at their leisure. Once the sensors are identified, the user can click on their icons and receive current as well as historical measurements collected by these sensors.

Current and historical sensor measurements are available from the *Life Under Your Feet website*, where measurements can either be viewed directly inside the web browser or downloaded as CSV files. Figure 12 presents a sample screenshot of Grazor covering the geographic area of one of our deployments.

In addition to Grazor, we also have a direct web-interface for performing arbitrary analysis by exposing the SQL schema and allow SQL queries directly to the database.

Fig. 12 Grazor User Interface from *Life Under Your Feet*

This "guru-interface" has proven invaluable for scientists using the Sloan Digital Sky Survey and has already been very useful to us. If there is some question you want to ask that is not built-in, this interface lets you ask that question.

4.3 Data Analysis

In addition to examining individual measurements and looking for unusual cases, scientists want a high level view of the measured quantities; they want to analyze aggregations and functions of the sensor data, cross-correlate them with external measurements, and perform these tasks using intuitive and easy-to-use tools. Some of the typical questions we expect these systems to answer are:

1. Display a physical quantity (average, min, max, standard deviation) for a particular time or time interval, for one sensor, for a subset of the sensors, for all sensors at a site, or for all sites. Show the results as a function of location, time, as well as a function of sensor subset ID or category.
2. Look for unusual patterns and outliers such as a node behaving differently from its neighbors or an unusual spike in measurements.
3. Look for extreme events, e.g., rainstorms, and show data in time-after-event coordinates.

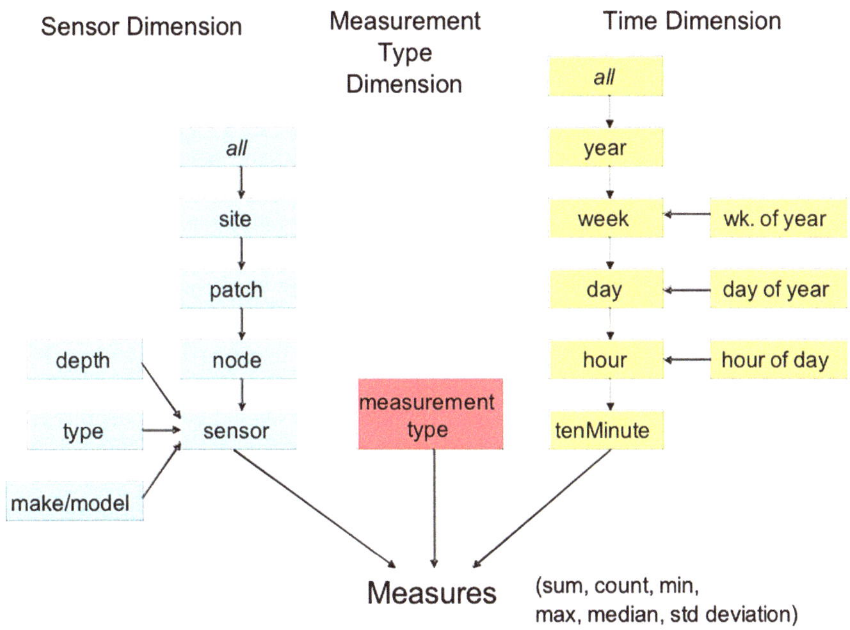

Fig. 13 Sensor data cube dimension model

4. Correlate measurements with external datasets (e.g., with weather data, data from CO_2 flux towers, or data from stream gauges).
5. Notify the user in real-time if the data has unexpected values, indicating that sensors might be damaged and need to be checked or replaced.

Queries 2–5 are standard relational database queries that fit the schema in Fig. 11 very nicely—indeed the database was designed for them. However, Query 1 is really the main application of the data analysis and calls for a specialized database design, called a *data cube*, that supports roll-up and drill-down data queries across many dimensions [14].

Figure 13 shows the unified dimension model for a data cube we built for the database shown in Fig. 11. It is built and maintained using the Business Intelligence Development Studio and OLAP features of SQL Server 2005. The cube provides access to all sensor measurements including air and soil temperature, soil water pressure and light flux averaged over 10-min measurement intervals, in addition to daily averages, minima and maxima of weather data including precipitation, cloud cover and wind.

The cube also defines calculations of average, min, max, median and standard deviation that can be applied to any type of sensor measurement over any selected spatio-temporal range. Analysis tools querying the cube can display these aggregates easily and quickly, as well as apply richer computations such as correlations that are supported by the multidimensional query language MDX [33]. Users can aggregate and pivot on a variety of attributes: position on the hillside, depth in the soil, under the shade versus in the open, etc.

The cube aggregates the `DataSeries` fact table around three dimensions (when, who, where)—Time (`DateTimes`), Location/Sensor (`Sensor`), and Measurement Type (`MeasurementType`) (see Fig. 13.) The Time dimension includes a hierarchy providing natural aggregation levels for measurement data at the resolution of year, season, week, day, hour and minute (to the grain of 10-min interval). Not only can data be summarized to any of these levels (e.g., average temperature by week), but these summarized data can then also be easily grouped by recurring cyclic attributes such as hour-of-day and week-of-year. The Location/Sensor dimension includes a geographic hierarchy permitting aggregation or slicing by site, patch, node or individual sensor, as well as a variety of positional or device-specific attributes (patch coordinates, node position, sensor manufacturer, etc.) This dimension itself is constructed by joining the relational database tables representing sensor, site, patch, and node.

The weather data available in the cube uses these dimensions as well, although at a different time and space grain. In the Location/Sensor and time dimensions, weather is available per-site and per-day respectively. By sharing the same dimensions as the sensor measurements, relationships between weather and measurement information can be readily analyzed and visualized side-by-side using the tools.

Data visualization, trending and correlation analysis is most effective when measurement data is available for every 10-min measurement interval of a sensor. While it is straightforward to handle large contiguous data gaps by eliminating a gap period from consideration, frequent gaps can interfere with calculations of daily or hourly averages. To avoid these problems, we plan to use interpolation techniques to fill any holes in the data prior to populating the cubes.

Figure 14 displays one example of the type of analysis enabled by the data cube. It displays the correlation between surface temperature and light intensity as a function of sensor ID, indicated by circles with different colors, averaged over the whole duration of the experiment. As daylight breaks, the temperature of the surface quickly rises (moving to the right), and reaches its maximum around 2–3 PM, since the deployment site has northern exposure. Then as dusk sets, temperature starts to decline reaching its minimum during the early morning hours. This example demonstrates the power of visualization to expose subtle data features and lead to deeper scientific insights as well as provide a decision tool for how the sensor-based experiment should be modified to better cover the scientists' needs.

5 Conclusion

A wireless sensor network is only the first component in an *end-to-end* system that transforms raw measurements to *scientifically significant* data and results. This end-to-end system includes calibration, interfaces with external data sources (e.g., weather data), databases, Web Services interfaces, analysis, and visualization tools.

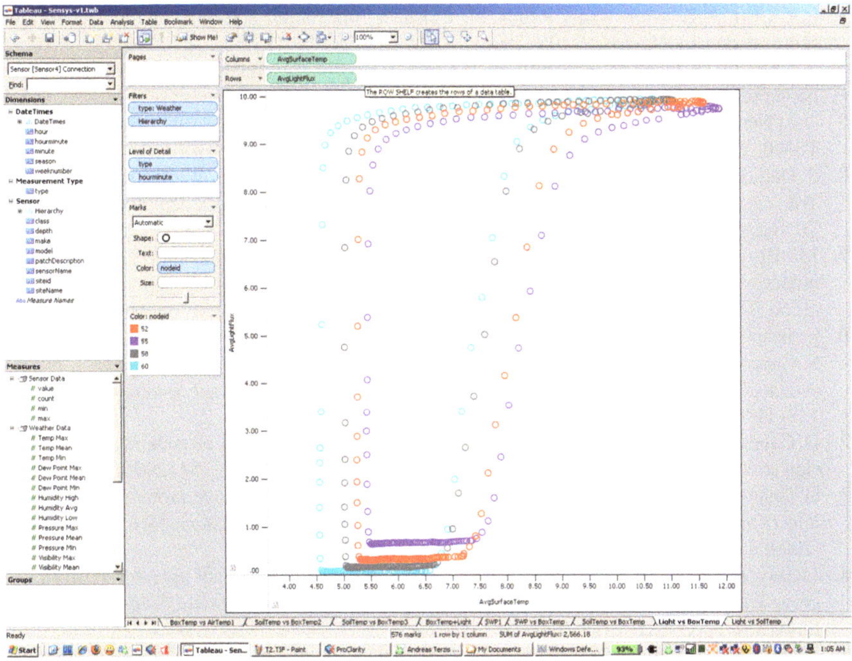

Fig. 14 Example of datacube analysis. Correlation between surface temperature and light intensity

The WSN community has focused its attention so far on routing algorithms, self-organization, and in-network processing among other things, environmental monitoring applications—sometimes derided as *academically dull applications*—require a different emphasis: reliable delivery of the majority (if not all) of the data and metadata, high quality measurements, and reliable operation over long deployment cycles. We believe that focusing on these problems will lead to interesting new avenues in WSN research.

References

1. M. Buettner, G. Yee, E. Anderson, R. Han, X-MAC: a short preamble MAC protocol for duty-cycled wireless sensor networks, in *Proceedings of the 4th ACM SenSys Conference* (2006)
2. N. Burri, P. von Rickenbach, R. Wattenhofer, Dozer: ultra-low power data gathering in sensor networks, in *Proceedings of the 6th IPSN Conference* (2007)
3. M. Chang, P. Bonnet, Meeting ecologists' requirements with adaptive data acquisition, in *Proceedings of the 8th ACM Conference on Embedded Networked Sensor Systems, SenSys '10* (ACM, New York, NY, USA, 2010), pp. 141–154. doi:10.1145/1869983.1869998. URL http://doi.acm.org/10.1145/1869983.1869998
4. M. Chang, C. Cornou, K. Madsen, P. Bonnet, Lessons from the Hogthrob deployments, in *WiDeploy* (2008)

5. B.R. Chen, K.K. Muniswamy-Reddy, M. Welsh, Ad-hoc multicast routing on resource-limited sensor nodes, in *REALMAN '06* (2006)
6. Y. Chen, O. Gnawali, M. Kazandjieva, P. Levis, J. Regehr, Surviving sensor network software faults, in *SIGOPS* (2009)
7. Dust Networks, Inc., Time Synchronized Mesh Protocol, http://www.dustnetworks.com/docs/TSMP_Whitepaper.pdf (2006)
8. P. Dutta, J. Hui, J. Jeong, S. Kim, C. Sharp, J. Taneja, G. Tolle, K. Whitehouse, D. Culler, Trio: enabling sustainable and scalable outdoor wireless sensor network deployments, in *IEEE SPOTS*, pp. 407–415 (2006)
9. J.E. Elson, L. Girod, D. Estrin, Fine-grained network time synchronization using reference broadcasts, in *Proceedings of the 5th Symposium on Operating Systems Design and Implementation (OSDI)*, pp. 147–163 (2002)
10. Federspiel Controls, http://www.federspielcontrols.com
11. S. Ganeriwal, R. Kumar, M.B. Srivastava, Timing-sync protocol for sensor networks, in *Proceedings of the 1st ACM Conference on Embedded Networked Sensor System (SenSys)*, pp. 138–149 (2003)
12. D. Ganesan, R. Cristescu, B. Beferull-Lozano, Power-efficient sensor placement and transmission structure for data gathering under distortion constraints, in *IPSN '04* (2004)
13. O. Gnawali, R. Fonseca, K. Jamieson, D. Moss, P. Levis, Collection tree protocol, in *Proceedings of the 7th ACM Conference on Embedded Networked Sensor Systems (SenSys)*, pp. 1–14 (2009)
14. J. Gray, A. Bosworth, A. Layman, H. Pirahesh, Data cube: a relational operator generalizing group-by, cross-tab and sub-totals, in *Proceeding of the 1996 International Conference in Data Engineering*, pp. 152–159 (1996)
15. J. Gupchup, D. Carlson, R. Musăloiu-E, A. Szalay, A. Terzis, Phoenix: an epidemic approach to time reconstruction, in *Proceedings of the 7th European Conference on Wireless Sensor Networks, EWSN'10* (Springer, Berlin, 2010), pp. 17–32. doi:10.1007/978-3-642-11917-0_2. URL http://dx.doi.org/10.1007/978-3-642-11917-0_2
16. J. Gupchup, R. Musaloiu-Elefteri, A.S. Szalay, A. Terzis, Sundial: using sunlight to reconstruct global timestamps, in *EWSN*, pp. 183–198 (2009)
17. S. Kim, R. Fonseca, P. Dutta, A. Tavakoli, P.L. David Culler, S. Shenker, I. Stoica, Flush: a reliable bulk transport protocol for multihop wireless networks, in *Proceedings of 5th ACM Sensys Conference* (2007)
18. J. Ko, J.H. Lim, Y. Chen, R. Musvaloiu-E, A. Terzis, G.M. Masson, T. Gao, W. Destler, L. Selavo, R.P. Dutton. MEDiSN: Medical emergency detection in sensor networks, in ACM Trans. Embed. Comput. Syst. 10(1), Article 11 (2010)
19. K. Langendoen, A. Baggio, O. Visser, Murphy loves potatoes: experiences from a pilot sensor network deployment in precision agriculture, in *Proceedings of the Parallel and Distributed Processing Symposium (IPDPS)* (2006)
20. H.K. Le, D. Henriksson, T. Abdelzaher, A control theory approach to throughput optimization in multi-channel collection sensor networks, in *IPSN '07* (2007)
21. H.K. Le, D. Henriksson, T. Abdelzaher, A practical multi-channel medium access control protocol for wireless sensor networks, in *IPSN '08* (2008)
22. P. Levis, D. Gay, V. Handziski, J.H. Hauer, B. Greenstein, M. Turon, J. Hui, K. Klues, R.S. Cory Sharp, J. Polastre, P. Buonadonna, L. Nachman, G. Tolle, D. Culler, A. Wolisz, T2: A Second Generation OS For Embedded Sensor Networks. Tech. Rep. TKN-05-007, Telecommunication Networks Group, Technische Universität Berlin (2005)
23. P. Levis, G. Tolle, *TEP 118 Dissemination*, http://www.tinyos.net/tinyos-2.x/doc/html/tep118.html
24. C.J.M. Liang, J. Liu, L. Luo, A. Terzis, F. Zhao, Racnet: a high-fidelity data center sensing network, in *Proceedings of the 7th ACM Conference on Embedded Networked Sensor Systems, SenSys '09* (ACM, New York, NY, USA, 2009), pp. 15–28. doi:10.1145/1644038.1644041. URL http://doi.acm.org/10.1145/1644038.1644041

25. Life Under Your Feet: Life Under Your Feet Schema Browser (2007), http://lifeunderyourfeet.org/en/help/browser/browser.asp
26. S. Lin, J. Zhang, G. Zhou, L. Gu, J.A. Stankovic, T. He, ATPC: adaptive transmission power control for wireless sensor networks, in *Proceedings of the 4th ACM Sensys Conference* (2006)
27. J. Liu, B. Priyantha, F. Zhao, C.J.M. Liang, Q. Wang, S. James, Towards fine-grained data center cooling monitoring using racnet, in *HotEmNets'08* (2008)
28. Y. Liu, Y. He, M. Li, J. Wang, K. Liu, L. Mo, W. Dong, Z. Yang, M. Xi, J. Zhao, X.Y. Li, Does wireless sensor network scale? A measurement study on greenorbs, in *INFOCOM, 2011 Proceedings IEEE*, pp. 873–881 (2011)
29. M. Lukac, P. Davis, R. Clayton, D. Estrin, Recovering temporal integrity with data driven time synchronization, in *IPSN*, pp. 61–72 (2009)
30. S. Madden, M.J. Franklin, J.M. Hellerstein, W. Hong, Tag: a tiny aggregation service for ad-hoc sensor networks, in *OSDI '02* (2002)
31. M. Marot, B. Kusy, G. Simon, A. Ledeczi, The flooding time synchronization protocol, in *Proceedings of the 2nd Conference on Embedded Networked Sensor Systems (SenSys)*, pp. 39–49 (2004)
32. A. Meliou, D. Chu, C. Guestrin, J. Hellerstein, W. Hong, Data gathering tours in sensor networks, in *Proceedings of IPSN* (2006)
33. Microsoft Corporation: Multidimensional Expressions (MDX) Reference, http://msdn2.microsoft.com/en-us/library/ms145506.aspx (2005)
34. R. Musaloiu-E., C.J. Liang, A. Terzis, Koala: ultra-low power data retrieval in wireless sensor networks, in *Proceedings of the 7th International Conference on Information Processing in Sensor Networks (IPSN)* (2008)
35. R. Musaloiu-E., C.J.M. Liang, A. Terzis, Koala: ultra-low power data retrieval in wireless sensor networks, in *IPSN '08: Proceedings of the 7th International Symposium on Information Processing in Sensor Networks* (2008)
36. J. Paek, R. Govindan, Rate-controlled reliable transport for sensor networks, in *Proceedings of the 5th ACM Sensys Conference* (2007)
37. J. Polastre, J. Hill, D. Culler, Versatile low power media access for wireless sensor networks, in *Proceedings of the 2nd ACM Sensys Conference* (2004)
38. S. Rangwala, R. Gummadi, R. Govindan, K. Psounis, Interference-aware fair rate control in wireless sensor networks, in *SIGCOMM '06* (2006)
39. J. Sallai, B. Kusy, Á. Lédeczi, P. Dutta, On the scalability of routing integrated time synchronization, in *EWSN*, vol. 3868 (Springer, Berlin, 2006), pp. 115–131, http://dblp.uni-trier.de/db/conf/ewsn/ewsn2006.html#SallaiKLD06
40. Sloan Digital Sky Survey: The Sloan Digital Sky Survey/SkyServer (2002), http://skyserver.sdss.org/
41. T. Stathopoulos, L. Girod, J. Heidemann, D. Estrin, Mote Herding for Tiered Wireless Sensor Networks. Tech. Rep. CENS-TR-58, University of California, Los Angeles, Center for Embedded Networked Computing (2005)
42. SynapSense Corporation: LiveImaging: Wireless Instrumentation Solutions, http://www.synapsense.com/ (2008)
43. R. Szewczyk, J. Polastre, A. Mainwaring, D. Culler, Lessons from a sensor network expedition, in *Proceedings of the 1st European Workshop on Wireless Sensor Networks (EWSN '04)* (2004)
44. Texas Instruments: 2.4 GHz IEEE 802.15.4/ZigBee-ready RF Transceiver, http://www.chipcon.com/files/CC2420_Data_Sheet_1_3.pdf (2006)
45. G. Tolle, J. Polastre, R. Szewczyk, N. Turner, K. Tu, P. Buonadonna, S. Burgess, D. Gay, W. Hong, T. Dawson, D. Culler, A macroscope in the redwoods, in *Proceedings of the Third ACM Conference on Embedded Networked Sensor Systems (SenSys)* (2005)
46. G. Werner-Allen, S. Dawson-Haggerty, M. Welsh, Lance: optimizing high-resolution signal collection in wireless sensor networks, in *SenSys '08* (2008)
47. G. Werner-Allen, K. Lorincz, J. Johnson, J. Lees, M. Welsh, Fidelity and yield in a volcano monitoring sensor network, in *Proceedings of the 7th USENIX Symposium on Operating Systems Design and Implementation (OSDI)* (2006)

48. A. Woo, T. Tong, D. Culler, Taming the underlying challenges in reliable multihop wireless sensor networks, in *Proceedings of ACM Sensys 2003* (2003)
49. Y. Wu, J. Stankovic, T. He, S. Lin, Realistic and efficient multi-channel communications in dense sensor networks, in *INFOCOM '08* (2008)
50. W. Ye, J. Heidemann, D. Estrin, An energy-efficient MAC protocol for wireless sensor networks, in *Proceedings of IEEE INFOCOM 2002* (2002)
51. J. Zhao, R. Govindan, Understanding packet delivery performance in dense wireless sensor networks, in *Proceedings of the ACM Sensys* (2003)
52. G. Zhou, C. Huang, T. Yan, T. He, J.A. Stankovic, T.F. Abdelzaher, MMSN: multi-frequency media access control for wireless sensor networks, in *INFOCOM '06* (2006)

Chapter 16
Data Gathering in Wireless Sensor Networks

Shouling Ji, Jing (Selena) He and Zhipeng Cai

Abstract Data gathering is one of the primary operations carried out in Wireless Sensor Networks (WSNs). It involves *data collection with aggregation* and *data collection without aggregation*, referred to as *data aggregation* and *data collection* respectively. In the last decade, many techniques for these two applications are proposed, with different focuses, such as accuracy, reliability, time complexity, and so on. This chapter reviews the state of the art of data aggregation and data collection techniques in order to present a comprehensive guidance on how to choose a more appropriate approach for different applications. The definitions of data aggregation and data collection are firstly introduced. Subsequently, the challenges of designing effective data aggregation and data collection methods are discussed. Then some typical data aggregation techniques and their classifications are presented. Particularly, a latest distributed data aggregation algorithm (DAS) is illustrated in details. For data collection, we begin with some new advances and then introduce several new tree-based and cell-based data collection algorithms. Finally, this chapter is ended by pointing out some possible future research directions.

1 Introduction

Wireless Sensor Networks (WSNs) have been successfully applied to a variety of fields. For all the applications, data gathering is one of the primary operations carried out in WSNs, where a base station collects all the data generated from each sensor through wireless communications. Data gathering is mainly for estimating network

---

S. Ji · Z. Cai (✉)
Department of Computer Science, Georgia State University, Atlanta, GA 30303, USA
e-mail: zcai@gsu.edu

J. He
Department of Computer Science, Kennesaw State University,
Kennesaw, GA 30144-5591, USA

H. M. Ammari (ed.), *The Art of Wireless Sensor Networks*,
Signals and Communication Technology, DOI: 10.1007/978-3-642-40009-4_16,
© Springer-Verlag Berlin Heidelberg 2014

Fig. 1 An example WSN

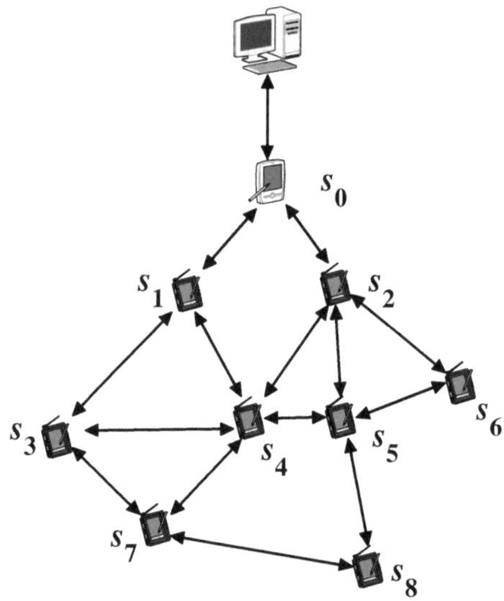

size, determining average system load, processing user queries, gathering interesting data, and so on. Much effort has been spent on designing effective data gathering approaches with different focuses such as energy-efficiency, network lifetime, delay bound, network throughput, energy-delay tradeoff, and so on [2, 3, 7, 15, 16, 19, 22, 24, 25, 33, 35–38, 41, 49, 51, 59, 62, 67, 69, 72, 73]. In this chapter, we introduce and summarize the recent advances of data gathering, and present some representative works.

In general, data gathering can be further classified as *data collection with aggregation* [2, 3, 15, 16, 22, 25, 35, 49, 59, 62, 67] and *data collection without aggregation* [7, 19, 24, 36–38, 41, 51, 69, 72, 73], referred to as *data aggregation* and *data collection* respectively. In data aggregation, specific aggregation functions are employed during the data gathering process, e.g., MAX, MIN, SUM, AVERAGE, and so on. In data collection, all the raw data produced at each node is gathered to the *sink* (base station) without any aggregation function. Figure 1 shows an example WSN consisting of one sink node denoted by s_0 and 8 sensor nodes denoted by s_i $(1 \leq i \leq 8)$. s_0 can communicate directly with a PC or some other user-operated devices in a wired or wireless manner. Each bidirectional edge between two nodes in Fig. 1 implies these two nodes have a one-hop communication link. For two nodes without a direct link, they can communicate via a multi-hop manner as long as the network is connected. For instance, in Fig. 1, s_1 and s_8 can communicate along the path $s_1 \leftrightarrow s_4 \leftrightarrow s_5 \leftrightarrow s_8$ or other paths. Now, suppose the WSN in Fig. 1 is used for monitoring the temperature of the deployed area. At a particular time instant t, assume the temperature values sensed by s_i $(1 \leq i \leq 8)$ are 48° at s_1, 58° at s_2, 60° at s_3, 57° at s_4, 61° at s_5, 58° at s_6, 47° at s_7, and 49° at s_8, respectively. Then, if we want to obtain the maximum temperature value of the monitored area at time t, we can conduct data aggregation in this network by applying the aggregation function MAX, which is used to obtain the maximum value of a candidate data set. Consequently, in this example, the data aggregation value is MAX$\{48°, 58°, 60°, 57°, 61°, 58°, 47°, 49°\} = 61°$. On the other hand, if we want to perform data collection over this network, e.g., to collect all the temperature values sensed by all the sensors at time t to the sink, all the 8 temperature values in the set $\{48°, 58°, 60°, 57°, 61°, 58°, 47°, 49°\}$ will be collected to the sink for further processing.

When dealing with data aggregation and data collection for WSNs, the constraints and limitations of WSNs on power supply, computation ability, and communication capacity introduce many challenges. We summarize some main challenges as follows.

- **Energy Efficiency**. Most sensor nodes are battery-powered, which implies the available power of a sensor node is very limited. Furthermore, in large-scale WSNs and WSNs which are deployed as monitoring and controlling systems where human intervention is not desirable or feasible, it is usually impossible to recharge a sensor node. Therefore, how to design energy-efficient data aggregation and data collection methods is a challenge. Especially for data collection, more traffic will be induced compared with data aggregation.
- **Timeliness and Real-time**. The limited communication capacity of each sensor node and the interference caused by wireless communication makes it more challenging to design effective data aggregation and collection algorithms for real-time applications. It is worth mentioning that more communication traffic will be induced for data collection compared with data aggregation, especially for the nodes close to the sink. Apparently, data are easily accumulated at the nodes close to the sink, which is called the data accumulation phenomenon. Thus, to resolve this problem and design elegant scheduling schemes is very important.
- **Scalability and Robustness**. WSNs tend to be large-scale networks and distributed systems. Some new nodes may join a network and some existing nodes may disappear at any time. This implies that the topological structure of a WSN varies over time. To deal with the dynamics, distributed algorithms with scalability are highly desired. On the other hand, it is difficult, sometimes impossible, to obtain the overall realtime network information to design optimal distributed algorithms. Therefore, designing optimal or sub-optimal distributed data aggregation and collection algorithms is challenging.
- **Time Synchronization and Distributed Solutions**. For large-scale WSNs consisting of vulnerable sensor nodes, it might be difficult and not realistic to achieve ideally strict time synchronization due to the unstable deployment environment, clock drifts, and technical limits.[1] Therefore, to comprehensively and profoundly

[1] Currently, some works have been proposed on the time synchronization issue for wireless sensor networks. For instance, in [23], a flooding-based time synchronization method, named Glossy, was designed. By testbed experiments (the testbed is a wireless sensor network consisting of 39–94 sensor nodes) and analysis, it can be shown that Glossy can achieve high-accurate network-wide time synchronization (at the millisecond level). Here, we emphasize that for large-scale wireless sensor networks deployed in an open/outdoor environment, e.g., GreenOrbs, a practical wireless sensor network consisting of 1000+ sensor nodes deployed in a forest [32], it might be difficult to achieve network-wide accurate time synchronization.

understand the performance of data aggregation and collection for practical WSNs, it is also important to investigate distributed data aggregation and collection algorithms for asynchronous WSNs.

- **Other Issues**. There exist many other challenges for data aggregation and collection in WSNs induced by node mobility, duty cycling, security issues, and so on, in different kinds of applications. To address these challenges, application-aware data aggregation and collection schemes are expected.

The rest of this chapter is organized as follows. In Sect. 2, the formal definition of data aggregation is provided. Then we review and summarize the existing data aggregation methods, followed by the discussion of a famous distributed data aggregation scheme. In Sect. 3, we go through some typical data collection algorithms. Finally, we conclude this chapter and point out some possible future research directions in Sect. 4.

2 Data Aggregation

In this section, we first provide some background knowledge of data aggregation. Subsequently, we review some existing data aggregation methods. Finally, we discuss one popular distributed data aggregation algorithm (DAS) [76], which is also the existing best algorithm with respect to minimizing the data aggregation latency.

2.1 Introduction of Data Aggregation

Data aggregation is an essential operation in WSNs and has many real applications. As introduced in [35], data aggregation can be used to determine network size by the COUNT aggregation function, which is an important parameter in many applications, e.g., Distributed Hash Tables (DHT) [56, 65], the configuration of a quorum in dynamic settings [1], and membership service in ad hoc networks [5]. There are many other applications of data aggregation by applying different aggregation functions such as MAX, MIN, AVERAGE, and SUM.

Generally, to accomplish a data aggregation task, we have to design a data aggregation schedule, which specifies how to carry out the aggregation task step by step in an interference-free manner. Now, we give the formal definition of an interference-free data aggregation schedule in randomly deployed WSNs under the protocol interference model.[2] For a WSN consisting of one single-radio sink node denoted by s_0 and n single-radio sensor nodes denoted by $s_1, s_2, \ldots, s_n$, the transmission range r and interference range r_I of all the nodes are assumed to satisfy $r_I = c_1 \cdot r$, where

[2] As indicated by some recent research, the protocol interference model may not be practical in realistic WSN applications. However, it has been employed since it simplifies the quantitative analysis process of a designed protocol which might provide some general guiding rules and insights on designing practical protocols for WSNs.

$c_1 \geq 1$ is a constant. The network time is assumed to be slotted. Since each node is equipped with a radio, within a time slot, a node s_i $(0 \leq i \leq n)$ can only work on a half-duplex communication mode, that is, a node can either send or receive data in a time slot but not both. Then, the topological structure of this WSN can be represented by a graph $G(V, E)$, where $V = \{s_i : 0 \leq i \leq n\}$ and E is the set of all the possible edges formed among all the nodes in V. For any two nodes $u, v \in V$, there is an edge between u and v if and only if $\|u - v\| \leq r$, where $\|u - v\|$ is the Euclidean distance between u and v. Under the protocol interference model, node u can successfully transmit data to node v only if (i) there is a link[3] between u and v ($(u, v) \in E$); and (ii) there is no other node w such that v is within the interference range of w ($\|w - v\| \leq r_I$) and w is trying to transmit data simultaneously with u.

Let X and Y be two subsets of V and $X \cap Y = \emptyset$. The data from the nodes in X can be aggregated to the nodes in Y if and only if all the nodes in X can simultaneously transmit data to some nodes in Y during a time slot in an interference-free manner. Then, an interference-free data aggregation schedule can be determined by a sequence of data sets $S_1, S_2, \ldots, S_\tau, \{s_0\}$. During the first time slot, the data from S_1 can be aggregated to the nodes in $V \backslash S_1$ in an interference-free manner. During the second time slot, the data from S_2 can be aggregated to the nodes in $V \backslash (S_1 \cup S_2)$ in an interference-free manner. This process continues until the τ-th time slot, during which the data from S_τ is aggregated to the sink s_0 in an interference-free manner. Therefore, a data aggregation schedule can be formally defined as follows.

Definition 1 ***[Data aggregation schedule].*** *A data aggregation schedule is a sequence of data sets $S_1, S_2, \ldots, S_\tau, S_{\tau+1} = \{s_0\}$ that satisfies the following conditions:*

- $\bigcup_{i=1}^{\tau+1} S_i = V$;
- $S_i \cap S_j = \emptyset$, *for* $1 \leq i \neq j \leq \tau + 1$;
- *All the data of the nodes from set S_k $(1 \leq k \leq \tau)$ can be aggregated to the nodes in $V \backslash \bigcup_{i=1}^{k} S_i$ during the k-th time slot, which implies the sink can obtain the final data aggregation value in the τ-th time slot.*

Currently, the study of data aggregation schedule in WSNs focuses on the following objectives: minimizing data aggregation delay, maximizing network throughput, maximizing network lifetime, and maximizing energy efficiency.

Note that, besides the protocol interference model, two other more realistic interference models are frequently exploited to capture wireless interference characteristics in WSNs, namely the physical interference model and the generalized physical interference model.[4] For the given definition of data aggregation, it can be

---

[3] We use *link* and *edge* interchangeably in this chapter.

[4] Under the physical interference model, a node u can successfully transmit data to another node v only if the Signal-to-Interference-plus-Noise Ratio (SINR) at v associated with u is no less than a threshold value. Thus, the data transmission under the physical interference model can also be viewed as a binary function, i.e., to be failed or to be successful. Under the generalized physical interference model, instead of modeling a data transmission as a binary function, the data receiving rate at v from u is a continuous function depending on the SINR value at v.

applied under the physical interference model and the generalized physical interference model directly. This is because that the definition specifies a proper data aggregation schedule from the interference-free scheduling perspective. For how to define interfering and interference-free, it is independent of data aggregation definition and only depends on different interference models. Similarly, in arbitrarily deployed WSNs, the formal definition of a data aggregation schedule can also be obtained with some according modifications.

2.2 Overview of Data Aggregation Schemes

In this subsection, we review some recent works on data aggregation. Some comprehensive surveys have been conducted in [2, 22, 35, 59, 67]. Furthermore, the research advances on secure data aggregation are summarized in [3, 25, 62]. In the following of this subsection, we summarize new advances on data aggregation techniques from three aspects: energy-efficiency, minimum-latency, and energy-delay tradeoff.

Energy-Efficient Data Aggregation. In [46], an energy-efficient data gathering method with reliability consideration is proposed. To conserve energy, three schemes are exploited. First, by adopting the network flow optimization techniques, the optimal strategy to balance the communication load among all the nodes is investigated. Second, to further balance the communication load, multiple communication trees, instead of a fixed tree, are constructed. In different task cycles of a periodic-task, one of the multiple trees is exploited. Finally, interference-free data transmission procedures are designed for data transmission. In [57], a new minimum spanning tree-based protocol, named Power Efficient Data gathering and Aggregation Protocol (PEDAP), as well as its power-aware version is proposed. PEDAP tries to prolong the lifetime of the last node in the system and meanwhile considering the lifetime of the first node.[5] On the other hand, with the cost of only slightly sacrificing the lifetime of the last node, the near optimal lifetime of the first node can be achieved in the power-aware version. In [20], the aggregation tree construction problem is investigated and an energy-aware distributed algorithm to generate a data aggregation tree based on the residual power of the nodes is proposed. In the sparse-node distribution situation, a multi-chain-based data aggregation scheme is proposed in [21], where a chain is a data transmission path connecting some or all of the network nodes. Since the performance of the scheme heavily depends on the construction of the chain, an energy-efficient chain construction algorithm is also proposed in this work. In [18], the use of approximation in data aggregation by exploiting small sketches is investigated. First, the authors generalized duplicate-insensitive sketches for approx-

[5] The network considered in [57] is a single-hop wireless sensor network where all the nodes can communicate directly. When considering the lifetime of the last node, it is defined as the time duration from the network being deployed to the time that the last node exhausts its energy. When considering the lifetime of the first node, it is defined as the time duration from the network being deployed to the time that a node exhausts its energy.

imating COUNT to deal with SUM. Subsequently, they proposed a method to use sketch to produce accurate results with low communication cost and computation cost. Considering the fact that it is not necessary to provide 100 % accuracy for most WSN applications, the authors in [63] introduced a data structure, named Quantile Digest or q-digest, which provides provable guarantees on approximation error and maximum resource consumption. Based on the experiments on a WSN involving 49 MICA 2 motes using a realistic traffic trace, a method to address reliable bursty convergecast scheme for WSNs is proposed [77]. In the proposed scheme, to improve channel utilization and reduce ack-loss, a window-less block acknowledgment scheme is designed; to alleviate retransmission-incurred channel contention, a differentiated contention control mechanism is implemented.

Minimum-Latency Data Aggregation. The Minimum-Latency Data Aggregation (MLDA) problem is studied in [10] for WSNs under the Unit Disk Graph (UDG) model and the protocol interference model. First, the authors proved that the MLDA problem is NP-hard even when all the sensors are grid-deployed by applying the result on orthogonal planar drawing and reduction from restricted planar 3-SAT. Then, a $(\Delta - 1)$-approximation algorithm for the MDAT problem is proposed, where $(\Delta - 1)$ equals the maximum number of sensors within the transmission range of any sensor. In [79], the ratio of $(\Delta - 1)$ proposed in [10] is improved, and a new approximated data aggregation algorithm with ratio $\frac{7\Delta}{\log_2 |S|} + c$ is proposed, where S is the set of sensors containing source data, and c is a constant. In [78], the challenges of fast and reliable data aggregation on the collision-prone CSMA MAC layer is studied. The authors considered two cases: first, the size of the packets produced by all the sensor nodes is much smaller than the maximum size of a data frame that can be transmitted in one time slot; second, some node does not have data for transmission and for those nodes which have, they may have lots of data that require more than one data packet. Latterly, Wan et al. [68] further improved the MLDA ratio bound under the UDG model and protocol interference model. In that work, they first designed three data aggregation algorithms with latency $15R + \Delta - 4$, $2R + O(\log R) + \Delta$, and $(1 + O(\log R/\sqrt[3]{R}))R + \Delta$, respectively, when the interference range is equal to the transmission range, where R and Δ are the range (network radius with respect to the base station) and maximum degree of the communication topology, respectively. Subsequently, they extended their results to the case that the interference range is larger than the transmission range, and proposed two aggregation algorithms. Recently, the authors of [76] and [74] studied distributed data aggregation algorithms for WSNs. Under the UDG model and protocol interference model, the proposed distributed data aggregation algorithm yields an aggregation schedule with latency $24D + 6\Delta + 16$, where D is the network diameter and Δ is the maximum node degree.

Energy-Delay Tradeoff Data Aggregation. In [50], two new schemes to minimize $energy \times delay$ for CDMA and non-CDMA WSNs are proposed. When the goal is to minimize the delay cost only, a binary combining scheme is designed, which yields a solution of latency $\log n$ and incurs a slight increase in energy cost. For

CDMA WSNs, a chain-based binary scheme with optimized performance in terms of $energy \times delay$ is proposed, where the chain-based binary scheme is a hierarchical binary-tree like data aggregation schedule. For non-CDMA WSNs, a chain-based 3-level hierarchy scheme is designed with good performance. In [75], the energy-latency tradeoff issue for WSNs is studied. The authors proposed some algorithms to minimize the overall energy dissipation of sensor nodes in the aggregation tree subject to the latency constraint. For the off-line problem, they designed a numerical optimal algorithm and a pseudo-polynomial time approximation algorithm based on dynamic programming. In [44], the authors studied the data aggregation problem for wireless ad hoc networks. First, they proposed a simple randomized distributed data aggregation algorithm with expected running time $O(\log n)$, where n is the number of nodes in the network. Subsequently, they investigated the trade-off between the energy and the latency of data aggregation. They showed that their proposed algorithm consumes at most $O(n \log n)$ times the optimal minimum energy. Recently, Li et al. [48] studied the MLDA schedule problem and proposed an energy-efficient distributed data aggregation scheduling algorithm based on a novel cluster-based aggregation tree. By theoretical analysis, they showed that the induced latency of their proposed algorithm is upper bounded by $4R' + 2\Delta - 2$, where Δ is the maximum degree and R' is the inferior network radius which is smaller than the network radius R. Furthermore, by simulations, they showed that the proposed algorithm has a comparable delay performance as that of the most recently published centralized algorithm, while consumes 78 % less energy.

2.3 DAS [76]

Since distributed data aggregation algorithms are preferred for WSNs, in this subsection, we discuss a recently published distributed data aggregation scheme DAS. DAS is a distributed data aggregation scheduling algorithm which tries to minimize the induced data aggregation latency [76].

DAS is proposed under the UDG model, which is an even simplified version of the protocol interference model. For simplicity, the transmission range and interference range of all the nodes are assumed to be equal and normalized to 1. The network time is assumed to be slotted and synchronized. DAS consists of two phases. In the first phase, a data aggregation tree is constructed. In the second phase, the distributed aggregation scheduling is performed.

In phase one, a Connected Dominating Set (CDS)-based tree is constructed by exploiting the existing methods [26–31, 68] and serves as the data aggregations tree. Let $G(V, E)$ be a unit-disk graph representing a WSN. Define the sink s_0 as the *center* of G. The *radius* of G with respect to s_0 is the maximum depth of the *Breadth-First-Search* (BFS) tree rooted at s_0. For a subset U of V, U is a *Dominating*

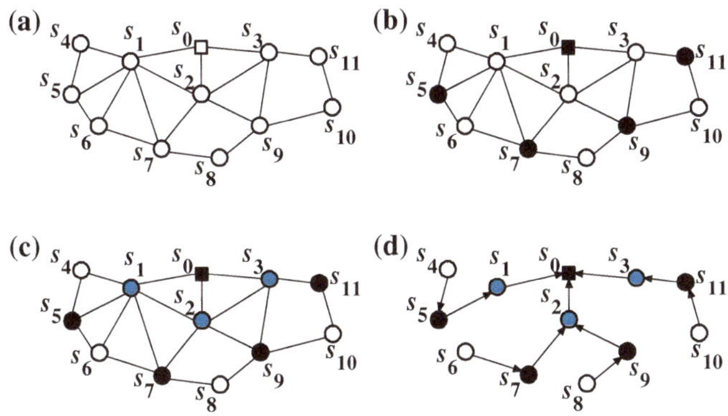

Fig. 2 The construction of a CDS-based data aggregation tree. s_0 is the sink. **a** Network topology, **b** black nodes as *dominators*, **c** blue nodes as *connectors* and **d** CDS-based data aggregation tree

Set (DS) of G if every node in V is either an element of U or adjacent[6] to at least one node in U. If the subgraph of G induced by U is connected, then U is called a *Connected Dominating Set* (CDS) of G. Since CDS can serve as a virtual backbone of a WSN, it receives a lot of attention recently [26–31]. We use the example shown in [43] to explain the aggregation tree construction process.

Taking the WSN shown in Fig. 2a as an example, a CDS-based data aggregation tree T can be built (shown in Fig. 2d) using the method proposed in [68]. Let G represent the network in Fig. 2a. T is rooted at sink s_0 and can be built according to the following steps. First, construct a BFS tree on G beginning at the sink and obtain a *Maximal Independent Set* (MIS) D according to the search sequence. As shown in Fig. 2b, the set of all the black nodes $\{s_0, s_5, s_7, s_9, s_{11}\}$ is an MIS of the network shown in Fig. 2a. Note that D is also a *Dominating Set* (DS) of G and an element in D is called a *dominator*. Clearly, every dominator is out of the communication range of any other dominator. Let G' be a graph on D in which two nodes in D are linked by an edge if and only if these two nodes have a common neighbor in G, e.g., s_0 and s_7. Obviously, sink s_0 is in G' and we also denote s_0 as the center of G'. Suppose that the radius of G' with respect to s_0 is L' and we denote the union of the dominators at level l $(0 \leq l \leq L')$ as set D_l. Note that, $D_0 = \{s_0\}$. Second, we choose nodes, also called *connectors*, to connect all the nodes in D to form a CDS. Let S_l $(0 \leq l \leq L')$ be the set of the nodes adjacent to at least one node in D_l and at least one node in D_{l+1} and compute a minimal cover $C_l \subseteq S_l$ for D_{l+1}. Let $C = \cup_0^{L'-1} C_l$ and therefore $D \cup C$ is a CDS of G. As shown in Fig. 2c, the blue nodes $\{s_1, s_2, s_3\}$ are connectors chosen to connect the dominators in $D_0 = \{s_0\}$ and $D_1 = \{s_5, s_7, s_9, s_{11}\}$. Meanwhile, the union of the dominators and connectors in Fig. 2c forms a CDS of the network shown in Fig. 2a. Finally, for any other node u,

[6] In this chapter, if we say two nodes u and v are adjacent/connected, we mean u and v are within the communication range of each other, i.e., $\|u - v\| \leq r$, where r is the transmission range of sensor nodes.

also called a *dominatee*, not belonging to $D \cup C$, choose the nearest dominator as u's parent node. In this way, the data aggregation tree T of G is obtained as shown in Fig. 2d. For the constructed data aggregation tree, it has many interesting properties, e.g., each connector node is adjacent to at most 5 dominators in which one of them is the parent node of that connector, and each dominator node is adjacent to at most 12 connectors in which one of them is the parent node of the dominator except for the sink node (the sink node is the root of the data aggregation tree which has no parent node).

The second phase is the aggregation scheduling phase. For each node $u \in V$, let $p(u)$ be u's parent node in the data aggregation tree T. Let $N(u)$ be the set of u's 1-hop neighbors except for $p(u)$ and $Ch(u)$ be the set of u's children, then u's *competitor set* is defined as $N(p(u)) \cup (\cup_{v \in N(u) \setminus Ch(u)} Ch(v)) \setminus \{u, p(u)\}$. In DAS, each node u has to maintain the following information:

- u's unique ID;
- u's competitor set, denoted by $M(u)$;
- u's earliest possible sending slot, K, initialized to 1;
- $R(u) = R_1(u) \cup R_2(u)$, where $R_1(u) = \{v | v \in M(u)$ and is ready to schedule$\}$, $R_2(u) = \{v | v \in M(u)$ and finished scheduling$\}$, and $R_1(u)$ and $R_2(u)$ are both initialized to be empty. For $\forall v \in R_1(u)$, v's earliest possible sending slot is maintained and for $\forall w \in R_2(u)$, w's schedule is maintained.
- The number x of u's children that have not been scheduled, which is initialized to the number of u's children.

Additionally, in DAS, at any time instant, each node is set to one of the following states: NOT-READY (NRY), READY (RY), WAIT0, WAIT1, SCHEDULE-COMPLETED (SC) and SLEEP. During the scheduling process, the behavior of each node is captured by the automaton shown in Fig. 3.

According to the automaton shown in Fig. 3, initially, all the leaf nodes on T are in the RY state, while all the other nodes are in the NRY state. Once a node is in the RY state, it has an aggregation value obtained by aggregating its own data and the data received from all of its children. Furthermore, each node in the RY state sends a MARK message containing its node ID and K to request for the feedback message from all its competitors and switches to the WAIT0 state. For each node, if it receives a MARK message, it replies a message back which contains its current state and the earliest possible sending slot. For a node u in the WAIT0 state, until it receives all the feedback messages, if its ID is larger than all the nodes in $R_1(u)$, it switches to state SC to call a function named FIX-SCH(u) to determine its actual schedule and sends a SCH-COMPLETE message containing its ID and K to the nodes in $R_1(u)$. The description of function FIX-SCH(u) is shown in Algorithm 1. Otherwise, if u's ID is not larger than the ID of every node in $R_1(u)$, it switches to state WAIT1 to wait for the event that its ID is larger than the nodes in $R_1(u)$. Until this event happens, u switches to state SC and conducts the same action mentioned above.

In DAS, each sensor node works according to the automaton shown in Fig. 3. Finally, the sink node can obtain the final data aggregation value. By theoretical analysis, it is shown that the time complexity of DAS is $O(n)$, the message complexity

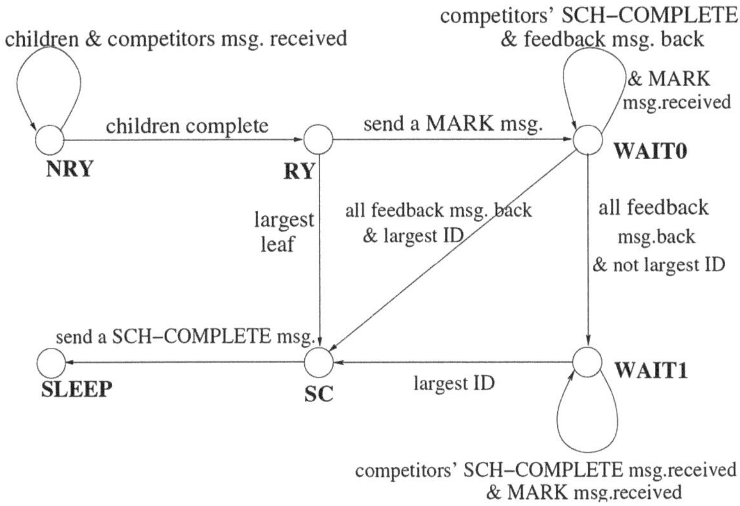

Fig. 3 The behavior of a node in DAS

Algorithm 1: FIX-SCH(u)

---

input : node u
output: u's actual schedule

$sch \leftarrow u.K$;
while *true* **do**
  if $\exists$ *node* $v \in R_2(u)$, $v.schedule = sch$ **then**
    $sch \leftarrow sch + 1$;
  else
    $u.schedule \leftarrow sch$;
    u sends a SCH-COMPLETE message and $u.schedule$ to $M(u) \cup \{p(u)\}$, then let u go to sleep and wake up in its time slot;

---

of DAS is $O(n\Delta)$, and the induced data aggregation delay is upper bounded by $24D + 6\Delta + 16$, where D is the network diameter and Δ is the maximum node degree.

3 Data Collection

In some applications, the base station needs to collect all the data from all the sensor nodes without aggregation [11, 12, 39, 43], referred to as data collection. In this operation, the union of all the data from all the sensor nodes at a particular time instant is called a *snapshot*, e.g., in the temperature-monitoring example shown in Fig. 1 (Sect. 1), the data set $\{48°, 58°, 60°, 57°, 61°, 58°, 47°, 49°\}$ produced by all the sensors at time t is a snapshot. The problem of collecting all the data of one snapshot is called *snapshot data collection*, and a snapshot data collection schedule can be defined formally as follows [39, 40, 43].

Definition 2 ***Snapshot Data Collection Schedule.*** *Suppose the network topology graph is represented by* $G = (V, E)$, *where* $V = \{s_0 = sink, s_1, s_2, \ldots, s_n\}$ *is the node and* E *is the set of all the possible formed links. A snapshot data collection schedule is a sequence of node transmission sets* $S_1, S_2, \ldots, S_\tau, S_{\tau+1} = \{s_0\}$ *that satisfies the following conditions*:

- $\cup_{i=1}^{\tau+1} S_i = V$;
- *During the* k*-th* ($1 \leq k \leq \tau$) *time slot, every node in set* S_k *can successfully transmits one data packet to some nodes in* $V \backslash S_k$;
- *At the end of the* τ*-th time slot, the sink* s_0 *can receive all the* n *data packets of a snapshot.*

Similarly, the problem of collecting multiple continuous snapshots is called *continuous data collection*, and a continuous data collection schedule is a schedule sequence of node transmission sets that gathers all the n data packets of each snapshot to the sink [39, 40, 43]. Furthermore, *data collection capacity*, which is defined as the achievable average data receiving rate at the sink during the data collection process [9, 11–13, 40, 43], is widely adopted as a performance measurement to measure how fast has data been collected to the sink.

Evidently, compared with data aggregation, data collection introduces more communication traffic, which implies a data collection network is more crowded and a data transmission suffers from more interference. Therefore, to design an effective data collection scheduling scheme is more challenging compared with data aggregation. In the following, recent research advances on data collection will be reviewed, followed by detailed discussion of several representative data collection methods, which yield the best result with respect to network capacity under the protocol interference model and the physical interference model, respectively.

3.1 Overview of Data Collection Schemes

Similar to data aggregation, data collection is another essential operation in WSNs. Therefore, much research has been conducted on this issue recently. In [69], recent data collection issues, challenges, and approaches in WSNs are summarized. In addition, the data collection methods in WSNs with mobile elements are surveyed in [24]. In the following, we will summarize the most recent data collection algorithms with different focuses.

Energy-Efficient Data Collection. In [71], Wang et al. studied the traffic-aware relay node deployment issue, which aims to maximize network lifetime. The authors developed optimal solutions for a simple case where there is only one source node with both single and multiple traffic flows. Subsequently, they showed that the general form of the deployment problem is NP-hard. Thus, they transformed the general problem into a generalized Euclidean Steiner Minimum Tree (ESMT) problem and developed a hybrid algorithm that can successfully return optimal results with all test cases which can be verified within acceptable time frames. However, the solution of ESMT is in the continuous domain and may yield fractional numbers of relay nodes. Therefore, they further developed algorithms for discrete relay node assignment, together with local adjustments that yield practical solutions. In [60], the authors investigated the delay and energy-efficiency tradeoffs for data collection in large-scale WSNs. They proposed efficient distributed algorithms for data collection with objectives of minimizing data collection delay, minimizing overall communication traffic, or minimizing total energy consumption. Furthermore, they theoretically proved that the proposed algorithms are either optimal or within constant factors of the optimum. To address the energy-efficient data collection challenges for WSNs, Liu et al. [52] designed a generic data collection framework on how to partition the sensors into clusters, how to dynamically maintain the clusters in response to environmental changes, how to make a schedule for the sensors in a cluster, how to explore temporal correlation, and how to restore the data in the sink with high fidelity.

Wang et al. [70] studied the approximate data collection issue for WSNs. They proposed an Approximate Data Collection (ADC) algorithm to partition the network into clusters, discover local data correlations on each cluster head, and perform a global approximate data collection at the sink according to model parameters uploaded by cluster heads. The adaptive data collection strategies for lifetime-constrained WSNs are studied in [66]. First, the authors implemented an off-line algorithm that can obtain the optimal data update strategy. Second, based on sensor readings, an adaptive strategy that makes data update decisions on the fly and meets network lifetime requirements is proposed. Furthermore, to cope with message losses, two methods, History and Expected, are also proposed for the adaptive strategy. Finally, in connection with the adaptive strategy for aggregate data collection, a scheme to assign the numbers of updates allowed to be sent by the sensor nodes based on their topological relations is designed.

Based on a distributed N-to-1 multi-path discovery protocol, a hybrid multi-path scheme (H-SPREAD), which aims to improve the security and reliability of data collection, is proposed in [53]. By combining with secret sharing, the end-to-end multi-path data dispersion in H-SPREAD enhances the security of the end-to-end data delivery. Furthermore, since H-SPREAD has multiple available data transmission paths, the reliability of data collection can be improved as well. In [61], Rothery et al. conducted an empirical study of data collection algorithms for WSNs. An important conclusion derived from that work is that the right choice of a data collection algorithm should come after understanding the operating environment of a WSN. In addition, this conclusion is shown to have significant impacts on the performance of a data collection scheme. By exploiting spatio-temporal load balancing, the authors in [47] designed a near-lifetime-optimal data collection strategy and an energy-efficient packet exchange mechanism for WSNs. In the designed strategy, instead of using a fixed communication topology, a set of communication topologies are constructed and can be applied in a duty-cycling manner with each topology for a data collection cycle. By integrating adaptively enabling/disabling the prediction scheme, Jiang et al. [42] proposed an energy-efficient framework for clustering-based data collection in WSNs. To efficiently implement the prediction techniques, they also designed an adaptive scheme.

Delay-Aware Data Collection. In [55], Luu et al. implemented an efficient scheduling for data collection through multi-path routing technique in WSNs. They also proposed an algorithm for deriving a super/lower bound on the shortest possible length of the data collection schedule generated by any algorithm. In [17], Cheng et al. designed a delay-aware data collection network structure for WSNs. With the objective of minimizing data collection delay, two network formation algorithms are designed to construct the proposed network structure in centralized and distributed manners, respectively. To reduce the redundancy of the data produced by all the sensors in a WSN, and thereby decreasing the communication traffic and energy consumption of a data collection task, Arici et al. [4] proposed a Pipelined In-Network COmpression (PINCO) scheme for data collection in WSNs. In PINCO, each group of data is a highly flexible structure such that compressed data can be recompressed without decompressing, which can further reduce communication traffic in a data collection task.

Data Collection with Mobile Elements. To improve data collection efficiency for WSNs and prolong network lifetime, mobile elements are usually considered. For instance, in [8, 34, 45, 58] and [6], the authors proposed to use mobile sinks or mobile nodes to improve data collection efficiency. In [58], the secure data collection problem using mobile data collectors is investigated for clustered WSNs. According to the tree-based key management scheme, three protocols for secure data collection in terms of different assumptions and constraints are proposed in [58], namely the Time Stamp Protocol (TSP), the Polynomial Points Sharing Protocol (PPSP), and the Secret Sharing Protocol (SSP), which are used to identify malicious Movement for Democratic Change and to maintain confidentiality of the collected data. The scalable data collection issue in WSNs with multiple mobile sinks is investigated in [45]. In this work, based on randomization and locally available information, a distributed data collection algorithm for multi-mobile-sink WSNs is designed and evaluated. Similar to [45], Chatzigiannakis et al.[8] also studied the data collection issue for WSNs with a mobile sink. They proposed four characteristic mobility patterns for the sink along with different data collection algorithms. By simulations, several important performance properties of each proposed protocol are examined. Another similar work exploiting node mobility for energy-efficient data collection in WSNs is [34]. In that work, the authors proposed a scheme to exploit mobile nodes in the monitoring area working as forwarding agents. They also showed an analytical model to understand the key performance metrics of data collection, e.g., data transfer, latency to the destination, and energy consumption. In [6], Borsetti et al. investigated the data collection issue for road-side sensor networks by using mobile nodes as data mules. In that work, they proposed a concept of Virtual Data Mule (VDM) and applied VDM to the case of data retrieval from road-side sensors through vehicular nodes. A federation problem was studied for data collection wireless sensor networks in the scenarios where the number of available mobile data collectors is less than the number of relay nodes required and more than the number of network segments (a network segment can be viewed as a connected network component in the network topological graph) in [64]. Specifically, an algorithm was presented to find an optimized travel routes for the mobile data collectors so that the average delay of the network can be minimized.

Data Collection Capacity. The delay-optimal data collection issue and the order-optimal data collection capacity problem are studied in [9, 11–14, 39, 40, 43]. In [12] and [13], the authors investigated the data collection capacity of arbitrary WSNs, which is a more general network deployment model compared with the traditional random WSNs. In that work, they first constructed a Breadth First Searching (BFS) tree as the data collection topology and then designed a data collection algorithm based on path scheduling. By theoretical analysis, they showed the proposed algorithm can achieve order-optimal data collection capacity of $\Theta(\frac{1}{8\alpha^2}W)$ for random deployed WSNs, where $\alpha > 1$ is a parameter in the protocol interference model, and W is the bandwidth of the wireless channel. For arbitrary WSNs, the proposed algorithm can achieve a data collection capacity of $\Omega(\frac{1}{\Delta^*}W)$, where Δ^* is a weighted-average value of the maximum interference among paths and branches of the BFS tree. In [39, 43], the authors studied the achievable continuous data collection capacity for dual-radio multi-channel WSNs. In that work, the authors first proposed a multi-path scheduling algorithm for snapshot data collection. By theoretical analysis, they showed that the multi-path scheduling algorithm has better performance than previous works, which yields a schedule plan of capacity lower bounded by $\frac{W}{\frac{3.63}{H}\rho^2+o(\rho)}$, where H is the number of available channels, W is the bandwidth of a channel, and ρ is the same parameter as α in [12] (which is actually a parameter in the protocol interference model). Subsequently, the authors proposed a novel continuous data collection algorithm for dual-radio multi-channel WSNs. By combining the pipeline scheduling technique and the compressive data gathering technique, the proposed continuous data collection algorithm speeds up the data collection process significantly.

Since the physical interference model is more accurate to describe wireless interference in WSNs, the authors in [11] and [40] studied the snapshot data collection capacity and continuous data collection capacity, respectively. By taking an elegant network partition method, two cell-based snapshot data collection algorithms are proposed, which are shown to be order-optimal. Additionally, the authors in [40] proposed a Segment-Based Pipeline Scheduling (SBPS) algorithm under the physical interference model. By analysis, SBPS can achieve a continuous data collection capacity of $\Omega(\sqrt{\frac{n}{\log n}} \cdot W)$ or $\Omega(\frac{n}{\log n} \cdot W)$, where n is the number of sensor nodes in a WSN. In [9] and [14], Chen et al. studied the data collection capacity problem for multi-sink WSNs. They showed that the achievable data collection capacity of their designed algorithm for a k-sink WSN is

$\Theta(k \cdot W)$, which implies multiple sinks can improve the achievable data collection capacity.

3.2 Tree-Based Data Collection

In [43] and [39], the authors studied the snapshot data collection and continuous data collection issues for randomly deployed dual-radio multi-channel WSNs under the protocol interference model. In this subsection, we discuss the methods proposed in those papers. First, we describe the network model employed in [43] and [39]. Subsequently, the multi-path scheduling algorithm for snapshot data collection is presented. The basic idea of the multi-path scheduling algorithm is trying to schedule multiple interference-free paths on the data collection tree concurrently to accelerate the data collection process. Third, the proposed pipeline scheduling algorithm for continuous data collection is given. In the pipeline scheduling algorithm, the data collection tree is partitioned into segments, and then a data collection pipeline can be formed on these segments to implement parallel transmission of multiple snapshots. Finally, some obtained theoretical results on the proposed algorithms are demonstrated.

Network Model. In [43] and [39], the considered network is a WSN consisting of n sensor nodes and one sink node. The transmission range of all the nodes is normalized to one and thus the network topology can be represented by a Unit Disk Graph (UDG). The interference range in the protocol interference model is assumed to be $\rho \geq 1$. For each node, the number of available orthogonal channels is H, denoted by λ_1, $\lambda_2, \ldots, \lambda_H$, respectively. In addition, the bandwidth of all the channels is same, each of which is W bits/second.

Multi-Path Scheduling for Snapshot Data Collection. To accomplish a data collection task, a Connected Dominating Set (CDS)-based routing tree is constructed first in [43] and [39] to serve as the communication structure. The construction process has been discussed in Sect. 2.3. For the snapshot data collection problem, a multi-path scheduling algorithm, which is based on a single path scheduling algorithm, is proposed. We will show the single-path scheduling algorithm first, and then we will review some multi-path scheduling algorithms.

In the single-path scheduling algorithm, the data collection on one path is scheduled. First, the links on the path are partitioned into two sets denoted by P_o and P_e, where P_o denotes the set of links on the path whose heads are dominators and whose tails have at least one packet for transmission, and P_e denotes the set of links on the path whose tails are dominators and have at least one data packet for transmission. For the path shown in Fig. 4a, s_0 and s_2 are dominators. Hence, $P_o = \{(s_3, s_2), (s_1, s_0)\}$ and $P_e = \{(s_2, s_1)\}$. Then, in the single-path scheduling algorithm, the data collection on a path will be scheduled by repeating the following two steps until all the data has been collected to the sink.

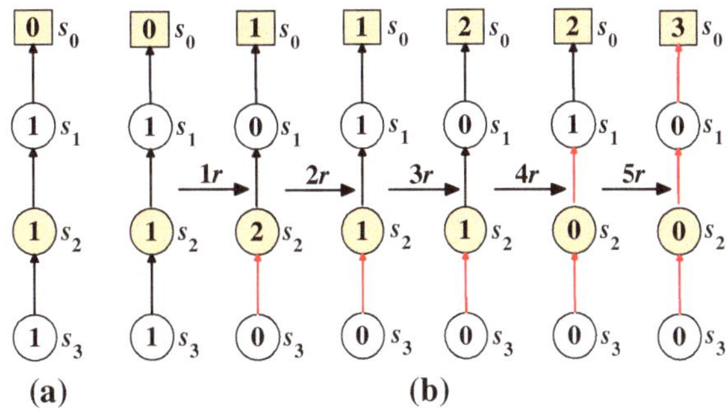

Fig. 4 **a** A single path and **b** its scheduling (r = round). The number within each node denotes the number of data packets at that node

- In an odd round, schedule every link in P_o once, i.e., assign a dedicated channel and one dedicated time slot to each link in P_o.
- In an even round, schedule every link in P_e once, i.e., assign a dedicated channel and one dedicated time slot to each link in P_e.

The scheduling process of the path in Fig. 4a is shown in Fig. 4b. During the first (odd) round, links (s_3, s_2) and (s_1, s_0) are scheduled and the packets at s_3 and s_1 are transmitted to their parent nodes. After the first round, s_3 has no packet to transmit. During the second (even) schedule, link (s_2, s_1) is scheduled and s_2 transmits one packet to its parent node. This process continues until all the packets on path P have been transmitted to s_0.

For the data collection task on a tree, the authors in [43] and [39] proposed some multi-path scheduling algorithms. Suppose that there are m leaf nodes in the data collection tree T, and the path from leaf node $s_i (1 \leq i \leq m)$ to the sink s_0 is denoted by P_i. Two paths P_i and P_j are said *intersecting* if they have at least one common node besides the sink node. The common node of two intersected paths having the largest number of hops from the sink is called an *intersecting point*. If path P_i intersects with other paths, the route from s_i to the nearest intersecting point of s_i is called a *sub-path*, denoted by F_i. Otherwise, F_i is actually P_i. The basic idea of the proposed *multi-path scheduling* algorithm is as follows: For the non-intersecting paths without wireless interference, schedule them by the single path scheduling algorithm concurrently. For the non-intersecting paths with wireless interference, schedule them according to a certain order. The path with the smallest subscript has the highest priority. For the intersecting paths, schedule them according to a certain order, e.g., if P_i is one of these intersecting paths with the smallest subscript, schedule P_i until there is no packets having to be transmitted on the sub-path F_i and then continue to schedule the next path. The multi-path scheduling algorithm can be further explained by the routing tree shown in Fig. 5a. In Fig. 5a, there are three leaf nodes s_1, s_2, and s_3 and their corresponding paths are P_1, P_2 and P_3, respectively. Path P_1 and path P_3 are not intersecting, therefore they are scheduled concurrently as long as they are interference-free. Since path P_1 and path P_2 are intersecting paths, path P_1 is

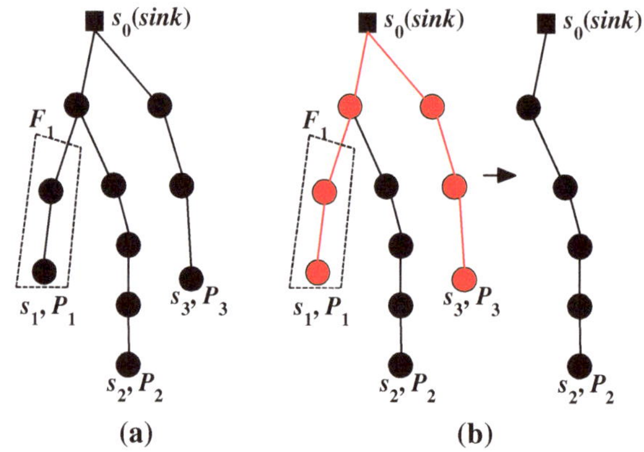

Fig. 5 **a** A routing tree and **b** its scheduling

scheduled first. After no packets on the sub-path F_1 of P_1 have to be transmitted, path P_2 is scheduled as shown in Fig. 5b.

Pipeline Scheduling for Continuous Data Collection. For the continuous data collection task in dual-radio multi-channel WSNs, the authors of [39, 43] proposed a pipeline scheduling algorithm, which is based on the Compressive Data Gathering (CDG) technique [54]. CDG is first proposed for snapshot data collection. The basic idea of CDG is to distribute the data collection load uniformly to all the nodes in the entire network.

In the proposed pipeline scheduling algorithm, the nodes in a data collection tree T are grouped by levels, denoted by D_e, $D_{L'}$, $C_{L'-1}$, $D_{L'-1}$, $C_{L'-2}$, ..., D_1, C_0, $D_0 = \{s_0 \text{ is the sink}\}$, in a bottom-up manner, where D_e is the set of all the dominatees, $D_i (0 \leq i \leq L')$ is the set of the dominators at the i-th level, and $C_i (0 \leq i \leq L' - 1)$ is the set of the connectors at the i-th level. Since every node has two radios, one radio can be dedicated to receiving data and the other is dedicated to transmiting data. Therefore, the nodes at every level can receive and transmit data simultaneously over different channels. Consequently, for a continuous data collection task consisting of N snapshots, the scheduling idea of the *pipeline scheduling* algorithm is as follows:

1. All the nodes in D_e transmit the packets of the j-th $(1 \leq j \leq N - 1)$ snapshot to their parent nodes in the CDG way. After all the packets of the j-th snapshot have been transmitted successfully, the nodes in D_e immediately transmit the packets of the $(j + 1)$-th snapshot in the CDG way.
2. After all the nodes in $D_l (1 \leq l \leq L')$ receive all the packets of the j-th snapshot from their child-level, they send the packets of the j-th snapshot to their parent nodes in the CDG way. After all the packets of the j-th snapshot have been transmitted successfully, the nodes in D_l immediately transmit the packets of the $(j + 1)$-th snapshot to their parent nodes in the CDG way, if they have received all the packets of the $(j + 1)$-th snapshot from their child-level.
3. After all the nodes in $C_l (0 \leq l \leq L' - 1)$ receive all the packets of the j-th snapshot from their child-level, they send the packets of the j-th snapshot to their

parent nodes in the CDG way. After all the packets of the j-th snapshot have been transmitted successfully, the nodes in C_l immediately transmit the packets of the $(j+1)$-th snapshot in the CDG way if they have received all the packets of the $(j+1)$-th snapshot from their child-level.

4. The sink restores the data of a snapshot in the CDG way after it receives all the packets of this snapshot.

Steps 1–4 provide the general frame of the pipeline scheduling scheme. Now, we discuss the idea to prevent radio confliction and channel interference in Steps 1–3. If two or more nodes have the same parent node, they are called *sibling* nodes. In Steps 1–3, radio confliction may arise if two or more sibling nodes send data to their parent node simultaneously even over different orthogonal channels. This is because every sensor only has one radio dedicated to receiving data. Suppose that there are at most Δ_e (resp., Δ_d and Δ_c) nodes in D_e (resp., $D_l (1 \leq l \leq L')$ and $C_l (1 \leq l \leq L' - 1)$) which have the same parent node. Usually, $\Delta_e < \Delta(T)$ except in one-hop WSNs, where any sensor is just one hop away from the sink, $\Delta_e = \Delta(T)$. Then, $\Delta_d \leq 4$ and $\Delta_c \leq 11$ (Note that $|C_0| \leq 12$). To avoid confliction, the nodes in D_e (resp., D_l and C_l) can be partitioned into Δ_e (resp., Δ_d and Δ_c) subsets to guarantee that each node belongs to one subset and no sibling nodes belong to the same subset. Then, when scheduling the nodes of each level, these subsets can be scheduled in a certain order. For the nodes in C_0, they can also be scheduled in a certain order.

The authors of [43] and [39] theoretically evaluated the proposed snapshot data collection algorithm and the continuous data collection algorithm. They showed that the achievable data collection capacity of the multi-path scheduling algorithm is $\Omega(\frac{1}{\frac{3.63}{H}\rho^2+o(\rho)} \cdot W)$, where W is the channel bandwidth, H is the number of the orthogonal channels, ρ is the ratio of the interference radius over the transmission radius of a node, and $o(\rho)$ is a linear equation of ρ, which is order-optimal. The achievable asymptotic network capacity of the pipeline scheduling algorithm in a long-run is $\frac{nW}{12M(\frac{3.63}{H}\rho^2+o(\rho))}$ when $\Delta_e \leq 12$, or $\frac{nW}{M\Delta_e(\frac{3.63}{H}\rho^2+o(\rho))}$ when $\Delta_e > 12$, where n is the number of the sensors, M is a constant value and usually $M \ll n$, and Δ_e is the maximum number of the leaf nodes having the same parent in the routing tree (i.e., data collection tree). A straightforward upper bound of a dual-radio WSN is $2W$, since a dual-radio sink can simultaneously receive two packets at most. Whereas, thanks to the benefit brought by the pipeline technique, the analysis shows that the pipeline scheduling algorithm can even achieve a capacity higher than $2W$.

3.3 Cell-Based Data Collection [40]

To capture the wireless interference in wireless networks, two interference models are frequently used, i.e., the protocol interference model and the physical interference model. The protocol interference model is a simplified model which is more convenient for analysis. On the other hand, under the physical interference model, a receiver can successfully receive data from the sender only if the Signal-to-Interference-plus-Noise Ratio (SINR) of the sender at the receiver is no less than a threshold (denoted by η) value. Compared with the protocol interference model, the physical interference model takes the received physical signal strength as a criterion, which is more accurate and reliable. Therefore, to describe the wireless interference more accurately in a data collection scheduling, the authors of [40] studied the snapshot data collection problem and the continuous data collection problem for WSNs under the physical interference model. In [40], the basic idea of the designed algorithms is to partition the network into cells. Subsequently, all the cells are further been partitioned into compatible cell sets where all the cells in a compatible cell can conduct data transmission concurrently and interference-freely. Based on the obtained compatible cell sets, a snapshot data collection algorithm and a continuous data collection algorithm are designed, which both schedule compatible cell sets sequentially and repeatedly until to finish the data collection task.

Network Model and Network Partition. The considered WSN in [40] consists of n sensors denoted by $s_1, s_2, \ldots, s_n$ and one sink deployed in a square area $A = cn$, where c is a constant. All the nodes are assumed to be independent and identically distributed (i.i.d.). The sink is assumed to be located at the top-right corner of the square. The communication radius of all the sensor nodes is assumed to be r and the bandwidth of the common available channel is W. The network time is slotted and during a time slot, a receiver s_j successfully receives the transmitted data from the sender s_i under the physical interference model if and only if the SINR of s_i at s_j is no less than a constant $\eta > 0$, i.e.,

$$SINR = \frac{P_i \cdot (\|s_i - s_j\|)^{-\alpha}}{N_0 + \sum\limits_{k \neq i} P_k(\|s_k - s_j\|)^{-\alpha}} \geq \eta, \quad (1)$$

where P_i is the transmission power of s_i, $\|s_i - s_j\|$ is the Euclidean distance between s_i and s_j, α is the path-loss exponent and usually $\alpha \in [3, 5]$, $N_0 > 0$ is a constant representing the background noise, and P_k is the transmission power of concurrent sender s_k $(1 \leq k \leq n, k \neq i)$.

Subsequently, the authors in [40] studied the geographical properties of the considered WSN. By partitioning the network into equal-size square cells as shown in Fig. 6, the authors derived the upper and lower bounds of the number of sensor nodes within each cell, as well as the average number of sensor nodes in each cell. Additionally, for the cells shown in Fig. 6, each of them is assigned a pair of coordinates (i, j) $(1 \leq i, j \leq \lambda)$, which indicates this cell is located at the i-th column and the j-th row. For convenience, the cell with coordinates (i, j) is denoted by $\kappa_{i,j}$.

Since the sink is located at the upper-right corner cell $\kappa_{\lambda,\lambda}$, for the sensors in cell $\kappa_{i,j}$, they will forward their data to the sensors located at cells $\kappa_{i+1,j}$, $\kappa_{i,j+1}$ or/and $\kappa_{i+1,j+1}$ as shown in Fig. 7, i.e., the sensors in each cell will forward their data to sensors in subsequent cells horizontally, vertically, or/and diagonally. Finally, the

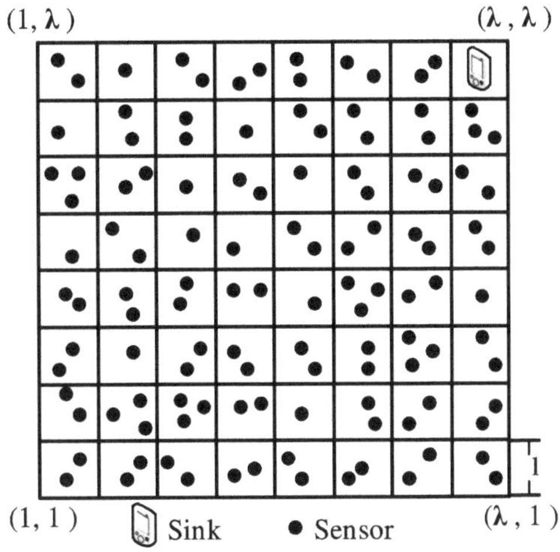

Fig. 6 Network partition

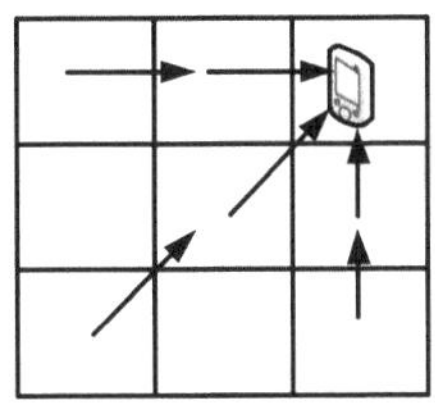

Fig. 7 Data transmission mode

data generated by all the sensors will be forwarded to the sink via this multi-hop fashion.

When data transmission is initialized between two neighboring cells, they may suffer interference caused by other concurrent data transmissions. To make all the concurrent data transmissions successful, the network is further partitioned into larger square zones with side length $R = \omega \cdot l$ (to avoid radio conflicts, $\omega > 2$, i.e., $R \geq 3l$) by another group of horizontal and vertical lines and these square zones are called *interference zones* as shown in Fig. 8. For each interference zone, it is assigned a pair of integer coordinates (i, j) $(1 \leq i, j \leq \lceil \sqrt{cn}/R \rceil)$ and interference zone (i, j) which is denoted by $o_{i,j}$. For a cell $\kappa_{i',j'}$ in an interference zone $o_{i,j}$, the *relative position* of $\kappa_{i',j'}$ in $o_{i,j}$ is defined as $(i' \cdot l - (i-1) \cdot R,\ j' \cdot l - (j-1) \cdot R)$. The cells having the same relative positions in different interference zones are called *compatible cells*. In Fig. 8, compatible cells having relative position (l, l) are highlighted. If two sensors are in different cells which are compatible cells, they can transmit data simultaneously without incurring any interference.

At any time, one sensor in each compatible cell can be selected to transmit data. All the selected sensors transmit data simultaneously. These transmitting sensors will not incur unacceptable interference to each other, since they are spread in different compatible cells. To this end, the authors also derived the necessary condition on the

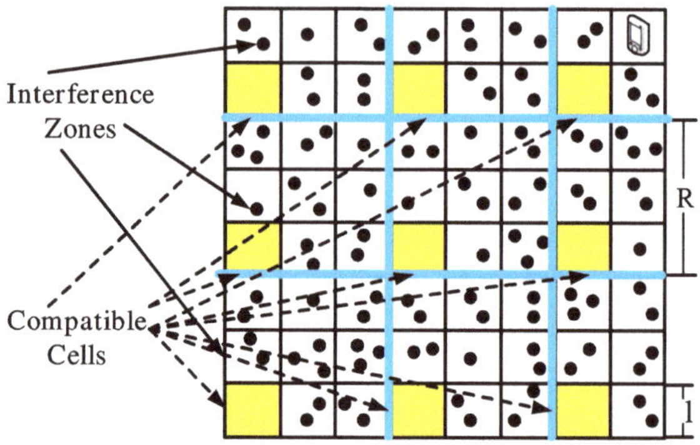

Fig. 8 Interference zones and compatible cells

value of R. By setting appropriate R value, all the compatible cells can conduct data transmissions simultaneously and successfully.

Based on the derived R value, the scheduling algorithms for snapshot data collection and continuous data collection are designed in [40], respectively.

Cell-based Path Scheduling for Snapshot Data Collection: For snapshot data collection, the authors of [40] proposed a Cell-Based Path Scheduling (CBPS) algorithm. First, by abstracting each cell $\kappa_{i,j}$ into a super-node $\nu_{i,j}$, a data collection tree can be constructed according to the following rules:

- For super-node $\nu_{\lambda,j}(1 \le j \le \lambda')$ (note that $\lambda' = \lambda - 1$), $\nu_{\lambda,j}$ transmits its data to $\nu_{\lambda,j+1}$, i.e., create a directed edge from $\nu_{\lambda,j}$ to $\nu_{\lambda,j+1}$;
- For super-node $\nu_{i,\lambda}(1 \le i \le \lambda')$, $\nu_{i,\lambda}$ transmits its data to $\nu_{i+1,\lambda}$, i.e., create a directed edge from $\nu_{i,\lambda}$ to $\nu_{i+1,\lambda}$;
- For super-node $\nu_{i,j}(1 \le i, j \le \lambda')$, $\nu_{i,j}$ transmits its data to $\nu_{i+1,j+1}$, i.e., create a directed edge from $\nu_{i,j}$ to $\nu_{i+1,j+1}$.

For the network shown in Fig. 8, a data collection tree can be constructed as shown in Fig. 9. In Fig. 9, p_i and p_i' denote the paths from the leaf super-nodes to the sink.

According to the example shown in Fig. 9, the basic idea of CBPS can be shown as follows, which consists of four steps.

Step 1: Schedule paths $p_1, p_2, \ldots, p_{\lambda'}$ until all the data packets on these paths have been transmitted to the super-nodes on p_v. When scheduling $p_1, p_2, \ldots, p_{\lambda'}$, it is obvious that they can be partitioned into at most ω groups $G_k(0 \le k \le \omega - 1)$ with each group consisting of mutual compatible paths. Thereafter, in the i-th super time slot (a super time slot equals to the maximum time needed to transmit the data of one cell to a neighboring cell), the paths in group $G_{(i-1)\%\omega}$ are scheduled. Taking the data collection tree shown in Fig. 9 as an example, $p_1, p_2, \ldots, p_7$ can be divided into three groups with $G_0 = \{p_1, p_4, p_7\}$, $G_1 = \{p_2, p_5\}$ and $G_2 = \{p_3, p_6\}$. Thereafter, G_0, G_1 and G_2 will be scheduled in a round-robin fashion. Within a group, the super-nodes on all the paths can also be divided into at most ω node-groups $g_k(0 \le k \le \omega - 1)$ with each node-group containing mutual compatible nodes. Then, in the j-th available super time slot for a particular node-group, the

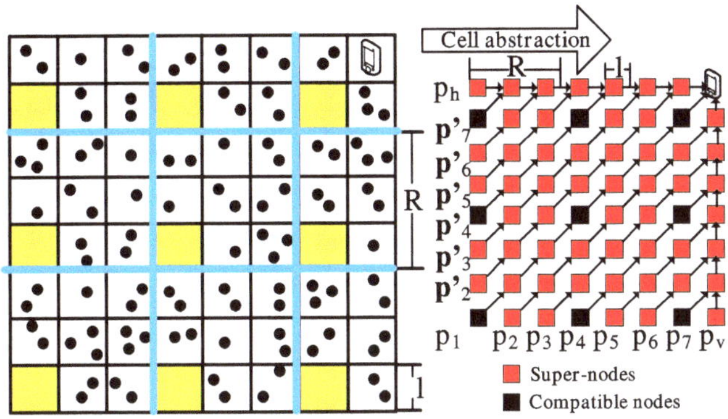

Fig. 9 Construction of a data collection tree

super-nodes in $g_{(j-1)\%\omega}$ are scheduled. For group G_0 in the previous example, the super-nodes on paths p_1, p_4 and p_7 can be divided into three node-groups and they can be scheduled in a round-robin manner in the available super time slots for G_0.

Step 2: Schedule paths $p'_2, p'_3, \ldots, p'_{\lambda'}$ until all the data packets on these paths have been transmitted to the super-nodes on p_h. This step can be done in a similar way as Step 1.

Step 3: Schedule path p_v until all the data packets have been transmitted to the sink. After Step 1, for any super-node $\nu_{\lambda,j}(1 \leq j \leq \lambda')$, it has the data of j super-nodes. Then, p_v can be abstracted to a *virtual tree* rooted at the sink, having λ' internal disjoint paths (except at the root) with lengths $1, 2, \ldots, \lambda'$ respectively by splitting super-node $\nu_{\lambda,j}$ into j *virtual nodes*. Now, in the virtual tree, every virtual node contains exactly the same data with a super-node as the result of the splitting. For instance, p_v in Fig. 9 is abstracted to a virtual tree shown in Fig. 10. Afterwards, we schedule each path of the resulting virtual tree by a similar path-scheduling method used in Step 1.

Step 4: Schedule path p_h until all the data packets have been transmitted to the sink. This step can be done in a similar way as Step 3.

After the above four steps of CBPS, the data of a snapshot can be collected by the sink.

Segment-Based Pipeline Scheduling for Continuous Data Collection. For continuous data collection, the authors of [40] proposed a Segment-Based Pipeline Scheduling (SBPS) algorithm, which is also based on the CDG technique [54].

Since a network is partitioned into interference zones $o_{i,j}(1 \leq i, j \leq \lceil\sqrt{cn}/R\rceil)$, a new term called *segment* can be defined based on interference zones. On the basis of interference zone $o_{i,i}(1 \leq i \leq \lceil\sqrt{cn}/R\rceil)$, the area consisting of interference zones $o_{j,i}(i \leq j \leq \lceil\sqrt{cn}/R\rceil)$ and $o_{i,j}(i \leq j \leq \lceil\sqrt{cn}/R\rceil)$ is called a *segment*, denoted by S_i. Taking the network shown in Fig. 11 as an example, there are three segments S_1 (consisting of interference zones $\{o_{1,1}, o_{2,1}, o_{3,1}, o_{1,2}, o_{1,3}\}$), S_2 (consisting of

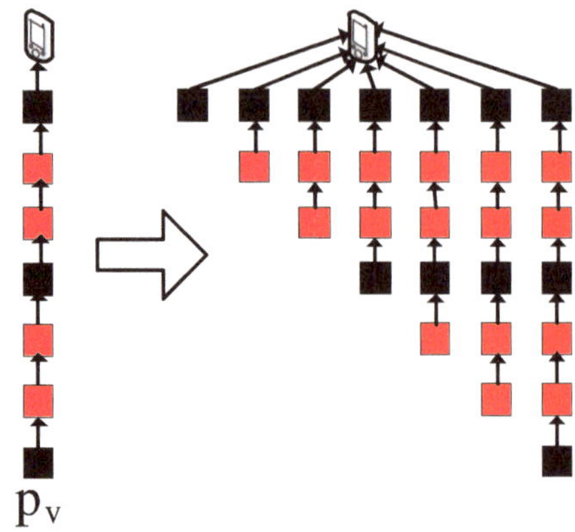

Fig. 10 A virtual tree

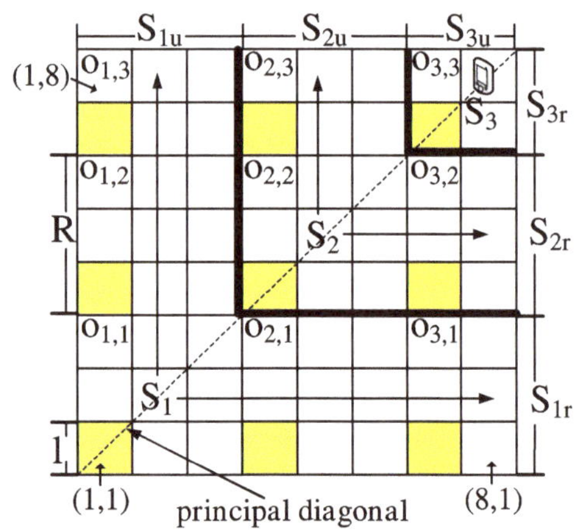

Fig. 11 Segments

interference zones $\{o_{2,2}, o_{3,2}, o_{2,3}\}$), and S_3 (consisting of interference zone $\{o_{3,3}\}$). Within a segment S_i, the area consisting of cells on and below the principal diagonal is denoted by S_{ir}, and the area consisting of the remaining cells is denoted by S_{iu}, i.e., $S_i = (S_{ir}, S_{iu})$. For instance, in the network shown in Fig. 11, $S_{1r} = \{\kappa_{i,1}(1 \leq i \leq \lambda), \kappa_{i,2}(2 \leq i \leq \lambda), \kappa_{i,3}(3 \leq i \leq \lambda)\}$, and $S_{1u} = \{\kappa_{1,j}(1 < j \leq \lambda), \kappa_{2,j}(2 < j \leq \lambda), \kappa_{3,j}(3 < j \leq \lambda)\}$.

For CDC, a similar routing structure as in the CBPS algorithm (note it does not imply the same scheduling) is employed, i.e., each cell is abstracted as a super-node and then a data collection tree is constructed following the same rules as in CBPS (cells and super-nodes are used interchangeably in the subsequent discussion). Unlike CBPS, a super-node here can compress its currently held data packets of a snapshot by the CDG technique.

Based on the defined segments and the constructed data collection tree, the general idea of SBPS can be presented in a hierarchy-level fashion as follows.

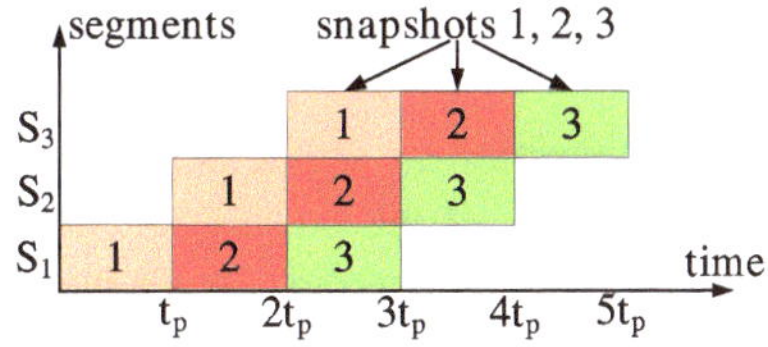

Fig. 12 Data collection pipeline of 3 segments

First, scheduling at the segment-level. At this level, each segment as a whole is considered. Since there is no intersection between any two segments, the data transmission on the segments can be pipelined (it can also be guaranteed that there is no wireless interference among segments in the next step), i.e., for each segment $S_i = (S_{ir}, S_{iu})$, S_i starts the data transmission of the $(k+1)$-th snapshot immediately after it transmits all the data of the k-th snapshot to segment S_{i+1}. Let $t_p = \max\{t(S_i)|1 \leq i \leq \lceil\sqrt{cn}/R\rceil$, $t(S_i)$ *is the number of time slots used by segment* S_i *to transmit all the data packets of a snapshot*$\}$. Then, a segment data transmission pipeline on all the segments is formed with each segment working with t_p time slots for every snapshot (here, a snapshot is an individual task in a traditional pipeline operation). For instance, Fig. 12 shows the data collection process of three snapshots by the segment data transmission pipeline formed from the network shown in Fig. 11.

Second, scheduling at the row/column-level, i.e., within a segment. For the k-th snapshot, within each segment $S_i = (S_{ir}, S_{iu})$, S_{ir} is scheduled first to transmit the data in the cells of S_{ir} to $S_{(i+1)r}$ row by row. Thereafter, S_{iu} is scheduled by a similar way to transmit the data in the cells of S_{iu} to $S_{(i+1)u}$ column by column. When schedule S_{ir}, the first row of cells of S_{ir}, i.e., the cells $\kappa_{j,i} (i \leq j \leq \lambda)$, are scheduled first, followed by the second row of cells, i.e., the cells $\kappa_{j,i+1} (i+1 \leq j \leq \lambda)$, and so on until the last row of cells of S_{ir}, i.e., the cells $\kappa_{j,i+\omega-1} (i+\omega-1 \leq j \leq \lambda)$, are scheduled. When scheduling each row, the cells on that row can be partitioned into ω compatible cell groups $g_1^i, g_2^i, \ldots, g_\omega^i$ with each group containing mutual compatible cells. Afterwards, $g_1^i, g_2^i, \ldots, g_\omega^i$ are scheduled in sequence. Note that the same approach can be followed when scheduling all the segments in the segment data transmission pipeline. Therefore, all the cells in $g_j^i (1 \leq j \leq \omega, 1 \leq i \leq \lceil\sqrt{cn}/R\rceil)$ are also mutual compatible cells. This implies all the segments can be scheduled without wireless interference. Afterwards, the cells in S_{iu} can be scheduled column by column in a similar way. Finally, S_i transmits all the data packets of the k-th snapshot to its subsequent segment S_{i+1}.

Third, scheduling at the cell-level, i.e., within each row/column. For every sensor in cell $\kappa_{i,j}$, it generates one data packet of the k-th snapshot. Furthermore, for the sensors in cells $\kappa_{i,1} (1 \leq i \leq \lambda)$ and $\kappa_{1,j} (1 \leq j \leq \lambda)$, they will not receive any data packets of the k-th snapshot according to the previous segment-level and row/column-level scheduling strategies (actually, this is true for any snapshot). Thus, the sensors in $\kappa_{i,1} (1 \leq i \leq \lambda)$ and $\kappa_{1,j} (1 \leq j \leq \lambda)$ transmit the packets of the k-th snapshot in the CDG way in their available time slots, i.e., for each sensor, it multiplies its data with M random coefficients respectively, and sends the new obtained M products to its parent node. For the sensors in

$\kappa_{i,j}(1 < i, j \leq \lambda)$, they will receive some data packets of the k-th snapshot (it is possible that some sensors do not have any children. In this case, they do the same operation as the sensors in $\kappa_{i,1}(1 \leq i \leq \lambda)$ and $\kappa_{1,j}(1 \leq j \leq \lambda)$). After they receive all the packets of the k-th snapshot from their children sensors, they combine their data and the received data in the same way as in CDG and transmit the obtained M data packets to their parent sensors, respectively. For the sink, it restores the data of a snapshot in the CDG way after it receives all the packets of that snapshot.

For the proposed CBPS and SBPS, the authors of [40] evaluate their performance through theoretical analysis and simulations. They proved that CBPS can achieve order-optimal data collection under the protocol interference model, and SBPS can accelerate the continuous data collection process significantly. Also, the simulation results show that CBPS and SBPS work well for WSNs under the physical interference model.

4 Conclusions and Future Work

In this chapter, the techniques of data aggregation and data collection in WSNs are discussed. First, the definitions of data aggregation and data collection are introduced, followed by the challenges of data aggregation and data collection. Subsequently, the recent advances of data aggregation techniques are summarized. A recently published distributed data aggregation algorithm DAS is then introduced in detail. DAS is fully distributed data aggregation algorithm which is more suitable for distributed wireless systems such as wireless sensor networks. Then different data collection algorithms are reviewed, with several state-of-the-art tree-based and cell-based data collection algorithms discussed in details.

In conclusion, for different data aggregation and collection applications, proper strategy choices in different scenarios are summarized in Tables 1 and 2.

Table 1 Data aggregation strategy choices in different scenarios

Scenarios	Choices
Energy-aware	[18, 20, 46, 57, 63, 77]
Latency-aware	[10, 68, 74, 76, 78, 79]
Energy-latency tradeoff	[44, 48, 50, 75]

Table 2 Data collection strategy choices in different scenarios

Scenarios	Choices
Energy-efficient	[42, 47, 52, 53, 60, 61, 66, 70, 71]
Delay-aware	[4, 17, 55]
Max-capacity/throughput	[9, 11–14, 39, 40, 43]
Mobile wireless networks	[6, 8, 34, 45, 58]

Furthermore, since WSNs are application-oriented networks, there are still many new challenges introduced by different applications involving data aggregation and data collection. Therefore, the following future research directions exist.

- Most of the existing works on data aggregation and data collection focus on snapshot data aggregation or snapshot data collection. Therefore, they may be inefficient for applications requiring continuous data aggregation or/and data collection. In continuous data aggregation and data collection, besides the challenges and constraints in snapshot data aggregation and data collection, more traffic and therefore more interference will be induced. Hence, designing effective continuous data aggregation and continuous data collection algorithms, with respect to energy-efficiency, delay-aware, energy-delay tradeoff, or/and other Quality-of-Service (QoS) objectives, is desirable.
- Almost all of the existing works on data aggregation are based on the assumption that data can be fully aggregated, where multiple data packets can be aggregated into one single data packet. However, for some applications, e.g., top-k queries, this assumption may not be valid. This is because, in these applications, instead of producing one data packet, k ($k \geq 1$) data packets may be produced after applying the data aggregation function. For convenience, this kind of data aggregation is referred to as *partial data aggregation*. Evidently, more traffic may be produced in partial data aggregation tasks compared with traditional fully data aggregation tasks, which implies most of the existing works do not work. Therefore, how to accomplish a partial data aggregation task is an interesting topic.
- Most of the works for data aggregation and data collection are for the single-radio single-channel case. However, with the fast development of hardware techniques, many multi-radio multi-channel sensor nodes are available nowadays. In multi-radio multi-channel WSNs, each sensor node is equipped with multiple radios and each radio can work on multiple orthogonal channels. Compared with traditional single-radio single-channel WSNs (which can only works on a half-duplex mode[7]), multi-radio multi-channel WSNs can achieve a higher network throughput, since a multi-radio node can works on a full-duplex manner and multi-channel improves the parallelism of the network. Therefore, to exploit the benefits brought by multiple radios and multiple channels, multi-radio multi-channel data aggregation and data collection algorithms deserve more research attention in the future.
- To capture the wireless interference in wireless networks, the generalized physical interference model is more accurate, but more complicated. However, most of the works are based on the protocol interference model and the physical interference model. Therefore, to be more practical, it is worth to study how to design data aggregation and data collection algorithms under the generalized physical interference model.

---

[7] In a wireless network, if a node works on the *half-duplex* mode, then, at any time, this node can either transmit data to some other node, or receive data from some other node, but not both. If a node works on the *full-duplex* mode, then, this node can transmit and receive data simultaneously without any confliction or interference.

- For simplicity, most of the works on data aggregation and data collection assumed that the networks are randomly deployed, explicitly or implicitly. Actually, a more general network deployment model is the arbitrary network model, where all the sensor nodes in the network are arbitrarily distributed, and thus two nodes even within the transmission range of each other may not conduct data transmissions directly because of the obstacles between them. Under the arbitrary network deployment model, how to implement energy-efficient data aggregation and collection algorithms, delay-optimal data aggregation and collection algorithms, as well as energy-delay tradeoff data aggregation and collection algorithms, is still an un-solved issue.
- Wireless networks, especially large-scale WSNs, tend to be distributed and asynchronous. Besides, they could be dynamic, i.e., at any time, some nodes may disappear due to energy depletion or damages, and some new nodes may join a network. Thus, centralized algorithms which require the overall network information may not be proper. Furthermore, it is difficult and not realistic to achieve strict overall time synchronization due to the unstable deployment environments, clock drift, and other technique limits. Therefore, designing distributed and asynchronous data aggregation and collection algorithms is another future research direction.

References

1. I. Abraham, D. Malkhi, Probabilistic quorums for dynamic systems. Distrib. Comput. **18**(2), 113–124 (2005)
2. K. Akkaya, M. Demirbas, R.S. Aygun, The impact of data aggregation on the performance of wireless sensor networks. Wireless Commun. Mob. Comput. **8**, 171–193 (2008)
3. H. Alzaid, E. Foo, and J.G. Nieto, Secure data aggregation in wireless sensor networks: a survey, in *CRPIT* (2008)
4. T. Arici, B. Gedik, Y. Altunbasak, and L. Liu, PINCO: a pipelined in-network compression scheme for data collection in wireless sensor networks, in *ICCCN* (2003)
5. Z. Bar-Yossef, R. Friedman, and G. Kliot, RaWMS—Random walk based lightweight membership service for wireless ad hoc networks. ACM Trans. Comput. Syst. **26**(2), 1–66 (2008)
6. D. Borsetti, C. Casetti, C.-F. Chiasserini, M. Fiore, J. María, Virtual data mules for data collection in road-side sensor networks, in *ACM MobiOpp* (2010)
7. Z. Cai, S. Ji, J. He, A.G. Bourgeois, Optimal distributed data collection for asynchronous cognitive radio networks, in *IEEE ICDCS* (2012)
8. I. Chatzigiannakis, A. Kinalis, S. Nikoletseas, Sink mobility protocols for data collection in wireless sensor netowrks, in *ACM MOBIWAC* (2006)
9. S. Chen, Y. Wang, X.-Y. Li, X. Shi, Order-optimal data collection in wireless sensor networks: delay and capacity, in *IEEE SECON* (2009)
10. X. Chen, X. Hu, J. Zhu, Minimum data aggregation time problem in wireless sensor networks, in *IEEE MSN* (2005)
11. S. Chen, Y. Wang, X.-Y. Li, X. Shi, Data collection capacity of random-deployed wireless sensor networks, in *IEEE GLOBECOM* (2009)
12. S. Chen, S. Tang, M. Huang, Y. Wang, Capacity of data collection in arbitrary wireless sensor networks, in *IEEE INFOCOM* (2010)
13. S. Chen, M. Huang, S. Tang, Y. Wang, Capacity of data collection in arbitrary wireless sensor networks, in *IEEE Transactions on Parallel and Distributed Systems* (2011)

14. S. Chen, Y. Wang, X.-Y. Li, X. Shi, Capacity of data collection in randomly-deployed wireless sensor networks. Wireless Netw. **17**, 305–318 (2011)
15. S. Cheng, J. Li, Sampling based (epsilon, delta)-approximate aggregation algorithm in sensor networks, in *ICDCS* (2009)
16. S. Cheng, J. Li, Z. Cai, O(ϵ)-approximation to physical world by sensor networks, in *IEEE INFOCOM* (2013)
17. C.-T. Cheng, C.K. Tse, F.C.M. Lau, A delay-aware data collection network structure for wireless sensor networks. IEEE Sens. J. **11**(3), 699–710 (2011)
18. J. Considine, F. Li, G. Kollios, J. Byers, Approximate aggregation techniques for sensor databases, in *ICDE* (2004)
19. M. Ding, X. Cheng, Robust event boundary detection in sensor networks—a mixture model based approach, in *IEEE INFOCOM* (2009)
20. M. Ding, X. Cheng, G. Xue, Aggregation tree construction in sensor networks, in *IEEE VTC* (2003)
21. K. Du, J. Wu, D. Zhou, Chain-based protocols for data broadcasting and gathering in the sensor networks, in *IEEE IPDPS* (2003)
22. E. Fasolo, M. Rossi, J. Widmer, M. Zorzi, In-network aggregation techniques for wireless sensor networks: a survey. IEEE Wirel. Commun. **14**(2), 70–87 (2007)
23. F. Ferrari, M. zimmerling, L. Thiele, O. Saukh, Efficient network flooding and time synchronization with glossy, in *IPSN* (2011)
24. M. D. Francesco, S.K. Das, Data collection in wireless sensor networks with mobile elements: a survey. ACM Trans. Sens. Netw. **8**(1), 1–31, 2011
25. J. Guo, J. Fang, X. Chen, Survey on secure data aggregation for wirless sensor networks, in *IEEE SOLI* (2011)
26. J. He, S. Ji, Y. Pan, Z. Cai, Approximation algorithms for load-balanced virtual backbone construction in wireless sensor networks, Theoretical Computer Science
27. J. He, S. Ji, M. Yan, Y. Pan, Y. Li, Load-balanced CDS construction in wireless sensor networks via genetic algorithm. Int. J. Sens. Netw. (2011)
28. J. He, S. Ji, M. Yan, Y. Pan, Y. Li, Genetic-algorithm-based construction of load-balanced CDSs in wireless sensor networks, in *MILCOM* (2011)
29. J. He, Z. Cai, S. Ji, R. Beyah, Y. Pan, A genetic algorithm for constructing a reliable MCDS in probabilistic wireless networks, in *WASA* (2011)
30. J. He, S. Ji, P. Fan, Y. Pan, Y. Li, Constructing a load-balanced virtual backbone in wireless sensor networks, in *ICNC* (2012)
31. J. He, S. Ji, Y. Pan, Z. Cai, Load-balanced virtual backbone construction for wireless sensor networks, in *COCOA* (2012)
32. http://www.greenorbs.org/
33. Q. Huang, Y. Zhang, Radial coordination for convergecast in wireless sensor networks, in *IEEE LCN* (2004)
34. S. Jain, R.C. Shah, S. Roy, Exploiting mobility for energy efficient data collection in wireless sensor networks. Mob. Netw. Appl. **11**, 327–339 (2006)
35. P. Jesus, C. Baquero, P.S. Almeida, A survey of distributed data aggregation algorithms, Technical report (2011)
36. S. Ji, Z. Cai, Distributed data collection and its capacity in asynchronous wireless sensor networks, in *INFOCOM* (2012)
37. S. Ji, Z. Cai, Distributed data collection in large-scale asynchronous wireless sensor networks under the generalized physical interference model. IEEE/ACM Trans. Netw. (in press)
38. S. Ji, R. Beyah, Z. Cai, Snapshot and continuous data collection in probabilistic wireless sensor networks. IEEE Trans. Mob. Comput. (in press)
39. S. Ji, Y. Li, X. Jia, Capacity of dual-radio multi-channel wireless sensor networks for continuous data collection, in *IEEE INFOCOM* (2011)
40. S. Ji, R. Beyah, Y. Li, Continuous data collection capacity of wireless sensor networks under physical interference model, in *IEEE MASS* (2011)

41. S. Ji, R. Beyah, Z. Cai, Snapshot/Continuous data collection capacity for large-scale probabilistic wireless sensor networks, in *INFOCOM* (2012)
42. H. Jiang, S. Jin, C. Wang, Prediction or not? an energy-efficient framework for clustering-based data collection in wireless sensor networks. IEEE Trans. Parallel Distrib. Syst. **22**(6), 1064–1071 (2011)
43. S. Ji, Z. Cai, Y. Li, X. Jia, Continuous data collection capacity of dual-radio multi-channel wireless sensor networks. IEEE Trans. Parallel Distrib. Syst. **23**(10), 1844–1855 (October 2012)
44. A. Kesselman, D. Kowalski, Fast distributed algorithm for convergecast in ad hoc geometric radio networks, in *IEEE WONS* (2005)
45. A. Kinalis, S. Nikoletseas, Scalable data collection protocols for wireless sensor networks with multiple mobile sinks, in *IEEE ANSS* (2007)
46. H. Lee, A. Keshavarzian, Towards energy-optimal and reliable data collection via collision-free scheduling in wireless sensor networks, in *IEEE INFOCOM* (2008)
47. H. Lee, A. Keshavarzian, H. Aghajan, Near-lifetime-optimal data collection in wireless sensor networks via spatio-temporal load balancing. ACM Trans. Sens. Netw. **6**(3) (2010)
48. Y. Li, L. Guo, S.K. Prasad, An energy-efficient distributed algorithm for minimum-latency aggregation scheduling in wireless sensor networks, in *IEEE ICDCS* (2011)
49. J. Li, S. Cheng, (ϵ, δ)-approximate aggregation algorithms in dynamic sensor networks. IEEE Trans. Parallel Distrib. Syst. **23**, 385–396 (2012)
50. S. Lindsey, C. Raghavendra, K.M. Sivalingam, Data gathering algorithms in sensor networks using energy metrics. IEEE Trans. Parallel Distrib. Syst. **13**(9), 924–935 (2002)
51. F. Liu, X. Cheng, D. Chen, Insider attacker detection in wireless sensor networks, in *IEEE INFOCOM* (2007)
52. C. Liu, K. Wu, J. Pei, An energy-efficient data collection framework for wireless sensor networks by exploiting spatiotemporal correlation. IEEE Trans. Parallel Distrib. Syst. **18**(7), 1010–1023 (2007)
53. W. Lou, Y. Kwon, H-SPREAD: a hybrid multipath scheme for secure and reliable data collection in wireless sensor networks. IEEE Trans. Veh. Technol. **55**(4), 1320–1330 (2006)
54. C. Luo, F. Wu, J. Sun, C.W. Chen, Compressive data gathering for large-scale wireless sensor networks, in *ACM MOBICOM* (2009)
55. H. V. Luu, X. Tang, An efficient scheduling algorithm for data collection through multi-path routing structures in wireless sensor networks, in *IEEE MSN* (2010)
56. G. Manku, Routing networks for distributed hash tables, in *PODC* (2003)
57. H. Ö. Tan, ï. Körpeoğlu, Power efficient data gathering and aggregation in wireless sensor networks, SIGMOD Record, vol. 32, No. 4, pp. 66–71 (2003)
58. A.S. Poornima, B.B. Amberker, Secure data collection using mobile data collector in clustered wireless sensor networks. IET Wireless Sens. Syst. **1**(2), 85–95 (2011)
59. R. Rajagopalan, P.K. Varshney, Data-aggregation techniques in sensor networks: a survey. IEEE Commun. Surv. Tutor. **8**(4), 48–63 (2006)
60. C. Ren, X. Mao, P. Xu, G. Dai, Z. Li, Delay and energy efficiency tradeoffs for data collections and aggregation in large scale wireless sensor networks, in *IEEE MASS* (2009)
61. S. Rothery, W. Hu, P. Corke, An empirical study of data collection protocols for wireless sensor networks, in *ACM RealWSN* (2008)
62. Y. Sang, H. Shen, Y. Inoguchi, Y. Tan, N. Xiong, Secure data aggregation in wireless sensor networks: a survey, in *IEEE PDCAT* (2006)
63. N. Shrivastava, C. Buragohain, D. Agrawal, S. Suri, Medians and beyond: new aggregation techniques for sensor networks, in *ACM Sensys* (2004)
64. J.L.V.M. Stanislaus, M. Younis, Delay-conscious federation of multiple wireless sensor network segments using mobile relays, in *IEEE VTC 2012-Fall* (2012)
65. I. Stoica, R. Morris, D. Karger, M. Kaashoek, H. Balakrishnan, Chord: a scalable peer-to-peer lookup service for internet applications, in *SIGCOMM* (2001)
66. X. Tang, J. Xu, Adaptive data collection strategies for lifetime-constrained wireless sensor networks. IEEE Trans. Parrellel Distrib. Syst. **19**(6), 721–734 (2008)

67. M. Thangaraj, P.P. Ponmalar, A survey on data aggregation techniques in wireless sensor networks. Int. J. Res. Rev. Wireless Sens. Netw. **1**(3), 36–42 (2011)
68. P.-J. Wan, S.C.-H. Huang, L. Wang, Z. Wan, X. Jia, Minimum-latency aggregation scheduling in multihop wireless networks, in *ACM MOBIHOC* (2009)
69. F. Wang, J. Liu, Networked wireless sensor data collection: issues, challenges, and approaches. IEEE Commun. Surv. Tutor. **13**(4) (2011)
70. C. Wang, H. Ma, Y. He, S. Xiong, Approximate data collection for wireless sensor networks, in *IEEE ICPADS* (2010)
71. F. Wang, D. Wang, J. Liu, Traffic-aware relay node deployment: maximizing lifetime for data collection wireless sensor networks. IEEE Trans. Parallel Distrib. Syst. **22**(8), 1415–1423 (2011)
72. W. Wu, X. Cheng, M. Ding, K. Xing, F. Liu, P. Deng, Localized outlying and boundary data detection in sensor networks. TKDE **19**(8), 1145–1157 (2007)
73. K. Xing, X. Cheng, Location-centric storage for on-demand warning in sensor networks, in *IEEE INFOCOM* (2005)
74. X. Xu, S. Wang, X. Mao, S. Tang, X. Li, An improved approximation algorithm for data aggregation in multi-hop wireless sensor networks, in *FOWANC* (2009)
75. Y. Yu, B. Krishnamachari, V.K. Prasanna, Energy-latency tradeoffs for data gathering in wireless sensor networks, in *IEEE INFOCOM* (2004)
76. B. Yu, J. Li, Y. Li, Distributed data aggregation scheduling in wireless sensor networks, in *IEEE INFOCOM* (2009)
77. H. Zhang, A. Arora, Y. Choi, M.G. Gouda, Reliable bursty convergecast in wireless sensor networks, in *ACM MOBIHOC* (2005)
78. Y. Zhang, S. Gandham, Q. Huang, Distributed minimal time convergecast scheduling for small or sparse data sources, in *RTSS* (2007)
79. J. Zhu, X. Hu, Improved algorithm for minimum data aggregation time problem in wireless sensor networks. J. Syst. Sci. Complexity **21**, 626–636 (2008)

Part VIII
Security

Chapter 17
Current Challenges and Approaches in Securing Communications for Sensors and Actuators

Zygmunt J. Haas, Lin Yang, Meng-Ling Liu, Qiao Li and Fangxin Li

Abstract Recent advances in MEMS hardware have enabled small-footprint and inexpensive sensors to be deployed in hard-to-access locations and to form wireless sensor networks (WSNs). WSNs are typically mission-oriented networks and offer appealing solutions to a range of practical problems. However, due to the characteristics of WSN, their design principles differ from other types of networks. For instance, the severe limitations of computational and energy resources in the network nodes restrict their ability to process and communicate information. These characteristics, particular to WSNs, dictate new security challenges and require new approaches to implementation of security protocols. In this chapter, we present some of the WSNs security challenges and discuss a number of selected solutions presented in the technical literature. The structure of the chapter is as follows. In Sect. 1, we provide background material on WSN security; in particular, we present the security goals, implementation constraints, potential attacks and defenses, and evaluation benchmarks. In Sect. 2, we discuss basic security challenges and approaches, including cryptography schemes, key management schemes, and attack detection and prevention mechanisms. Then, in Sects. 3, 4, and 5, we discuss secure routing, secure localization, and secure data aggregation, respectively. Finally, we conclude the survey in Sect. 6.

1 Background

1.1 Security Goals

The field of Information Security defines numerous goals with respect to protecting information, with the following being considered the top three: *confidentiality*, *integrity*, and *availability* [1]. (These three components, being the core principles

Z. J. Haas (✉) · L. Yang · M-L. Liu · Q. Li · F. Li
Wireless Networks Laboratory (WNL), Cornell University, Ithaca, NY 14853, USA
e-mail: haas@ece.cornell.edu
URL:wnl.ece.cornell.edu

H. M. Ammari (ed.), *The Art of Wireless Sensor Networks*,
Signals and Communication Technology, DOI: 10.1007/978-3-642-40009-4_17,
© Springer-Verlag Berlin Heidelberg 2014

of information security, are often referred to as the *CIA triad* in the *Information Assurance* field [2].) Information confidentiality is defined as the concealment of information, so that the information can only be available and disclosed to the intended parties. Integrity of information refers to its trustworthiness, so that an adversary cannot create, modify, or destroy information, including protection against injection of fraudulent, duplicate, or old (expired) information (e.g., the *replay* attack). Information availability denotes the ability of timely and reliable access to the information. The operation of averting information availability is called the *denial of service* attack. Note that ensuring the above elements of information security requires protection of multiple hardware and software components of the system, with the network being just one such a component.

Other major elements of information security include:

- *Authentication*—the ability of a party to verify the identity of the other communicating party or parties.
- *Non-repudiation*—the ability to assert that a particular party generated the information.
- *Authorization*—the ability to restrict information access only to permitted parties.
- *Authenticity*—the ability to verify that the information has been created or modified by the declared party.
- *Privacy*—the ability to conceal the meta-data about the information-generating entity (e.g., identity of the entity, its location) that generated the transmitted data (to be distinguished from confidentiality).

As for most networks, WSNs are also expected to support a subset of the above information security goals, a set which depends on the particular WSN application. However, the new aspect of WSN security relates to the approaches through which the above information security goals are implemented. In particular, meeting the above security goals in WSNs are especially challenging, due to the hardware limitations of the sensor nodes (energy, computation, storage), as well as due to the operational modes of the network (e.g., unattended operation, large number of nodes, limited node lifetime, fixed deployment, unknown and/or changing propagation conditions). Furthermore, due to the fact that in some applications cost is an important consideration, individual network nodes tend to be inexpensive and, thus, less reliable. However, overall the network needs to support some required reliability level. Furthermore, the cost consideration often prevents reliance on more expensive hardware (e.g., tamper-resistant modules), which could otherwise reduce the complexity of some security-related protocols.

1.2 Implementation Constraints

The particular features of the sensor nodes, especially their limitations, affect the design of the security protocols. For instance, typically, nodes in a WSN are inexpensive sensing devices with quite limited computational capabilities, as compared with nodes in traditional communication networks. Consequently, WSN's nodes are unable to execute security protocols designed for such other networks. Furthermore, the wireless communication channel and the operational characteristics of WSNs introduce additional implementation challenges.

Hardware Constraints

The complexity of a software and the performance requirements of its underlying application dictate the amount of hardware resources needed for the software execution, including the size of memory/storage, the code space, the CPU clock rate, and the energy source. Sensors are usually limited in physical size, which in turn limits its ability to store data and code. Most contemporary sensor devices have RAM sizes ranging from 1 to 10 KB, and program memory less than 1 MB [3]. For example, a popular sensor type, the TelosB, uses the TI MSP430F1611 processor, which is a 16-bit, 8-MHz RISC CPU, with only 10 KB RAM, 48 KB program memory, and 1024 KB flash storage. Moreover, the code space required to support the OS alone is on the order of several kilobytes. For example, the de-facto operating system for wireless sensors, the TinyOS, requires about 4 KB of memory. Therefore, the implementation of security protocol must be very space-conscious to fit the code in the limited available memory of a sensor node.

Power Constraints

Power constraints are always a major concern in WSNs, because the small physical size of the sensor node limits the battery size and, thus, its energy capacity. Furthermore, in many application scenarios in which sensor nodes operate unattended, once a sensor node is deployed it is impossible to replace its battery. Although recharging is possible if an energy scavenging mechanism is implemented in the nodes, often the amount of energy collected in this way is limited and inadequate to fully support by itself the node's energy needs. In contrast to the above power limitation, security related operations can be especially power demanding. Walters et al. summarized in Ref. [4] three major sources of power consumptions due to security related operations in WSN:

- the processing required for security functions, such as encryption, decryption, data signing, and signature verification;
- the energy required to transmit the security-related data or overhead, such as initialization vectors for encryption and decryption; and
- the energy required to maintain security parameters in a secure manner, such as storage of cryptographic keys.

Physical-Level Constraints

In general, wireless communication medium makes mounting security attacks much easier, as compared to attacks in wired networks. For example, the attacker can passively eavesdrop on the network's radio frequency range to steal messages in transit, or it can inject malicious messages into the network at will [5]. However, WSN are particularly vulnerable due to several of their attributes. The large number of deployed nodes and the lack of tight binding among the network nodes render it easier for an attacker to compromise a small number of nodes, while remaining undetected. Moreover, due to the simplicity of their hardware, sensor nodes will typically lack sophisticated protection schemes against physical-layer attack, such as jamming. Furthermore, in many WSNs, the senor nodes are unreliable and prone to hardware malfunction or depletion of energy, so that compromising a sensor node could be misinterpreted as a failure, rather than a security breach. To provide the required level of security, WSN security protocol design must take these physical-level constraints into consideration.

1.3 Potential Attacks and Defenses

Wood and Stankovic in Ref. [6] exploited layered network architecture to analyze security issues and to improve robustness. The network layers architecture is divided into physical, data link, network, and transport layers. Each layer is susceptible to different types of attacks, and security attacks can exploit interaction among layers or cut across multiple layers. Table 1 lists the layers of a typical sensor network, and describes each layer's security vulnerabilities and possible defenses.

Table 1 Sensor network layers and denial-of-service defenses (From [6])

Network layer	Attacks	Defenses
Physical	Jamming	Spread-spectrum, priority message, lower duty cycle, region mapping, mode change
	Tampering	Tamper-proofing, hiding
Data Link	Collision	Error-correcting code, collision detection
	Exhaustion	Rate limitation
	Unfairness	Small frames
Network	Neglect and greed	Redundancy, probing
	Homing	Encryption
	Misdirection	Authorization, monitoring
	Black holes	Authorization, monitoring
Transport	Flooding	Client puzzles
	Desynchronization	Authorization

There are two major types of attacks at the physical layer: jamming and tampering [6]. Jamming refers to interfering with the transmissions on the radio frequencies that the network's nodes are using, such that an adversary can disrupt the entire network of N nodes with only k randomly distributed jamming nodes, where $k << N$. The standard defense against jamming involves various forms of spread-spectrum communication. However, such defense may not be available for sensor nodes, because sensor devices are typically assumed to be low-cost, low-power devices. Other defenses include switching to a lower duty cycle to outlive an adversary or mapping the jammed region and rerouting traffic. Tampering refers to physically compromising the nodes in the network. Tamper protection falls into two categories: passive and active [7]. Passive mechanisms do not require energy and include technologies that protect a circuit from being detected (e.g., tamper-proofing, protective coat). Active tamper protections involve special hardware circuits, which consumes more energy. Therefore, it would be more appropriate for sensor nodes to employ passive techniques.

The data link layer is susceptible to three major attacks: collisions, exhaustion, and unfairness [6]. Error-correcting codes can be used to alleviate some of the effects of collisions. However, they cannot completely solve the problem, as the adversary can still corrupt more data than could be corrected by the network. Collision detection is another way to deal with a collision attack, but it cannot completely defend against collision attacks, because proper transmission still need cooperation among nodes, while subverted nodes could intentionally and repeatedly deny channel access. Exhaustion refers to attacks that deplete the energy source of the network nodes, thus jeopardizing the availability of the network. Existing MAC techniques, such as random back-offs and scheduled access, are only intended to solve the problem of random collisions. When collisions are intentional, these techniques become largely ineffective. A possible solution is inclusion of a rate limiting mechanism in MAC admission control, so that the network would simply ignore the excessive requests generated by an attacker without responding to such requests, and thus avoiding further increase in the traffic volume. However, rate limiting feature has its disadvantages; for example, it reduces the overall network capacity and it limits the maximum data rate of individual users even when the network is underutilized. Unfairness can be caused by selective intermittent application of the attacks mentioned above, by abusing a cooperative MAC-layer priority scheme, or by monopolizing the channel. One defense against this threat is to use small frames, so that an individual node can capture the channel only for a short time period. However, if the messages typically transmitted by the network nodes are long, then splitting the messages into smaller frames incurs additional framing and channel-access overheads.

The network layer is subject to four types of attacks: *neglect and greed*, *homing*, *misdirection*, and *black holes* [6]. A malicious node is *neglectful* when it arbitrarily neglects to route some messages. This node is also *greedy* if it gives undue priority to its own messages. A solution to this type of problem is to use multiple (alternate) routing paths or send redundant messages. Location-based network protocols that rely on geographic forwarding expose the network to *homing* attacks, in which an adversary observes the traffic to obtain the location of critical nodes. Once found, these nodes are subject to being attacked. One solution to this problem is to encrypt the header, so that an adversary cannot learn the location of the critical nodes from reading the header. This solution assumes a secure key management scheme, such that all neighbors share cryptographic keys, and so a passive adversary cannot learn the source or destination of the messages from the headers. *Misdirection* is a more powerful attack, which forwards messages along a wrong path, perhaps by fabrication malicious route advertisements. This attack can target either the sender or an arbitrary victim. (The defense to this attack is similar to that for a black-holes attack, which is discussed next.) A *Black-holes* attack is an even more effective attack against distance-vector-based routing protocols. In this attack, compromised nodes advertise zero-cost routes, thus forming *routing black holes* within the network [8]. This causes excessive messages to be routed through the compromised nodes, and therefore causes intensive bandwidth contention around malicious nodes. It also causes the neighbors of malicious nodes to quickly exhaust their energy supplies due to excessive routing, and therefore could potentially create partitions in the network. Authorization and monitoring are ways to defend against misdirection and black-holes attack [6]. In an authorization-based solution, only authorized nodes (i.e., nodes with valid public/private key pairs) are allowed to exchange routing information. Nodes may use public key infrastructure to sign and verify routing messages, thus ensuring the confidentiality and integrity of routing information. Zero-cost routes in black holes attack are thus eliminated, as an adversarial node does not have the valid public/private key pairs to generate encrypted routing information. Such a scheme requires a reliable certification authority for authentication. Since a centralized certification authority can become a single point of failure, distributed certification authority schemes have been proposed. For example, Zhou and Haas in Ref. [9] proposed a distributed certification authority scheme by distributing the certification function among n servers and, by using threshold cryptography, ensures the certification authority is compromised only if at least t servers are compromised. Monitoring-based solution relies on nodes monitoring their neighbors to ensure their proper routing behavior. Nodes then select routing paths utilizing nodes that exhibit long-term proper routing behavior. It is assumed in monitoring-based solution that nodes with a longer period of proper routing behavior are more trustworthy, and therefore routing information from these nodes are less likely to be inaccurate. This defends misdirection and black-holes attack, as compromised nodes can be quickly detected by their neighbors, and the compromised nodes are then not selected again for routing paths.

The transport layer can be threatened by flooding and desynchronization attacks [6]. A naïve solution to *flooding* is to limit the number of allowed connections, but this would also degrade the overall network throughput, as there would be fewer connections available for each node. A better solution is to have the server node ask the client node to solve some computationally expensive puzzle upon requesting a connection, so that if the client is a compromised node and repeatedly requests establishment of connections, the client would deplete its power while repeatedly solving the puzzles. *Desynchronization* disrupts end-to-end connections. In this attack, the adversary disrupts the communication between two nodes by forging messages that carry sequence numbers or other control messages. If the adversary is successful, the communicating nodes will waste energy by excessively executing the synchronization-recovery protocol without exchanging any useful information. A counter to this attack is authentication of all exchanged packets, assuming that the adversary cannot forge the authentication mechanism.

1.4 Evaluation Benchmarks

There are various benchmarks for evaluating whether a security scheme is appropriate for WSNs. Some of these benchmarks are [10]:

- Resiliency—the network capability to continue to offer sufficient security services while some of the network nodes are compromised.
- Resistance—the network capability to avert an attacker from fully controlling the network.
- Scalability—the capability to support security in very large networks (on the order of hundreds to thousands of nodes).
- Self-organization and flexibility—the capability to adaptively restructure the security scheme as a result of network changes.
- Robustness—the capability of the network to continue to operate in spite of irregularities (e.g., security attacks, hardware failure, etc). (Robustness is a more generalized notion than resiliency, as resiliency is focused on providing security services while under attack, whereas robustness considers providing general network services while under attack.)
- Energy efficiency—the degree to which the network lifetime is maximized by preserving energy.
- Assurance—the confidence that the major elements of information security (e.g., confidentiality, integrity, availability, etc.) are adequately met [2].

2 Basic Security Challenges and Approaches

2.1 Cryptography Schemes

Cryptography is the basic element in any security system. It deals with encrypting the message in order to achieve secure communication in the presence of third party adversaries. There are two types of encryption methods: symmetric key cryptography, which uses the same key for encryption and decryption, and asymmetric (public) key cryptography, which uses different keys for encryption and decryption. Compared to asymmetric key cryptography schemes, symmetric key cryptography schemes have the advantage of lower computational overhead because they involve relatively simply bit-wise operations (e.g., XOR) that can be directly implemented in hardware, but they require more complex key distribution and key management schemes [11].

There has been extensive work done on the evaluation of different cryptography schemes. In Ref. [12], Ganesan et al. investigated the performance of five different symmetric key cryptography schemes (RC4, IDEA, RC5, MD5, and SHA1), over six different hardware platforms that use 8/16/32-bit word size (Atmega 103, Atmega 128, M16C/10, SA-1110, PXA250, and UltraSparc2). Their experiments indicate that: (1) the cycle overhead (i.e., the number of clock cycles to perform a cryptographic operation on a hardware platform) is mostly uniform within each word-size class (8/16/32 bit), but there are differences among the three word-size classes; (2) the impact of caches (i.e., the additional clock cycles due to cache misses in memory fetch) is negligible; and (3) hashing techniques require almost an order of a magnitude higher clock cycle overhead than symmetric key encryption techniques. Table 2 shows the execution times for the various encryption algorithms on the various platforms. Figure 1 shows the byte overhead for the various algorithms and platforms. They also derived a model to assess the computational overhead of embedded architectures for encryption protocols, in general.

Although public key cryptography schemes have received much less attention in WSN security due to their expensive computational overhead, there are several studies that discuss the possibilities of incorporating public key cryptography schemes into WSN. In Ref. [13], Gaubatz et al. showed that special purpose ultra-low power hardware implementations of public key algorithms can be used in sensor nodes. They implemented three different public key cryptography schemes (Rabin's Scheme

Table 2 Execution times (in μs) for algorithms, platforms, and plaintext sizes (in bytes) (From [12])

Algorithm	Size	Action	Atmega 103	Atmega 128	M16C/10	Strong ARM	Xscale (400)	Xscale (200)	Sparc (440)
MD5	0	Digest	5863	1466	1083	46	26	53	23
	1–26	Digest	5890	1473	1075	46	26	53	23
	62–80	Digest	10888	2722	2011	74	45	90	39
SHA-1	1	Digest	15249	3812	2651	69	12	102	27
	3	Digest	15781	3945	5303	69	12.3	103	27
	56	Digest	14543	3636	7955	133	25.8	205	55
	64	Digest	31107	7777	10907	145	25.7	207	56
RC5	16	Init	9641	2410	2074	41	45	91	28
		Enc	1651	413	197	3	3	6	2
		Dec	1636	409	202	3	3	7	2
IDEA	16	Init enc	1523	381	727	26	15.54	47	11
		Init enc	9417	2354	1927	76	25.16	69	36
		Enc	2555	325	596	16	3.24	17	9
		Dec	2614	325	597	16	3.27	17	9
RC4		Init	1886	472	2455	155	66.8	216	96
		Enc	344	86	123	10	5	9	4

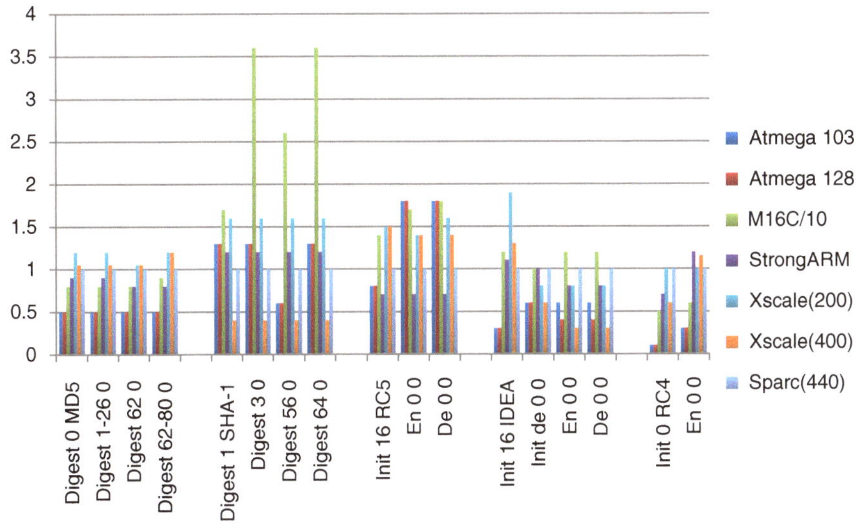

Fig. 1 Normalized overhead for algorithms, platforms and, plaintext sizes (in bytes) (From [12])

[14], NtruEncrypt and NtruSign [15], and Elliptic Curve Cryptography [16, 17]) in a sensor node that is embedded with a custom-designed low-power co-processor to handle all the computation-intensive tasks. They concluded that the use of public key cryptography can reduce the amount of traffic overhead due to key management in WSN, and that the computational cost is within acceptable limits and sufficiently fast on their special-purpose hardware. Wander et al. in Ref. [18] also performed a series of experiments to quantify the energy costs of authentication and key exchange based on ECC [16, 17] and RSA [19] public key cryptography schemes on an 8-bit microcontroller platform. Their studies indicate that authentication and key exchange protocols using optimized software implementations of public key cryptography are quite viable on small wireless devices. They also recommend ECC over RSA for larger energy savings.

Another cryptography technique is watermarking. In Ref. [20], Koushanfar and Potkonjak proposed the first watermarking approach for protecting data and information generated in wireless embedded sensor networks. They considered a sensor network application in which sensor nodes collect data (the data-acquisition phase) to solve nonlinear optimization problem (the data-processing phase). Node signatures are embedded during the data-acquisition and data-processing phases, without compromising the quality of the recorded data or the results of data processing. They conducted two experiments—acoustic atomic trilateration and light source determination—to study the trade-offs between the security protection and the watermarking overhead and to study the situations where such watermarking scheme is the most effective.

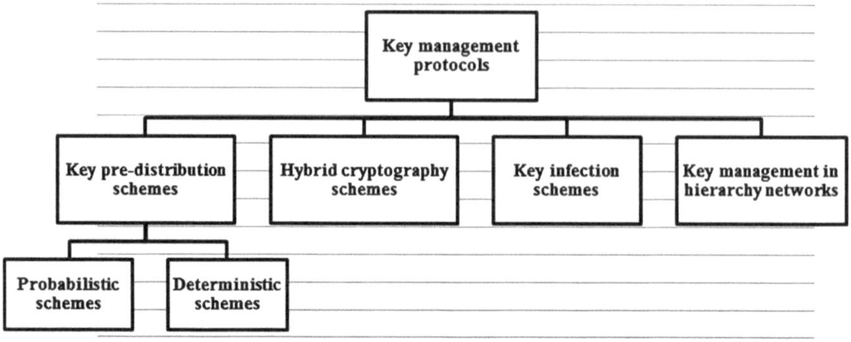

Fig. 2 Taxonomy of key management protocols

2.2 Key Management Schemes

Key management deals with distributing and storing encryption and decryption keys to implement secure communication. A trivial solution to key management is to use a global key for all the sensor nodes. However, in this scheme, if any node in the network is compromised, then the adversary obtains the global key and the whole network security is defeated. Another trivial solution is to have each node store $N-1$ different keys (where N is the total number of nodes in the network), with each key corresponding to a different node in the network. However, this solution is too complex, as a sensor node's limited memory may be insufficient to store the $N-1$ keys, especially for a large network. Clearly, key management is an important yet challenging task, in particular for encryption schemes that use symmetric key cryptography.

There have been extensive research works done in the area of key management schemes. Here, we discuss four categories of key management schemes—key pre-distribution schemes, hybrid cryptography schemes, key infection schemes, and key management in hierarchy networks. Figure 2 shows the taxonomy of key management.

2.2.1 Key Pre-distribution Schemes

In key pre-distribution schemes, sensor nodes store some initial keys before the nodes are deployed [21]. Key pre-distribution schemes are further divided into probabilistic schemes and deterministic schemes.

For probabilistic schemes, the existence of one or more common predistribution keys between intermediate nodes is not certain, but is instead guaranteed only probabilistically. Eschenauer and Gligor in Ref. [23] proposed one of the earlier probabilistic schemes. In their scheme, a ring of keys is distributed to each sensor node before node deployment. Each key ring consists of a randomly chosen k keys from a large pool of P keys, which is generated offline. A pair of nodes can communicate if they share any key among their key rings. Although a pair of nodes may not always have a shared key, if a path between them exists, they can use that path to exchange a key that establishes a direct link. An enhancement over this scheme is proposed by Chan et al. in Ref. [22], in which a q-composite random key pre-distribution scheme is proposed. This scheme requires q keys ($q > 1$) instead of just one common key among the key rings of a pair of communicating nodes. The authors showed that the q-composite key scheme strengthens the network's resilience against node capture when the number of captured nodes is small. Figure 3 shows how the fraction of additional communications that the attacker can compromise varies with the number of nodes captured by the attacker. As a point of reference, comparing the two cases of $q = 1$ and $q = 2$, in terms of the amount of additional compromised communications in a network with 50 compromised nodes is 9.52, and 4.74 %, respectively. The disadvantage of the q-composite keys schemes is that a larger portion of the network is revealed to the adversary as larger number of nodes becomes compromised. This scheme thus trades off the protection against an unlikely large-scale network attack in order to significantly improve the strength of the random key pre-distribution scheme against smaller-scale attacks. Another probabilistic scheme is GKMPSN proposed by Zhu and Zhang in Ref. [24]. GKMPSN is a centralized group key distribution scheme, in which a network controller broadcasts new group keys, as well as node revocation information (i.e., information that identifies a compromised node), to all the nodes whenever a compromised node is detected. Prior to the deployment of the network, each node stores a random set of keys out of a common large key pool. The group re-keying operation then takes two steps. In the first step, the pre-deployed random keys at each node are used to create secure channels between nodes in order to deliver new keying materials to legitimate nodes. In the second step, each node uses the received keying materials to update both the group key and the pre-deployed keys that are invalidated by the compromised nodes. GKMPSN has an attractive property

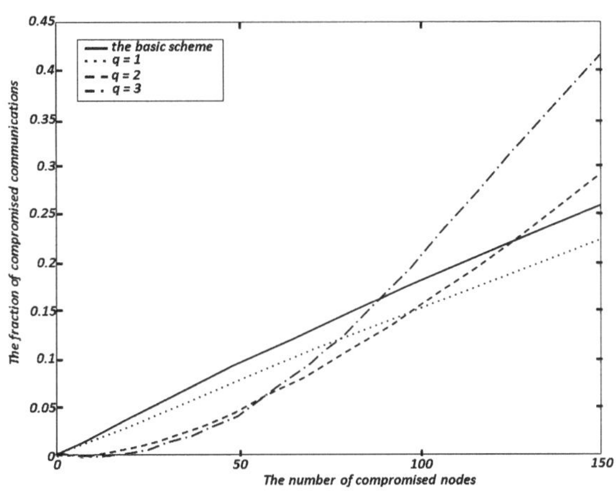

Fig. 3 Probability that a random communication link between two randomly chosen nodes can be decrypted by an adversary, as a function of the number of nodes captured by the adversary (excluding the two communicating nodes). Key ring size is 200, and probability of successful key-setup with a neighbor is 0.33 (From Ref. [22])

of partial statelessness, in which a node can decode the current group key, even if the node missed a few previous group re-keying operations. This is an attractive feature as: (1) typically packet losses are high in WSN due to unreliable communication, and (2) the scheme facilitates new nodes joining the network after initial network deployment.

For deterministic schemes, any two intermediate nodes are guaranteed to share one or more predistributed keys. An example of a deterministic scheme is LEAP (Localized Encryption and Authentication Protocol) proposed by Zhu et al. [25]. LEAP is motivated by the observation that different types of messages have different security requirements and that a single keying mechanism is not suitable for meeting these different security requirements. LEAP supports the establishment of four types of keys for each sensor node—an individual key shared with the sink node, a pairwise key shared with another sensor node, a cluster key shared with multiple neighboring nodes, and a group key that is shared by all the nodes in the network. The individual key allows a node to securely send sensor readings to the sink node. The pairwise key prevents a compromised node's attack, because once a compromised node is detected, its neighbors will typically immediately revoke the pair-wise keys shared with that compromised node. However, even if the compromised node is not detected for some time, its damage is only limited to its near neighbor, as the pair-wise keys are only shared between one-hop neighbors. The cluster key is used for secure local broadcast, for example routing control information. The group key authenticates the sink node to the sensor nodes to facilitate secure network-wise operations, such as key refreshments. Blom in Ref. [26] proposed another pair-wise deterministic key distribution scheme. The scheme can defend against up t compromised nodes. In pre-distribution phase the sink node generates a $(t+1)$-by-N matrix $\mathcal{G}$ over some finite field GF(q), where N is the total number of nodes in the network and $(t+1)$ is the codeword size. The matrix is known to all the nodes in the network. Then the sink node creates a $(t+1)$-by-$(t+1)$ symmetric matrix $\mathcal{D}$ over GF(q). This allows the sink node to compute a matrix $\mathcal{A} = (\mathcal{D}\mathcal{G})^T$ with the property that $\mathcal{K} = \mathcal{A}\mathcal{G}$ is a symmetric matrix (since $\mathcal{A}\mathcal{G} = (\mathcal{D}\mathcal{G})^T\,\mathcal{G} = \mathcal{G}^T\mathcal{D}^T\mathcal{G} = \mathcal{G}^T\mathcal{D}\mathcal{G} = \mathcal{G}^T\mathcal{A}^T = (\mathcal{A}\mathcal{G})^T$). Each node i in the network is assigned with a public column vector $\mathcal{G}^{(i)} =$ ith column of $\mathcal{G}$, and a private row vector $\mathcal{A}(i) =$ ith row of $\mathcal{A}$. Then, for nodes i and j to establish a pairwise key, they can exchange $\mathcal{A}^{(i)}$ and $\mathcal{A}^{(j)}$, and compute their pairwise key $K_{ij} = K_{ji} = \mathcal{G}^{(j)}\mathcal{A}\,(i) = \mathcal{G}^{(i)}\mathcal{A}\,(j)$. Assuming there are $m < t+1$ compromised nodes, then these nodes know m rows and, due to symmetry, m columns of $\mathcal{K}$. A node needs to know at least $(t+1)$ elements in a codeword in order to acquire information about other element in the codeword of other nodes. Therefore, if there are less than $(t+1)$ cooperating nodes, no information about the unknown key is revealed. It has been proven in that the Blom scheme securely protects the pairwise key if any $t+1$ columns of $\mathcal{G}$ are linearly independent.

2.2.2 Hybrid Cryptography Schemes

Hybrid cryptography schemes use computationally expensive asymmetric key cryptography at the sink nodes and computationally cheaper symmetric key cryptography at all other sensor nodes. An example of hybrid scheme is proposed by Huang et al. in Ref. [27], which is based on a combination of elliptic curve cryptography and symmetric key operations. This scheme reduces the high cost of elliptic curve random point scalar multiplications at the sensor side and replaces them with low cost and efficient symmetric key-based operations. On the other hand, it authenticates the two identities based on elliptic curve implicit certificates to avoid the typical key management problem in pure symmetric key-based protocols.

2.2.3 Key Infection Schemes

In key infection schemes, keys are sent in plaintext and thus are not secure. However, these schemes assume that the number of adversaries at key establishment phase is very small. For example, Anderson et al. in Ref. [28] proposed a scheme in which each node bootstraps itself by broadcasting an initial key in the clear. Nodes then exchange keys and build up trust structures as they perform network and resource discovery. The scheme assumes that the adversary can only monitor a small proportion of the communications during deployment phase (i.e., initial key setup phase), but is fully capable of launching attacks after the deployment phase. This is often a realistic assumption, because the initial deployment duration is on the order of seconds, while the overall lifetime of the network can be up to years. Despite the apparent insecurity of this scheme, the proposed scheme uses *multipath secrecy amplification* and *multi-hop key propagation* to enhance the security of the network, such that at most a fixed proportion of communications links can be eavesdropped. *Multipath secrecy amplification* combines keys propagated along different paths to update pairwise keys. For example, consider three nodes W_1, W_2, and W_3, with pairwise keys k_{12}, k_{13}, and k_{23}. Suppose W_1 wants to update k_{12} to k'_{12}. W_1 can ask W_3 to send the key-update request to W_2. The request from W_1 to W_3 is encrypted using k_{13}, and the request from W_3 to W_2 is encrypted using k_{23}. Therefore, if k_{12} is the only key being compromised, the adversary cannot update to a new k'_{12} as long as neither k_{23} nor k_{13} is compromised. Table 3 compares the ratio of the compromised links of the two schemes: the basic key infection scheme and the security amplification scheme (SA), showing the improvement of the latter scheme. In the table, α is a varying density of the adversarial nodes (referred to as "black dust"), assuming values of 1 %, 2 %, and 3 %, and d is the average number of neighbors of a node. *Multi-hop key propagation* uses intermediate nodes to temporarily store the pairwise key update information for two nodes at the ends of a path. For example, if W_1 links to W_2, W_2 links to W_3, and W_1 wants to update k_{13} to k'_{13}, then W_1 and W_3 can invoke W_2's help to set up a new key that W_2 immediately forgets, so a potential node compromising W_2 in the future does not reveal k'_{13}. Multi-hop key propagation supports end-to-end, rather than link-level cryptography, which helps energy efficiency as sink-to-node

Table 3 Improvement of secrecy amplification (SA) over the basic key infection scheme (From Ref. [28])

d	$\alpha = 1\,\%$		$\alpha = 2\,\%$		$\alpha = 3\,\%$	
	Basic (%)	SA (%)	Basic (%)	SA (%)	Basic (%)	SA (%)
2	1.20	0.97	2.29	2.00	3.38	2.93
3	1.81	1.37	3.44	2.67	5.42	3.93
4	2.30	1.80	4.45	3.71	6.50	5.55
5	2.93	2.37	5.73	4.68	8.73	6.75

communications can be encrypted using end-to-end keys rather than translated at intermediate nodes. With the assumption of limited adversaries at the initial key deployment phase, and with the enhancement using multipath secrecy amplification and multi-hop key propagation, the simulation showed that the key infection scheme is almost as secure as using pre-loaded initial keys.

2.2.4 Key Management in Hierarchy Networks

Some key management schemes take advantage of the fact that nodes are often categorized into different types, such as sink nodes, gateway nodes, and sensor nodes, and different types of nodes have different computational resources. In Ref. [29], Jolly et al. present a key management scheme in a clustered sensor network. The method uses pre-deployed symmetric keying, in which sensor nodes store a minimum number of keys that they share with other nodes. Gateway nodes store a larger number of keys, and the sink nodes have no restrictions and store all the keys in the network. Their simulation showed that the energy consumption overhead for the key management is remarkably low and they report an order of magnitude of energy saving. Chorzempa et al., in Ref. [30] proposed another hierarchical key management scheme for WSNs. The scheme is called SECK (Survivable and Efficient Clustered Keying), with three tiers of nodes. The bottom tier consists of low-end sensor nodes, which are clustered. Each cluster is managed by a second-tier cluster head to perform data aggregation and forwarding. At the top tier there is a globally trusted sink node. After initial network deployment, the low-end sensor nodes undergo a location training phase to establish clusters, and cluster coordinate system is used in low-end node recovery procedure. Clusters are then used for establishing and updating administrative keys. A session key between a pair of nodes can then be obtained from administrative keys. Simulations suggested that the scheme is resilient against multiple node captures, and can efficiently recluster and salvage compromised nodes. Figure 4 depicts the comparison of the resiliency against node capture of the SECK scheme and the basic probabilistic pairwise scheme of Eschenauer and Gligor [23], while showing that the resiliency of both schemes is comparable. Realizing that the Eschenauer and Gligor scheme is considered as having good resiliency, one can conclude that the SECK scheme has overall good resiliency property as well.

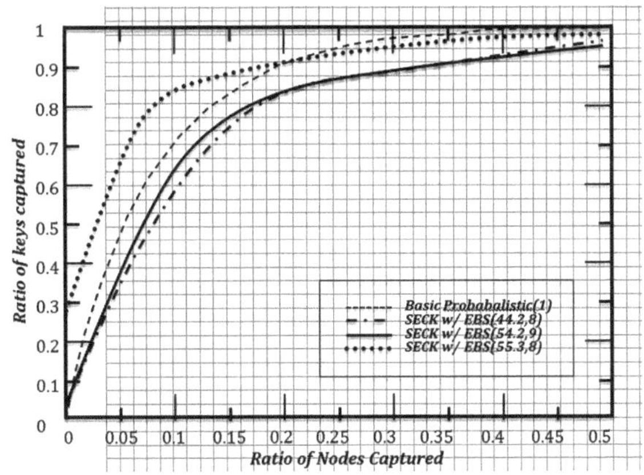

Fig. 4 The ratio of keys captured versus the ratio of the captured nodes (From Ref. [23])

2.3 Attack Detection and Prevention

Section 1.3 briefly discussed several basic attacks and defenses in WSNs. This section extends the previous discussion, and focuses on attack detection and prevention mechanisms for two well-known attacks in WSNs: the Sybil attack and the Wormhole attack. Another, a more complex issue, compromised node detection, is discussed here as well.

2.3.1 Sybil Attack Detection and Prevention

Newsome et al. systematically analyzed the Sybil attack and its defensive measures in Ref. [31]. In the Sybil attack, a node illegitimately claims multiple identities. This attack can be exceedingly detrimental to many important functions of WSN. Figure 5 demonstrates Sybil attack, where an adversary node ‘AD’ is present with multiple identities. ‘AD’ appears as node ‘F’ to ‘A’, as node ‘C’ to ‘B’, and as node ‘A’ to ‘D’, so when ‘A’ wants to communicate with ‘F’, it sends the message to the adversarial node ‘AD’.

For distributed storage for WSNs, the Sybil attack can defeat replication and fragmentation mechanisms in a distributed hash table, such as GHT [32]. Thus, the system may not realize that while it replicates or fragment data across a number of nodes, in fact, it is storing data on a number of Sybil identities which were created by the same malicious node. For routing, the Sybil attack can defeat multipath or dispersity routing protocols, such that seemingly disjoint paths could, in fact, traverse through a single malicious node which represents several Sybil identities. In geographic routing protocols, a Sybil node could appear as being present in more than one place at the same time. For data aggregation, the Sybil attack can have one malicious node contribute to the aggregate many times to alter the aggregate reading. For voting, the Sybil attack allows a malicious node to vote multiple times to control

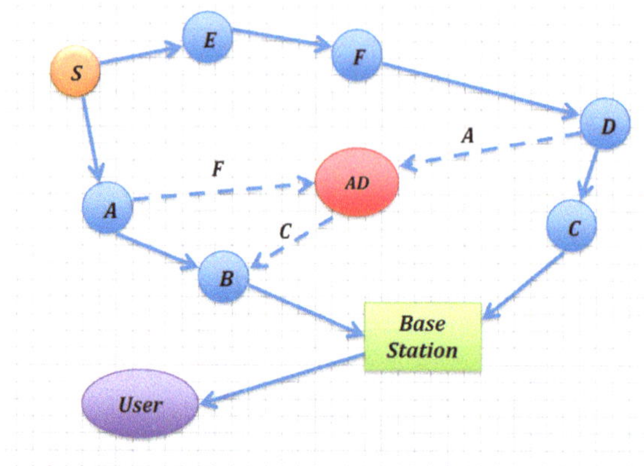

Fig. 5 Sybil attack

outcome of a vote, such as in a blackmail attack. For fair resource allocation, a Sybil node can claim multiple identities and therefore obtain more network resources. For misbehavior detection, Sybil nodes could "spread the blame" by making it appear that the level of misbehaving of the Sybil identities is large enough for the system to take an action. Defenses against the Sybil attack include radio resource testing, random key predistribution, position verification, and registration [31]. In radio resource testing [31], it is assumed that any sensor node has only one radio, and that a radio is incapable of simultaneously sending or receiving on more than one channel. When a node *A* wants to test whether any of its neighbors are Sybil nodes (i.e., a node with multiple identities), the node *A* can assign to each of its neighbors a different channel to broadcast some messages. The node *A* can then choose randomly a channel to listen to. Due to the assumption that each node has only one radio, if *A* does not hear anything on a chosen channel, then the node *A* can be suspect that the node being assigned to that channel is a Sybil node. Figure 6 shows the probability of not detecting the presence of some Sybil nodes using this method.

In random key predistribution [31], each node is assigned a subset of a large set of keys, such that any two nodes share at least a secret key for communication, and no two nodes are assigned with the same subset of keys. As a result, the sensor node can be uniquely identified by the subset of keys that it possesses. A network is able to verify the identity of a node by the keys that the node possesses, referred to as *key validation*. To launch a successful Sybil attack, the attacker is challenged to find the exact subset of keys of a node to steal the node's identity. The estimated probability of a randomly created Sybil identity being effective is depicted in Fig. 7 as a function of the number of compromised nodes. As a point of reference, for an attack to succeed, the attacker needs to compromise at least 150 nodes (in the full validation case).

Position verification is another approach to defend against Sybil attacks [31]. This approach is only applicable in static WSNs (i.e., where sensor nodes are not mobile). In this approach, the network verifies the positions of each node. Sybil nodes can be detected because the Sybil nodes advertised by a single malicious node now all have

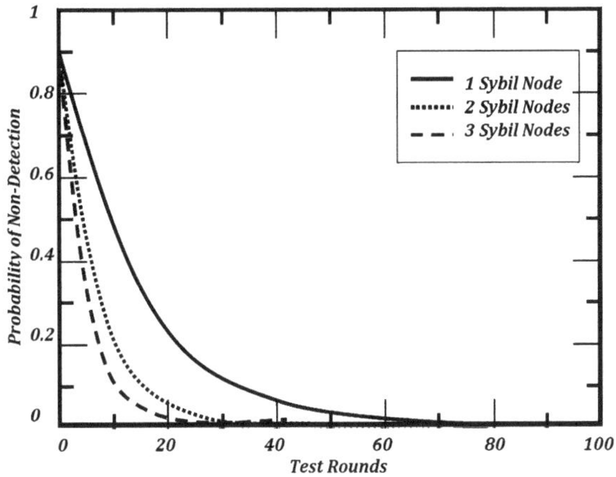

Fig. 6 Probability of no Sybil nodes being detected while using the radio defense in the case of a channel being assigned to every neighbor (From Ref. [31])

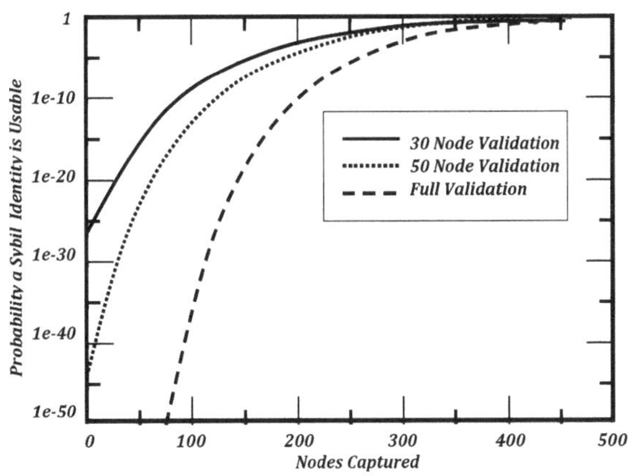

Fig. 7 Probability of a randomly created Sybil node being effective in the key pool scheme as a function of the number of compromised nodes (From Ref. [31])

the same physical position, which would raise an alarm, as the assumption is that a single physical location could be associated with at most one node.

Registration is another potential solution against Sybil attacks [31]. In this approach, there is a trusted central authority that manages the network. The central authority keeps a list of trusted nodes and deployment of nodes. In order to detect a Sybil attack, an entity would poll the network; the results would then be compared with the information about the known deployment. To prevent attack, any node could query the central authority for checking the trusted node list. However, this scheme requires that the central authority must be able to store the trusted node list and deployment information securely.

2.3.2 Wormhole Attack Detection and Prevention

In the wormhole attack, attacker records packets at one location in the network, tunnels them to another location, and retransmits the packets at the other location, making it appear as the two parts of the tunnel are in close proximity to each other. Figure 8 demonstrates the wormhole attack, where 'WH' is the adversary node which creates a tunnel between nodes 'E' and 'I'. These two nodes are at most at the distance of two hops from each other.

The wormhole attack can form a serious threat in wireless networks. For example, the wormhole attacker can gain unauthorized access, disrupt routing, or perform a Denial-of-Service attack. In Ref. [33], Hu et al. described the wormhole attack and proposed a mechanism called *packet leashes* for detecting and defending against wormhole attacks. A leash is an additional part of a packet that limits the packet's maximum allowed transmission distance. The authors introduced two types of leashes—geographic leash, which limits the distance a packet travels, and temporal leash, which limits the time a packet lives (and hence limits the travelling distance, as packet's speed is limited by its speed of propagation). To construct a geographical leash, each node is assumed to know its own position, and all nodes in the network are assumed to have loosely synchronized clocks. Consider two nodes: source, s, and receiver, r. Let ps, pr, ts, and tr denote positions and times of nodes s and r, respectively. If v is an upper bound on the velocity of any node, and Δ is the maximum clock difference between any two nodes, then upon receiving the packet from source s, the receiver can compute an upper bound on the distance between the sender and itself, as $d_{sr} \leq ||ps - pr|| + 2v\,(tr - ts + \Delta) + \delta$, where δ is the maximum relative error in location information between any two nodes. To use temporal leashes, the packet includes an expiration time, after which the receiver does not accept the packet. The expiration time is based on the allowed maximum transmission distance and the speed of light. A specific protocol, called TIK (TESLA with Instant Key

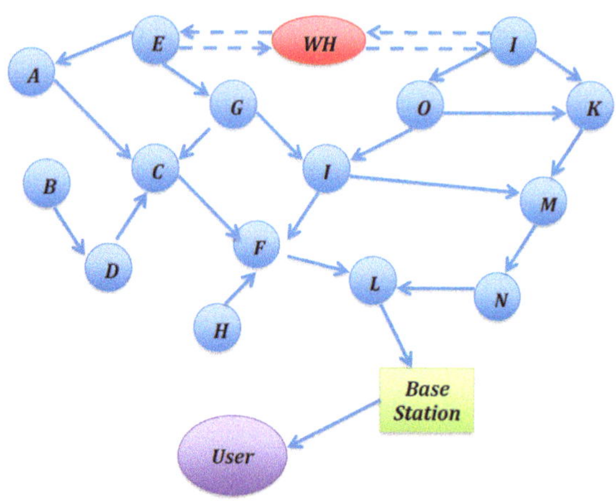

Fig. 8 The Wormhole Attack

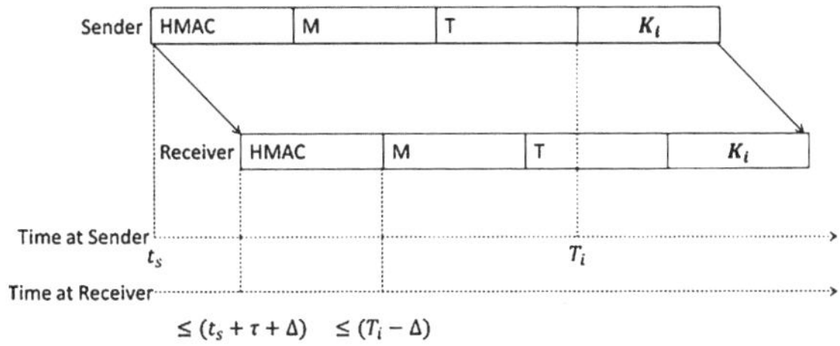

Fig. 9 Timing of a packet in transmission of the TIK protocol (From Ref. [33])

disclosure), is also presented in Ref. [33] to implement temporal leashes. TIK consists of three states: *sender setup*, *receiver bootstrapping*, and *sending and verifying authenticated packets*. In the *sender setup* phase, the sender uses a pseudo-random function F and a secret master key X to derive a series of keys $K_0, K_1, \ldots, K_w$, where $K_i = F_X(i)$. The pseudo-random function is assumed to be secure in the sense that it is computationally intractable for an attacker to find the master secret key X, even if all the keys K_0, K_1,..., K_w are known. In addition, without the secret master key X, it is computationally intractable for an attacker to derive a key K_i that the sender has not yet disclosed. Each K_i expires after some time interval I, which is selected by the sender. In the *receiver bootstrapping* phase, the receiver synchronizes with the source to agree on the initial time T_0 and the time interval I. Finally, in the *sending and verifying authenticated packets* phase, if the sender sends a packet P at time T_i, then the sender also sends a message authentication code (HMAC) of P generated using some undisclosed key K_{i+j}, in which j is large enough such that P arrives at the receiver before time T_{i+j} and K_{i+j} has not yet been disclosed. The receiver can wait until time T_{i+j} for the source to release the key K_{i+j}, and verify the HMAC for P. The timing diagram of a TIK packet is shown in Fig. 9. The protocol assumes that all the clocks are synchronized within the maximum timing error of Δ. Upon receipt of the HMAC value and based on the time as the time of disclosure of the key , the receiver confirms that the corresponding key was not yet sent by the sender. After all the verifications of the protocol were successfully completed, the receiver accepts the packet.

2.3.3 Compromised Node Detection

Besides attack detection and prevention, compromised node detection is another important problem in WSN security. Compromised node detection is usually implemented by software or hardware code-testing schemes. However, for WSNs, hardware-based code-testing schemes are often not feasible for lightweight sensor nodes. Software-based code-testing is more promising, because it requires neither dedicated hardware nor physical access to the device. Software-based approaches are usually based on a challenge-response scheme, where the verifier (usually the sink node) challenges a prover (a target device) to compute a checksum of its memory [34]. Examples of software-based code-testing schemes are SWATT [35] and SCUBA [36].

SWATT (SoftWare-based ATTestation for Embedded Devices) [35] by Seshadri et al. is a software-based attestation technique to verify the memory contents of embedded devices. For each sensor device, SWATT adds an external verifier that is physically distinct from the device. The verifier and the device then run the challenge-response protocol of SWATT. To ensure that the device can return the correct answer only if its memory contents are correct, the verification procedure uses a pseudo-random memory traversal, in which the verifier sends to the device a randomly generated seed for the device to generate a pseudorandom starting memory address for the verifier to access. The verifier then traverses the memory randomly, and iteratively updates a checksum of the memory contents. Since the verifier's memory traversal is random, the attacker cannot predict which memory location is accessed, and therefore after a certain number of iterations of memory accesses, the verifier can eventually detect whether the memory is maliciously altered.

SCUBA (Secure Code Update By Attestation) [36] by Seshadri et al. is another example of software-based code-testing scheme. SCUBA enables a sensor network to detect compromised nodes. Once a compromised node is detected, SCUBA allows the network to either repair the compromised node through code updates, or revoke the compromised node. SCUBA is based on ICE (Indisputable Code Execution), which is a challenge-response-based protocol that ensures that a remote sensor node does not execute a malicious executable. To achieve this, each sensor node is installed with a special executable called the ICE verification function. The ICE verification function is responsible for checking the integrity of an executable on the sensor node, and also for setting up an execution environment to provide atomic execution for this executable (i.e., when an executable is executed in this execution environment, no other executable can interrupt this execution). To ensure that the ICE verification function itself is untampered, the ICE verification function is implemented as a self-checksumming code, which is a sequence of instructions that compute a checksum over themselves, such that the checksum would be wrong or the computation would be slower if the sequence of instructions were modified. The SCUBA protocol then works as follows. To invoke a sensor node's executable X, the invoker node (e.g., a sink node) first sends a "check integrity and execute" request to the sensor node. The ICE verification function on the sensor node then checks for the integrity of X, and executes X if the integrity checking passed. After finishing the execution of X, the sensor node sends the execution result, together with the checksum returned by the ICE verification function, back to the invoker. The invoker can detect whether the sensor node's executable is compromised by reviewing the ICE checksum.

3 Secure Routing

Many WSN routing protocols are based on traditional ad hoc network routing protocols. The original focus of ad hoc routing protocols was on performance, but security issues were extensively studied as research on ad hoc routing protocols matured (e.g., [37–42]). Of course, secure routing is also an important requirement

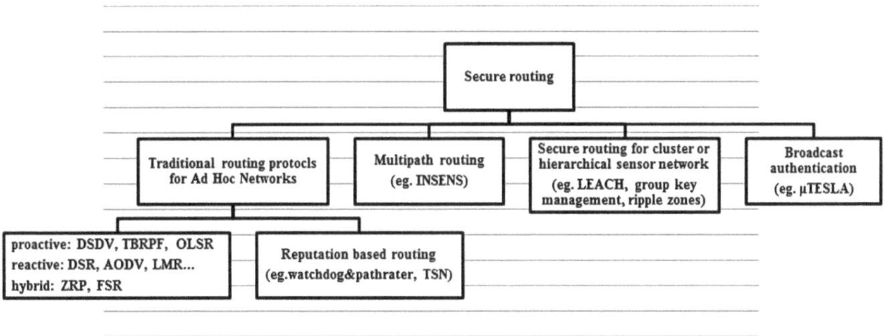

Fig. 10 Taxonomy of secure routing

for WSN applications. This section examines existing approaches to secure routing for WSNs. Figure 10 shows the taxonomy of secure routing. A comprehensive discussion of many of the attacks on routing protocols is also presented in Ref. [43] by Karlof et al.

3.1 Traditional Routing Protocols for Ad Hoc and Sensor Networks

Routing protocols for ad hoc and sensor networks can be broadly classified into three categories: proactive, reactive, and hybrid [44]. In proactive routing protocols, nodes periodically exchange routing information, so typically correct routes are almost always known at the time a routing request is placed. Examples of proactive routing protocols include DSDV [45], TBRPF [46], and OLSR [47]. In reactive routing protocols, nodes exchange routing information only when a communication request is pending. Examples of reactive routing protocols include DSR [48], AODV [49], LMR [50], ABR [51] and TORA [52]. Hybrid routing protocols are a mixture of proactive and reactive routing protocols. ZRP [53] and FSR [54] are examples of hybrid routing protocols. Proactive routing protocols have lower latency, since routing information is consistently maintained; however such protocols are wasteful in communication overhead when the traffic activity is small and, especially, when the network is highly mobile. In contrast, reactive routing protocols incur smaller communication overhead at the expense of larger delays. Thus, reactive protocols may be more practical for low-activity and mobile ad hoc networks, while proactive protocols may better fit highly-active and more static networks. Similarly, for mobile sensor networks which support applications that require infrequent communications, reactive protocols might be a better choice. Hybrid routing protocols typically outperforms proactive and reactive routing protocols, because they use a combination of the two [55], and often optimize their performance based on the network conditions [53, 56].

The original designs of many of the ad hoc network routing protocols are based on performance metrics (e.g., energy efficiency) rather than on security provisions, but there have been numerous works that extend these original ad hoc network routing protocols to improve security. Zapata in Ref. [37] noted that ad hoc networks protocols are being designed without security in mind. He proposed the secure ad hoc on-demand distance vector (SAODV) routing protocol to address the problem of securing a MANET network. SAODV assumes that each node has a signature key pair from a suitable asymmetric cryptosystem. Further, each ad hoc node is capable of securely verifying the association between the address of a given ad hoc node and the public key of that node. AODV uses two mechanisms to secure its messages: Digital signatures and Hash chains. Digital signatures are used by AODV to authenticate the non-mutable fields of a message. Hash chains are used in AODV to secure the mutable part of a message, which is hop count information. On the other hand, for route error messages, a node uses digital signatures to sign the whole message, and any neighbor of the node that receives such a route error message authenticates the signature.

Papadimitratos and Haas [39, 57] proposed the Secure Message Transmission (SMT) protocol, which is based on the notion of information dispersion. SMT assumes that there is another underlying protocol capable of discovering routes in the network (e.g., SRP [7]), although the routes may contain malicious nodes. SMT adds redundancy and partitions the information into fragments, while transmitting the fragments across multiple routes, so that even if some of the fragments are lost (i.e., those that are sent over the routes with malicious nodes), the remaining fragments suffice to reconstruct the original transmission. While SMP transmits data simultaneously over multiple routes, a modification of SMT, the Secure Single Path (SSP) protocol, transmits data over multiple routes in an alternative manner. The salient feature of these protocols is that they do not rely on trustworthiness of nodes in the network (with the exception, of course, of the source and the destination nodes). Indeed, the protocols can deliver highly reliable and low-delay communication even when a large fraction of the network nodes act maliciously. The protocols are, in particular, useful for reliable and secure real-time communications, when retransmissions may not be an option. As a point of reference, SMT can deliver 93 % of messages without retransmissions, even when 50 % of the nodes randomly drop packets. Figure 11 depicts an example of the SMT operation.

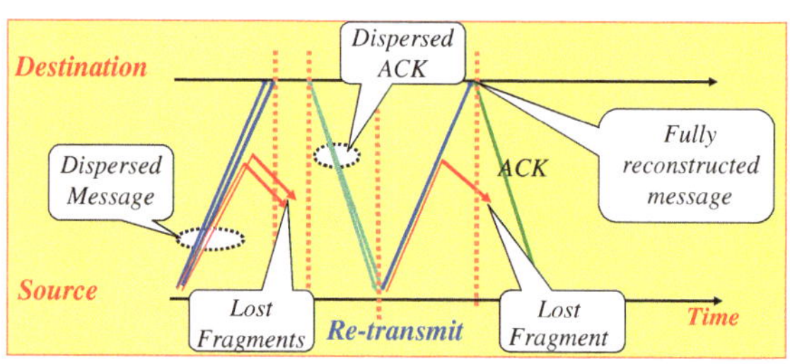

Fig. 11 SMT transmission of a single message through dispersion

Ariadne [40], by Hu et al., is an on-demand secure routing protocol for ad hoc networks. Ariadne's goal is to prevent attacks by tampering with uncompromised routes (i.e., routes formed by uncompromised nodes), and also prevent many types of DoS attacks. The operation of the Ariadne protocol is based on the target node authenticating the Route Requests. This is accomplished by the initiator including a MAC, which is calculated over the unique data in the Route Request, and using key K_{sd}. Ariadne uses three alternative mechanisms for route data authentication, which are: TESLA protocol, Digital signatures, and standard MACS. Additionally, per-hop hashing technique is used to confirm that there is no missing node in the list of nodes in the request. The design of the protocol is based on a reactive routing protocol DSR [48] and a broadcast authentication protocol TESLA [58, 59]. The operation of DSR is divided into *Route Discovery* and *Route Maintenance*. In *Route Discovery*, the source node *S* broadcasts a ROUTE_REQUEST message containing the identifier for the destination node *D* (which is referred to as the "target"). There are two situations in which discarding of a ROUTE_REQUEST occurs. One such a situation is when the node's address is already listed in the route's record. Discarding the request avoids a request propagating around a loop. The other such a situation involves discarding the request upon determination that the host has recently seen a copy of the same request, one carrying the same initiator address and the same request id. This guarantees that a later copy of the request that arrived at this node by a different route is removed. If the request is not discarded, the node appends its own identifier to the ROUTE_REQUEST message and re-broadcasts the message. When ROUTE_REQUEST reaches the target *D*, *D* replies with a ROUTE_REPLY message, which contains the routing information, back to the source node *S*. Node *S* then uses the route in ROUTE_REPLY message to forward data to the node *D*. *Route Maintenance* is a mechanism used to detect broken links on an established route. If any of the intermediate hop transmission along the path fails, the node unable to make the next hop transmission returns a ROUTE_ERROR message back to *S*, and *S* repeats the *Route Discovery* phase. TESLA uses a secret key to generate message authentication code (MAC) for messages to ensure broadcast authentication. The secret key should be kept secret by the message originator, so that no other node can forge the MAC, however, the receiving nodes need the secret key for verification. Instead of using a computationally expensive asymmetric cryptography scheme such as RSA [19], TESLA achieves this asymmetry from loose time synchronization and delayed key disclosure. In TESLA, the sender chooses a random initial key K_N, and generates a one-way key chain by repeatedly computing a one-way hash function H on its starting value: $K_i = H(K_{i+1}) = H^{N-i}(K_N)$. To compute any previous key K_j from a key K_i in which $j < i$, a node can compute $K_j = H^{i-j}(K_i)$ Then, at time slot t_i, the key K_i is used to generate the MAC. At the next time slot t_{i+1}, K_i is published, so that the receiving nodes can verify the MAC using K_i. If the source node has additional broadcast messages to send, K_{i+1} is used to generate the MAC for the new messages at time t_{i+1} Similar to DSR, Ariadne also has a *Route Discovery* phase and a *Route Maintenance* phase. To support secure routing, the ROUTE_REQUST, ROUTE_REPLY, and ROUTE_ERROR messages are all authenticated using the TESLA scheme described above.

Marti et al. in Ref. [38] proposed a reputation-based secure routing for ad hoc networks. In reputation-based routing, the next-hop of a path is chosen based on reliability of links and reputation of nodes. Marti et al. used a watchdog that identifies misbehaving nodes and a Path-rating scheme that helps routing protocols to avoid these nodes; Watchdog and Pathrater are the two mechanisms used to detect and mitigate routing misbehavior. These mechanisms are implemented on top of source routing protocols. Detection of misbehaving nodes is done by the Watchdog by keeping a buffer of packets that were recently sent. The Watchdog then attempts to verify whether a packet has, indeed, been forwarded by the next node by overhearing the transmissions of the neighboring nodes. The Watchdog removes the packet from its buffer when it determines that the packet has been forwarded by the next node. If after a timeout the packet is still in the buffer, a failure count of the node that was responsible for forwarding on the packets is incremented by the Watchdog. If the count surpasses a particular threshold, the node is considered a misbehaving node. The Pathrater is run by every network node. The most likely reliable route is chosen by taking into the consideration the knowledge of misbehaving nodes and the data about links' reliability.

The algorithm of the Pathrater assigns ratings to nodes in five steps: firstly, when the Pathrater becomes aware of a node in the network, the Pathrater assigns the node the rating of 0.5 (every node assigns itself the value of 1.0.) Secondly, at periodic time intervals (of 200 ms), the Pathrater increments the ratings of nodes on all actively used paths by the value of 0.01, with the maximum rating value being 0.8. Thirdly, when the Pathrater detect a misbehaving during packet forwarding, it decrements a misbehaving node's rating by 0.05. Fourthly, negative path values suggest that there is one or more suspected misbehaving nodes on the path. Of course, the goal is to choose the path with the highest ratings.

The Grudger Protocol by Buchegger and Boudec in Ref. [41] is also a reputation-based secure routing for ad hoc networks. Detecting and isolating misbehaving nodes becomes possible by utilizing the Grudger Protocol. Trust relationships and routing decisions are based on behavior of other nodes, which is gathered through experience, observation, or reports. It is intended to be implemented and run on top of any existing ad hoc routing protocols, such as DSR or AODV. Each node of the Grudger protocol consists of four components: monitor, reputation system, path manager, and trust manager. The monitor detects deviance by watching its neighborhood. This is accomplished by listening to the transmissions of the next node to verify that it forwards the packet. As a result, nonconformities can be detected. The trust manager plays an important role in three aspects: firstly, using trust function it calculates the trust levels of nodes and manages trust levels in a trust table; secondly, the incoming ALARM messages are filtered based on the trust level of the reporting nodes and maintaining information about received alarms in an alarm table; thirdly, forwarding the ALARM messages according to the friends list. The reputation system is in charge of maintaining a table of the ratings of the nodes. A rating is determined based on a function that includes the node's own experience, the observations, and the reported experience. The path manager re-ranks paths based on a security metric, deletes paths which contain malicious nodes, ignores route requests generated by

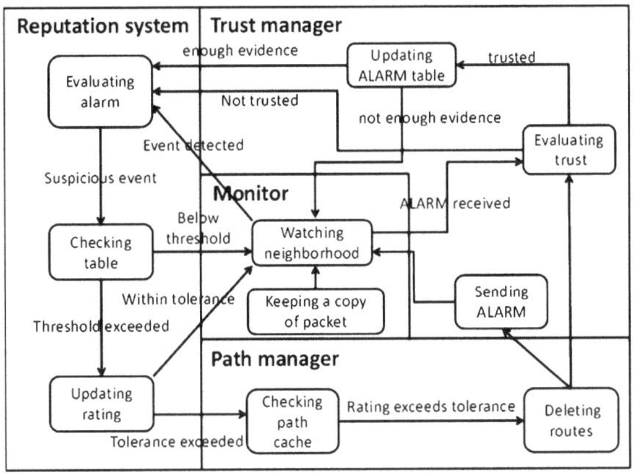

Fig. 12 The relationship between monitor, reputation system, path manager, and trust manager (Based on Ref. [41])

malicious node, and ignores requests for a route which contains a malicious node in the source route (while alerting the source node).

Figure 12 describes the relationship between the four components of the Grudger Protocol (the monitor, the reputation system, the path manager, and the trust manager). The operation is explained as follows in four steps. Step 1: When the monitor detected a suspicious event, such information is passed on to its reputation system. Step 2: The reputation system determines whether the event happened more often than some predefined threshold, and if so, the rating of the node that caused the event is adjusted by the reputation system. Step 3: If the resulting rating of the node is too high, then the path manager removes all the routes that contain this node from the cache of the paths. The trust manager then sends out an ALARM message. Step 4: Upon receipt of such an ALARM message by a monitor component from a node that is (at least) partially trusted, the monitor passes such a received ALARM message on to the trust manager, and the ALARM table in the trust manager is updated. Depending on the level of evidence, the information about the node reported in the ALAM will be passed on to the reputation system.

Despite the effectiveness of those secure routing protocols for ad hoc networks, they may not always be directly applied to WSNs. This is because WSNs usually have a directional data flow, from the data collector nodes towards the sink nodes; whereas in ad hoc networks, data flows are more uniform among nodes. Routing protocols designed for ad hoc networks do not take this directional data flow characteristic into considerations. Therefore, depending on the situation WSNs may need their own versions of such secure routing protocols.

3.2 Multipath Routing

Multipath routing protocols take advantage of the redundancy of sensor nodes in the network. They are robust against limited number of compromised nodes, at the expenses of larger communication overhead. INSENS (Intrusion-tolerant Routing Protocol for Wireless Sensor Networks), proposed by Deng et al. in Ref. [60], is an example of a multipath routing protocol. The main goal of INSENS is to support operation in spite of the harm caused by an intruder who was able to compromised sensor nodes with the intention to inject, modify, or block packets. INSENS assumes that after the initial deployment, sensor nodes can have only bounded mobility. INSENS is effective against DoS attacks and false routing information spreading. In this scheme, each node shares a secret key only with the sink node and not with other nodes. To defend against DoS attacks, broadcast is only permitted by the sink node. To defend against false routing information spreading, initially the sink node computes a multi-hop, multi-path data forwarding tree in three rounds. In round 1, the sink node broadcasts route request message to all sensor nodes. In round 2, the sensor nodes reply back to the sink node with their local topological information. In round 3, the sink node computes a routing table, and then securely unicasts it to each sensor node in a breadth-first manner. This forms a data forwarding tree rooted at the sink node. Data forwarding then proceeds according to this data forwarding tree. Multipath routing in INSENS enhances intrusion tolerance, so that even if an intruder compromises a node or a path, alternate forwarding paths still exist on the data forwarding tree. Bidirectional verification is used to protect against the rushing attack. Nodes joining and leaving a network are supervised by secure maintenance mechanisms.

In Fig. 13, multiple routes are derived between each source and destination. The intent is that these paths should be as independent as possible; i.e., that the paths share in common minimum number of nodes and links. In the best case, only the

Fig. 13 Selection of paths in the multipath routing policy (From Ref. [60])

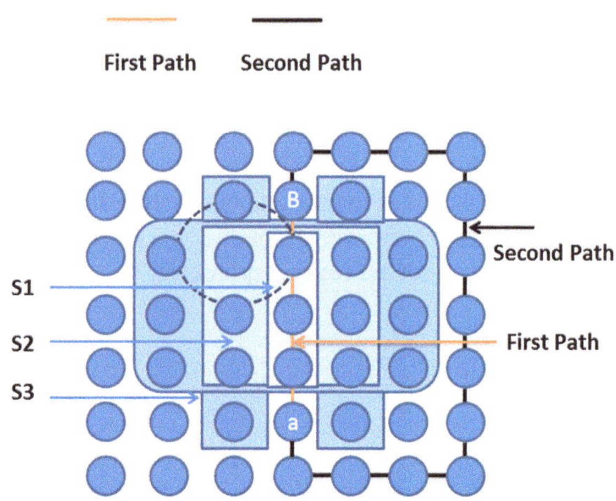

source node and the destination node are common between two paths. In fact, the second path should exclude the nodes on the first path (area S1 in Fig. 13), their neighbors (area S2), and the neighbors of their neighbors (area S3). One or more intruders along some paths can jeopardize the delivery of some of the copies of a message. However, as long as there is at least one path that is not affected by an intruder, the destination will receive a correct copy of the message.

3.3 Secure Routing for Cluster or Hierarchical Sensor Networks

Many WSN architectures are "cluster-based." In such architectures, each cluster has a cluster head and many subordinates, and the cluster head is very close (one-hop or only few-hops away) from all subordinates in its cluster. Subordinates collect data and send the data to the cluster head, and then the cluster head determines routing path and transmits aggregated data. LEACH in Ref. [61], proposed by Heinzelman et al., is one of the first cluster-based routing protocols that significantly reduces energy consumption. LEACH is a self-organizing, adaptive clustering protocol that uses randomization to distribute the energy load evenly among the sensor nodes. After sensor node deployment, nodes cluster themselves and elect one cluster head for each cluster. Since cluster heads typically perform more intensive processing, they are more prone to faster battery drainage and reduced lifetime. To reduce this problem, LEACH randomly rotates the cluster-head position. Furthermore, to reduce energy and to enhance system lifetime, the transmissions to the cluster head are compressed using local data fusion. The cluster head selection process is done probabilistically: each sensor node elects itself to be a local cluster head with certain probability. A cluster head broadcasts its status (e.g., remaining energy, location information, etc.) to other sensors. The non-cluster head nodes then join a cluster by choosing the cluster head that requires the minimum communication energy. After all the nodes are arranged into clusters, every custer-head generates a schedule to be used by the nodes that belong to the cluster-head. The energy dissipated in the sensor can be minimized by turning off the radios of nodes, when the nodes are not transmitting. The LEACH protocol also equalizes the energy used by the nodes, so that nodes are depleted of energy at about equal rate, thus allowing maintaining a more uniform coverage of the environment.

Operation of the LEACH algorithm is based on rounds, and the algorithm comprises the following phases: (a) the advertisement phase, (b) the cluster set-up phase, (c) the schedule creation phase, and (d) the data transmission phase.

There are two steps in the advertisement phase; the cluster-heads are chosen in the first step by having each node n, select a random number between 0 and 1. If this number is less than the threshold $T(n)$, then the node serves as a cluster-head in the current round. The threshold function is set as follows:

$$T(n) = \begin{cases} \frac{P}{1-P\cdot(r \bmod P^{-1})} & \text{if } n \in G \\ 0 & \text{otherwise,} \end{cases}$$

where P is the intended percentage of cluster heads, r is the number of the current round, and G represents the set of nodes that have not served as cluster-heads in the last $1/P$ rounds. The second step in the advertisement phase consists of forming the clusters: transmitting with the same power, the cluster-heads transmit their advertisement using the CSMA (Carrier Sense Multiple Access) protocol. Each non-cluster-head node selects its cluster-head (and, thus, the cluster) for this round based on the measured signal strength of the received advertisement transmissions. During the cluster set-up phase, a non-cluster-head node transmits its selection to the cluster-head and, thus, becomes a member of the cluster. The cluster-head node generates a TDMA schedule, which is based on the number of the nodes in its cluster, and which indicates to each node when a node can transmit. In the data transmission phase, the non-cluster nodes transmit to the cluster head based on the TDMA schedule. When all the data from the non-cluster nodes have been received by the cluster head, the cluster head compresses the data into a single signal, which the cluster head then transmits to the base station. To reduce the interference between the transmissions of the nodes in different clusters which are in close proximity one to another, the clusters use different CDMA (Code Division Multiple Access) codes to communicate. The cluster head send the information about the choice of a particular spreading code to the nodes in its cluster. Using the particular spreading code, the cluster head can then extract the information sent by its nodes, while reducing the interference caused by transmission of nodes in other clusters.

Despite the energy-efficiency characteristic of LEACH, Karlof et al. in Ref. [43] pointed out that such cluster-based protocols are susceptible to the Sybil attack, in which a compromised node can claim multiple identities and advertises itself as multiple cluster heads. Tubaishat et al. proposed in Ref. [62] a cluster-based routing protocol, the Secure Routing Protocol for Sensor Networks (SRPSN). The goal of the SRPSN protocol is to protect the data packet in the sensor networks from different types of attacks. It uses a group key management scheme, which contains group communication policies, group membership requirements and an algorithm for generating a distributed group key for secure communication. This secure routing protocol stores the routing table in a cache. The protocol uses hierarchical architecture and highly efficient symmetric cryptographic operations. In the group key management scheme, the key computation starts from the initiator node by using its partial key. The group key is computed by a leader using the partial keys which are contributed by every sensor node in a group; i.e., one can consider this as a "bottom up approach," as the accumulation of the partial keys is done from a leaf nodes up to the parent nodes. A modified multiparty Diffie–Hellman protocol [63] is used for computing group key, while the updating of group keys is done using the concept of key trees. The routing information messages (e.g., route request and route reply messages) are then encrypted using the group key.

Fig. 14 Ripple-zone-based WSAN routing (From Ref. [64])

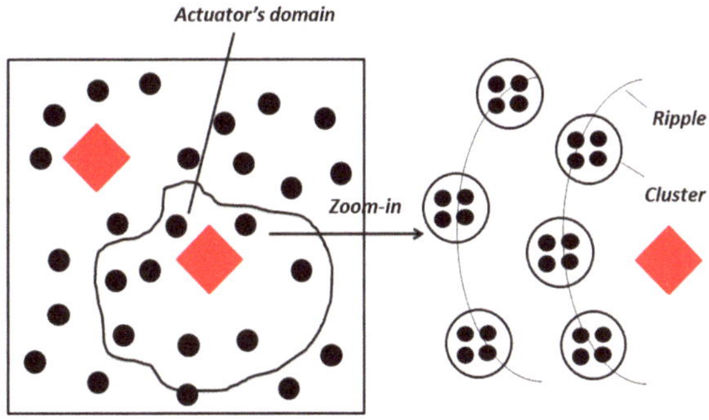

Cluster-based secure routing protocols are also used in wireless sensor and actuator networks (WSANs), which is a special type of WSNs that consists of both, low-power sensor nodes that form the traditional WSN and high-power actuator nodes. Therefore, while WSNs are mostly concerned only with sensors-to-sensors communications, WSANs have to consider four types of communications: sensors-to-sensors, sensors-to-actuators, actuators-to-sensors, and actuators-to-actuators. Furthermore, the natures of the four types of communications are different. For example, sensors-to-sensors communication is usually many-to-one (sensors to the sink) or many-to-many (sensors to sinks), whereas actuators-to-actuators communication is usually peer-to-peer. Hu et al., in Ref. [64] proposed a secure routing protocol based on WSAN's hierarchical network architecture. A scalable and energy-efficient routing architecture, referred to as Ripple-zone (RZ), is employed to implement WSAN security. A multiple-key management scheme, together with the Ripple-zone routing architecture, improves the security of in-network processing, for example, of the data aggregation operation. The scheme uses a Member Recognition Protocol (MRP) to allow actuators and sensors to self-organize themselves into separate domains, with each actuator as the domain center. As shown in Fig. 14, within each domain, sensor nodes are grouped into ripple zones around the domain center actuator, such that nodes in a ripple zone all have the same number of hops to the actuator. Within each ripple zone, sensor nodes are further clustered, and each cluster elects a sensor node as the cluster head or "master." A "master" is responsible to accumulate data from the sensors in its zone. The "master" then transmits the data to the "master" in the next "ripple," which is located closer to the actuator. Each node (sensor or actuator) shares a global key and a pairwise key with the sink node, which are updated periodically. In the high-level (among actuators), two types of keys exist: session key (SK), which is used to secure data packet transmission, and a backbone key (BK), which is used to secure control packets that include session key re-keying information. Figure 15 shows the relationship between these two keys. To protect against attacks, the session keys (SKs) are periodically re-keyed, while the refreshing of the backbone key (BK) is event-triggered, based on events such as actuator insertion, node death, or node compromise. A chain of session keys is generated at the sink node by continuously applying a known one-way hash function, and the key

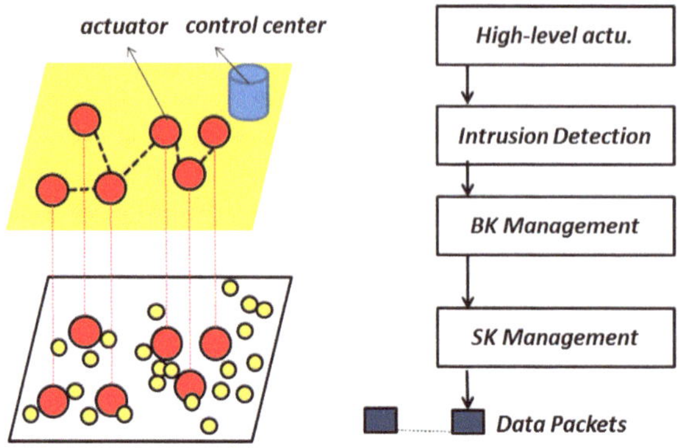

Fig. 15 Backbone Key (BK) and Session Key (SK) (From Ref. [64])

chain is sent to the actuators. An actuator keeps a buffer to store the key chain in order to tolerate multiple key losses. To support different security levels for different types of messages, multiple types of keys are introduced in the low level: Master-to-Actuator Key (MAK), Inter-Master Pairwise Key (MPK), Sensor-to-master Pair-wise Key (SPK), Zone Key (ZK), and Ripple Key (RK). A MAK is shared between each master and its domain actuator and is used for direct master-to-actuator secure communication. MPK is used occasionally to establish secure channels between two masters that belong to two actuator domains. SPK is shared between a master and each of the sensors in its zone. ZK is used for data aggregation and also for propagation of a query message to the whole cluster and is shared among all sensors in the same cluster. RK is used to achieve hop-to-hop security in an actuator domain. This multi-key management scheme allows for the establishment of a secure routing protocol based on ripple-zone, in which messages are routed in a hop-by-hop manner across ripple zones.

3.4 Broadcast Authentication

Broadcast is a fundamental operation in many networks, and it is an essential component in many routing protocols. Perrig et al. in Ref. [65] proposed μTESLA, an authenticated broadcast protocol for the SPINS (Security Protocols for Sensor Networks). In general, authentication operation based on asymmetric cryptography is too computationally intensive for WSN nodes. μTESLA overcomes this problem through the concept of delaying the disclosure of symmetric keys. To send an authenticated packet, the sink node computes a MAC (message authentication code) on the packet with a key that is secret at that point in time. Upon receiving the packet, the node temporarily stores the packet in a buffer and waits for the key disclosure from the sink node. At the time of key disclosure, the sink node broadcasts the verification key to all the receivers. When a node receives the disclosed key, it can verify the key. If the key is correct, the node can use it to authenticate the packet. Each

MAC key K_i is a key of a key chain, generated by a public one-way function F, such that $K_i = F(K_{i+1})$ where the subscript denotes the time interval.

4 Secure Localization Schemes

Depending on the application, localization can be an essential service in a WSN. For example, a location-based routing protocol, GPSR (geographic routing protocol) proposed by Karp et al. in Ref. [66], relies on accurate position of sensor nodes to perform routing. Localization is a well-studied topic, but almost all localization systems operate in a non-adversarial setting [67]. Secure localization only recently emerged as an active area of research. Secure localization schemes can be categorized into two groups—beacon-based and non-beacon-based.

4.1 Beacon-Based Schemes

In the beacon-based schemes, some nodes in the WSN (referred to as *beacon nodes*) are equipped with GPS hardware. These beacon nodes can correctly identify their own location via GPS signals. The beacon nodes can then help the non-beacon nodes to obtain their location information. The localization schemes can be categorized into two types of schemes: "range dependent" and "range-independent." In the range-dependent schemes, the calculation of a node's location is based on the estimates of distances and angles to reference points, where the locations (coordinates) of the reference points are known. Such estimates are usually obtained by one of the following ways: received signal strength, Time of Arrival (ToA), Time Difference of Arrival (TDoA), and Angle of Arrival (AoA) [67].

On the other hand, the range-independent localization schemes do not rely on the nodes performing time, angle, or power measurements. For example, Lazos et al. in Ref. [68] proposed a range-independent localization algorithm called SeRLoc that is beacon-based. SeRLoc is a distributed algorithm based on a two-tier network architecture that allows sensors to passively determine their location without interacting with other sensors. There are two types of nodes: sensor nodes equipped with omnidirectional antennas, and locator nodes equipped with multi-directional antennas and GPS. Locator nodes first obtain their accurate location via GPS, and then each locator transmits beacons with their individual coordinates and coverage areas. Each sensor node collects location information from all the locator nodes that it can receive and then, using this information, assembles a search area of its own location. After receiving enough beacons from different locators, the sensor estimates its location as the center of gravity of the overlapping region of the coverage areas. After analytically evaluating the probability of sensor displacement due to security threats in WSNs, such as the wormhole attack, the Sybil attack, and compromise of network

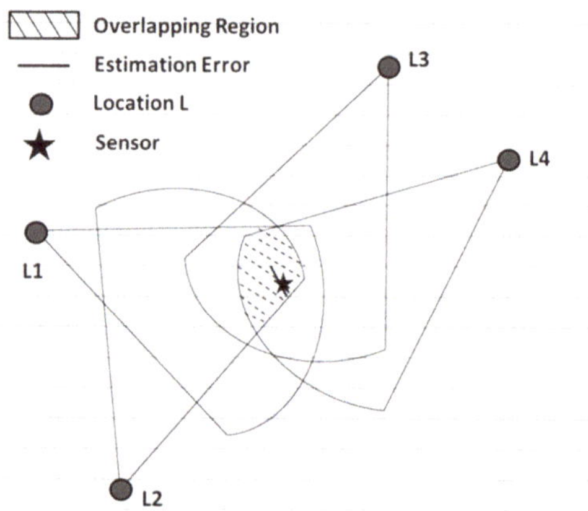

Fig. 16 A sensor estimates its location as a Center of Gravity based on the beacons from locators L1, L2, L3, and L4 (Based on Ref. [68])

entities, they showed that SeRLoc provides accurate location estimation even in the presence of these threats. See Fig. 16 for additional details.

Li et al. in Ref. [70] proposed a robust statistical method for secure localization using triangulation. In triangulation, a sensor node gathers a collection of $\{(x, y, d)\}$ values, where d is an estimated distance from the sensor node to a beacon node at location (x, y). In the ideal case, these $\{(x, y, d)\}$ values map out to a parabolic surface $d^2(x, y) = (x - x_0)^2 + (y - y_0)^2$. Thus, to estimate its location, the sensor node can simply solve for a Least Square problem from the gathered data set $\{(x, y, d)\}$. However, in the presence of adversaries, some of the (x, y, d) values can be outliers. Therefore, instead of using Least Square, the authors proposed to use Least Median Square [69] for achieving robustness in localization. Unlike Least Square, which minimizes the sum of the residue squares, Least Median Square minimizes the

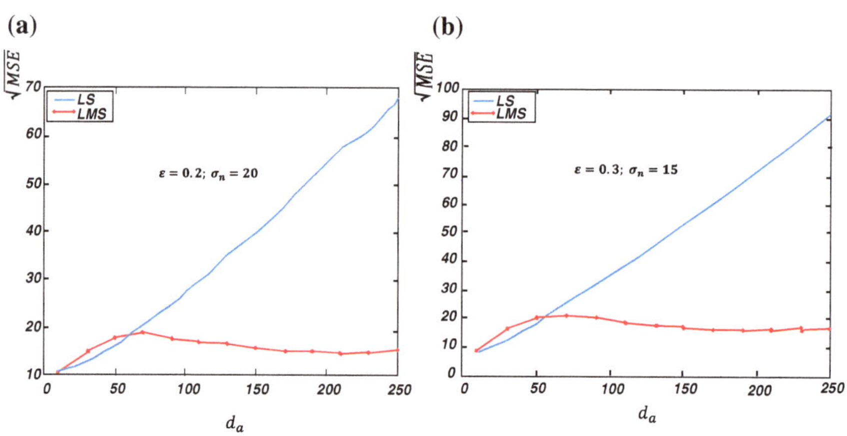

Fig. 17 The performance comparison between LS and LMS for localization (From Ref. [69])

median of the residue squares. As a result, outliners have a much smaller effect on the optimization cost function, which makes the location estimation more robust. Contamination ratio (ϵ) is the fraction of the samples that are outliers and the noise level is assumed to be σ_n. Figure 17 shows the square root of mean square error (MSE) as a function of distance $d_a = \sqrt{(x_a - x_0)^2 + (y_a - y_0)^2}$ (measurement of the strength of the attack). In particular, the performances at two pairs of ϵ and σ_n values are presented in the Fig. 17a: $(\epsilon, \sigma_n) = (0.2, 20)$ and Fig. 17b: $(\epsilon, \sigma_n) = (0.3, 15)$. The results demonstrate that the estimation error of ordinary LS increases with d_a, which is caused by the non-robustness of LS to outliers. On the other hand, the estimation error of LMS exhibits a different behavior; it first increases until reaching a maximum (which occurs at a critical value of d_a), then the estimation error slightly decreases, and finally stabilizes. These results could be interpreted as saying that, if LMS is used for localization, the adversary does not gain by mounting a too powerful attack.

4.2 Non-Beacon-Based Schemes

Since equipping sensor nodes with GPS hardware can be costly, in some practical environments beacon-based localization schemes may not be feasible. In non-beacon-based schemes, a node calculates the position of another node of interest by making an estimation based on the known locations of existing nodes. Non-beacon-based schemes are less accurate, but are also less expensive to implement than beacon-based schemes [67].

Fang et al. in Ref. [71] proposed a non-beacon-based localization scheme. The scheme is based on the following observation: in practice, it is quite common for sensor nodes to be deployed in groups. The locations of the groups (deployment points) are pre-determined prior to deployment and are stored in each sensor's memory. Sensors from the same group can be placed in locations which follow some *a priori* known spatial probability distribution, for example, a two-dimensional Gaussian distribution. With this prior deployment knowledge, sensors can estimate their locations by observing the group memberships of its neighbors. The scheme modeled the localization problem as a statistical estimation problem and used the Maximum Likelihood Estimation method to estimate the location.

5 Secure Data Aggregation

WSN applications that involve extensive amount of data processing typically do not have all the processing done at the central sink node, but instead some processing could be done by the network. Such in-network processing via data aggregation in large-scale sensor networks has been shown to improve scalability, eliminate information redundancy, and increase the lifetime of the network, but the drawback is that data aggregation renders the security problem more difficult [72].

In a large-scale data processing network, sensor nodes can be classified into two groups. Most of the nodes are data collector nodes that are only responsible for collecting sensor measurements. The other nodes are data aggregator nodes that perform aggregation functions upon receiving data from collector nodes. The massive data processing performed by the collector–aggregator architecture can significantly reduce the communication overhead in the network. However, from a security perspective, there are two types of potential threats: the first one is that aggregators can receive false data from collectors; the second one is that the sink node receives false data from compromised aggregators [10]. Secure data aggregation schemes are developed to overcome these two threats. These schemes can be classified into plaintext-based schemes and cipher-based schemes.

5.1 Plaintext-Based Schemes

In the plaintext-based secure data aggregation schemes, intermediate nodes in the path can read the data in transit. Hu et al. in Ref. [73] proposed such an example. In this scheme, each node A is initialized before deployment with a symmetric secret key, K_{AS}, shared with the sink node. Time is slotted, so that as time progresses, a sequence of temporary encryption keys will be generated for node A. For example, in time slot i, the temporary encryption key for A would be $K_{Ai} = E\,(K_{AS}, i)$. After each time slot i, the temporary encryption key K_{Ai} will be revealed to all sensor nodes. The data aggregation proceeds as follows. Consider the following sequence of connected nodes: $A \to B \to C \to S$, as in Fig. 18. At time slot i, node A transmits its reading R_A, identifier A, and a message authentication code $MAC\,(K_{Ai}, R_A)$ to its next hop node B. Node A will hold the data until time slot $i + 1$, when K_{Ai} is revealed. This

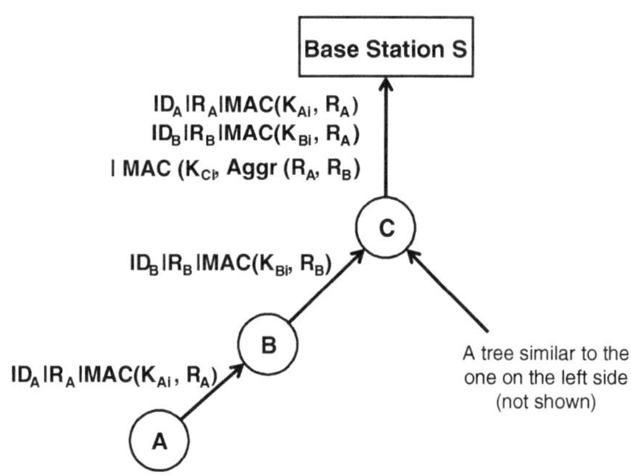

Fig. 18 Example Sensor Network. (Based on Ref. [73])

same sequence of operations is done at node B (i.e., at the time of slot i, B sends $\{R_B, B, MAC(K_{Bi}, R_B)\}$ to C). At the time of the slot $i+1$, K_{Ai} and are revealed to all the nodes. Therefore node B can verify the integrity of R_A, and if R_A is verified, then B forwards A's message $\{R_A, A, MAC(K_{Ai}, R_A)\}$ to C. Similarly, C can verify R_A's and R_B's integrity using K_{Ai} and K_{Bi}, respectively. If the verification test at C passed, C can perform aggregation over R_A and R_B. This data aggregation scheme is therefore a delayed aggregation—aggregation is performed not at the immediate next hop, but at a later hop. As a result, the intermediate node has to forward both its own data and the received data to the last node, and therefore an additional transmission cost is incurred. However, delayed aggregation benefits data integrity—an adversary who obtains key material from a compromised node cannot tamper with many sensor readings.

5.2 Cipher-Based Schemes

In cipher-based secure data aggregation schemes, intermediate nodes on the path cannot read the data in transit. One implementation of such a scheme is Concealed Data Aggregation (CDA), proposed by Girao, et al. in Ref. [74]. CDA is based on a concept called *privacy homomorphism* (PH) proposed by Domingo–Ferrer in Ref. [75]. PH is a particular encryption transformation with additive and multiplicative homomorphic properties, so that direct computation over encrypted data is possible. Suppose Q and R are two rings, where "+" and "*" are the corresponding addition and multiplication operations for both rings. Let K denote the set of keys, E to be an encryption function ($E : K \times Q \rightarrow R$), and D to be a corresponding decryption function ($D : K \times R \rightarrow Q$) Then PH ensures that, for all $a, b \in Q$ and $k \in K$, we have $a + b = D_k(E_k(a) + E_k(b))$ (i.e., homomorphic addition), and $a * b = D_k(E_k(a) * E_k(b))$ (i.e., homomorphic multiplication). CDA uses PH to encrypt aggregated data along the path. The additive and multiplicative homomorphism properties allow processing data while the data is encrypted, without the necessity to decrypt the data at each intermediate node. This allows preservation of data confidentiality and integrity while the data is routed within the network.

6 Conclusion

In this chapter, we discussed several important aspects of WSN security, including cryptography schemes, key management schemes, secure routing protocols, secure localization, and secure data aggregation.

Cryptography schemes are classified into public key cryptography and symmetric key cryptography. Public key cryptography schemes are more computationally demanding, but require less care in key distribution and management. Due to limited resources at the sensor nodes, public key cryptography schemes are often considered infeasible for WSNs, although recent results showed that some public key cryptography schemes can be implemented in WSNs by choosing appropriate algorithms, parameters, etc. However, achieving energy-efficient public key cryptography schemes still need further research. For symmetric key cryptography schemes, efficient key management schemes need to be designed.

We discussed four categories of key management schemes—key pre-distribution schemes, hybrid cryptography schemes, key infection schemes, and key management in hierarchical networks. Although key management has been an active research area in the past decade, there are still certain open problems in this area. Current key management schemes are mostly concerned with static WSNs, and key management schemes for mobile WSNs still lack appropriate solutions. Most key management schemes require trustworthy sink nodes, which may not be a valid assumption in many applications; therefore new schemes need to be designed to secure the sink node.

Currently there are many secure routing algorithms for WSNs, and many of them are derivatives of secure ad hoc network routing algorithms. We reviewed several ad hoc network routing protocols, then surveyed two categories of secure routing protocols specifically designed for WSNs—multipath routing and cluster-based routing. Although many secure routing algorithms can prevent or detect node compromise to some extent, there is still a window of vulnerability in which a compromised node can go unnoticed and false routing information may be spread. Designing secure routing protocols to minimize this window of vulnerability is another important research area. Yet another consideration for future secure routing research is to expand the evaluation metrics. Current evaluations of secure routing are mostly focused on security metrics; other metrics such as QoS need to be considered in addition to security.

Secure localization is divided into two categories: beacon-based and non-beacon-based schemes. However, both types of schemes are only suitable for static WSNs. Mobile WSNs secure localization still need further investigation.

Secure data aggregation schemes include plaintext-based and cipher-based schemes. Data aggregation schemes usually assume aggregators as more powerful sensor nodes than data collector sensor nodes. Therefore it is desirable to design secure data aggregation schemes that can be applied in a homogeneous WSN, where all the sensor nodes have equal capabilities. Another potential research direction in secure data aggregation is to investigate the tradeoffs between security and energy efficiency gains.

A somewhat newer topic related to WSN that has not been covered in this chapter is that of security of *Internet of Things* (*IoT*) networks. The reader is referred to references [76–81].

Acknowledgments This work was sponsored in part by the NSF grants numbers ANI-0329905, CNS-1040689, ECCS-1308208, CNS-1352880, CNS-0626751, and by the AFOSR contract number FA9550-09-1-0121/Z806001.

References

1. M. Bishop, *Computer Security: Art and Science* (Addison-Wesley, Boston, 2003)
2. G. Stoneburner, C. Hayden, A. Feringa, *Engineering Principles for Information Technology Security*, NIST Special Publication (Rev A, June 2004), pp. 800–827
3. M. Healy et al., *Wireless Sensor Hardware: A Review* (IEEE Sensors, Lecce, 2008
4. J.P. Walters et al., in *Wireless Sensor Network Security: A Survey Security in Distributed, Grid, and Pervasive Computing*, ed. by Y. Xiao (CRC Press, Boca Raton, 2006)
5. E. Shi et al., Designing secure sensor networks. IEEE Commun. Mag. **11**(6), 38–43 (2004)
6. A. Wood et al., Denial of service in sensor networks. IEEE Comput. **35**(10), 54–62 (2002)
7. P. Papadimitratos, Z.J. Haas, Secure routing for mobile ad hoc networks, in *Proceedings of SCS CNDS* (San Antonio, 2002) 27–31 Jan 2002, pp. 193–204
8. J. Jeong, G.Y. Lee, Z.J. Haas, Prevention of black-hole attack using one-way hash chain scheme in ad hoc networks, in *International Conference on Information Networking* (Estoril, 2007) pp. 22–25
9. L. Zhou, Z.J. Haas, Securing Ad Hoc networks. IEEE Netw. **13**(6), 24–30 (1999)
10. X. Chen et al., Sensor network security: a survey. IEEE Commun. Surv. Tutorials **11**(2), 52–73 (2009)
11. S.A. Camtepe et al., Key distribution mechanisms for wireless sensor networks: a survey. Computer Science Department at RPI Techical report TR-05-07 (2005)
12. P. Ganesan et al., Analyzing and modeling encryption overhead for sensor network nodes, in *Proceedings of 2nd ACM International Conference on Wireless Sensor Networks Applications* (2003), pp. 151–159
13. G. Gaubatz et al., State of the art in ultra-low power public-key cryptography for wireless sensor networks, in *Proceedings of 3rd IEEE International Conference on Pervasive Computing and Communications Workshops* (2005), pp. 146–150
14. M. Rabin, *Digitalized Signatures and Public-Key Functions as Intractable as Factorization* (MIT Laboratory for Computer Science, Cambridge, 1979)
15. J. Hoffstein, J. Pipher, J. Silverman, NTRU: a ring based public key cryptosystem, in *Algorithmic Number Theory (ANTS III)* (Portland, 1998)
16. N. Koblitz, Elliptic curve cryptosystem. Math. Comput. **48** (177), 203–209. JSTOR 2007884
17. V. Miller, Use of elliptic curves in cryptography. CRYPTO **85**, 417–426 (1985)
18. A. Wander et al., Energy analysis for public-key cryptography for wireless sensor networks, in *IEEE PerCom'05* (Pisa, 2005)
19. R.L. Rivest, A. Shamir, L.M. Adleman, A method for obtaining digital signatures and public-key cryptosystems. Commun. ACM **21**(2), 120–126 (1978)
20. F. Koushanfar, M. Potkonjak, Watermarking techniques for sensor networks: foundation and applications, in *Security in Sensor Networks*, ed. by Y. Xiao (Auerbach Publications, Boca Raton, 2006)
21. J. Jeong, Z.J. Haas, Predeployed secure key distribution mechanism in sensor networks: current state-of-the-art and a new approach using time information. IEEE Wirel. Commun. 42–51 (2008)
22. H. Chan et al., Random key predistribution schemes for sensor networks, in *Proceedings of IEEE Symposium Security Privacy* (2003), pp. 197–203
23. L. Eschenauer et al., A key-management scheme for distributed sensor networks, in *Proceedings of Conference on Computer and Communications Security* (2002), pp. 41–47
24. S. Zhu, W. Zhang, Group key management in sensor networks, in *Security in Sensor Networks*, ed. by Y. Xiao (Auerbach Publications, Boca Raton, 2007)
25. S. Zhu et al., LEAP: efficient security mechanisms for large-scale distributed sensor networks, in *Proceedings of 10th ACM Conference on Computer and Communications Security* (2003), pp. 62–72
26. R. Blom, An optimal class of symmetric key generation systems, advances in cryptology, in *Proceedings of EUROCRYPT84*, LNCS, Vol. 209 (1984), pp. 335–338

27. Q. Huang et al. Fast authenticated key establishment protocols for self-organizing sensor networks, in *Proceedings of 2nd ACM International Conference on Wireless Sensor Networks Applications* (2003), pp. 141–150
28. R. Anderson, et al., Key infection: smart trust for smart dust, in *Proceedings of 12th IEEE International Conference on Network Protocols (ICNP)* (2004)
29. G. Jolly et al., A low-energy key management protocol for wireless sensor networks, in *Proceedings of 8th International Symposium Computers and Communications (ISCC)*, Vol. 1 (2003), pp. 335–340
30. M. Chorzempa et al., SECK: Survivable and efficient clustered keying for wireless sensor networks, in *Proceedings of IEEE Workshop on Information Assurance in Wireless Sensor Networks*(Phoenix, 2005), pp. 453–458
31. J. Newsome et al., The Sybil attack in sensor networks: analysis and defenses, in *Proceedings of 3rd International Symposium on Information Processing in Sensor Networks* (2004), pp. 259–268
32. S. Ratnasamy et al., GHT: a geographic hash table for data-centric storage, in *WSNA* (2002)
33. Y. Hu et al., Packet leashes: a defense against wormhole attacks in wireless ad hoc networks, in *Proceedings of IEEE INFOCOM* (2003)
34. C. Castelluccia, et al., On the difficulty of software-based attestation of embedded devices, in *CCS'09* (2009), pp. 9–13
35. A. Seshadri, et al., SWATT: software-based attestation for embedded devices, in *Proceedings of IEEE Symposium Security Privacy* (2004), pp. 272–282
36. A. Seshadri, et al., SCUBA: Secure code update by attestation in sensor networks, in *Proceedings of 5th ACM Workshop Wireless Security* (2006), pp. 85–94
37. M.G. Zapata, Secure Ad-Hoc on-demand distance vector routing. Mobile Comput. Commun. Rev. **6**(3), 106–107 (2002)
38. S. Marti, et al., Mitigating routing misbehavior in mobile ad hoc networks, in *Proceedings of 6th Annual International Conference Mobile Computing Networking* (2000), pp. 255–265
39. P. Papadimitratos, Z.J. Haas, Securing data communication in mobile Ad Hoc networks. JSAC **24**(2), 343–356 (2006). Special issue on Security in Wireless Ad Hoc Networks
40. Y. Hu, A. Perrig, Ariadne: a secure on-demand routing protocol for ad hoc networks, in *ACM MOBICOM* (2002), pp. 12–23
41. S. Buchegger, J.L. Boudec, Nodes bearing grudges: towards routing security, fairness, and robustness in mobile ad hoc networks, in *Proceedings of the 10th Euromicro Workshop on Parallel, Distributed and Network-based Processing* (IEEE Computer Society, Canary Islands, 2002), pp. 403–410
42. M.C. Wong et al., Security issues in ad hoc networks, in *Security in Sensor Networks*, ed. by Y. Xiao (Auerbach Publications, Boca Raton, 2007)
43. C. Karlof et al., Secure routing in wireless sensor networks: attacks and countermeasures. AdHoc Netw. J. Spec. Issue Sensor Netw. Appl. Protoc. **1**(2–3), 293–315 (2003).
44. Y. Wang, Y. Tseng, Attacks and defenses of routing mechanisms in Ad Hoc and sensor networks, in *Security in Sensor Networks*, ed. by Y. Xiao (Auerbach Publications, Boca Raton, 2007)
45. C.E. Perkins, P. Bhagwat, Highly dynamic destination-sequenced distance-vector routing (DSDV) for mobile computers, in *ACM Conference on Communications Architectures, Protocols and Applications*, Vol. 1 (1994), pp. 234–244
46. B. Bellur, R.G. Ogier, A reliable, efficient topology broadcast protocol for dynamic networks. in *IEEE INFOCOM*, Vol. 1 (1999), pp. 178–186
47. P. Jacquet et al., Optimized link state routing protocol for ad hoc networks. in *IEEE International Multi Topic Conference*, Vol. 1 (2001), pp. 62–68
48. D.B. Johnson, D.A. Malts, Dynamic source routing in Ad Hoc wireless networks, in *Mobile Comput.*, ed. by T. Imielinski, H. Korth (Kluwer Academic Publishers, Norwell, 1996), pp. 153–181
49. C.E. Perkins, E.M. Royer, Ad-Hoc on-demand distance vector routing, in *IEEE Workshop on Mobile Computing Systems and Applications*, Vol. 1 (1999), pp. 90–100

50. M.S. Corson, A. Ephremides, A distributed routing algorithm for mobile wireless networks. Wirel. Netw. **1**(1), 61–81 (1995)
51. C.K. Toh, Associativity-based routing for ad hoc mobile networks. Wirel. Pers. Commun. **4**(2), 103–139 (1997)
52. V.D. Park, M.S. Corson, A highly adaptive distributed routing algorithm for mobile wireless networks. in *IEEE INFOCOM*, Vol. 1 (1997), pp. 1405–1413
53. Z.J. Haas, M.R. Pearlman, The performance of query control schemes for the zone routing protocol. IEEE/ACM Trans. Netw. **9**(4), 427–438 (2001). doi:10.1109/90.944341
54. G. Pei, M. Gerla, T.W. Chen, Fisheye state routing: a routing scheme for Ad Hoc wireless networks. in *IEEE International Conference on Communications*, Vol. 1 (2000), pp. 18–22
55. P. Samar, M.R. Pearlman, Z.J. Haas, Independent zone routing: an adaptive hybrid routing framework for Ad Hoc wireless networks. ACM/IEEE Trans. Netw. **12**(4), 595–608 (2004)
56. M.R. Pearlman, Z.J. Haas, Determining the optimal configuration for the zone routing protocol. IEEE J. Sel. Areas Commun. **17**(8), 1395–1414 (1999). doi:10.1109/49.779922
57. P. Papadimitratos, Z.J. Haas, Secure message transmission in mobile Ad Hoc networks. Elsevier Ad Hoc Netw. J. **1**(1), 193–209 (2003)
58. A. Perrig, R. Canetti, D. Song and J.D. Tygar, Efficient and secure source authentication for multicast, in *Proceedings of the Network and Distributed System Security Symposium, NDSS'01* (2001), pp. 35–46
59. A. Perrig, R. Canetti, J.D. Tygar, D. Song, Efficient authentication and signing of multicast streams over lossy channels, in *Proceedings of the IEEE Symposium on Security and Privacy* (2000), pp. 56–73
60. J. Deng et al., INSENS: intrusion-tolerant routing in wireless sensor networks. Comput. Commun. **29**, 216–230 (2006)
61. W.R. Heinzelman et al., Energy-efficient communication protocol for wireless microsensor networks, in *33rd Annual Hawaii International Conference on System Sciences* (2000), pp. 3005–3014
62. M. Tubaishat et al., A secure hierarchical model for sensor network. ACM SIGMOD Rec. **33**, 7–13 (2004)
63. W. Diffie, M.E. Hellman, Privacy and authentication: an introduction to cryptography. Proc. IEEE **67**(3), 397–427 (1979)
64. F. Hu et al., Scalable security in wireless sensor and actuator networks, in *Security in Sensor Networks*, ed. by Y. Xiao (Auerbach Publications, Boca Raton, 2007)
65. A. Perrig et al., SPINS: security protocols for sensor networks. Wirel. Netw. **8**, 521–534 (2002)
66. B. Karp et al., GPSR: greedy perimeter stateless routing for wireless networks, in *Proceedings of the 6th Annual International Conference on Mobile Computing and Networking* (ACM Press, New York, 2000), pp. 243–254
67. K. Ravichandran, K.M. Sivalingam, Secure localization in sensor networks, in *Security in Sensor Networks*, ed. by Y. Xiao (Auerbach Publications, Boca Raton, 2007)
68. L. Lazos et al., SeRLoc: robust localization for wireless sensor networks, in *Proceedings of 3rd ACM Workshop Wireless Security* (2004), pp. 21–30
69. P. Rousseeuw, A. Leroy, *Robust Regression and Outlier Detection* (Wiley-Interscience, New York, 2003)
70. Z. Li, W. Trappe, Y. Zhang, B. Nath, Robust statistical methods for securing wireless localization in sensor networks, in *Proceedings of 4th International Symposium Information Processing in Sensor Networks* (2005)
71. L. Fang et al., A beacon-less location discovery scheme for wireless sensor networks, in *Proceedings of IEEE INFOCOM* (2005)
72. T. Dimitriou, I. Krontiris, Secure in-network processing in sensor networks, in *Security in Sensor Networks*, ed. by Y. Xiao (Auerbach Publications, Boca Raton, 2007)
73. L. Hu et al., Secure aggregation for wireless networks, in *Proceedings of Symposium Applications Internet Workshops* (2003), pp. 384–391
74. J. Girao et al., CDA: concealed data aggregation in wireless sensor networks, in *Proceedings of ACM WiSe* (2004)

75. J. Domingo-Ferrer, A provable secure additive and multiplicative privacy homomorphism, in *Proceedings of Information Security Conference* (2002), pp. 471–483
76. L. Atzori et al., The internet of things: a survey. Comput. Netw. **54**, 2787–2805 (2010)
77. R. Roman et al., Integrating wireless sensor networks and the internet: a security analysis. Internet Res. **19**(2), 246–259 (2009)
78. R. Roman et al., Do wireless sensor networks need to be completely integrated into the internet? in *Furture Internet of People, Things and Services (IoPTS) eco-Systems* (Brussels, 2009)
79. R. Hummen et al., A security protocol adaptation layer for the IP-based internet of things, *Interconnecting Smart Objects with the Internet Workshop* (2011)
80. T. Zahariadis et al., *Securing Wireless Sensor Networks Towards a Trusted Internet of Things* (IoS Press, 2009), pp. 47-56. ISBN: 978-1-60750-007-0
81. R. Roman et al., Key management systems for sensor networks in the context of the internet of things. Comput. Electr. Eng. **37**(2), 147–159 (2011)

Chapter 18
Privacy Enhancing Technologies for Wireless Sensor Networks

Chi-Yin Chow, Wenjian Xu and Tian He

Abstract Since wireless sensor networks (WSNs) are vulnerable to malicious attacks due to their characteristics, privacy is a critical issue in many WSN applications. In this chapter, we discuss existing privacy enhancing technologies designed for protecting system privacy, data privacy and context privacy in wireless sensor networks (WSNs). The privacy-preserving techniques for the system privacy hide the information about the location of source nodes and the location of receiver nodes. The data privacy techniques mainly protect the privacy of data content and in-network data aggregation. The context privacy refers to location privacy of users and the temporal privacy of events. For each of these three kinds of privacy in WSNs, we describe its threats and illustrate its existing privacy-preserving techniques. More importantly, we make comparisons between different techniques and indicate their strengths and weaknesses. We also discuss possible improvement, thus highlighting some research trends in this area.

1 Introduction

Privacy is a critical issue when applying theoretical research in wireless sensor networks (WSNs) to scientific, civilian and military applications [35, 45], e.g., environmental sensing, smart transportation and enemy intrusion detection.

C.-Y. Chow · W. Xu
Department of Computer Science, City University of Hong Kong, Kowloon, Hong Kong
e-mail: chiychow@cityu.edu.hk

W. Xu
e-mail: wenjianxu2@student.cityu.edu.hk

T. He (✉)
Department of Computer Science and Engineering, University of Minnesota,Minneapolis, MN, USA
e-mail: tianhe@cs.umn.edu

H. M. Ammari (ed.), *The Art of Wireless Sensor Networks*,
Signals and Communication Technology, DOI: 10.1007/978-3-642-40009-4_18,
© Springer-Verlag Berlin Heidelberg 2014

WSNs are vulnerable to privacy breaches because they possess the following characteristics:

- **Wireless communication.** Wireless sensors need to communicate with each other through wireless communication. Wireless communication signals are easy to be tracked or eavesdropped by adversaries. We show later in this chapter that, in some kinds of applications, privacy breaches take place when adversaries are able to track or eavesdrop wireless communication signals even if the content of the transmitted data is protected securely.
- **Open environments.** WSNs are usually deployed in open environments to provide sensing and monitoring services. Such open environments could cause privacy concerns because malicious people can easily approach the system area or even physically access the sensor.
- **Large-scale networks.** The number of sensor nodes in a WSN is often large, so that protecting every node from being compromised by adversaries is difficult. Thus, the privacy enhancing technology designed for the WSN should be able to deal with a situation that the network contains some compromise sensor nodes which can be controlled by adversaries.
- **Limited capacity.** In general, wireless sensors have scarce resources, e.g., limited computational power, constrained battery power, and scarce storage space. As a result, existing privacy enhancing technologies designed for the Internet or wireless networks are not applicable to WSNs.

Due to these limitations, it is very challenging to design secure privacy-preserving techniques for WSNs. It is essential for researchers to study existing privacy enhancing technologies for WSNs, investigate their strengths and weaknesses, and identify new privacy breaches to improve them. To help researchers to understand the state-of-the-art privacy enhancing technologies for WSNs, we category them into three main types of privacy, namely, *system privacy*, *data privacy*, and *context privacy*. These three kinds of privacy are defined as follows:

1. **System privacy** is the ability of a system to protect the information about the setting of its WSN (e.g., the location information of its base station) and the communication information among its network components (e.g., the source node of data).
2. **Data privacy** is the ability of a system to preserve the data content through the course of transmission or in-network aggregation.
3. **Context privacy** is the ability of a system to protect the user location monitored by sensor nodes, or the time when an event is detected by sensor nodes.

For each of these three kinds of privacy, we discuss its threats and then highlight its existing privacy-preserving techniques.

The rest of this chapter is organized as follows: Sect. 2 presents an overview of this chapter. Sections 3, 4, and 5 describe the threat models and solutions of system privacy, data privacy, and context privacy, respectively. Section 6 concludes this chapter and discusses future research directions.

2 Overview of This Chapter

Various WSN applications require different privacy protection techniques. For example, in an event detection application, the location information of source sensor nodes is the sensitive information and can be inferred by adversaries through wireless communication signal analysis even without knowing the transmitted data content. Such event detection applications require system privacy protection. In a data collection application, its sensor node's readings are sensitive and should be protected during the course of transmission. Thus, data collection applications need data privacy protection. In a location monitoring application, the location information of monitored individuals is sensitive and should be protected. Location monitoring applications call for context privacy protection. From these three applications, we can see that different types of WSN applications have their own definition of sensitive information and they require different privacy protection techniques. In other words, exiting privacy enhancing technologies for WSNs are application-oriented because many of them are designed for a particular WSN application.

In this chapter, we identify three main types of privacy for existing WSN applications, namely, system privacy, data privacy, and context privacy. For each type of privacy, we first discuss its threat model and then describe its protection techniques, as depicted in Fig. 1. For system privacy, we mainly discuss privacy-preserving techniques designed to protect the source node of data and the location information of base stations. For data privacy, we survey privacy-preserving techniques for protecting data content during transmission or in-network aggregate processing. For context privacy, we focus on protection techniques designed for preserving the location privacy of people monitored by sensor nodes and the temporal information of events detected by sensor nodes.

3 System Privacy

Tracking wireless communication signals in a WSN could reveal different kinds of sensitive information such as which sensor node generates a reading, which sensor node is the destination of a packet and which sensor nodes are near a base station (or a

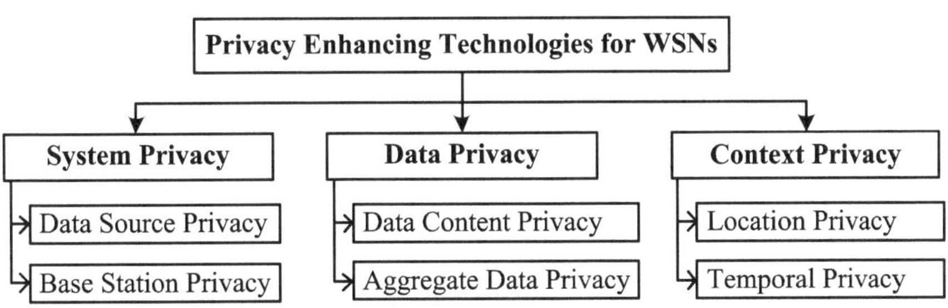

Fig. 1 Our taxonomy of privacy enhancing technologies for wireless sensor networks (WSNs)

sink node). For example, sensors deployed for detecting and monitoring a particular type of endangered animal in a forest (e.g., giant pandas [57]) can be utilized by hunters to locate the monitored animals. After a sensor node detects a target animal, it sends a report to the base station. A hunter can eavesdrop wireless communication signals to trace from the base station back to the source node, even without capturing and analyzing the transmitted data content. This kind of privacy breach would cause serious consequences and is difficult to be detected. More powerful adversaries can also compromise a sensor node and capture packets to learn the routing path from the source sensor node to the base station.

Under the system privacy, we identify two main types of privacy issues, namely, *data source privacy*, and *base station privacy*. Adversaries may locate the data source or the base station by analyzing wireless communication signals. Privacy-preserving techniques designed for protecting data source and base station privacy may prevent such malicious attacks.

3.1 Data Source Privacy

Consider our example where a WSN is deployed for detecting and monitoring a particular type of endangered animal in a forest. After a sensor node detects a target animal, it generates a report about the animal's activities and send the report to the base station through a multi-hop routing protocol [2, 3]. Suppose an adversary is equipped with devices such as antenna and spectrum analyzers that can be used to capture wireless signals between sensor nodes and measure the angle of arrival of signals. And all sensor nodes keep silent until they detect a target animal in their sensing area. After a sensor node detects a target animal, it sends a report to the base station. The adversary would capture wireless communication signals along the data transmission path from the source sensor node to the base station, and then he can carry out a traffic backtracking attack to find the source sensor node by capturing wireless signals for each hop, analyzing their direction, and then identifying the sender of each hop, as illustrated in Fig. 2. In practice, the adversary could use traffic backtracking to locate the endangered animals and bring danger to them, even though he is not able to read the report content. In this section, we discuss existing privacy enhancing techniques for two main types of attacks: local eavesdropping (Sect. 3.1.1) and global eavesdropping (Sect. 3.1.3).

3.1.1 Local Eavesdropping

Phantom routing is a routing protocol designed to prevent the traffic backtracking attack [32, 46]. The main idea of the phantom routing protocol is to select a phantom source node at a location far away from the real one. The protocol consists of two main steps:

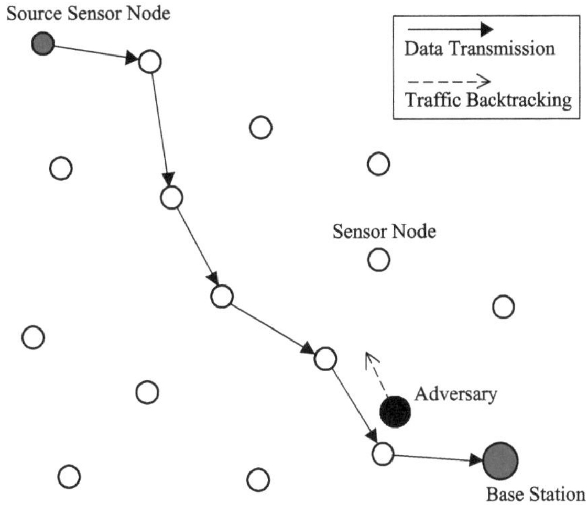

Fig. 2 The traffic backtracking attack in a WSN

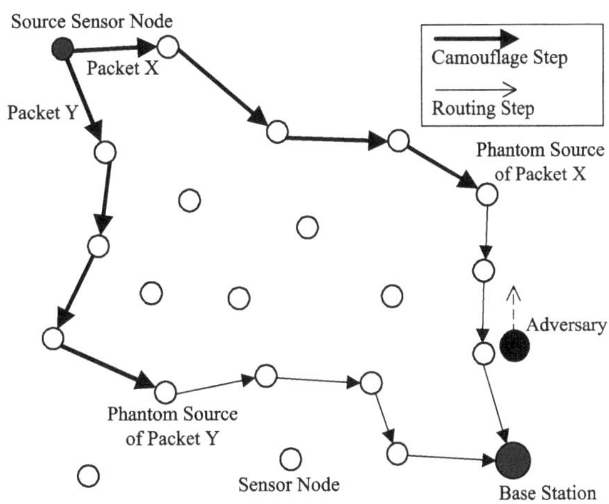

Fig. 3 The phantom routing paths of two data packets from the real source node to the base station

1. **Camouflage step.** This step ensures that a phantom source sensor node is located far away from the real source node. For example, when a sensor node wants to send out a packet, the packet is forwarded to the other node with hop distance h.
2. **Routing step.** This step makes sure that the packet can be delivered to the base station. After forwarding the data packet h hops, a conventional routing protocol (e.g., flooding [39], probabilistic broadcast [19], or single-path routing [28, 33]) is used to route the packet to the base station.

Figure 3 shows an example of the phantom routing where $h = 4$ and a source sensor node sends out two packets X and Y. During the camouflage step (i.e., the first four hops), the routing paths of the packets X and Y (indicated by bold arrows) from the real source sensor node are different. After forwarding the packets 4 hops, they are forwarded to the base station through a single-path routing protocol. These camouflage paths make an adversary difficult to backtrack to the source sensor node in a given time period. The basic idea is that the adversary can trace back to a certain intermediate node by capturing the transmission signal of one of these two packets, but he may never catch any other packet from the real source sensor node at that intermediate node because the routing paths of other packets do not pass through that node. In this way, the phantom routing could direct the adversary to a place far away from the real source sensor node, and therefore it cuts off the backtracking of the adversary.

We next discuss several techniques that have been proposed for the camouflage step.

1. **Random walk** [32]. For each of the first h hops, the sensor node randomly chooses a neighbor node to forward the packet. Since such a random walk may visit a sensor node more than once and the node at the h-th hop (i.e., the phantom source) may remain clustered around the actual source node, the pure random walk is inefficient for leading the phantom source node to be far away from the actual source node [32, 46].
2. **Neighbor-grouping-based directed walk** [32]. When a sensor node forwards a packet, it divides its neighbor nodes into two sets S and S'. The node first randomly selects one of these two sets and then randomly forwards the packet to a neighbor node in the selected set. For example, the neighbor nodes can be divided into two sets based on their hop distance to the base station, i.e., S includes all the neighbor nodes with the hop distance smaller than or the same as the sensor node's hop distance and S' includes all the nodes with the hop distance larger than the sensor node's hop distance.
3. **Greedy random walk** [58]. Before a sensor node forwards a packet, it uses a Bloom filter [5] to store the identifiers of itself and its neighbor nodes and then forwards the packet with the Bloom filter. After a neighbor node receives the packet, it randomly selects a neighbor node whose identifier is not stored in the Bloom filter as the next hop, adds the identifiers of its neighbor nodes in the Bloom filter, and then forwards the packet with the updated Bloom filter.
4. **Minimum distance** [37]. When a sensor node wants to send out a packet, it randomly selects the location of a phantom source node such that the distance between the sensor node and the phantom source node is at least certain distance d_{min}. Figure 4 illustrates this technique in our example, where both packets X and Y are first forwarded to their phantom source nodes that are far away from the source sensor node by at least d_{min}. Then, the phantom source nodes route X and Y using a single-path routing protocol. However, the sensor node may not have the actual location of every node in the system, if an intermediate node that is able to tell that the selected phantom source node does not exist, it becomes the phantom source node. The actual source node can also select all the intermediate nodes on the routing path to a selected phantom node using either an angle-based or a quadrant-based approach.

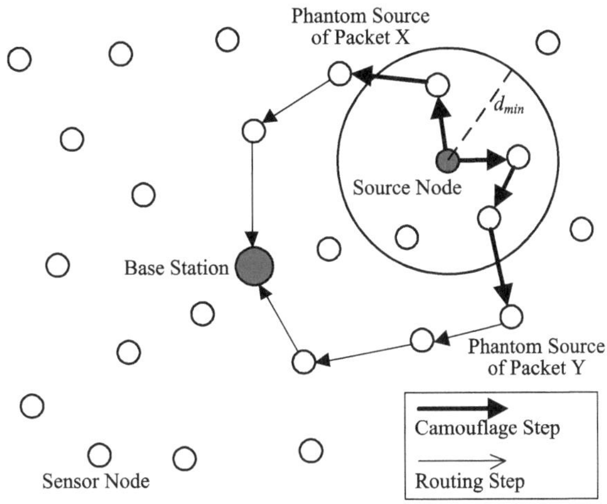

Fig. 4 The minimum distance technique in the camouflage step of phantom routing

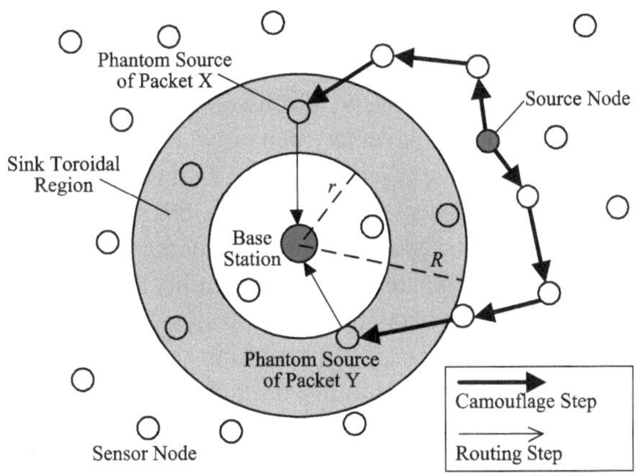

Fig. 5 The sink toroidal region (STaR) technique in the camouflage step of phantom routing

5. **Sink toroidal region** [38]. In this technique, a phantom source node is always selected inside a sink toroidal region (STaR) around a base station or a sink. The STaR is defined by two circles centered at the base station with radius r and R, where $r < R$. The intersection area between these two areas constitutes the STaR. A source sensor node randomly selects a location within the STaR as the phantom source, and then forwards its packet to it, as in the minimum distance technique. The STaR area should be large enough such that it is not practical for an adversary to monitor the entire STaR. Figure 5 depicts STaR in our example, where the shaded donut shape is the STaR. Packets X and Y are first forwarded to their randomly selected phantom source nodes inside the STaR and then a single-path routing protocol is used to forward them to the base station.

3.1.2 Discussion

All these techniques share the same goal: randomly finding a phantom source node that is far away from the actual source node. The first three techniques require an intermediate node to make its local decision to find a phantom node, while the last two techniques allow the source sensor node to determine the phantom node, and even all the intermediate nodes in the routing path between itself and the phantom node in the minimum distance technique. The last two techniques are more reliable, energy-efficient, and secure, because the sensor node has more control over the selection of its phantom node. However, if the sensor node does not have a global view of the system, it may not be able to use the minimum distance or STaR technique. It would be useful to extend them to allow intermediate nodes to make their local decision based on the source sensor node's requirements and its performance can be the same as the centralized decision scenario.

3.1.3 Global Eavesdropping

In the previous section, we discussed the privacy-preserving techniques designed for a scenario that an adversary can only eavesdrop on a limited portion of the network at a time, i.e., a *local eavesdropper*. In this section, we consider a stronger adversary called a *global eavesdropper* [41, 51, 59] who is capable to deploy his sensor nodes or compromise sensor nodes to monitor the communication traffic in the entire WSN. With a global view of the network traffic, the adversary can easily infer the location of a source node using the traffic analysis attack.

The basic approach of preventing the traffic analysis attack from global eavesdroppers is to inject dummy traffic into a WSN to make adversaries confused and thus unable to distinguish a real data source from a set of dummy data sources. There are three main kinds of techniques to inject dummy traffic into a WSN:

1. **Periodic collection** [41]. Every sensor node independently and periodically sends out data packets regardless of whether it has real data packets to send out or only dummy data packets. More specifically, each sensor node has a timer that fires at a constant rate. When the timer fires, if the node has a packet in its buffer, it forwards the packet; otherwise, it sends a dummy packet with a random payload. In this way, the adversary is not able to distinguish a real source sensor node from other dummy sensor nodes because they are all sending out data packets in the same manner. The drawback of this solution is that a network with more sensor nodes incurs higher communication overhead.
2. **Source simulation** [41]. Although the periodic collection technique provides optimal source-location privacy, if the constant rate is small, the system delay may be very high; if the rate is large, the system may have too much dummy traffic and suffer from high power consumption. The source simulation technique artificially creates multiple fake traces in the network to hide the traffic generated by real objects. For example, if a WSN is designed to monitor pandas in a forest,

their historical habits and behaviors are studied to create fake traces for pandas. Before deployment, we randomly select a set of sensor nodes and initialize a unique token in each of them. These tokens are transmitted between sensor nodes in the network to simulate the behavior of pandas. Thus, the adversary who wants to trace pandas using the traffic analysis attack probably finds a virtual one.

3. **Probabilistic dummy generation** [51]. The objective of this technique is also to reduce the amount of dummy traffic and latency of the periodic collection technique. This technique uses the exponential distribution to control the rate of dummy generation by creating a sequence of dummy packets such that the time intervals between two consecutive messages follow the predefined exponential distribution. For a packet generated by a real event, the packet is delayed as long as its transmission time also follows the predefined exponential distribution. The whole concept of this technique is based on a statistical property: if two probabilistic distributions are both exponential distributions with very close means, they are statistically indistinguishable from each other. Experimental results show that this technique effectively reduces communication overhead while it can provide the same level of privacy protection as the periodic collection technique.

3.1.4 Discussion

The basic idea of these dummy traffic injecting techniques is to use dummy traffic to hide the real data source sensor nodes. There is a tradeoff between the communication overhead (e.g., bandwidth and power consumption) and the privacy protection [29]. K. Mehta et al. provide a method to estimate a lower bound on the communication overhead needed to achieve a certain level of privacy protection in the network [41]. In addition, some scientists study how to reduce communication overhead for dummy traffic without sacrificing any privacy protection. For example, Y. Yang et al. suggest placing some proxy sensor nodes in a WSN to filter out dummy packets and drop them after a certain number of hops of transmission [59].

There is a new research direction for source data privacy in unattended wireless sensor networks (UWSNs), where critical sensor nodes replicate their readings in a certain number of randomly selected nodes d-hop away from the critical sensor nodes. Recent study has found that there is a tradeoff between source-location privacy and data survivability (i.e., the number of data replicas) [7]. When an adversary finds the data source node, he can destroy the critical node in the system.

3.2 Base Station Privacy

In some WSN applications, such as security monitoring systems, the physical location of a base station (or a sink node) is considered as sensitive information. The main reason is that revealing the physical location of a base station to the adversary may give him a chance to make either physical or denial-of-service attacks to the base station in order to disable the WSN [15]. However, many routing protocols would reveal obvious traffic patterns in the network. The sensor nodes near the base station forward a greater number of data packets than other sensor nodes that are far away from the base station [16]. Based on such unbalanced traffic, it is easy for the adversary to infer the base station's physical location. The main idea of hiding the location information of the base station is similar to that of the phantom routing and dummy traffic injection techniques, but it focuses on how to make the adversary difficult to distinguish a sink node (or a base station) from other sensor nodes. We describe four techniques for protecting the base station privacy.

3.2.1 Multi-Parent Routing

In the multi-parent routing [16], each sensor node has multiple parent nodes. Each sensor node s finds its multiple parents based on their hop distance to the base station. The base station broadcasts a beacon message with a *level* field that is initially set to one. When s receives the beacon message, the value of *level* indicates s's hop distance to the base station. s next increases *level* by one and rebroadcasts the beacon message with the increased *level* value to its neighbors. After a certain time period, s selects all neighbor nodes whose *level* value is less than s's *level* value as its parent nodes. Figure 6 depicts an example, where a WSN consists of seven sensor nodes s_1 to s_7 and each sensor node already finds its *level* value. s_3 have four neighbor nodes (i.e., s_1, s_2, s_6, and s_7), but only the *level* values of s_1 and s_2 are less than s_3's *level* value. Thus, s_1 and s_2 are s_3's multiple parent nodes. Each sensor node erases its *level* value after all the sensor nodes have found their multiple parent nodes. When a sensor node wants to send a data packet, it randomly selects one of its multiple parent nodes to send the packet. Although this technique is similar to the *phantom routing* technique, it aims at spreading out traffic evenly in the whole WSN to make the adversary difficult to infer the physical location of the base station instead of finding an intermediate phantom node. Multi-parent routing often works with random walk and fractal propagation, as will be discussed later, to enhance the privacy-protecting performance.

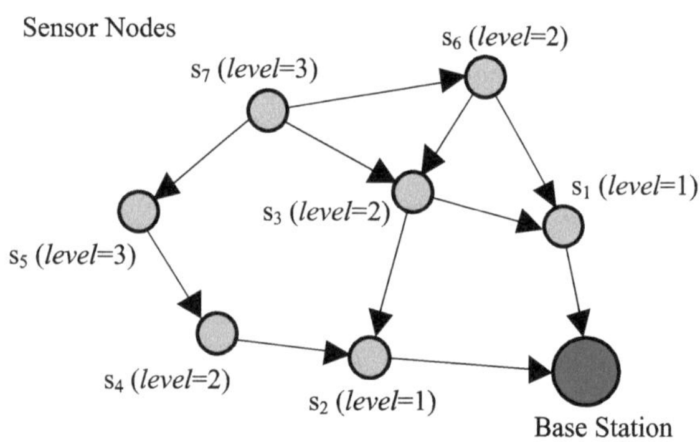

Fig. 6 Multi-parent routing

3.2.2 Random Walk

Although a random walk has been proposed for reliable data transmission [56], it can also be used to protect the base station privacy [16]. The basic idea is that a sensor node has two ways to forward a packet: (1) the node forwards the packet to one of its parent nodes with equal probability or (2) it forwards the packet of one of its neighbors with equal probability. The node uses the first way with probability p or the second one with probability $1 - p$.

3.2.3 Fractal Propagation

The multi-parent routing and random walk techniques can avoid forwarding packets to the base station through the shortest or static path. The fractal propagation technique can further spread out fake packets [16]. When a sensor node overhears that its neighbor node forwards a data packet, it also generates a fake data packet following a predefined probability distribution with a system parameter k and forwards it to a randomly selected neighbor node. After the neighbor node receives the fake packet, it reduces the value of k by one and forwards it to one of its neighbor nodes with the updated k value. When a node receives the fake packet with $k = 0$, it simply drops the packet. As a result, this technique spreads out the communication traffic of a data packet in the network. This technique works due to the fact that the normal nodes can distinguish a fake packet from a real one, while the adversary cannot distinguish since it does not know the encryption key. If we use this technique with the multi-parent routing and random walk techniques, a full picture of the real and fake communication flows triggered by a data packet transmission looks like a fractal shape, as depicted in Fig. 7.

3.2.4 Hiding Traffic Direction Information

The goal of hiding traffic direction information is to make the directions of both incoming and outgoing traffic at a node uniformly distributed [30]. The basic idea is that each sensor node divides its neighbor nodes into two disjoint lists: *closer* and *further lists*, where the closer list contains the neighbor nodes that are closer to a receiver and the further list consists of the neighbor nodes that are further way from the receiver. The distance between two sensor nodes can be measured by their hop distance or their Euclidean distance. When a sensor node forwards a data packet, it selects the next hop from the further list with probability p, and from the closer list with probability $1 - p$. p is used to balance a performance trade-off between communication overhead (i.e., latency and power consumption) and privacy. For example, if p is smaller, the node selects more next hops in the closer list. Thus, the routing paths are shorter and the power consumption is lower, but the receiver's location privacy is weaker. On the other hand, if p is larger, more next hops are elected from the further list. Although the privacy protection is stronger, the routing paths

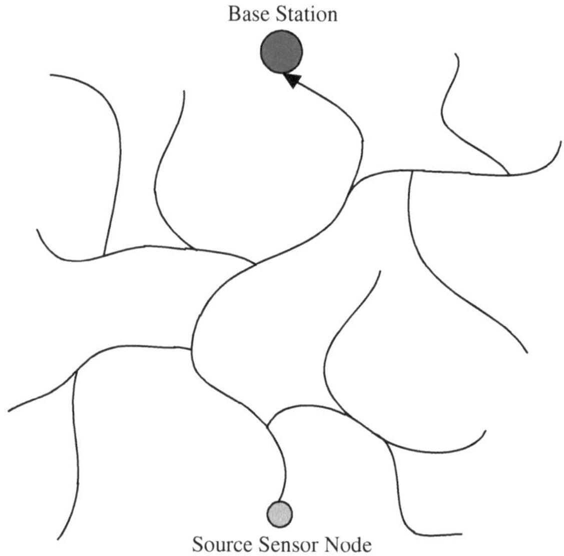

Fig. 7 The real and fake communication traffic trigged by a data packet transmission using the fractal propagation technique with the multi-parent routing and random walk techniques

are longer and the power efficiency is lower. To further protect the receiver-location privacy, fake packets can also be injected with this technique to smooth out the traffic. Whenever a sensor node forwards a packet, it also sends a fake packet to a randomly chosen neighbor in the further list, and the fake packet will be further forwarded away by h hops ($h \geq 2$). Experimental results show that this technique with fake packet injection deliveries more packets than the phantom routing technique [32] and the fractal propagation technique [16], when $p \geq 0.4$.

3.2.5 Discussion

In general, the techniques designed for protecting the base-station- or receiver-location privacy are based on random forwarding and fake packet injection. Some of these techniques (i.e., [16, 30]) require that each sensor node knows its hop distance from a sink through a broadcast message from the sink. An intelligent fake packet injection technique has been designed to protect the base-station privacy during such a topology discovery period [36]. Most of existing techniques for protecting base station privacy only consider stationary base stations. It is interesting to consider a scenario where a base station is able to move to sensor nodes to collect data. There are only a few attempts to tackle this scenario. For example, data forwarded to random nodes and mobile base stations move in the network following some random paths to collect data from sensor nodes [43]. Actually, the mobility of the base station can enhance the privacy of the base station itself [1]. Specifically, the base station relocates itself within the WSN periodically, which obfuscates previous traffic analysis conducted by the adversary. Thus, the privacy of the base station is protected. Nevertheless, more research efforts are needed to tackle the privacy issue of mobile base stations with stationary sensor nodes or even with mobile sensor nodes.

4 Data Privacy

Wireless sensor nodes are usually deployed to monitor surroundings by providing a variety of readings. In many WSN applications, sensor readings may be sensitive information, e.g., a log of identifications of border-crossing vehicles. Such readings must be protected from malicious attacks. This section describes existing privacy-preserving techniques for two main types of data privacy: *data content privacy* and *aggregate data privacy*.

4.1 Data Content Privacy

Many research efforts have mainly focused on how to design encryption and authentication mechanisms for WSNs that consider the computational and power constraints of wireless sensor nodes. Since these techniques are more related to security issues in WSNs, we briefly highlight two well-known suites of security protocols. Interested readers are referred to two survey papers [8, 50].

1. **SPINS** [47]. SPINS consists of two components, a secure network encryption protocol (SNEP) and a micro version of timed efficient streaming loss-tolerant authentication protocol (μTESLA). SNEP provides secure channels between a senor node and a base station for data confidentiality, data authentication for two-party communication, data integrity through data authentication, and guarantee for data freshness (i.e., no adversary replays old data). μTESLA provides authenticated broadcast communication.
2. **Localized Encryption and Authentication Protocol (LEAP)** [60]. LEAP supports the management of four types of keys to provide different security requirements for different kinds of data communications on sensor nodes. (1) Every sensor node has a unique *individual key* that is shared with the base station for their secure communication; (2) the sensor nodes in the same group have a shared *group key* for building a secure broadcast channel from the base station to the whole group; (3) every sensor node shares a *cluster key* with its neighbors for securing local broadcast messages; and (4) every node also shares a *pairwise shared key* with each of its neighbor nodes for secure communication with authentication. In addition, LEAP provides an efficient protocol for establishing and updating these keys, as well as an authentication mechanism for them.

There are two key limitations of security protocols in WSNs [44, 53]. (1) *Constrained resources*. Since sensor nodes usually have limited battery and computational power, only simple and fast methods of cryptography can be used in WSNs.

Wireless communication is a major cause of power consumption, so unnecessary information exchange should be avoided. (2) *Unsecure key storage.* Since sensor nodes are usually deployed in an open area, they have no secure storage for their secret keys. Recent results have shown that pairing-based cryptography (PBC) is suitable for constrained sensors [44, 53]. Based on PBC, authenticated identity-based non-interactive security protocols can be designed for WSNs. Experimental results have shown that PBC is feasible on 8-, 16-, and 32-bit sensor processors and various types of sensor platforms, e.g., MICA2/MICAz, TelosB/Tmote Sky, and Imote2. Therefore, it is important for the sensor network security community to restudy how PBC can be used to enhance the existing suites of security protocols.

4.2 Aggregate Data Privacy

One of the important functions of WSN applications is the support for in-network data processing, which means sensor nodes can collaborate with each other to provide services or answer queries without a centralized server. End-to-end encryption and authentication techniques are not applicable to in-network data processing because intermediate nodes cannot access any by-passing data. In practice, many WSN applications only need to provide aggregate statistics such as SUM, AVERAGE, MIN, or MAX of sensor readings in a certain region or within a certain time period [40]. These applications can employ in-network data aggregation to reduce the amount of raw sensor readings to be reported, so sensor and network resources can be saved. Data aggregation techniques often assume that all sensor nodes in the WSN are trustworthy [49]. However, this assumption may not be realistic because sensor nodes can be compromised by the adversary who wants to steal sensor readings. If the raw sensor readings are passing through and being aggregated on these compromised nodes, this would cause privacy leakage.

Before presenting privacy-preserving data aggregation techniques, we summarize their desired characteristics as follows:

1. **Privacy.** The data generated by a sensor node should be only known to itself. Furthermore, a privacy-preserving data aggregation technique should be able to handle attacks and collusion among compromised nodes, since it is possible that some nodes may collude to uncover the private data of other nodes.
2. **Efficiency.** Data aggregation reduces the amount of traffic in a WSN, thus saving bandwidth and power usage. However, a privacy-preserving data aggregation technique introduces additional computational and communication overhead to sensor nodes. A good technique should minimize such kinds of overhead.
3. **Accuracy.** A privacy-preserving data aggregation technique should not reduce the accuracy of aggregate values.

Different WSN applications make different trade-offs among these performance metrics. The rest of this section discuss four privacy-preserving schemes for data aggregation, namely, cluster-based private data aggregation (CPDA), slice-mix-aggregate (SMART), secret perturbation, and k-indistinguishable privacy-preserving data aggregation (KIPDA).

4.2.1 CPDA: Cluster-Based Private Data Aggregation

CPDA [25] is designed to support privacy-preserving SUM aggregation in WSNs. In CPDA, sensor nodes are randomly grouped into clusters. For each cluster, algebraic properties of polynomials are used to calculate an aggregate SUM. CPDA guarantees that the data of individual node is not exposed to other nodes. Finally, the intermediate aggregate values in each cluster are further aggregated along a routing path to a base station. In general, CPDA consists of three main steps:

1. **Formation of clusters.** The first step in CPDA is to construct clusters to perform intermediate aggregations. Every cluster consists of one cluster head (CH) and many cluster members. The CH is responsible for calculating intermediate aggregations and reporting their results to a base station. Figure 8 depicts a cluster with three sensor nodes s_0, s_1, and s_2, where s_0 is the CH.
2. **Intermediate aggregation.** This step is based on a random key distribution mechanism proposed in [18]. Consider a cluster with one head and n members, where s_0 is the CH and $s_1, \ldots, s_n$ are other sensor nodes in the cluster. Each node s_i $(0 \leq i \leq n)$ sends a seed A_i to other members in its cluster. As depicted in Fig. 8a, s_0 sends A_0 to s_1 and s_2, s_1 sends A_1 to s_0 and s_2, and s_2 sends A_2 to s_0 and s_1. For each node s_j $(0 \leq j \leq n)$, each node s_i $(0 \leq i \leq n)$ perturbs its own private reading into V_j^i based on s_j's seed and n random numbers generated by s_i, and it then sends $E(V_j^i)$ in an encrypted form to s_j, where $i \neq j$. As illustrated in Fig. 8b, s_0 sends $E(V_1^0)$ and $E(V_2^0)$ to s_1 and s_2, respectively; s_1 sends $E(V_0^1)$ and $E(V_2^1)$ to s_0 and s_2, respectively; and s_2 sends $E(V_0^2)$ and $E(V_1^2)$ to s_0 and s_1, respectively. Each node s_j $(1 \leq j \leq n)$, except the CH, next adds all received V_j^i $(0 \leq i \leq n)$ and its V_j^j together to compute a sum F_j, and then send F_j to its CH. In the running example, s_1 and s_2 send F_1 and F_2 to the CH s_0, respectively. Finally, after the CH receives F_j from each member s_j, it is

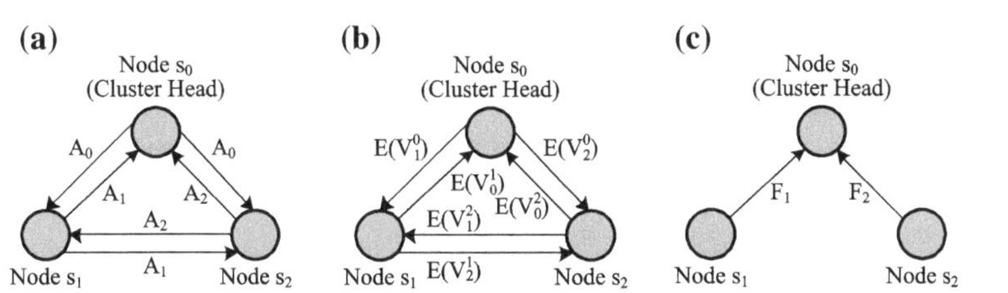

Fig. 8 An example of CPDA within a cluster of three sensor nodes. **a** Broadcast seeds. **b** Send encrypted perturbed values. **c** Send assembled values to CH

able to compute the sum of the original readings from all the nodes in the cluster without compromising the privacy of individual nodes' data value.
3. **Cluster data aggregation.** The cluster head reports its intermediate aggregate sum to the base station. The base station computes an aggregate SUM by summing up all collected intermediate aggregate sums.

CPDA guarantees that if less than $(n - 1)$ nodes collude in a cluster of size n, the individual sensor readings in the cluster cannot be disclosed. Therefore, larger average cluster size contributes to better privacy-preservation performance, but it also incurs more computational overhead to compute the intermediate aggregation value. As a result, there is a design tradeoff between the privacy protection and computation efficiency.

Although CPDA can provide private data aggregation, it cannot guarantee data integrity. If an adversary changes the intermediate aggregate result in some clusters, the aggregate result would deviate from the actual one dramatically [26]. CPDA has been extended to iCPDA [27] that can guarantee data integrity through additional piggybacks. In iCPDA, every node in a cluster collects necessary information to calculate an intermediate aggregate result within the cluster. Hence, all the nodes in the cluster can figure out the intermediate aggregated value in the cluster, enabling them to monitor their CH and detect data pollution attacks. Experimental results showed that the communication overhead of iCPDA is a little bit higher than CPDA due to the extra message exchange.

4.2.2 Slice-Mix-Aggregate (SMART)

The basic idea of SMART [25] is to slice readings and use the associative property of addition to compute aggregate SUM. In general, SMART consists of three main steps:

1. **Slicing readings.** Suppose a WSN has N sensor nodes. Each sensor node s_i randomly selects a set of m peers within a certain hop distance. After s_i gets a private reading d_i, d_i is sliced into m pieces. s_i randomly keeps one piece, and then each of the remaining $m - 1$ pieces is randomly sent in an encrypted form to a distinct one of the selected m peers. Let d_{ij} be a piece of d_i that is sent from s_i to another sensor node s_j, and hence, $d_i = \sum_{j=1}^{N} d_{ij}$, where $d_{ij} = 0$ if s_j does not receive any piece of d_i.
2. **Mixing slices.** When a sensor node s_j receives k encrypted pieces within a certain time interval, s_j decrypts each piece and sums up all received pieces to compute a mixed value $r_j = \sum_{i=1}^{N} d_{ij}$, where $d_{ij} = 0$ if s_i is not one of the senders of the k pieces. Then, s_j sends r_j to the query server.
3. **Aggregation.** After the query server receives the mixed values from all the sensor nodes within a certain time interval, it sums them up to get the final result $\sum_{j=1}^{N} r_j = \sum_{i=1}^{N} \sum_{j=1}^{N} d_{ij} = \sum_{i=1}^{N} d_i$, where $r_j = 0$ is the query server does not receive any mixed value from node s_j and $d_i = 0$ if node s_i does not report any reading within the time interval.

Unlike CPDA, SMART does not need to form any clusters, so it can reduce the computational cost of cluster-wise data aggregation.

4.2.3 Secret Perturbation

The basic idea of secret perturbation [20] is that each sensor node s_i has a pre-assigned secret S_i, which is only known to s_i and the base station. s_i does not send its original reading d_i to the base station. Instead, s_i only sends a perturbed version of d_i, i.e., $\widehat{d_i} = d_i + S_i$, to the base station. Each intermediate node between s_i and the base station receiving a perturbed value can perform an additive aggregation function on it with other perturbed values sent from other nodes to get a single perturbed value. Since the base station knows the secret of each sensor node, it can subtract the perturbations from a perturbed value to get an actual reading. In general, the *basic secret perturbation scheme* consists of three main steps to compute aggregate SUM:

1. **System initialization.** Given N sensor nodes and the range of each sensor reading $[0, d_{max}]$, the base station selects an integer L, a prime number q and a secure hash function $hash(x)$ such that $max\{2^{L-1}, 2N\} < q < 2^L$, $d_{max} < 2^L$ and $0 \leq hash(x) < 2^L$. It also assigns each sensor node s_i with two secret numbers.
2. **Perturbed aggregation.** When a sensor node s_i receives a query, it gets a reading d_i, uses $hash(x)$ to calculate a perturbed reading value $\hat{d_i}$ and an auxiliary reading value $\hat{A}_i$, and initializes a list of IDs of reporting sensor nodes $list_i = \{i\}$. If s_i is a leaf node or it has no downstream node that reports data, s_i simply sends $\langle \hat{d}_i, \hat{A}_i, list_i \rangle$. On the other hand, if the other node s_j receives $\langle \hat{d}_{i_k}, \hat{A}_{i_k}, list_{i_k} \rangle$ from its downstream nodes, where $k = 0, 1, \ldots, m$ and $m < N$, it computes its own $\hat{d}_j$, $\hat{A}_j$ and $list_j$, and then performs an additive aggregation $\langle \hat{d}_j, \hat{A}_j, list_j \rangle$ on all the received $\langle \hat{d}_{i_k}, \hat{A}_{i_k}, list_{i_k} \rangle$.
3. **Retrieving the original aggregation at a base station.** After the base station s_0 receives the perturbed tuples $\langle \hat{d}_{i_k}, \hat{A}_{i_k}, list_{i_k} \rangle$ from its downstream nodes, where $k = 0, 1, \ldots, m$, $m \leq N$, and the ID of the base station is 0, s_0 computes its $\hat{d}_0$, $\hat{A}_0$ and $list_0$. s_o is then able to find the sum of its reading and the original readings of all the received tuples.

Taiming Feng et al. have extended the *basic secret perturbation scheme* to the *fully-reporting secret perturbation scheme* [20] that does not require sensor nodes to report their IDs by requesting every sensor node to report an actual or dummy reading in the perturbed form. They further proposed the *adaptive secret perturbation scheme* to minimize communication overhead and avoid reporting node IDs.

4.2.4 *k*-Indistinguishable Privacy-Preserving Data Aggregation (KIPDA)

The KIPDA scheme [21] is designed for MAX/MIN aggregation in WSNs based on the concept of k-anonymity [52]. The basic idea of KIPDA is that each sensor node reports its actual reading along with $k - 1$ other restricted or unrestricted camouflage values such that the actual reading is indistinguishable among the k values. A base station is able to find an exact aggregate MAX/MIN result based on k-indistinguishable data reported from sensor nodes. We present the four main steps in KIPDA with a running example with $k = 7$ for aggregate MAX, as depicted in Fig. 9.

1. **System setup.** The base station decides *a set of global real value positions* for a vector with k elements. Each position is assigned to each sensor node to indicate which position in a reported vector should store its actual reading. For each sensor node, the remaining unassigned positions are divided into two disjoint sets: (1) a set of unrestricted positions, where a random camouflage number is generated for each position, and (2) a set of restricted positions, where a camouflage value that is required to be less than or equal to an actual reading for aggregate MAX while a camouflage value that is required to be larger than or equal to an actual reading for aggregate MIN. In our example (Fig. 9), we assume that k is seven, the size of each set of unrestricted positions is two, and the size of each set of restricted positions is four.
2. **Filling camouflage values.** After a sensor node receives a query, it puts its actual reading at its real value position assigned by the base station in a vector. Then, it generates a camouflage value at every unrestricted or restricted position in the vector. Finally, it sends the vector to the base station through its parent node. In the running example, sensor node 1 puts is actual reading (i.e., 23) at the first position in a vector (indicated by an underline) and generates two unrestricted camouflage values 30 and 21 that are put at the fourth and sixth positions in the vector (indicated by shaded cells), respectively, as Fig. 9. Since the query computes aggregate MAX, node 1 generates four camouflage values that are less than the actual reading (i.e., 23). In this example, node 1 puts 18, 15, 20, and 8 at

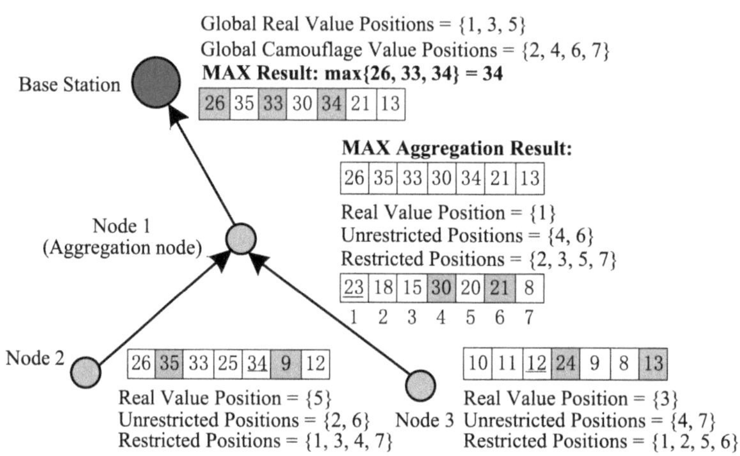

Fig. 9 An example of processing aggregate MAX using KIPDA with $k = 7$

the second, third, fifth, and seventh positions in the vector, respectively. Similarly, nodes 2 and 3 generate their vectors. Since node 1 is the parent node of nodes 2 and 3, nodes 2 and 3 send their vectors to node 1.

3. **Aggregation at an intermediate node.** After a sensor node receives vectors from its children nodes, it computes an intermediate aggregation result vector. For each position in the result vector, the node selects the maximum value at the same position among all the received vectors. The node next sends the result vector to its parent node. For example (Fig. 9), node 1 receives the vectors from nodes 2 and 3. The value at the first position of the result vector is $\max\{23, 26, 10\} = 26$. Similarly, the value at the second position of the result vector is $\max\{18, 35, 11\} = 35$ and so on. After node 1 computes the intermediate result vector $\langle 26, 35, 33, 30, 34, 21, 13\rangle$, the vector is sent to the base station through its parent node.
4. **Retrieving the original aggregation at a base station.** After the base station receives an aggregate vector, it selects the maximum value among the values at the global real value positions in the vector as the query answer. For example (Fig. 9), the base station receives an aggregate vector $\langle \underline{26}, 35, \underline{33}, 30, \underline{34}, 21, 13\rangle$ from node 1. The base station selects the maximum values at the first, third and fifth positions in the vector. Thus, the final aggregate MAX is $\max\{26, 33, 34\} = 34$.

Since the aggregate process of KIPDA does not require end-to-end or link-level encryption, it achieves higher efficiency than the encryption-based data aggregation technique in terms of power consumption and latency. In terms of robustness, KIPDA can tolerate up to a large number of compromised sensor nodes or communication link. However, it is not easy to apply KIPDA to other aggregation functions other than MAX and MIN.

4.2.5 Discussion

All the private data aggregation techniques discussed in this section can provide precise aggregate results if there is no packet loss. In terms of privacy protection, the secret perturbation technique always prevents the adversary from finding out an individual sensor's data or an aggregate result, regardless of the number of compromised sensor nodes. Contrarily, CPDA, SMART and KIPDA can only tolerate up to a certain threshold number of compromised sensor nodes or communication link. In terms of efficiency, experimental results [20] showed that the secret perturbation technique consumes less bandwidth than CPDA and SMART. KIPDA is more efficient than conventional encryption-based data aggregation techniques which incur high latency due to the decryption and re-encryption operations. Finally, CPDA, SMART and secret perturbation techniques are specially designed for aggregate SUM, while KIPDA is tailored for aggregate MAX/MIN. A WSN application should select the best private data aggregation technique based on its requirements for privacy protection and system efficiency.

5 Context Privacy

In this section, we focus on two kinds of context detected by sensor nodes, namely, *location privacy* and *temporal privacy*, in Sects. 5.1 and 5.2, respectively. Location privacy protection techniques are designed to anonymize the location information of people monitored by sensor nodes. Privacy-enhancing techniques for temporal privacy are designed to hide the time when a query is issued by a user or an event is detected by a sensor node.

5.1 Location Privacy

Sensor-based location systems have been proposed to support indoor positioning and monitoring, e.g., [24, 48, 55]. Such indoor monitoring systems can provide many services: (1) *Location-based queries.* They can answer queries like "how many customers on the second floor" and "which shop is the densest one during lunch hours." (2) *Security and control.* The system alerts the administrative staff when it detects that someone enters an office at midnight or the number of people in a room is larger than a system-specified limit. (3) *Resource management.* When the system has detected no people in an certain area for a certain period of time, it turns off some building facilities (e.g., lights, escalators and elevators) to save energy. However, similar to GPS, sensor-based location systems would threaten the user privacy. For example, if an adversary knows the location of a person's office, the adversary can easily determine whether the person is in his or her office by monitoring the sensing information, e.g., the number of people in a sensing area, reported from the sensor deployed in his or her office. Once the adversary identifies an individual's location, the adversary can track the user's movements by monitoring location updates from other sensor nodes [11, 22].

Cricket [48] is a privacy-aware sensor-based location system. Cricket has two strategies to protect the user's location privacy: (1) It deploys sensors in a system area. Every user wears a tag that receives signals from multiple sensors to detect the user's location. This decentralized positioning approach does not require any centralized processing on user location information or store user locations. (2) If a user concerns about location privacy, he or she can turn off the tag, and thus, no monitoring system can know the user's location. However, Cricket [48] is not suitable for location monitoring systems. The main reason is that if many people do not report location information to a location monitoring system, the system cannot provide any meaningful services.

Note that location privacy is different from data source privacy described in Sect. 3.1. Location privacy protection techniques aim at protecting the privacy of individuals' location information collected by source sensor nodes, while the objective of data source privacy protection techniques is to protect the privacy of the location of source sensor nodes themselves. We discuss four privacy enhancing techniques for indoor sensor-based location monitoring systems, and then describe how a privacy-preserving monitoring system provides location-based services without compromising the user location privacy.

5.1.1 Pseudonyms

To protect the privacy of individuals' location information collected by sensors while taking advantage of location-based services, users' true identity should be hidden from the applications receiving their location information. An even-driven middleware has been designed to act as a proxy server between the user and the application to help the user hide his/her real identity [4]. As depicted in Fig. 10, after a user registers his/her interest in a particular location-based service (LBS) with the middleware, the LBS provider receives event callbacks from the middleware when the user enters or exits a certain system-specified area. For example, a shopping mall application is configured to enable an LBS "sending e-coupons to users entering the shopping mall." This application should register certain areas in front of the shopping mall's entrances and wait for the event callback. When a registered user enters the application's registered area, the application sends relevant e-coupons to the user. Since the users' real identity can be anonymized by the middleware, they can enjoy LBS without revealing their real identities. However, it is still risky for a user to use a long-term pseudonym, even if different LBS providers give out different pseudonyms to the same user to avoid collusion. This is because an adversary could identify a user by following the "footsteps" of a pseudonym to or from some places which are strongly associated with the user's real identity (e.g., a residential house). One possible solution is to frequently change a user's pseudonym, but it may significantly degrade the quality of LBS. We discuss a mix-zone approach that can balance between the user location privacy and the quality of services.

Fig. 10 The middleware model for pseudonyms

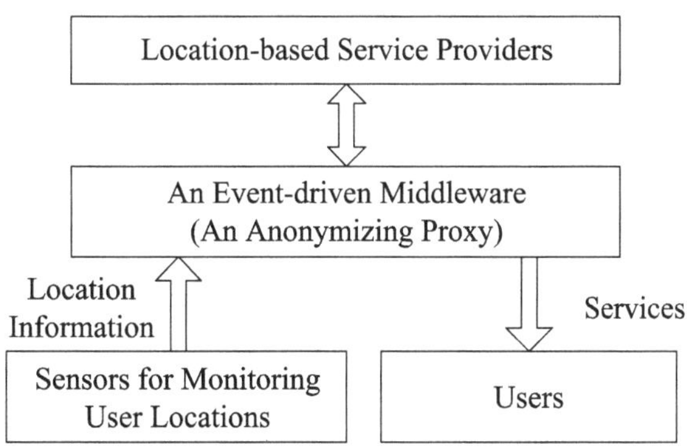

5.1.2 Mix Zones

The idea of *mix zones* [4] is derived from the concept of *mix nodes* designed for anonymous communication systems [6]. A *mix zone* is defined as a spatial region with a system-specified maximum number of users who have not registered with any application callback, while an *application zone* is defined as an area where LBS applications could register for event callbacks. An example is depicted in Fig. 11, where there is one mix zone, which is represented by a shaded area, and three application zones, which are represented by white rectangles, an art gallery (A), a book store (B), and a coffee shop (C). Let's use this simple example to illustrate the basic idea of mix zones. Suppose that two users Alice and Bob enter the mix zone from B at the same time, and their identities are mixed; after a certain period, two users exit from the mix zone and appear in C. We cannot infer that these users are Alice and/or Bob or other users located in the mix zone before Alice and Bob entering it. However, if a mix zone has a diameter much larger than the distance the user can reach during one location update period, it might not be able to protect users adequately. For the same example depicted in Fig. 11, A is much closer to B than C. If two users enter the mix zone from A and C at the same time and a user appears in B at the next location update time, an adversary may tell that the user entering B from the mix zone is not the one who emerged at C before. Furthermore, if there is nobody in the mix zone at this time, the user in B can only be the one from A, thus revealing A's identity. To address this privacy issue, two metrics are proposed to measure the level of privacy protection, namely, *anonymity set* and *entropy*. A user can specify the minimum size of his/her anonymity set that is the number of people visiting the mix zone during the same location update period. The user is not willing to reveal his/her location information to any application until the mix zone finds a qualified anonymity set. The entropy is used to measure the level of uncertainty that a hostile adversary knows about a user's location information based on the user's historical movement data.

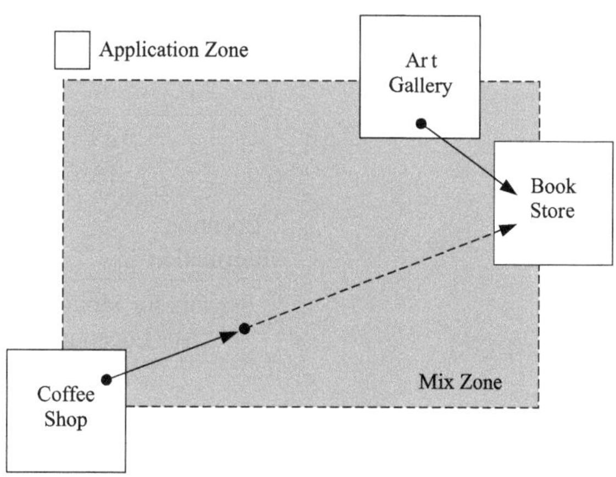

Fig. 11 A sample mix zone for three application zones

(a)

1101	1100	1011	0110
Building ID	Floor ID	Room ID	Node ID

(b)

1101	1100	1011	XXXX
Building ID	Floor ID	Room ID	Node ID

(c)

1101	1100	XXXX	XXXX
Building ID	Floor ID	Room ID	Node ID

Fig. 12 Examples of node ID cloaking in the hierarchical location perturbation algorithm. **a** Sensor node detected at least *k* subjects. **b** At least *k* subjects can only be found in the room level. **c** At least *k* subjects can only be found in the room level

5.1.3 Hierarchical Location Perturbation

Sensors could be deployed inside buildings to track the locations of individuals, which is used for adaptive computing services. However, the location privacy of individuals may be breached due to the vulnerability of WSN. The basic idea of the hierarchical location perturbation algorithm [22] is to provide less spatial accuracy and perturb the count of subjects in a monitored area. In general, the algorithm consists of two main steps:

Step 1: Defining a hierarchical structure. The system area is partitioned into several physical hierarchies, e.g., rooms, floor, and building. Each sensor node is assigned with a unique hierarchical ID. Figure 12 depicts an example of a hierarchical structure, where the highest level is a building, the second level contains all the floors in the building, the third level contains all the rooms on each floor, and the lowest level contains all the sensor nodes in each room. If each node in the hierarchical structure does not have more than 16 child nodes, the ID of each level can be represented by four bits, as depicted in Fig. 12a.

In each hierarchy, a sensor node is selected as a leader by using a distributed leader election protocol. Each leader keeps track the number of subjects monitored in its corresponding hierarchy. Also, each node knows a system-specified anonymity level k, i.e., the monitoring system can only receive the subject count information of a monitored area containing at least k subjects.

Step 2: Location perturbation. When a sensor node detects the number of subjects that is equal to or larger than k, it cloaks the number of subjects by randomly rounding up or down the subject count to the nearest multiple of k, and sends an exact node ID to the leader in the upper level. Otherwise, the sensor sends the exact number of detected subjects with a cloaked node ID to the leader in the upper level. The leader repeats this step until at least k subjects in a higher level can be found. In our example, if a sensor node can detect at least k subjects, it can report its exact

ID with a cloaked subject count to the leader in the upper level (i.e., the room level) (Fig. 12a). Otherwise, it sends the exact subject count with its cloaked ID "XXXX" to the leader in the room level. Figure 12b shows the case that the leader in the room level can detect at least k subjects. If this is not the case, it further sends the exact sum of the subject counts of its child nodes with its cloaked ID to the leader in the floor level. Figure 12c shows that the leader in the floor level can find at least k subjects, so it sends its exact ID with cloaked room and node IDs and the cloaked subject count to the leader in the building level.

5.1.4 Spatial Cloaking

TinyCasper [9, 11] is a privacy-preserving location monitoring system designed for WSNs. The basic idea of TinyCasper is that sensor nodes can communicate with each other and work together to find k-anonymized aggregate locations that are reported to a server. A k-anonymized aggregate location is defined as (A, N), where A a monitored area that contains at least k people and N is the number of people detected inside A. Figure 13a depicts an example where 17 sensors are deployed in the system, a nonzero number beside a sensor node indicates the number of people detected inside its sensing area, the location of six people are represented by circles, and $k = 3$. After the sensor nodes communicate with other peers, every node with a nonzero people count blurs its sensing area into a cloaked area that contains at least $k = 3$ people. Then, the sensor node reports the cloaked area A along with the number of people N located in A as an aggregate location, i.e., (A, N), to the server. In our example, sensor nodes s_1, s_5, s_6, s_7, and s_{15} detect a nonzero people count, they communicate with nearby peers to blur their sensing area. For example, s_7 needs to communicate with nodes s_5 and s_6 to find three people, and then it blurs its sensing area into a cloaked area that contains the sensing areas of nodes s_5, s_6 and s_7 (represented by the left-bottom shaded rectangle in Fig. 13b. s_7 reports the cloaked area with three people as an aggregate location to the server. In this example, the server receives two three-anonymized aggregate locations (represented by the two shaded rectangles) (Fig. 13b).

Two in-networks spatial cloaking algorithms have been proposed for TinyCasper [9, 11]. (1) The *resource-aware* algorithm employs a greedy approach for sensor nodes to determine their aggregate locations. Its objective is to minimize the communication and computational overhead. (2) The *quality-aware* algorithm aims at enabling sensor nodes to determine their aggregate locations with the minimal cloaked area, in order to maximize the usability of the aggregate locations that are used by the server to provide location-based monitoring services. Spatial cloaking techniques have also been extended to support mobile devices, e.g., smartphones and mobile sensors, through peer-to-peer communication [12, 13, 42].

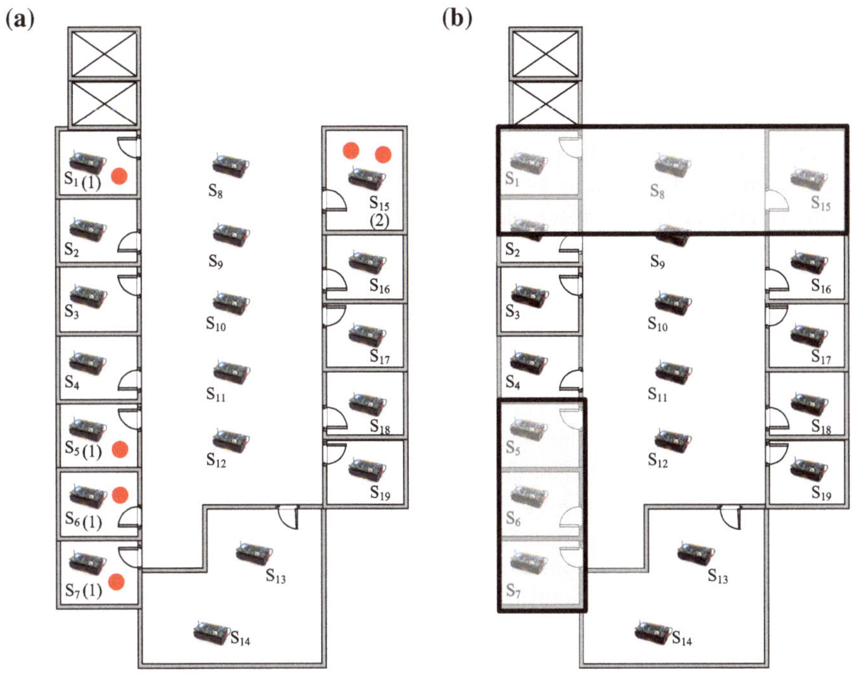

Fig. 13 Example of spatial cloaking in a sensor-based location monitoring system ($k = 3$). **a** Sensor deployment. **b** Two k-anonymized aggregate locations

5.1.5 Discussion

Using pseudonyms may not be secure because an adversary could link the old and new pseudonyms to infer a user's read identity. The effectiveness of the mix zone greatly depends on its size, the user population, the sensing resolution, and the user movement speed. The hierarchical location perturbation algorithm does not consider sensor nodes deployed at the same level in the predefined hierarchy, e.g., the sensor nodes in the same room. When the subject count of a sensor node does not satisfy a certain number, the algorithm goes up to its parent. The algorithm trends to generate cloaked areas larger than necessary, such large aggregate locations would degrade their usability in location monitoring services. On the other hand, the spatial cloaking algorithms do not reply on any hierarchical structure and they can support both indoor and open environments. In addition, since they aim at maximizing the usability of their cloaked areas by minimizing their area, they can be used to provide better location monitoring services. We discuss how a WSN can provide location monitoring services based on cloaked areas in next section.

5.1.6 Privacy-Preserving Location Monitoring Services

As discussed in Sects 5.1.3 and 5.1.4, the privacy-preserving monitoring system can only report aggregate locations for the server to provide monitoring services. To this end, a privacy-preserving aggregate query processor is designed to employ a spatio-temporal histogram to estimate the subject distribution based on aggregate locations reported from sensor nodes [10]. The spatio-temporal histogram partitions the system area into disjoint equal-sized cells that is stored as a two-dimensional array. Each element $M[i, j]$ is an estimator that represents the estimated number of subjects located in its corresponding cell area, where i and j indicate the i-th row and j-th column in the array, respectively. In general, the privacy-preserving aggregate query processing has two main steps.

Step 1: Histogram updates. The server is able to detect the total number of people N_{total} in the system area. Initially, N_{total} is uniformly distributed among all estimators in the histogram. When the server receives an aggregate location (A, N), the basic approach calculates the sum S of the estimators of the cells intersecting the aggregate location area A and uniformly distributes the reported people count N among the cells intersecting A. The difference between S and N is uniformly distributed among the cells outside A. Several optimization techniques have been proposed for an adaptive spatio-temporal histogram, in which the server updates the histogram based on various spatial, temporal, and historical factors.

Figure 14 depicts an example of a basic histogram where the system area is divided into 25 cells and $N_{total} = 100$. Initially, N_{total} is uniformly distributed among the 25 cells, so the value of each estimator is 4 (Fig. 14a). Figure 14b shows an aggregate location with $N = 20$ people and its cloaked area A is represented by a bold rectangle. The sum of the four estimators of the four cells intersecting A is $4 \times 4 = 16$. Since 20 is uniformly distributed among the four cells intersecting A, each estimator in these four cells is set to $20/4 = 5$. The difference $16 - 20 = -4$ is evenly distributed

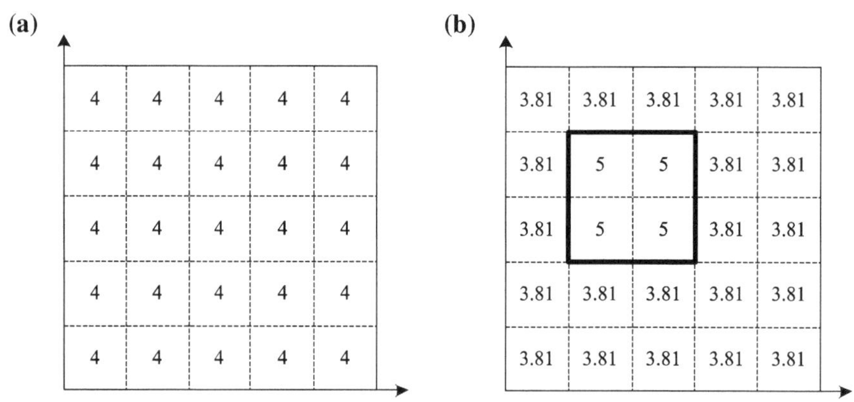

Fig. 14 An update on the basic spatio-temporal histogram. **a** Initial histogram. **b** An aggregate location with 20 people

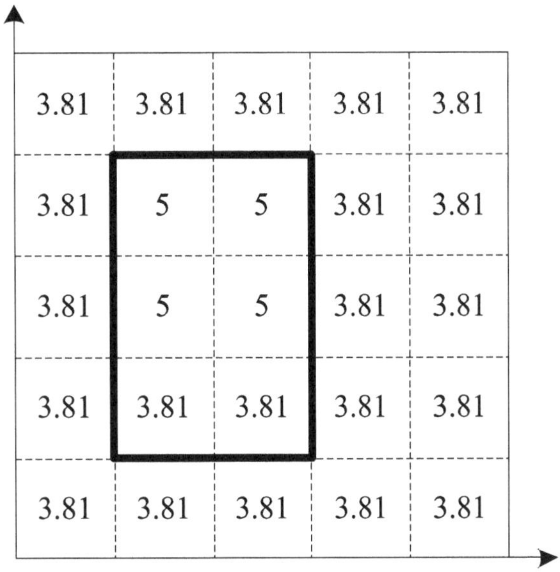

Fig. 15 Privacy-preserving aggregate query processing

among the estimators of the cells outside A, so each of those estimators is set to $4 + (-4/21) = 3.81$.

Step 2: Aggregate query processing. The query processor is able to answer aggregate queries. Given an aggregate query with a query range, the query processor calculates the sum of the estimators of the cells intersecting the query range and returns it as a query answer. Figure 15 depicts an example of an aggregate query with a spatial query range which is represented by a bold rectangle. The query answer is the sum of the estimators of the cells intersecting the spatial query range, i.e., $5 \times 4 + 3.81 \times 2 = 27.62$.

Recently, a stochastic flow model [54] has been designed to provide location monitoring services in an indoor environment based on the sensor node's actual subject count. Since the flow model considers the network topology, it may provide more accurate location monitoring services. It would be very useful to apply the flow model to TinyCasper.

5.2 Temporal Privacy

In this section, we discuss two privacy-preserving techniques for temporal privacy in WSNs, namely, *adaptive delaying* [31] and *probabilistic sampling* [23]. The adaptive delaying technique is mainly designed for protecting the event time or the packet transmission time, while the probabilistic sampling technique aims at protecting the user's query patterns and the unusual event time.

5.2.1 Adaptive Delaying

In some WSN applications, the time when an event is detected or data is transmitted by a sensor node is considered as sensitive information. An adversary would be able to infer temporal information by merely capturing the arrival time of data packets at some sensor nodes. Packet delaying techniques have been applied to anonymous network communication. Most of these early techniques rely on the concept of pool mixes [14], where the pool mix waits until it buffers a certain number of packets before taking mixing action to protect their anonymity. Thus, the pool mix delays packets before forwarding them. The concept of pool mixes has been extended to delay individual incoming packets according a probabilistic distribution before sending them out [17, 34].

P. Kamat et al. [31] have further used packet delaying to protect the temporal privacy of detected events and transmitted packets. Suppose a source node S detects an event and generates a packet at some time X and sends the packet in an encrypted form to a destination node R, and a compromised node E monitors traffic arriving at R. The adaptive delaying technique can protect temporal privacy in three network scenarios:

1. **Two-party single-packet network.** S obfuscates the time X by storing the packet in its local buffer for a random time period Y before transmitting it to R. E witnesses the packet arrives at a time $Z = X + Y$, while R can decrypt the packet to know the actual event detection or packet generation time X.
2. **Two-party multiple-packet network.** Suppose S generates a stream of n packets $p_1, p_2, \ldots, p_n$ at times $X_1, X_2, \ldots, X_n$ and delays them by $Y_1, Y_2, \ldots, Y_n$, respectively, before sending them to R. E observes the packets at $Z_1, Z_2, \ldots, Z_n$, where $Z_i = X_i + Y_i$ $(1 \leq i \leq n)$. Similar to the first scenario, R can decrypt each packet and know its actual event detection or packet generation time. If the system needs to maintain packet ordering, Y_j has to be at least the time until all previous packets $p_1, p_2, \ldots, p_i$ (where $1 \leq i < j \leq n$) were transmitted. Thus, there is an ordering of $Z_1 < Z_2 < \ldots < Z_n$. Otherwise, Y_j can be independent of each other and independent of the event detection or packet generation times.
3. **Multihop network.** Suppose S sends a stream of packets to R through m intermediate nodes $S \rightarrow N_1 \rightarrow N_2 \rightarrow \ldots \rightarrow N_m \rightarrow R$. The delay of a packet p_i is $Y_i = Y_{0,i} + Y_{1,i} + \ldots + Y_{m,i}$, where $Y_{k,i}$ denotes the delay time of i-th packet at k-th intermediate node and $Y_{0,i}$ denotes the delay time of i-th packet at S. Consequently, E observes the arrival of each packet p_i at $Z_i = X_i + Y_i$.

Delaying packets protects temporal privacy, but it increases burden on the buffer at an intermediate node, especially the nodes close to the base station. When a buffer is full, a replacement strategy is needed to replace a victim packet in the buffer for a newly received packet. The victim packet is transmitted immediately instead of dropping it. Four replacement strategies have been designed to choose a victim packet from a buffer as follows:

1. **Longest delayed first (LDF).** The packet has been delayed in the buffer for the longest time is selected as the victim packet.

2. **Shortest delay time first (SDTF).** The packet has been delayed in the buffer for the shortest time is selected as the victim packet.
3. **Longest remaining delay first (LRDF).** The packet with the longest remaining delay time in the buffer is selected as the victim packet.
4. **Shortest remaining delay first (SRDF).** The packet with the shortest remaining delay time in the buffer is selected as the victim packet.

A rated-controlled adaptive delaying mechanism is used to adjust the delay process based on the employed replacement strategy. In the adversary model of simulations, the adversary estimates the creation time of each packet. The total estimation error for m packets is calculated as the *mean square error*, i.e., $\sum_{i=1}^{m}(x_i' - x_i)^2/m$, where x_i is the true creation time for packet i and x_i' is estimated by the adversary. The larger this value is, the better the temporal privacy is protected. At the same time, tolerable end-to-end delivery latency for each packet should be maintained. Simulation results show that LRDF is the best replacement strategy in terms of both temporal privacy and latency.

5.2.2 Probabilistic Sampling

Data can be collected from sensor nodes in an on-demand or a periodic manner. The on-demand manner is easy for an adversary to infer when a user issues a query or a sensor node detects an event. Although the periodic manner can protect the temporal privacy of queries issued by users and events detected by sensor nodes, its power consumption is very high if the data collection time interval is short. However, it may not satisfy a user-specified deadline if the time interval is long. To this end, a probabilistic sampling technique is designed [23]. The basic idea is to carefully report data at random times to blend user requests and events with the routine traffic, thus making user requests and events indistinguishable to an attacker. In this technique, the day is divided into time intervals, *TimeInterval*, and each time interval contains the same number of equal-sized time slots. Figure 16 depicts an example, where the time slot is one second and the day is divided into 86,400 time slots. Each time interval is five seconds, so it contains five time slots and the day has 17,280 time intervals. This technique has three important policies:

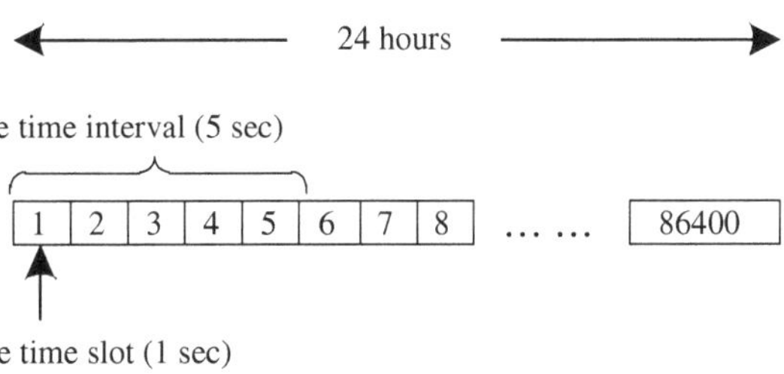

Fig. 16 Probabilistic sampling

1. A time slot is a short time period, in which at most one data report or query can transmit or issue, respectively. The base station and the sensor node have the same probability P (e.g., $1/(\mathit{TimeInterval} \times 2)$) of issuing a query or initiating a data report, respectively.
2. A user should only be able to issue one query every time interval.
3. When there are too many time slots between two reports from a sensor node, P should be increased until the sensor node can generate a data report. After that, P is reset to its original value.

With the careful design of the generation probability of automatic data reports, the privacy of user queries and unusual events in the network is protected. Experimental results show that the probabilistic sampling technique can effectively reduce the chance that an adversary can identify whether a user issues a query in a time interval.

5.2.3 Discussion

The adaptive delaying technique [31] does not consider any user- or system-specified deadline for a query or data collection. It would be more interesting if this technique can be extended to be time-constrained. In addition, it is important to investigate whether such a time constraint degrades its privacy protection. On the other hand, although the probabilistic sampling technique [23] considers user-specified deadlines, it is not clear how to adjust the probability P to meet the deadline and maximize the accuracy of a query answer. More sophisticated models are needed to answer these questions.

6 Conclusion and Future Directions

In this chapter, we highlighted the existing privacy enhancing technologies for wireless sensor networks (WSNs). We categorized these technologies into three main types of privacy, namely, *system privacy*, *data privacy* and *context privacy*. For each type of privacy, we presented its major privacy-preserving techniques. In the context privacy, the user location privacy is a new type of privacy in WSNs. There are three main open privacy issues in privacy-preserving location monitoring services. (1) Existing solutions only aim at protecting snapshot location privacy. It is important to study continuous location monitoring privacy to avoid tracking a target user's location by analyzing consecutive snapshots of aggregate locations reported from sensor nodes. (2) The state-of-the-art privacy-preserving aggregate query processor can only support range queries. It is essential to design more advanced query processors to support more complex analysis, e.g., data mining. (3) There is a tradeoff between user privacy and data accuracy. For example, a higher anonymity level would lead to a larger aggregate location area that degrades the user location accuracy. It would be very interesting to derive a theoretical model to balance such a tradeoff and find an appropriate privacy protection level for a desired level of accuracy.

References

1. U. Acharya, M. Younis, Increasing base-station anonymity in wireless sensor networks. Ad Hoc Netw. **8**(8), 791–809 (2010)
2. K. Akkaya, M. Younis, A survey on routing protocols for wireless sensor networks. Ad Hoc Netw. **3**(3), 325–349 (2005)
3. J.N. Al-Karaki, A.E. Kamal, Routing techniques in wireless sensor networks: A survey. IEEE Wirel. Commun. **11**(6), 6–28 (2004)
4. A.R. Beresford, F. Stajano, Location privacy in pervasive computing. IEEE Pervasive Comput. **2**(1), 46–55 (2003)
5. B.H. Bloom, Space/time trade-offs in hash coding with allowable errors. Commun. ACM **13**(7), 422–426 (1970)
6. D.L. Chaum, Untraceable electronic mail, return addresses and digital pseudonyms. Commun. ACM **24**(2), 84–88 (1981)
7. C.W. Chen, Y.R. Tsai, Location privacy in unattended wireless sensor networks upon the requirement of data survivability. IEEE J. Sel. Areas Commun. **29**(7), 1480–1490 (2011)
8. X. Chen, K. Makki, K. Yen, N. Pissinou, Sensor network security: A survey. IEEE Commun. Surv. Tutor. **11**(2), 52–73 (2009)
9. C.Y. Chow, M.F. Mokbel, T. He, TinyCasper: A privacy-preserving aggregate location monitoring system in wireless sensor networks, in *Proceedings of the ACM SIGMOD International Conference on Management of Data*, 2008
10. C.Y. Chow, M.F. Mokbel, T. He, Aggregate location monitoring for wireless sensor networks: A histogram-based approach, in *Proceedings of the IEEE International Conference on Mobile Data Management*, 2009
11. C.Y. Chow, M.F. Mokbel, T. He, A privacy-preserving location monitoring system for wireless sensor networks. IEEE Trans. Mob. Comput. **10**(1), 94–107 (2011)
12. C.Y. Chow, M.F. Mokbel, X. Liu, A peer-to-peer spatial cloaking algorithm for anonymous location-based services, in *Proceedings of the ACM SIGSPATIAL International Conference on Advances in Geographic Information Systems*, 2006
13. C.Y. Chow, M.F. Mokbel, X. Liu, Spatial cloaking for anonymous location-based services in mobile peer-to-peer environments. GeoInformatica **15**(2), 351–380 (2011)
14. G. Danezis, The traffic analysis of continuous-time mixes, in *Proceedings of the International Conference on Privacy Enhancing Technologies*, 2005
15. J. Deng, R. Han, S. Mishra, Intrusion tolerance and anti-traffic analysis strategies for wireless sensor networks, in *Proceedings of the IEEE/IFIP International Conference on Dependable Systems and Networks*, 2004
16. J. Deng, R. Han, S. Mishra, Decorrelating wireless sensor network traffic to inhibit traffic analysis attacks. Pervasive Mob. Comput. **2**(2), 159–186 (2006)
17. C. Díaz, B. Preneel, Taxonomy of mixes and dummy traffic, in *Proceedings of the International Information Security Workshops*, 2004
18. L. Eschenauer, V.D. Gligor, A key-management scheme for distributed sensor networks, in *Proceedings of the 9th ACM conference on Computer and communications, security* (2002)
19. P. Eugster, R. Guerraoui, S. Handurukande, P. Kouznetsov, A. Kermarrec, Lightweight probabilistic broadcast. ACM Trans. Comput. Syst. **21**(4), 341–374 (2003)
20. T. Feng, C. Wang, W. Zhang, L. Ruan, Confidentiality protection for distributed sensor data aggregation, in *Proceedings of the IEEE INFOCOM*, 2008
21. M.M. Groat, W. Hey, S. Forrest, KIPDA: k-indistinguishable privacy-preserving data aggregation in wireless sensor networks, in *Proceedings of the IEEE INFOCOM*, 2011

22. M. Gruteser, G. Schelle, A. Jain, R. Han, D. Grunwald, Privacy-aware location sensor networks, in *Proceedings of the USENIX International Workshop on Hot Topics in Operating Systems*, 2003
23. J. Guerreiro, E.C.H. Ngai, C. Rohner, Privacy-aware probabilistic sampling for data collection in wireless sensor networks, in *Proceedings of the International Wireless Communications and Mobile Computing Conference*, 2011
24. A. Harter, A. Hopper, P. Steggles, A. Ward, P. Webster, The anatomy of a context-aware application. Wireless Netw. **8**(2–3), 187–197 (2002)
25. W. He, X. Liu, H. Nguyen, K. Nahrstedt, T. Abdelzaher, PDA: Privacy-preserving data aggregation in wireless sensor networks, in *Proceedings of the IEEE INFOCOM*, 2007
26. W. He, X. Liu, H. Nguyen, K. Nahrstedt, T. Abdelzaher, iPDA: an integrity-protecting private data aggregation scheme for wireless sensor networks, in *Proceedings of the IEEE Military Communications Conference*, 2008
27. W. He, X. Liu, H. Nguyen, K. Nahrstedt, T. Abdelzaher, A cluster-based protocol to enforce integrity and preserve privacy in data aggregation, in *Proceedings of the IEEE International Conference on Distributed Computing Systems Workshops*, 2009
28. C. Intanagonwiwat, R. Govindan, D. Estrin, J. Heidemann, F. Silva, Directed diffusion for wireless sensor networking. IEEE/ACM Trans. Netw. **11**(1), 2–16 (2003)
29. A. Jhumka, M. Leeke, S. Shrestha, On the use of fake sources for source location privacy: Trade-offs between energy and privacy. Comput. J. **54**(6), 860–874 (2011)
30. Y. Jian, S. Chen, Z. Zhang, L. Zhang, Protecting receiver-location privacy in wireless sensor networks, in *Proceedings of the IEEE INFOCOM*, 2007
31. P. Kamat, W. Xu, W. Trappe, Y. Zhang, Temporal privacy in wireless sensor networks: Theory and practice. ACM Trans. Sens. Netw. **5**(4), 1–24 (2009)
32. P. Kamat, Y. Zhang, W. Trappe, C. Ozturk, Enhancing source-location privacy in sensor network routing, in *Proceedings of the IEEE International Conference on Distributed Computing Systems*, 2005
33. B. Karp, H.T. Kung, Gpsr: Greedy perimeter stateless routing for wireless networks, in *Proceedings of the ACM International Conference on Mobile Computing and Networking*, 2000
34. D. Kesdogan, J. Egner, R. Büschkes, Stop-and-Go-MIXes providing probabilistic anonymity in an open system, in *Proceedings of the International Workshop on Information Hiding*, 1998
35. N. Li, N. Zhang, S.K. Das, B. Thuraisingham, Privacy preservation in wireless sensor networks: A state-of-the-art survey. Ad Hoc Netw. **7**(8), 1501–1514 (2009)
36. X. Li, X. Wang, N. Zheng, Z. Wan, M. Gu, Enhanced location privacy protection of base station in wireless sensor networks, in *Proceedings of the International Conference on Mobile Ad-hoc and Sensor Networks*, 2009
37. Y. Li, J. Ren, Source-location privacy through dynamic routing in wireless sensor networks, in *Proceedings of the IEEE INFOCOM*, 2010
38. L. Lightfoot, Y. Li, J. Ren, Preserving source-location privacy in wireless sensor network using STaR routing, in *Proceedings of the IEEE Global Communications Conference*, 2010
39. H. Lim, C. Kim, Flooding in wireless ad hoc networks. Comput. Commun. **24**(3), 353–363 (2001)
40. S. Madden, M.J. Franklin, J.M. Hellerstein, W. Hong, TAG: A tiny aggregation service for ad-hoc sensor networks. ACM SIGOPS Oper. Syst., Rev. **36**(SI), 131–146 (2002)
41. K. Mehta, D. Liu, M. Wright, Protecting location privacy in sensor networks against a global eavesdropper. IEEE Trans. Mob. Comput. **11**(2), 320–336 (2012)
42. M.F. Mokbel, C.Y. Chow, Challenges in preserving location privacy in peer-to-peer environments, in *Proceedings of the International Workshop on Information Processing over Evolving Networks*, 2006
43. E.C.H. Ngai, I. Rodhe, On providing location privacy for mobile sinks in wireless sensor networks, in *Proceedings of the ACM International Conference on Modeling, Analysis and Simulation of Wireless and Mobile Systems*, 2009
44. L.B. Oliveira, D.F. Aranha, C.P.L. Gouvêa, M. Scott, D.F. Câmara, J. López, R. Dahab, TinyPBC: pairings for authenticated identity-based non-interactive key distribution in sensor networks. Comput. Commun. **34**(3), 485–493 (2011)

45. N. Oualha, A. Olivereau, Sensor and data privacy in industrial wireless sensor networks, in *Proceedings of the International Conference on Network Architectures and Information Systems Security*, 2011
46. C. Ozturk, Y. Zhang, W. Trappe, Source-location privacy in energy-constrained sensor network routing, in *Proceedings of the ACM International Workshop on Security of Ad hoc and Sensor Networks*, 2004
47. A. Perrig, R. Szewczyk, V. Wen, D. Culler, J.D. Tygar, SPINS: Security protocols for sensor networks. Wireless Netw. **8**(5), 521–534 (2002)
48. N.B. Priyantha, A. Chakraborty, H. Balakrishnan, The cricket location-support system, in *Proceedings of the ACM International Conference on Mobile Computing and Networking*, 2000
49. R. Rajagopalan, P. Varshney, Data-aggregation techniques in sensor networks: A survey. IEEE Commun. Surv. Tutor. **8**(4), 48–63 (2006)
50. J. Sen, A survey on wireless sensor network security. Int. J. Commun. Netw. Inf. Secur. **1**(2), 55 (2009)
51. M. Shao, Y. Yang, S. Zhu, G. Cao, Towards statistically strong source anonymity for sensor networks, in *Proceedings of the IEEE INFOCOM*, 2008
52. L. Sweeney, k-anonymity: A model for protecting privacy. Int. J. Uncertainty, Fuzziness Knowl. Based Syst. **10**(5), 557–570 (2002)
53. P. Szczechowiak, A. Kargl, M. Scott, M. Collier, On the application of pairing based cryptography to wireless sensor networks, in *Proceedings of the ACM International Conference on Wireless Network Security*, 2009
54. S. Wang, X.S. Wang, Y. Huang, Tracking the dynamic distribution of people in indoor space with noisy partitioning sensors, in *Proceedings of the IEEE International Conference on Mobile Data Management*, 2012
55. R. Want, A. Hopper, V. Falcao, J. Gibbons, The active badge location system. ACM Trans. Inf. Syst. **10**(1), 91–102 (1992)
56. A. Woo, T. Tong, D. Culler, Taming the underlying challenges of reliable multihop routing in sensor networks, in *Proceedings of the ACM International Conference on Embedded Networked Sensor Systems*, 2003
57. World Wildlife Fund: http://wwf.panda.org [Accessed: April 24, 2012]
58. Y. Xi, L. Schwiebert, W. Shi, Preserving source location privacy in monitoring-based wireless sensor networks, in *Proceedings of the IEEE International Symposium on Parallel and Distributed Processing*, 2006
59. Y. Yang, M. Shao, S. Zhu, B. Urgaonkar, G. Cao, Towards event source unobservability with minimum network traffic in sensor networks, in *Proceedings of the ACM International Conference on Wireless Network Security*, 2008
60. S. Zhu, S. Setia, S. Jajodia, LEAP: Efficient security mechanisms for large-scale distributed sensor networks, in *Proceedings of the ACM International Conference on Computer and Communications Security*, 2003

Part IX
Middleware

Chapter 19
Middleware Platforms: State of the Art, New Issues, and Future Trends

Flávia C. Delicato, Paulo F. Pires and Albert Y. Zomaya

Abstract In this chapter we examine the rapid advances that have occurred recently in the wireless sensor networks (WSNs) domain and argue that intelligent middleware is needed to tackle the challenges brought by these changes. We present an overview on existing design approaches for WSN middleware, as well as the most common middleware services and programming abstractions. We describe key features that must be incorporated in middleware for the current generation wireless sensor networks and conclude the chapter with a discussion on new issues and future trends in the design of WSN middleware.

1 Introduction

Wireless Sensor Networks (WSNs) technology [1] experienced a major breakthrough in the last decade, attracting increased attention from the scientific community and the industry. Wireless Sensor Network nodes work collaboratively, extracting environmental data, performing some simple processing and transmitting them to one or more exit points of the network called sink nodes, to be analyzed and further processed. The computational capacity, albeit limited, of sensor nodes, enables them to realize the so-called *in-network processing*. Thus, WSN nodes are not merely passive devices to collect data, but are also capable of performing operations such

F. C. Delicato (✉) · P. F. Pires
Federal University of Rio de Janeiro, Cidade Universitaria, Ilha do Fundao, RJ-Brazil
e-mail: fdelicato@gmail.com

P. F. Pires
e-mail: paulo.f.pires@gmail.com

A. Y. Zomaya
University of Sydney, Sydney, NSW, 2006, Australia
e-mail: albert.zomaya@sydney.edu.au

H. M. Ammari (ed.), *The Art of Wireless Sensor Networks*,
Signals and Communication Technology, DOI: 10.1007/978-3-642-40009-4_19,
© Springer-Verlag Berlin Heidelberg 2014

as event detection, data fusion and data aggregation [2]. Such WSN feature brings potential advantages such as, for instance, rapid response to critical situations.

Considering the wide range of environmental variables that can be collected by sensor nodes, there are a large number of applications that can benefit from the use of WSN, including fire detection [3], object detection and tracking [4] security monitoring [5], environmental monitoring, structural health monitoring (like bridges, buildings, tunnels and dams) [6, 7], monitoring of natural disasters (such as landslides [8]), just to name a few. WSN applications differ greatly in their data delivery models, QoS, connectivity and mobility requirements, thus demanding different organizations and features from the underlying network. In Chap. 6 of this Book there is a comprehensive survey on WSN applications, including a multidimensional taxonomy addressing the most relevant application features and requirements.

The increasing abundance of WSN applications and the heterogeneity of wireless sensor technologies currently available bring forth different challenges and require changes in how these networks should be designed and used. Several important issues in the early years of WSN research, as routing, topology control and media access control, already have relatively suitable and efficient solutions, but the new usage scenarios of these networks have opened a whole new range of challenges to be overcome.

One of the changes the WSN field has witnessed recently regards the scope of such networks. Early work on WSN had focused on simple applications of data acquisition, and in most cases, considered a single application over a network. The research focus was mostly on providing algorithms and protocols for efficient power management, routing, localization, and other low level issues. The design of network protocols and applications were usually tightly coupled, and the application logic along with underlying protocols were bundled in a monolithic code to be deployed in the nodes. Such works assumed *application-specific* WSNs (a *single application scope).* This strong coupling aimed to provide a high efficiency in terms of energy consumption. However, the design strategies were often *ad hoc* and imposed direct interaction of the application with the underlying embedded operating system, or even with hardware components of sensor nodes. Furthermore, such design approach generated rigid systems with WSNs built specifically for a single target application, with no reuse of software artefacts and of the sensing/communication infrastructure.

Recent work is considering that, given the cost of building and deploying the WSN infrastructure, the potentially long operational lifetime of the WSN and the wide range of applications that could use it, the same network can be shared by different applications. Therefore, the current trend is to design WSNs with long lifetimes, meeting the requirements of multiple applications from the same or different domains, which execute concurrently in the network nodes. Such approach gives rise to the vision of *shared sensor networks* [9] (*multiple applications scope*), in which the lack of flexibility previously adopted in the WSN design is undesirable.

Finally, the new generation of WSNs, that has gained momentum nowadays, is supposed to have an *Internet-scale* scope. Unlike previous application-specific or shared WSNs, in this new scenario the interface among different networks and applications is highly relevant, thus interoperability becomes an issue. This kind of WSN is related to the new paradigm of "the Internet of things" [10, 11].

Apart of the WSN scope, the development of applications for such networks has always been a challenge. First of all, WSN programming has traditionally been an error-prone task since it requires programming individual nodes, using low-level abstractions provided by the sensor operating system and interfacing with the hardware and network protocols [12]. Second, developers of WSN applications are usually experts in their knowledge field (as civil engineer, biology, geology, etc), not in networks. Even when network experts are available to be in charge of dealing with programming issues, the requirements of the applications that will run in the WSN need to be taken in account to efficiently program the nodes; therefore, the domain experts always need to be involved. So, there is a gap between the high-level requirements from WSN applications (the *application level knowledge*), and the complexity of operations in the underlying network infrastructure (the *network level knowledge*). A potential solution to bridging this gap and removing (or alleviating) the hindrances and challenges posed by the WSN application development and execution is to adopt a middleware platform.

In distributed systems, middleware is usually defined as software that lies between the operating system and applications running on each node of the system. In general, middleware is supposed to hide the internal operation and heterogeneity of the underlying system, providing standard interfaces, abstractions and a set of services. By hiding the inherent complexity of distribution, the use of middleware in traditional distributed computing facilitates the work of application developers. Tasks such as concurrency control, transactions, data replication, security, and other infrastructure services are examples of services performed by a middleware.

WSN systems could also benefit from the use of a middleware layer that offers a generic execution environment for applications, providing a high level abstraction of the network infrastructure functionality. In short, WSN middleware has to provide support for the development, management, deployment and execution of sensing-based application. Among other functions, the middleware can decide the best protocols to be used according to the application requirements, coordinate the operation of sensors to achieve the application goals and intelligently manage the use of network resources. In order to efficiently provide the quality of service required by applications, it is often necessary to interact with the lower levels of protocols or even with hardware components. The middleware can perform this interaction for the benefit of the application. All these functions facilitate the tasks of the application developers and of the network managers.

However, despite the well-known advantages of using middleware platforms in traditional distributed systems, only recently the researchers began to consider its adoption in the WSN design [13–17]. Considering the simplicity of the first applications and the application-specific nature of early WSNs, the overhead of adding a middleware layer in the network design was often not worth. But to accommodate the new scenarios of shared and Internet-scale WSN, the development of WSNs will require systematic design approaches based on high level and preferable standardized abstractions. Thus, the support of middleware becomes a crucial need in order to provide: (i) appropriate **system abstractions**, so that the application developer can focus on the application logic without having to deal with the lower level implementation details, (ii) standard and reusable **services** for several applications, so that developers can deploy and execute the application without worrying about complex, error-prone and tedious functions, (iii) runtime environment able to manage the execution of multiple applications, (iv) mechanisms for network infrastructure management and adaptation to allow the efficient use of WSN resources. It should also support interoperability with external networks, as the Internet, or enterprise systems.

Since WSNs have several particular features and constraints, conventional middleware technologies are not suitable for these networks. Instead, a middleware specifically tailored to WSN is required, and the design and implementation of a successful WSN middleware is not trivial. It needs to deal with challenges dictated by the WSN features on one hand and the applications requirements on the other.

Considering the WSN scarce resources, a middleware must be robust, fault tolerant, lightweight and with short storage requirements. Regarding the design, middleware platforms used in traditional distributed systems (based on fixed and wired connected devices) have been built adhering to the "black box" metaphor, that is, the inherent complexity of the distribution should be hidden from the users and developers, so that the system appear as a single integrated computational entity. In other words, the distribution becomes transparent. However, completely hiding the network operation details from applications can be inefficient in WSNs. WSN applications need to detect and react to changes in the environment (regarding connectivity, bandwidth and available energy), as well as to changes in the application requirements. The application ability to modify the network behaviour can be crucial to improve the overall system performance. Therefore, the middleware should act as a broker between applications and the WSN, translating application requirements into WSN configuration parameters. Due to the dynamics of WSN environments, the middleware should supply mechanisms that allow the application to monitor the network state through a high level interface. From the monitoring of the execution context, the application may decide to dynamically change some behaviour of the network, or aid the middleware to perform some adaptation. Therefore, WSN middleware should provide context-awareness, instead of transparency, for the applications.

Independently on the design approach or specific purpose, a complete WSN middleware solution should include three major components: programming abstractions, system services and runtime support. It should be mentioned that it is not necessary for a specific WSN middleware to include all these components. Also, functions of several components may be combined together and implemented as a single component. Moreover, it is desirable to include some mechanism for QoS provision and management. QoS issues in WSN are complex and often have a *crosslayer* nature. Thus a complete QoS solution requires not only middleware support but also the participation of all levels of the WSN protocol stack. Runtime support supplies the underlying execution environment of applications and can be considered as an extension of the embedded operating system which provides functions of scheduling of tasks, inter-process communication (IPC), memory control, and power control in terms of voltage scaling and component activation/inactivation. The need for runtime support in WSN middleware derives from the facts that the hardware and firmware of the sensor nodes may not always provide enough support for the implementation of the middleware services. Runtime support of WSN middleware is often embodied as a virtual machine over a specific embedded operating system [18].

In this chapter we will start by summarizing the requirements a WSN middleware should meetl (Sect. 2) and then will focus on currently existent programming abstractions (Sect. 3) and middleware services (Sect. 4) for WSN. Section 5 sheds light on new issues and future trends in the design of WSN in general and middleware concerns in particular. Section 6 concludes the chapter presenting final remarks.

2 Requirements and Challenges of Middleware for WSN

The main purpose of WSN middleware is to support the development, maintenance, deployment and execution of sensing-based applications. This support includes mechanisms to formulate high-level sensing tasks, communicate such tasks to the WSN, distribute them to individual sensor nodes, coordinate nodes to perform the tasks and report the final result back to the task issuer (applications and/or final users). All mechanisms provided by a middleware platform should respect the special features of WSN, mainly the energy efficiency, robustness, dynamic execution context and scalability. In this Section we will first depict WSN constraints and specific features that dictate middleware requirements, and then outline some desired design principles and characteristic the middleware should have to meet such requirements.

2.1 WSN Features

Wireless sensor networks are a category of ad hoc networks, sharing several characteristics of these networks but having additional requirements and constraints. Following we summarize the main WSN features that pose challenges in the design of a WSM middleware.

Resource Constraint. The limited resources of WSN nodes have always been an issue in the design and operation of these networks. The main restrictions of sensor nodes concern their processing power, memory capacity, bandwidth and energy. Due to the processing and memory limitations, any software solution for WSN should be lightweight, with a low computational load. Such feature often prevents the adoption of sophisticated techniques for the implementation of services in the nodes. Moreover, resource limitation make it challenging to meet QoS requirements as response time, availability, bandwidth allocation, data reliability, among other application specific needs, as a long operational lifetime (for long running applications).

In spite of the fact that recent advances on technology indicate that limited processor and memory are temporary constraints in WSNs that tend to disappear with fast developing fabrication techniques [19], the energy constraint, on the other hand, remains as a critical issue that needs to be tackled so that WSNs can be widely employed. Therefore, it is an important requirement of WSN to have mechanisms to manage the energy consumption in sensor nodes in an intelligent way.

Dynamic Execution Context. WSNs are subject to unpredictable changes in their execution context. Besides the intrinsic variations in the network connectivity, typical in wireless links, the availability of sensors in the network is also highly variable due to node failures, energy depletion and mobility. The network topology is also frequently changed by topology control protocols [20] that decide when and which sensors should be turned off to save power. The required sensing tasks, that dictate the network behaviour, can also change since the emergent WSN are assumed to be used by different applications (concurrently or not). These characteristics demonstrate the high degree of dynamics in the execution context of WSNs. Therefore, it is crucial that such networks are endowed with mechanisms that provide context-awareness and adaptation.

Data-Centricity. Sensor nodes are used to collect data from its surroundings. Since applications are interested in the data collected by a sensor, and express their interests in terms of type of data and QoS requirements, the scheme for node addressing in a WSN should be based in attributes that describe the node capabilities to provide data (instead of using a scheme based on any global and unique network address). Moreover, individual readings of a single node are often not relevant to the application, which is more interested in the merged or synthesized information gathered from different nodes. Considering the typical spatial density in a WSN, there is often a redundancy in the data collected for different nodes inside a given region of interest. Since in WSN the data communication in general consumes more energy than data processing, individual measurements should be aggregated as near to the data source as possible so that only the resulting, relevant information is transmitted to sink nodes, thus saving transmission energy. Therefore, in-network processing should be performed in every node of the WSN, aiming at saving energy of nodes, while reducing data redundancy,thus increasing the relevance of the information reported back to the application.

Scalability and Heterogeneity. A particular feature of WSNs is that they may consist of hundreds to thousands of nodes. Moreover, in the emergent WSN environment, it is likely that the devices, communication links, protocols and software components will be heterogeneous in terms of capabilities, programming abstractions, models and development tools. Such heterogeneous nature of WSNs, with a large number of sensors of different hardware platforms running applications developed by different teams will require the use of an abstraction layer that provides a common way to program, access the networks and extract the sensing data. The high number of nodes will require mechanism to coordinate them in the execution of the required sensing tasks.

Additional Features. Another unique feature of WSN is the use of application knowledge to configure and optimize the network operation. Such knowledge should be used in the initial decision making about WSN parameters and behaviour as well as to trigger dynamic adaptation whenever is needed after the application execution.

An additional feature concerns the concept of time and location of the sensed event. As WSNs monitor real-world data, information of time and space are important, and consist in key elements to properly merge individual sensor readings. Finally, since WSN can be deployed for sensitive applications like military surveillance and forecasting systems, security requirements such as confidentiality, authentication, integrity, freshness, and availability should be considered. Since most existing algorithms and security models are not suitable for WSNs, solutions specifically tailored for these networks must be designed that take into account the node resource constraints.

2.2 WSN Middleware Requirements

All the aforementioned WSN features should be taken into account when designing a successful WSN middleware and specifying its services. The following subsections describe the main requirements such WSN middleware should met.

2.2.1 Management of WSN Resources

The WSN scarcity of resources dictates that the middleware imposes a low computational load in the sensor nodes. Moreover, middleware should provide services to manage and optimize the network resource usage in an efficient way while meeting the application requirements. One example of such services is the selection of the set of nodes that will participate in a given sensing task, always taking into account the tradeoff between meeting the application requirements and optimizing energy consumption. Depending on the desired granularity of control, the middleware may also be in charge of (i) selecting the voltage levels used to process each task allocated to a sensor, using some mechanism of dynamic voltage scaling (DVS), (ii) dynamically changing the data modulation scheme by adopting a dynamic modulation scaling (DMS) technique, or (iii) dynamically adapting the data sampling rate of sensors [21], always with the final aim of optimizing energy consumption. Another useful service is to implement data fusion techniques to merge sensor readings of individual sensors into a high-level result to be sent to the application, thus saving transmission energy.

2.2.2 Management of Dynamic Execution Contexts

Considering the WSN highly dynamic environment, WSN middleware should provide services to flexibly adapt the system to the changes without decreasing its overall performance. Furthermore, the middleware should report failures and take measures to circumvent problems until the detected failures can be fixed. To deal with the dynamic execution context while providing a reasonable quality of service to its users, the WSN behaviour cannot rely on static parameters defined at the design time. The middleware should make decisions on the initial WSN configuration and then be able to change its operation dynamically, with or without the participation of applications. The execution context must be continually monitored and applications need to be "aware" of such context [6] and actively participate in the decision making. To this end, the middleware must expose to applications part of the knowledge about context, allowing them to participate in decisions about the infrastructure and in the activation of adaption strategies whenever it is necessary. The authors in [6] argue that the use of computational reflection [19] provides a powerful way for building middleware platforms that enable the development of context aware applications. WSN middleware can use reflective principles to interact with the infrastructure of the underlying network, keep updated context information in their data structures and make them available to applications.

Application knowledge should be used in the middleware decisions at several levels, including support to routing strategies and data aggregation functions. Traditional middleware are general purpose platforms, being designed to accommodate a wide variety of applications without needing application knowledge. WSN middleware, on the other hand, has to provide mechanisms for injecting application knowledge into the WSN and exploit such knowledge for optimization purposes. However, since WSNs can support a wide class of applications, tradeoffs need to be explored between the degree of generality and the application specificity of the middleware. A promising approach is to embed the particular features of an application in some type of specification or configuration file, building an application *profile*, as is done for example in [22]. The information embedded in the application profile can be interpreted by the middleware and used to guide its operations and the network behaviour.

2.2.3 Management of WSN Heterogeneity

The middleware should provide proper abstractions and mechanisms to deal with the heterogeneity of sensor nodes [23]. It should encapsulate the different communication protocols, software and hardware technology, providing the application with a high-level interface to interact to.

An additional feature concerns the concepts of time and location of sensed events. Since WSNs monitor real world data, time and spatial information are relevant, being key elements for fusing individual sensor readings. Therefore, support for time synchronization and location management should be integrated into a WSN middleware. Security solutions can be also provided as middleware services for sensitive applications. All services provided by the middleware should respect the particular characteristics of WSNs, especially robustness, scalability and efficiency in terms of energy. Finally, it is important to note that the scope of middleware for WSN is not restricted to the sensor network alone, but also encompasses external networks connected to the WSN (such as Internet) as well as the applications issuing queries or tasks to the WSN.

2.2.4 Summarization of WSN Middleware Requirements

To conclude this Section we summarize the aforementioned WSN requirements in a set of design principles for building WSN middleware systems. In [24] the authors proposed principles to be adopted in the design of a WSNs middleware. These principles and some additional ones that we consider relevant for a successful WSN middleware are described below:

- The middleware must provide mechanisms for both data-centric and in-network processing.
- Application knowledge should be used to optimize the network operation. Therefore, it is important to integrate knowledge of the application level to the services provided by the middleware.
- The middleware should be light in terms of communication and computing requirements. This requirement calls for using simple and efficient heuristics that generate sub-optimal solutions for decisions taken by the middleware.
- Due to the limited resources, it is very likely that the performance requirements of all running applications cannot be simultaneously met. Thus, the middleware should intelligently negotiate QoS of various applications against each other.
- All the solutions should preferably be based on the use of localized algorithms [25, 26]. Such algorithms provide robustness and scalability to the system.
- The middleware design should incorporate context awareness and adaptive properties.

3 Programming Abstractions

As we previously mentioned, a complete WSN middleware solution should include three major components: programming abstractions, system services and runtime support, besides QoS mechanisms. Programming abstractions are the underpinning of WSN middleware. They provide the high-level programming interfaces to the application developer, separating the programming of WSN applications from the operation in the underlying WSN infrastructure. They also provide the basis for developing the middleware services. Such abstractions, being supported by a suitable programming model, compiler and runtime support, unburden developers from handling low-level mechanisms such as messaging and routing protocols, power management, among others.

Existing programming abstractions encompass three aspects, discussed in the next two subsections: *abstraction level* (Sect. 3.1), *programming paradigm, and interface type* (Sect. 3.2). The decision of adopting a particular abstraction level and selecting an appropriate programming paradigm (and corresponding type of interface) depends on the specific application requirements and the underlying WSN infrastructure. Section 3.3 presents a brief comparison among commonly adopted programming paradigms.

3.1 Programming Abstractions: Abstraction Levels

Abstraction level refers to how the application developer views (and interact to) the WSN system. Basically, we can identify two different levels for programming abstractions: node-centric or local behaviour and macroprogramming or global/system behaviour.

Node-centric level abstracts the WSN as a distributed system consisting of a collection of sensor nodes, and the developer is provided with support to program the individual nodes [27, 28]. Therefore, the developer has to translate the global application behaviour in terms of local actions on each node, and individually program the nodes using the corresponding programming model. Examples of middleware platforms that consider such scope for programming WSN applications are Hood [29], Abstract Regions [7], Logical Neighborhoods [30], and Virtual Nodes [31].

On the other hand, macroprogramming [32] or *global level* abstracts the WSN as a single virtual system and allows the programmer to build a unique centralized program (global behaviour) that is then further decomposed (sometimes automatically or semi- automatically) into subprograms to execute on local nodes (node level behaviour) [33, 34]. This approach relieves application developers from directly dealing with concerns at each network node. Regiment [35], Kairos [33, 36] and ATaG [37] are significant examples of this type of abstraction.

Roughly speaking, node level abstraction promotes the development of applications in a more flexible and energy efficiency way, and generates smaller communication and interpretation overhead. On the other hand, system level abstractions are easier to use since nodal behaviours can be generated automatically so that the developer can concentrate on the network-wide tasks, without directly handling the collaboration among sensor nodes to perform the assigned tasks [18].

3.2 Programming Abstraction: Paradigms and Interface Types

Programming paradigm refers to the model adopted to program the applications. It defines the respective type of *interface* to be used in the middleware platform. The suitability of a specific approach is often dependent on the application requirements. One of the main features of WSN applications that affects the choice of the middleware programming approach is the required data delivery model. Data delivery in WSN can be continuous, event-driven, or query-based. Correspondingly, for different applications, WSN middleware may use different programming paradigms, such as database, mobile agent, and Publish/Subscribe (Pub/Sub). For example, the data base paradigm is often more suitable for query-based data collection, while the Pub/Sub paradigm can be a good choice for event-driven applications.

The next subsections briefly present existent programming paradigms for WSN middleware. It is important to notice that some of these approaches are orthogonal to each other and a same middleware platform can fit in more than one.

For instance, a middleware can adopt an event-based and a component-based and/or an application-driven programming approach at the same time.

3.2.1 Programming Paradigm Based on Virtual Machines

This approach allows developers to write application code in separate, small modules that are injected and distributed throughout the network using specialized algorithms, which aim at minimizing the overall energy consumption and resource usage. The Virtual Machine (VM) in each node then interprets the injected modules. Examples of this approach include Maté [38], ASVM [39] and DAViM [40]. Maté consists in a byte code interpreter that runs on TinyOS [41]. Code is broken into capsules of 24 byte-long instructions, and the set of capsules representing a program are injected into the network. Maté's components include the VM, the network, the logger, the hardware and the boot/scheduler. A synchronous model is adopted that, according to Maté's developers, makes application-level programming simpler and less prone to bugs than dealing with asynchronous event notifications. On the other hand, the use of such synchronous model makes Maté not so suitable for event-based WSN applications.

3.2.2 Programming Paradigm Based on Database

In the database approach, the whole network is abstracted as a virtual database system. This approach provides an easy-to-use and high-level interface that allows user to issue queries to the WSN in order to extract the sensing data of interest. We can consider that the database approaches were the first attempt to propose a middleware layer for WSNs. Examples include Cougar [42], TinyDB [9], SINA [43]. The first two proposals are based on pure database systems, which essentially provide a distributed database solution tailored to resource-constrained WSNs, focusing on efficient query routing and processing. In Cougar system [42] the processing capabilities of the sensors are exploited to perform part of query processing within the network, instead of centralizing such processing only in the WSN sink node. In SINA an SQL-like query language is used to express queries, but the middleware also provides additional functions which are out of the scope of traditional database systems. It provides support for scripting that enables the management of sensor hardware access, communication and event handling.

3.2.3 Event-Based Programming Paradigm

Another programming approach to WSN middleware is based on the concept of events. In such approach, the application specifies its interest in given state changes of the environment (basic events). Upon detecting an event, a sensor node sends an *event notification* towards interested applications. The application can also specify certain patterns of events (composite events), so that the application is only notified if events occur that match these patterns [12]. In [15], a sophisticated set of event operators for describing event patterns in WSN was proposed. Mires [44] is another example of middleware using an event-based approach, in which a publish/subscribe solution was designed and implemented to run on TinyOS using nesC language. It adopts a component-based programming model using active messages to implement its publish/subscribe-based communication infrastructure. Other programming abstraction that provides publish-subscribe mechanisms is the message-oriented approach. The strength of this paradigm, in the same way as the event-based one, lies in its support for asynchronous communication that promotes energy savings in the WSN. Both are very suitable for WSN applications that required and event-driven data delivery model.

3.2.4 Application-Driven Programming Paradigm

The application driven approach allows developer to fine-tune the network operation, configuration and management according to application requirements. In spite of the fact that such approach brings benefits regarding the WSN performance, it may provide a specialized middleware that is tightly coupled with the application, instead of a general purpose WSN middleware. MiLAN is an example of middleware [45] that adopts this approach. MILAN receives a description of the application requirements through a standard API and chooses the best network configuration according to such requirements, while seeking to maximize the overall network lifetime. MILAN incorporates changes in the application interests based on updated context information and manages the network conditions over time. In its first version, MILAN adopts a fully centralized approach, suitable for small scale WSNs, as those used to monitor patients in hospitals.

3.2.5 Component-Based Programming Paradigm

Component-Based Software Engineering (CBSE) is a modern methodology that proposes software construction by plugging software components [46]. Based on component interoperability, this programming approach aims at building more flexible and adaptable software. Recently, some middleware proposals based on CBSE have emerged in the WSN field. In [47] RUNES is introduced, a reconfigurable component-based middleware for networked embedded systems. RUNEs architecture encompasses two layers: a foundation layer, called the middleware kernel, which is the runtime realization of a simple but well-defined software component model, and a top layer of component frameworks that offer a configurable and extensible set of middleware and application services [12]. Other example is MWSAN [48], a real-time component-based middleware for WSANs, which provides a set of high level services for sensors and actuators. Besides considering the real-time as a major concern, it also takes into account issues such as the network configuration and quality of service (QoS) parameters.

3.2.6 Programming Paradigm Based on Tuple Spaces

The need for coordination in WSNs has attracted the attention of the Coordination paradigm community [49]. Coordination [50] is a programming paradigm "whose goal is to separate the definition of the individual behaviour of application components from the mechanics of their interaction." This goal is usually achieved by using either *message passing* or *data sharing* as a model for interaction. Publish-subscribe, previously described, is an example of the former, where coordination occurs only through the exchange of messages (*events*) among publishers and subscribers. While message passing, in its pure form, is inherently *stateless*, data sharing enables coordination among components by manipulating the (distributed) *state* of the system [50]. Linda [51] can be considered one of the most representative coordination languages and it brought attention of the community for the tuple space abstraction. Linda is based on a shared memory model where data is represented by elementary data structures called tuples, and the memory is a multiset of tuples called a *tuple space*. The tuple space model has several features that make it suitable for using in wireless networks environments in general. First, only a small set of operations is needed to manipulate the tuple space, and therefore to enable distributed component interaction [50]. Second, in this model, the coordination among processes in is decoupled in both time and space, i.e., tuples can be exchanged among producers and consumers without being simultaneously available, and without mutual knowledge of their identity or location. Such decoupling is crucial in the presence of wireless connectivity, considering the inherent fluctuations in the radio communication. Moreover, tuple spaces can be straightforwardly used to represent the context perceived by the coordinating components. On the other hand, this beneficial decoupling is achieved at the costs of the global accessibility to all components and persistence, two features not feasible or efficient in WSN environments [50]. Therefore, to benefit from the advantages of the tuple space model, middleware for WSN needs to adapt to the resource constrains of sensor nodes. Examples of WSN middleware that builds on the concept of tuple spaces are TinyLime [52] and TeenyLime [53]. In TinyLime, sensor nodes are sparsely distributed in an environment, not necessarily able to communicate with each other, and a set of mobile base stations (laptops) move through the space accessing the data of nearby sensors. Each base station owns a tuple space and federated tuple spaces can be established in order to communicate and synchronize several base stations and some client hosts. TinyLime exploits the benefits of using tuple spaces while circumventing its potential drawbacks for the WSN context.

3.2.7 Programming Paradigm Based on Mobile Agents

In the mobile agent based computing paradigm, a task-specific executable code traverses the relevant sources to acquire sensing data. Mobile agents can be used to reduce the communication cost in the network, by moving the processing function to the data rather than bringing the data to a central node [12]. Examples of proposals that adopt mobile agents in WSNs are Agilla [54], MAWSN [55] and actorNet [56]. Agilla could also be included in the tuple space middleware approach, since agents coordinate through tuple spaces. Agilla aids the fast deployment of adaptive applications in WSNs. It allows the developer to create and inject mobile agents, which can migrate across the WSN performing application-specific tasks. Mobile agents can intelligently move or clone themselves to target locations in response to changing conditions in the environment. Each node maintains a local tuple space, and different agents can coordinate through local or remote operations on these tuple spaces. This flow of code and state has the potential to transform a WSN into a shared, general-purpose computing platform able of running several autonomous applications at a time.

3.3 WSN Middleware Programming Paradigms

The middleware programming abstraction paradigms aforementioned provide mainly communication services allowing high-level applications to issue (synchronous or asynchronous) queries on WSN generated data. Each paradigm has its own features and consequently its own advantages and disadvantages. Table 1 summarizes the pros and cons of the described programming paradigms.

A current issue with existing programming paradigms for WSN middleware is that they only provide limited flexibility and interoperability in terms of interaction with external applications and end-users. Updating the interfaces and providing automated machine-to-machine (M2M) interactions in these types of solutions is often constrained due to restricted interaction models and pre-defined interfaces and operations [57]. Another promising and recent solution for middleware programming paradigm is adopting a service oriented approach [58, 59], according to which middleware components are provided as *services*. Chapter 21 addresses service oriented WSN middleware.

4 Middleware Services

Services embody the functionalities and consist in the core of a WSN middleware. They are exposed to the application developer through the abstraction programming interface, and provide the support for application deployment, execution, as well as sensor and network management. During deployment, the functionalities of WSN middleware can be distributed to the sensor nodes, sink nodes, and applications.

Table 1 Programming paradigm comparison

Programming Abstraction	Pros	Cons
Virtual machine	• It minimizes the overall energy consumption and resource usage (energy efficiency is high) • It has high scalability • It allows software updates at runtime, thus adapting the WSN behavior to different application requirements and execution contexts (adaptability is high)	• The execution of instructions can introduce high overhead on nodes • In some existing proposals, for instance Maté, code updates disturb the entire system, since it is required to disseminate a complete TinyOS code image with the modified virtual machine in the network • Code updates are installed in all WSN nodes—there is no control where they will be installed • Programming the applications is not easy (usability is poor)
Database	• It provides an easy-to-use and high-level interface to the final users (usability is good) • It is energy efficient	• It lacks time space relations between events • Real time applications are not supported • Only approximate results are provided • It assumes that sensor nodes are largely homogeneous (heterogeneity support is poor) • Existing implementations do not support complex sensor types, as surveillance cameras with image processing • The use of SQL-like languages is not suitable for WSN since sensor data capture observations and not facts. Moreover, since SQL is query-oriented it does not properly support the task-oriented programming required by WSN
Event-based	• It allows a good decoupling of event producers and subscribers, and supports additional operations related to time and spatial conditions • It is suitable for event-based WSN applications	• The implementation of such paradigm can be complex • It assumes that all sensors report accurate data and that the set of sensors is homogeneous (heterogeneity support is poor)

(continued)

Table 1 (continued)

Application driven	• It allows developer to fine-tune the network operation, configuration and management according to application requirements • Usability is good • Scalability is high	• It lacks support for OS and hardware heterogeneity • High coupling with the target application (domain specific middleware rather than a generic middleware)
Component-based	• It provides high support for heterogeneity, allowing applications run on various types of devices • It scales well • It allows building flexible and adaptable software (adaptability is high)	• The application programming is not easy (poor usability)
Tuple space approach	• It allows high decoupling among data sources and consumers	• It is resource-intensive • It does not scale well • It does not support adaptability
Mobile agent based	• It allows resource optimization, by reducing the communication cost in the network (energy efficiency is high) • It scales well • The migration of agents enables adapting network to different applications with different requirements (adaptability is high) • The use of multiple agents allows multiple users from different applications share the network simultaneously (resource sharing is good)	• It is a platform dependent solution • Programming the applications is not easy (usability is poor)

The distributed middleware components located in different nodes of the network communicate with each other to achieve common goals in benefit of the user/manager.

Middleware services can be classified into two broad categories: *common or generic services* and *domain or application-specific services*. Common services are the basic services shared by all WSN applications. They assist the handling of the application-level information and the management of WSN infrastructure. Domain services facilitate the development of applications in a specific domain. They make use of the common services and add application-specific functionalities.

4.1 Common Services of WSN Middleware

The functionalities provided by common services of a WSN middleware include: code and data management, resource discovery and management; and integration. The following subsections detail each of these services.

4.1.1 Code Management Service

Code Management encompasses services for code deployment, i.e., allocation and migration of code to sensor nodes. Code allocation determines a set of sensor nodes, on which the execution of code representing the sensing tasks is be activated. Code migration transfers the code on a sensor node to another one [60]. Strategies for code allocation can be context aware and allow adaptation to the current context. For instance, in MiLAN [45] application-level QoS is applied to the control of the code allocation. The approach enables adapting the application operations according to the current application requirements, which can be adjusted depending on the output of the application itself. Code migration can be implemented at not only the middleware layer but also in the underlying embedded operating systems, as in BerthaOS [61] and MagnetOS [62]. However, because the operating system of WSN nodes often does not support code interpretation, code migration implemented at the OS level is error prone and subject to malicious attacks. Current implementation techniques for code migration at the middleware level include the use of mobile code (e.g., TCL script in Sensorware [63], SQTL scripts in SINA [43]) and mobile Java object (e.g., TinyLime [52]).

4.1.2 Data Management Service

This service is responsible for ***data acquisition, data processing, data synchronization,*** and ***data storage*** functionalities. As previously mentioned, WSN applications are data centric, where data refers mainly to the sensed data, however it can also refer to the network infrastructure information of interest by the applications (for instance to participate in optimization decisions). Approaches to implement the data management services strongly depend on the application data delivery model.

Data acquisition is an essential service for WSN applications, responsible for delivering the relevant and accurate data required by the application. For the event-based data delivery model, data acquisition support is focused on the event definition, event register/cancel, event detection and event delivery. The application specifies its interest in specific state changes of the data. Upon detecting such an event, the middleware will help sending event notification to interested applications. TinyDB [9], DSware [64], Mires [44], and Impala [28] all support event-based data acquisition. For query-based applications, data acquisition support is focused on the query processing model and methods. Middleware for query-based data model usually uses a declarative interface, with global level abstraction and database programming model, as for example TinyDB, Cougar and SensorWare.

Regarding the support for **data processing** in WSNs we can identify two different approaches, with several degrees between them. In a totally ***centralized processing***, sensing data are collected and sent to a central node for processing. WSN nodes are passive devices, only in charge of achieving the environmental measurements, while all processing is performed in the central (sink) node. In a fully ***distributed processing***, every sensor node is involved with data sensing, processing and routing, and is aware of the final decision to be sent to the application. Such approach fully exploits the computational capacity of WSN nodes performing the so-called in-network processing. In the middle of these two extremes, we have ***partially distributed processing***. One possible approach for ***partially distributed processing*** consists in raw data being collected and pre-processed in the sensor nodes to obtain partial results which are then sent to the sink node for further processing in order to achieve the final result to be sent to the application. Other possible approach consists in the final results being obtained through both in-network processing by individual sensor nodes and information exchange between the sensors, and between the sensor and the sink node. Considering that the communication cost is higher than the computation cost at a sensor node, WSN middleware should support in-network distributed data processing service, mostly by implementing data fusion/aggregation techniques [2]. Another service related to data processing is *time synchronization* aimed to ensure the proper synchronization between different sensor nodes. Several WSN applications require precise time synchronization among the readings on different sensor nodes. Achieving time synchronization is an important function of a middleware platform. Several time synchronization algorithms specific for WSN have been proposed in the last years [65]. All of them could be provided as a middleware service.

Regarding **data storage** support in WSNs three approaches can be identified [66, 67]. External storage stores the data in the sink node—external to the WSN. Local storage stores the data where it is generated, reducing communication but increasing the cost to achieve (query) data. Data centric storage provides a trade-off between the previous two approaches and it is the most popular approach implemented in existing WSN middleware [68]. In data centric-storage sensing data is stored according to its event type at designated WSN nodes. As a consequence, users can send queries for a particular data type directly to the node that stores such data type rather than flooding the network with queries, thus saving communication energy.

4.1.3 Resource Discovery Service

The ubiquity and the wireless nature of sensor devices require network architectures to support *ad hoc* configuration. A key technology of true *ad hoc* networks is service discovery, functionality by which "services" (functionalities provided by nodes) can be described, advertised, and discovered by other devices or applications. There are typically two levels of resource discovery in WSN: an internal level is used by sink nodes to discover the characteristics of every sensor node in the network; an external discovery level is used by applications to find out which WSN supplies the required services, and how to invoke them. Considering the dynamic environment of WSN, such service is required for discovering sensor nodes newly joined to the network and for detecting nodes which are becoming unavailable either as a result of mobility or energy depletion. After an application starts executing, the service also provides the application with the underlying network information to achieve context-awareness and to support adaptive resource management services.

The ***resource discovery*** service can return, among other useful information: (i) the data type (s) that a discovered node can provide, (ii) the modes a sensor can operate, (iii) the node transmission power level, (iv) the supported data sensing and data sending rates, (v) the node residual energy level, (vi) information about the network topology, network protocols, and (vi) neighbours and locations of the discovered nodes.

Compared to the resource discovery in traditional networks [16] which involves identifying and locating services and resources in the system, resource discovery services in WSN are more difficult to implement due to the lack of unique node ID and the lack of a generic service specification, and because the services need to be provided in a power-aware way. Most of the current service discovery and capability description mechanisms in WSN are based on *ad hoc* representation schemes and lack from standardization. Some WSN middleware platforms (as MiLAN, for instance), adopt service discovery protocols from traditional computer network solutions, e.g., SLP [69] and Bluetooth SDP [52]. Other systems, e.g., TinyLime, developed novel solutions to implement the resource discovery service, specific tailored for WSNs.

A crucial requirement for the future, widely accessed WSNs is interoperability under unpredictable conditions, i.e., networks which were not designed for specific, predefined purposes, should be able to be accessed by different applications, which dynamically discover their functionality and take advantage of it. Crucial tasks involved in the dynamic utilization of WSN services are service discovery and description. Service oriented approaches enhanced with semantic technologies, as for instance, the adoption of ontology language are promising solutions to describe the characteristics of WSN devices, their sensing capabilities, and specific information of applications accessing the WSN, in a standardized way. In [70] the authors describe a WSN middleware that exploits such technologies.

4.1.4 Resource Management Service

This is the service responsible for managing the node (e.g., energy, memory, sensing devices, communication module) and network (e.g., topology, routing, system time) resources. Resource management allows the WSN to be self-configured, self-organized and self- healing. Resource management services are usually in charge of resource configuration at setup time and resource adaptation at runtime, and they are essential to ensure the QoS properties. Resource management at the OS layer is platform-dependent, thus changes at this level might affect resource requirements of the different applications running in a sensor node. On the other hand, application-level resource management imposes an extra burden on the application, and adaptation mechanisms developed at this level cannot be reused. In contrast, resource management at the middleware layer has more flexibility. Examples of resource management services are: clustering [71, 72], task scheduling and allocation [73, 74], active node selection. These services are supported by finer granular services such as power level management, transmission level management, among others.

4.1.5 Integration Service

For wide scope applications, WSN needs to be integrated into other existing network infrastructures, such as the Internet, grid, cloud, and database systems. Therefore, a service is required to integrate WSN and its running applications into external systems. Since a WSN is a "close" network, it is not easy to implement the integration service at the lower layers (e.g., OS or MAC layer), thus middleware should provide this service [75]. In a WSN middleware, integration is related to both task coordination and data sharing, and can be implemented at the application level or data level [18]. Application level integration is more related to task coordination, where the applications are running in both the WSNs and the external system. Data level integration, on the other hand, is more related to data sharing, where only the data provided by the WSNs are used in the external system. A promising solution to achieve integration at both levels is the service oriented approach. It allows implementing WSN integration based on standardized and open architecture technologies as Web services. It provides a common information and communication format to facilitate the integration.

4.1.6 Additional Services

Besides the aforementioned services, there are several others that can be included as common services in a WSN as, for instance, naming, location and security. The naming service is used whenever a unique local identification for each node of the network is required (with a smaller size in bytes than the serial number, or MAC address). The localization service is useful when some nodes of the network do not have GPS receiver. In these cases, a localization algorithm [76] can be implemented as a service provided by the middleware, and executed on demand. Security is the service in charge of meeting the requirements of confidentiality, authentication, integrity, freshness, and availability whenever the application data or code is sensitive.

4.2 Domain Specific Services of WSN Middleware

Domain services facilitate the development of applications in a specific domain. They make use of the common services and leverage them by adding application-specific functions. Since there is a wide range of application domains in WSN, a comprehensive listing of this type of service is difficult. Significant examples are EnviroTrack [77], a WSN middleware that supports environmental Target tracking, and Impala, a middleware designed for the ZetbraBet project, a wildlife monitoring project. Impala has two layers: the upper layer contains the application specific protocols and functions, and the lower layer contains the common services such as code management. There are also several proposals for WSN-SHM middleware [6, 7, 66, 78], designed for developing structural health monitoring applications. Such applications, besides other common requirements of WSN application, require high frequency sampling and are highly resource consuming, thus posing additional challenges for the middleware to tackle.

5 New Issues and Future Trends

The most notable changes required in the WSN design to accommodate the new application scenarios and technologies relate to the following factors: (i) sharing of the WSN infrastructure, (ii) interoperability between different WSNs, (iii) access to sensor generated data, and (iv) use of WSN in applications considered critical. We will discuss these changes in the following sections.

5.1 WSN Infrastructure Sharing

The first change is directly related to the "application-specific" feature typical of WSN. Since the emergence of these networks, they were usually designed as systems dedicated to a single target application. With such approach, there was always a large coupling between the requirements of the target application and the underlying network, including node hardware and software platforms. By adopting this design strategy, WSNs could be built by taking advantage of the knowledge at the application level to optimize the other levels of network protocols in order to achieve maximum energy efficiency. Applications developed for a particular sensor platform needed to be completely rewritten to be used on other platforms, with virtually no reuse of both software artefacts and the deployed sensors. This scenario tends to change considerably.

In many commercial and/or large scale WSN deployments nowadays, the return-of-investment (ROI) is a crucial factor to determine the successfulness of those deployments. On one hand, as it is always the case in networks composed of resource constrained devices, the system lifetime is the fundamental requirement to be taken into account in the design of a WSN system. In this sense, the ROI is considered beneficial from a longer system lifetime. On the other hand, the *utilization* of a WSN plays an increasingly important role in current WSN design, and such utilization may be very low in networks specifically designed for a single application. For instance, a WSN designed for a smart building [67, 79] may provide the following four services: temperature and humidity monitoring, security alarms, light control and structural health monitoring. Obviously, in this case, each service requires a set of sensors (dependent on the service type) to perform the associated sensing tasks. In the traditional design, the smart building project would require four service-specific WSNs to fulfil the needs of the applications. However, with such design the utilization of each individual WSN would be low, and the maintenance cost high. Therefore, one direct consequence of this type of design is the decrease of the WSN ROI.

Another issue related to the aforementioned change in the design approach of WSN was raised in works such as [80], which discuss the potential use of WSN for sustainable development.[1] According to the authors, such potential will remain largely underexploited as these networks are designed primarily for application contexts relatively rich in resources. The cost of WSNs is one of several barriers that prevents them from being exploited for the sustainable development of applications. Many components of WSN are becoming less expensive (the processing boards, for example), but the sensing units remain relatively expensive components, depending on their type. As stated in [44], "successful projects based on technology development depend upon shared technology, due to the excessive cost of individual devices/personal." However, as mentioned, the traditional design approach of most WSN is based on long-term deployments, owned by a single user and focused on a single application, a paradigm that does not promote sharing, and consequently using WSN for sustainable development.

For these reasons, and considering the physical networks' infrastructure cost and the fact that the same set of sensors can be useful for a variety of applications, the design trend of WSNs starts shifting from energy efficient application-specific approach towards an energy efficient multiple applications approach [23, 81, 82]. Compared to the traditional design approach, this new one enables multiple applications to share the resource of sensors and networking, increasing the use of the system so that the total cost is reduced and ROI increased, and promoting the sustainable development of WSN. Following the same trend of sharing data/infrastructure, one can still imagine a scenario in which multiple applications not only benefit from an installed network infrastructure, but also from multiple sensor networks, producing a shared multifunctional system.

[1] Sustainable development can be defined as a process of development that "meets the needs of the present without compromising the ability of futures generations to meet their own needs" and whose interdependent underpinnings are "the social and economic development, as well as environmental protection."

However, this new trend hinders the issue of obtaining energy efficiency in the WSN and poses new challenges regarding simultaneously meeting requirements of different applications. In fact, this new trend of designing shared WSN systems has an impact on all levels, from network design and development of their applications, to the use of data produced by it, affecting the other changes mentioned above, required for new WSN scenarios. In particular, the desired sharing creates new challenges when it comes to promoting the interconnection and interoperability of WSN and to achieve the desired energy efficiency, while meeting the requirements of different applications.

5.2 WSN Interoperability

Regarding the interoperability and interconnection of networks, WSN nodes typically can be produced by different manufacturers using different technologies, hardware, software, languages, APIs, programming and communication models. Currently, there is not a *de facto* standard for sensor programming languages and neither for accessing sensor generated data. This feature poses a challenge to be surpassed to enable new scenarios, especially regarding the shared use of WSN. In this scenario, there must be interoperability between different WSN, possibly between different applications, and between WSN and external networks as the Internet. Moreover, the way how data produced by these networks are extracted and accessed should also be reviewed to enable the extraction and easy access to the wide range of data available for the end user. Thus, the adoption of a uniform API, protocols and standardized languages is necessary to allow such interoperability in an environment with high degree of heterogeneity.

Concerning data access, having WSN originally designed to operate as standalone networks, sensor readings were usually disseminated using proprietary data formats, toward the sink node located at the edges of these networks. In the sink nodes the target application was in charge of performing more complex processing whenever required, and any translations or format conversions. However, in the new shared scenario, with several applications developed by different users from different domains, there is the need of interconnecting these networks to external networks, enterprise networks and the Internet. Moreover, WSN applications and the produced sensed data become part of corporate systems in a M2M interaction. Integrating a WSN into the information system of an enterprise at a high abstraction level is a recent need already being tackled by several authors [83]. The use of proprietary data formats and protocols is not suitable in these scenarios. Therefore, the adoption of common APIs, standardized formats and protocols is a need.

In the context of WSN interconnection with other networks, a recent outspread of the advances in embedded systems has enabled the integration in the Internet of physical objects of everyday life, such as household appliances, industrial machinery, vehicles, sensors and actuators. The paradigm of connecting everyday objects (the so called smart objects or smart things), in the Internet is known as the Internet of Things—IoT [11]. IoT aims to connect the physical (analog) environments to the Internet (digital), promoting a large number of new possibilities for application domains. Recently, a new paradigm of application development inspired by the idea of IoT, known as Web of Things (WoT) [84–86] has emerged. This new paradigm is based on the use of protocols and standards widely accepted and already in use in traditional Web, such as HTTP (Hypertext Transfer Protocol) and URIs (Uniform Resource Identifier). The purpose of the WoT is to leverage the vision of connectivity between the physical and digital worlds, enabling the current Web to include everyday life objects, which will be handled in the same way as any other Web resource. In the WoT paradigm, the interfaces of a smart object is built according to REST principles (Representational State Transfer) [87], which allow exposing the services of these objects as resources in a ROA approach (resource -Oriented Architecture) [88]. These interfaces provided by objects that are to be integrated as Web resources are called RESTful APIs, and the resources provided by the APIs are called RESTful resources. In the WoT context, the sensors of a WSN can be considered as smart objects [10] and the data provided by them can be made available as a resource through RESTful APIs. A number of approaches have been proposed [89–91] to enable building REST-based WSN, and the required interconnection of WSN with the Internet.

5.3 WSN for Critical Applications

A final point to consider with respect to new issues in the WSN context that affect the middleware usage concerns the fact that the adoption of these networks in critical applications can bring benefit still underexploited. According to the European Commission [92], critical application scenarios consist of "[....] networks, facilities, services and physical and IT assets that, if suffering interruption or are damaged, will have a serious impact on health, security or well being economic of citizens. [....]". The number of critical scenarios where WSN can be used is remarkably high, as can be observed in the studies reported in [1]. However, its widespread use in such settings is still limited by a lack of confidence in the resilience of these networks. Resilience can be defined as [93]: "the persistence or reliability when facing 'changes'." This perspective leads to new requirements, especially regarding the fault tolerance of such networks, such as the ability to accommodate unforeseen environmental disturbances. The use of WSN for critical applications such as SHM, monitoring of natural disasters, control of factories and plants, Smart Grids, among others, introduces specific reliability requirements in addition to the usual WSN requirements. Among these, the authors in [7] reported: (i) reliable synchronization of the sensor collected measurements, (ii) reliable delivery of a significant amount of measurements, and (iii) minimization of human intervention on the network (i.e., autonomy of the network). Such requirements must be satisfied in WSN systems so that they can be used in applications considered critical. A middleware platform consists of the most appropriate component of a WSN system to implement the features that meet such requirements.

6 Final Remarks

WSN middleware aims at supporting the development, management, deployment and execution of sensing-based applications. It can also contribute to dealing with the heterogeneity of nodes in a same WSN, promoting interoperability among networks and applications and reuse of software and hardware artefacts. In order to achieve its goals, WSN middleware provides an interface and a set of services. Depending on the application features and needs different services will be required. In Chap. 6 of this Book, the authors classify WSN applications according to different dimensions. The authors identified that most of the researched applications require some form of mechanism for data storage (persistency), data processing (as filtering, fusion, decision making, event defection), as well as additional, sometimes complex services as time synchronization and localization. All these services and mechanisms can be provided by a middleware layer, instead of being addressed repeatedly and in an ad hoc manner by each application developer. Even if the application is simple enough so that it does not require middleware services, the provided middleware API aids in the application programming, shielding the developer from dealing with the low level languages typical of WSN environments. Therefore, WSN middleware is useful for most of WSN application classes and scenarios. Only in situations where the application requirements are overly specific, the WSN scope is by design (or due to some policy/administrative issue) restricted to a single application, the network nodes are homogeneous and highly constrained (thus making the resource optimization the main issue) and no interoperability, portability and/or reuse are desirable, the use of a middleware layer and its inherent overhead could be discarded.

In this chapter we presented an overview of the concept of WSN middleware, its main features and benefits of usage, and discussed existing solutions. It was not our goal to exhaust the subject, since it is a broad and interdisciplinary research field. For some comprehensive and recent surveys on WSN middleware the reader is referred to [17, 18, 59]. Our goal in this chapter was to motivate for the necessity of using middleware when building WSNs and developing applications for these networks, as well as shedding light in several aspects to be considered when designing a middleware specific tailored for WSNs. Moreover, we presented the emergent scenarios of WSN and discussed how, along with the rapid evolution in this area, the adoption of a middleware platform will become even more important to increase the popularity of WSN-based applications [41, 58, 86].

Acknowledgments This work was partially supported by Brazilian agencies FAPERJ and CNPq. We want to express our gratitude to Professor Habib M. Ammari who kindly invited us to contribute to this Book.

References

1. I.F. Akyildiz et al., Wireless sensor and actor networks: research challenges. Ad Hoc Netw. J. (Elsevier) **2**(4), 351–367 (2004)
2. E.F. Nakamura, A.A.F. Loureiro, A.C. Frery, Information fusion for wireless sensor networks: methods, models, and classifications. ACM Comput. Surv. **39**, 9/1–9/55 (2007)
3. P. Levis, D. Culler, The firecracker protocol, in *Proceedings of the 11th ACM SIGOPS European, Workshop*, Sept 2004
4. L. Evers, P. Havinga, Supply chain management automation using wireless sensor networks. *IEEE International Conference on Mobile Adhoc and Sensor Systems Conference*, pp. 1–3, 2007
5. T. He, S. Krishnamurthy, J.A. Stankovic, T. Abdelzaher, L. Luo, R. Stoleru, T. Yan, L. Gu, J. Hui, B. Krogh, Energy-efficient surveillance system using wireless sensor networks, in *Proceedings of the 2nd International Conference on Mobile Systems, Applications, and Services*, pp. 270–283. ACM, New York, NY, USA, 2004
6. L. Capra, W. Emmerich, C. Mascolo, Carisma: context-aware reflective middleware system for mobile applications. IEEE Trans. Softw. Eng. **29**(10), 929–945 (2003)
7. M. Cinque, D. Cotroneo, G. De Caro, M. Pelella, Reliability requirements of wireless sensor networks for dynamic structural monitoring, *International Conference on Dependable Systems and Networks (DSN)*, 2006
8. A. Rosi, M. Berti, N. Bicocchi, G. Castelli, M. Mamei, A. Corsini, F. Zambonelli, Landslide monitoring with sensor networks: experiences and lessons learnt from a real-world deployment. Int. J. Wirel. Sens. Netw. **10**(3), 111–122 (2011)
9. S. Bhattacharya et al., Multi-application deployment in shared sensor networks based on quality of monitoring, in *Proceedings of the 2010 16th IEEE Real-Time and Embedded Technology and Applications, Symposium*, 2010
10. F.C. Delicato, P.F. Pires, L. Pirmez, T. Batista, Wireless sensor networks as a service, in *Proceedings of the 17th IEEE International Conference and Workshops on the Engineering of Computer-Based Systems (ECBS '10)*, IEEE Computer Society, Washington, DC, USA, 2010
11. O. Vermesan et al., Internet of things strategic research roadmap. *Aerospace Technologies and Applications for Dual Use*, 2008
12. B. Rubio, M. Diaz, J.M. Troya, Programming approaches and challenges for wireless sensor networks, in *Proceedings of the 2nd International Conference Systems and Networks Communications*, France, p. 36, 25–31 Aug 2007
13. F.C. Delicato, L. Fuentes, N. Gamez, P. Pires, Variabilities of wireless and actuators sensor network middleware for ambient assisted living, in *Proceedings of 10th International Work-Conference on Artificial Neural Networks, IWANN 2009 Workshops*, vol. 5518, pp. 851–858, Salamanca, Spain, 2009
14. F.C. Delicato et al., A flexible middleware system for wireless sensor networks, in *Proceedings of the ACM/IFIP/USENIX International Middleware Conference*, Rio de Janeiro, July 2003
15. E. Souto, G. Guimaraes, G. Vasconcelos, M. Vieira, N. Rosa, C. Ferraz, A message-oriented middleware for sensor networks, in *Proceedings of the 2nd International Workshop on Middleware for Pervasive and Ad-Hoc Computing (MPAC 2004)*, pp. 127–134, Toronto, Canada, October 2004
16. H.A. Duran-Limon, G.S Blair, G. Coulson. Adaptive resource management in middleware: a survey. IEEE Distrib. Syst. Online **5**(7), 1541–4922 (2004)
17. K. Henricksen, R. Robinson. A survey of middleware for sensor networks: state-of-the-art and future directions, in *Proceedings of the International Workshop on Middleware for Sensor Networks*, pp. 60–65, Melbourne, Australia, 2006
18. M.M. Wang, J.N. Cao, J. Li et al., Middleware for wireless sensor networks: a survey. J. Comput. Sci. Technol. **23**(3), 305–326 (2008)
19. B.C. Smith, Reflection and semantics in a procedural programming language. Ph.D. thesis, MIT, USA, 1982

20. P. Santi, Topology control in wireless Ad Hoc and sensor networks. ACM Comput. Surv. **37**(2), 164–194 (2005)
21. C. Yeh, Dynamic reconfiguration techniques for wireless sensor networks. Masters theses, University of Massachusetts (2008), http://scholarworks.umass.edu/theses/119
22. F.C. Delicato et al., Reflective middleware for wireless sensor networks, in *Proceedings of the 20th ACM Symposium on Applied Computing*, USA, March 2005
23. K. Römer, O. Kasten, F. Mattern, Middleware challenges for wireless sensor networks. ACM Mobile Comput. Commun. Rev. **6**(2), 59–61 2002
24. Y. Yu, B. Krishnamachari, V.K. Prasana, Issues in designing middleware for wireless sensor networks. Special issue on middleware technologies for future communication networks. IEEE Netw. Mag. **18**(1), 15–21 (2004)
25. S. Meguerdichian et al., Localized algorithms in wireless ad-hoc networks: location discovery and sensor exposure, in *Proceedings of the 2001 ACM Symposium on Mobile Ad Hoc Networking and Computing*, pp. 106–116, USA, Oct 2001
26. H. Qi, P.T. Kuruganti, Y. Xu, The development of localized algorithms in wireless sensor networks. Invited paper, Sensors 2002 vol. 2, pp. 286–293, 2002
27. P. Levis, D. Culler, Mate: a tiny virtual machine for sensor networks, in *Proceedings of the 10th International Conference Architectural Support for Programming Languages and Operating Systems (ASPLOS-X)*, pp. 85–95. ACM Press, San Jose, USA, 2002
28. T. Liu, M. Martonosi, Impala: a middleware system for managing autonomic, parallel sensor systems, in *Proceedings of PPoPP'03*, pp. 107–118, San Diego, California, USA, June 2003
29. W. Whitehouse, C. Sharp, E. Brewer, D. Culler, Hood: a neighborhood abstraction for sensor networks. in *Proceedings of the 2nd International Conference on Mobile Systems, Applications and Services*, pp. 99–110, New York, USA, 2004
30. L. Mottola, G.P. Picco, Logical neighborhoods: a programming abstraction for wireless sensor networks, in *Proceedings of the 2nd International Conference on Distributed Computing in Sensor Systems*, San Francisco, CA, USA, June 2006
31. P. Ciciriello, L. Mottola, G. Picco, Building virtual sensors and actuators over logical neighborhoods, in *Proceedings of the 1st International Workshop on Middleware for Wireless Sensor Networks*, pp. 19–24, Melbourne, Australia, 2006
32. R. Gummadi, O. Gnawali, R. Govidan, MacroProgramming wireless sensor networks using Kairos, in *Proceedings of the International Conference on Distributed Computing in Sensor Systems*, vol. 3560, pp. 126–140. Springer, LNCS, 2005
33. R. Gummadi et al., Macro-programming wireless sensor networks using kairos, in *Proceedings of the International Conference on Distributed Computing in Sensor Systems (DCOSS 05)*, pp. 126–140. Springer, Marina del Rey, USA, LNCS 3560, 2005
34. M. Welsh, G. Mainland, Programming sensor networks using abstract regions, in *Proceedings of the 1st Usenix/ACM Symposium Networked Systems Design and Implementation (NSDI 04)*, pp. 29–42, San Francisco, CA, March 2004
35. R. Newton, M. Welsh, Regions streams: functional macroprogramming for sensor networks, in *Proceedings of the 1st International Workshop on Data Management for Sensor, Networks (DMSN'04)*, pp. 78–87, 2004
36. T.C. Rodrigues et al., Model-driven development of wireless sensor network applications, in *Proceedings of the 9th IEEE/IFIP International Conference on Embedded and Ubiquitous Computing*, Melbourne, 2011
37. A. Bakshi, V.K. Prasanna, J. Reich, D. Larner, The abstract task graph: a methodology for architecture-independent programming of networked sensor systems. in *Proceedings of the International Workshop on End-to-End Sense-and-Respond Systems (EESR'05)*, pp. 19–24, 2005
38. P. Levis, D. Culler, Mat'e: a tiny virtual machine for sensor networks, in *Proceedings of the 10th International Conference on Architectural Support for Programming Languages and Operating Systems (ASPLOS-X'02)*, San Jose, CA, USA, Oct 2002
39. P. Levis, D. Gay, D. Culler. Active sensor networks, in *Proceedings of the 2nd International Symposium on Networked Systems Design and Implementation (NSDI'05)*, pp. 29–42, San Francisco, CA, USA, March 2005

40. S. Michiels, W. Horr'e, W. Joosen, P. Verbaeten, DAViM: a dynamically adaptable virtual machine for sensor networks, in *Proceedings of the 1st International Workshop on Middleware for Sensor Networks*, Melbourne, Australia, 2006
41. TinyOS communities. TinyOS specification, http://www.tinyos.net
42. P. Bonnet, J. Gehrke, P. Seshadri, Towards sensor database systems, in *Proceedings of the 2th International Conference on Mobile Data Management (MDM'01)*, vol. 1987, pp. 3–14. Springer, LNCS, 2001
43. C. Curino, M. Giani, M. Giorgetta, A. Giusti, A.L. Murphy, G.P. Picco, TinyLime: bridging mobile and sensor networks through middleware, in *Proceedings of the 3rd IEEE International Conference on Pervasive Computing and Communications (PerCom 2005)*, pp. 61–72, Hawaii, USA, March 2005
44. E. Brewer, M. Demmer, B. Du, M. Ho, M. Kam, S. Nedevschi, J. Pal, R. Patra, S. Surana, K. Fall, The case for technology in developing regions, in *IEEE Computer*, June 2005
45. W. Heinzelman, A. Murphy, H. Carvalho, M. Perillo, Middleware to support sensor network applications. IEEE Netw. **1**(18), 6–114 (2004)
46. G. Heineman, W. Councill, *Component-Based Software Engineering: Putting the Pieces Together* (Addison-Wesley, Reading, 2001)
47. P. Costa, G. Coulson, C. Mascolo, L. Mottola, G.P. Picco, S. Zachariadis, Reconfigurable component-based middleware for networked embedded systems. Int. J. Wirel. Inf. Netw. **14**(2), 149–162 (2007)
48. J. Barbar'an, M. D'ıaz, I. Esteve, D. Garrido, L. Llopis, B. Rubio, A real-time component-oriented middleware for wireless sensor and actor networks, in *Proceedings of the IEEE International Conference on Complex, Intelligent and Software Intensive Systems*, pp. 3–10, Vienna, Austria, April 2007
49. N. Carriero, D. Gelernter, Coordination languages and their significance. Commun. ACM **35**(2), 97–107 (1992)
50. P. Costa, L. Mottola, A.L. Murphy, G.P. Picco, Tuple space middleware for wireless networks, in *Invited Chapter in Middleware for Network Eccentric and Mobile Applications* ed. by B. Garbinato, H. Miranda, L. Rodrigues. Springer Press, 2008
51. D. Gelernter, Generative communication in Linda. ACM Trans. Program. Lang. Syst. **7**(1), 80–112 (1985)
52. Bluetooth Special Interest Group, Bluetooth specification, version 1.1, Feb 2001, http://en.wikipedia.org/wiki/Bluetooth
53. P. Costa, L. Mottola, A.L. Murphy, G.P. Picco, TeenyLIME: transiently shared tuple space middleware for wireless sensor networks, in *Proceedings of the 1st International Workshop on Middleware for Wireless Sensor Networks (MidSens 2006)*, pp. 43–48, Melbourne, Australia, Nov 2006
54. C.-L. Fok, G.-C. Roman, C. Lu, Rapid development and flexible deployment of adaptative wireless sensor network applications, in *Proceedings of the 25th International Conference on Distributed Computing Systems*, pp. 653–662. IEEE Computer Society Press, Columbus, Ohio, USA, June 2005
55. M. Chen, T. Kwon, Y. Yuan, V. Leung, Mobile agent based wireless sensor networks. J. Comput. **1**(1), 14–21 (2006)
56. Y. Kwon, S. Sundresh, K. Mechitov, G. Agha, ActorNet: an actor platform for wireless sensor networks, in *Proceedings of the IEEE International Joint Conference on Autonomous Agents and Multiagent Systems*, Hakodate, Japan, May 2006
57. H. Abangar, P. Barnaghi, K. Moessner, R. Tafazolli, A. Nnaemego, K. Balaskandan, A service oriented middleware architecture for wireless sensor networks, in *Proceedings of Future Network and Mobile Summit*, Florence, Italy, June 2010
58. OASIS, Reference architecture foundation for service oriented architecture 1.0, committee specification draft 03, July 2011
59. J. Al-Jaroodi, N. Mohamed, Service-oriented middleware: a survey. J. Netw. Comput. Appl. **35**(1), 211–220 (2012)

60. M. Kuorilehto, M. HÄannikÄainen, T. D. HÄamÄalÄainen, A survey of application distribution in wireless sensor networks. EURASIP J. Wirel. Commun. Networking **38**(5), 774–788 (2005)
61. J. Lifton, D. Seetharam, M. Broxton, P.J. Pushpin, Computing system overview: a platform for distributed, embedded, ubiquitous sensor networks, in *Proceedings of the 1st International Conference on Pervasive Computing*, Switzerland, 2002
62. R. Barr et al., On the need for system-level support for ad hoc and sensor networks. Operating Syst. Rev. **36**(2), 15 (2002)
63. A. Boulis, C.-C. Han, M. B. Srivastava, Design and implementation of a framework for efficient and programmable sensor networks, in *Proceedings of the First International Conference on Mobile Systems, Applications, and Services (MobiSys 03)*, SAN Francisco, CA, USA, pp. 187–200, 5–8 May, 2003
64. S. Li, S. Son, J. Stankovic, Event detection services using data service middleware in distributed sensor networks, in *Proceedings the 2nd International Workshop Information Processing in Sensor Networks*, pp. 502–517, Palo Alto, California, USA, 22–23 April 2003
65. J.E. Elson, Time synchronization in wireless sensor networks. PhD thesis, University of California, Los Angeles, 2003
66. K. Chintalapudi, J. Paek, O. Gnawali, T.S. Fu, K. Dantu, J. Carey, R. Govindan, E. Johnson, S. Masri, Structural damage detection and localization using NETSHM, in *Proceedings of the 5th International Conference on Information Processing in Sensor Networks*, pp. 475–482, USA, April 2006
67. K.-C. Tsai, J.-T. Sung, M.-H. Jin, An environment sensor fusion application on smart building skins, in *Proceedings of the IEEE International Conference on Sensor Networks, Ubiquitous, and Trustworthy Computing (SUTC '08)*, pp. 291–295. IEEE Computer Society, Washington, DC, USA, 2008
68. S. Shenker et al., Data-centric storage in sensornets, in proceedings of the 1st ACM SIGCOMM workshop on hot topics in networks. ACM SIGCOMM Comput. Commun. Rev. **33**(1), 137–142 (2003)
69. E. Guttman, C. Perkins, J. Veizades, M. Day, Service location protocol, version 2, IETF, RFC 2608, June 1999
70. F.C. Delicato et al., Exploiting web technologies to build autonomic wireless sensor networks, in *Proceedings of IFIP 19th World Computer Congress, TC-6, 8th IFIP/IEEE Conference on Mobile and Wireless Communications Network*, vol. 211, pp. 99–114, Santiago, 2006
71. A.R. Rocha et al., WSNs clustering based on semantic neighbourhood relationships. Comput. Netw. **56**(5), 1627–1645 (2012)
72. Y.-C. Chang, Z.-S. Lin, J.-L. Chen, Cluster based self-organization management protocols for wireless sensor networks. IEEE Trans. Consum. Electron. **52**(1), 75–80 (2006)
73. Voinescu, A., Tudose, D.S., N. Tapus, Task scheduling in wireless sensor networks, in *Proceedings of the Sixth International Conference on Networking and Services*, Cancun, Mexico, 07–13 March 2010
74. Y. Yu, V.K. Prasanna, Energy-balanced task allocation for collaborative processing in wireless sensor networks. ACM/Kluwer J. Mobile Netw. Appl. **10**(1–2), 115–131 (2005)
75. K. Hwang, J.I.N. Park, D. Eom, A design and implementation of wireless sensor gateway for efficient querying and managing through world wide web. IEEE Trans. Consum. Electron. **49**, 1090–1097 (2003)
76. W. Heinzelman, A. Murphy, H. Carvalho et al., Middleware to support sensor network applications. IEEE Netw. Mag. **18**(1), 6–14 (2004)
77. T. Abdelzaher, B. Blum, Q. Cao, D. Evans, J. George, S. George, T. He, L. Luo, S. Son, R. Stoleru, J. Stankovic, A. Wood, EnviroTrack: towards an environmental computing paradigm for distributed sensor networks, in *Proceedings of the 24th International Conference Distributed Computing Systems*, Tokyo, Japan, pp. 582–589, 23–26 March 2004
78. D. Musiani, K. Lin, T. Simunic Rosing, Active sensing platform for wireless structural health monitoring, in *Proceedings of the 5th International Conference on Information Processing in Sensor Networks*, Massachusetts, 25–27 April 2007

79. F. Shu, M.N. Halgamuge, W. Chen, Building automation system using wireless sensor networks: radio characteristics and energy efficient communication protocols. Special issues: network on building monitoring: from theory to real application. Electron. J. Struct. Eng. 66–73 (2009)
80. N. Ramanathan et al., Gaurav Sukhatme designing wireless sensor networks as a shared resource for sustainable development. *International Conference on Information and Communication Technologies and Development*, May 2006
81. D. Guinard et al., Sharing using social networks in a composable web of things, in *2010 8th IEEE International Conference on Pervasive Computing and Communications Workshops (PERCOM Workshops)*, pp. 702–707, 2010
82. H.B. Lim et al., The national weather sensor grid: a large-scale cyber-sensor infrastructure for environmental monitoring. Int. J. Sens. Netw. **7**, 19–36 (2010)
83. I.K. Samaras, J.V. Gialelis, G.D. Hassapis, Integrating wireless sensor networks into enterprise information systems by using web services. *Third International Conference on Sensor Technologies and Applications*, pp. 580–587. IEEE Press, 2009
84. D. Guinard, V. Trifa, Towards the web of things: web mashups for embedded devices, in *Proceedings of Workshop on Mashups, Enterprise Mashups and Lightweight Composition on the Web, WWW Conferences*, Spain, 2009
85. D. Guinard, Towards opportunistic applications in a web of things, in *IEEE International Conference on Pervasive Computing and Communications Workshops*, 2010
86. D. Guinard, V. Trifa, T. Pham, O. Liechti, Towards physical mashups in the web of things, *In Proceedings of IEEE Sixth International Conference on Networked Sensing Systems*, Pittsburgh, USA, June, 2009
87. R.T. Fielding, Architectural styles and the design of network-based software architectures. Doctoral dissertation, University of California, Irvine, 2000
88. X. Guo, J. Shen, Z. Yin, On software development based on SOA and ROA, in *Control and Decision Conference (CCDC)*, pp. 1032–1035. Publishing Press, 2010
89. A. Dunkels, T. Voigt, J. Alonso, Making TCP/IP viable for wireless sensor networks, in *Proceedings of the First European Workshop on Wireless Sensor Networks (EWSN 2004)*, Berlin, Germany, Jan 2004
90. A. Dunkels, T. Voigt, J. Alonso, H. Ritter, J. Schiller, Connecting wireless sensornets with TCP/IP networks, in *Proceedings of the Second International Conference on Wired/Wireless Internet Communications (WWIC2004)*, Frankfurt (Oder), Germany, Feb 2004
91. J.W. Hui, D.E. Culler, IP is dead, long live IP for wireless sensor networks, in *Proceedings of the 6th ACM Conference on Embedded Network Sensor Systems*, pp. 15–28. ACM, Raleigh, North Carolina, USA, 2008
92. K. Chintalapudi, T. Fu, J. Paek, N. Kothari, S. Rangwala, J. Caffrey, R. Govindan, E. Johnson, S. Masri, Monitoring civil structures with a wireless sensor network. IEEE Internet Comput. **10**(2), 26–34 (2006)
93. J.C. Laprie, From dependability to resilience, *38th IEEE/IFIP International Conference On Dependable Systems and Networks*, Anchorage, Alaska, June 2008

Chapter 20
Service-Oriented Middleware: Overview and Illustrative Example

Flávia C. Delicato, Paulo F. Pires and Albert Y. Zomaya

Abstract In this chapter we present an emerging approach to develop systems for WSN, named Service-Oriented Middleware (SOM), in which the WSN is logically viewed as a service provider for consumer applications. SOM provides abstractions for the complex underlying WSN through a set of generic and/or application-specific services. Services as data aggregation, adaptation, security, self-organization, resource management as well as other advanced services can be designed, implemented, and integrated in an SOM framework to provide a flexible and easy environment to develop effective WSN applications. Moreover, the intrinsically decoupled nature of the various components involved in a service-oriented architecture promotes interoperability between service providers and consumers. The adoption of service oriented approach provides WSN users with a unified protocol to access and communicate with the WSN components and developers with a flexible programing model to build efficient and scalable WSN systems. Besides presenting the basic concepts of SOM development and discussing the potential benefits of such approach in building WSM middleware systems, in this chapter we also present a concrete example of a WSN SOM to further illustrate the described features.

F. C. Delicato (✉) · P. F. Pires
Federal University of Rio de Janeiro, Rio de Janeiro, Brazil
e-mail: fdelicato@gmail.com

P. F. Pires
e-mail: paulo.f.pires@gmail.com

A. Y. Zomaya
University of Sydney, Sydney, Australia
e-mail: albert.zomaya@sydney.edu.au

H. M. Ammari (ed.), *The Art of Wireless Sensor Networks*,
Signals and Communication Technology, DOI: 10.1007/978-3-642-40009-4_20,

© Springer-Verlag Berlin Heidelberg 2014

1 Introduction

Wireless sensor networks (WSNs) operate in a highly dynamic and distributed environment. Such networks are becoming increasingly heterogeneous regarding the devices, software technologies and hardware components employed, and they potentially serve the needs of applications from multiple domains. All these features, combined with the fact that developers generally are not networking experts, make the development of applications for sensor networks a challenging task. There are usually integration, scalability, reliability, security, usability, QoS, and operational issues to be considered when building WSN applications. Tacking these challenges is crucial to enable WSN to reach its full potential as a widely distributed sensing infrastructure.

In traditional distributed computing, the use of middleware [1–4] facilitates the work of application developers, unburdening them to deal with the inherent complexity of distribution. Activities such as concurrency control, transactions, data replication, security, and other infrastructure services are performed by a middleware layer placed between applications and the operating system. WSNs can also benefit from the usage of a middleware layer that provides a runtime environment for generic applications and abstracts the complexity of the underlying network infrastructure. Among other functions, WSN middleware can decide the most suitable protocols to be used according to the application requirements; coordinate the nodes' operation in the accomplishment of the application goals; and intelligently manage the use of network resources. To efficiently provide the quality of services required by WSN applications, it is often necessary to interact with the lower levels of the network protocol stack, or even with the hardware devices. The middleware layer can execute such interactions for the benefit of final users or applications. By performing all these functions, a WSN middleware facilitates the task of application developers and network managers.

The design of new generation WSNs often incorporates a middleware layer as a solution to face the challenges involved in building and executing WSN applications. There are different approaches for WSN middleware, concerning the provided abstraction programing to build applications and the provided services to support applications' execution. Chapter 19 of this book presents a comprehensive survey on existing proposals for WSN middleware. All the surveyed approaches provide important functions for different applications such as efficient software configuration and data aggregation. However, more advanced functions are required in the middleware layer as, for instance, the capability to adapt the network behaviour to the ever changing execution context of WSNs and the provision of several levels of interoperability (among different networks, applications and with the Internet). Future WSNs are envisioned as shared systems [5–7], composed of an underlying sensing and communication infrastructure, possibly belonging to multiple owners, providing services to a wide range of applications, from several groups of users. To accommodate the features of this emergent scenario, a new architectural approach is needed, in which application logics and requirements are separated from the data gathering and dissemination functions. In such architecture, the components should be loosely coupled, and accessed through well-defined interfaces. To achieve energy efficiency, applications should be able to dynamically change the network behaviour. However, these changes should be expressed in a high abstraction level, through a common, preferably standardized protocol. Such features will allow the design of WSN to be independent from the applications that will use them.

One of the most recent approaches that provide high flexibility for adding new and advanced functions to WSN middleware is the service-oriented paradigm. Service-oriented middleware (SOM) logically views the WSN as a service provider for consumer applications [8]. The SOM provides abstractions for the complex underlying WSN through a set of generic and/or application-specific services. Services such as data aggregation, adaptation, security, self-organization, reliability, management as well as other advanced services can be designed, implemented, and integrated in an SOM framework to provide a flexible and easy environment to develop effective WSN applications. Moreover, the intrinsically decoupled nature of the various components involved in a service-oriented architecture promotes interoperability between service providers and consumers. The adoption of service oriented approach provides WSN users with a unified protocol to access and communicate with the WSN components and developers with a flexible programing model to build efficient and scalable WSN systems.

In this chapter we will first discuss (Sect. 2) basic concepts of the service oriented approach and the main advantages of adopting such approach in WSN middleware. In Sect. 3, as an aid to fully understanding WSN SOM, we present in details an illustrative example of a middleware system built on the service paradigm. Section 4 concludes the chapter.

2 Overview on Service-Oriented Middleware for WSN

The service oriented paradigm is a promising approach for designing distributed systems which has been considered a compelling candidate to deal with the heterogeneity in the Internet. In recent years it has been adopted in the design of WSN middleware. Service-oriented middleware (SOM) logically views WSNs as service providers for client applications. The distinctive feature of WSN SOM is that the complex underlying WSN is managed through a set of services required by WSN applications which are organized following the Service Oriented Architecture (SOA [9]). SOA is independent of any specific technology and focuses on defining services as autonomous and heterogeneous computational components running on different platforms [10]. The following subsections present basic concepts related to service oriented computing and SOA. We also discuss the application of SOA principles in the building of WSN middleware.

2.1 Background Concepts: SOC and SOA

Service-Oriented Computing (SOC) is the computing paradigm that uses services as underpinning elements for developing distributed applications. Services are self-describing, platform-agnostic computational elements that support fast, low-cost composition of applications [11, 12]. Services may range from simple function requests to complex business processes. SOC enables software vendors to present a software application *as a service*. By adopting SOC, organizations are able to expose their core competencies programmatically over a network (Inter or intra-net) using standard (XML-based) languages and protocols. Services are exposed, accessed and implemented via a self-describing interface-based on open standards [12]. By adopting standardized protocols and languages to describe services and their interactions, SOC facilitates integrating services provided by various vendors, and allows building business applications from the composition of other existing services. SOC successfully addresses problems related to the integration of heterogeneous applications in a distributed environment. Such feature has motivated the use of SOC as a promising programming model for handling node heterogeneity in the Internet and, more recently, in the WSN domain.

According to [12] in order to meet their design requirements, services should be:

- Technology neutral: they must be invocable through standardized lowest common denominator technologies that are available to almost all IT environments. This implies that the invocation mechanisms (protocols, descriptions and discovery mechanisms) should comply with widely accepted standards.
- Loosely coupled: they must not require knowledge or any internal structures or conventions (context) at the client or service side.

Services shoud also support location transparency, meaning that they should have their definitions and location information stored in a repository (the service registry) and be accessible by clients independently of their location.

To build the service model, SOC relies on the Service Oriented Architecture (SOA) [10]. SOA is an architectural design pattern that enables applications to be developed using loosely coupled and interoperable services. The main building block in SOA is the *service*, an entity describable and discoverable through well-defined interfaces. All participants in a SOA revolve around the service, interacting and performing different functions according to their established roles. SOA specifies three roles: the *service provider*, the *service discovery agency*, and the *service requestor* (client). The interactions among them involve the *publish*, *find* and *bind* operations [13] (Fig. 1). Clients are software agents that request the execution of a service. Providers are software agents that provide the service. Agents can be simultaneously both service clients and providers. Providers are responsible for publishing a description of the service(s) they provide. Clients must be able to find the description(s) of the services they require and to bind to them. In a typical service-based scenario a service provider hosts a network accessible software module (an implementation of a given service). The service provider defines a service description and publishes it to a discovery

Fig. 1 SOA roles and operations

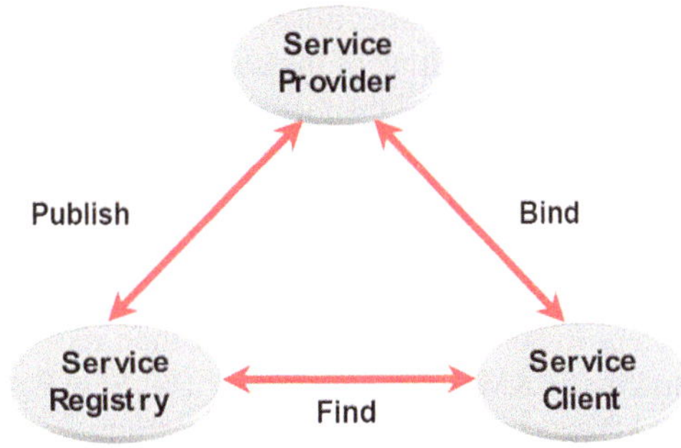

agency through which the service is made discoverable. The service requestor uses a find operation to retrieve the service description typically from a discovery agency and uses the service description to bind with the service provider and invoke the service or interact with the service implementation.

Among other features, according to [12], the SOA approach enables that: (i) multiple applications executing in various platforms can effectively communicate with each other; (ii) software services are provided to end-user applications through published/discoverable interfaces, and (iii) services are loosely coupled, thus reducing dependencies and facilitating reuse.

2.2 Implementation Technologies

SOA only defines the fundaments of service orientation; as a consequence it is necessary the use of a technology that implements SOA principles. The most disseminated technology for service implementation is Web Services technology. Web services can be defined as modular programs, generally independent and self-describing, that can be discovered and invoked across the Internet or an enterprise intranet. In the same way as component-based middleware systems, Web services expose an interface that can be reused without concerning on how the service is implemented. Unlike current component-based middleware [1, 14–16], Web services are not accessed via protocols dependent on a specific object-model. Instead, Web services are accessed via ubiquitous Web protocols and data formats, such as Hypertext Transfer Protocol (HTTP [17]) and XML [18], which are vendor independent.

The *Web Services Description Language* (WSDL) [19, 20] is an XML language to describe the interface of a Web service enabling a program to understand how it can interact with it. Each Web service publishes its interface as a WSDL document that completely specifies the service interface so that clients and client tools can automatically bind to the Web service. A WSDL document defines services as collections of network endpoints or ports [19]. Besides, messages and port types are defined. *Messages* are abstract descriptions of the data being exchanged, and *port types* are abstract collections of operations. In WSDL, there is a separation between the *abstract* definition of messages and their *concrete* network implementation. This allows the reuse of abstract definitions of *messages* and *port types*. The concrete protocol and data format specification for a particular port type defines a reusable *binding*. A *port* is specified by associating a network address with a reusable binding. A *service* is defined as a collection of ports.

SOAP protocol extends XML so that computer programs can easily pass parameters to server applications and then receive and understand the returned semi-structured XML data document. The SOAP specification has four parts [21]. The SOAP *envelope* construct defines a framework for expressing the content of a message. The SOAP *binding framework* defines an abstract framework for exchanging SOAP envelopes between peers using an underlying protocol for transport. The SOAP *encoding rules* defines a serialization mechanism that can be used to exchange instances of application-defined data, arrays, and composite types. SOAP standard communication model is asynchronous, however, it can be mapped to represent more complex communication models such as RPC-like (solicit and response) or broadcast communication.

Since the Web services technology uses XML as the encoding system, data is easily exchanged between computing systems with incompatible architectures and incompatible data formats. WSDL completely describes the Web service interface, while SOAP completely describes parameters, data types and exceptions included in a message being exchanged between Web services. The XML language is very flexible to model services descriptions and messages but it is also rather verbose. For most applications the verbosity of XML is not a problem. WSN applications, however, are constrained by the sensor resources, that are limited in processing power and memory capacities, and, most importantly, have a very slow communications channel available. Therefore, a more compact mechanism for representing the data is needed. One example of such a mechanism is the WAP Binary XML Content Format (WBXML [22]). This format defines a compact binary representation for XML [18], intended to reduce the size of XML documents for transmission and to simplify parsing them. WBXML was designed to be used as part of the WAP protocol [22]. The binary XML content format was designed to allow more effective use of XML data on narrowband communication channels with no loss of functionality or semantic information.

Recently, the REpresentational State Transfer (REST) [23] was introduced as a lighter alternative architectural style to the WS-* technologies. RESTful Web services are built based solely on current standard Web mechanisms; any entity on the Web is identified as a resource at some URI that can be accessed through standard HTTP operations. The maind advantage of REST is its universality and the uniform service interface that leads to a potentially lightweight implementation, since it does not imposes any overhead to interact with a service as it occurs in the case of the SOAP protocol. However, the simplicity of the REST architecture currently provides support to only basic distributed interaction and coordination [24], leaving many open issues that have been tackled by SOC, such as dealing with security, distributed transactions, services composition, and many other features needed to built a complete solution to common problems of a distributed environment.

2.3 SOM for WSN

As we have seen in Chap. 19 of this Book, WSN have requirements and features that hinder the development of applications for these environments. Among these, we can highlight the limited resources of nodes, the high dynamics and heterogeneity of the network. For WSN to realize their full potential of usage by industry and scientific communities, new programming models and frameworks are required to deal with the challenges posed by these environments, and also to allow the applications to make use of resources available in the network at each moment, adapting themselves to the contextual changes. The adoption of a SOM can be a successful design decision to face the challenges involved in building and executing WSN applications.

Briefly stated, Service-Oriented Middleware (SOM) supports the service-oriented interaction pattern through the provision of proper functionalities for deploying, publishing/discovering and accessing services at runtime. SOM also often provides support to realize more complex composite services by integrating simpler ones [25].

By separating the software of a WSN in service consumers and service providers which interact in a flexible way through publish, discovery and bind operations, SOM provides an automated and elegant solution for [26]:

- Facilitating application development from the discovery and composition of services provided by networks of possibly heterogeneous nodes;
- Facilitating coordination of applications in terms of consumption of the services provided by the underlying network infrastructure;
- Facilitating the adaptation of applications, in an energy-efficient manner for the network, in order to handle the highly dynamic execution context.

Recent efforts have targeted the use of SOM for WSN to provide high flexibility for adding new functions to WSN middleware. Unlike other approaches, new functions can be added to the middleware simply by adding new services. These services can be designed, implemented, and easily integrated in the SOM to provide a flexible and easy environment to develop WSN applications.

There are several proposals for WSN SOM presented in the literature. For a comprehensive survey on the subject the reader is referred to [8]. Most of current approaches [27, 28] consider the WSN as a *data provider* for client applications that run outside the network and access the provided resources through a base station (or sink node) that acts as the gateway between the WSN nodes and external systems. Such approaches promote interoperability among heterogeneous nodes but do not leverage the ability of nodes to support in-network processing, thus not fully exploiting their collaboration and self-adaptation capacities. If the WSN nodes are considered not only as data providers but as providers of *services*, including simple processing, decision making and implementation of adaptive behaviour, the use of SOC can bring many additional benefits to the design and implementation of applications, especially concerning the network efficiency and flexibility. An example of this approach is Servilla [29]. Servilla is a service provisioning framework for heterogeneous WSN. Using Servilla, an application can dynamically discover and bind to local and remote services, facilitating in-network collaboration between heterogeneous nodes and achieving high levels of efficiency and flexibility. In Servilla, applications are composed as platform-independent tasks and platform-specific capabilities are exposed as services. Tasks search for services that match their requirements, and dynamically bind to them when they are available. A specialized service description language is adopted that enables tasks to selectively bind to services that exploit the capabilities of whatever hardware is available at a particular time and place [29]. In contrast to Servilla, that adopts a service description language specifically tailored for the system, in the next Section we describe a WSN SOM that is completely built on standardized Web services technologies and protocols. The described middleware will be used as a tool to support the learning of the paradigm presented in this Chapter, in a *learning-by-example* approach.

3 Illustrative Example of WSN SOM

As an illustrative example of a WSN service oriented middleware, in this section we detail the work presented in [30–33]. In the proposed system, the middleware components are organized following the SOA roles and their implementation is based on the Web Service technologies. From an external point of view, applications are *service requestors* and sink nodes are *service providers*. A sink node exposes the descriptions of the services provided by the WSN as a whole and offers access to these services. From an internal point of view, sinks are the *service requestors* and sensor nodes are the *service providers*. Sensors send the descriptions of their services to sink nodes, which keep a repository of the service descriptors of each type of sensor existing in the network. The interfaces of the provided services are physically described by WSDL documents and XML Schema [34] and the messages exchanged between the external and internal components of the system are implemented as SOAP or XML messages.

The main components of the middleware are the **communication module** and **middleware services** (Fig. 2). The communication module is composed of a message router, a set of handlers, a set of XML drivers and a SOAP Proxy. The communication module in sink nodes is based on the SOAP protocol. However, to avoid the overhead incurred by the adoption of SOAP, inside the network all data messages are represented in XML language, encapsulated through a specific, compact format.

SOAP Proxy. The communication between the WSN and external applications occurs through sink nodes. SOAP proxies are programs that translate function calls made in the application programming language to SOAP messages. Conversely,

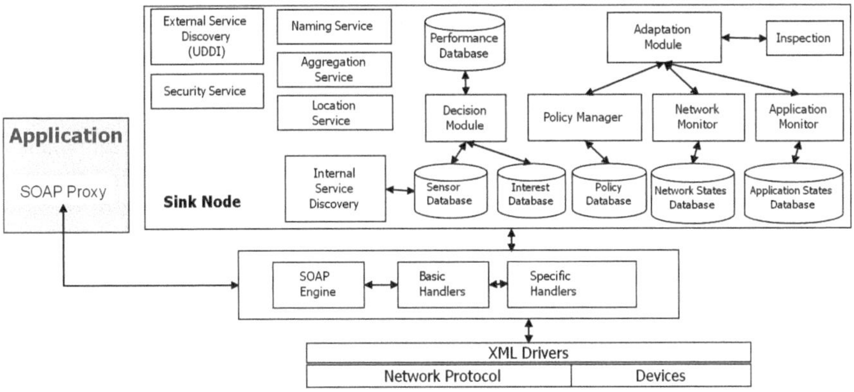

Fig. 2 Middleware Architecture [33]

SOAP reply messages are converted to data and function calls in the application language. SOAP proxies are automatically created from WSDL documents describing the WSN provided services.

XML Drivers. As the middleware can be used in the top of multiple network infrastructures, an abstraction layer, implemented as XML drivers, is provided to represent the underlying behaviour of protocols and devices. XML drivers consist of software modules that translate the middleware messages and commands (defined in XML language) to the protocol language and vice-versa.

Message Router. This component coordinates the flow of SOAP/XML messages through the several handlers of the communication module. Handlers represent the message processing logic and act as dispatchers to the several middleware services. There are a set of basic handlers and a set of specific handlers. Basic handlers are responsible for parsing and composing XML messages, header processing and data type conversions. Specific handlers are defined for each implemented middleware service.

Middleware Services. The basic service provided by a WSN is the delivery of data collected by sensors for client applications. Such delivery depends on the discovery of sensing capabilities available in the network nodes, on the request for data according to the application needs and on how communication occurs between source nodes (data producers) and applications (data consumers). The proposed middleware provides the application with an abstraction of this delivery service so that it can be configured according to the different requirements of each application. To support data delivery, the middleware provides a **discovery service**, based on Web services protocols. Besides the abstraction of the basic service of data delivery, the middleware offers a **data aggregation service** as well as **resource management services**, which encompass (i) a **decision service** to facilitate the application development and (ii) an **inspection and adaptation service** to provide the dynamic behaviour of the system. The decision service is responsible for taking the necessary configuration decisions to meet a particular set of application requirements. Examples of decisions supported by this service are the choice of the best routing protocol and the selection of the subset of nodes to remain active during the execution of the required sensing tasks [35]. The inspection and adaptation service [36] is in charge of monitoring the system state and activating adaptation policies whenever it is necessary.

The middleware services are implemented as modular software components. Therefore, depending on the computational resources of the sensor nodes, as well as on the specifc requirements of the running applications, a different set of services can be installed. Additional generic services as naming, location and security, can be implemented by third parties and incorporated to the middleware architecture. The modular architecture and the use of XML in the internal middleware communication allow the service components to be implemented in any programming language and easily "plugged" in the middleware. This feature provides the middleware architecture with flexibility and extensibility properties, required to accommodate new requirements of emerging applications and WSN scenarios. Section 3.2 describes an extension implemented with the purpose of augmenting the middleware adaptation capability by enhancing the services's descrpttion with semantic information. The next sections present the main steps of the middleware operation.

3.1 System Operation

The basic operation of the described middleware consists of a series of phases that are interleaved with the WSN operation itself.

Service Discovery. Before system start-up, it is necessary to discover the services provided by the WSN. Two levels of service discovery are required: external and internal; each one managed by a specific middleware component. Internal discovery (Fig. 3) starts with an initial configuration phase, during which nodes exchange XML messages describing their services. Configuration messages include: node identifier, timestamp, sensor types, geographical location, residual energy, maximum & minimum confidence degrees, and maximum & minimum data rates. Configuration messages have to reach at least one sink node in the network. A multi-hop protocol is used to route such messages. This phase takes place only once during system start-up, and the protocol being used is not so far the optimized solution for a given application. Sink nodes store the contents of received configuration messages in a local XML-based repository.

The external discovery level is used by applications to find out which WSN supplies the required services, and how to invoke them. The use of SOAP and XML, both part of the Web Services architecture [37], makes the UDDI standard [38] to be a natural choice for service discovery protocol. UDDI is a protocol for communicating with registries. UDDI defines an infrastructure based on XML for the service discovery, and uses SOAP as the protocol to invoke these services. After using UDDI to find the WSN that meets its requirements, the application obtains through the sink node the WSDL document [19] that describes the interface of the services provided by the network.

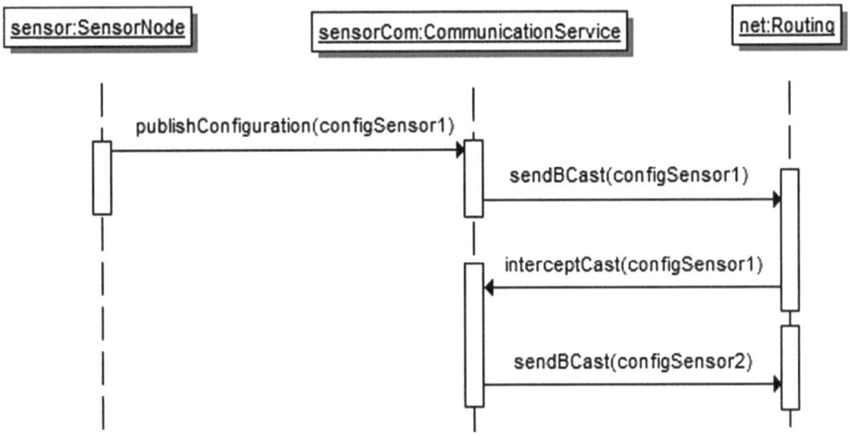

Fig. 3 UML sequence diagram of Internal Discovery Phase

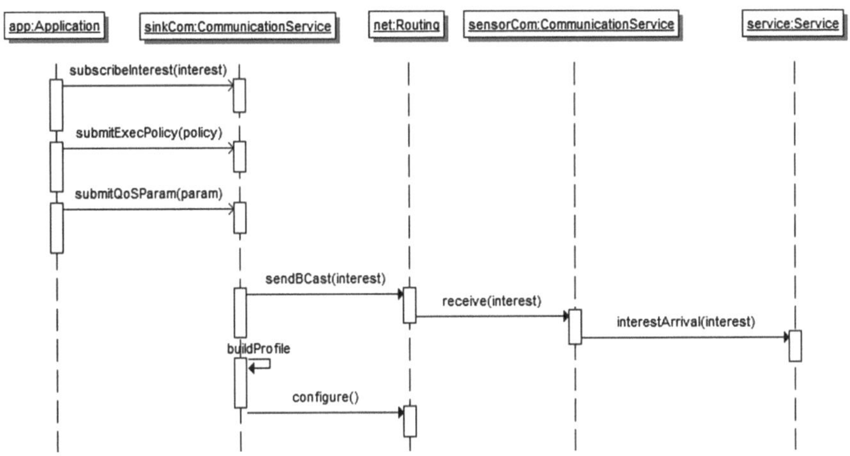

Fig. 4 UML sequence diagram of Interest Submission Phase

Submission of Application Interests and Requirements. The WSDL document provided by the sink node enables the application to learn about the services available in the WSN and how to invoke them. The application can submit its interests (Fig. 4) using different types of SOAP messages for interest advertising [31]. A SOAP message for interest advertising contains the sensor type, the coordinates of the geographical target area, the data acquisition duration and rate, data aggregation functions to be applied on sensed data, among other information. Besides interests, applications can send, for each requested service, a list of execution policies, indicating the QoS parameters to be respected in each execution context, during the delivery of the service.

Network Configuration. The middleware implements an automated decision process to select the best protocol/topology to be used in the network for a given application. Such process is a service provided by the decision service. A decision algorithm [33] is responsible for choosing the best data dissemination strategy, protocol and network topology. Choices are based on the results of previous experiments and simulations. The middleware keeps a database with past information of different configurations and application requirements. Such information is constantly refined when new values are reported by the current applications. After the middleware using the information on the advertising interest SOAP message to discern about network configuration, such message is propagated to the sensor nodes using the underlying data dissemination protocol.

Creation of Application Profile. The application profile is an XML-based data structure which contains: (i) the description of the application requirements in terms of its sensing interests, and (ii) the execution policies for each requested service, i.e., the QoS parameters to be met in each possible execution context. The initial profile is built from information extracted from the SOAP message of interest advertising. The middleware inserts its decisions about protocol and topology configuration in the application profile. Furthermore, the different adaptation policies applied during the execution of the application can be included in the profile.

In the described system, the QoS requirements for the basic service of data delivery can be defined in terms of three parameters: delay, accuracy and data rate. QoS requirements for the aggregation service are defined according to two parameters: the aggregation degree (ratio of the number of received messages to the number of sent messages) and the aggregation delay (delay incurred by the time a sensor node should wait for incoming messages to be aggregated before their transmission). Other services require different sets of QoS parameters.

The execution state is represented by a set of application and system parameters. Application parameters concern the information known by the current application and are defined in terms of the values of sensing data about the monitored phenomenon [39]. System parameters regard information known by the middleware and include: battery level; range/power/transmission rate; and geographic location of nodes.

After creating the application profile, the middleware establishes the current network task and controls the network/sensors as a function of the application QoS and network performance. The middleware interacts with the infrastructure of the subjacent network to keep the context always up to date. Whenever a change in the context occurs, the system verifies in the active application profiles if QoS requirements are being met. If some requirement is not being satisfied, the middleware applies an adaptation policy.

Data Dissemination. When a sensor node generates data, the data is transferred to the communication module, where an XML driver converts it into an XML representation. A specific handler verifies if the data attributes match some aggregation service request. If there is a match, the data is dispatched to the respective aggregation service. The aggregated data is forwarded to the dissemination protocol, as a new XML message of data advertisement. When receiving a packet from another sensor, the dissemination protocol verifies (through a field in the packet header) if the packet contains an application message (data or interest advertising) or an infrastructure (control) message. If it is a control packet, the message is processed by the protocol of data dissemination itself. In case of an application message, it is transferred to the communication module. A basic handler is responsible for verifying the message type (interest or data) and for dispatching it to the specific handler, which process the message and forwarding to the respective services indicated in the message.

System Inspection. The middleware inspection capacity allows an application to request information on the current execution context whenever it wants (reification process [40]). The request is submitted as a SOAP message. From the analysis of the provided information, the application may decide to modify the system behaviour, changing some previously registered QoS parameter or execution policy. The adaptation module keeps a table to register the parameters that each application requests to monitor. To capture any sensible change, monitoring components existent in the sensor nodes periodically check the values of requested parameters.

Adaptation Policies. Adaptation policies are pre-registered in the system as sets of actions to be performed when the QoS requirements established by an application are not being fulfilled, for a given execution context. There are aggressive or conservative policies, and their main goal is to balance the application QoS requirements with the network state, aiming to extend the WSN global lifetime. Examples of defined adaptation policies are: (i) increase the data reliability (data accuracy); (ii) decrease the energy consumption; (iii) increase the available bandwidth. A policy of decreasing the energy consumption may be implemented by two actions: decreasing the data rate and turning off some sensors (considered as providing redundant information to the application). The system, however, must verify if the action to be performed will not damage the minimum level of QoS requested by running applications.

3.2 Semantic Extension of the WSN Middleware

The pervasiveness and the wireless nature of sensor devices require WSN architectures to support *ad hoc* configuration. A key technology of true *ad hoc* networks is service discovery, functionality by which "services" offered by nodes can be described, advertised, and discovered by other devices or applications. Current service discovery and capability description mechanisms are based on *ad hoc* representation schemes and rely on standardization. A crucial requirement for the future, widely accessed WSNs is interoperability under unpredictable conditions, i.e., networks which were not designed for specific, predefined purposes, should be able to be accessed by unforeseen applications, which dynamically discover their functionality and take advantage of it. To enable fully automatized services discover in a highly distributed and heterogeneous systems, semantic information is necessary to augment the syntactic information provided by WSDL and XML schema. An approach that has been successfully adopted to increase the degree of automation in the Web services environment is the Semantic Web [41]. Aiming to provide semantically enriched Web services descriptions, the Semantic Web approach proposes using service ontologies, such as the Ontology Language for Web Services (OWL-S) [42], for augmenting service description. As OWL-S definitions are interpretable by software agents through the employment of inference techniques, Web services described by these ontologies can be automatically discovered and composed.

The tasks involved in the dynamic utilization of WSN services include service discovery and description. The description of a service encompasses information about the sensing task and QoS parameters. Thus, an ontology language is useful to describe the characteristics of WSN devices, their sensing capabilities, and specific information of applications accessing the WSN. In order to endow the proposed middleware with semantic capabilities, new mechanisms were included in its original architecture to acquire, to reason about and to adapt the WSN behavior according to context information. Such capability allows sensor nodes to maintain consistent contextual knowledge and change their behavior according to it. Such knowledge is achieved through sharing context information among the different entities of a WSN system, namely, sensor nodes, applications and infrastructure components. Sensors monitor and periodically send contextual information (for instance the current status of the device and the network connectivity). Applications inspect the current context and eventually change previously stated execution policies. Service components must guarantee that defined QoS parameters are met, and that the current execution context is a valid one. In order to meet this goal, the extended version of the described middleware adopts a set of common ontologies to support the communication among the different entities that comprise the WSN domain. Such ontology also facilitates sharing knowledge among devices from different vendors and among different WSNs.

Enhanced Communication Module. In the implemented extension, the communication module includes, besides a SOAP proxy and XML drivers, a knowledge base and a reasoning engine. The **knowledge base** corresponds to the ontology database. It contains the adopted ontology model, that is, the definitions of the classes and properties created for describing sensor features, execution contexts and policies, application queries and tasks. The full database is implemented only in sink nodes. Sensor nodes keep a sub-part of the ontology definitions needed for representing their own capabilities and information about execution contexts. The **reasoning engine** is a software module responsible for reasoning with ontology knowledge, that is, static knowledge derived from the underlying ontology model. Its function is to decide whether WSN nodes can meet the requirements of a submitted application task.

Enhanced Services. The new extension enhanced the adaptation capabilities of the proposed middleware. In such version, the middleware inspection **and adaptation service** were implemented as two independent modules: (i) the **inspection module**, that allows the application to inspect the network behavior at runtime, exposing a representation of the current execution state; and (ii) the **monitoring and adaptation module**, responsible for monitoring the states of the network and running application(s) and for activating adaptation policies whenever it is necessary or requested. The **monitoring and adaptation module** accesses the local ontology database and, similarly to the communication module, it contains a reasoning engine. This engine is responsible for reasoning with both ontology knowledge and contextual knowledge. Contextual knowledge is a dynamic knowledge that is inferred from situational information reported by sensor nodes. Once that information is available, the module verifies if the WSN execution context at every given moment represents a valid state. Otherwise, a predefined adaptation policy is triggered to repair the network state.

WSN Ontology. We detected three situations in which it would be worthwhile to add semantics in the context of accessing and using WSNs:

- To discover networks that potentially meet the interests of an application, given a high level description of the requested services. The goal here is to find the address (URL) of the access point (sink) of such WSNs, through which applications are able to access and use the WSN services. In this case, the UDDI protocol, used for discovering WSNs, should be added with semantic capacities. Addressing such situation is out of the scope of our work.
- Once a specific WSN has been chosen, to determine if the sensing task requested by the application can be fully accomplished by such network, given the task detailed description, including QoS requirements.
- Once the required task has been initiated, to share knowledge on the execution context, allowing (i) sensor nodes to send information about the network and the application current states; (ii) applications to monitor such state; and (ii) service components to verify if a given execution state is valid and the eventual need of triggering adaptation mechanisms in case of violation of QoS parameters.

We designed a WSN ontology to capture the most relevant features of sensors, execution context and application requirements for the purposes of service discovery and context monitoring. Therefore, we created classes and properties to describe concepts related to the descriptions of sensor node capabilities, application tasks, policies, and execution contexts. For the purpose of service discovery, we defined: (i) three main classes for describing WSN features (*WSN*, *SensorNode* and *SensorField*); and (ii) four main classes for describing application requirements (*Task*, *Query*, *QoSParameters* and *SensorType*). For the purpose of describing execution policies and contexts, and verifying if the current state fits in a valid policy, the main classes created are: *ExecutionContext*, *ApplicationState*, *NetworkState*, *ExecutionPolicy* and *CurrentState*. The defined ontologies are concisely depicted in Figs. 5 and 6. The middleware reasoning engines have a set of rules that allows reasoning based on such ontologies.

The ontologies designed for the WSN domain were defined by using the OWL/RDF language [43]. The Web Ontology Language (OWL) is intended to provide a language to describe the classes and relations between them that are inherent in Web documents and applications. The OWL language can be used (i) to formalize a domain by defining classes and properties of those classes; (ii) to define individuals and assert properties about them, and (iii) to reason about these classes and individuals to the degree permitted by the formal semantics of the OWL language. We will not present the OWL files containing the complete description of ontologies, for lacking of space. The OWL-DL [43] version of the language was used for representing the

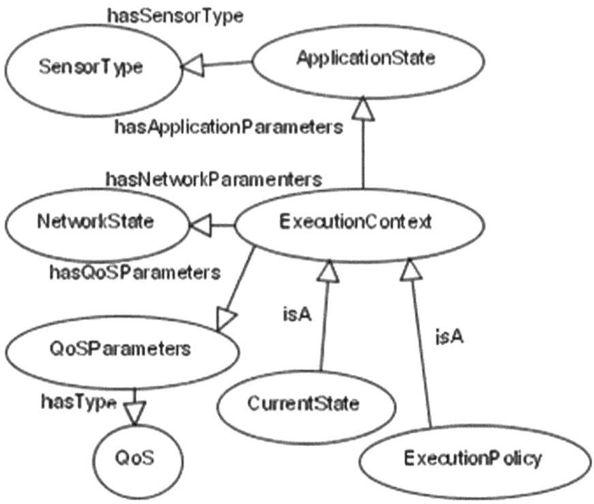

Fig. 5 Main ontologies for execution policies and contexts

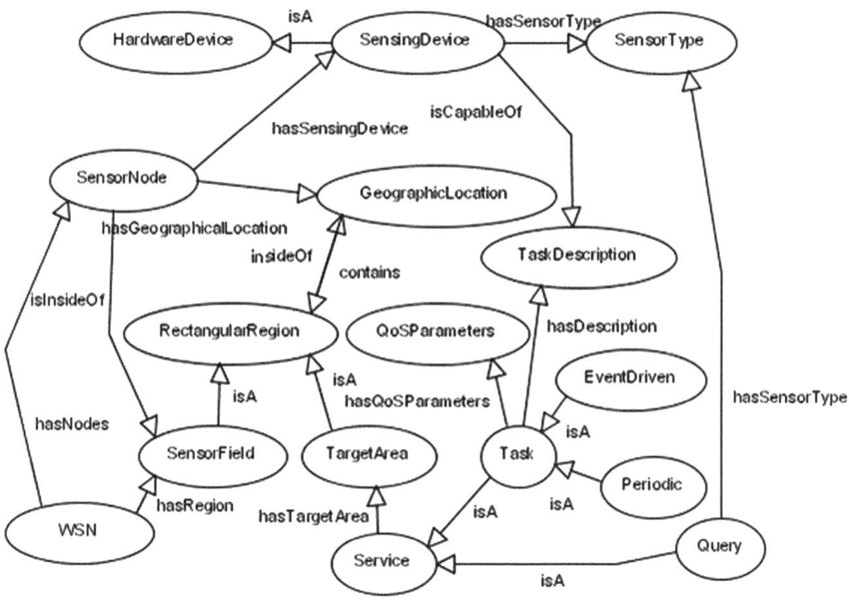

Fig. 6 Main ontologies for tasks and WSN descriptions

ontologies stored on the sink node knowledge base and the OWL-Lite [43] version for ontologies on the sensor node database.

4 Conclusion

Sensor network applications often run on networks composed of hundreds or thousands of nodes spread over a wide geographical area. Sensor nodes have very scarce computational and energy resources, communicate via highly volatile wireless links, and often are exposed to environmental factors such as storms, critical temperatures and humidity, among others. In addition, sensor nodes can have different capacities, be manufactured and operated by different vendors, and be accessed by multiple clients requiring different functionalities from the network. When applying the service oriented paradigm to the WSN system design, the WSN is seen *as a service*, with different granularities: the whole network can be the service provider for client applications; groups of nodes can be providers of composite services; each sensor node can be a provider of simple services; or even a single sensing unit can provide an atomic service for other nodes or applications. The adoption of a service-oriented middleware provides flexibility in designing WSN applications, as it builds on well-known standards for data representation and packaging, for describing the functionality provided by the WSN (as a service), thus facilitating the search for available services that can be invoked to meet application needs. Furthermore, the use of SOM can facilitate coordination of the nodes in performing the sensing tasks, contributing to the efficient use of network resources. The use of SOC has been proven success on the Internet. Its adoption in WSN is still in its infancy but has the potential to leverage the widespread use of these networks, especially in emerging scenarios in which more and more sensor networks will be built as shared infrastructures with multiple owners and serving multiple users. In these scenarios, issues such as interoperability and ubiquitous access to sensor generated data will be even more crucial.

Acknowledgments This work was partially supported by the following Brazilian funding agencies: FAPERJ (**Fundação Carlos Chagas Filho de Amparo à Pesquisa do Estado do Rio de Janeiro**), CNPq (National Council for Scientific and Technological Development), under processes 311363/2011-3 and 470586/2011-7 for Flávia C. Delicato; 480359/2009-1 and 311515/2009-6 for Paulo F. Pires and RNP. We also want to express our gratitude to Professor Habib M. Ammari who kindly invited us to contribute to this Book.

References

1. Microsoft Corporation. Distributed Component Object Model Protocol-DCOM/1.0, draft (Nov 1996), http://www.microsoft.com/Com/resources/comdocs.asp. Accessed June 2012
2. OMG (Object Management Group). *The Common Object Request Broker: Architecture and Specification*. Revision 2.0. (July 1995)
3. OMG Common Object Request Broker Architecture: Core Specification. V. 3.0.3, http://www.omg.org/technology/documents/formal/corba_2.htm. Accessed March 2012
4. SUN Microsystems, Enterprise JavaBeans Specification 2.0, http://java.sun.com/products/ejb/docs.html. Accessed June 2012

5. S. Bhattacharya, A. Saifullah, C. Lu, G.-C. Roman, Multi-application deployment in shared sensor networks based on quality of monitoring. 16th IEEE real-time and embedded technology and applications symposium (2010), pp. 259–268
6. C. Efstratiou, I. Leontiadis, C. Mascolo, J. Crowcroft, A shared sensor network infrastructure, in *Proceedings of the 8th ACM Conference on Embedded Networked Sensor Systems (SenSys '10)*. ACM, (New York, 2010) pp. 367–368. doi: 10.1145/1869983.1870026
7. F. Wu, Y. Kao, Y. Tseng, From wireless sensor networks towards cyber physical systems, Pervasive Mob. Comput. **7**(4), pp. 1574–11 (2011). ISSN 397–413
8. J. Al-Jaroodi, N. Mohamed, Service-oriented middleware: a survey. J. Netw. Comput. Appl. **35**(1), 211–220 (2012)
9. S. Graham et al., Building Web Services with Java: Making Sense of XML, SOAP, WSDL, and UDDI (Sams Publishing, Indianapolis, 2002)
10. OASIS Reference Architecture Foundation for Service Oriented Architecture 1.0, Committee Specification Draft 03, (July 2011)
11. M. P. Papazoglou, P. Traverso, S. Dustdar, F. Leymann, Service-oriented computing: state of the art and research challenges. IEEE Comput. **40**(11), 38–45, IEEE Computer Society, 2007
12. M.P. Papazoglou, Service-oriented computing: concepts, characteristics and directions. Keynote for the 4th international conference on web information systems engineering (WISE 2003), December 10–12, 2003, IEEE CS
13. M. Champion, C. Ferris, E. Newcomer, D. Orchard, Web Services Architecture W3C Working Draft, www.w3.org/TR/ws-arch/, Nov 2002
14. Microsoft Corporation and Digital Equipment Corporation, The Component Object Model Specification, http://www.opengroup.org/pubs/catalog/ax01.htm, Accessed October 1995
15. Microsoft Corporation. Net Framework, http://www.microsoft.com/net. Accessed January 2012
16. SUN Microsystems, Enterprise JavaBeans Specification 2.0. Sun Microsystems, http://java.sun.com/products/ejb/docs.html Accessed August 2001
17. Ietf, RFC 2616, Hypertext Transfer Protocol—HTTP/1.1, http://www.ietf.org/rfc/rfc2616.txt, Accessed June 1999
18. W3C (World Wide Web Consortium) Recommendation, Extensible Markup Language (XML) 1.0 (Second Edition), http://www.w3.org/TR/REC-xml, Accessed October 2000
19. W3C (World Wide Web Consortium) Note, Web Services Description Language (WSDL) Version 2.0 Part 0: Primer, http://www.w3.org/TR/2007/REC-wsdl20-primer-20070626/, Accessed June 2007
20. W3C (World Wide Web Consortium) Note. Web Services Description Language (WSDL) 1.1, http://www.w3.org/TR/2001/NOTE-wsdl-20010315, Accessed March 2001
21. W3C (World Wide Web Consortium), SOAP Version 1.2 Part 0: Primer (Second Edition), http://www.w3.org/TR/2007/REC-soap12-part0-20070427/, Accessed April 2007
22. Wireless Application Protocol Forum: What is WAP and WAP Forum? http://www.wapforum.org /what/index.htm
23. R.T. Fielding, R.N. Taylor, Principled design of the modern Web architecture. ACM Trans. Internet Technol. **2**, 115–150 (2002)
24. C. Pautasso, O. Zimmermann, F. Leymann, Restful web services vs. "big" web services: making the right architectural decision, in 17th international conference on World Wide Web (WWW) (2008)
25. V. Issarny et ai., Service-oriented middleware for the future internet: state of the art and research directions. J. Internet Serv. Appl. **2**(1), 23–45 (2011)
26. C.L. Fok, G.C. Roman, C. Lu, Adaptive service provisioning for enhanced energy efficiency and flexibility in wireless sensor networks, Elesvier Science of Computer Programming, Special Issue on Best Papers of Coordination'10, Available online 29 December 2011, ISSN 0167–6423
27. E. Avilés-López, J. García-Macías, Tinysoa: a service-oriented architecture for wireless sensor networks, Service Oriented Computing and Applications. doi:10.1007/s11761-009-0043-x.

28. N.B. Priyantha, A. Kansal, M. Goraczko, F. Zhao, Tiny web services: design and implementation of interoperable and evolvable sensor networks, in *SenSys '08* (2008), pp. 253–266. doi:10.1145/1460412.1460438
29. C.L. Fok, G.C. Roman, C Lu, Servilla: a flexible service provisioning middleware for heterogeneous sensor networks. Sci. Comput. Program. **77**(6), 663–684 (2012). ISSN 0167–6423
30. F.C. Delicato et al., A service approach for architecting application independent wireless sensor networks. Cluster Comput. **8**(2–3), 211–221 (2005). ISSN: 1386–7857
31. F.C. Delicato et al., A Flexible Middleware System for Wireless Sensor Networks, in *Proceedings of the ACM/IFIP/USENIX International Middleware Conference*, Rio de Janeiro, July 2003
32. F.C. Delicato et al., Reflective Middleware for Wireless Sensor Networks, in *Proceedings of the 20th ACM Symposium on Applied Computing (SAC2005)* (USA, 2005), pp. 1155–1159
33. F. Delicato, P. Pires J. Rezende, L. Pirmez, Service Oriented Middleware for Wireless Sensor Networks. Technical Report NCE04/04, http://www.nce.ufrj.br/labnet/research/networksensors/publications.htm, (2004)
34. W3C (World Wide Web Consortium) Recommendation 28 October 2004 [Online], XML Schema Part 0: Primer Second Edition, http://www.w3.org/TR/xmlschema-0/. Último acesso: 24/05/2005
35. F.C. Delicato et al., Application-Driven Node Management in Multihop Wireless Sensor Networks, in *Proceedings of the 4th IEEE International Conference on Networking*, Reunion Island, April 2005
36. F. C. Delicato et al., An efficient heuristic for selecting active nodes in wireless sensor networks. Comput. Netw. **50**, 3701–3720, Elsevier Science, ISSN: 1389–1286, 2006
37. Web Services Architecture, W3C Recommendation 2004 [cited 2009 August]; Available from: http://www.w3.org/TR/ws-arch/
38. OASIS Spec Technical Committee Draft, UDDI Version 3.0.2, http://uddi.org/pubs/uddi_v3.htm 2004
39. W. Heinzelman, A. Murphy, H. Carvalho et al., Middleware to support sensor network applications. IEEE Netw. Mag. Spec. Issue **18**(1), 6–14 (2004)
40. L. Capra, W. Emmerich, C. Mascolo, Carisma: context-aware reflective mIddleware system for mobile applications. IEEE Trans. Software Eng. **29**(10), 929–945 (2003)
41. T. Berners-Lee, J. Hendler, O. Lassila, The Semantic Web. Scientific American Magazine (May 17, 2001)
42. OWL-S Coalition, OWL-S 1.1 Release, http://www.daml.org/services/owl-s/1.1/
43. W3C Recommendation, OWL Web Ontology Language (February 10, 2004); http://www.w3.org/TR/owl-guide/

Part X
Sensor Technology, Standards, and Operating Systems

Chapter 21
System Architecture and Operating Systems

Yanjun Yao, Lipeng Wan and Qing Cao

Abstract The emergence of resource constrained embedded systems such as sensor networks have introduced unique challenges for the design and implementation of operating systems. In OS designs for these systems, only partial functionality is required compared to conventional ones, as their code is running on a much more restricted and homogeneous platform. In fact, as illustrated by microcontrollers, most hardware platforms in wireless sensor networks (WSNs) simply do not have the required resources to support a full-fledged operating system. Instead, operating systems for WSNs should adapt to their unique properties, which motivate the design and development of a range of unique operating systems for WSNs in recent years. In this chapter, we systematically survey these operating systems, compare them in their unique designs, and provide our insights on their strengths and weaknesses. We hope that such an approach is helpful for the reader to get a clear view of recent developments of wireless sensor network operating systems.

1 Introduction

The traditional roles of an operating system include the management and protection of system resources for different users, in addition to providing programming and execution support for concurrent applications. Residing between the hardware and applications, operating systems are known for their complexity: the Linux operating

Y. Yao (✉) · L. Wan · Q. Cao
Department of Electrical Engineering and Computer Science, University of Tennessee, Knoxville, TN, USA
e-mail: yyao9@utk.edu

L. Wan
e-mail: lwan1@utk.edu

Q. Cao
e-mail: cao@utk.edu

H. M. Ammari (ed.), *The Art of Wireless Sensor Networks*,
Signals and Communication Technology, DOI: 10.1007/978-3-642-40009-4_21,
© Springer-Verlag Berlin Heidelberg 2014

system as of 2012 contains around 15 million lines of code [1], whereas Windows XP reportedly has 45 million [2]. Such complexity is necessary to implement the wide range of tasks that operating systems are involved in, such as task scheduling, memory protection, storage management, among others.

For resource constrained embedded systems such as sensor networks, however, only partial functionality is required for their operating systems compared to conventional ones, as their code is running on much more restricted and homogeneous platforms. In fact, as represented by microcontrollers, most hardware platforms in WSNs simply do not have the required resources to support a full-fledged operating system. Furthermore, operating systems for WSNs should support their unique requirements, such as energy management, sensor sampling, and low-power communication. These requirements motivate the design and development of a wide range of unique operating systems for WSNs, which will be our main topics of discussion in this chapter.

Specifically, this chapter is organized as follows. In the remaining of this section, we describe the critical tasks performed by sensor network operating systems. In the next section, we describe a variety of hardware platforms. In Sects. 3–5, we describe three representative operating systems for sensor networks, including TinyOS [3], Contiki [4], and LiteOS [5]. We briefly describe several more operating systems, and compare them from different perspectives in Sect. 6. Finally, we conclude in Sect. 7.

1.1 Kernel Scheduler

The operating system kernel serves the central functionality of scheduling tasks, through which it selects the next job to be admitted into the system. There are multiple types of schedulers, such as first-in-first-out (FIFO), shortest-job-first (SJF), priority-based scheduling, round-robin scheduling, and multilevel queue scheduling. Each of these scheduling policies makes different tradeoffs between CPU overhead, throughput, turnaround time, and response time. Detailed treatment on this topic can be found in textbooks such as [6].

For WSNs, the types of commonly found scheduling policies are more limited due to resource constraints. Different operating systems can be categorized according to (i) what types of entities are being scheduled, and (ii) what are the scheduling policies used. For example, the TinyOS operating system uses a streamlined FIFO scheduler for tasks, where these tasks are special types of function pointers stored in a task queue. These tasks are envisioned to be long-running functions that are posted by triggering events. Different from TinyOS, another operating system, SOS [7], focuses on messages, which are asynchronous and behave like tasks in TinyOS. These messages serve as communication mechanisms between system modules. SOS maintains a high priority queue for time-critical messages, such as those from ADC interrupts and a limited set of timers, and a low priority queue for other, non-time-critical messages. Finally, some operating systems provide scheduling for multiple types of entities. The LiteOS operating system, for example, provides scheduling support for both threads and tasks, and allows different policies to be implemented for each entity: a priority-driven scheduler is provided to support tasks, while a round-robin scheduler is provided to allow multiple threads to share the CPU without starvation.

Which type of scheduling is the best? We don't believe that there is a specific answer to this question. The different implementations of scheduling, however, do have a significant impact on the way programs are written and executed. We leave the detailed discussions on the impact of scheduling on programming models to our descriptions of individual operating systems.

1.2 Programming Model

In this section, we describe those programming models offered by operating systems, that is, how developers write programs based on different system level APIs. One of the fundamental problems faced by programming models is how they handle concurrency. Concurrency is challenging in sensor networks. This is because applications written for them are fundamentally asynchronous: a radio message event, for example, could come in at any time. One of the simplest ways to handle this is by polling the radio to decide whether a packet has arrived. Such a simple model suffers, however, from excessive CPU usage and the risks of missing data when the packets are being processed. Clearly, better concurrency models are needed to build efficient and reliable applications.

We can categorize existing operating systems for sensor networks into two broad domains in terms of their concurrency models: event driven and thread driven. Some operating systems provide compatible APIs for both types of models, such as the 2.x versions of TinyOS. In this chapter, we primarily consider the 1.x version of TinyOS to illustrate the idea of event-driven programming model.

1.2.1 Event-Driven Programming Model

The concept of the event-driven programming model is conceptually simple: the system handles an event using interrupts, where an event can be triggered by both hardware and software sources. For example, they could be an event indicating data availability from a sensor, a packet arrival event, or a timer firing event. Such an event is then handled by a short sequence of instructions that only carries out the very basic activities, for example, storing the content of the packet or a sensor's value into a local buffer. The actual processing of these data is not done in these event handler routines, but is handled separately and decoupled from the actual occurrences of events. These handling routines are usually referred to as long-running processing tasks. Unlike events, which can interrupt at any time, tasks are only handled in a specific order (FIFO in TinyOS), which means that later tasks cannot preempt the earlier tasks.

The reasoning behind such design choices is profound: using short and simple event handling sequences can avoid their interference with normal code execution. Event handlers typically do not interrupt each other (as this would complicate stack handling procedures), but are simply executed one after each other. As a result, this event-based programming model creates two different "contexts": one for the time-critical event handlers and one for normal instruction flows. Interestingly, programmers usually find such a programming methodology quite complicated, especially for larger scale applications: reasoning about the application usually involves formal methods in finite state machines (FSM), which are implicitly embedded in event-driven applications.

Although difficult to write, the event-based programming model comes with the advantage that it saves context switch overhead: only one thread of execution is continually running with its own stack. Therefore, such models typically have lower memory footprint and less runtime overhead compared to thread-based models [5].

1.2.2 Thread-Driven Programming Model

Because of the complexity of adopting the event-based programming model, most modern operating systems support concurrent execution of multiple processes or threads on a single CPU. On sensor networks, such a model was originally challenged: as the early papers on TinyOS illustrated [3], the concern of overhead in context switches makes this model less preferable on resource constrained devices, hence motivating the event-driven design. Later work, however, challenged this argument by showing that multiple threads can indeed be created on the resource-constrained platforms with only a few kilo bytes of RAM space. Representative systems following this philosophy include the Mantis operating system in 2007 [8], the LiteOS operating system in 2008 [5], and the TinyOS 2.x version in 2009 [9]. The critical advantage of thread-based model is that it greatly simplifies software development: using threads, the programmer can easily reason about the execution sequences of the programs, leading to more logical organizations of software components.

1.3 Protocol Stack

One conventional way for developing communication protocol stacks, e.g., the protocol stack of the Internet, is to provide a layered architecture of protocols, where one protocol resides on top of another to provide desired functionalities. The key advantage of this approach is that it allows each layer to be developed separately and independently of its higher or lower layer protocols. Therefore, as long as one layer's interface remains the same, its implementation can be flexible. The layered approach has been shown to be highly effective in conventional wired network environments, where two models are widely popular: the OSI seven-layer model and the TCP/IP four-layer model.

However, for WSNs, such a layered approach is less preferable. One primary reason is the need for cross-layer optimization to achieve the best performance under resource constraints. For example, in a sensor network that collects real-time sensor data, its sampling period should not be fixed due to frequently changing radio channel conditions. Instead, the sampling period should be adjusted dynamically according to the real-time link quality. In this case, the application layer (data collection) exploits the routing layer (link quality) information to achieve the best packet delivery performance.

Because of their differences with conventional networks, protocol stacks for sensor networks tend to use separate models. In the TinyOS operating system, a component-based model is proposed. Specifically, TinyOS adopts a holistic approach, where a separate programming language called nesC [10] is developed to facilitate this component-based model. These components are *wired* together to fulfill the global functionality, and they interact with each other over clearly-defined interfaces. The main difference compared to the layered architecture is that such a component-based model no longer maintains hierarchical relations between components. Therefore, an application layer component can easily invoke the interface provided by a routing layer component. Another advantage of this design is that such a component based architecture fits well with the event-driven programming model of TinyOS: functions provided by hardware or software can all be conveniently designed and implemented as self-contained components to improve system maintainability and modularity.

1.4 Storage Management

The storage space is a precious resource in sensor networks mostly due to energy consumption concerns. Storage management has conventionally been considered as one important goal of operating systems, as demonstrated by the design of various file systems and storage services [11, 12]. On sensor nodes, typically flash storage is provided to store sensor data temporarily before they are transmitted back to the base station. Therefore, the management of storage space determines how data are managed during the execution time of the system.

The operating system for sensor networks provides varying interfaces for storage management. Earlier operating systems such as TinyOS, Contiki and SOS provide direct access to the on-board flash memory by allowing users to read and write specific blocks directly. Empirical studies, however, found such an approach to be too low-level for most practical needs of applications. Later operating systems such as Mantis and LiteOS provide reliable file system interfaces that are similar to Unix. Such design choices make it much easier to develop large-scale storage-intensive applications.

Perhaps one of the interesting trends in sensor networks is that storage management can also be implemented as totally separate components compared to the host operating systems. For example, the ELF file system [13] was implemented on the TinyOS operating system, yet reflects a separation of concerns by serving as an extension to the original functionalities of the host OS. Some other storage-oriented services, such as EnviroMic [12], focuses on additional storage issues such as data redundancy. Finally, some other work presents support for data management over additional hardware, such as flash cards, whose capacity may scale up to a few gigabytes [14].

1.5 Additional Requirements

In this section, we briefly introduce a few more cross-cutting requirements in sensor networks. These requirements include memory management, energy conservation, real-time support, reliability, and portability.

1.5.1 Memory Management

On resource-constrained sensor nodes, memory space is very limited. Therefore, most sensor network operating systems do not provide dedicated memory management. TinyOS, for example, assumes that everything is statically allocated at compile time, and does not support dynamic memory allocation. The same design principle is followed in several following operating systems such as Contiki and Mantis. Some operating systems believe otherwise, such as SOS and LiteOS, which provide dynamic memory allocation in the form of memory pools. In SOS, such support is needed to facilitate its dynamic messaging system, while in LiteOS, a Unix-like **malloc** API is provided for user applications, which is especially useful on motes such as IRIS nodes [15] where more RAM space is available.

1.5.2 Energy Conservation

Most operating systems for sensor networks mention energy conservation as one of the fundamental goals in their system designs. However, in our survey of representative operating systems, we find that energy conservation has usually been associated with the application layer or the networking layer, but not with the operating system itself. This explains why only a few primitives on energy conservation are provided in the OS, such as turning the CPU into low-power mode, or turning the radio into sleeping mode, etc. Indeed, if the OS decides to take aggressive energy conservation measures, the applications layer will be affected no matter it is intended or not. Therefore, most operating systems allow application layers to develop their own energy saving protocols (e.g., the tripwire system in VigilNet [16]), while the kernel only provides basic primitives through its APIs.

1.5.3 Real-Time Support

Real-time support has been less considered by operating system designs. On the other hand, many sensor network applications such as surveillance and environmental monitoring are time-sensitive in nature. To support such applications, one operating system, Nano-RK [17], provides a reservation-based real-time operating system for use in WSNs. Specifically, Nano-RK supports fixed-priority preemptive multitasking for guaranteeing that task deadlines are met, along with support for CPU and network bandwidth reservations. During runtime, tasks can specify their resource demands, so that the operating system can provide timely, guaranteed and controlled access to CPU cycles and network packets in resource-constrained embedded sensor environments.

1.5.4 Reliability

Due to the difficulty of debugging WSNs, reliability has recently emerged as a critical problem. How to address reliability in the operating system layer is still an active area of research. Interestingly, although the original papers on the initial versions of the current operating systems did not mention much on reliability, follow-up publications filled these gaps by addressing specifically system reliability. For example, on top of TinyOS alone, the following debugging and reliability tools have been provided: EnviroLog [18], Neutron [19], t-kernel [19], among others.

1.5.5 Portability

Portability refers to the ability of the operating systems to be easily ported across different platforms. We provide a survey of the different types of hardware in the next section. Note that, however, not all platforms are supported by one single operating system. Among the operating systems we surveyed, TinyOS and Contiki are the two that support the widest range of devices. On the other hand, we are optimistic on the portability of the remaining operating systems, as we consider this to be mostly an engineering problem: given that most of these operating systems are written in C or variants of C, porting them across system platforms will not be too challenging.

2 Hardware Platforms and Architecture

In this section, we concentrate on introducing the hardware components of a sensor node, and the characteristics of different commercial motes.

2.1 Sensor Node Components

A sensor node, also called mote, is an autonomous device working in WSNs. It is capable of processing information, gathering sensed data and communicating with other sensor nodes. A typical sensor node consists of the following hardware components:

- **Microcontroller**: With the characteristics of being low-cost and low-power, a microcontroller performs data processing and controls the functionality of other components in a sensor node.
- **Transceiver**: A transceiver is composed of both a transmitter and a receiver. The transceiver is in charge of sending and receiving packets to or from other sensor nodes in the network. It usually has multiple operational states, such as transmit, receive, idle, and sleep, where each state consumes different amount of energy.
- **Sensor Boards**: To measure the physical conditions, sensor boards are either integrated into or installed separately on the microcontroller boards. Usually, a sensor node is equipped with passive sensors that sense the environment without affecting them. Within the coverage area of sensors, the observed event can usually be reliably and accurately reported.
- **Power Source**: The power source provides energy to the sensor node. The most common power source is battery. Some sensor nodes are able to gather additional energy at runtime from sources such as solar energy, temperature differences, or vibrations. When connected to the PC, sensor nodes can draw power through USB connectors.
- **External Memory**: Some sensor nodes are equipped with flash memories, which are used to expand their storage capacity. The external memory can be used to store application related data or programs.

2.2 Commercial Sensor Nodes

Since the concept of SmartDust was proposed a decade ago [20], many companies have developed different sensor nodes. In this section, we present an overview of several commercial sensor nodes as shown in Table 1. These nodes include Mica [21], Mica2 [22], MicaZ [23], EPIC [24], IRIS [15], Sun SPOT [25], LOTUS [26], TelosB [27], Cricket [28], Waspmote [29], and Imote2 [30].

2.3 Comparison of Sensor Nodes

In this section, we compare the characteristics of microcontrollers, transceivers, sensors and memories on the above-mentioned sensors.

Table 1 Properties of popular commercial sensor nodes

Sensor node	Microcontroller	Transceiver	Storage	Vendor	Year
Mica	Atmel ATmega103	RFM TR1000 radio	512 KB	Crossbow	2001
Mica2	Atmel Atmega128L	Chipcon CC1000	512 KB	Crossbow	2002
MicaZ	Atmel Atmega128L	TI CC2420 802.15.4 radio	512 KB	Crossbow	2002
EPIC	TI MSP430	TI CC2420 802.15.4 radio	2 MB	UC Berkeley	2007
IRIS	Atmel ATmega1281	Atmel AT86RF23 radio	512 KB	Crossbow	2007
Sun SPOT	Atmel AT91RM920T	TI CC2420 802.15.4 radio	None	Oracle	2006
LOTUS	Cortex M3 10–100 MHz	RF231 radio	64 MB	MEMSIC	2011
TelosB	TI MSP430	TI CC2420 802.15.4 radio	1024 KB	Crossbow	2004
Cricket	Atmel ATmega128L	Chipcon CC1000	512 KB	MEMSIC	2007
Waspmote	Atmel ATmega1281	ZigBee/802.15.4/ DigiMesh/RF, 2.4 GHz/868/900 MHz	2 GB	Libelium	2009
Imote2	Intel PXA271 XScale	TI CC2420 802.15.4 radio	32 MB	Crossbow	2008

2.3.1 Microcontroller and Memory

Table 2 reviews the microcontroller specifications for each model mentioned in the previous section. The speed of the microcontroller is closely related to the sizes of bus, clock, RAM, EEPROM and flash. Most of the motes are equipped with a microcontroller consisting of 8-bit bus, and 8 MHz clock, and its memory contains 4 KB of RAM, 4 KB of EEPROM, and 128 KB of flash. These limited resources are sufficient to deal with ordinary tasks, while more resources are provided for special ones.

Most of the microcontrollers, which operate on only one frequency, do not support dynamic voltage scaling (DVS). As adjusting the frequency of the processor is one of the most effective methods of optimizing power consumption, microcontrollers operate on multiple CPU frequencies are developed. For example, the Intel PXA271 XScale supports switching between four CPU frequencies as 13, 104, 208, and 416 MHz to balance performance and power efficiency.

Table 2 Microcontroller specification comparison

Microcontroller	Bus	Clock (MHz)	RAM	EEPROM	Flash
ATmega103	8-bit	4	4 KB	4 KB	128 KB
Atmega128L	8-bit	8	4 KB	512 KB	128 KB
TI MSP430 microcontroller	16-bit	4–8	10 KB	None	48 KB
ATmega1281	8-bit	8	8 KB	4 KB	128 KB
Atmel AT91RM9200	32-bit	180	512 KB	None	4 MB
Cortex M3	32-bit	10–100	64 KB SRAM	None	512 KB + 64 MB
Intel PXA271 XScale	16/32-bit	13–416	32 MB	None	32 MB

Table 3 Radio specification comparison

Radio model	Frequency band (MHz)	Outdoor range	Current draw
CC1000	300–1000	500–1000 ft	Transmit: 7.4 mA, receive: 10.4 mA, idle: 74 μA, sleep: 0.2 μA
CC2420	2400–2483	75–100 m	Transmit: 11–17.4 mA, receive: 19.7 mA, idle: 20 μA, sleep: 1 μA
CC2480	2405–2480	>300 m	Transmit: 27 mA, receive: 27 mA, sleep: 0.3–190 μA
Atmel AT86RF230	2400–2483.5	100 m	Transmit: 16.5 mA, receive: 15.5 mA, sleep: 20 μA

2.3.2 Radio Transceivers

The radio transceiver properties of each mote are as shown in Table 3. Most of the motes are equipped with 802.15.4 transceivers with slightly different characteristics on frequency band and outdoor range. One important difference is that these models support different number of power states, and each of the power states consumes various amount of power.

2.3.3 Sensor Support

There are several types of sensors for detecting humidity, temperature, light, pressure, acceleration/seismic, acoustic, magnetic, video, vibration and other types of environment conditions. Motes are equipped with sensors in two ways, on-board sensors or plug-in sensors. Most of motes, such as Mica, Mica2, MicaZ, IRIS and so on, provide no on-board sensors. Instead, an expansion connector is provided for installing external sensor boards. As shown in Table 4, just a few sensor nodes, such as TelosB, provide on-board sensors.

Table 4 Sensor specification comparison

Sensor node name	Sensors
Mica/Sun SPOT	Light, temperature, acceleration/seismic sensors
MicaZ/Mica2/EPIC/	Expansion connector for light, temperature, barometric pressure, acceleration/seismic, acoustic, magnetic and other sensor boards
IRIS/LOTUS/Cricket	Expansion connector for light, temperature, barometric pressure, acceleration/seismic, acoustic, magnetic and other sensor boards
TelosB	Optional integrated temperature, light and humidity sensor
Waspmote	Expansion connector for gas, temperature, liquid level, weight, pressure, humidity, luminosity, accelerometer, soil moisture, solar radiation

3 The TinyOS Operating System

3.1 Introduction

TinyOS is a well-known and widely-used open source operating system designed for wireless sensor networks. It started as a project at UC Berkeley as part of the DARPA NEST program. Since it was made available to public in 2000, TinyOS has attracted thousands of academic and commercial developers and users worldwide.

3.2 System Overview

There are four requirements of wireless sensor networks that motivate the design of TinyOS:

- **Limited Resources**: Generally, the sensor nodes or motes have very limited hardware resources to achieve the goals of small size, low cost, and low power consumption.
- **Reactive Concurrency**: In wireless sensor networks, a sensor node must be able to respond to many types of events, such as sending and receiving packets. Therefore, a concurrency management that reduces potential bugs while satisfying resource and timing constraints is required.
- **Flexibility**: The variety of hardware and applications in sensor networks implies that the operating system for sensor nodes must be flexible. In addition, the operating system for sensor nodes should also support fine-grained modularity so that it can be customized and reused.
- **Low-Power Operations**: To ensure prolonged operation lifetime, it is crucial to improve the energy efficiency of motes. Therefore, the low power operation is an important goal in the operating system design.

To achieve these four requirements, the design of TinyOS concentrates on two basic principles: the event-driven programming model and the component-based system architecture. Specifically, TinyOS supports an event-driven concurrency model which consists of split-phase interfaces, asynchronous events, and tasks. TinyOS is implemented in the nesC programming language, a dialect of C, which supports the TinyOS concurrency model, as well as mechanisms for combining different software components together to form robust network embedded systems. The nesC programming language is designed to allow application developers to easily build different components and link them together to construct complete, concurrent systems. This allows developers and users to enjoy enough flexibility to customize the system based on hardware and applications.

TinyOS is composed of a tiny scheduler and a graph of components. Each component is an independent computational unit that has one or more interfaces. These interfaces, which are bi-directional, provide the only point of access to the component. There are three computational abstractions for components: commands, events, and tasks.

Commands and events are designed for inter-component communication, while tasks are used to demonstrate intra-component concurrency. Typically, a command is a request sent to a component to perform some service, such as reading data from a sensor, while an event signals the completion of that service. Hardware interrupts or packet arrivals can also signal the events asynchronously. The command returns immediately while the event signals the completion of the service some time later.

Commands and events may post a task, which is a function that will be executed by TinyOS scheduler at a later time, rather than being executed immediately. Tasks are atomic with respect to each other and run to completion, though they can be preempted by events. Tasks can be used to call lower level commands, signal higher level events, and schedule other tasks which are within a component. The run-to-completion execution model of tasks makes it possible to allocate a single stack that is assigned to the currently executing task, which allows tasks to be much more lightweight compared to threads. A non-preemptive, FIFO scheduling policy is used in the standard TinyOS task scheduler.

3.3 Event-Based Concurrency Model

In wireless sensor network applications, fine-grained concurrency is needed, because events can arrive at any time and must interact with ongoing computations. There are two approaches to solve this problem: (1) queuing the events on their arrival so that they could be executed later, as in most message-passing systems [31], and (2) executing an event handler immediately, also referred to as the active message approach [32]. TinyOS chooses the latter approach because some of those events are time critical.

The execution model of TinyOS is event-driven which consists of run-to-completion tasks which represent the ongoing computation, and interrupt handlers that are invoked asynchronously by hardware. A program uses the `post` operator to submit a task to the scheduler for execution, and all tasks run to completion and the other tasks will not preempt the running task. Waiting tasks in the task queue will be executed in the order of their arrivals. In other words, tasks are atomic with respect to each other. However, tasks are not atomic with respect to interrupt handlers, or the commands and events invoked by the interrupt handlers.

On one hand, the non-preemptive TinyOS concurrency model can provide programmers a simple way to deal with the race conditions, deadlocks and other concurrency-related issues. On the other hand, it also can cause some serious problems in many applications when long running tasks are involved. In these problems, even priority tasks may be delayed, which will reduce the responsiveness of the system significantly. The concurrency model of TinyOS makes TinyOS most suitable for developing simple, non-time-critical applications. Programmers may even can split large tasks into smaller pieces to handle the non-preemptive tasks problem, so that the maximum delay for incoming priority tasks will be reduced. However, as the computational capabilities of sensor nodes increase and their applications become more complex, this concurrency model may not be suitable in some scenarios.

3.4 Component-Based System Architecture

Another important concept of TinyOS's programming model is the so called components which encapsulate a set of specific services, which are specified by interfaces. A general component structure is shown in Fig. 1. The TinyOS component contains four interrelated parts: a set of command handlers, a set of event handlers, a fixed-size frame, and a set of simple tasks. Tasks, command and event handlers run in the context of the frame and operate on its state.

A component has two classes of interfaces: one consists of those the component provides and the other consists of those the component uses. Figure 2 shows a simplified form of the TimerM component (TimerM component is a part of the TinyOS timer service), which provides the StdControl and Timer interfaces and uses a Clock interface. Interfaces, which are bi-directional, contain both commands and events. The providers of an interface implement the commands while the users of an interface implement the events. For instance, as shown in Fig. 3, the timer interface defines `start()`, `stop()` commands and a `fired()` event. The `start()` and `stop()` commands are implemented in the component that provides this interface, while the `fired()` event is implemented in the component that uses this interface.

In nesC, there are two types of components: modules and configurations. As shown in Fig. 2, modules are written in a dialect of C and used to call and implement commands and events. A module declares private state variables and data buffers,

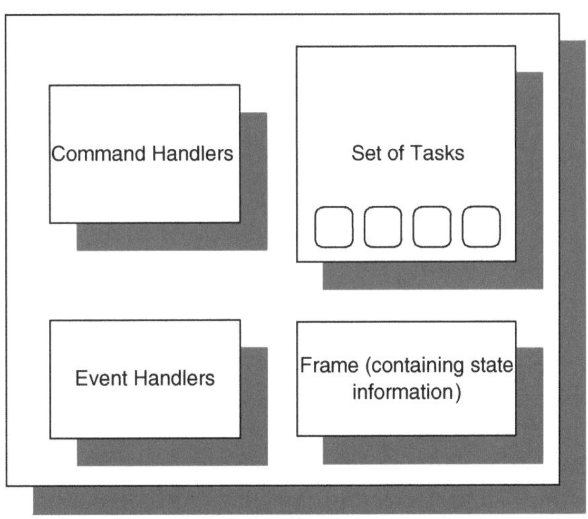

Fig. 1 The structure of a TinyOS component [33]

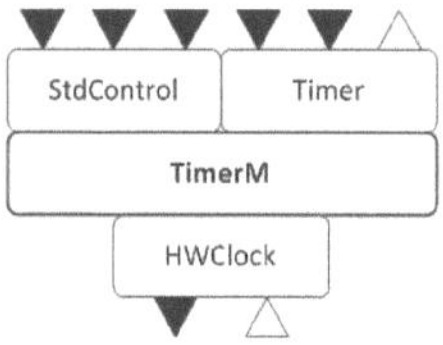


```
module TimerM{
    provides {
       interface StdControl;
       interface Timer[uint8_t id];
    }
    uses interface Clock;
}
Implementation {
    ... a dialect of c ...
}
```

Fig. 2 The TimerM component [3]

```
interface StdControl {
  command result_t init();
  command result_t start();
  command result_t stop();
}

interface Timer {
  command result_t start(char type, uint32_t interval);
  command result_t stop();
  event result_t fired();
}

interface Clock {
  command result_t setRate(char interval, char scale);
  event result_t fire();
}

Interface SendMsg {
  command result_t send(uint16_t address,
                        uint8_t length,
                        TOS_MsgPtr msg);
  event result_t sendDone(TOS_MsgPtr msg,
                          result_t success);
}
```

Fig. 3 TinyOS interface example [3]

which can only be referenced by itself. Configurations are used to wire different components together through interfaces of these components. For instance, as shown in Fig. 4, the TinyOS timer service is a configuration (TimerC) that ties the timer module (TimerM) and the hardware clock component (HWClock) together. Multiple components can be aggregated together into a single "supercomponent" through configurations.

nesC not only provides the component programming model, but also imposes some limitations on C to improve the efficiency and robustness of the source code. First, the function pointers are prohibited in nesC which allows the compiler to know the call graph of a program precisely. The benefit of this is cross-component optimizations for entire call paths become possible, which can help remove the

Fig. 4 TimerC configuration [3]

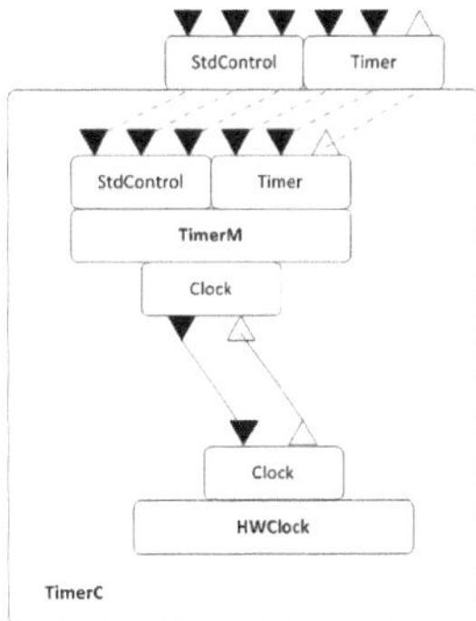


```
configuration TimerC {
   provides {
      interface StdControl;
      interface Timer[uint8_t id];
   }
}
Implementation {
   components TimerM, HWClock;

      StdControl = TimerM.StdControl;
      Timer = TimerM.Timer;

      TimerM.Clk -> HWClock.Clock;
}
```

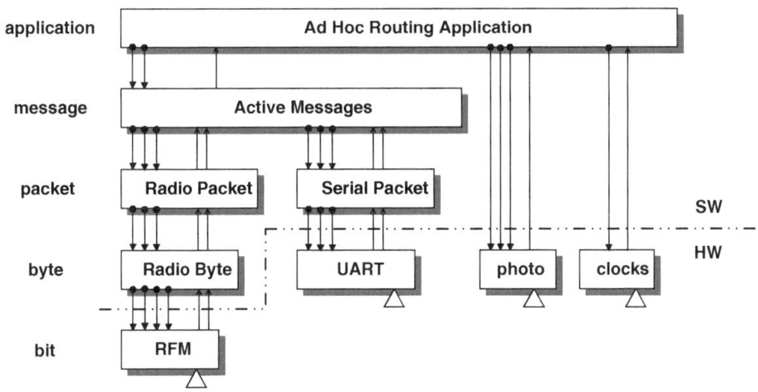

Fig. 5 Ad hoc networking application component graph [32]

overhead of cross-module calls and the inline code for small components into its callers. Second, nesC does not support dynamic memory allocation, which prevents memory fragmentation and runtime allocation failures.

3.5 Networking Architecture

The communication and networking architecture of TinyOS is based on Active Messages (AM), which is a simple, extensible paradigm for message-based communications. Each Active Message consists of two parts: the name of a user-level handler which will be invoked on a target node and a data payload which will be passed as arguments. Specifically, in TinyOS, the packets of the active message are 36 bytes long and the handler ID is 1 byte. As an active message is received, the node dispatches the message to one or more handlers that are registered to receive messages of that type. Figure 5 shows the complete component graph of an ad hoc networking application which is based on the active message model in TinyOS.

The components underlying the packet level is used to transmit the block of bytes out over the radio. The interface of the packet level component provides a mechanism

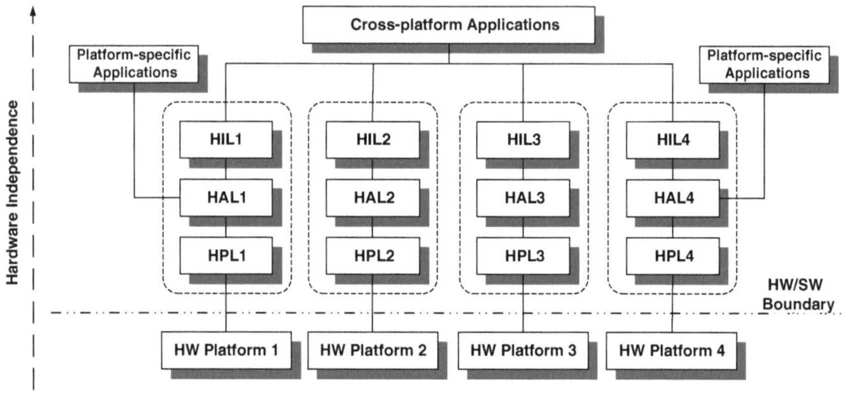

Fig. 6 Hardware abstraction architecture for TinyOS 2.x [34]

to initiate the transmission, and when a transmission or a reception is complete two events will be fired. In Fig. 5 we can see AM not only provides an unreliable, single-hop datagram protocol, but also a unified communication interface to the radio and the serial port. The application level protocols which are built on top of the AM interface provide multi-hop communication, large ADUs, or other features.

3.6 New Features in TinyOS 2.x

TinyOS 2.x can be regarded as a clean slate redesign and re-implementation of TinyOS. The motivation for developing TinyOS 2.x is to eliminate several limitations in TinyOS 1.x which make 1.x hard to meet nowadays requirements and uses. For instance, the fundamental limitations of the structure and interfaces defined in TinyOS 1.x, cause the following problems: components are tightly coupled, interactions are hard to find, and it is difficult for a newcomer to learn sensor network programming quickly. The main modifications and new features of TinyOS 2.x lie in the following aspects.

3.6.1 Hardware Abstraction

To implement the hardware abstractions, TinyOS 2.x follows a three-level abstraction hierarchy, called the HAA (Hardware Abstraction Architecture) shown in Fig. 6.

At the bottom of the HAA is the HPL (Hardware Presentation Layer), which is a thin software layer on top of the raw hardware. HPL is used to present hardware such as IO pins or registers as nesC interfaces. Generally there is no state in the HPL besides the hardware itself, which has no variables.

At the middle of the HAA is the HAL (Hardware Abstraction Layer), which is built on top of the HPL. HAL provides higher-level abstractions that are easier to use than the HPL. Though HAL provides high-level abstractions of hardware, it still provides the full functionality of the underlying hardware.

At the top of the HAA is the HIL (Hardware Independent Layer), which is built on top of the HAL. HIL provides abstractions that are hardware independent, which means that the HIL usually does not provide all of the functionality that the HAL does.

3.6.2 Scheduler

TinyOS 2.x scheduler also has a non-preemptive FIFO policy, but tasks in 2.x operate slightly differently compare to those in 1.x. In TinyOS 2.x, tasks can have their own reserved slots in the task queue, and a task can only be posted once. A post fails if and only if the task has already been posted. If a component needs to post a task multiple times, an internal state variable can be set in the component, which means when the task executes, it can repost itself.

This slight change in semantics significantly simplifies the component code, because a component can just post the task instead of testing to see if a task has already been posted. Components do not need to try to recover from failed posts with retries.

3.6.3 Multi-Thread Support

TinyOS 1.x does not provide any multi-thread support, so that application development should strictly follow the event driven programming model. Since version 2.1, TinyOS provides support for multi-thread and these TinyOS threads are called TOS Threads [35]. The TOS thread package, which is backward compatible with existing TinyOS code, provides a thread-based programming model which is compatible with an event-driven kernel. In TinyOS, though application-level threads cannot preempt tasks and interrupt handlers, they can preempt other application-level threads. The TinyOS scheduler is run by a high priority kernel thread which is dedicated to this task. Control Blocks (TCB) are dynamically allocated by TOS threads Thread with space for a fixed size stack that does not grow overtime. Context switches and system calls of TOS thread introduce an overhead which is less than 0.92 % [35].

3.6.4 Booting/Initialization

TinyOS 2.x has a different boot sequence than TinyOS 1.x. The interface StdControl in TinyOS 1.x has been split into two interfaces `Init` and `StdControl`. The latter one only has two commands `start` and `stop`. In TinyOS 1.x, components will be powered up and started at boot if they are wired to the boot sequence. However, in TinyOS 2.x this case will not happen, because the boot sequence only initializes components. When boot sequence finishes initializing the scheduler, hardware, and software, it will signal the `Boot.booted` event. This event will be handled by the top-level application component which will start services accordingly.

3.6.5 Virtualization

The concept of a "generic" or instantiable component, which is introduced by nesC 1.2, allows TinyOS 2.x to have reusable data structures, like bit vectors and queues, which can make application development more simple and efficient. Furthermore, with the help of generic configurations, services can encapsulate complex wiring relationships for clients that need them. This means many basic TinyOS services now can be virtualized. A program can instantiate a service component that provides the needed interface rather than wiring to a component with a parameterized interface (e.g., GenericComm or TimerC in 1.x).

3.7 Implementation and Hardware Support

TinyOS and its applications are very small, which are suitable for those platforms where hardware resources are often limited. TinyOS can run on a wide range of hardware platforms, including EPIC, Imote2, Shimmer, IRIS, Kmote (Telos Rev B), Micaz, Mica2, Mica2Dot, NXTMOTE (TinyOS on LEGO MINDSTORMS NXT), Mulle, TMote Sky (Telos Rev B), TinyNode, Zolertia Z1, UCMini, among others. Supported microcontrollers include Atmel AT90L-series, Atmel ATmega-series, Texas Instruments MSP-series and Intel XScale PXA271. The details of these hardware platforms can be found in Sect. 2.

4 Contiki: A Lightweight and Flexible Operating System

4.1 Introduction

Developed by the Swedish Institute of Computer Science, Contiki is a lightweight and flexible operating system for WSNs. Contiki provides dynamic loading and unloading of individual components and services. Although it implements an event-driven kernel, it also supports preemptive multi-threading, which is implemented as a library that is linked only with programs that explicitly require multi-threading support.

In the following subsections, we present the system overview, kernel architecture, and key features of the Contiki operating system.

4.2 System Overview

Contiki is an operating system developed for sensor motes with constrained resources. A running Contiki system is composed of four components: the kernel, the libraries, the program loader, and a set of processes. As shown in Fig. 7, the memory of Contiki is separated into two parts at the compile time: the core and the loaded program. The Contiki kernel, the program loader, the language run-time, support libraries, and a communication stack with device drivers for the communication hardware are saved in the core. On the other hand, programs are loaded into the system by program loaders.

Prior to the deployment of the system, the core is compiled into a single binary image and stored in the memory. After the deployment, the core is not modified, unless a special boot loader is used to overwrite or patch the core. In the program loader, programs binaries may be obtained either by using the communication stack, or by using external storage such as EEPROM. In most cases, programs loaded into the system are first stored in EEPROM before being programmed into the code memory.

Saved in the core, the kernel controls the communication between processes. This functionality is not implemented by providing a hardware abstraction layer as is the case with TinyOS, but by letting device drivers and applications communicate directly in the hardware. A process, which is controlled by kernel, can be either an application program or a service. As mentioned above, the programs are loaded by the program loader. On the other hand, a service is a set of application processes working together to implement a specific functionality. Both the application programs and services can be dynamically replaced in-time. A process is defined by an event handler function and an optional poll handler function, where the event handler is an asynchronous callback subroutine that handles inputs received in a program, and a poll handler specifies an action when a process is polled. The state of the process is kept in its private memory and the kernel only keeps a pointer to the process state.

More specifically, the kernel provides a lightweight event scheduler and a polling mechanism. The event scheduler is in charge of dispatching events to run processes and periodically calls processes' polling handlers, which specifies the action of the

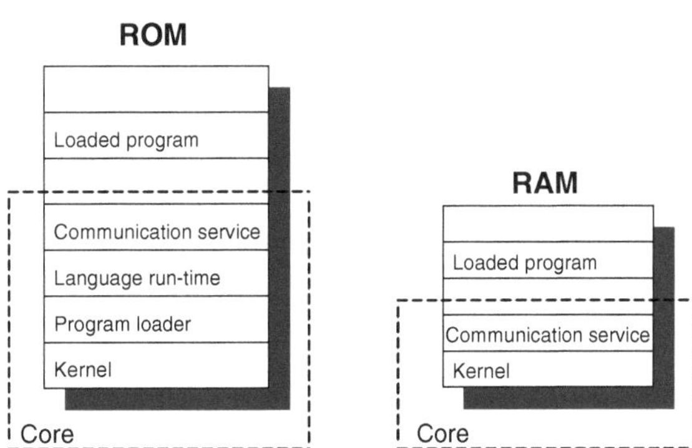

Fig. 7 Contiki as partitioned into core and loaded programs [4]

polled process. Once an event handler is scheduled, the kernel cannot preempt it. In that case, event handlers must run to completion, if no internal mechanisms are used to achieve preemption. On the other hand, the polling mechanism specifies high priority events that are scheduled in-between each asynchronous event. Polling mechanism is used by processes that operate near the hardware to check for status updates of hardware devices. All processes that implement a poll handler are called in order of their priority, when a poll is scheduled.

Contiki uses a two-level scheduling hierarchy to implement the event preemption. First, in Contiki, all event scheduling is done at a single level and events cannot preempt each other, unless there is an interrupt. An interrupt can be implemented by using hardware interrupts or underlying real-time execution support. In Contiki, the interrupts are never disabled. In that case, Contiki does not allow events to be posted by interrupt handlers to avoid race conditions in the event handler. Instead, a polling flag, which provides interrupt handlers with a way to request immediate polling, is provided in the kernel to request a poll event.

The events, which trigger the execution of programs, can be dispatched by the kernel or through the polling mechanism. There are two types of events supported in Contiki kernel, asynchronous events and synchronous events. Asynchronous events are a form of deferred procedure calls, which are enqueued by the kernel and are dispatched to the target process some time later. The use of asynchronous events reduce stack space requirement, because the stack is rewound between each invocation of event handler. However, in order to implement an immediate schedule of target processes, synchronous events are supported, too.

4.3 Key Features

As we mentioned above, Contiki is a lightweight event-driven operating systems that supports both preemptive multi-threading and dynamic loading and unloading of individual components and services. In this section, we will introduce how Contiki implements those features in detail.

4.3.1 Easy Event-Driven Programming

In general, Contiki is an event-driven operating system. This is because event-driven systems do not need to allocate memory for per-thread stacks, which leads to lower memory requirements. As an operating system designed for sensor motes with constrained resource, it is important to keep the system to be lightweight.

However, as we discussed earlier, event-based programming is typically complicated, as an event-driven model does not support blocking wait, an abstraction that is usually desired to express the program logic flows. In that case, programmers of such systems frequently need to use state machines to implement control flow for high-level logic that cannot be expressed as a single event handler. To cope with this problem, a novel programming abstraction called Protothreads [36] is proposed for Contiki. Protothreads, which provide a conditional blocking wait operation, can be used to reduce the number of explicit state machines in event-driven programs, and make it possible to write event-driven programs in a thread-like style with a memory overhead of only two bytes per protothread.

From the view of implementation, in Contiki operating system, processes are implemented as protothreads running on top of the event-driven kernel. The protothreads mechanism does not specify how to invoke or schedule a protothread. Instead of that, the system using the protothread defines all these. In general, the protothread of a process is invoked when the process receives an event, such as a message from another process, a timer, or a sensor input. The blocking of the thread is executed while the process receives an event using the protothread conditional blocking statements. We can treat protothreads as a combination of events and threads. From threads, blocking wait semantics have been inherited by protothreads. From events, protothreads have inherited the stacklessness and the low memory overhead. The linear sequencing of statements in event-driven programs is supported by the blocking wait semantics. As a protothread does not require its own stack, protothreads are very lightweight compare to traditional threads. Besides, all protothreads run on the same stack, and context switching is done by stack rewinding. These two features make protothread work better on memory constrained systems.

More specifically speaking, protothreads can be used to replace state machines. The steps of doing this are as following. In general, the control flow of a state machine can be decomposed to three primitive patterns: sequences, iterations and selections as shown in Fig. 8. All these three patterns can be easily mapped to protothreads as shown in Fig. 9. In that case, to rewrite an event driven state machine with protothreads, we just need to first find the patterns in the state machine, and then translate the patterns to corresponding protothreads.

Based on the pseudocode in Fig. 9, we have already had a peak on how to program on protothread. Let us have look at the details on this in the following part. In general, there are a set of APIs been provided by Protothreads:

1. **PT_BEGIN(pt)**: begins the protothread.
2. **PT_END(pt)**: ends the protothread.
3. **PT_INIT(pt)**: initializes the protothread.
4. **PT_EXIT(pt)**: exits from protothread.

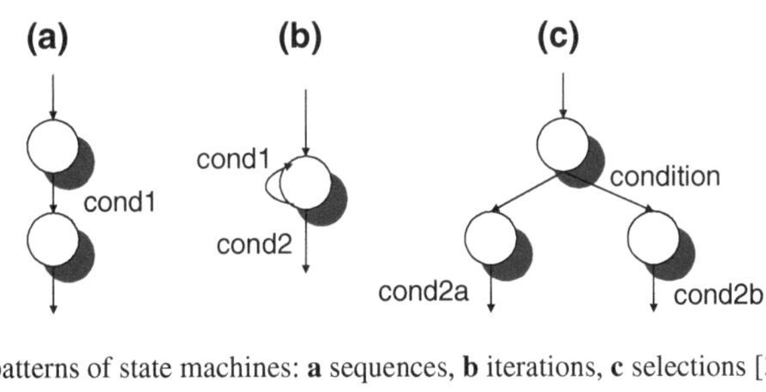

Fig. 8 Three patterns of state machines: **a** sequences, **b** iterations, **c** selections [36]

```
a_sequence:
    PT_BEGIN
    (*...*)
    PT_WAIT_UNTIL(cond1)
    (*...*)
    PT_END
```

```
an_iteration:
    PT_BEGIN
    (*...*)
    while (cond1)
        PT_WAIT_UNTIL (cond1
                       or cond2)
    (*...*)
    PT_END
```

```
a_selection:
    PT_BEGIN
    (*...*)
    if(condition)
        PT_WAIT_UNTIL(cond2a)
    else
        PT_WAIT_UNTIL(cond2b)
    (*...*)
    PT_END
```

Fig. 9 Pseudocode of protothreads for different state machine transitions [36]

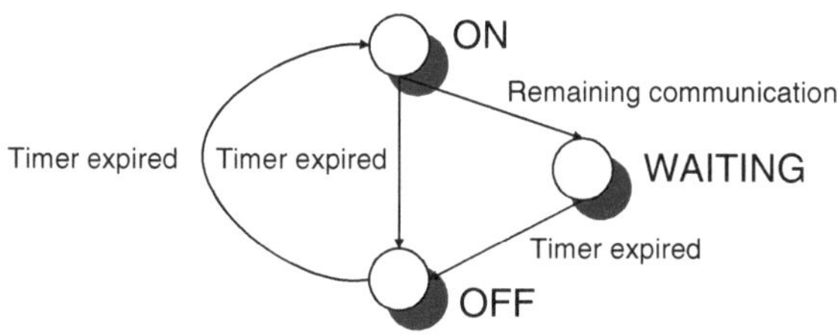

Fig. 10 State machine of the hypothetical MAC protocol [36]

5. **PT_WAIT_UNTIL(pt, condition)**: the operation takes a conditional statement and blocks the protothread until the statement evaluates to be true.
6. **PT_YIELD(pt)**: performs a single unconditional blocking wait that temporarily blocks the protothread until the next time the protothread is invoked.
7. **PT_SPAWN(pt)**: initializes a child protothread and blocks the current protothread until the child protothread is either ended with PT_END or exited with PT_EXIT.

The functionality of each API is as following. The beginning and end of a protothread are declared with PT_BEGIN and PT_END statements. Other protothread statements, such as PT_WAIT_UNTIL, must be placed between PT_BEGIN and PT_END. A protothread can exit prematurely with a PT_EXIT statement. A protothread is stackless, because instead of having a history of function invocations, all protothreads in a system run on the same stack that can be rewound. In that case, protothread can only block at the top level of the function. In other words, it is impossible for regular function called from a protothread to block inside the called function, unless an explicit PT_WIAT_UNTIL() statement is used. As a result of that, the programmer should always be aware of which statements may block the thread.

Let us take the code for hypothetical MAC protocol as an example to see how to translate the code of state machines to protothreads. In general, the hypothetical MAC works as following: turn on the radio at t_0; wait until $t = t_0 + t_{awake}$; turn radio off, if all communication has completed; if communication has not completed, wait until it has completed or $t = t_0 + t_{awake} + t_{wait_{max}}$; turn radio off, and wait until $t = t_0 + t_{awake} + t_{sleep}$; start all over again.

```
State:{ON,WAITING, OFF}

radio_wake_eventhandler:
  if (state=ON)
    if (expired(timer))
      timer ← t_sleep
    if (not communication_complete())
      state ← WAITING
      wait_timer ← t_wait_max
    else
      radio_off()
      state ← OFF
  elseif (state=WAITING)
    if (communication_complete() or
        expired(wait_timer))
      state ← OFF
      radio_off()
  elseif (state=OFF)
    if (expired(timer))
      radio_on()
      state ← ON
      timer ← t_awake
```

```
radio_wake_protothread:
  PT_BEGIN
  while (true)
    radio_on()
    timer ← t_awake
    PT_WAIT_UNTIL(expired(timer))
    timer ← t_sleep
    if (not communication_complete())
      wait_timer ← t_wait_max
      PT_WAIT_UNTIL(
                communication_complete()
                   or expired(wait_timer))
    radio_off()
    PT_WAIT_UNTIL(expired(timer))
  PT_END
```

Fig. 11 Pseudocode of the hypothetical MAC protocol with states (*left*) and Protothreads (*right*) [36]

The above steps can be generalized as the state machine shown in Fig. 10. On the other hand, the pseudocode for the state machine and the protothreads are shown in Fig. 11. From the pseudocode, we can see that protothreads not only make the programming simple, but also considerably decrease the code size.

4.3.2 Preemptive Multi-Threading

As mentioned above, Contiki is an event-driven operating system that supports preemptive multi-threading. In this section, we will show how preemptive multi-threading is implemented in details.

In general, preemptive multi-threading is implemented on top of the event-based kernel as a library that can be linked on demand. More specifically, only applications that explicitly require a multi-threaded model will be linked with this library.

The multi-threading in Contiki is designed as following: each thread requires a separate stack, and threads execute on their own stack until they explicitly yield or are preempted. In that case, the multi-threading library is divided into two parts: a platform independent part that provides interfaces to the event-based kernel, and a platform specific part which implements the stack switching and preemption primitives between threads. The preemption is implemented by using a timer interrupt that saves the processor registers onto the stack and switches back to the kernel stack.

As a loadable library, a set of stack management functions are provided. We hide the details of the internal code of each function, and only show the APIs of the library here as following:

1. **mt_yield()**: yield from the running thread.
2. **mt_post(id, event, dataptr)**: post an event from the running thread.
3. **mt_wait(event, dataptr)**: wait for an event to be posted to the running thread.
4. **mt_exit()**: exit the running thread.
5. **mt_start(thread, functionptr, dataptr)**: start a thread with a specified function call.
6. **mt_exec(thread)**: execute the specified thread until it yields or is preempted.

In the APIs above, `mt_yield()`, `mt_post()`, `mt_wait()`, and `mt_exit()` can be called from a running thread. On the other hand, `mt_start()` and `mt_exec()` are called to set up and run a thread. The actual scheduling of a thread is performed in `mt_exec()`, which needs to be called from an event handler.

4.3.3 Run-Time Dynamic Linking

The third feature of Contiki is dynamic loading and unloading of individual components and services. This is because of the following reason. In a variety range of situations, it is necessary to update system software for sensor networks. In most of these scenarios, just a small part of the system needs to be modified. Without support for dynamic loading and unloading of system components, it is costly to perform those changes. In that case, being able to dynamically reprogram parts of the sensor network system, instead of reloading the whole system image, helps shorten the development cycle time distinctly.

The Contiki system uses loadable modules to perform dynamic reprogramming. With loadable modules, only parts of system need to be modified when a single program is changed. It works as following: the native machine code of the program that is loaded into the system, which holds references to functions or variables in the system, is contained in a loadable module. The references must be resolved to the physical address of the functions or variables before the machine code can be executed. The process of resolving the references is called linking. There are two ways to link modules, the pre-linked modules and the dynamically linked modules. The first way is done when the module is compiled. It contains the absolute physical addresses of the referenced functions or variables. On the other hand, the second way is done when the module is loaded, and it contains the symbolic names of all system core functions or variables that are referenced in the module.

The original Contiki [4] uses pre-linked binary modules for dynamic loading. Compare to dynamically linked modules, pre-linked modules have two benefits: first, pre-linked modules are smaller than dynamically linked modules, hence less information needs to be transmitted. Second, the process of loading a pre-linked module into the system is less complex than the process of linking a dynamically linked module. In that case, the pre-linked method has smaller run-time overhead than dynamically linked method. However, the pre-linked module needs to keep the absolute addresses of all functions and variables that are referenced by the module, which means that these physical addresses are hard-coded. Therefore, the pre-linked

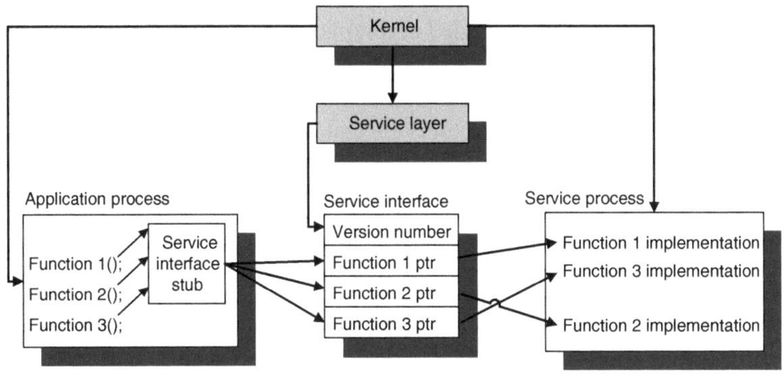

Fig. 12 How an application process calls a service [4]

module can only be loaded into the system with the exact same physical addresses as the original system.

In the later versions of Contiki, a dynamic linker is proposed and implemented [37]. The dynamic linker is designed to link, relocate, and load the standard object files, such as Executable and Linkable Format (ELF), Compact ELF (CELF), and so on. It takes four steps to link, relocate and load the object files: first, the dynamic linker parses the object files and extracts relevant information about where in the object files the code, data, symbol table, and relocation entries are stored. Secondly, memory for the code and data is allocated in flash ROM and RAM, respectively. After that, the code and data segments are linked and relocated to their respective memory locations. Finally, the code is written to flash ROM and the data to RAM.

4.3.4 Callable Services

To make the dynamic loading and unloading of a set of applications easier, Contiki proposed the fourth feature, callable services. In Contiki, the functionalities that the application processes may call are implemented as services. A service can be seen as a form of shared library. The most typical services include communication protocol stacks, sensor device drivers, and higher level functionalities such as sensor data handling algorithms. In Contiki, services can be dynamically replaced and dynamically linked at run-time.

As shown in Fig. 12, services are managed by a *service layer*, which sits next to the kernel. The service layer keeps track of running services and finds installed services. Services can be identified by strings of text that describe them. In that case, the installed services are searched with those strings. A typical service consists of a service interface and a process that implements the interface. The service interface consists of a version number and a function table, which holds function pointers to the functions implementing the interface.

Services can be dynamically loaded and replaced at run-time. When a service is to be replaced, the kernel informs the running version of the service by posting a special event to the service process. With the help of the kernel, a pointer to the new service process is provided, and the service can produce a state description that is then passed to the new process. Besides that, a version number of the new service is tagged to the service state description, so that only one compatible version of the same service will be loaded. Meanwhile, the old service removes itself from the system.

4.4 Network Protocols

Besides the basic services provided by the operating system, two extended components on network protocols have been proposed for Contiki. In this section, we introduce them in details.

4.4.1 ContikiRPL

RPL (Routing Protocol for Low-power and Lossy Networks) is a standard protocol for IPv6 routing over Low-power, Lossy Networks (LLNs). Optimized for many-to-one traffic pattern, RPL is designed for networks with significantly higher packet loss rates than traditional IP routing protocols. The network topology of RPL is a Destination-Oriented Directed Acyclic Graph (DODAG) rooted at a border router, where all nodes in the network maintain a route to the router.

In Contiki, a variant of RPL, called ContikiRPL [38], is proposed. ContikiRPL sets up forwarding tables only and leaves the actual packet forwarding to other network layers, instead of making forwarding decisions per packet. ContikiRPL gets link cost estimated by an external neighbor information module, and recomputes the path cost to the sink via the updated link, and checks if there is a need to change preferred parents on the forwarding table. The link cost is computed as an exponentially weighted moving average function over the number of link layer transmissions, and the neighbor with smallest cost will be kept in the forwarding table.

4.4.2 ContikiMAC

Contiki group proposed a MAC layer protocol called ContikiMAC [39], which uses a power-efficient wake-up mechanism with a set of timing constraints to allow devices to keep their transceivers off. In general, ContikiMAC is a simple asynchronous protocol, which needs no signaling messages and additional packet headers.

ContikiMAC is a radio duty cycling protocol that uses periodical wake-ups to listen for packet transmissions from the neighbors. For a sender, it repeatedly sends a packet, until it gets the acknowledgment from the receiver. For a receiver, it periodically wakes up to check if there is a packet sent to it. If the receiver detects a transmission to it, it should stay awake to receive the full packet and transmit a link layer acknowledgment. For other nodes that detect the packet, they go to sleep immediately. The sender can learn the wake-up time of a receiver based on the time it gets the acknowledgment. In that case, more power can be saved while the sender keeps on sleeping until the time receiver wakes up.

4.5 Implementation and Hardware Support

The code size of Contiki, which is tightly related with the provided services, is larger than the code size of TinyOS. As a set of services are provided, Contiki's event kernel is significantly larger than that of TinyOS. Compared to TinyOS, whose event kernel only provides a FIFO event queue scheduler, the Contiki kernel is more complex, as both FIFO events and poll handlers with priorities are supported. On the other hand, the dynamic loading and unloading property of Contiki requires more run-time code than TinyOS.

Contiki is ported onto a number of architectures, such as Texas Instruments MSP430, Atmel AVR, Hitachi SH3 and the Zilog Z80. In that case, it can run in a large range of hardware platforms, such as Mica, Mica2, MicaZ, EPIC, IRIS, Sun SPOT, LOTUS, TelosB, Cricket, and Waspmote mentioned in Sect. 2.

5 The LiteOS Operating System

5.1 Introduction

The LiteOS operating system is a multi-threaded operating system that provides Unix-like abstractions for WSNs. Aiming to be an easy-to-use platform, LiteOS offers a number of novel features, including: (1) a hierarchical file system and a wireless shell interface for user interaction using UNIX-like commands; (2) kernel support for dynamic loading and native execution of multithreaded applications; and (3) online debugging, dynamic memory, and file system assisted communication stacks. LiteOS also supports software updates through a separation between the kernel and user applications, which are bridged through a suite of system calls.

The key contribution of LiteOS is that it presents a familiar, Unix-like abstraction for WSNs by leveraging the likely existing knowledge that common system programmers already have: Unix, threads, and C. By mapping sensor networks to file directories, it allows applying user-friendly operations, such as file directory commands, to sensor networks, therefore reducing the learning curve for operating and programming sensor networks.

5.2 System Overview

In this section, we first present an overview of the LiteOS operating system, and then present its three subsystems.

5.3 Architectural Overview

Figure 13 shows the overall architecture of the LiteOS operating system, partitioned into three subsystems: LiteShell, LiteFS, and the kernel. Implemented on the base station PC side, the LiteShell subsystem interacts with sensor nodes only when a user is present. Therefore, LiteShell and LiteFS are connected with a dashed line in this figure.

LiteOS provides a wireless *node mounting mechanism* (to use a UNIX term) through a file system called LiteFS. Much like connecting a USB drive, a LiteOS node mounts itself wirelessly to the root filesystem of a nearby base station. Moreover, analogously to connecting a USB device (which implies that the device has to be less than a USB-cable-length away), the wireless mount currently supports devices within wireless range. The mount mechanism comes handy, for example, in the lab, when a developer might want to interact temporarily with a set of nodes on a table-top before deployment. Ideally, a network mount would allow mounting a device as long as a network path existed either via the Internet or via multi-hop wireless communication through the sensor network. Once mounted, a LiteOS node looks like a *file directory* from the base station. A sensor network, therefore, maps into a higher level directory composed of node-mapped directories. The shell, called LiteShell, supports UNIX commands, such as copy (cp), executed on such directories.

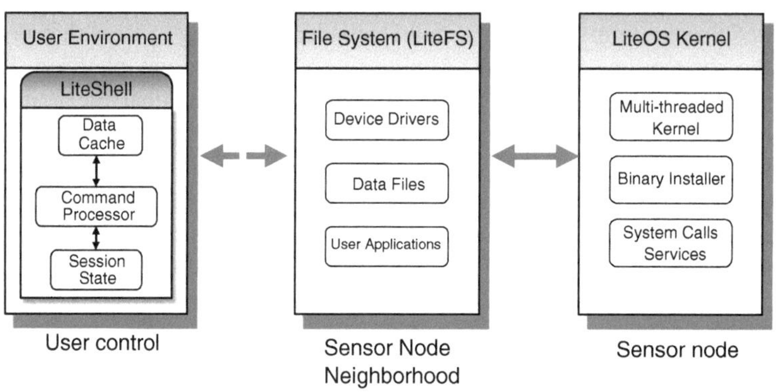

Fig. 13 LiteOS operating system architecture [5]

5.4 *Key Features*

5.4.1 LiteShell Subsystem

The LiteShell subsystem provides Unix-like command-line interface to sensor nodes. This shell runs on the base station PC side. Therefore, it is a front-end that interacts with the user. The motes do not maintain command-specific state, and only respond to translated messages (represented by compressed tokens) from the shell, which are sufficiently simple to parse. Such an asymmetric design choice not only significantly reduces the code footprint of the LiteOS kernel that runs on motes, but also allows us to easily extend the shell with more complicated commands, such as authentication and security.

Currently, LiteOS supports the list of commands listed in Table 5. They fall into five categories: file commands, process commands, debugging commands, environment commands, and device commands.

File Operation Commands

File commands generally maintain their Unix meanings. For example, the **ls** command lists directory contents. It supports a **–l** option to display detailed file information, such as type, size, and protection. To reduce system overhead, LiteOS does not provide any time synchronization service, which is not needed by every application. Hence, there is no time information listed. As an example, a **ls –l** command may return the following:

```
$ ls -l
    Name     Type    Size     Protection
    usrfile file     100      rwxrwxrwx
    usrdir  dir      ---      rwxrwx---
```

In this example, there are two files in the current directory (a directory is also a file): **usrfile** and **usrdir**. LiteOS enforces a simple multilevel access control scheme. All users are classified into three levels, from 0 to 2, and 2 is the highest level. Each level is represented by three bits, stored on sensor nodes. For instance, the **usrdir** directory can be read or written by users with levels 1 and 2.

Once sensor nodes are mounted, a user uses the above commands to navigate the different directories (nodes) as if they are local. Some common tasks can be greatly

Table 5 The list of LiteOS shell commands

Command list	
File commands	ls, cd, cp, rm, mkdir, touch, pwd, du, chmod
Process commands	ps, kill, exec
Debugging commands	debug, list, print, set breakpoint, continue, snapshot, restore
Environment commands	history, who, man, echo
Device commands	./DEVICENAME

simplified. For example, by using the **cp** command, a user can either copy a file from the base to a node to achieve wireless download, or from a node to the base to retrieve data results. The remaining file operation commands are intuitive. Since LiteFS supports a hierarchical file system, it provides **mkdir**, **rm** and **cd** commands.

Process Operation Commands

LiteOS has a multithreaded kernel to run applications as threads concurrently. LiteShell provides three commands to control thread behavior: **ps**, **exec**, and **kill**. We illustrate these commands through an application called **Blink**, which blinks LEDs periodically. Suppose that this application has been compiled into a binary file called Blink.lhex,[1] and is located under the C drive of the user's laptop. To install it on a node named **node101** (that maps to a directory with the same name) in a sensor network named **sn01**, the user may use the following commands:

```
$ pwd
   Current directory is  /sn01/node101/apps
$ cp /c/Blink.lhex Blink.lhex
    Copy complete
$ exec Blink.lhex
    File Blink.lhex successfully started
$ ps
    Name   State
    Blink  Sleep
```

As illustrated in this example, we first copy an application, **Blink.lhex**, into the **/apps** directory, so that it is stored in the LiteFS file system. We then use the **exec** command to start this application. The implementation of **exec** is as follows. The processor architecture of Atmega128 follows the Harvard architecture, which provides a separate program space (flash) from its data space (RAM). Only instructions that have been programmed into the program space can be executed. Hence, LiteOS reprograms part of the flash to run the application.

Once the Blink application is started, the user may view its thread information using the **ps** command. Finally, the **kill** command is used to terminate threads.

Debugging Commands

We now describe the debugging commands. Eight commands are provided, including those for setting up the debugging environment (**debug**), watching and setting variables (**list**, **print**, and **set**), adding/removing breakpoints (**breakpoint** and **continue**), and application checkpoints (**snapshot** and **restore**). Note that all debugging commands keep information on the front-end, i.e., the PC side. In fact, there is no debugging state stored on the mote, which means that there is no limit on the maximum number of variables (or the size of variables) can be watched, or how many breakpoints can be added. We now describe these commands in more detail.

The user first invokes the **debug** command to initiate the environment. This command takes the source code directory of the application as its parameter. For example, if supplied with the kernel source code location, it allows debugging the kernel itself.

[1] LiteOS uses a revised version of the Intel hex format, called lhex, to store binary applications. lhex stands for LiteOS Hex.

Once invoked, this command parses the source code as well as the generated assembly to gather necessary information, such as memory locations of variables. Such information is then used by other debugging commands for diagnosis purposes. For instance, it is used by the command **list** to display the current variables and their sizes, commands **print** and **set** to watch and change variable values, and commands **breakpoint** and **continue** to add and remove breakpoints. Once a breakpoint is added, the command **ps** tells whether a thread has reached the breakpoint.

We now explain the commands **snapshot** and **restore**. **Snapshot** allows adding a checkpoint to an active thread, by exporting all its memory information, including variable values, stack, and the program counter, to an external file. **Restore**, on the other hand, allows importing such memory information from a previously generated file, essentially restoring a thread to a previous state. Combined use of these two commands allows replaying part of an application by *rewinding* it, and is particularly useful for locating unexpected bugs.

Environment Commands

The next four commands support environment management: **history** for displaying previously used commands, **who** for showing the current user, **man** for command references, and **echo** for displaying strings. The meanings of these commands are similar to their Unix counterparts.

Device Commands

The LiteOS shell provides an easy way to interact with the sensors. Every time the file system is initialized, a directory **dev** is created, which contains files that map to actual device drivers. On MicaZ, the **dev** directory contains the following files:

```
$ls
led, light, temp, magnet, accel, radio
```

In this directory, led refers to the LED device. There are four sensors, light, temperature, magnetic, and accelerator, respectively. There is also the radio device, which sends and receives packets. An example of reading 100 data samples from the light sensor at a frequency of 50 ms is written as follows, where the first parameter is the frequency and the second parameter is the number of readings (Fig. 14).

```
./light 50 100
```

5.4.2 LiteFS Subsystem

We now describe the LiteFS subsystem. The interfaces of LiteFS provide support for both file and directory operations. The APIs of LiteFS are listed in Table 6.

While most of these APIs resemble those declared in "stdio.h" in C, some of them are customized for sensor networks. For instance, two functions, **fcheckEEPROM** and **fcheckFlash**, are unique in that they return the available space on EEPROM and the data flash, respectively. Another feature of LiteFS is that it supports simple search-by-filename using the **fsearch** API, where all files whose names match a

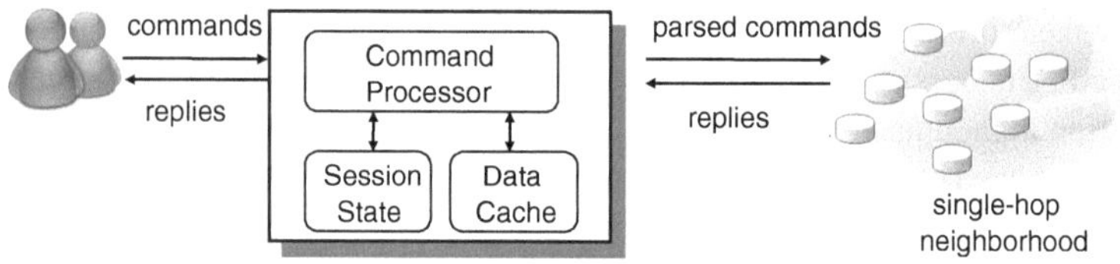

Fig. 14 The implementation of LiteShell [5]

Table 6 LiteFS API list

API usage	API interface
Open file	FILE* fopen(const char *pathname, const char *mode);
Close file	int fclose(FILE *fp);
Seek file	int fseek(FILE *fp, int offset, int position);
Test file/directory	int fexist(char *pathname);
Create directory file	int fcreatedir(char *pathname);
Delete file/directory	int fdelete(char *pathname);
Read from file	int fread(FILE *fp, void *buffer, int nBytes);
Write to file	int fwrite(FILE *fp, void *buffer, int nBytes);
Move file/directory	int fmove(char *source, char *target);
Copy file/directory	int fcopy(char *source, char *target);
Format file system	void formatSystem();
Change current directory	void fchangedir(char *path);
Get current directory	void fcurrentdir(char *buffer, int size);
Check EEPROM usage	int fcheckEEPROM();
Check flash usage	int fcheckFlash();
Search by name	void fsearch(char *addrlist, int *size, char *string);
Get file/directory info	void finfonode(char *buffer, int addr);

query string are returned. These APIs can be exploited in two ways; either by using shell commands interactively, or by using application development libraries.

Implementation of LiteFS

Figure 15 shows the architecture of LiteFS, which is partitioned into three modules. It uses RAM to keep opened files and the allocation information of EEPROM and the data flash in the first module, uses EEPROM to keep hierarchical directory information in the second, and uses the data flash to store files in the third. Just like Unix, files in LiteFS represent different entities, such as data, application binaries, and device drivers. A variety of device driver files, including radio, sensor, and LED, are supported. Their read/write operations are mapped to real hardware operations. For example, writing a message to the radio file (either through the shell by a user or through a system call by an application) maps to broadcasting this message.

In RAM, our current version of LiteOS supports eight file handles (this number is adjustable according to application needs), where each handle occupies eight bytes. Hence, at most eight files can be opened simultaneously. LiteFS uses two bit vectors

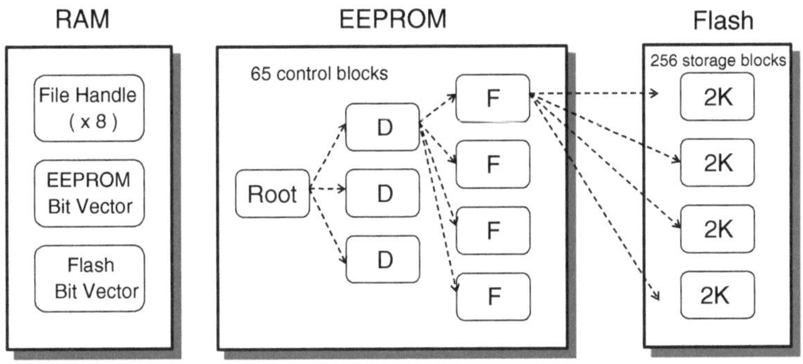

Fig. 15 LiteFS architecture [5]

to keep track of EEPROM/flash allocation, one with 8 bytes for EEPROM, the other with 32 bytes for the serial flash. A total of 104 bytes of RAM are used to support these bit vectors.

In EEPROM, each file is represented as a 32-byte control block. LiteFS currently uses 2080 bytes of the 4096 bytes available in EEPROM to store hierarchical directories, while the remaining EEPROM is available for other needs. These 2080 bytes are partitioned into 65 blocks. The first block is the root block, which is initialized every time the file system is formatted. The other 64 blocks are either directory blocks (specified with D) or file blocks (specified with F) according to application needs. Just like Unix, files represent data, binary applications, and device drivers.

5.5 Programming Model

In this section, we describe the programming model provided by LiteOS. In TinyOS/-nesC, the view is that an application is driven by events, which can optionally post long-running tasks to a FIFO scheduler. Such a programming model typically requires the use of finite state machines (FSMs) to reason with large-scale applications. In contrast to the event based programming model adopted by TinyOS, LiteOS uses threads to maintain execution contexts. Threads, however, do not completely eliminate the use of events for efficiency reasons. Observe that there are two types of events: internally generated events, and externally generated events. For instance, a **sendDone** event always follows a packet sending operation, and is therefore internally generated. A radio receive event, however, is triggered by external events, and may not be predictable. LiteOS treats these two types of events separately. Generally, internally generated events are implicitly handled by using threads, where events like sendDone are no longer visible. It is the externally generated events that deserve special attention.

We illustrate how externally generated events are handled in LiteOS using the radio receive event as an example. LiteOS provides two solutions to handle this event. The first is to create a new thread using the **createThread** system call, which blocks until the message arrives. Consider a reliable communication example. Suppose that the application thread creates a child thread to listen to possible acknowledgements. If such a message is received before **T_timeout**, the child thread wakes up its parent thread using the **wakeupThread** method. Otherwise, if its parent thread wakes up without receiving an acknowledgement, this child thread is terminated by its parent thread using the **terminateThread** method.

While this thread-based approach is usually sufficient for handling externally generated events, it introduces the overhead of creating and terminating threads. This is typically not a problem because it wastes a few hundred CPU cycles, less than 0.1 ms. For computationally intensive applications, however, user applications want to reduce overhead. LiteOS provides another primitive for this purpose: callback functions.

A callback function registers to an event, which could range from radio events to detections of targets, and is invoked when such an event occurs. For instance, the function **registerRadioEvent** tells that a message has been received. Its prototype is defined as follows:

```
void registerRadioEvent(uint8_t port, uint8_t *msg,
uint8_t length, void (*callback)(void));
```

This interface requires the user thread to provide a buffer space for receiving incoming messages. After the kernel copies the message to this buffer, it invokes the callback function. Based on this mechanism, the previous reliable communication example can be implemented as follows:

Part I: Application

```
1 bool wakeup = FALSE;
2 uint8_t currentThread;
3 currentThread = getCurrentThreadIndex();
4 registerRadioEvent(MYPORT, msg, length, packetReceived);
5 sleepThread(T_timeout);
6 unregisterRadioEvent(MYPORT);
7 if (wakeup == TRUE) {...}
8 else {...}
```

Part II: Callback function

```
9 void packetReceived()
10 {
11 _atomic_start();
12 wakeup = TRUE;
13 wakeupThread(currentThread);
14 _atomic_end();
15 }
```

We briefly explain this example. First, the thread listens on MYPORT, and allocates a buffer for receiving incoming packets (line 4). This thread then sleeps (line 5). When it wakes up, it checks if it has been woken by the callback function (line 7), or by a timeout event (line 8). The thread then handles these two cases separately.

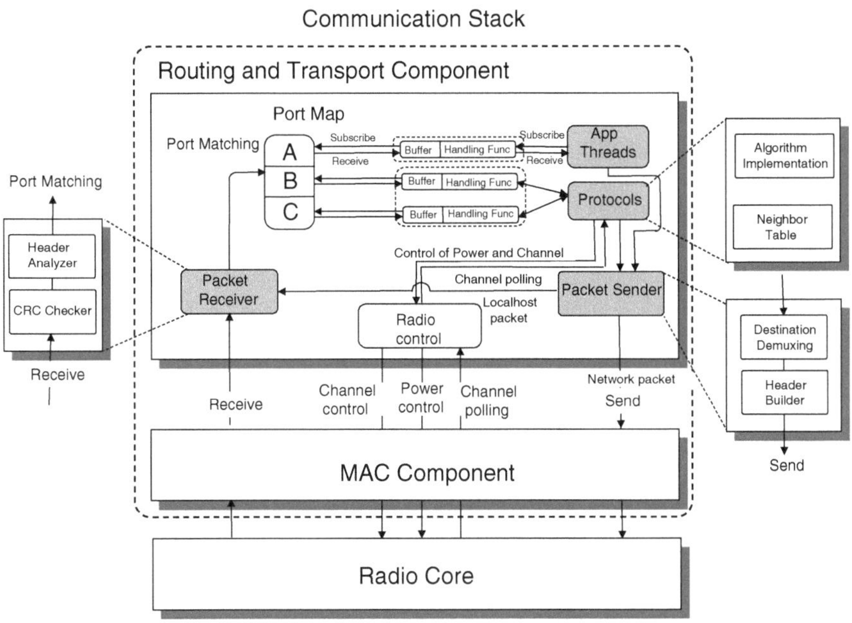

Fig. 16 The architecture of communication stack [5]

5.6 Communication Architecture

In this section, we describe the communication stack of LiteOS. The design of flexible and powerful communication stacks has been a critical challenge to sensor network applications. In LiteOS, we treat communication stacks, such as routing and MAC protocols, as threads. Hence, different communication protocols can be exchanged easily, and dynamically loaded just like any user application.

To use a protocol, an application organizes the header of packets in such a way that the communication protocol can correctly parse it. Packets are then sent to the port which the protocol is listening on. For example, a flooding protocol uses a different listening port compared to a multi-hop routing protocol. In the following example, our multi-hop routing protocol listens on port 10, and the application sends a packet of 16 bytes through this protocol using the following code sample:

```
//10 is the port number, 0 means local node protocol
//16 is the message length, and msg is the message pointer
 radioSend(10, 0, 16, msg);
```

Our implementation of this multi-hop protocol (geographic forwarding) maintains a neighbor table of 12 entries, and contains around 300 lines of code. Its binary image consumes 1436 bytes of program flash, and 260 bytes of RAM, including stack. Its overhead is so low that it can co-exist with other protocols, such as flooding.

Figure 16 shows the communication stack that serves as the foundation for the routing layer protocols. In this figure, both the receiving (on the left) and the sending (on the right) operations are illustrated. When the sender intends to deliver packets, it puts the destination address as well as the port number for the destination node into the packet header. The packet is then delivered to the MAC component and broadcasted over the radio. When the packet is received by a neighbor, its CRC field is first checked for integrity. If this packet is sent to the current node, its port number is matched against each process that is listening to incoming packets. The thread that has a match in port number is considered the right thread for the incoming packet. The contents of the packet are then copied into the internal buffer that is provided by the thread, which is in turn woken up to handle the incoming packet. Note that this communication stack is similar to the port-based socket model in Unix in their way of handling packets.

5.7 Implementation and Hardware Support

The code size of LiteOS is comparable to that of Contiki, though considerably larger than TinyOS. This is because LiteOS employs more complicated scheduling algorithms and implementation details. LiteOS has been ported onto a number of architectures, such as MicaZ and IRIS motes. LiteOS can also run on the Avrora simulator [40] without any modification.

6 Comparison of Different Operating Systems

Besides the three operating systems we described in this chapter, there are a few other operating systems for WSNs. In this section, we first briefly describe these operating systems, and then we compare them.

6.1 The SOS Operating System

SOS [7] is developed by the University of California, Los Angeles. Similar to Contiki and LiteOS, it also supports dynamically loadable modules. Specifically, SOS consists of a list of modules and a common kernel, which implements messaging, dynamic memory, and module loading and unloading, among other services. Please note that in SOS, modules are not processes: they are scheduled in a cooperative manner, and there is no memory protection between modules. On the other hand, the system protects against common module bugs using techniques such as typed entry points, watchdog timers, and primitive resource garbage collection. The latest version of SOS, version 2.0.1, is no longer under active development because of the graduation of the core developers.

6.2 The Mantis Operating System

Mantis OS [8] is developed by the University of Colorado, and aims to develop a more traditional OS for sensor networks. It implements a conventional preemptive time-sliced multi-threading on sensor nodes. Its programming model is primarily thread-driven, therefore, it also provides support for a set of concurrency control primitives, including semaphores. Its power-efficient scheduler helps save energy when all threads are sleeping and then sleeps the microcontroller for a duration deduced from each thread's sleep time. Mantis OS also supports advanced sensor OS features such as multimodal prototyping, dynamic reprogramming, and remote shells. As of today, it is no longer under active development after the 1.0 beta version.

6.3 The Nano-RK Operating System

Nano-RK [17] is developed to support real-time sensor network applications such as environment surveillance and monitoring. Developed by the Carnegie Mellon University, Nano-RK implements a reservation-based real-time scheduler. Nano-RK supports fixed-priority preemptive multitasking for guaranteeing that task deadlines are met, along with the support for CPU and network bandwidth reservations. Tasks can specify their resource demands, and the operating system provides timely, guaranteed and controlled access to CPU cycles and network packets in resource-constrained embedded sensor environments. Nano-RK also introduces the concept of virtual energy reservations that allows the OS to enforce energy budgets associated with a sensing task by controlling resource accesses. Nano-RK has been implemented on the Atmel ATMEGA128 processor with the Chipcon CC2420 802.15.4 transceiver chip. The latest version of Nano-RK supports various wireless link layer protocols such as U-Connect [41].

6.4 The RETOS Operating System

RETOS [42] is developed in Yonsei University, Korea. It has four distinct objectives, which are to provide (1) a multithreaded programming interface, (2) system resiliency, (3) kernel extensibility with dynamic reconfiguration, and (4) WSN-oriented network abstraction. RETOS is a multithreaded operating system, hence it provides the commonly used thread model of programming interface to developers. It uses various implementation techniques to optimize the performance and resource usage of multithreading. RETOS also provides software solutions to separate kernel from user applications, and supports their robust execution on MMU-less hardware. The RETOS kernel can be dynamically reconfigured, via a loadable kernel framework, so an application-optimized and resource-efficient kernel is constructed.

Finally, the networking architecture in RETOS is designed with a layering concept to provide WSN-specific network abstractions. The latest version of RETOS is version 1.2, and is no longer under active development.

6.5 The Enix Operating System

Enix is a lightweight dynamic operating system for tightly constrained platforms for WSNs. Enix provides a cooperative threading model, which is applicable to event-based WSN applications with little run-time overhead. Virtual memory is supported with the assistance of the compiler, so that the sensor platforms can execute code larger than the physical code memory they have. To enable firmware updates for deployed sensor nodes, Enix supports remote reprogramming. The commonly used libraries and the main logical structure are separated; each sensor device has a copy of the dynamic loading library in the Micro-SD card, therefore, only the main functions and user-defined sub-routines need to be updated. A lightweight, efficient file system named EcoFS is also included in Enix. Enix was last released in August, 2010, but is no longer under active development.

6.6 Comparison of Sensor Network Operating Systems

In this section, we compare the features of different operating systems in different angles. Specifically, Fig. 17 shows our comparison results. One general principle is that different operating systems have their own pros and cons, and users should take these features into account when making decisions on which operating system to choose.

Specifically, we have the following observations on the comparison of OS in sensor networks.

First, we observe that sensor network operating systems are far from maturation. Even for the most mature one, TinyOS, its impact is still limited mostly within academia. Indeed, the lack of serious industry participation seems to explain why some of the operating systems are no longer in active development. For example, interestingly, the SOS website explained that the reason that SOS is no longer developed is that its core developers have graduated. In the future, sustained effort from industry or community seems to be very much needed to broaden the impact of sensor network operating systems.

Second, we observe that a set of benchmark requirements will be highly beneficial for future development of sensor network OS software. Just like standard benchmarks exist for measuring the pros and cons of CPU chips, such benchmarks could significantly increase the usability of sensor network operating systems. We hope that the academic community can work together in the future to decide which

	TinyOS	Contiki	LiteOS	SOS	Mantis	RETOS	Nano-RK	Enix
Current license	New BSD	BSD	BSD	Modified BSD	Modified BSD	Not available	GNU	Not available
User shell interface	No (application specific shell such as SimpleCmd exists)	Yes(on-the-mote shell is provided)	Yes (wireless remote shell)	No	Yes (terminal)	No	No	No
Real-time support	No	No	No	No	No	No	Yes	Yes
File system	Single level (ELF, Matchbox)	Coffee file system	LiteFS, Unix-like hierarchical	No	No	No	No	EcoFS
Thread support	Yes (in 2.x)	Yes (protothreads)	Yes	No	Yes	Yes	Yes	Yes
Event-based programming	Yes (by default)	Yes	Yes	Yes	No	Yes	No	No
Debugging	Yes (through various research projects)	Yes	Yes (built-in GDB like debugging)	No	No	Yes (static analysis and dynamic analysis)	No	Yes
Wireless reprogramming	Yes (Deluge)	Yes	Yes	Yes (module level)	One node only	Yes	Yes	Yes
Dynamic memory	No	Yes (malloc interface)	Yes (malloc interface)	Yes	No	No	Yes	No
First publication/release date	2002 (version 1.0)	2004	2008	2005	2005	2007	2005	2010
Latest release	Version 2.1.1, April, 2010	Version 2.5 September, 2011	Version 2.1, October, 2011	Version 2.0.1 April, 2008	Version 1.0 beta Oct, 2007	Version 1.2 September 2007	R877 July 2009	Release 2, August 2008
Platform support	Atmega128 and MSP430	MSP430 and Atmega128	Atmega128	Atmega128	Atmega128, MSP430	MSP430	Atmega128	Nordic nRF24LE1
Simulator	TOSSIM, PowerTossim	Cooja	Through AVRORA	Source level Simulator/ Through AVRORA	Through AVRORA	No	Through AVRORA	No

Fig. 17 Comparison of sensor network operating systems

properties and features are crucial for sensor network OS designs, and make sure that they are well implemented.

A third observation is that there is a general lack of good simulators for OS-independent simulation. For example, the only one that we are aware of on the AVR platform is the Avrora. Without such simulation support, it is extremely difficult to carry out large-scale simulations without the use of large-scale testbed.

7 Conclusion Remarks

The difficulty of building a comprehensive operating system on resource-constrained sensor nodes is underscored by the fact that only a few have been developed in the past decade of research on WSNs. The support from operating systems, on the other hand, is crucial to facilitate the design, implementation, and testing of wireless sensor network software. In this chapter, we systematically presented the design considerations of operating systems for WSNs, the major components of three representative OS systems, and compared the current OSes in multiple angles and aspects. We hope our descriptions so far can help the reader become familiar with the basic concepts and knowledge of operating systems for WSNs.

Currently, operating system research and development for WSNS remains an active research area, and we predict that we will continue to see new operating systems emerge in this area driven by new hardware and sensor network applications. Indeed, there are many open problems that need to be further investigated. For example, how to choose the right programming models based on application needs will continue to be a central topic of explorations in the field of OS research for WSNs. Various trade-offs among overhead, energy efficiency, system reliability, flexibility, and user interface designs will certainly shape the design choices in future operating systems for sensor networks in the days to come.

Acknowledgments The work in this book chapter is in part supported by the NSF grant CNS-0953238 and CNS-1117384.

References

1. J. Corbet, G. Kroah-Hartman, A. McPherson, Linux Kernel Development: How Fast It Is Going, Who Is Doing It, What They Are Doing, and Who Is Sponsoring It. White Paper (online) (2012). http://go.linuxfoundation.org/who-writes-linux-2012
2. A history of Windows. Microsoft (online) (10/12/2012). http://windows.microsoft.com/en-US/windows/history
3. J. Hill, R. Szewczyk, A. Woo, S. Hollar, D. Culler, K. Pister, *System architecture directions for networked sensors, in Proceedings of the International Conference on Architectural Support for Programming Languages and Operating Systems* (ASPLOS), New York, NY, USA, 2000)

4. A. Dunkels, B. Gronvall, T. Voigt, Contiki–a lightweight and flexible operating system for tiny networked sensors, in Proceedings of the IEEE Workshop on Embedded Networked Sensors (EmNets) (IEEE (Comput. Soc.), Washington, DC, USA, 2004), pp. 455–462.
5. Q. Cao, T. Abdelzaher, J. Stankovic, T. He, The LiteOS operating system: towards Unix-like abstractions for wireless sensor networks, in Proceedings of the International Conference on Information Processing in Sensor Networks (IPSN), 2008, pp. 233–244.
6. A. Silberschatz, P.B. Galvin, G. Gagne, *Operating System Concepts* (Wiley, New York, 2005)
7. C.c. Han, R. Kumar, R. Shea, E. Kohler, M. Srivastava, A dynamic operating system for sensor nodes, in Proceedings of the International Conference on Mobile Systems, Applications, and Services (MobiSys), Seattle, Washington, DC, 2005, pp. 163–176.
8. S. Bhatti, J. Carlson, H. Dai, J. Deng, J. Rose, A. Sheth, B. Shucker, C. Gruenwald, A. Torgerson, R. Han, *MANTIS OS : an embedded multithreaded operating system for wireless micro sensor platforms, in ACM/Kluwer Mobile Networks & Applications* (MONET), Special Issue on Wireless Sensor Networks, August, Secaucus, NJ, USA, 2005)
9. P. Levis, D. Gay, V. Handziski, J.h. Hauer, B. Greenstein, M. Turon, J. Hui, K. Klues, C. Sharp, R. Szewczyk, J. Polastre, P. Buonadonna, L. Nachman, G. Tolle, D. Culler, A. Wolisz, T2: a second generation OS for embedded sensor networks. Technical Report TKN-05-007, Telecommunication Networks Group, Technische Universität Berlin (2005).
10. D. Gay, P. Levis, R. von Behren, M. Welsh, E. Brewer, D. Culler, *The nesC language: a holistic approach to networked embedded systems, in Proceedings of the SIGPLAN Conference on Programming Language Design and Implementation* (PLDI), San Diego, CA, USA, 2003)
11. N. Tsiftes, A. Dunkels, *A database in every sensor, in Proceedings of the ACM Conference on Embedded Networked Sensor Systems* (SenSys), Seattle, Washington, DC, 2011)
12. L. Luo, Q. Cao, C. Huang, T. Abdelzaher, J. Stankovic, M. Ward, *EnviroMic: towards cooperative storage and retrieval in audio sensor networks, in Proceedings of the International Conference on Distributed Computing Systems (ICDCS)* (ON, Toronto, 2007), pp. 1–22
13. H. Dai, M. Neufeld, R. Han, *ELF: an efficient log-structured flash file system for micro sensor nodes, in Proceedings of the ACM Conference on Embedded Networked Sensor Systems* (SenSys), Baltimore, MD, USA, 2004)
14. G. Mathur, P. Desnoyers, D. Ganesan, P. Shenoy, *Ultra-low power data storage for sensor networks, in Proceedings of the International Conference on Information Processing in Sensor Networks (IPSN)* (ACM, New York, NY, USA, 2006), pp. 1–8
15. IRIS sensor nodes, Memsic website (online) (10/11/2012), http://www.memsic.com/products/wireless-sensor-networks/wireless-modules.html
16. T. He, S. Krishnamurthy, L. Luo, T. Yan, L. Gu, R. Stoleru, G. Zhou, Q. Cao, P. Vicaire, J. Stankovic, T. Abdelzaher, J. Hui, B. Krogh, VigilNet: an integrated sensor network system for energy-efficient surveillance. ACM Trans. Sensor Netw. **2**(1), 1–38 (2006)
17. A. Eswaran, A. Rowe, R. Rajkumar, *Nano-RK: an energy-aware resource-centric RTOS for sensor networks, in Proceedings of the IEEE Real-Time Systems Symposium (RTSS)* (IEEE Computer Society, Washington, DC, USA, 2005)
18. L. Luo, T. He, G. Zhou, L. Gu, T. Abdelzaher, J. Stankovic, Achieving repeatability of asynchronous events in wireless sensor networks with EnviroLog, in Proceedings of the IEEE International Conference on Computer Communications (INFOCOM) (2007).
19. Y. Chen, O. Gnawali, M. Kazandjieva, P. Levis, J. Regehr, *Surviving sensor network software faults, in Proceedings of the ACM Symposium on Operating Systems Principles (SOSP)* (ACM Press, New York, NY, USA, 2009), p. 235
20. J. Kahn, R.H. Katz, K. Pister, *Next century challenges: mobile networking for Smart Dust, in Proceedings of the ACM/IEEE International Conference on Mobile Computing and Networking* (Mobicom), Seattle, Washington, DC, USA, 1999)
21. Mica sensor node datasheet (2001), http://stomach.v2.nl/docs/Hardware/DataSheets/Sensors/MICA_data_sheet.pdf
22. Mica2 sensor node datasheet (2002), https://www.eol.ucar.edu/rtf/facilities/isa/internal/CrossBow/DataSheets/mica2.pdf

23. MicaZ sensor node (2002), http://www.memsic.com/products/wireless-sensor-networks/wireless-modules.html
24. EPIC sensor node (2007), http://www.eecs.berkeley.edu/ prabal/projects/epic/
25. Sun SPOT sensor node datasheet (2006), http://www.sunspotworld.com/docs/Yellow/eSPOT8ds.pdf
26. LOTUS sensor node (2011), http://www.memsic.com/products/wireless-sensor-networks/wireless-modules.html
27. TelosB sensor node (2004), http://www.memsic.com/products/wireless-sensor-networks/wireless-modules.html
28. Cricket sensor node (2007), http://www.memsic.com/products/wireless-sensor-networks/wireless-modules.html
29. Wasp mote website (2009), http://www.libelium.com/products/waspmote
30. L. Nachman, J. Huang, J. Shahabdeen, R. Adler, R. Kling, Imote2: serious computation at the edge, in Wireless Communications and Mobile Computing Conference, IWCMC '08. International, Crete Island **2008**, 1118–1123 (2008)
31. T. Stanley, T. Close, M.S. Miller, Causeway: a message-oriented distributed debugger. Technical Report HPL-2009-78, HP Laboratories (2009).
32. P. Buonadonna, J. Hill, D. Culler, Active Message Communication for Tiny Networked Sensors (2001), http://ww.w.tinyos.net/papers/ammote.pdf
33. L.f.u. Innsbruck, M. Lang, TinyOS. Habitat (2006).
34. V. Handziski, J. Polastre, J.H. Hauer, C. Sharp, A. Wolisz, D. Culler, The Hardware Abstraction Architecture of TinyOS 2.x. Tech. Rep., Telecommunication Networks Group Technische Universität Berlin, Computer Science Department University of California, Berkeley (2010).
35. P. Levis, A. Terzis, R. Govindan, *TOSThreads: thread-safe and non-invasive preemption in TinyOS, in Proceedings of the ACM Conference on Embedded Networked Sensor Systems* (SenSys), Berkeley, CA, 2009)
36. A. Dunkels, O. Schmidt, T. Voigt, M. Ali, *Protothreads: simplifying event-driven programming of memory-constrained embedded systems, in Proceedings of the ACM Conference on Embedded Networked Sensor Systems (SenSys)* (Boulder, CO., USA, 2006)
37. A. Dunkels, N. Finne, J. Eriksson, T. Voigt, *Run-time dynamic linking for reprogramming wireless sensor networks, in Proceedings of the 4th International Conference on Embedded Networked Sensor Systems (Sensys) (ACM* (Boulder, CO., USA, 2006), pp. 15–28
38. N. Tsiftes, J. Eriksson, A. Dunkels, *Poster abstract: low-power wireless IPv6 routing with ContikiRPL, in Proceedings of the International Conference on Information Processing in Sensor Networks (IPSN)* (Stockholm, Sweden, 2010), pp. 4–5
39. A. Dunkels, The ContikiMAC Radio Duty Cycling Protocol. Technical Report T2011:13, Swedish Institute of Computer Science. Swedish Institute of Computer Science (2011).
40. B.L. Titzer, D.K. Lee, J. Palsberg, *Avrora: scalable sensor network simulation with precise timing, in Proceedings of the International Conference on Information Processing in Sensor Networks* (IPSN), Los Angeles, CA, 2005)
41. A. Kandhalu, K. Lakshmanan, R.R. Rajkumar, *U-Connect: a low-latency energy-efficient asynchronous neighbor discovery protocol, in Proceedings of the International Conference on Information Processing in Sensor Networks* (IPSN), Stockholm, Sweden, 2010)
42. H. Cha, S. Choi, I. Jung, H. Kim, H. Shin, J. Yoo, C. Yoon, *RETOS: resilient, expandable, and threaded operating system for wireless sensor networks, in Proceedings of the International Conference on Information Processing in Sensor Networks (IPSN)* (IEEE, Cambridge, MA, USA, 2007), pp. 148–157

Chapter 22
Programming Languages, Network Simulators, and Tools

Dilan Sahin and Habib M. Ammari

Abstract Wireless sensor networks offer many advantages in different application areas with its ease of deployment, low-cost, low-power capabilities. With the increased interest to the wireless sensor networks, the research community have started to carry out network simulations to better analyze the network's behavior and performance since they provide significant reduction in cost and simulate the different types of sceneries in tolerable time intervals. This chapter introduces the network simulators, i.e., NS-2, OMNET++, J-Sim, OPNET and TOSSIM, respectively with some of the major network programming languages, e.g., nesC and Mate.

1 Introduction

Wireless sensor networks (WSNs) offer many advantages in different application areas with its ease of deployment, low-cost, low-power capabilities [1]. WSN enables sensing, monitoring and controlling of the physical environment to produce high quality and easy accessible information for further analysis purposes. However, it has some inadequacies to fully perform its tasks due to the nature of wireless sensor nodes. Wireless sensor nodes have limitations in their computational capacities since they are low-cost and low-power. Hence, to provide reliable, secure and long-term wireless sensor network communications, a lot of protocols have been developed to efficiently use the resources, accurately route the sensor packets, and reliably provide wireless communications. To better analyze the performance of WSN pro-

D. Sahin
University of Michigan-Dearborn, Dearborn, MI, USA
e-mail: dsahin@umd.umich.edu

H. M. Ammari (✉)
WiSeMAN Research Lab, Department of Computer and Information Science,
University of Michigan-Dearborn, Dearborn, MI 48128, USA
e-mail: hammari@umd.umich.edu

H. M. Ammari (ed.), *The Art of Wireless Sensor Networks*,
Signals and Communication Technology, DOI: 10.1007/978-3-642-40009-4_22,
© Springer-Verlag Berlin Heidelberg 2014

tocols, preliminary test phases on large scales, and under complex and changing conditions should be conducted. Since, creating real test beds, is costly and in some cases impossible, reliable network simulations are used to test the functionalities and performances of the network protocols in changing environment conditions and specific network requirements. Hence, with the increased interest to the wireless sensor networks, the research community have started to carry out network simulations to better analyze the network's behavior and performance. The complex configurations of sensor networks, some difficulties to provide the appropriate network devices and transmission media and some other reasons lead the researchers to prefer using the network simulation tools to better meet the network requirements and to have a chance to test the different types of scenarios in tolerable time intervals.

The network simulator programs are always the software tools that some of them commercially available for free-of-charge, while some do not. Network simulators use specific programming languages, while some of them even combine a couple of programming languages to model the networks. Most of the network simulators consist of some basic modules, such as, wireless medium, physical environment, network protocols, events and applications. *Wireless medium* is modeled by medium module which enables nodes to broadcast signals and use specific propagation models while *environment module* is responsible to model the physical phenomenon which is related to temperature, humidity, sound and light etc. The network protocol modules cover *physical protocol* which is the lowest layer of the protocol stack and responsible of some basic networking stuff, such as, sending and receiving data packets, carrier sensing and adjusting communication channels, *MAC protocol* which is generally implemented in software running on node's processor and responsible for setting protocol parameters and adjusting energy consumption, *routing protocol* which provides routing packets between sensor nodes with special algorithms to meet the requirements of WSN, *application layer* which acts as the interface between lower protocol layers and WSN applications. Here are some of the advantages of network simulators.

- Low-cost: Network simulators provide significant reduction in cost. Preparing real network test beds takes too much time to build and increases the cost. However, network simulators can simulate the different types of sceneries in tolerable time intervals. On the other hand, they provide a pre-test environment before investing money to costly network equipment.
- Ease of usability: With the "add" command, adding sensor nodes or other network equipment to a network is easy in network simulators. Most of the network simulators provide user friendly interfaces to the user.
- Customization: Some network models may not be modeled by simulators, however, it is possible for users to create their own network models.
- Scalability: It is possible to create wireless networks with hundreds of sensor nodes in tolerable time intervals and test the reliability of the communication between them in network simulators.

There is a wide range of network simulators, however not all of them are suitable to simulate WSN scenarios. In this chapter, four basic network simulation programs, e.g., NS-2, OMNET++, J-Sim and OPNET, are detailed investigated with the programming languages used to model the networks. A comprehensive comparison is presented. Furthermore, a brief description of other network simulation tools is also introduced. Furthermore, a brief comparison of other network simulators, e.g., TOSSIM, QualNet, Atarraya, GLoMoSim, JiST, are presented. Some of the simulators have some limitations to conduct some special simulations, for instance, some of them do not have any support for energy modeling of the sensor nodes or some of them have some limitations in supporting a scalable network. While energy consumption concept is very critical to create reliable simulations in some situations, some of them require scalability feature from a simulator. Here are some of the shortcomings/opportunities of network simulators to have a general understanding before giving details.

- Comparing to NS-2, J-Sim can simulate larger number of network nodes, around 500 network nodes. NS-2 is better for small-scale networks since, large-scale networks require so much memory and CPU time. However, J-Sim is not initially developed for WSN, hence, the execution time is quite longer than the execution time of NS-2.
- A detailed documentation and a regularly updated manual simulations are provided by NS-2. J-Sim and OMNET++ network simulators cannot provide such a detailed and updated documentation.
- The disadvantage of NS-2 is that NS-2 needs to be re-compiled every time when there is a change in the user code.
- While J-Sim network simulator can work on all platforms, OMNET++ and NS-2 network simulators can run on Windows and Unix systems.
- The architecture of TOSSIM and NS-2 have some similarities, however, they have different aspects network simulation, while NS-2 focuses on simulation of the sensor nodes at the packet level, TOSSIM focuses on simulating TinyOS sensor nodes the bit level.
- While NS-2 advantages of large user community and on-going development, J-Sim and OMNET++ lacks of comprehensive documentations.
- OMNET++ is very rich in terms of providing several features. NS-2 is the same as OMNET++ in this field, hence becoming familiar with these two simulators may take some time of the users, according to be familiar to J-Sim network simulator.
- In terms of GUI features, OMNET++ supports more tools and functionalities than J-Sim supports.
- J-Sim provides a rich support for wireless sensor networks, while NS-2 enables simulation of some common wireless sensor network protocols.
- TOSSIM's execution model does not capture CPU time and it does not model the energy consumption.
- OMNET++ has a better design engine for simulations than NS-2 platform.
- According to a study conducted by Vienna University of Technology [2], OMNET++ has a better performance than NS-2, and it enables the creation of the simulation from the scratch with good speed optimizations. Furthermore,

OMNET++'s programming interface is clearer and more open than NS-2's programming interface since NS-2 puts some restrictions to some of its parts.
- NS-2 supports OTcl script interface which provides fast coding strategies for modules.

The remainder of this chapter is organized as follows. In Sect. 2, the importance of wireless sensor network modeling concept is introduced with different models, e.g., wireless channel modeling, energy consumption modeling, MAC modeling, and routing modeling. Section 3, on the other hand, gives a quick review about the network simulators before presenting them in detailed. Sections 4, 5, 6, 7 and 8, introduce the network simulators, i.e., NS-2, OMNET++, J-Sim, OPNET and TOSSIM, respectively. Some of the major network programming languages, e.g., nesC and Mate, are presented in Sect. 9, while Sect. 10 gives a general comparison about the network simulators. Section 11 gives a brief description about some short-comings of network simulators, while Sect. 12 presents briefly some other network simulation tools.

2 Wireless Sensor Network Modeling

2.1 Wireless Channel Modeling

The key to the successful design of reliable wireless communication systems is how realistically the radio propagation channel is modeled. An accurate analysis of a radio propagation model may enable valuable insights for wireless sensor network optimization and performance. There are three important radio propagation models that most of the network simulation tools have implemented, e.g., free space model, two-way ground reflection model and log-normal shadowing model. Free space model, the simplest propagation model, assumes that wireless communication range is the perfect circle, and the wireless channel has the ideal propagation conditions in which there is a direct line-of-sight path between transmitter and receiver. Two-way ground reflection model, on the other hand, takes into account both propagation distance from transmitter to receiver and ground reflection approximations of the signal. Hence, it gives more accurate results than free-space model for long distance communications. The more detailed model, log normal shadowing model, takes into account both fading and distance affects in the surrounding of transmitters and receivers, hence, it provides more accurate estimations about the received power than the other two propagation models.

The radio propagation models in simulation environments must reduce the computations and calculations to a minimum level to enable reasonable time intervals. Most of the network simulators have implemented these three radio propagation channel models to give more precise assumptions about wireless data communications. While free space and two-way ground are easy to implement in network simulation environments, to implement log-normal shadowing model, it is needed some other information rather than distance, e.g., interferences and fading effects.

This chapter presents how each network simulator has implemented these models in the following sections.

2.2 Energy Consumption Modeling

In WSN, energy consumption minimization is one of the most important tasks to increase the network lifetime. Hence, the knowledge of power consumption characteristics of sensor nodes is a great value to design accurate power saving strategies. Each network simulation tool has applied different approaches to this critical subject. In next sections, detailed analysis will be provided for each network simulator.

2.3 MAC Modeling

Wireless sensor nodes mostly operate in harsh environmental conditions which changes the network conditions, such as, network topology may change due to many different reasons or network size and network density may change. Hence, MAC modeling should provide valuable insights to prevent the adverse effects of these changes on network performance, and network lifetime. There are two important MAC schemes, e.g., time division multiple access (TDMA), and code division multiple access (CDMA) in wireless sensor networks. TDMA is preferred when the energy consumption is an issue since it decreases the energy consumption of the sensor nodes by letting them sleep during their inactive states. CDMA, on the other hand, is a contention-based MAC scheme which means that nodes are responsible to find a free-channel by sensing the medium. However, nodes have to be awake for longer times and spend more power. In the following sections, detailed analysis will be provided for each network simulator.

2.4 Routing Modeling

The most important thing to be considered while designing routing protocols should be their energy saving features. Furthermore, they should transmit the data reliably from one point to another. In the literature, there have been many routing techniques implemented for WSN. Each network simulator applies different methods to meet the specific requirements of the networks. In the following sections, detailed analysis will be provided for each network simulator.

3 A Quick View on Network Simulators

3.1 NS-2

- **General:** NS-2 is very popular discrete-event simulation tool for WSN, and intentionally developed for academic research, hence, it is open source and easily extendible.
- **Support:** NS-2 supports IEEE 802.11 and IEEE 802.15.4 wireless MAC. Furthermore, it provides energy-saving strategies for IEEE 802.15.4 wireless MAC.
- **Routing:** NS-2 supports ad hoc routing protocols for wireless IP network.
- **Language:** C/C++, OTcl languages are used to develop NS-2, and OTcl sits in the center of simulation and controls and creates the simulation environment.
- **Scalability:** NS-2 does not scale well for large sensor networks since it does not have an application model, hence, some public extensions are integrated to do some specific tasks.

3.2 OMNET++

- **General:** OMNET++ is an open, extensible component-based discrete-event simulation tool, and developed by a user community in academia.
- **Support:** OMNET++ supports wired/wireless IP communications networks, and wireless sensor networks with some extensions. The main support was for IP networks, hence, it does not provide efficient energy models or protocols for WSN.
- **Language:** C++ language is used for OMNET++ simulations.
- **Wireless Channel Modeling:** Castalia is a modular and extensible simulation environment which supports wireless channel and radio modeling for WSN.
- **Scalability:** OMNET++ scales well for large sensor networks.

3.3 J-Sim

- **General:** J-Sim an extensible, open-source, discrete event network simulation environment.
- **Routing:** AODV routing algorithm is one of the basic algorithms implemented in J-Sim.
- **Language:** J-Sim is a java-based tool and also it offers higher-level Tcl-based programming interface.
- **Wireless Channel Modeling:** J-Sim only supports free-space and two-way ground propagation models.

3.4 OPNET

- **General:** OPNET is a discrete-event, general purpose network simulator which was intentionally developed for military purposes. OPNET has strong libraries for fixed networks, on the other hand, with some extensions, it also supports WSN simulations for ZigBee compatible.
- **Wireless Channel Modeling:** OPNET is very successful at accurate modeling of the radio transmission since it can detaily model in different characteristics of physical-transceivers and antennas. Furthermore, it can also model some 3D outdoor scenarios.

3.5 TOSSIM

- **General:** TOSSIM is a discrete-event simulator which is released as a simulator tool for TinyOS operating system for embedded sensor nodes.
- **Energy Consumption Modeling:** The PowerTOSSIM is an extension of TOSSIM to be the enabler of an accurate energy consumption modeling.
- **Language:** TOSSIM is programmed based on a component-based programming model, supported by nesC programming language.
- **Real-Time Modeling:** TOSSIM may not be accurate in real-time modeling since its simulation code may not run in a real node as it runs perfectly in simulation.
- **Scalability:** The probabilistic bit error model makes TOSSIM very scalable.

4 NS-2

NS-2 which is an open source, discrete event network simulation tool, is one of the most popular simulation tools in the research community. It is the second version of Network Simulator (NS) which has been supported by central research and development organization for the Department of Defense DARPA (Defense Advanced Research Projects Agency) through the Virtual InterNetwork Testbed (VINT) project with the joint effort of University of California at Berkeley, University of Southern California's Information Sciences Institute (USC/ISI), Lawrence Berkeley National Laboratory (LBNL), Xerox Palo Alto Research Center (PARC) and the National Science Foundation (NSF) in early 1989 [3].

NS-2 is an object-oriented simulator written in C++ and OTcl programming languages to simulate a variety of IP networks. TCP, routing, and multicast protocols over wired and wireless networks are some of its simulation areas. It does not only focus on the typical network simulations, it also performs resource allocation, real-time communication, energy issues in ad hoc networks, transport protocols in wireless sensor networks, and control strategies for wireless robots [4].

ADVANTAGES OF Ns-2	DISADVANTAGES OF Ns-2
Provide slarge number of available Models and realistic mobility models	Need to be recompiled every time a change occurs
A detailed documentation and a regularly updated manual simulations	Not so well structured software architecture
Supports powerful and flexible scripting and simulation setup	It is difficult to analyze and understand the code.
A set of randomized mobility model is supported	Not perform well for large-scale networks
Includes an energy model and it allows user to easily generate traffic and movement patterns.	Do not support hierarchical modeling

Fig. 1 Advantages and disadvantages of NS-2

NS-2 supports simulation of TCP, routing and multi-cast protocols over wired and wireless networks, while support for wireless sensor networks, wireless LAN protocols were added in later versions.

The architecture of NS-2 is very similar to Open Systems Interconnection (OSI) Reference Model. Hence, a packet needs to go to the network layer, link layer, MAC layer and physical layer. Local Area Network (LAN), Wireless LAN (WLAN), and satellite networks are supported on the lower layers. On the network layer, NS-2 supports static, dynamic, unicast, and multicast routing with several queueing techniques, such as First In, First Out (FIFO) algorithms or stochastic fair queuing. On the transport layer, NS-2 supports TCP, User Datagram Protocol (UDP), and Real-time Transport Protocol (RTP). Basically, NS-2 was developed to simulate wired networks, however, with the later extensions, it can simulate IEEE 802.11 and several other extensions for Bluetooth [3]. An energy model and a set of randomized mobility models are also integrated to NS-2 to enable the users to easily generate traffic and movement patterns and provide realistic.

The extensibility and object-oriented features of NS-2 have made it very popular, however, there are also some short-comings and limitations in terms of customization and easy-to-use. NS-2 does not support an application model which is very important for sensor networks to create interactions between network and application levels.

A detailed documentation and a regularly updated manual simulations are provided by NS-2. Some of the advantages and disadvantages of NS-2 is represented by Fig. 1. NS-2 has three analyzing tools, network animator (Nam), trace file analyzer (Trace Graph), and visualization and analysis tool for wireless simulations (INSpect).

- **Nam:** Nam is the standard GUI of NS-2 to visualize the simulation results, network topology, packet flows, queue lengths and packet drops. Nam is developed to be able read large animation data and be used in different network visualization tools [3]. Before visualizing the simulation, Nam generates a trace file which holds topology information, e.g., nodes, links, and packet traces. Hence, the user can generate topology configurations, layout information, and packet traces using tracing events in ns. After the trace file generation, it is read by nam, and a topology is created with a pop up window. Nam provides many controls to the user. For instance, the user can jump to any point in time by using a time line and time actuator. Nam also provides a service to convert Nam animations to animated gifs and MPEG movies. Nam also enables the creation of simple scenarios.

- **Trace Graph:** Trace Graph' responsibility is to graphically represent trace files of NS-2. Throughputs, round trip times, or node dependent information, e.g., the number of forwarded packets can all be visualized with 200 different 2- and 3-dimensional graphs provided by Trace Graph [5]. However, Trace graph is not part of NS-2 bundle.
- **INSpect:** The main feature of INSpect is to visualize the wireless environments, connectivity graphs, communication ranges, and coordinates of static and mobile nodes [5]. INSpect is originally based on Nam visualization tool.

4.1 WSN Modeling in NS-2

- **Propagation Modeling:** NS-2 simulator implements three different radio propagation models, e.g., free space, two-way ground and log-normal shadowing propagation models. Basically, the receiving power P_r is calculated for data transmissions between sensor nodes with the preferred propagation model, and later, based on the value, channel model decides what action to take. There are two important cases in which receiver takes an action: in the first case, P_r is greater that the receiving threshold which means that there is enough power to allow reception at receiver side, in the second case, P_r is less than the receiving threshold, but greater than the carrier sense threshold, the receiving node drops the packet. Most importantly, NS-2 implements the most accurate radio propagation model, log-normal shadowing, and it is up to the user to pick the most suitable propagation model in terms of the requirements or network characteristics.
- **Energy Consumption Modeling:** Energy consumption model is one of the most important components of NS-2 in implementing a reliable wireless sensor network. Sensors are all battery-powered, and to provide precise assumptions during simulations, energy consumption of sensors should be carefully taken into account in energy consumption models. NS-2 supports an *EnergyModel* which is implemented as a node attribute. The initial energy value given at the beginning of the simulation changes during the each packet reception and transmission. When the all energy source of the node is consumed, the other nodes are notified to take the necessary actions. Since MAC protocol controls the radio modules which consume the most energy, the choice of the MAC protocol is also an important parameter to control the energy consumption levels of sensor nodes.

- **MAC Modeling:** IEEE 802.11 MAC protocol, S-MAC, T-MAC, TDMA are implemented in NS-2 simulation environment. IEEE 802.11 MAC protocol is very efficient in terms of bandwidth and latency. However, the energy control function of IEEE 802.11 cannot meet the demands of sensor network applications [6]. Hence, it is not preferred in wireless sensor network applications if energy consumption is an important criteria. On the other hand, S-MAC is more energy efficient than IEEE 802.11 MAC. S-MAC enables efficient energy consumption for sensor nodes by offering them a tunable periodic active/sleep cycle for sensor nodes. However, it is not efficient in network bandwidth and latency. Hence, T-MAC is introduced which is capable of controlling the frame size dynamically. TDMA, on the other hand, is very efficient in terms of energy preservation, since it provides active/sleep cycles for each node.
- **Routing Modeling:** Destination-Sequenced Distance Vector (DSDV), Dynamic Source Routing (DSR), Ad hoc On-Demand Distance Vector (AODV), Temporally-Ordered Routing Algorithm (TORA) and Protocol for Unified Multi-casting Through Announcements (PUMA) are some of the most important routing algorithms implemented in NS-2. In the literature, there are many researches about NS-2 simulator and implementation and comparison of these routing algorithms and many more for other routing algorithms, e.g., optimized link state routing (OLSR), dynamic MANET On Demand (DYMO), fish-eye state routing (FSR) [7–14].

4.2 NS-2 Simulation and Programming Languages

NS-2 uses two languages, C++ to define the internal mechanism of the simulation objects, and Object-oriented Tool Command Language (OTcl) to set up the simulation by configuring the objects while scheduling discrete events. C++ is preferred for its run-time speed to efficiently manipulate bytes, packet headers and implement algorithms to run over large data sets, while OTcl is preferred for its great performance to be changed very quickly and interactively [3]. NS-2 provides a class hierarch in C++ which can be refereed as the compiled hierarchy and a similar class hierarchy with the OTcl which can be defined as the interpreted hierarchy. The root of the hierarchies is the TclObject class in which the methods automatically establish the interpreted class hierarchy [3]. New simulator objects are created through the interpreted hierarchy and mirrored by another object in the compiled hierarchy. In the general class hierarchy, TclObject is the superclass of other classes in the interpreted and compiled hierarchies, as depicted in Fig. 3. NsObject is the subclass of TclObject, as well as, the super class of all basic network component objects, e.g., nodes and links. Connecter and classifier are the subclasses of NsObject. Node is a compound object of NS-2 with a node entry object and classifiers. Unicast node which consist of an address classifier and a port classifier, and multicast node which contains a classifier to classify multicast packets from unicast packets, and to perform multicast routing, are two different types of nodes in NS-2. Link is another compound object that is used between nodes as duplex-link or simplex-links. They can be recognized as OTcl objects which aggregates the delay, queuing, and possibly loss modules. Event scheduler is another concept of NS-2 network simulator. A real-time and non-real-time schedulers exist in NS-2 to help the interaction of the simulator with a real live network, capture live packets just like a common node and also inject packets into the live network, and handle the events generated by network objects at a scheduled time, respectively. Since, NS-2 is an event driven simulator, schedulers have quite important roles. There are four types of schedulers that each of them is implemented with a specific data structure. The basic task of a scheduler is to select the next earliest event and execute it, then select other earliest event to do the same task. Here are some of the types of the schedulers.

- **The List Scheduler:** Simple linked-list data structure is used for implementation of the list scheduler. The elements in the list is hold according to a time-order, hence, for adding or deleting an event, a scanning process is needed to be performed to find the appropriate entry.
- **The Heap Scheduler:** A heap data structure is used for implementing the heap scheduler. The heap schedular is the superior version of the list scheduler with large number of events [3].
- **The Calendar Queue Scheduler:** The Calendar Queue Scheduler is implemented based on a different data structure. The events on the same month/day of multiple years can be recorded in one day.
- **The Real-Time Scheduler:** The main focus area of the real-time scheduler is to synchronize the execution of events with real-time. It is the subclass of the list scheduler, and the real-time capability is still under development. This scheduler can work well with relatively slow network traffic data rates.

In NS-2 simulations, the following classes are the fundamental parts for performing simulations.

- **TclClass:** TclClass is the virtual class that enables two important functionalities to its subclasses; one is constructing the interpreted class hierarchy to mirror the compiled class hierarchy and another is providing methods to instantiate new TclObjects.
- **TclCommand:** TclCommand provides a mechanism to export simple commands to the interpreter which can be executed within a global context by the interpreter.
- **EmbeddedTcl:** The development of functionality is allowed in either compiled code, or via the interpreter code. The Tcl scripts can be loaded and evaluated by the objects of EmbeddedTcl class. NS-2 can be easily extended by adding OTcl code via scripts. Hence, the scripts must be integrated to NS-2. For this purpose, the scripts are converted into EmbeddedTcl objects.
- **InstVar:** InstVar class focuses on the methods and mechanisms to bind a C++ member variable in the compiled shadow object to a specified OTcl instance variable in the equivalent interpreted object [3]. When a class is established within the interpreter, a variable must be bound and interpreter can operate on an object in that class. Hence, whenever a TclObject is created within the interpreter, method

execution context is set up with the interpreter and a compiled shadow object of the interpreted TclObject is generated. After these steps, the constructor of the compiled object has chance to bind its member variables of the object to interpreted instance variables in the context of the newly created interpreted object [3].
- **Simulator:** Simulator class enables a set of interfaces to configure a simulation and to choose the type of event scheduler to drive the simulation. As mentioned before, user needs to write a simulation script to start the simulation, hence, the corresponding script starts with generating an instance of the *simulation class*, and then, call the other methods to create nodes, topologies, and other aspects of the simulation. When a new simulation class object is created, some operations are performed, such as, initializing the packet format, creating a scheduler etc. The event scheduler starts to run the simulation in an event-driven manner.

4.3 NS-2 Fundamental Components

As mentioned before, to start a simulation, an instance of Simulator class, topology, network nodes should be created. Here are some of the basic elements for NS-2 simulations.

4.3.1 Node

Node is a stand-alone class In OTcl while most of its components are TclObjects. The simple structure of a node, which is depicted in Fig. 2, consists of two TclObjects as mentioned before, an address classifier and a port classifier. Their main tasks are to distribute incoming packets to the proper agent or outgoing link. NS-2 nodes have address or id field, a list of neighbors, a list of agents, and a routing module. By default, nodes are created for unicast simulations. To differentiate a multicast routing from unicast routing, it should be looked at the highest bit of the address which is 0 to represent the unicast routing, or other than 0 to represent the multicast routing.

4.3.2 Classifier

When a node receives a packet, the packet's fields, e.g., its destination address, its source address should be examined, and then, the values should be mapped to an outgoing interface object which is the next downstream recipient of the corresponding packet. This whole task is performed by the classifier object. Hence, it is not wrong to define a node as a collection of classifiers. A simple node has one address classifier and a port classifier, however, when there is a need to add more functionality to the node, more classifiers can be added. Different types of classifiers are used for different purposed by each node. According to a logical criteria, a packet is matched and a reference is obtained to another simulation object based on the match results

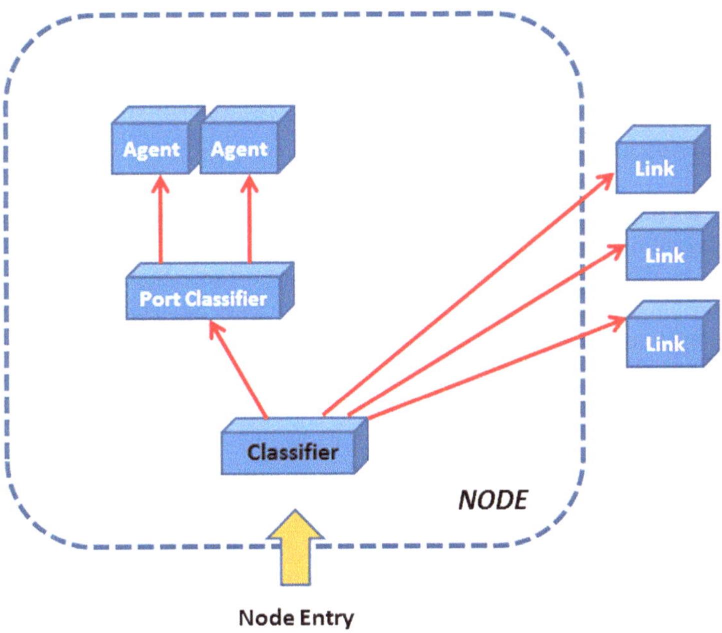

Fig. 2 Simple structure of a node

by the classifier [3]. A table of simulation objects indexed by slot number is included to each classifier to associate the slot number with a received packet and forward it to the object referenced by that slot. There are four types of classifiers.

- **Address Classifiers:** Address classifiers are used for unicast packet forwarding. A bitwise shift and mask operation to a packet's destination address are performed to generate slot numbers [3].
- **Multicast Classifiers:** Multicast classifiers are used to classify packets according to the source and destination addresses. A table mapping is performed for mapping source/group pairs to slot numbers.
- **MultiPath Classifier:** MultiPath classifier is used to provide equal cost multipath forwarding, in case of a node has multiple equal cost routes to the same destination, wants to use all of them simultaneously.
- **Hash Classifier:** Hash classifier is used to classify a packet as a member of a particular flow by using hash tables to assign packets to flows [3]. When flow-level information is needed, hash classifiers are used.

The organization of these classifiers and construct a bridge to them is very important and can be handled by class inheritance concept [3]. However, this method has some complications, hence, object composition concept is introduced. A set of interfaces for classifier access and organization is needed to allow individual routing modules which perform implementation of their own classifiers to insert their classifiers into the node, to allow route computation blocks to populate routes to classifiers in all routing modules, and to provide a single point of management for existing routing modules [3].

4.3.3 Routing Module

In NS-2 routing, three functions carry important roles, *Routing agent* is used to exchange routing packet with neighbors, *route logic* is used to perform information gathering by the routing agents for the actual route computation, and lastly, *classifiers* to enable computed routing table to manage packet forwarding [3]. In the new routing implementation phase, not all of the three functional blocks should be implemented, same classifiers can be used. Routing module is required to manages all these three function blocks, when a new routing protocol which has its own classifier, is implemented. The main functionalities of the routing module are,

- Initialize its connection to a node, and tears the connection down. Usually, it informs the about its interest level for route updates and transport agent attachments, and creates its classifiers and install them in the node.
- To be informed by the node, if it is interested in knowing routing updates,
- to be informed by the node, if it is interested in learning about transport agent attachment and detachment in a node.

4.3.4 Links

Links are other fundamental elements of the NS-2 simulator. They are used to connect the network nodes in the topology. While nodes are composed of classifiers, links are composed of connectors. Link is a stand-alone class as node class in OTcl. Two nodes are connected together with a point to point link via a simple link class with specified bandwidth and delay characteristics [3]. There are five important elements that define a a link.

- **Head:** Head is the entry point to the link which points to the first element in the link.
- **Queue:** Queue is the reference to the main queue element of the link. One queue is owned per simple link. Multiple queue elements may be owned by other more complex types of links [3].
- **Link:** Links is recognized as a reference to the element which actually takes the responsibility to model the link, according to the delay and bandwidth characteristics.
- **Ttl:** Ttl is referred as the reference to the element which makes adjustments to the ttl in each packet.
- **Drophead:** Drophead is the reference to an object which is the head of a queue of elements that process link drops [3].

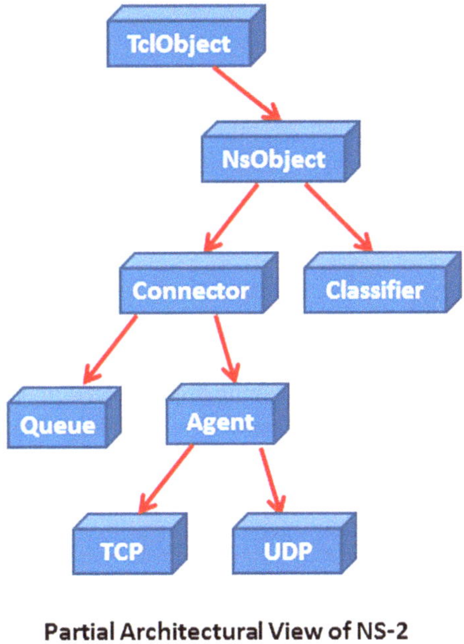

Fig. 3 A partial architecture of NS-2

4.3.5 Connectors

The only responsibility of connectors is to generate data for one recipient. A packet is received by a connector which will perform some functions on it, deliver it to the neighbors and then drop it. Here are some of the different types of connectors.

- **Network Interface:** The main task of network interface connector is to label packets with incoming interface identifier which is used by some multicast routing protocols.
- **DynaLink:** DynaLink connector can be referred as an object which allows/disallows traffic depending on whether the link is up or down. It is inserted to the link just before the simulation is started, hence, represents the head of the link.
- **DelayLink:** DelayLink connector is referred as the object which can model the link's delay and bandwidth characteristics. DelayLink manages the schedules the receive events for the downstream object for each packet received at the appropriate time for that packet in case of the link is not dynamic.
- **Queues:** Queues connector maintains the modeling of the output buffers integrated to a link in a real router in a network.
- **TTLChecker:** The main task of the TTLChecker connector is to decrement the ttl in each packet that it receives. In case of a ttl has a positive value, the packet is forwarded to the next element on the link [3] (Fig. 3).

4.3.6 Agents

The end-points where network packets are generated or consumed, or are in the implementation of protocols at various other layers can be recognized as the agents [3]. Agent class has partly implementation in C++ and partly implementation in OTcl. Packet generation and reception functionalities are both supported by Agents. At different layer, agents are used for protocol implementations.

4.3.7 Timers

As agents, timers can be implemented in both C++ and OTcl. Timers are based on an abstract base class in C++. Mostly, they are used by agents.

4.3.8 Packets

The structure of a packet is defined in the packet class which provides member functions to handle a free list for objects of this type [3]. Packet class has a pointer to an array of characters where the packet headers are hold and a pointer to the packet data. The packet structure has some fields to hold some special information about the packet; *time stamp field* is used for measurement of queuing delay at switch nodes, *ptype_ field* is used for identification of packet types, *uid_ field* is used for scheduling packet arrivals, *size_ field* is used for packet's size information etc.

4.3.9 OTcl

OTcl is the Tcl script language with Object-oriented extensions developed at MIT. NS-2 uses OTcl to create and configure a network. Hence, NS-2's configuration part is written in OTcl. The user needs to write a Tcl script to start the NS-2 simulation. A Tcl script has the proper information to run the simulation, such as definitions of the network topology and protocols, generation of network traffic and other events, commands to generate trace files. After the network topology is defined by the network objects and the plumbing functions in the library, the event scheduler should be invoked to manage the traffic sources to start or stop transmitting packets [15]. Then, The OTcl script is analyzed and C++ libraries are used to implement the data.

4.3.10 C++

C++ programming language is very suitable to run a large simulation, hence, C++ is the preferred language to implement the main part of the simulator. The user do not need to write a C++ code to start the simulation, if there is a need to add new protocols and models, C++ code must be written. The C++ codes need to be

ADVANTAGES OF OMNET++	DISADVANTAGES OF OMNET++
Supports a powerful GUI	Do not support a great variety of protocols
Easy debugging and tracing	The mobility extension is not complete
Accurate modeling of the physical phenomena	Poor analysis of typical measures
Proffesional technical support	Leaves a significant amount of backround work for users to test their own protocols
Growing customer base	Simulation results reporting is not powerful

Fig. 4 Advantages and disadvantages of OMNET++

compiled and linked to an executable file hence, changing in C++ code may take time. The user may use three different C++ programming styles; *Basic C++* which is the simplest style with basic instructions, but is not flexible enough, *C++ coding with input arguments* which addresses the flexibility problem by taking the system parameters as input parameters not to compile the program again, *C++ coding with configuration files* which address the length problem of large number of inputs of the second type by saving all the system parameters in a configuration file. With the introduction of the third style, the previous problems are solved.

5 OMNET++

The discrete event simulation environment, OMNET++ which is stands for objective modular network testbed in C++, has been developed by Andrs Varga to focus on the traffic modeling of telecommunications network, the simulation environment of communication networks, complex IT systems, queuing networks and hardware architectures. OMNET++ is an extensible, open-source, component-based C++ simulation library with integrated development and graphical runtime environment which supports real-time simulations, alternative programming languages (C#, Java), database integration and network emulation [16]. Rather than being a network simulator, OMNET++ provides an appropriate environment with basic machinery and tools for developers to write simulations. Some of the advantages and disadvantages of OMNET++ is presented in Fig. 4. OMNET++ runs on Linux, Unix systems, Windows OS platforms. There are many model frameworks, e.g., INET framework, Castalia, MIXIM, OverSim, developed independently from OMNET++ simulation framework to support computer networks, queuing networks and hardware architectures [16].

OMNET++ provides many advantages to programmers with its public-source, component-based, modular and open-architecture environment and high expendability and modularization features. OMNET++ eases the discrete event-driven application and complex software systematic performance assessment simulations [17].

OMNET++ includes different facilities to perform the important tasks. Here are some of them.

- **Simulation Kernel Library:** Simulation kernel library has some definitions of objects which are used for scenario definition and topology creation.
- **The Simulation IDE:** IDE provides a dual-mode, e.g., graphical and source, round-trip editor for NED files [16].
- **Network Description (NED):** NED is the language of OMNET++ that enables users to declare simple modules, and connect and assemble them into compound modules [16].
- **Graphical Network Editor (GNED):** GNED enables the graphical topology build and file creation in NED language with the use of modules, gates and connections. GNED provides easy definition of scenarios. The reusability of models is easier since GNED enables creation of topology templates. Furthermore, the NED files can be exported into XML format and XML files can be imported according to the NED document type definition to for topology creation [15].

OMNET++ has a different perspective in representing the network elements in a model. The component architecture concept is introduced in OMNET++ framework which gives a flexibility to the designer to map the network devices or protocols into model components that can be recognized as *modules*. Hence, OMNET++ consists of a hierarchically nested modules. Logical structure of the system is depends on the user, hence, the depth of the module nesting is not limited. Message passing is the basic way for modules to communicate with each other. Events, jobs, packets, commands or other entities depending on the model domain can be recognized as messages transferred between modules. Furthermore, modules can be connected to each other via gateways.

5.1 OMNET++ MODEL

OMNET++ modules are gathered together to form the OMNET++ model. There are two types of modules, *simple modules* which are written in C++ with the usage of simulation class library, and *compound modules* which are formed by a group of simple modules. Routing tables, user agents, traffic sources and sinks, protocol entities, e.g., TCP, 802.11 interface card represent the simple modules in a computer network while network nodes, e.g., routers, hosts represent the compound modules. Model reuse concept is adopted in OMNET++, a model type as a building block can be recognized as a simple or a compound model as depicted in Fig. 5. Hence, a model can be split up into many simple modules or a compound module can behave as a simple module with modifications of its functionality. Simple and compound

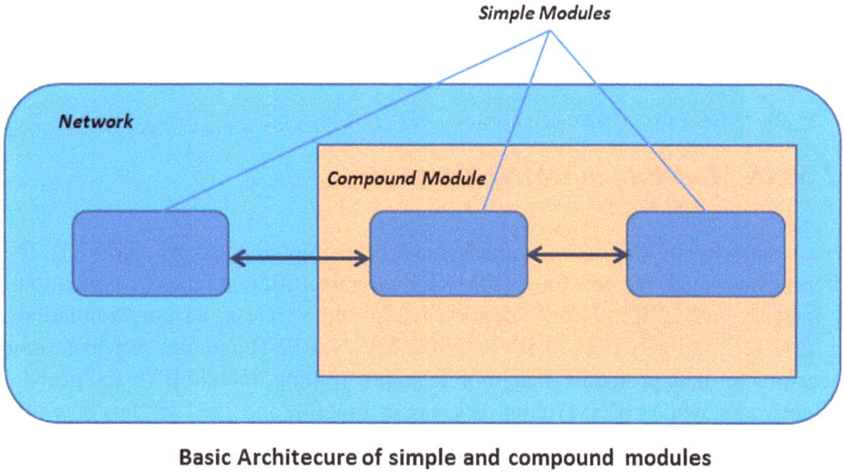

Fig. 5 Basic architecture of simple and compound modules

modules are the instances of module types that serve as components of more complex module types. Hierarchically nested modules form OMNET++ models which can be also recognized as networks. System module represents the top level model and contains submodules. All OMNET++ modules can be recognized as submodules or sub-submodules of the system module [16]. Moreover, different module types can be saved in separate files and a group of these types can create component libraries. OMNET++ models have some important components to perform its special tasks, here are some of them.

- **Gate:** Gates which are the input, output and input interfaces of modules, have important roles in OMNET++, for instance, messages are send via output gates and received via input gates by simple modules.
- **Connection:** An input gate and an output gate or two input gates can be linked by a connection. Two submodules, a submodule with parent and two gates of the parent module can be linked by a connection. Within a single module of hierarchy, connections are created and not allowed to span through the hierarchical levels not to hinder the model reuse. Propagation delay, data rate, bit error rate etc. can be assigned to the connections.
- **Messages:** Messages enable modules to communicate with each other. Frames or packets in a computer network, jobs or customers in a queueing network can represent the messages. Furthermore, complex data structures can be contained in messages. Messages can be sent directly or via gateways and connections to the simple modules.
- **Parameters:** Modules can have parameters, e.g., string, integer, double, boolean, and XML element tree, to transmit the configuration data to simple modules and define the model topology [16]. These parameters can be encapsulated into channel

objects. Moreover, parameters can also define the number of submodules and number of gates within a compound module.

5.2 WSN Modeling in OMNET++

A vast majority of simulation models have been implemented by OMNET. The different simulation frameworks of OMNET supports different types of simulations, for instance, INET framework supports wired and wireless, ad hoc simulations, e.g., 802.11, Ethernet, TCP, IP, IPv6, OSPF, MPLS, RSVP, and also provides some extensions for new protocols, e.g., mobile ad hoc routing, mobile IPv6 and peer-to-peer networks, while MiXiM framework supports mobile and fixed wireless networks with detailed models of radio wave propagation, interference estimation, and wireless MAC protocols, e.g., ZigBee. Furthermore, Castalia provides researchers to test their protocols in realistic wireless channel and radio models.

- **Propagation Modeling:** The wireless channel access is one of the most important modules in OMNET++. Hence, three important radio propagation models, e.g., free-space, two-way ground and log-normal shadowing, have been implemented in OMNET++. For instance, Kuntz et. al implemented log-normal shadowing model in [18] to describe the real conditions sufficiently. The free-space propagation model is already the default propagation model of INET framework of OMNET++, and, two-way ground model is also supported [19].
- **Energy Consumption Modeling:** In OMNET++, a battery module has been implemented to control the real-time energy levels of each node. The total time that is spent by the radio module in different states is calculated to find the actual consumed energy with the given radio power models. In [20], Chen et. al. proposed an energy model in which the energy consumption at the radio is measured by counting the total time that the radio has spent in each state of the simulation.
- **MAC Modeling:** The Ad hoc operation is supported by both OMNET++'s INET and Mobility frameworks. Furthermore, in different versions of OMNET++, different MAC protocols have been implemented. IEEE 802.15.4 CSMA and L-MAC have been implemented in OMNET++, MiXiM 2.1 version, while S-MAC and T-MAC have been implemented in OMNET++ 3.2, MAC Sim 0.2.2 versions [16]. However, IEEE 802.15 MAC is the core module of the implementation.
- **Routing Modeling:** Routing modeling supports star and tree-cluster topologies and also helps forming the cluster-tree PANs.
- **Traffic Modeling:** Chen et al. developed traffic module as a traffic generator to run the simulation [20]. Hence, traffic module provides traffic generation with multiple types, e.g., CBR, on-off and exponential. Furthermore, it acts like a traffic producer and also as a traffic sink to collect and analyze the data packets.

5.3 OMNET++ NED Language

The domain specific language (DSL) of OMNET++ can be recognized as NED language which stands for *network description*. NED language is chosen from a variety of other languages, e.g., XML, Tcl, Python, since it is simple and supported with a graphical editor. NED enables the declaration of simple modules *which describe the interface of the modules, such as, gates and parameters*, or definitions of compound modules *which consist of the declaration of the module's external interface, e.g.,gates and parameters, and the definition of submodules and their interconnection*, or definitions of networks *which are compound modules that are recognized as self-contained simulation model* [16]. Here are some of the specific features of NED language.

- **Hierarchical:** To deal with complexity, especially in very large projects, hierarchical representation is a very common concept. Hence, single complex modules can be broken down into smaller modules and can be used as compound modules.
- **Inheritance:** Sub-classifying is another method that is adopted by OMNET++. Modules and channels can be subclassed to create and inheritance overall the system. Hence, some adjustments can be made to the derived modules and channels, such as adding new parameters, gates and connections.
- **Inner Types:** To prevent namespace pollution, channel and module types used locally by a compound module can be defined within a compound module.
- **Packages:** Package usage is preferred by NED language to prevent the name clashes.
- **Meta-Data Annotations:** Modules, channel types, parameters, gates and submodules can be annotated by adding properties to store graphic attributes, e.g., position, icon, to declare measurement units, to denote the C++ namespace, to mark gates which remain unconnected. For instance, a module's graphical representation can be specified as meta-data annotations.
- **Component Based:** The reusability of simple modules and compound modules enable the reduction in code copying and existence of component libraries.

5.4 Network Simulation Frameworks

The following network simulation frameworks were developed for OMNET++ and some characteristics of these frameworks is presented in Fig. 6.

- **INET Framework:** INET framework built upon OMNET++ is an open-source communication network simulation package that supports wireless and mobile simulations while containing important protocol implementations such as IPv4, IPv6, TCP, UDP, etc. The module concept is adopted from OMNET++ and, while routers, hosts, switches are represented by OMNET++ compound modules; protocols, applications and other functional units are represented by OMNET++

INET	MiXiM	Castalia	OverSim
For Wireless and Mobile Simulations	For Wireless and Mobile Simulations	For WSN and Body Area Networks	Peer-to-peer Networks
Supports several Internet protocols: UDP, TCP, SCTP, IP, IPv6, Ethernet	Supports communication protocols especially (MAC) level	Adopts Log-normal shadowing	Contains P2P systems and overlay protocols.
Has emulation capabilities	Supports a powerful GUI	Physical process modeling, several MAC/Routing protocols implemented	Supports central module for gathering and processing statistics.

Fig. 6 Some characteristics of network simulation frameworks of OMNET++

simple modules [21]. INET hierarchical packets consist of modules and they are organized according to OSI layers. While some modules implement protocols, manage communication modules or autoconfiguration of a network, some modules are just responsible for holding the data, such as, routing tables. Network interfaces, e.g., ethernet, IEEE 802.11, are generally recognized as compound modules and they consist of a queue, MAC and other simple modules.

- **Castalia:** The specialty of Castalia framework is networks of low-power embedded devices, body area networks and wireless sensor networks. Castalia fully supports advanced tests of distributed algorithms and protocols in realistic wireless channel since it adopts log-normal shadowing model to accurately model the wireless channel by taking into account different parameters, e.g., scattering, fading, noise, interference [22]. Castalia is a highly parametric and can simulate a wide range of platforms. Castalia supports advanced channel model based on empirically measured data, advanced radio model based on real radios for low-power communication and extended sensing modeling provisions [22].
- **MIXIM:** MIXIM is an OMNET++ network simulation framework that supports mobile and fixed wireless networks and provides a detailed models wireless channel, wireless MAC protocols, e.g., ZigBee, wireless connectivity, mobility models etc [23]. A user friendly, graphical representation of wireless mobile networks is provided with advanced support of debugging and complex wireless scenarios.
- **OverSim:** OverSim is an overlay and peer-to-peer network simulation framework based on INET framework. OverSim has many structured, e.g., Chord, Kademlia, Pastry, Bamboo, Koorde, and Broose and unstructured P2P systems and overlay

protocols to facilitate the implementation of additional protocols and to make them more comparable [24]. The overlay, underlay and all network packets are detaily represented by the graphical interface of OMNET++. A P2P network architecture can be modeled by OverSim's modular architecture, and easy exchange and extendibility of modules and code reuse are all available in OverSim [24].

5.5 OMNET++ Programming

OMNET++ simulation tool provides modeling and simulation of different kinds of networks. The user can easily describe the network topology, and GNED editor enables easy definition of network scenarios with a graphical and script-based description of the topology with modules, gates and connections, and furthermore, GNED editor enables exporting NED files to XML which is the enabler language for OMNET++ to interface with other systems, for instance, a network management application may import the network topologies saved in a SQL database into OMNET++ [15]. OMNET++'s C++ classes describe the simulation objects, and simple modules are the basic components of OMNET++ programming and they are implemented as C++ classes, derived from *cSimpleModule* library class. Simple modules' main task is message sending which is represented with *cMessage* class. The events can be generated and read by each module which can be recognized with discrete events. Discrete event simulation (DES) is one of the other fundamental concept of OMNET++ programming. Computer networks are usually recognized as DES systems in which state changes happen at discrete instances in time, hence, between two consecutive events, no state change takes place [16]. In a computer network, start of a packet transmission and end of a packet transmission can be recognized as two consecutive events that no state change happens between them. *Arrival time* term refer the time when events take place. Messages are used to represent the events, and are sent from one module to another. The message's destination module is the place where the event will occur and the time when the event will occur is the arrival time of the message [16].

The other components of OMNET++ programming are compound modules and channels. Channel objects may represent the connections in simulation. Channel objects can be programmed by C++ users and represent the channel behavior, e.g., propagation and transmission time modeling, error modeling etc. All of the components are represented with C++ *cComponent* while the abstract module class *cModule* and abstract channel class *cChannel* are the subclasses of *cComponent*. Figure 7 depicts other classes and the inheritance relationship of them.

Furthermore, a set of user interfaces are used to perform some specific tasks, such as, managing state estimations, visualizing of simulation results, enabling users to start or stop the simulation execution. These user interfaces are places in different simulation libraries, and they are connected to the simulation kernel or to the models through interfaces. There are two types of OMNET++ user interfaces; e.g., *Tkenv*

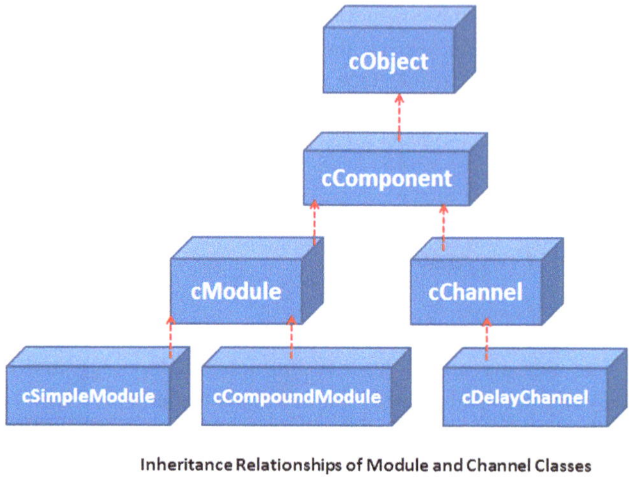

Fig. 7 Inheritance relationship of components of OMNET++

which is based on graphical, windowing user interface, and *Cmdenv* which is the command-line user interface [15].

- **Tknenv:** Interactive execution of the simulation, tracing and debugging are all supported by Tknenv user interface which provides a detailed visualization of simulation state, separate window for output of each module, event-by-event execution, labeled breakpoints and detail report of the entire simulation etc.
- **Cmdenv:** Cmdenv which is a small, fast user interface, is designed for batch execution and can run on UNIX or Windows. Three data files are used to write the simulation results; *the output vector files* which can be read by Matlab, Octave and Plove tool, *the output scalar files* which can be visualized by Scalar tool and *user own files* [15].

Here are some of the other fundamental elements of OMNET++ programming.

5.5.1 Library Classes

The fundamental parts of OMNET++ simulation library consist of the parts of the component model, e.g., modules, channels, gates, objects etc. The network topology can be extracted from a model by the *topology discovery* class of the library [16]. Furthermore, random number generation and several distributions are all available. Some of the distributions are log-normal, Poisson, Bernoulli, uniform etc. Moreover, simulation library also includes several statistical classes.

5.5.2 Network Packets

In network simulations, the representation of network packets is very important. Packets which are derived from *cPacket*, are C++ classes and they are subclasses of *cMessage*. The packet's length, the pointer to the encapsulated packet, error flag which indicates if the packet is corrupted, are all included in *cPacket*'s field.

5.5.3 Wireless Packet Transmission

In wireless packet transmission, packets are directly sent to the corresponding nodes. Channel controller module is dedicated to keep track of the range and frequency information of nodes. Channel controller also provides the modeling of the wireless channel and the radio reception in destination node's side. In this level, the model frameworks, e.g., MiXiM, Castalia etc., decide which propagation, interference and reception model to implement [16].

5.5.4 Simulation Signals

Simulation signals are the notification mechanism of OMNET++ simulation library, that allows communication between components [16]. Components emit the signals and propagate them on the module hierarchy up to the roof. There is a listener concept that are notified when a signal is emitted. Listeners have the ability to receive signals from all of the components during the whole simulation. Listeners help the modules to get notified of the events via the signals. Components register listeners for the specific events that they want to be aware of. Furthermore, signals can be used for emitting variables to be recorded as simulation results [16].

6 J-Sim

J-Sim is an extensible, open-source, discrete event network simulation environment build entirely in JAVA programming language which is a wide-spread, well-known and easy-to-learn programming language. J-Sim enables a flexible simulation of distributed, parallel, and discrete event systems and has a wide range of application areas which have discrete-time character [25]. J-Sim is a real-time process driven network simulator, developed by a team at the Distributed Realtime Computing Laboratory (DRCL) of the Ohio State University and has been sponsored by the National Science Foundation (NSF), DARPA's Information Technology Office, Air Force Office of Scientific Research's Multidisciplinary University Research Initiative (AFORS MURI), the Ohio State University and the University of Illinois at Urbana-Champaign [26]. J-Sim is inspired by the first widely used simulation language, *Simula* which was developed in the 1960s by Ole-Johan Dahl and Kristen Nygaard [25]. The advantages

ADVANTAGES OF J-Sim	DISADVANTAGES OF J-Sim
Platform Independent	The execution time is longer
Fully supports object oriented mechanism	Not firstly developed to simulate WSN
Persistence	Adding new protocols is very difficult
Supports both pseudo-real-time and virtual time simulations	Adding new node components is very difficult
Includes specific WSN packages with both battery and power model	Only, IEEE 802.11 MAC protocol is supported

Fig. 8 Advantages and disadvantages of J-Sim network simulator

and disadvantages of J-Sim is depicted in figure Java is not the only language used in J-Sim; for the description and implementation of models, Java programming language is preferred, while a script language is preferred for construction, configuration and control the simulation at run-time [15]. Tcl, Perl or Python script languages are all supported by J-Sim, however, the J-Sim implementation is based on Tcl script language. J-Sim is implemented on top of the Autonomous Component Architecture (ACA) where components are written in Java. Here are some of the fundamental parts of J-Sim (Fig. 8).

- **Component:** In J-Sim, every element is recognized as a component, such as a node, a link, or a protocol. Each component can be a single component or they can be grouped to form other components as depicted in Fig. 9. Each component make connections to each other through *ports* where they exchange the data flows. The ports can be connected as one-to-one, one-to-many or many-to-many. A component can be realized as a *box* that sends output flows and receive input flows with the help of the ports. J-Sim has the ability to disable the components to see the consequences of a failure in a simulated network.
- **Composite:** Components can come together to form other components. These components can be recognized as composite that consist of several inner components.
- **Contract:** Contract can be realized as the behavior of a component. There are two types of contracts; *port contract* which is specific to a port of the component and focuses on the communication pattern of the components that are connected to it, and *component contract* which defines the input-output relation of a component to better describe the arrival time, and the arrival place of data packets and the data process [15].

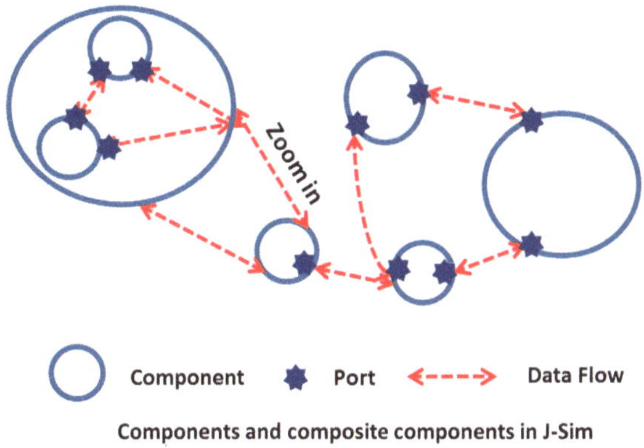

Fig. 9 Components and composite components of J-Sim

A network is a composite component which consists of nodes, links and smaller networks, while a node is a composite component which consists of applications and protocol modules. ACA mimics the integrated circuit design and its software architecture is so modular that contract binding is separated from component binding at system integration time. Hence, J-Sim has a loosely-coupled component architecture that components can be designed, implemented and tested individually [25]. With ACA, J-Sim provides an extendible network simulation and enables its components easy plug-and-play notion to the simulator environment. A generalized packet-switched internetworking framework (INET) implemented on top of ACA based on some features gathered from different layer of protocol stack.

6.1 WSN Simulation in J-Sim

J-Sim provides an extensible component architecture and framework for wireless sensor network simulations with the great support for sensor, sink nodes, wireless communication channels and power and mobility models. In [27], the wireless simulation framework consists of sensor nodes which are responsible to detect the signals generated by target nodes, and send them to the sink nodes via wireless channel. Since, wireless communication between sensor nodes is handled between wireless channels, to correctly simulate the wireless sensor networks, a channel component is implemented to simulate channels. Furthermore, two different propagation models, e.g., sensor propagation model and wireless propagation model, are included to differentiate the signal propagation between sensor node to sink nodes and sensor nodes to target nodes. The other modules for energy consumption, routing or MAC layers are introduced in the followings.

- **Propagation Modeling:** While free-space and two-way ground propagation models are available in J-Sim simulator, log-normal shadowing model is not supported. However, J-Sim supports a different propagation model, irregular terrain model, which is based on the electromagnetic theory and make predictions of the median attenuation of a radio signal as a function of distance [28].

- **Energy Consumption Modeling:** J-Sim's energy consumption model is implemented in WirelessPhy component to control the energy consumption of each sensor node. When a node consumes all of its energy, its the interface card is called as "dead," and no further communications can be done with this node.
- **MAC Modeling:** IEEE 802.11 protocol is the only MAC protocol implemented in the Mac_802_11 component for J-Sim wireless extension. IEEE 802.11 MAC frames are transmitted between wireless interface card and the Mac_802_11 component when their reception and decode process are successfully accomplished.
- **Routing Modeling:** AODV routing algorithm is one of the basic algorithms implemented in J-Sim. Greedy Perimeter Stateless Routing has been also implemented in J-Sim network simulator tool [29].
- **WirelessPhy Model:** Some features of the physical layer of a wireless card are simulated in this model. The receiving signal strength of each frame is measured and calculated to determine if it can be correctly decoded.

6.2 INET Platform

INET platform is component-based architecture which is dedicated to network simulation and provides modeling of all kinds of infrastructures. INET platform is fully integrated with J-Sim network simulator. INET defines the basic components of hierarchical networks, such as network, node, link etc., protocol component which serves as the base class for protocol module implementations, and finally contract classes for the core service layer [25]. Hierarchical networks in random depth are supported by INET, for instance, an internetwork with networks, nodes and links. Address allocation mechanism is adopted by INET to address the network nodes at different levels of the network hierarchy [25]. A typical INET node contains a core service layer which provides services which are defined in terms of contracts, such as the data forwarding/delivery service, the identity service, the routing service, the interface/neighbor service and the packet filter configuration service [25].

- **Data Forwarding/Delivery Service:** The downward contract and the upward contract come together to form a data forwarding service. The downward contract focuses on defining the packet sending service provided by the core service layer to upper layer protocols. An upper layer protocol must clarify some of the information before sending a packet, for instance, the destination address, the maximum number of hops before the packet is discarded (TTL), the router alert flag which indicates if the corresponding packet should be prevented by intermediate routers, and the type of service (ToS) that packet will receive. The upward contract focuses on describing the service to deliver a packet to an upper layer protocol. The prerequisite information should be provided by core service layer, such as the sender address, the incoming interface on which the packet arrives, and the type of service (ToS).

- **Node Identity Service:** A list of identities, or addresses, of the nodes should be provided in the core service layer. INET defines a unicast address as an identity that belongs to only one node in the network, and a multicast address as another identity that can be used by one or more nodes to be identified in the network. Node identity service provides identity lookup and identity configuration contracts for upper layers to configure the node identities. In identity lookup, *default identity lookup*, *identity inquiry* and *all identities inquiry* services are provided for upper layer protocols to look up the node identities. In identity configuration service, service for configuring the identities of a node is supported.
- **Routing Service:** A routing table is used at each node's core service layer. A routing table contains the information source and destination addresses, proper interfaces for each packet to be relayed. A routing table can be configured by an upper layer with route lookup, route configuration and route request service contracts. *Route lookup contract* specifies the source address, the destination address, and the incoming interface on which the packet arrived, *route configuration* describes the services for configuring the routing table, such as, adding, removing and modifying a routing table entry, *route request* is defined when a packet does not have the route by the core service layer [25].
- **Interface/Neighbor Services:** There exist a database to hold the information about the interfaces of the node which includes information about the local address, the addresses of the neighbors within an interface, and the bandwidth etc. Upper layer protocols can be notified by the neighbor events with the help of core layer service.
- **Packet Filter Configuration Services:** The extensible part of the core service layer can be recognized as the packet filters which is a part of each node. When a packet is forwarded to an interface, the packet may meet a couple of packet filters.

The core service layer of INET is composed of five different components, e.g., Packet Dispatcher, Identity, Routing Table, Hello, and PacketFilterSwitch to better reuse the components as depicted in Fig. 10 [25].

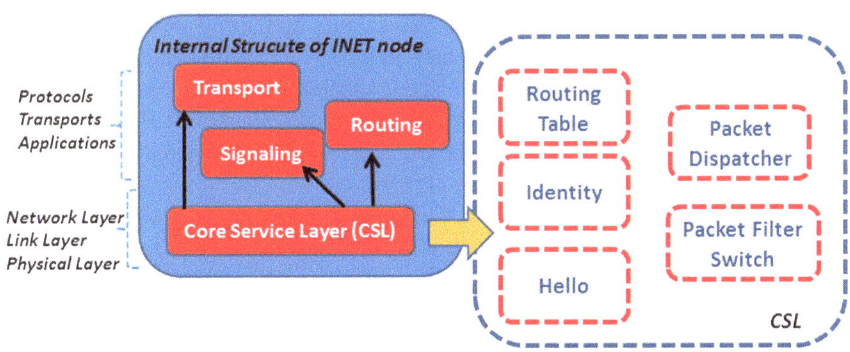

Fig. 10 Components of an INET node [25]

- **Packet Dispatcher Component:** The data sending/delivery services to the upper layer protocols are provided by packet dispatcher component. The incoming packets are forwarded to a proper number of output ports which are connected to an upper layer protocol or a lower layer component. The identity lookup and route lookup services are invoked and TLL fields of the packets are checked to decide whether discard a packet or not.
- **Identity Component:** Identities of the nodes are used to provide the identity services to the upper layer protocols and other components in the core service layer.
- **Routing Table Component:** The routing table entries are used to provide the routing services to the upper layer protocols, e.g., the routing protocols.
- **Hello Component:** The interface and neighbor information of the nodes are used to provide interface/neighbor services to both the upper layer protocols and other components in the core service layer by hello component.
- **Packet Filter Switch Component:** Packet filter switch component works between an upper layer protocol and the components in the packet filter banks. The configuration requests are switched for packet filter components.

6.2.1 J-Sim Library

J-Sim library provides the model creation in a flexible way. J-Sim library is an easy-to-use, and not a complex simulation model that can be built with minimum coding. J-Sim library is based on five important packages; *Queue package* which consists of four different types of queues including FIFO, LIFO, Priority, and Temporal queues, *Statistic package* which provides data collection and analysis, *Variate package* which includes random number generators and many commonly-used random variate generators, *Process package* which provides implementation of the process interaction approach, and lastly, *Event package* which provides event scheduling approach [30].

- **Queue Package:** Queue class is an abstract base class and derive FIFO Queue (First-In, First-Out queue), LIFO Queue (Last-In, First-Out queue), PriorityQueue, and TemporalQueue classes. The subclasses derived from the Queue classes need to implement the methods of the Queue classes since they are all abstract and empty to indicate the queue to be empty. Each queue has a infinite capacity, however, the user can set limitations while contracting queue objects. There are some basic methods of queue classes; *length method* to return the number of elements in the queue, *clear method* to remove all the elements in the queue, *enqueuemethod* to insert a single element into the queue etc [30].
- **Statistic Package:** The main purpose of the statistic package is to collect statistical information with its special classes. It is an abstract class as the Queue class and consist of special methods to analyze statistical data; *SampleSat* class to collect sample statistical data, *TimeStat* class to collect time-persistent statistics, *Statistic* class to calculate minimums, maximums, means, variances, standard deviations, root mean squares and confidence interval half widths [30].

- **Variate Package:** A wide variety of random variates is provided by variate package. Discrete random variates, e.g., Bernoulli, Binomial, DiscreteProb, Geometric, HyperGeometric, NegativeBinomial, Poisson, and Randi, and continuous random variates, e.g., Beta, Cauchy, ChiSquare, Erlang, Exponential, F Distribution, Gamma, HyperExponential, LogNormal, Normal, StudentT, Triangular, Uniform, and Weibull, are all supported by J-Sim's variate package.
- **Process Package:** Process package provides creation of simulation models using the process interaction paradigm of simulation. The classes in this package are easy-to-use with the help of automatic code generation, hence, designing a model do not need any additional effort. Process package has some special classes to do some specific tasks; *VirtualScheduler* class to implement virtual time simulation, *Clock* class to keep the time of the simulation model for both pseudo-real-time and virtual-time simulations, *SimObject* to represent a single simulation entity and encompass its basic functionality, and *Model* class to encapsulate the basic features of a simulation model applet.
- **Event Package:** Event Package is used to build event-scheduling simulation models which consists of the Event classes, Entity and Scheduler to implement event scheduling.

6.2.2 J-Sim Simulation

Pseudo-real time simulation and virtual time simulation are both provided by J-Sim as depicted in Figs. 11 and 12, respectively. Virtual time simulation is used in case of a need for fast simulation in which the user do not want to be limited by the speed of a real time clock. Hence, the virtual clock is used instead of a real time clock. In pseudo-real time simulations, the user wants to see results of the simulation with animations. This type of simulation can also be used for testing the validity of a model where each entity is implemented as a Java thread. It is easy to switch between these two alternatives. Virtual Scheduler Class is a virtual time simulation manager which schedules of entities that generate events based on the time when the entity causes the event. Virtual Scheduler Class has some similar methods with Java's Thread methods, e.g., *vStart*, *vStop*, *vSuspend*, *vResume* etc.

J-Sim builds many classes on top of this. Here are some of the fundamental classes of J-Sim.

- **SimObject Class:** An instance of *SimObject* class can represent a single simulation entity. *SimObject* can extend the Java Thread class and enable each entity in a J-Sim model as a separate entity.
- **DynamicNode Class:** *DynamicNode* is an abstract class which encapsulates the features that appear as Server, Facility, Signal, Source and Sink.
- **Server Class:** Server class can be recognized as service provider which creates a number of service units and provides servers to *SimObject*s requesting service.

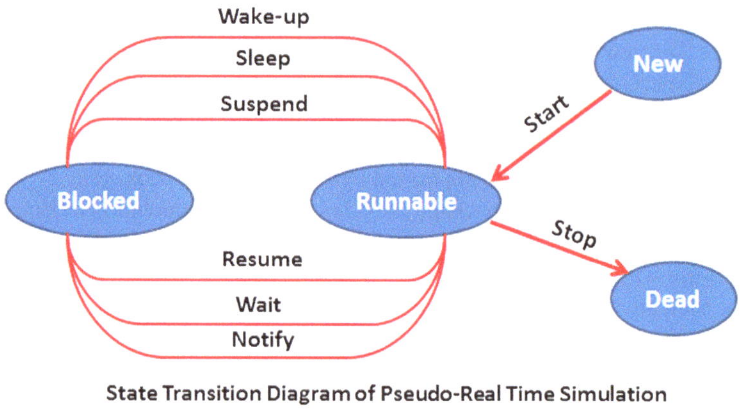

Fig. 11 State transition diagram of pseudo-real time simulation

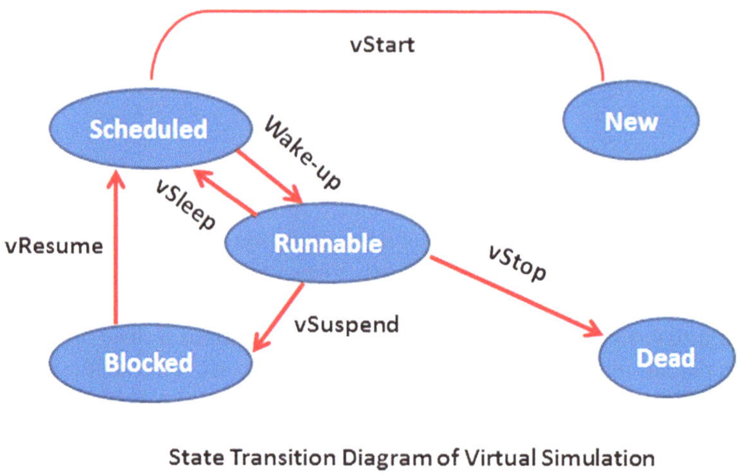

Fig. 12 State transition diagram of virtual time simulation

- **Facility Class:** This class is derived from Server class, hence it is very similar to server class. A queue is encapsulated as a private data member. A facility provides service to simulation entities via the *request* method.
- **Signal Class:** The behaviors of servers and increasing or decreasing number of service units are all affected by signal class. A traffic light can be recognized as a signal class in a simulation.
- **Source Class:** Source class can be recognized as the creator of *SimObjects* depending on the defined parameters, e.g., inter-arrival time, number of entities. Since simulation model will request creation of different types of entities, initially, source class is designed as an abstract class. To provide the required entity, the source class should be extended and *makeEntity* abstract method should be implemented.

Furthermore, the lifetime of the Source class is implemented to periodically create an entity to the inter-arrival time distribution.

- **Sink Class:** Sink class is developed as the opposite of the source class which destroys the *SimObjects*. When *SimObjects* complete their lifetimes, they directly go to the sink class to be terminated with the *capture method*.
- **Transport Class:** Edges of the simulation model graph that connect two nodes are generated by the transport class. Simulation entities travel along the transport class and move to the transport class with *move* method.
- **Model Class:** Model class enables multiple models to run simultaneously in a separate window frame. Furthermore, it provides statistical summary of the results when the simulation is over.

J-Sim simulations can run in two different modes; console mode and graphics mode. In the console mode, *JSimulation* is one of the fundamental classes of J-Sim which is inherited from *Object* and has the ability to execute the simulation. Hence, in every J-Sim simulation program, an instance of this class should take place. Every simulation object has some special tasks to perform. Here are some of them [25].

- **Information About Processes:** Simulation object need to update the processes in its list when a new process is created or killed. Every instance of *JSimulation* has a queue of *JSimProcessInfo* elements where information about the simulation processes are stored. *JSimProcessInfo* object is created and inserted into the queue that is called as *infoQueue*, whenever a new process is added to the simulation. *deleteProcess*() and *deleteAllProcesses*() are the methods used to delete the processes from the queue.
- **The Calendar:** A calendar of events as instances of *JSim Calendar* in simulation object list, should be updated when a process is activated, suspended or canceled. *JSim Calendar* represents a queue of *JSimCalendarItem* elements that each of them holds the time information of one event about when the event is scheduled and the number of processes which will be run at that time. Some methods of calendar perform some specific tasks, *addEntry*() to add a new element to the queue, *deleteEntries*() to delete the first or all events from the calendar, and *getFirstProcessTime*() to return the time of the first event in the calendar. The mentioned methods cannot be used by processes. *JSimulation* suggests some methods, e.g., *addEntryToCalendar*() and *deleteEntriesInCalendar*() to use them.
- **Simulation Time:** The simulation time should be kept and should be increased from its initial value, 0, as every step is executed. *JSimSimulation*'s *step*() method is the only place where the simulation time is changed. While it cannot be set by any other methods, it can be read by every method.
- The simulation object needs to know switching between console mode and graphics mode. To switch from console mode to graphics mode, the user need to call a method named *runGUI*(). Each simulation can create its own simulation mode as an instance of *JSimGUIMainWindow* class (Tables 1, 2).

Table 1 General comparison 1 of network simulators

Network simulator	Graphical support	Language support	Interface	Scalability	Available modules
OMNET	Yes	C++	NED	Large	Wired, wireless, ad hoc, and wireless sensor networks
NS-2	No	C++	OTcl	Small	Wired, wireless, ad hoc, and wireless sensor networks
J-Sim	Yes	Java	Tcl	Large	Wired, wireless, ad hoc, and wireless sensor networks

Table 2 General comparison 2 of network simulators

Network simulator	Extendible	Simulation technique	Documentation	Licence	Initial concentration focus
OMNET	Yes	Discrete event simulation	Medium	Open source	Wired network
NS-2	Yes	Discrete event simulation	Excellent	Open source	Wired network
J-Sim	Yes	Discrete event simulation	Lacks a comprehensive manual	Open source	Wired network

6.2.3 Tcl Language

Tcl is one of the most widely accepted scripting language in the world. Tcl language is used for configuration of simulation scenarios which requires a certain learning overhead. gEditor helps the configuration of Tcl files. The main contribution of Tcl is to enable topology setup and a limited means of simulation control. J-Sim offers higher-level Tcl-based programming interface. Tcl operations are all commands and written in prefix notation. Redefinition and overridden concepts are adopted. Tcl gains extendibility feature via Java in J-Sim. String handling, file system access, several native data types, unicode support, threading are some of features that supported by Tcl language.

6.2.4 Java Programming Language

J-Sim is a java-based tool. Java is widely supported by many communities and one of the most widely-spread and well-known programming language. Furthermore, its runtime environments and compilers are free of charge. Java provides many advantages for J-Sim. For instance, Java pre-compiled code can be interpreted in the target environment, hence, both source texts and pre-compiled code are portable. This important feature of Java is an important indication that Java is platform-independent and enables users to use the tool in all possible computing environments and user's simulations can be ported from one platform to another with minimum effort.

Java is an object-oriented language with the concepts of classes, instances, encapsulation, inheritance and polymorphism which provides the basic classes to be extended both in functionality and in data content. Java enables several processes to run concurrently, provides a safe method of their synchronization with the usc of *Thread* concept. With *Thread* class, instances can run parallel with other instances.

7 OPNET

OPNET is another discrete event network simulator written in C++. It is one of the most widely used network simulation programs. The downside of OPNET is not commercially available for free-of-charge. However, It can be used free of charge by the users applying to the University program of the simulator. The advantage of Opnet among some of the others is that OPNET provides modeling of different network-specific hardware, e.g., physical-link transceivers and antennas [31]. OPNET supports a various number of protocols, e.g., MAC protocols of IEEE 802.11a/b/g and Bluetooth technologies. OPNET also provides a user graphical interface for modeling, making graphs and animating the simulator results. Windows XP/2K, Linux and Solaris platforms support the OPNET network simulator.

As NS-2, OPNET also provides a comprehensive manual for technical support. Furthermore, it enables to collect some special kind of statistics, e.g., global statistics, node statistics, attribute statistics and animation statistic.

OPNET supports four simulation technologies as depicted in Fig. 13 [31];

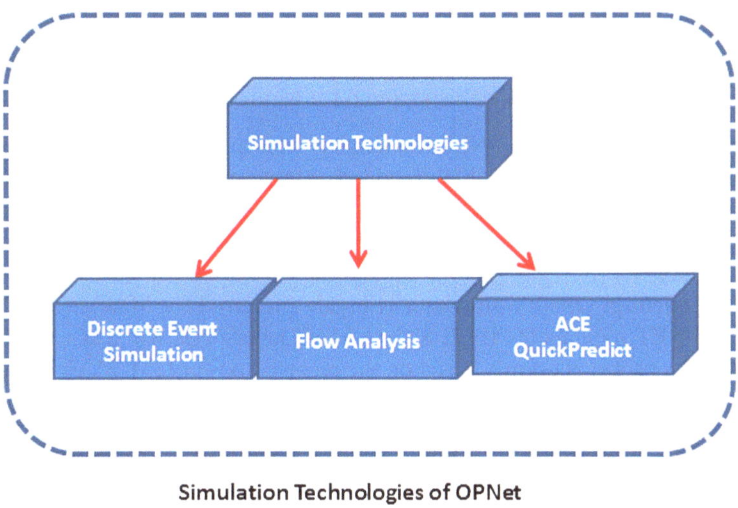

Fig. 13 OPNET simulation technologies

- **Discrete Event Simulation:** Highly detailed models which provide explicit simulation packets and control messages are supported with discrete event simulation feature. Discrete event simulation feature enables high-fidelity results, the simulation run times may take quite longer.
- **Hybrid Simulation:** Two different modeling techniques, e.g., analytical and discrete, are combined to achieve accurate and detailed results for specific flows. Background and explicit network traffics place important roles for hybrid simulation feature. Background traffic can be recognized as the network's ambient load at an abstract level, while explicit traffic models are used to detaily represent the selected network application flows. The runtime of the executions are quite faster compared to the discrete event simulation technology.
- **Flow Analysis:** Analytical techniques and algorithms are preferred by flow analysis technology to model the steady-state network behavior. Instead of modeling individual protocol messages or packets, Flow analysis technology prefers studying the routing and reachability features of the network in steady state, and in scenarios with one or more failed devices. As hybrid simulation technology, Flow analysis technique manages execution times faster than discrete event simulation technique.
- **Application Characterization Environment (ACE) Quick Predict:** Analytical technique is the preferred solution to monitor the impact on application response time of changing network parameters, e.g., bandwidth, latency, utilization, packet loss. OPNET ACE fully supports this technique.

7.1 WSN Simulation OPNET

OPNET enables simulation of different types of WSN. Terrain, mobility, path-loss, hybrid or analytical models are all supported for wireless network simulations.

- **Propagation Modeling:** OPNET implements free-space and longley-rice propagation models. Furthermore, it allows developers to create their own propagation models as well. Longley-rice model takes into account many parameters to measure the signal strength, such as, vertical, horizontal, polarization, earth's curvature and conductivity [32].
- **MAC Modeling:** OPNET simulator is highly preferred for MAC layer simulations. OPNET implements IEEE 802.11 MAC layer in its wireless module and decides if a packet is correct based on the received power [33]. MAC layer uses different metrics, e.g., delay, bandwidth, energy consumed per bit to measure the communication performance.
- **Routing Modeling:** There are no special routing protocols implemented in OPNET.

7.2 OPNET Simulation

A project-and-scenario approach is preferred to model the networks.

- **Project:** Project can be recognized as the collection of network-related scenarios, and each of them tried to address a particular aspect of the network design. All the projects can have only one scenario.
- **Scenario:** Scenario can be recognized as the single instance of a network. Scenario represents a unique configuration of the network, e.g., topology, protocols, applications, traffic, and simulation settings.

After that, there are three steps that need to be performed for the network simulation.

- **Specifying Data Collection:** One of the most important decision that must be decided before starting the simulation should be the form of the simulation result, e.g., application-specific, statistics, behavioral characterizations, visual animations, time-dependent series of values, parametric relationships or application-specific visualization [34]. Hence, in the first step the form of the simulation should be decided.
- **Construction of the Simulation:** The execution of the OPNET simulation program can be managed by using an executable file in the host computer's filesystem.
- **Execution the Simulation:** The last step is the simulation execution. If the results are not satisfying, additional changes can be made to the model's specification and additional simulations can be executed. Internal and external execution are some of the options for running the network simulations.

During these important steps, there are some sub-steps that should be made to start a simulation. Project editor is used to construct and edit the topology of a network model. To construct a network topology, the objects from an Object Palette can be dragged to the Project Editor workspace, or the existing network topologies can be loaded and built or from an external system, a network topology can be imported. Furthermore, OPNET provides a wide range of models from its libraries. For instance, a standard library contains subnetworks, nodes, links, LANs, clouds and utility objects. Several simulation scenarios can be modeled, and it is also possible to communicate with other simulators. OPNET supports the hierarchical modeling and describes a network as a collection of sub-models representing sub-networks or nodes. Manual creation of simulation topologies is available. OPNET supports a variety of different protocols while it enables the users to implement their own models. Stochastic models can be implemented since network nodes can be configured by just setting their parameters. Each node has transmission and reception modules as their protocol layer or physical resource to provide connections to the links. Message forwarding is the basic way for modules to communicate [31]. The simulation of wireless sensor networks is supported with the extension, named, the OPNET Modeler Wireless Suite. Full protocol stack modeling is supported with this extension, hence, all aspects of wireless transmission, RF propagation, transmitter, receiver characteristics, interference, node mobility etc., are modeled. MANET, IEEE

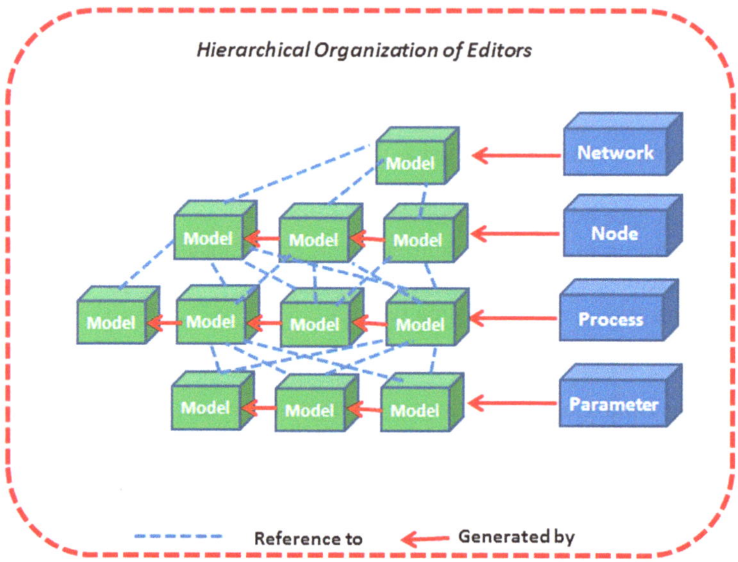

Fig. 14 Hierarchical organization of editors

802.11, 3G, Ultra Wide Band, IEEE 802.16, Bluetooth are some of the supported wireless technologies by OPNET.

As mentioned above, OPNET provides hierarchical modeling in which it provides three different editors to model the systems as depicted in Fig. 14. These are network, node, process editors that help OPNET to gain a hierarchical fashion.

- **Network Model:** A physical topology of a communications network can be specified to position the network entities, e.g., nodes, links. The node's behavior can be customized with a set of parameters or characteristics that are attached within each model. The network nodes can behave as fixed or mobile nodes, and duplex or simplex links can connect them. Radio links are also supported to provide mobile communications. It is possible to modify the links to simulate the actual communication channels. Subnetwork concept is introduces to decrease the complexity level of the network. This concept introduces the abstraction concept. Each subnetwork can contain multiple other subnetworks while the lowest level subnetwork consists of nodes and links. The communication between different subnetworks can be managed by communication links.
- **Node Model:** Node editor provides the required specifications for the network devices which are connected to the network. Node models can be recognized as interconnected modules. Packet generators, point-to-point transmitters and radio receivers can be introduces as the first group of node models which all have predefined characteristics and parameters. Processors and queues can be recognized as the second group of models which have high programable characteristics. All the models can be connected to each other with packet streams and static wires. A block structured data flow diagram is always used to represent a node.

- **Process Model:** Process editors construct the process models whose main responsibility is to introduce the logic flow and behavior of processor and queue modules. Furthermore, process models can also define the functionality of each programmable block in a node model. Proto-C language which contains the state transition diagrams, a library of kernel procedures, and the standard C programming language can define the process models [34]. A powerful state-transition diagram is preferred to support any specification of any type of protocol, resource, application, algorithm, or queuing policy. Child process can be generated to do the sub-tasks.

To gain more insights about the network elements, a brief description of each element is itemized as the followings:

- **Node:** Node is a basic network device with a wide range of possible capabilities, e.g., router, switch, hub, workstation, server, firewall, etc.
- **Link:** Link can be recognized as the physical media and properties, such as, line rate in bits per second, delay, likelihood of data corruption, etc. Line segments or a series of line segments with arrowheads represent the links.
- **LANs:** A LAN infrastructure is abstracted by the LAN object which enables the reduction of the amount of configuration required to do to represent an internetwork of LANs, and the amount of memory needed to run the simulation [31].
- **Cloud:** The WAN infrastructure is abstracted into one object by the cloud object which provides high-level characteristics, such as, packet latency and discard ratio.
- **Utility Objects:** Utility objects manage logical functions in the network, such as, configuration of network resources, scheduling special events, etc.

7.2.1 OPNET Probe Editor

There are some techniques to collect the simulation data. The simulation data can be automatic animation, customized programmed animation predefined and user defined statistics. Probe editor can be used to decide the type of outputted data. A probe can be assigned for each source of data which is desired by the user. Many probes can come together to provide a group to be used when the simulation is executed. There are many types of probes. Here are some of them.

- **Statistic Probe:** Statistic probe is a kind of probe that is applied to predefined, standard statistics and monitoring characteristics, e.g., bit error rates or throughput.
- **Custom Animation Probe:** This type of probe is also supported by process and link models. The characteristics of animation is defined.
- **Automatic Animation Probe:** Generation of animation sequences of the simulation is provided with the automatic animation probe.
- **Coupled Statistic Probe:** The output data is generated as the data generated by the statistic probe. However, additionally, a primary module and a coupled module should be defined. Some data is generated in primary module while some of them is generated in coupled module.

7.2.2 OPNET Analysis Tool

OPNET Analysis Tool provides the data visualization with the help of graphs. As mentioned before, the simulation results can be represented by different forms, e.g., f numerical data, animation, messages printed in the console window, to generation of ASCII or binary files, and even live interactions with other programs [34]. Rectangular areas in the graphs provided by analysis tool, are recognized as the analysis panels which can be produced with several different operations. A plotting area, with two numbered axes, which is generally referred to as the abscissa axis (horizontal), and the ordinate axis (vertical) is included in an analysis tool. Hence, OPNET Analysis Tool enables the user to extract the data from the output files to display them in various forms. Several mechanisms for numerically processing the data and generating new data sets for plotting purposes, e.g., computing probability density functions and cumulative distribution functions, histograms, are also introduced by the Analysis Tool.

7.2.3 OPNET Filter Tool

The data which is represented by OPNET Analysis Tool can also be operated by OPNET Filter Tool which is represented as the block diagram consisting of interconnected filter elements. These can be either built-in numeric processing elements, or references to other filter models [34]. The hierarchial model is introduced in filter tools which consist of other filters. The discrete and ordered sets of numeric data can be called as the vector o which the filters can also operate.

8 TOSSIM

TOSSIM is another discrete event simulator to ease the development of sensor network applications. First of all, we need to have a closer look at the TinyOS to better understand TOSSIM. TinyOS is the most popular event-driven, component-based operating system for use with embedded sensor nodes [35] and TOSSIM enables the simulation of these nodes by abstracting the hardware capability and software modeling [36]. TOSSIM can fully simulate the TinyOS network stack from the bit level, allowing experimentation with low-level protocols in addition to top-level application systems. TinyViz is the graphical user interface of TOSSIM which can visualize and interact with running simulations. The independent components are linked together to enable building an application-specific OS into each application. It is written with one of the most popular programming language, nesC, and it has the abilities to access and control the radio communication and hardware parts of sensor nodes. TOSSIM is actually a TinyOS simulator and utilizes the TinyOS's hierarchical model with the replacement of lower level hardware components with software emulated ones [37]. It can be compiled directly from TinyOS code which provides reducing the gap between real environment and the simulator. TOSSIM enable thousands of nodes to run simultaneously with high levels of reliability with the help of great design of TinyOS since it supports mote-based applications which are small in size and have their own private and static frames which reduce the simulation overhead [37]. Here are some of the advantages of TOSSIM.

- In scalability point of view, TOSSIM can handle wide range of networks with thousands of nodes, e.g., a network with 850 nodes, and simulate many of the sensor nodes at once even they have limited memory and CPU resources with high reliability levels.
- TOSSIM can highly perform well at capturing the behaviors of motes by providing wide range of experiments.
- TOSSIM supports a seamless PC-based TinyOS tool-chain which help developers to perform seamless transitions between running applications on motes and in simulation and also provides detailed visualizations of running simulations with GUI support.
- TOSSIM provides a bridging between algorithms and implementation to give a chance to developers to test their code on real hardware.

8.1 TOSSIM Simulation

For wireless sensor network, TOSSIM uses a very simple and powerful abstraction in which a WSN is represented as a directed graph where each vertex is a sensor node while each edge is a bit error probability. Hence, with this abstraction, many problems, i.e., hidden terminal problem, problems during data transmission, can be captured easily.

8.1.1 Radio Models

TOSSIM supports two basic radio models, i.e., simple which places all nodes in a single cell and lossy which places nodes in a directed graph. Simple radio model is used for testing single-hop algorithms and provides every bit to be transmitted and received without any errors. Lossy radio model enables TOSSIM to model asymmetric links and capture many reasons of packet loss and noise in the network.

8.1.2 Network Monitoring and Visualization

The SerialForwarder is the standard TinyOS interface tool which enables user to interact with the simulated network. TOSSIM provides two modes, i.e., serial port mode and the snooping mode to connect to SerialForwarder. On the other hand, message interface generator is another tool to generate necessary Java classes for TinyOS packets.

8.2 WSN Simulation of TOSSIM

The basic idea of TOSSIM is to provide a flexible, simple and efficient environment for the needs of users, hence, it does not provide any modeling about the real world. However, it enables outside tools to manage the necessary modeling according to the needs of simulation.

- **Propagation Modeling:** TOSSIM does not provide a radio propagation modeling for WSN, however, it supports a radio abstraction of bit errors between two nodes in the network which provides an easy modeling for asymmetric links. Furthermore, TOSSIM enables an external program managing the required radio propagation modeling and mapping it to the bit errors between nodes.
- **Energy Consumption Modeling:** TOSSIM, itself, does not provide a proper energy modeling, or energy-consumption estimation modeling for wireless sensor nodes since it has instantaneous transitions from an event to another with no tracking of execution time. Hence, to add proper energy-consumption estimation features to it, PowerTOSSIM is written as an extension, event-driven simulation environment to TOSSIM to handle accurate and per-node estimations of power consumptions. It provides a detailed energy model for each type of sensor nodes, for instance, the first model was developed for Mica2 which is the third generation multi-channel radio transceiver used for low power WSN, while the second model was developed for MicaZ which offers a 2.4 GHz, IEEE/ZigBee 802.15.4, with PowerTOSSIM-z [37, 38] (Table 3).

9 Other Major Programming Languages for Wireless Sensor Networks

WSN consist of thousands of sensor nodes which are easy-to prone to failure, hence, even they are not physically reachable, they should be reprogrammable for the changing needs with the lowest energy-consumption [39]. Here are some of the major programming languages for WSN.

9.1 nesC

nesC is a very popular low-level network programming language used for wireless sensor networks to support special needs of sensor nodes by providing event-driven execution, component-oriented application design, and a flexible concurrency model [40]. It is an extension of C language and supports almost the same data types, structures and set of operators. nesC provides many advantages in terms of reduction in code size, simplification of application development, reliability and resource

Table 3 Modeling suitability comparison of network simulators [15]

Modeling suitability	OMNET++	J-Sim	NS-2	OPNET
Creation new models	Yes	Yes	Ycs	Not specified
Extendibility to existing and predefined models	Yes	Yes	Yes	Yes
Generic models support	Yes	Yes		Not specified
Distributed on a platform network	Parallel simulation is possible			Yes
Hosting operating systems	Windows, Linux, Unix	All platforms are supported	Windows, Unix, free Bsd	Windows, Linux, Sun solaris
Hosting heterogeneous modeling formalisms	Stochastic and deterministic	Stochastic and deterministic	Stochastic and deterministic	Stochastic and deterministic
Support for hierarchical modeling	Yes	Yes		Yes
Easy integration with other tools	Yes	Yes	Yes	Yes
Continuous and discrete variable representation	Yes	Yes	Yes	Yes
Stochastic modeling	Yes	Yes	Yes	Yes

consumption, for instance, it introduces whole-program analysis which includes data-race detection in reliability point of view and aggressive function inlining in resource consumption point of view [40]. It has a strong relationship with TinyOS, the micro-based event-driven operating system for sensor nodes. nesC language has been used for development of WSN applications on TinyOS platform and. The holistic system design of nesC has been realized as the key focus of it. This design provides mote applications to be tied to hardware and gives a chance to each mote to run a single application at one time. Hence, all resources are known and designing flexible decompositions is easy since hardware and software boundaries vary on the application and hardware platform. Furthermore, nesC programming language has two important modular concepts, the interface and component. While components are assembled to form the whole program, interfaces are responsible to declare some *events* which imply the completion of an operation and *commands* which indicate the requests for that operation. There are two types of components, i.e., *configuration*which indicates the number of components that are wired to one another through interfaces and *module* which is responsible to implement the commands that have been identified by the interfaces or events. Here are some of the advantages of nesC language.

- nesC programming language supports a strong component model which supports event-driven systems, a flexible hardware/software boundary and provides bidirectional interfaces and efficient implementations.
- nesC programming language provides applications to show synchronous behaviors with very limited resources since it's compiler can detect most of the data races at compiler time.
- nesC programming language performs very well in terms of improving reliability and reducing code size.
- Whole program inlining, dead-code elimination, static component instantiation are some of the other advantages that nesC provides.

Here are some of the shortcomings of nesC language.

- nesC is a static programming language which does not support dynamic memory allocation or function pointer.
- nesC has complex syntax and semantics according to other networking languages. Hence, it is not easy to establish formal models from nesC programs.
- Modeling nesC programs on TinyOS may be complicated since it is also required to model the hardware operations on motes.

Apart from all the advantages and disadvantages, nesC should accomplish some goals which are related to the nature of WSN. One of the most important goal of nesC is to provide long-lived applications for WSN for reliable data communications since sensor nodes may have to collect data for long-periods without human interaction. On the other hand, nesC has to deal with the limited physical resources that sensor nodes have.

9.2 Mate

Mate is the first virtual machine for sensor networks and implemented on top of the TinyOS. It is basically a byte-code interpreter to run on sensor nodes and designed to meet the specific requirements of sensor nodes which are limited in processing and energy-consumption. Mate can quickly and easily manage the installation of wide range of programs with the lowest energy-consumption and network traffic since it provides capsules of 24 instructions to fit in a packet and forward themselves via the network [39]. Mostly single byte instructions, bytecode, form its instructions set and arithmetic operations, variable and stack manipulations are encoded as instructions. Mate has two stacks, *operand stack* which provides the environment to most instructions to operate on, and return address stack. Mate has three execution contexts, i.e., clock timers, message send requests and message receptions, that can run simultaneously on instruction granularity [39].

Here are some of the highlighted advantages of Mate.

- Mate has a small, robust and simple programming model which provides memory, execution and energy-consumption efficiency for sensor nodes.
- Mate provides secure operations and prevents malicious capsules.
- Mate hides race conditions and asynchrony of TinyOS which prevents the need for managing the message buffers.
- Mate supports both a built-in ad hoc routing algorithm and mechanisms for writing new ones.

10 General Comparison of Network Simulators from WSN Perspective with Example Scenarios

- **IEEE 802.15.4 Simulation Point of View:** Most of the existing network simulation tools were originally designed for wired networks, and then extended for wireless sensor networks. Hence, realistic wireless sensor network simulations cannot be conducted. The energy consumption of sensor nodes is one of the most important criteria that must be taken into account while designing a simulation. However, most of the network simulators cannot support this feature in a reliable manner. For instance, NS-2 was developed for IP networks, however, later it is extended for IEEE 802.11 wireless networks. IEEE 802.11 MAC, on the other hand, is not efficient for the battery-powered wireless sensor nodes, since it introduces additional overheads. Hence, NS-2 cannot provide an efficient and reliable simulation environment for wireless sensor network simulations, e.g., ZigBee. However, in the literature, there has been some researches conducted to develop a reliable model for IEEE 802.15.4 protocols for WSN simulations with OPNET simulator [41, 42]. In [41], the OPNET modeler simulator is used to create a reliable simulation tool for IEEE 802.15.4 slotted CSMA/CA mechanism with the support of deterministic real-time traffic, and cluster-tree topology and hierarchical tree routing mechanisms. J-Sim is not suitable to simulate IEEE 802.15.4 networks, since it only implements the IEEE 802.11 MAC.
- **IEEE 802.11 Simulation Point of View:** NS-2 network simulator was originally developed for simulation of TCP/IP, multicast protocols, and wireless sensor network support has been added later. However, the extensibility feature of NS-2 enables the integration of new protocols. J-Sim, on the other hand, provides an energy modeling for sensor networks. With its component based architecture model, it enables the sensor network to scale well (Table 4).

Table 4 Opportunities and challenges of network simulators in WSN point of view

Simulation suitability	Opportunities	Challenges	Support of MAC protocols
OMNET++	Supports a graphical network editor	Needs other frameworks to run WSN simulations	Aloha, CSMA (MiXiM)
J-Sim	Supports energy modeling, component-based architecture	Radio energy consumption is not supported	IEEE 802.11
NS-2	Visualization is available and very popular	Customization, and application model is not enabled	IEEE 802.11, a single-hop TDMA protocol
OPNET	Supports very large-scale WSN	Introduces complexity in propagation models	IEEE 802.15.4
TOSSIM	Provides accurate and reliable simulations	Energy consumption model is not supported	–

11 Shortcomings of Network Simulators

- **NS-2:** NS-2 is one of the most complex network simulators. It needs advanced skills and a long-learning process to create reliable and meaningful simulations. On the other hand, the customization is not available and application model which is a very important component for sensor networks to make interactions with network protocol level is not included in NS-2.
- **TOSSIM:** The energy consumption is the most important measurement of sensor nodes in WSN. However, TOSSIM does not model the energy consumption, hence it is, alone, not preferable simulator in most cases. On the other hand, TOSSIM makes assumptions very simplified which may lead a perfectly running simulation not run on a real mote, furthermore, the TOSSIM simulation results should not be recognized as the final-point of evaluations.
- **J-Sim:** J-Sim is another complex simulator tools, and the downside of it is that only IEEE 802.11 MAC protocol has been implemented so far. Hence, it is not a preferred simulator tool for realistic WSN simulations. Furthermore, it introduces some additional overhead and some inefficiencies regarding to the Java programming language.
- **OPNET:** OPNET is one of the most preferred simulator for WSN, however, it does not support any special routing protocols.
- **OMNET++:** OMNET++, by itself, is not a wireless sensor network simulator. Castalia, for instance, is a WSN simulator built on top of OMNET++. It has some limitations, for example, it is not a sensor specific platform (Table 5).

Table 5 Simulation suitability comparison of network simulators [15]

Simulation suitability	OMNET++	J-Sim	NS-2	OPNET
Fast simulation capabilities	Not specified			Yes
Exchanging information with other simulators				Yes
Real network/system interactions	Yes	Yes	Yes	Yes
Trace files of the simulation results creations	Yes	Yes	Yes	Yes
External information as input	Yes	in output files	Yes	Yes
GUI for model creation/management	C++ editor for component behaviour	component behavior in Java	Yes	Yes
GUI for scenario creation/management	GNED for topology	Yes	Limited	Yes
Design, archive and modify scenarios	Yes	Yes	Yes	Yes
Network visualization tools and network data base systems	Yes	A plotter only	Yes	Yes
Analysis tools for state estimations of the network, results analysis, etc.	Yes	A plotter only	Yes	Yes
Disturbances/corruptions generations inside the simulated networks	Defined in C++ code	Yes	Yes	Yes
Simulation of SCADA system components	Yes	Yes	Yes	Yes
Simulation of electricity and telecommunication networks	Generic modeling	Generic modeling	Telecom	Telecom

12 Other Network Simulators

There are many other network simulators which are briefly covered in this section. Here are some them.

- **GloMoSim:** GloMoSim is another popular network simulator platform which focuses on simulation of mobile wireless networks [43]. GloMoSim is written in Persec programming language which is the extension of C programming language for parallel programming. The user needs to know how to write in Persec to add new protocols or methods. As NS-2, GloMoSim realizes the OSI concept. GloMoSim is very good at simulating IP networks, however, its performance is not good enough for other types of networks. GloMoSim has been updated as a commercial product, QualNet.
- **QualNet:** QualNet is a commercial product based on GloMoSim network simulator to simulate ad hoc networks [44]. Persec programming language is used

for developing the QualNet. The main focus area of QualNet is wireless sensor networks, hence, it provides several set of tools for network simulation and modeling.

- **Atarraya:** Atarraya is another event-driven simulation platform for teaching and researching topology control algorithms and protocols for wireless sensor networks. Atarraya enables creation and testing of old and new topology control protocols with the help of its graphical interface to illustrate how these protocols work. Several topology construction and several topology maintenance algorithms and protocols, deployment generation using different distributions and support for different energy models are all the features that Atarraya supports [45].
- **JiST:** JiST which referees to the *Java in Simulation Time*, is another network simulator for mobile ad hoc networks [46]. The official development of JiST is no longer supported by its original author, however, Ulm University has released some enhancements and improvements.

Acknowledgments This work is funded by WiSeMAN Research Lab Department of Computer and Information Science College of Engineering and Computer Science University of Michigan-Dearborn. The authors gratefully acknowledge the insightful comments of the anonymous reviewers which helped improve the quality and presentation of the paper significantly. This work is partially supported by the US National Science Foundation (NSF) grants 0917089 and 1054935.

References

1. S. Misra, M. Reisslein, G. Xue, A survey of multimedia streaming in wireless sensor networks. IEEE Commun. Surv. Tutorials **10**, 1553–1877 (2008)
2. A Comparison Study of Network Simulators, Online: https://ti.tuwien.ac.at/ecs/people/albeseder/simcomp/simcomp
3. NS-2 Official WebSite, Online: http://www.isi.edu/nsnam/ns/
4. S. Ivanov, A. Herms, G. Lukas, Experimental validation of the NS-2 wireless model using simulation, emulation, and real network in *Proceedings of the 4th Workshop on Mobile Ad-Hoc Networks (WMAN'07)*, pp. 433–444. VDE Verlag, (2007)
5. K.M. Reineck, Evaluation and comparison of network simulation tools. Master thesis (2008)
6. H. Tseng, S.-H. Yang, P.-Y. Chuang, H.-K. Wu, G.-H. Chen, An energy consumption analytic model for a wireless sensor MAC protocol. *2004 IEEE 60th Vehicular Technology Conference, 2004. VTC2004-Fall*, vol. 6, pp. 4533–4537, 26–29 Sept 2004
7. I. Vijaya, A.K. Rath, Simulation and performance evaluation of AODV, DSDV and DSR in TCP and UDP environment. *3rd International Conference on Electronics Computer Technology (ICECT), 2011*, vol. 6, pp. 42–47, 8–10 April 2011
8. S. Wasiq, W. Arshad, N. Javaid, A. Bibi, Performance evaluation of DSDV, OLSR and DYMO using 802.11 and 802.lip MAC-protocols. *IEEE 14th, International Multitopic Conference (INMIC), 2011*, pp. 357–361, 22–24 Dec 2011
9. J. Ma, The study on multi-path DSDV in Ad Hoc. *IEEE 3rd International Conference on Communication Software and Networks (ICCSN), 2011*, pp. 299–303, 27–29 May 2011
10. N. Javaid, A. Bibi, A. Javaid, S.A. Malik, Modeling routing overhead generated by wireless proactive routing protocols. *2011 IEEE GLOBECOM Workshops (GC Wkshps)*, pp. 1072–1076, 5–9 Dec 2011
11. L. Shrivastava, S.S. Bhadauria, G.S. Tomar, Performance evaluation of routing protocols in MANET with different traffic loads. *2011 International Conference on Communication Systems and Network Technologies (CSNT)*, pp. 13–16, 3–5 June 2011

12. K. Han, G. Pei, B. Ravindran, C. Hyeonjoong, E.D. Jensen, RTRD: real-time and reliable data delivery in Ad Hoc networks. *WCNC 2008. IEEE Wireless Communications and Networking Conference, 2008*, pp. 2253–2258, 31 March, 3 April 2008
13. S. Sagar, N. Javaid, Z.A. Khan, J. Saqib, A. Bibi, S.H. Bouk, Analysis and modeling experiment performance parameters of routing protocols in MANETs and VANETs. *2012 IEEE 11th International Conference on Trust, Security and Privacy in Computing and Communications (TrustCom)*, pp. 1867–1871, 25–27 June 2012
14. S.S. Kushwah, G.S. Tomar, Investigation of effects of mobility on routing protocols in MANET. *2011 International Conference on Ubiquitous Computing and Multimedia Applications (UCMA)*, pp. 82–84, 13–15 April 2011
15. White Paper, IRRIIS integrated risk reduction of information-based infrastructure systems. List of available and suitable, simulation components (2006)
16. OMNETT++ Community Website, Online: http://www.omnetpp.org/ Last visit: 28 June 2012
17. Q. Xue, X. Ren, Research of routing protocols simulation for wireless sensor networks based on OMNeT++. *2012 International Conference on Quality, Reliability, Risk, Maintenance, and Safety Engineering (ICQR2MSE)*, pp. 79–82, 15–18 June 2012
18. A. Kuntz, F. Schmidt-Eisenlohr, O. Graute, H. Hartenstein, M. Zitterbart, Introducing probabilistic radio propagation models in OMNeT++ mobility framework and cross validation check with NS-2. in *Proceedings of the 1st International Conference on Simulation tools and Techniques for Communications, Networks and Systems and Workshops (Simutools '08)*, Article 72, p. 7. ICST (Institute for Computer Sciences, Social-Informatics and Telecommunications Engineering), ICST, Brussels, Belgium, Belgium, 2008
19. J.-C. Maureira, P. Uribe, O. Dalle, T. Asahi, J. Amaya, Component based approach using OMNeT++ for train communication modeling. *2009 9th International Conference on Intelligent Transport Systems Telecommunications, (ITST)*, pp. 441–446, 20–22 Oct 2009
20. F. Chen et al., An energy model for simulation studies of wireless sensor networks using OMNeT++
21. INET Framework Manual, Online: http://inet.omnetpp.org/doc/inet-manual-DRAFT.pdf, Last visit: 28 June 2012
22. Castalia User Manual, Online: http://castalia.npc.nicta.com.au/
23. MIXIM Manual, Online: http://mixim.sourceforge.net/index.html
24. OverSim User Manual, Online: http://www.oversim.org/
25. J-Sim Community Website, Online: http://www.J-Sim.zcu.cz/
26. M. Malowidzki, Network simulators: a developer's perspective. Military Communication Institute, Zegrze
27. A. Sobeih, J.C. Hou, L.-C. Kung, N. Li, H. Zhang, W.-P. Chen, H.-Y. Tyan, H. Lim, J-Sim: a simulation and emulation environment for wireless sensor networks. IEEE Wirel. Commun. **13**(4), 104–119 (2006)
28. K. Sarabandi, I. Koh, G. Liang, H. Bertoni, Propagation modeling for FCS. in *Proceedings of the IEEE Military Communications Conference (IEEE MILCOM'01)* Oct 2001
29. F. Kuhn, R. Wattenhofer, Y. Zhang, A. Zollinger, Geometric Ad-Hoc routing: of theory and practice. in *Proceedings of the ACM Symposium on Principles of Distributed Computing (ACM PODC'03)*, July 2003
30. J.A. Miller et al., JSIM: a Java-based simulation and animation environment. Department of Computer Science, University of Georgia, Athens.
31. OPNet Official Website, Online: http://www.opnet.com/solutions/network_rd/modeler.html
32. P.L. Rice, A.G. Longley, K.A. Norton, A.P. Barsis, Tech note 101: transmission loss predictions for tropospheric communication circuits, vol. I, II. U.S. Government Printing Office, Washington, Jan 1967
33. A. Daga, D.K. Borah, G.R. Lovelace, P. De Leon, Physical layer effects on MAC layer performance of IEEE 802.11 a and b WLAN on the Martian surface. *IEEE Aerospace Conference*, p. 8, 2006
34. X. Chang, Network simulations with OPNET. *Proceedings of the 1999 Winter Simulation Conference*

35. A.I. McInnes, Using CSP to model and analyze TinyOS applications. *16th Annual IEEE International Conference and Workshop on the Engineering of Computer Based Systems ECBS 2009*, pp. 79–88, 14–16 April 2009
36. TOSSIM Official WebSite, Online: http://www.cs.berkeley.edu/pal/research/tossim.html
37. P., Enrico et al., PowerTOSSIM z: realistic energy modelling for wireless sensor network environments. *Proceedings of the 3nd ACM Workshop on Performance Monitoring and Measurement of Heterogeneous Wireless and Wired Networks*. ACM 2008
38. V. Shnayder, M.Hempstead, B. Chen, G.W. Allen, M.Welsh, Simulating the power consumption of large scale sensor network applications. Division of Engineering and Applied Sciences, Harvard University (2003)
39. P. Levis, D. Culler, Mate: a tiny virtual machine for sensor networks. SIGOPS Oper. Syst. Rev. **365**, 85–95 (2002)
40. D. Gay, P. Levis, R.v. Behren, M. Welsh, E. Brewer, D. Culler, The nesC language: a holistic approach to networked embedded systems. in *PLDI*, p. 111, 2003
41. P. Jurcik, A. Kouba, M. Alves, E. Tovar, Z. Hanzalek, A simulation model for the IEEE 802.15.4 protocol: delay/throughput evaluation of the GTS mechanism. in *Proceedings of 15th IEEE International Symposium on Modeling, Analysis, and Simulation of Computer and Telecommunication Systems (MASCOTS07)*, Istanbul (Turkey), Oct 2007
42. A. Kouba, M. Alves, B. Nefzi, Y.Q. Song, Improving the IEEE 802.15.4 slotted CSMA/CA MAC for time-critical events in wireless sensor networks. in *Proceedings of the Workshop of Real-Time Networks (RTN 2006), Satellite Workshop to (ECRTS 2006)* July 2006
43. Glomosim Official Website, Online, http://pcl.cs.ucla.edu/projects/glomosim/
44. QualNet Official Website, Online, http://www.scalable-networks.com/content/
45. Atarraya Official WebSite, Online, http://www.csee.usf.edu/ labrador/Atarraya/
46. E. Weingartner, H.V. Lehn, K. Wehrle, A performance comparison of recent network simulators. Distributed Systems Group RWTH Aachen University Aachen, Germany

Chapter 23
Network Architectures and Standards

Dilan Sahin and Habib M. Ammari

Abstract Wireless sensor network is composed of a collection of sensor nodes that sense the physical phenomena for further analysis purposes. ZigBee, WirelessHART, 6LoWPAN and ISA.100.11a are some of the wireless sensor network technologies that are presented throughout this chapter with the details of their network structure, protocol layers, key characteristics and application areas.

1 Introduction

There has been tremendous interest in wireless sensor networks (WSNs) technologies from academia, industry and technology developers due to its seamless, energy efficient, reliable, low-power and low-cost features. Basically, a WSN consists of a collection of sensor nodes spread over a geographical area to sense the physical environment and send the information to the collection points for further data analysis purposes [69]. Sensor nodes have limited computational capabilities, memory resources and bandwidth to be low-power and prolong the sensor network lifetime. WSN brings comfort to daily lives with advance applications, e.g., advanced control to healthcare systems, remote monitoring and control to security systems and advanced control and monitoring capabilities to automation industry, power grid, military services [29, 41, 62]. To be able to support these applications and address their specific needs, different protocols were proposed for WSNs.

D. Sahin
University of Michigan-Dearborn, Dearborn, MI, USA
e-mail: dsahin@umd.umich.edu

H. M. Ammari (✉)
WiSeMAN Research Lab, Department of Computer and Information Science, University of Michigan-Dearborn, Dearborn, MI48128, USA
e-mail: hammari@umd.umich.edu

H. M. Ammari (ed.), *The Art of Wireless Sensor Networks*,
Signals and Communication Technology, DOI: 10.1007/978-3-642-40009-4_23,

© Springer-Verlag Berlin Heidelberg 2014

This chapter introduces the key characteristics, protocol structures, application areas and network architectures of ZigBee, WirelessHART, 6LoWPAN and ISA.100.11a technologies, and finally gives a brief comparison in terms of network topology, reliability, data rate, security, power consumption and scalability features.

ZigBee provides an effective, low-power, low cost connectivity for a large number of devices into a single network. It has been an advantageous technology for most of the application areas, e.g., house automation, medical care, smart buildings, industrial control, since it overcomes the limits of wire connections [86]. However, the current trend has been moving to IP-based wireless networking technologies, such as, 6LoWPAN, which form a strong bridge between Internet and wireless embedded devices [41].

6LoWPAN is low-cost, low-power WSN technology that interconnects low-power wireless area networks to IP networks. However, switching to a new technology will bring its own obstacles, since, most of the wireless sensors have been deployed over years.

WirelessHART, on the other hand, is the first open wireless technology in industrial processing applications.

As mentioned before, there are a variety of different WSN-based applications that each of them has different, and unique requirements, however, the overall requirements can be generally itemized as the followings.

- **Energy Consumption Efficiency:** Energy consumption efficiency is one of the most critical requirements for most of the WSN applications, since the sensor nodes are battery powered, and they need to survive for long period of times unattended [40]. Hence, they have to consume the energy wisely to increase the life-time of sensor network and provide reliable and continuous wireless data communications since battery recharging/replacementmay not be feasible in large-scale deployments. There have been tremendous efforts towards minimizing the energy consumption of sensor nodes [43]. The most popular approach towards this problem is switching the mode of a sensor node to the sleeping mode, since a sensor node may spend as much energy in the idle node as in receiving and transmitting mode [6, 81]. Furthermore, the size of the sensor transmission radius, the type of operating system, sensor' event detection delay, the preferred routing algorithms, network topology, participation density are some of the other parameters that affect the energy consumption of a sensor node [3, 6, 21]. Dutta et al. proposed clustering techniques to provide optimal energy consumption among sensor nodes [11], while Unterassinger et al. proposed an energy management unit by employing several state machines to control different energy consumption domains by turning on/off depending on the operating mode of the wireless sensor node [63].
- **Security:** Security is another critical requirement that must be addressed for the continuity of the reliable WSN communication, since sensor nodes are vulnerable to various types of malicious attacks and threats, and in most applications, e.g., disaster recovery, bio-weapons, military, the wireless data is security fragile [4]. The data transmission between sensor nodes are placed over unreliable wireless channel, and in most scenarios, they are unattended. Furthermore, they are limited

in memory, processing and computational power. All these factors introduce major obstacles in the implementation process of traditional efficient security techniques, e.g., cryptography, authentication and key management, for instance, applying encryption techniques may require the transmission of additional data bits, increase the delay and packet loss [38, 54]. There has been various research on various security schemes proposed or implemented so far for wireless sensor networks. Holohan et al. proposed a new security paradigm for WSN which provides a mechanism to combat with the difficulties in securing resource constrained devices on distributed ad hoc networks by establishing primary-key infrastructure concepts [24]. Furthermore, Karlof et al. focused on secure routing in WSN [31], while Wood et al. introduced a mapping protocol, JAM, which detects the faulty region and prevents the data routing among that region [80].

- **Self-Healing:** Self-healing ability is another important requirement for WSN, since, sensor nodes operate unattended and autonomously in harsh environmental conditions which may easily lead to node failures or sensor nodes' battery depletion or changes in topology. Hence, the sensor nodes should easily adjust to the new network conditions and heal itself for the continuity of quality of service (QoS) and reliable data communications [68]. In academia, there have been many studies in this special area, for instance, Dutta et al. developed a new self-healing key distribution scheme with revocation capability to recover and increase the scalability of WSN and efficiency in storage and communication overhead over a lossy wireless network [10], while Han et al. proposed another self-healing distribution scheme to help group users to recover from the loss of broadcast messages over wireless network [19]. On the other hand, Qui et al. proposed an efficient solution for especially ZigBee networks to recover from any node failures or communication breakdowns [68].

2 IEEE 802.15.4 Protocol

IEEE 802.15.4 standard is the leading standard for most of the wireless sensor network technologies. It is a low-cost, low-power consumption and low-data rate wireless standard which has been increasingly replacing the wired technologies in existing applications. It specifies the physical layer and media access control layer for wireless technologies, e.g., ZigBee, WirelessHART, ISA100.11a etc, however it does not support upper layers or networking methods [15]. The standard provides enough flexibility in terms of distance and positions of devices in the network to meet the special requirements of WSN applications [9].

2.1 Protocol Structure of IEEE 802.15.4

- **Physical Layer:** Physical layer specifies the radio characteristics and responsible from data modulation, sending and receiving data. It provides 27 channels in 868 MHz (available in Europe), 915 MHz (available in US), 2.4 GHz (available world wide) bands [49]. The standard supports two physical layers, direct sequence spread spectrum (DSSS) technique and frequency-hopping spread spectrum (FHSS) to increase the reliability of the communications.
- **MAC Layer:** MAC layer has a key role in defining how efficiently and low-cost a communication is carried on [15]. The basic functionality is to transfer MAC frames via the physical channel with very low duty cycles which is appealing for WSN applications since the energy consumption is one of the main concerns. IEEE 802.15.4 MAC layer specifies two medium access modes; beacon-enabled mode (coordinated mode) and nonbeacon-enabled mode (uncoordinated mode) [9]. Beacon frames are periodically broadcasted by the network coordinator for the synchronization of equipment in the network in beacon-enabled communication. This type of mode is preferred when the network uses star topology. The frames can be transmitted from a device to a coordinator in which a device uses carrier sensing multiple access with collision avoidance (CSMA-CA) or the transmission can be made from a coordinator to a device [49]. Sending beacon frames in nonbeacon-enabled communication becomes randomly according to the channel state, if channel is busy, it is waited for a random period of time, if channel is idle, the data is transmitted [85]. MAC layer defines four frame structures; *beacon frame* used by network coordinator for beacon transmissions, *response frame* is the simplest frame among the others which is used for the confirmation of successful frame receptions, *data frame* is used for almost all data transmissions, and *command frame* is a simple frame that is responsible for network connection, disconnection and valid channel access.

2.2 Network Architecture of IEEE 802.15.4

Two types of devices, full-function device (FFD) and reduced-function device (RFD), have been defined in the standard. FFD is basically capable of performing all the network functionalities, while RFD is limited to perform a set of simple functions, such as, measuring some network parameters, e.g., light, humidity, temperature [9]. Each network is controlled by a network coordinator which is responsible for setting up the network and maintaining all the necessary functionalities. FFD can take the responsibilities of a network coordinator while RFD just has the abilities to connect with network coordinator and join/disjoint the network. Star and mesh networking topologies are supported by the standard depending on the application requirements.

3 ZigBee Wireless Standard

ZigBee is a standardized, short-distance, low cost, low power wireless technology based on IEEE 802.15.4 protocol, designed to meet the increasing demands for low-powered and efficient wireless networking between sensory and control network applications. ZigBee is proposed by the ZigBee Alliance which is a group of companies providing innovative, reliable and easy-to-use ZigBee standards [25]. ZigBee is a self-organizing and self-healing wireless mesh network technology, supporting more than 64,000 devices on a single network, operating at 2.4 GHz Industrial, Scientific, and Medical (ISM) non-licensed frequency band with 16 channels for global use, 915 MHz with 10 channels for North America and 868 MHz with one channel for EU countries [26]. ZigBee technology allows 250 Kbs data rate at 2.4 GHz, 40 Kbs data rate at 915 Mhz and 20 Kbs at 868 Mhz, and also transmission distances vary between 10–100 m, depending on the environmental characteristics [26].

Two implementation options are offered for ZigBee Specification; ZigBee and ZigBee PRO feature sets. While ZigBee feature set is designed for smaller networks, ZigBee PRO feature set improves the capabilities of ZigBee feature set to support larger networks by providing the same reliability and stability features [27].

ZigBee standard has entered the wireless technologies market with extraordinary control, scalability, security, ease-of-use, low-cost and low-power features. ZigBee offers many advantages in wireless communication over other wireless technologies as presented as foll

- **Reliable:** ZigBee is considered as a reliable technology since it applies end-to-end acknowledgment (ACK) and retransmission mechanisms with filtering duplicate packets [34]. The use of link quality metrics is an important indicator for the reliability assignment of the technology. ZigBee benefits the link quality estimator offered by IEEE 802.15.4 based on the bit error rate (BER) estimate which is implemented in many radio chips. ZigBee star topology is the least complex topology which can lead to simplicity and increased reliability, on the other hand, peer-to-peer topologies may increase the reliability since they include multiple paths to the ZigBee coordinator and ZigBee end devices.
- **Interference Avoidance:** Most of the technologies used in home electrical devices, such as Bluetooth, Wi-Fi, wireless USB and microwave ovens, share the same, 2.4 GHz ISM band with ZigBee. Hence, the interference problem is unavoidable in such situations. ZigBee uses different spread-spectrum techniques to protect itself from multipath and narrowband interference [17]. Indeed, IEEE 802.15.4 suggests various channels in the 915 MHz and 2.4 GHz bands for ZigBee. ZigBee can select the least interfered channel to fight against the interference. Moreover, ZigBee coordinator have the capability to reunite the ZigBee network in a different channel in case of any interference situations [17]. Furthermore, IEEE 802.15.4 MAC layer is based on CSMA/CA(Carrier sense multiple access with collision avoidance) that the sensor node will listen the channel before transmitting the packet.

- **Global Implementation:** ZigBee protocol has a global acceptance in almost all countries since it adopts IEEE 802.15.4 PHY layer with 16 channels for global use, 915 MHz with 10 channels for North America and 868 MHz with one channel for EU countries [26].
- **Low Cost:** ZigBee-based devices and platforms are cost-effective with simplicity and the underlying flexibility of 802.15.4 protocol. The system maintenance, flexibility, and battery life should be also considered in the overall system cost [67].
- **Security:** ZigBee has strong the encryption and authentication mechanisms and they are applied at application, network and MAC layers. While MAC layer manages its own security processing, other layers can decide which security level to use. ZigBee adopts the advanced encryption standard (AES) block cipher with counter mode (CTR) and cipher block chaining message authentication code with powerful 16-bit cyclic redundancy check [35]. Master, link and network keys are used to secure the transmitted frames, e.g., routing messages, network join requests. ZigBee network shares a network key as a common key, and all the network devices use it to secure the network frames. On the other hand, link key is a secret session key which are unique for two communicating ZigBee devices. Master key is responsible for generating link keys.
- **Long Battery Life:** ZigBee nodes are very efficient in battery life times, ranging from a few months to many years [5]. ZigBee standard adopts sleeping mode strategy to prolong the network life time. While ZigBee routers and coordinators are always awake, ZigBee end devices may sleep for the majority of time and send data for a small period of time [65]. During the sleep time, ZigBee coordinator and ZigBee routers hold the data and when the ZigBee end devices are awake, they send a request to their parent nodes if there are any packets to be sent.
- **Scalable:** IEEE 802.15.4 PHY and MAC layers that support handling large number of devices in a network are adopted for ZigBee protocol, hence, ZigBee network can be highly expandable; 255 ZigBee nodes(one is the master node and others are slave nodes) can be contained in a ZigBee network. In case of the network coordinator interconnects each other, e.g., star topology, the size of the whole network can be up to 65,000 nodes [35].

3.1 The Protocol Structure of ZigBee

The ZigBee protocol layering is based on the standard Open Systems Interconnection (OSI) seven-layer model. ZigBee benefits for a powerful physical radio specified by IEEE 802.15.4-2003 that defines the characteristics of ZigBee's physical layer (868/915 MHz or 2.4 GHz) and medium access control(MAC) layer [1, 37]. Network layer and application layer are specified by ZigBee Alliance with some technical shortcomings, e.g., address allocation, scalability, management tools, routing mechanisms, and interoperability with the Internet [39]. Here is the protocol structure of ZigBee as depicted in Fig. 1.

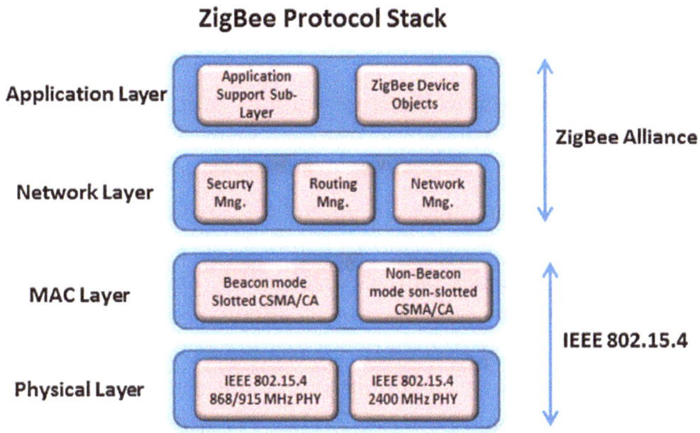

Fig. 1 ZigBee protocol stack

- **Physical Layer:** The low-cost feature of ZigBee comes with the design principles of physical layer; hence, ZigBee technology has simple and high level of integrations.
- **MAC Layer:** Multiple topologies and handling large number of devices are supported in ZigBee networks with the special design principles of MAC layer. MAC layer provides low power management capabilities, the usage of RFD devices, data framing, frame detection, medium access and error control for ZigBee technology.
- **Network Layer:** ZigBee network layer is one of the key layers in protocol stack, it provides mechanisms to ensure normal operations for MAC layer and application layer. Network layer is responsible for applying security frames, controlling nodes for joining or leaving the network and managing discovery and maintenance of routes between devices [1]. When a node wants to join the network, network layer starts the network discovery procedure to find a parent node and establish parent-child relationship [5].
- **Application Layer:** ZigBee Alliance provides application layer that consists of application support sub-layer (APS), ZigBee device objects (ZDO) and application framework [46]. APS acts as an interface between network layer and application layer, and it is responsible for binding devices based on their services and needs, transmitting messages and providing security services between them. ZDO defines the roles of the devices in the network, responses to binding requests and provides secure relationship between devices. Application framework provides the guidelines to be able to build a profile onto the ZigBee stack. It enables specifications to the frame formats for transporting data and standard data types for ZigBee profiles.

3.2 The Network Architecture of ZigBee

ZigBee network layer can support different network topologies, e.g., star topology, cluster tree topology and mesh network. For time-critical applications and long-life time networks, *star topology* is preferred as the least resource demanding topology in which all the end devices have direct communication with the network coordinator; for larger physical environments, *cluster tree network* is preferred which is a hybrid star/mesh topology and provides increased reliability and long network life time; *mesh network* with alternate route flexibility allows peer-to-peer communication between network devices, and decentralized routing is established throughout the network to provide self-healing capability in case of a node failure [7, 36]. Each ZigBee wireless network should have one ZigBee Network Coordinator to initialize the network and assigning network parameters. ZigBee coordinator is responsible for data collection, address assignment and multi-hop delivery in a tree topology [17]. ZigBee network can have more than one ZigBee router to extend the network and many end devices for data acquisition and control. Furthermore, ZigBee technology provides two different physical device types; reduced function device (RFD) and full function device (FFD). FFD can define full MAC layer functions that enable them to act as a network coordinator by sending beacons, synchronizing network equipment etc., RFD can only act as a simple network end device. Here are the brief descriptions of ZigBee network components. A general view of the network is depicted in Fig. 6.

- **ZigBee Network Coordinator (ZNC):** ZNC has many responsibilities, such as starting and configuring the ZigBee network, assigning a radio frequency channel, network ID and other network parameters. Furthermore, ZNC supports associations, designs trust center, manages network nodes and stores the information of network nodes by maintaining the overall network knowledge. Each ZigBee network has only one ZNC.
- **ZigBee Router:** ZigBee router provides multi-hopping for routing the messages between network nodes and it is also capable of accepting a join request to the network as a ZNC. Every ZigBee network except a ZigBee star network supports ZigBee routers and more than one ZigBee router can exist in ZigBee networks.
- **ZigBee Trust Center (ZTC):** For ZigBee security architecture, ZTC is the fundamental component and all other ZigBee devices trust ZTC for its services, e.g., trust management, device management and network management. ZTC stores and distributed keys to ZigBee devices.
- **ZigBee Gateway:** ZigBee gateway is responsible for serving as a bridge between ZigBee network and other network. A ZigBee network and a wired network should not be directly connected to each other. Hence, a ZigBee gateway is needed to separate them for security concerns and better traffic management.
- **ZigBee End Devices:** The simplest devices in ZigBee networks are ZigBee end devices with minimum memory size equipment and high energy savings. Their job is only communication with each other, requesting data from ZNC and searching for available networks.

ZigBee devices have two addresses, a 16 bit short network address and a 64 bit IEEE extended address. The parent coordinator or router device assigns a 16 bit network address to each node dynamically while joining to the ZigBee network. Each node uses this 16 bit network address for data transmissions and data routing. The 64 bit address is unique for each device and it stays the same when the node is manufactured.

ZigBee standard provides an automatic formation and self-organizing network, hence a sensor node should serve as a the coordinator of the network to initiate the network formation, set security levels of the network and be the network manager. The corresponding sensor node is the ZigBee coordinator. Hence, ZigBee coordinator as an FFD device takes the maximum control of the network. To form a ZigBee network, the network should contain at least one FFD to act as a network coordinator, and one FFD or RFD to serve as a ZigBee end device.

ZigBee network establishment contains two basic steps, e.g., network initialization and sensor nodes' joining to the network. The network coordinator is selected before the finding the appropriate channel for the network communications. Assigning a network identifier is the coordinator's responsibility. ZigBee coordinator decides the depth of the network, the number of children of a ZigBee router and number of child routers of a router before initializing the network. After the network establishment, other ZigBee devices can join to the network. The new sensor node starts to search for the network. It scans the available channels to find the operating networks and decides which one it should join. After searching for the network, the new node should select a parent node, and it prefers the parent node with the highest signal. The new node is ready for asking the network coordinator for joining the network. The network coordinator accepts the new node to the network, if the new node is a permitted device and has enough address space.

ZigBee standard supports two routing schemes, mesh routing which is based on Adhoc on demand vector (AODV) routing [47] and tree routing which is based on cluster tree routing algorithm [22]. Network layer uses AODV. In AODV routing, ZigBee router which is ready to transmit a packet from source to destination, prepares a routing table entry for the route. A route discovery process is took place on-demand in which the source node broadcasts a request and destination node replies that request. When the routing table entries are formed, the route can be used for data transmissions. In this type of routing, FDD devices act as ZigBee routers. In tree routing, only ZigBee routers transmit packets in a simplified routing notion.

3.3 ZigBee and Sensor Network Applications

ZigBee provides more reliable, energy efficient and cheaper communications in low transmission rate and high capacity networks than other wireless networking technologies, e.g., Wi-Fi, Bluetooth [7]. Hence, ZigBee technology can be a proper choice for the applications that need short-range and low data rate communications, and low complexity and low power consumptions.

- **Personal Health Care:** Healthcare expenditures has been increasing with growing population, increased living standards, longer life-spans, sophisticated high-tech treatments in mostly all over the world. Hence, to reduce the costs, new medical devices based on ZigBee technology can be manufactured. For instance, with wireless low-power communication capability, ZigBee is a promising technology for the adoption of Assistive Technology that help disabled or elderly to maintain their independence and mobility [76]. The strong mesh networking capability provides accurate equipment location services in hospital, medical center environments. Furthermore, chronic disease monitoring, personal wellness monitoring and personal fitness monitoring are other application areas that ZigBee technology focuses on.
- **Home Automation:** Home automation is the biggest market for ZigBee technology. The idea of an interconnected home creates a new ecosystem to maintain cost-effective, high-comfort, easy-to-use smart homes. There exist a diverse home automation applications ranging from security, light and climate control systems, environment and energy management systems to MP3s and DVD movie selection via a ZigBee wireless remote and TV interface, media management, audio servers that ease the lives of human-beings with self-organizing, simple set-up and maintenance features. There are a lot of companies that develop wireless home automation products based on ZigBee technology; automatic discovery of IPbased control devices, auto creation of programming and user interface for integration of new products are some of the features of Control4, on the other hand, Eatons Home Heartbeat enables the concept of home awareness by monitoring the activities of home appliances and alerting homeowners to take actions [75]. ZigBee wireless technology may be applied to meter reading systems to enable monitoring center reliable and low-cost data analysis and measurement for accurate electricity consumption values.
- **ZigBee Smart Energy:** With growing population, there is a significant increase in energy demand. Most of the consumers are not aware of the decreased energy resources, and how electric utilities are having difficulties to meet the energy demand and finding new ways to make consumers more conscious about the situation and integrating new energy resources to the power grid. Electric utility companies, government agencies, technology providers are looking smarter ways to reduce the energy consumption. ZigBee Smart Energy Profile aims to decrease the energy consumption with advances, ZigBee-based, green products. ZigBee Smart Energy Profile offers demand response and load control support by providing multiple control methods including temperature set points and offsets, criticality levels and targeting specific groups of devices including HVACs, water heaters, lighting, electric vehicles, and generation systems; real-time pricing support for multiple commodities including electric, gas, water, and thermal [87].

4 WirelessHART

WirelessHART is the first open industrial wireless communication standard which was designed as the wireless extension to the Highway Addressable Remote Transducer (HART) protocol [13, 59, 60]. Compared to the HART protocol which is widely used in the industrial process automation with thousands of HART networks and HART devices, wirelessHART has many advantages in the field of industrial control. WirelessHART brings simplicity, robustness, lower installation and maintenance costs and more flexible configuration to the industrial automation and control applications [60]. WirelessHART is a robust wireless networking technology operating at 2.4 GHz Industrial, Scientific, and Medical (ISM) non-licensed frequency band as several other wireless technologies. WirelessHART is a mesh network and built on the IEEE 802.15.4 physical layer and adds its own time-synchronized *data link layer* which is responsible from formatting the data packets, detection and correction of bit errors; *network layer* which defines the routing, topology control, end-to-end security and session management; *transport layer* which provides the end-to-end transmission reliability and flow control; *application layer* which defines various device commands, responses, data types and status reporting [32, 59].

WirelessHART was officially released in September 2007 by HART Communications Foundation (HCF) as the first open wireless communication standard designed to meet the unique requirements of wireless networks operating in process plants [44]. WirelessHART is a secure, highly reliable and TDMA-based wireless mesh networking technology which is interoperable with wireless device types of different manufacturers, and backward compatible with widely-used HART technology in the process industry [60, 61].

- **Interference:** WirelessHART uses frequency hopping spread spectrum (FHSS) and blacklisting methods to overcome the interference within the communication channels in industrial environment [59]. FHSS provides WirelessHART to be able to hop across the 16 channels defined in the IEEE802.15.4 standard to prevent interference while blacklisting feature prevents the usage of certain channels. Furthermore, clear channel assessment (CCA) mechanism is adopted as an optional choice before a message transmission.
- **Simple:** The simplicity is the main building block of WirelessHART standard. Some of the features of WirelessHART, such as self-organizing and self-healing wireless mesh network, reduced installation and wiring costs, easy-adjustments to changes in plant infrastructure, enable users to easily implement and gain benefits of WirelessHART standard [44].
- **Reliable:** Reliability is one of the most important requirement in the field of industrial control. However, the quality of data communications in industrial environments is under threat of the changing conditions, diverse sources of radio-frequency and electromagnetic interference. WirelessHART uses channel hopping, time division multiple access (TDMA), and direct sequence spread spectrum coding (DSSSC) technologies to overcome interference with other overlapping networks; easily finds alternate paths and adjusts communication paths to provide

optimal performance [44]. Moreover, all the field devices have routing capabilities and easily form a mesh network.

- **Secure:** WirelessHART is a secure protocol since robust, always-on, multi-tiered security measurements are employed to provide end-to-end, per-hop, and peer-to-peer security on data transmissions. WirelessHART provides security at both network level and MAC level by using four different security keys, e.g., public key, join key, network key and session key. Public key is responsible for helping the network manager to authenticate the new nodes; join key which used by sensor nodes side and is unique for each sensor node, sends joining request packets to the networks; all the network devices can share network key which used in MAC layer; and lastly, session key is used by network manager to encrypt critical data packets and it is unique between two devices [2]. Furthermore, for secure encryption of messages, an industry standard, advanced Encryption Standard (AES-128) block cipher in Counter with CBC-MAC Mode is used [51]. Moreover, data link layer is also responsible from security control by providing per-hop security between two neighboring wireless devices.
- **Interoperable:** WirelessHART is not a completely new protocol, it is the wireless extension of the HART protocol for industrial process automation and control protocol, released as the latest version HART 7.2. To be able to take advantage from widely used HART protocol, wirelessHART is designed as backward compatible with HART protocol even they have different physical and data-link layers with advances mechanisms for the integration of HART devices with WirelessHART networks [51].

4.1 The Protocol Structure of WirelessHART

WirelessHART uses a simplified version of OSI-7 layer reference model and implements the IEEE 802.15.4 physical layer as its first layer up to 250 Kbps. WirelessHART have unique specifications for data link layer and network layer, and uses the same transport and application layer of HART protocol. WirelessHART is the extension of widespread HART protocol and provides greater flexibility and scalability of wireless networking and meet the specific requirements of process automation applications that have a minimum cycle time on the order of seconds. A general view of the network is depicted in Fig. 6 [13]. Here is the protocol structure of WirelessHART as depicted in Fig. 2.

- **Physical Layer:** WirelessHART physical layer is the interface to the medium and responsible from transmitting and receiving raw data packets with additional control mechanisms to select operating channels and make channel assessments [48].
- **Data Link Layer (DLL):** The operation of industrial process automation should meet the strict timing requirements to provide a complete solution for real-time process control applications [60]. Hence, WirelessHART has a time-synchronized

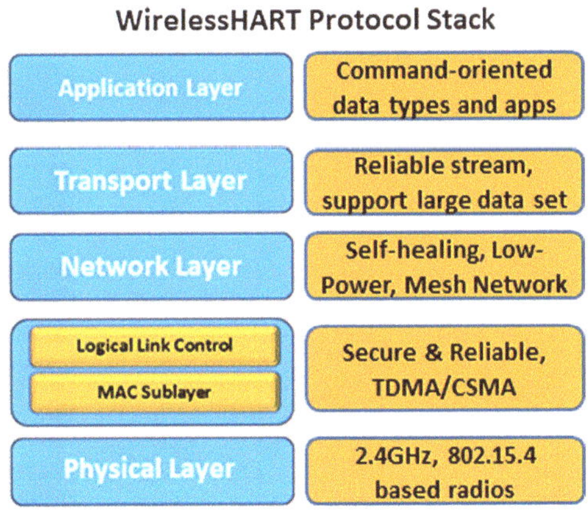

Fig. 2 WirelessHART protocol stack

data link layer which has a strict 10ms time slot and utilizes TDMA technology for collision-free and deterministic and coordinated communications [59]. Other responsibilities of DLL are handling acknowledgement frames, frequency hopping, channel blacklisting and security control [48]. DLL is divided into two sub layers, a logical link control (LLC) and a medium access layer (MAC). TDMA and slow frequency-hopping approaches are used by MAC layer and all the network nodes share the same sense of time or aligned to the same superframe structure which is formed by 100 timeslots per second and repeated continuously [13]. Timeslots are established to each WirelessHART field device to be able to provide multiple links. The timeslots are all grouped into superframes (10 ms). TDMA needs accurate internal slot timing to be able to provide proper interaction between sender and the receiver. Hence, superframes are generated periodically, sent and stored in each WirelessHART device. Proper number of superframes are gruoped and each group forms a network cycle, hence, each field device receives one time slot for data transmission within each network cycle. Some network devices which have more duties that the others may have more than one time slot to finish its task (Fig. 3).

- **Network Layer:** The basic responsibilities of network layer is to route packets from initial source to final destination, maintain the routing tables and providing secure, end-to-end communications. In addition, a central network manager is introduced in network layer to maintain the full knowledge of the network topology, manage the routing, handling the assignments of routing tables of every device in the joining the network. There are two types of routing protocols in network layer; *graph routing* which is an ID-based routing protocol where all the network devices are configured with graph information, including specification of neighbors to forward the packets; *source routing* which is the supplement of graph routing, used for network diagnostics and maintains a fixed path between source and destination.
- **Transport Layer:** Transport layer operates at the end communication points and is responsible from end-to-end packet delivery across several devices in the network.

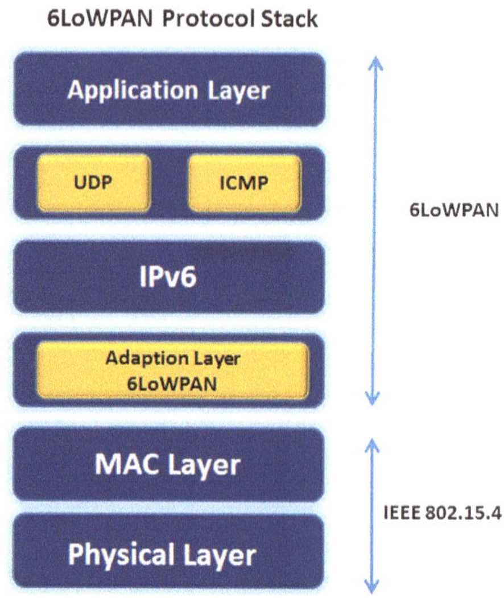

Fig. 3 6LoWPAN protocol stack

Transport Layer supports both (REQUEST/RESPONSE) acknowledgments and unacknowledged transactions.

- **Application Layer:** WirelessHART inherits application layer from application layer of HART protocol which is responsible from necessary communication services for object-to-object communication between distributed field applications and defines severe device commands, responses to data types and status reporting [48, 59]. *Universal commands* can be recognized as a collection of commands that must be supported by all wirelessHART devices; *Common practice commands* are optional and can be applied to a wide range of devices; *Device specific commands* are formed according to the needs of the field device by the manufacturers.

4.2 The Network Architecture of WirelessHART

To enable wireless networking reliable and successful in industrial applications, it must have self-healing and self-organizing features. It must be scalable, redundant enough to extend network range and provide real-time access and control to the applications. Wireless mesh network is capable of providing redundancy and self-healing features to the network since each node has a connection to at least one other node and all the nodes can find their neighbors and establish paths for data forwarding. The components of WirelessHART form a wireless mesh network that is managed by a network manager. Network manager is responsible for forming and monitoring the mesh network, and controlling the new devices to join the network. Star and hybrid (mesh/star) network topologies are also used by wirelessHART.

WirelessHART network consists of wireless devices; field devices, adapters, access points, routers, and wired devices; gateways, network and security manager, and plant automation hosts [51]. The popularity of using industrial field devices are increasing since they are less-expensive, easily-installed and more flexible than wired devices and they reduce modification and maintenance time [42]. These devices are also capable of finding their neighbors and establishing paths towards them. A mesh network is established between wireless devices where each device acts like a router. Mesh network provides some advantages to the WirelessHART network. Since, all the sensor nodes are connected to each other, the network range can be increased. Furthermore, since there are a lot of alternative paths between sensor nodes, in case of a node failure, the packets can be rerouted by alternative paths and reliability of the network is increased. Here is the brief introduction to the wirelessHART components:

- **Field Devices:** Field devices are the distributed actual sensors which are connected to the process or plant equipment and capable of routing and forwarding packets. Field devices can be connected to the plant equipment via other wireless network or they can be directly connected to the WirelessHART network.
- **Adapters:** Adapters are the enablers of the integration of wired HART devices into the wirelessHART network. One or more HART devices can be connected to a WirelessHART network via adapters. In case of point-to-point HART networks, only one adapter can be used to connect a HART device to the network.
- **Routers:** Routers are the special kinds of field devices, however, they do not interfere with process until they are needed for the improvement of the wireless connectivity.
- **Gateway:** Gateways are recognized as the bridges that provide connection between WirelessHART network and process plant. Each network has one gateway which consists of a virtual gateway and one or more access points. Virtual gateways have direct communications with network manager.
- **Access Points:** The actual physical connection to the WirelessHART network is provided by access points. Each access point has a unique ID.
- **Network Manager:** Network manager is responsible for the overall configuration and maintenance of WirelessHART network. It collects information from field devices via the gateway to determine the network health, routes to be set up, manage dedicated and shared resources. Network manager updates the routing information and communication schedule when new sensor nodes join to the WirelessHART network. Furthermore, network manager provides TDMA schedule and manages time slots in TDMA to be placed on different frequencies and to be allocated hop-by-hop based while other stations are all allowed to sleep.
- **Security Manager:** Security manager is responsible of monitoring the security status of the network, preventing attacks, generating session, joint and network keys with the incorporation of network manager.

All the WirelessHART devices need to have routing capabilities. Hence, all them are treated as the same in terms of network initialization, installation, network capability etc. With these features, WirelessHART is a simple and self-organizing network. WirelessHART uses two different mechanisms for message routing, e.g., graph routing and source routing. In graph routing, pre-determined paths are used for routing messages from source to destination. To prevent path redundancy, the graph route is divided into different several route paths between source and destination. On the other hand, source routing does not use path diversity and prefers ad hoc created routes for routing the packets. To be able to initiate the WirelessHART network, network and security managers, an access point, a gateway, some field devices are the fundamental components to be ready. After that, the network Id and a joint key should be assigned to the network as the first step for initializing the network. While the network manager settings are done through a PC connected to it, WirelessHART device settings must be made via a HART port. After these steps, WirelessHART network is ready to operate.

4.3 WirelessHART and Industrial Applications

Industry environments have harsh and unique characteristics that impose technical challenges to wireless communication technologies, e.g., strict timing requirements, high security concerns, RF interference, dynamic topologies and variable link capacities. The publicly available ZigBee and Bluetooth standards exist to be used in industrial applications, however, their capabilities cannot meet the unique requirements of industrial applications even ZigBee and WirelessHART use the same physical layer of IEEE 802.15.4. HART Communication Foundation and its member companies made some efforts to create a wireless alternative add-on to the existing wired HART technology to provide a complete solution for industrial applications and provide end-to-end reliable and secure communication with a wireless technology. With the WirelessHART open standard, field devices have a wireless interface and form a self-organizing and self-healing wireless mesh network with easy-installation and maintenance efforts. WirelessHART technology has a lot of advantages for process automation and control applications, since the data on plant operations can be securely and reliably collected, stored, analyzed and monitored, physical interventions for monitoring and maintenance of plants or expensive-wire installations are reduced.

5 6LoWPAN

Low-power wireless personal area networks (LoWPANs) is a promising technology for embedded applications, e.g., monitoring and control applications that require severe sensor nodes to cover a large geographic area with low-cost, low-power consumption and low-computation capabilities. However, LoWPANs may not meet the reliability, security requirements all the time because of short-range, low-power features, the wireless sensor node failures and sleep duty cycles. Hence, Internet Engineering Task Force (IETF) 6LoWPAN working group released 6LoWPAN technology to overcome these problems by enabling IPv6 to work with low-power, low-rate wireless embedded applications and networks with the integration of an adoption layer, new packet format and address management [56, 82]. The idea of providing wireless internet connectivity at low data rates with a low duty cycle for low-capacity devices was very promising to scale the wireless networks and guarantee the end-to-end communication capability. The previous assumptions about extending IP to LoWPANs were not encouraging due to the high resource-consuming feature and larger packet sizes (at least 1280 bytes in length) of IP technology. To overcome this issue, 6LoWPAN's adaption layer creates smaller packets to fit into IEEE 802.15.4 frame size (128 bytes) with its header compression functionality. Physical and data link layer of 6LoWPAN were adopted from IEEE 802.15.4 protocol. Here are some of the key characteristics of 6LoWPAN technology [58].

- **Scalability:** Most of the wireless sensor network technologies cannot overcome the scalability problem when the network size has become bigger. 6LoWPAN overcomes this issue by adopting an adaption layer with its header compression functionality; hence, connecting to other IP-based networks is easy without additional gateways.
- **Mobility:** The wide-spread IP-mobility technology can be used easily for the mobility feature of 6LoWPAN technology [58].
- **Manageability:** Network management is very important for 6LoWPAN standard to be able to scale well while it has limited display and input capabilities. Hence, 6LoWPAN networks can benefit from the advances of IP technology to ease the network management functionality. The already used tools for managing, commissioning and diagnosing, and simple network management protocol (SNMP) ease the management capability of 6LoWPAN technology [56].
- **Interference:** 6LoWPAN is based on low-power, low-throughput IEEE 802.15.4 standard, hence, the wireless communications are more prone to link failures and interference with other technologies which share the same communication band.
- **Security:** Guaranteeing the end-to-end security at 6LoWPAN networks is still an open-research issue [18]. There have been some researches towards this specific area. Granjal et al. proposed new compressed 6LoWPAN security headers to protect the IPv6 communications on IEEE 802.15.4 WSNs. Furthermore, Raza et al. proposed a system which adds IPsec (defines an authentication header and an encapsulating security payload) support to 6LoWPAN network to provide end-to-end secure communication between IP enabled sensor networks and traditional internet [50].

The future trend in wireless technologies seems to be towards IP-based WSN which enables devices in a network to directly connect to other IP-based networks without the need for translation gateways or proxies, hence, the network scalability, large-scale network management, reliable data communications, reduction in latency and data loss while application-layer protocol translations, can be accomplished successfully [39]. 6LoWPAN is the enabler technology for this future trend.

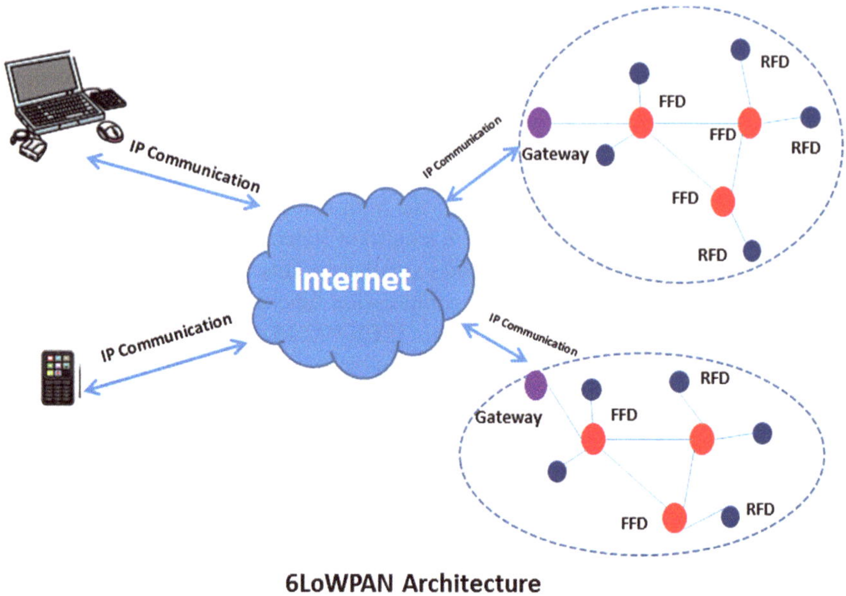

Fig. 4 6LoWPAN architecture

5.1 The Protocol Structure of 6LoWPAN

Some wireless sensor protocols do not have an IP network layer protocol, however, future WSNs require an internet connection between thousands of nodes and networks. IETF defined IPv6 over LoWPAN to provide the integration of TCP/IP into WSN. Hence, integrating an adaption layer above the IEEE 802.15.4 link layer enables a sensor node IP communication capabilities with low-power, short-range, low-cost and low-bit rate features. 6LoWPAN protocol stack is almost identical with a simple IP stack with some differences. IEEE 802.15.4 standards specifies the physical and MAC layer of 6LoWPAN protocol. 6LoWPAN implements the IPv6 with LoWPAN adaption layer as its network layer. The transmission control protocol (TCP) and the user datagram protocol (UDP) are implemented as 6LoWPAN's transport protocol. Furthermore, The Internet control message protocol v6 (ICMPv6) is used as 6LoWPAN's transport protocol which is responsible for control messaging, e.g., ICMP echo, ICMP destination unreachable and Neighbor Discovery messages. Here is the protocol structure of 6LoWPAN as depicted in Fig. 4.

- **Physical Layer:** 6LoWPAN adopts the IEEE 802.15.4 physical layer with a data rate of 250 Kbps and operating frequency of 2.4 GHz. 27 channels are defined in PHY layer and they are all allocated into different frequency band with different data rates, 868–868.6 MHz for European, 902–928 MHz for North America and 2400 MHz for worldwide.

- **Data Link Layer:** 6LoWPAN adopts the IEEE 802.15.4 MAC layer. The main responsibilities of data link layer are beacon generation and synchronization, network association, providing channel access via carrier sense multiple access with collision avoidance (CSMA/CA) mechanism. 6LoWPAN's data link layer should provide framing, unicast transmission, strong error-checking and addressing capabilities. To be able to distinguish the nodes on a link, and generate IPv6 addresses, addressing should be applied by data link layer. 6LoWPAN link layer should provide different services for different types of links; for instance, a broadcast service should be enabled for multi-access links. 6LoWPAN's data link layer should provide strong error-checking mechanism for secure and reliable communications since UDP and ICMP transport protocols have a simple 16-bit checksum mechanism. Four MAC frames are supported by the standard, such as beacon frame, data frame, acknowledgment (ACK) frame and MAC command frame, however, beacon frame is used only in star topology while other frames are used in both star and mesh technologies. ACK frame and MAC command frame handle MAC peer entity control transfers and confirm the successful frame receptions.
- **Adoption Layer:** Adoption Layer is the main layer of 6LoWPAN technology. IPv6 adoption for network layer does not meet the MTU (Maximum Transferable Unit) specifications. At least a MTU of 1280 bytes which is ten times of the one specified for 802.15.4 networks and 40-bytes length IPv6 header which is a huge overhead for 6LoWPAN technology are required for IPv6 protocol. Hence, an adoption layer is needed to take place between network and data link layers to provide packet fragmentation, header compression and support for data link layer forwarding of IP packets [29]. In *header compression*, IPv6 header fields are eliminated to reduce transmission overhead, in *fragmentation process*, packets are fragmented into multiple link layer frames to provide the minimum MTU requirement when they cannot fit into MAC frame payload size. Adaption layer supports four basic header types, e.g, dispatch header, mesh header, fragmentation header and HC1 header (IPv6 header compression header). To be able to provide a full routing functionality with the headers, additional routing header is required to be encapsulated in a packet.
- **Network Layer:** 6LoWPAN adopts IPv6 protocol as its network layer. The network layer of an IP-based wireless technology should be responsive, adaptive and energy-efficient to overcome interference, link failures, dynamic link qualities, and asymmetric links of IEEE 802.15.4 radio environment [29].
- **Transport Layer:** The header compression and fragmentation features of 6LoWPAN enables it to adopt the commonly used transport protocols, e.g., UDP and ICMv6.

5.2 The Network Architecture of 6LoWPAN

6LoWPAN consists of many low-power wireless area networks (LoWPANs) that can be also called IPv6 stub networks which are responsible from sending and receiving

Table 1 Comparison of wireless technologies

Feature	WirelessHART	ZigBee	6LoWPAN	ISA.100.11a
Security	AES-128 plus data link layer security	AES-128 plus application layer security	AES-128	AES-128, Join Key, Network ID, End to end security
Scalability	80–100 field devices supported by per gateway	65,535 nodes are supported by per coordinator	High	High
Reliability	Channel hopping, TDMA, DSSSC	ACK, retransmissions, link quality estimators	Simple 16-bit checksum mechanism	Channel hopping, TDMA, CSMA, hybrid channel blacklisting
Power consumption	Low	Low	Medium	Low
Network management	High	Medium	High with SNMP	High
Data rate	250 Kbps	250 Kbps	20–250 Kbps	Low data rates
Application types	Industrial app.	Home automation, building automation	Health care, industrial monitoring	Industrial process and control app.

IP packets. LoWPANs consists of nodes which can act as a host or as a router, and can be identified by a unique IPv6. LoWPANs nodes can send and receive IPv6 packets but they are limited in processing capabilities. In the 6LoWPAN architecture, there are 3 different LoWPANs; *Simple LoWPANs* that consists of multiple hosts, routers and one LoWPAN Edge Router to connect other IP networks with a backhaul/backbone link, *Extended LoWPANs* that includes many simple LoWPANs with multiple edge routers with a backbone link, and finally, *Ad hoc LoWPANs* that do not have any internet connection and operate without an infrastructure [56].

6LoWPAN supports both star and mesh network topology. In star topology, FFD acts a coordinator and has direct connection with each node. Furthermore, the network coordinator performs device communication based on guaranteed time slots hence, the use of collision avoidance techniques are avoided. In mesh topology, extra mesh header is needed by the 6LoWPAN adaptation layer to keep the originator and final destination address while the IEEE 802.15.4 MAC layer carries the source and destination node address that change at each hop [12]. Furthermore, in a mesh topology, intermediate nodes must provide multi-hop routing services to compensate appropriate number of routing packets, computation capabilities and energy consumption (Tables 1, 2 and 3).

Table 2 Application types of wireless technologies

WSN Technology	Application types
WirelessHART	Process automation and control applications
ZigBee	ZigBee smart energy, home automation, personal healthcare
6LoWPAN	Industrial monitoring, health care, disaster management
Wi-Fi Direct	One-to-many infrastructure connection, high affinity with legacy Wi-Fi, software AP
ISA.100.11a	Reliable monitoring and alerting, asset management, predictive maintenance, condition monitoring

Table 3 General Comparison of wireless technologies [1]

Protocol layers	WirelessHART	ZigBee	6LoWPAN	ISA.100.11a
PHY Layer	IEEE 802.15.4-2006 with 2.4 GHz Radio only	IEEE 802.15.4-2003 with 868/915 MHz or 2.4 GHz Radio	IEEE 802.15.4 with 2.4 GHz Radio	IEEE 802.15.4, 2.4 GHz-band
Network topology	Star, mesh	Star, tree, mesh	Mesh	Star, mesh
Application areas	Industrial process monitoring and control	Home automation, consumer devices	Environment monitoring and control	Industrial process monitoring and control
Network routing	Graph routing	AODV and tree routing	Ad hoc on-demand distance vector routing and dynamic MANET on-demand routing	Graph routing

In 6LoWPAN routing, when a sensor node wants to send a packet to an IP-enabled device outside of the 6LoWPAN network, the packet is firstly sent to the FFD in the same network. Then, FFD will relay the packet hop by hop manner to the 6LoWPAN gateway which will make a connection to 6LoWPAN with IPv6 domain and send the packet to the destination by using its IP address. Basically, there are three types of 6LoWPAN routing based on AODV routing protocol, such as Adhoc on demand distance vector routing (LOAD), dynamic MANET on demand (DYMO-low) and hierarchical routing (HiLow). However, LOAD routing is the most preferred one among them. LOAD routing is the simplified version of AODV and operates on top of the adaption layer. LOAD supports a mesh network topology and provides multi-hop routing between IEEE 802.15.4 devices for establishing routes in a 6LoWPAN network. In the route discovery, LOAD broadcasts messages to spread the route request messages. After the route discovery, LOAD provides data structures and manages local connections, and uses *routing tables* to store the required information, such as destination, next hop node, and *route request table* to store temporary route information which is used while route discovery process. DYMO-low routing do not force the data packets to be fragmented since route request messages act as IEEE 802.15.4 broadcast messages. DYMO-low routing uses 16 bit sequence numbers to provide loop free transmissions. The main purposed to use HiLow routing is to increase the network scalability since HiLow uses 16 bit short address for memory optimization and larger scalability.

5.3 The Application Areas of 6LoWPAN

6LoWPAN is the enabler technology for embedded devices to connect to Internet, or low-power, low-throughput wireless networks to be connected together to scale and form large network infrastructures with mobility features. Hence, there are lots of application areas that can benefit from the scalability, reliability and security features of 6LoWPAN technology. Some of the 6LoWPAN-based applications are briefly explained.

- **Industrial Monitoring:** Wireless technologies in industrial environments should be robust enough to provide secure and reliable data communications. Process monitoring and control, machine surveillance, supply chain management and asset tracking, and storage monitoring applications can benefit from scalability, energy efficiency, and safety features of IP-based low-power WPAN technology [53].
- **Health Care:** Healthcare applications are suffering environment delay or data loss which can cause the death of a person. Hence, real-time and reliable data transmissions are the basic requirements of healthcare applications. Other wireless technologies may fail in providing these requirements, however, 6LoWPAN can be a proper choice with its mobility, IP-connectivity features to provide reliable and real-time data communications for simple wearable remote controls for tele-assistance or intermediate systems with wearable sensors monitoring systems [53].
- **Disaster Management:** Disaster management applications are delay-sensitive and cannot tolerate false alarms. Hence, the wireless technology should provide real-time and secure communications to be able to detect malicious activities on time [53]. 6LoWPAN technology can provide the special requirements of disaster management systems with its interconnectivity feature.

6 ISA.100.11a

The ISA.100 committee which is a part of the international society of automation (ISA), was founded in 2005 to initiate the establishment of standards, recommendation of practices and technical reports and definition of technologies and procedures for implementing wireless systems in the automation and control environment, with an initial focus on the field level [14, 20]. In 2009, ISA.100 committee approved ISA.100.11a standard (ISA.100.11a: Wireless Systems for Industrial Automation: Process Control and Related Applications) which is an open, multi-functional wireless networking technology standard to provide reliable, robust and secure data communications for non-critical, alerting, supervisory control, open-loop control, and closed-loop control industrial applications [14, 48]. ISA.100.11a has a wide range of application coverage and easily connects to different kinds of communication networks. Low data rate wireless connectivity is supported with increased security and system management levels and the applications on the order of 100ms are fully supported. Here are some of the key characteristics of ISA.100.11a;

- **Robust:** ISA.100.11a protocol is very robust since it uses three different diversity types, *space diversity* to utilize multi-paths for data forwarding with mesh networking capability, *frequency diversity* to enable slow-frequency hopping, and *time diversity* to handle a retry mechanism [48]. These diversity provides the protocol to be robust with the combination of the performance of PHY layer.
- **Scalability:** ISA.100.11a network is capable of handling thousand of devices, however there is a practical limitations not to adversely increase the data traffic of the network and power consumption of sensor nodes. Hence, devices are allowed in an ISA.100.11a network in the order of 50–100 [48].
- **Security:** Security is one of the fundamental requirements of industrial and process automation systems. ISA100.11a uses integrity checks and optional encryption at data link layer. Furthermore, a security mechanism is also provided in transport layer. 128 bits keys are used for both transport and data link layers [84]. A shared global key, a private symmetric key or certificate are the prerequisites for a sensor node to have to be able to join a ISA100.11a network [20].
- **Power Consumption:** The sensor nodes are battery powered, hence this condition puts some limitations for the continuity of reliable and efficient data communications for ISA.100.11a standard. To overcome this limitation factor, some techniques are adopted [64]. For noncritical control applications, synchronizing sampling mechanism is implemented. This mechanism provides reduction of reporting rates since the transmissions take place when the rate of change of the measured data exceeds a certain threshold. Another technique is adaptive transmission power control. This control provides ISA.100.11a field devices to dynamically select a transmit power level, hence, power optimization can be accommodated. The last technique preferred mostly in star network topologies is to use of non-routing transmitters which is very effective in prolonging the network lifetime. Overall, ISA.100.11a is based on IEEE 802.15.4 which has short active periods to provide significant power consumption reduction, and furthermore, the usage of DSSS technique have a significant effect on reduction of power consumption since it employs short settling times in its channel filters and higher frequency references in its channel spacing [57, 66].
- **Reliability:** To provide reliability throughout the network, the interference avoidance mechanisms should be developed. ISA.100.11a uses some mechanisms to overcome the interference and increase the communications reliability [64]. The

first technique which is implemented is DSSS modulation technique. DSSS divides the ISA.100.11a signal into small fragments to spread over the available frequency channels and look like noise to the other wireless technologies with in the range [48, 64]. Furthermore, the adoption of IEEE 802.15.4 PHY provides increased reliability levels in data communication since IEEE 802.15.4 is a proven technology with its coexistence capability within the congested areas. Dynamic power control capability of ISA.100.11a also increase the reliability level of data communications since this mechanism prolongs the network lifetime.

- **Interoperability:** ISA.100.11a uses IEEE 802.15.4 standard for its radio communications. Since IEEE 802.15.4 is a worldwide standard, this feature gives a chance to ISA.100.11a for large-scale implementation for process automation applications. Furthermore, ISA.100.11a provides smooth communication with the existing application protocols, e.g., HART, Modbus, Profibus etc. ISA.100.11a adopts 6LoWPAN protocol for its network and transport layers. Hence, interoperability can be managed with internet hosts and sensor nodes in other WSN networks with the IPv6 compatibility [20].
- **Security:** ISA.100.11a enables secure data communication throughout the network. Security manager provides secure data communications, moreover, IEEE 802.15.4 standards' security mechanisms, e.g., AES-128 bits, message authentication codes, access control list enables security, and also the message integrity codes and unique symmetric keys of 128 bits of the link and transport layers enable security throughout the ISA.100.11a network.

6.1 The Protocol Structure of ISA.100.11a

The protocol structure of ISA.100.11a is based on the simplified version of seven layered open systems interconnection (OSI) reference model and ISA.100.11a adopts IEEE 802.15.4 PHY and MAC layers specification as most of the other wireless technologies. Moreover, the network and transport layers of ISA.100.11a are based on 6LoWPAN, IPv6 and UDP standards. Here is the protocol structure of ISA.100.11a as depicted in Fig. 5.

- **Physical Layer:** Physical Layer is the interface to the physical medium which is responsible for performing data transmission and reception, and control mechanisms for channel selection and energy detection [48]. ISA.100.11a uses IEEE 802.15.4 PHY standard with minor adjustments and operated on 2.4 GHz band, using Channels 11–26 with 2.5 MHz bandwidth usage [48]. DSSS divides the ISA.100.11a signal into small fragments to easily spread over available channels while FHSS provides random channel changes on a packet level to minimize the load on one channel.
- **Data Link Layer (DLL):** The main purpose of data link layer is to provide radio channel access and radio synchronization while handling the acknowledgment frames, security controls and defining the data format. Data link layer is divided

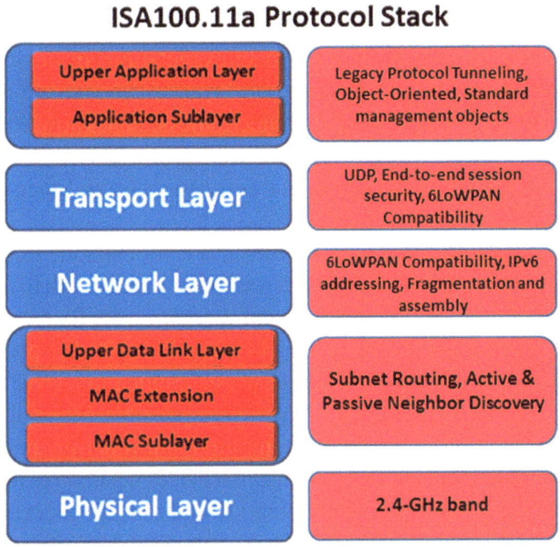

Fig. 5 ISA100.11a protocol stack

into a MAC sublayer, a MAC and an upper DLL. While MAC layer is a subset of IEEE 802.15.4, MAC extension has some different features not supported by IEEE 802.15.4 [48]. MAC sublayer sends and receives the individual frames, and also its security mechanism defends the attackers outside of the system, while MAC extension focuses on the changes to CSMA-CA mechanisms and upper DLL provides channel hopping mechanism, mesh routing and TDMA. Furthermore, DLL implements graph routing and TDMA features, and handles most of the work of forwarding data messages between sensor nodes which are called as a DL subnet, since they are connected to a single local mesh or a personal area network. DLL provides an abstraction to the higher layers since it prevents the messages, which are received with in a DL subnet or border router, to be used by IP layers. Since, ISA100.11a supports fully mesh wireless networking, DLL adopts mesh routing capabilities.

- **Network Layer:** The basic functionalities of the network layer is to handle routing, addressing and maintain the quality of service requirements. Network layer provides mesh to mesh routing and most of the data routing and processing is performed in the data link layer, hence, this situation enables to minimize the network layer overhead and increase the payload of the network layer. In this situation, network layer uses basic header format. Furthermore, the network layer of ISA.100.11a is compatible with 6LoWPAN protocol. Hence, full (IPv6) header format is adopted in case of potential future use of 6LoWPAN networks as the backbone network. ISA100.11a routing takes place at two levels, e.g., mesh level and backbone level, and when switching between these two levels, an appropriate address translation between 16 bit DL subnet address and 128 bit backbone

address should be performed. Hence, different requirements and resources should be adopted to be able to meet expectations of flexible routing.
- **Transport Layer:** Transport layer of ISA100.11a is also compatible with 6LoW-PAN, hence manages connectionless, unacknowledge service with optional security that extends user datagram protocol over IPv6 protocol. Hence, advanced integrity checks, additional authentication and encryption mechanisms are adopted [48].
- **Application Layer:** Application layer enhance ISA.100.11a application environment by providing the integration of the network to other host control systems with appropriate field devices and gateways. Application layer also supports object oriented modeling concepts [14]. The ISA.100.11a protocol divides application layer into 2 sublayers, e.g., the upper application layer and the application sublayer. the upper application layer is responsible for application processes and handles input/output hardware, computational function and protocol tunneling, while the application sublayer manages services needed for upper application layer, such as, object-oriented communication and routing packets [48].

6.2 The Network Architecture of ISA.100.11a

ISA.100.11a network consist of all the required components to be able to manage the network resources, route the network data traffic and easily integrate with other systems. Moreover, there is a backbone network that can be defined as the wired network where different ISA100.11a devices can be connected. Generally, ISA100.11a has two basic types of devices, e.g., backbone devices and field devices. Field devices may have routing or non-routing capabilities. For instance, a handled can be recognized as a non-routing device which can be attached to a full-function device for monitoring or data transmission purposes. On the other hand, backbone devices are recognized as full-function devices which have continues power sources. Backbone can be configured as a data network such as, industrial Ethernet, IEEE 802.11, or any other network within the facility. Backbone router, gateway, system manager, security manager, routing device, non-routing device are some of the components of ISA100.11a network. ISA.100.11a network may have non or more than one backbone routers or gateways. A general view of the network is depicted in Fig. 6. Here are some of the ISA100.11a network devices.

- **Gateway:** Gateway is one of the most important field device in the network. It acts as an interface between wireless and plant network [48].
- **System Manager:** System Manager is responsible for controlling the network, network devices, network resources and communications. After joining to the network, to be able to optimize the network topology, it starts to assign resources and provides a list of the appropriate neighbor nodes. For this purpose, system manager should be aware of connectivity level of the network with the actual measurements of link qualities [2].

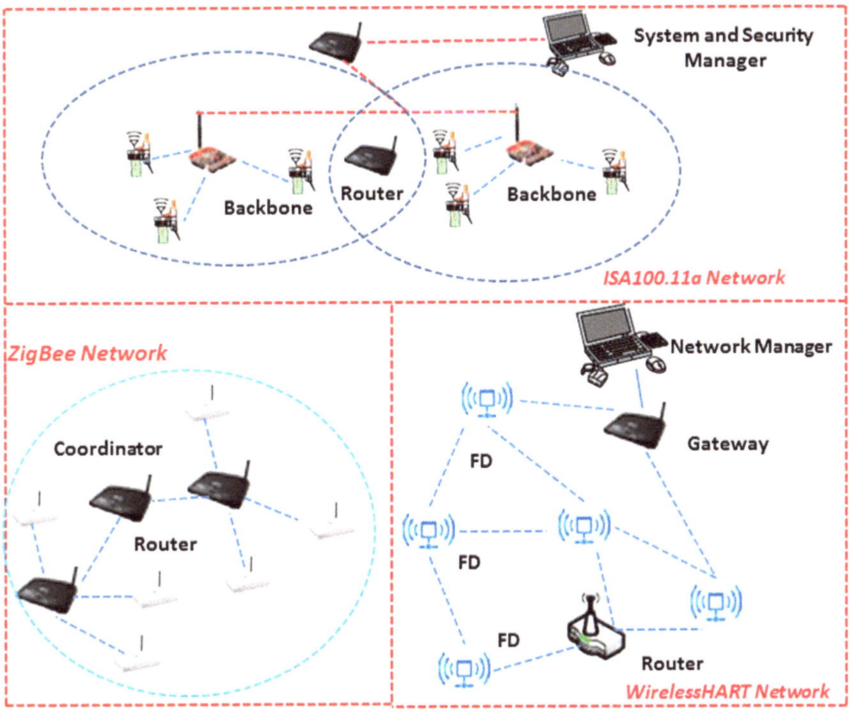

Fig. 6 General view of ZigBee, wirelessHART and ISA100.11a wireless technologies

- **Security Manager:** Security manager provides key management services to enable secure data communications with the cooperation of system manager, authenticates devices. To be able to establish a communication with a neighbor node, the corresponding node should want a new session key from security manager. Hence, it can guarantees that only authenticated devices can communicate with each other.
- **Routing Device:** Routing device is capable of routing data to the other devices in ISA.100.11a network.
- **Non-routing (I/O) Device:** Non-routing device is not capable of routing data, it provides sensory data to the other devices or it uses data from other devices.
- **Backbone Router:** Backbone router is responsible for establishing connection with other networks by routing data to the backbone network or from backbone network. It encapsulates the data that is arrived to the ISA100.11a devices.

ISA.100.11a employs three types of network, e.g., star, mesh and star-mesh topologies according to the roles of devices in the network. For instance, in a ISA.100.11a network, routing role is separated from sensor and actuator roles. With this separation, ISA.100.11a field devices are named as the end devices with no routing capability or router nodes with routing capabilities.

6.3 ISA.100.11a and Industrial Applications

Reliable monitoring and alerting, asset management, predictive maintenance, condition monitoring, factory automation, location services and logistics are all the application areas which have specific requirements and performance characteristics that can be covered by ISA.100.11a protocol. ISA.100.11a protocol effectively supports deterministic timing requirement-based control applications and most of the industrial plant's field device applications, e.g., emergency action which is an always critical application, closed loop regularity control which is often critical, closed loop supervisory control which is usually non-critical, etc.

7 Hardware Point of View of WSN Technologies

7.1 ZigBee from Hardware Point of View

There are many certified semiconductor companies, e.g., Freescale, Digi International, Ember, that have been providing successful designs for ZigBee products. Here are some of the hardware products of each companies.

- **XBee and XBee-PRO:** XBee and XBee PRO are ZigBee embedded RF modules developed by Digi International to provide low-cost, low-power wireless connectivity in ZigBee mesh networks [70]. XBee modules provide easy-deployment process for users. They can work in variety of protocols and frequencies.
- **MC1319x:** FreeScale semiconductor company provides a ZigBee ready platform with MC1319x family of transceivers for easy-to-use, low-power, low-cost applications through ZigBee networks [72]. For instance MC13191 RF transceiver can operate in 2.4 GHz frequency band and supports wireless applications such as, wireless security systems, remote controls and patient monitoring.
- **EM250:** Ember company developed a single-chip solution, EM250, to operate at 2.4 GHz frequency band and support low-cost, low-power ZigBee applications, e.g., building automation and control, home automation and control and asset tracking [71].

7.2 WirelessHART and ISA.100.11a from Hardware Point of View

- **Nivis:** Nivis company offers hardware solutions, e.g., radio modules, gateways, access points, system manager and security manager, for WirelessHART and ISA.100.11a standards to provide interoperable and flexible industrial automation and utility management applications [74]. Nivis solutions provide various flexible network topologies based on the requirements of each specific WirelessHART or ISA.100.11a application.

- **XYR 6000:** Honeywell OneWireless XYR 6000 transmitters provide a seamless integration with ISA.100.11a standard [73]. The system is flexible at connecting to any plant system and supporting existing protocols with high reliability and end-to-end security operations.

7.3 6LoWPAN from Hardware Point of View

- **SensiNode:** SensiNode company provides a complete hardware and software solution to support low-power wireless networking with IP technology [55].
- **Hitachi:** Hitachi company provides 6LoWPAN solutions on Renesas Technology motes with inter-operation of Arch Rock System [23, 52].

8 Other Existing Wireless Sensor Technologies

8.1 Bluetooth

Bluetooth is based on IEEE 802.15.1 standard and designed for short-range wireless communications between devices, e.g., mouse, keyboards, printers and joy-sticks [35]. FHSS is used as the radio technology and data is transmitted in range of 2.4 GHz. Piconet and scatternet are defined as the connectivity topologies. In piconet topology, a bluetooth device serves as a master and forms the wireless network while one or more bluetooth devices act as slaves. Slaves an communicate only with master device while master can perform communications as in point-to-point or point-to-multi point manner. On the other hand, scatternet is kind of a form where a collection of bluetooth piconets overlap in time and space [35].

8.2 Z-Wave

Z-Wave is wide-spread, short-distance, low power RF communication technology which focuses on wireless residential command, control and data exchange applications [33]. Z-Wave is developed by Zen Sys for automation in residential and light commercial environments, operates in the 900 MHz ISM bands and allows transmission at 9.6 and 40 kb/s data rates [17]. Two types of devices, controller and slave, can be defined with Z-Wave; the controllers poll or send commands to the slaves, the slaves are like monitoring sensors, which reply to the controllers or execute the commands [17].

8.3 Wireless M-Bus

M-Bus (Meter-Bus) is a European standard designed to full fill the need of remote reading of all types of smart meters, such as electric and gas meters [30]. Wireless M-Bus is the wireless type of the M-Bus technology which has low-cost, low-power consumption and robustness characteristics.

8.4 Wavenis

Wavenis is a two-way wireless connectivity platform by Coronis Systems and promoted by Wavenis Open Standard Alliance for control and monitoring applications which requires small amounts of data with low-data rates and low-power management. Wavenis provides 4.8 and 100 kb/s data rates and operates mainly in the ISM bands, 433, 868 and 915 MHz bands in Asia, Europe, and the United States, respectively [17]. Wavenis-enabled devices can be used for in-home comfort, building and industrial automation, smart metering, alarms and security, access control, medical and other UHF-active long-range RFID applications.

8.5 Wi-Fi

Wireless Fidelity, "Wi-Fi," based on the IEEE 802.11 standard has been gaining popularity in HAN applications [28]. Wi-Fi utilizes the unlicensed spectrum in the 2.4 and 5 GHz frequency bands with maximum raw data rate of 54 Mbps, and provides free wireless communication at high data rates [8]. Wi-Fi is used for two primary purposes in [16]; Wi-Fi is described as the technology standard for in-home multimedia applications and on the other hand, it is seen as the alternative choice to ZigBee which provide communication path between home automation system and Wi-Fi enabled devices.

8.6 Wifi-Direct

Wi-Fi Alliance has announced a new, breakthrough specification, "Wi-Fi Direct" [77, 78], which allows peer-to-peer wireless connectivity between devices to transfer, share, print or display the content. The Wi-Fi Direct specification which was previously code-named "Wi-Fi peer-to-peer" can be implemented to wide range applications, e.g., digital televisions, printers, cameras, gaming devices, smart phones, PCs, keyboards, headphones providing one-to-one or one-to-many connections. The exitance of Wi-Fi access point for device connections is not a mandatory require-

Table 4 Basic characteristics of wireless technologies

WSN technology	Basic characteristics
WirelessHART	Backward compatible, easy-to-use, simple communications, interoperable
ZigBee	Low-cost, long-battery life, usage of DSSS coding, small scale network support
6LoWPAN	Small packet size, low bandwidth, low-power, low-cos, large-scale network support
ISA.100.11a	Highly reliable message delivery, not backward compatible, costly, uses frequency diversity

ment anymore with Wi-Fi Direct technology. The aim of Wi-Fi Direct is to enable any-time, any-where connections for Wi-Fi Direct-Certified devices. Here are some of the key characteristics of Wi-Fi Direct technology.

9 Conclusion and Comparison of Wireless Technologies

There is a wide range of WSN-based applications that differ in reliability, security, scalability and bandwidth requirements. The proposed technologies address different requirements of monitoring and control, health care, home automation, and consumer electronics applications. The communication technology choice should be analyzed carefully to meet the demands of these specific applications. Here is a brief comparison of ZigBee, WirelessHART, ISA.100.11a, 6LoWPAN and Wi-Fi Direct technologies. Table 4 also gives a brief description of characteristics of wireless sensor network technologies.

- **Scalability and Network Management:** All of the mentioned wireless technologies are scalable, however some of them may have limitations in scalability and network management. 6LoWPAN is a competitive alternative to ZigBee, since it uses Internet Protocol version 6 (IPv6) for the network layer which is more promising when scalability is concerned [39]. 6LoWPAN uses the same IEEE 802.15.4 protocol as ZigBee for its physical and data link layer as depicted in Table 3, however it differs from ZigBee in its network layer which provides great end-to-end network communications. For direct communication with Internet, a router is needed to collect the data from ZigBee network and convert the ZigBee protocol to IP which causes ZigBee technology to degrade its performance on network management and routing capabilities [39]. Furthermore, 6LoWPAN can provide more powerful tools, e.g., SNMP (Simple Network Management Protocol) for monitoring and analyzing the network in case of the network extension [39]. On the other hand, Wi-Fi Direct can connect directly to an IP network and form a scalable network, however, Wi-Fi Direct do not have all the functionalities of a Wi-Fi Access Point. Hence, Wi-Fi Direct cannot provide a wide range of wireless network without the help of access points or wireless routers. The management of

Wi-Fi Direct devices in a wireless network is another issue that is hard to accomplish. ISA.100.11a adopts 6LoWPAN for its network layer, however, the network scalability is limited until it starts to affect the real-time and deterministic behavior of ISA.100.11a network. Hence, not to make the sampling rate slow, non-routing transmitters and backbone mesh networks are used [64].

- **Flexibility:** ISA.100.11a provides great flexibility to adopt to various applications, while WirelessHART puts some limitations to the field devices to behave the same for different vendors which prevents flexibility to respond specific requirements of different applications [48].
- **Reliability:** All of the mentioned wireless technologies are reliable in data transmissions, however, the harsh environmental conditions, wireless link quality and interference may degrade the performances of some them. For instance, WirelessHART technology is specifically designed for industrial process control applications and addresses some deficiencies of ZigBee technology in industrial applications. Industry applications have some stringent timing and high security requirements and harsh environmental conditions which poses great interferences and obstacles that are hardly addressable by ZigBee technology [59]. On the other hand, ZigBee has different spread-spectrum techniques to fight against the interference. Hence, ISA.100.11a and WirelessHART technologies are appropriate since ISA.100.11a use DSSS, channel hopping, channel blacklisting techniques to overcome interference in industrial environment. However, WirelessHART may have some coexistence with IEEE 802.11 protocol.
- **Energy Efficiency:** The underlying radio standard of ZigBee, WirelessHART, ISA.100.11a and 6LoWPAN technologies is IEEE 802.15.4. Hence, the common characteristic of these technologies is energy efficiency, and they are preferred according to the requirements of specific wireless sensor network applications.
- **Interoperability:** All of the mentioned wireless technologies are interoperable. WirelessHART is backward compatible with widely-used HART technology in the process industry; Wi-Fi Direct is backward compatible with widely used Wi-Fi technology; 6LoWPAN can connect to IP-based technologies that already exist without additional routers or proxies and can be applied to any low-power, low-rate wireless radio, while ZigBee can only communicate between 15.4 nodes due to limitation to a single radio standard. ISA.100.11a adopts 6LoWPAN as its network layer and can connect to other IP-based networks easily.
- **Higher Data Rates:** Wi-Fi Direct has 802.11n higher data rates (300 Mbps+), hence, Wi-Fi Direct can be a proper choice for the applications that need higher data rates, e.g, multi-player gaming, screen sharing, file sending. On the other hand, ZigBee, WirelessHART and 6LoWPAN support low data rates (250 Kbps) as depicted in Table 5.
- **Protocol Structure:** ZigBee, WirelessHART, ISA.100.11a, 6LoWPAN use IEEE 802.15.4 at the physical layer, while Wi-Fi uses IEEE 802.11 protocol.
- **Frequency and Channel Configurations:** ZigBee, WirelessHART, ISA.100.11a, 6LoWPAN are based on IEEE 802.15.4 standard while Wi-Fi Direct is based on IEEE 802.11 standard. 14 channels are defined in 2.4 Ghz band for IEEE 802.11. IEEE 802.11 uses non overlapping channels, e.g., 1, 6 and 11 to maximize the uti-

Table 5 Wireless sensor nodes

Sensor node	RF transceiver	Data memory	Features
IMote 2.0	ZigBee compliant radio	32 MB SRAM	TinyOS support
KMote	2.4 GHz IEEE 802.15.4 Chipcon wireless transceiver	10 KB RAM	TinyOS and SOS support
MicaZ	IEEE 802.15.4/ZigBee compliant radio	4 KB RAM	TinyOS, SOS and MantisOS support
Arago systems WiSMote Dev	CC2520	16 Kbytes RAM	Contiki and 6LoWPan supported
T-Mote sky	2.4 GHz IEEE 802.15.4 Chipcon wireless transceiver	10 KB RAM	Contiki, TinyOS, SOS and MantisOS support
INDriya_CS_03A14	IEEE 802.15.4 compliant XBee radios	4 KB RAM	IPv6 network supportive stacks for internetworking

lization of the frequency band, hence, IEEE 802.15.4 can accomplish interference free operations only in Channels 15, 20 and 26 [48] which provides flexibility for the technologies that adopt IEEE 802.15.4 as their PHY layer, e.g., ZigBee, WirelessHART, ISA.100.11a, 6LoWPAN .

9.1 Comparison of WSN Technologies for Different Application Areas

- **Building Automation:** The basic expected requirements from WSN technologies in building automation systems is them to be highly interoperable with other devices and provide low-power, and low-cost seamless data communications backbone since they are realized by a large number of sensor nodes distributed in 3D space with many specific subsystems, such as, lighting, electricity, HVAC, fire, security systems [69]. Among the wireless sensor network technologies mentioned throughout the chapter, ZigBee and Bluetooth seem to be the proper wireless technologies which can meet the specific demands of building automation systems. ZigBee technology is designed for periodic data delivery which increases the battery performance of sensor nodes, hence, it seems a proper technology choice for building automation systems [69]. On the other hand, for simple point-to-point communications, Bluetooth which is designed for continuous data transmissions, may be preferred. In the academia, there are lots of researches, experiments conducted for the performance analysis of ZigBee technology in buildings, for instance, Yang et al. monitored the performance of a fire hazard system based on the ZigBee ad hoc system which lead satisfactory results. Park et al. proposed a building automation model over ZigBee to adopt it as a wireless data link layer protocol in proposed system, and results have shown that the system is appropriate to be applied to building automation systems with tolerable response times, e.g., 10 ms [45]. Furthermore, Yong et al. conducted a research and implementation on ZigBee networking for building automation systems which resulted in low-cost, high-security, simple structure ZigBee-based system [83].

- **Industrial Applications:** The devices for industrial processing and control applications are robust, hence, there are some strict requirements for wireless sensor technologies to provide reliable data transmissions between them [79]. ZigBee, Bluetooth, WiFi are some of the technologies which are not accepted by the industry due to the some shortfalls, such that, ZigBee cannot provide the required QoS support for handling latency and message flow determinism for industrial applications which will cause other protocols, e.g., HART, Modbus, Profibus not to be supported by ZigBee, and Bluetooth, on the other hand, cannot scale well for large process control systems. ZigBee, on the other hand, can only utilize DSSS, hence, its performance may easily degrade in case of continues of noise. However, WirelessHART can handle this obstacle by applying channel hopping and channel blacklisting. Hence, WirelessHART and ISA100.11a technologies have been adopted for industrial applications due to their reliability, and low-cost features. However, WirelessHART does not support multiple protocols as ISA100.11a does. The transmission of HART messages are the only information specified and supported by WirelessHART. Hence, ISA100.11a is more preferable technology in this area since it can scale well and increase the network life-span. A general view of wireless sensor network applications is provided in Table 2.
- **Home Automation:** The realization of WSN technologies provided a new dimension to the process of the home automation systems, with the mobility of nodes, easy installation and deployment costs [62]. Many wireless sensor network technologies may fill the needs of home automation systems, however, some of them are advantageous over others, such that, 6LoWPAN offers low-cost, low-deployment, and most importantly, adaptability features to the existing technologies. Bluetooth is not suitable for home automation systems since it is limited in number of sensor nodes and its energy consumption is higher than other competitor wireless technologies, e.g., ZigBee, while WirelessHART is more suitable for industrial applications [62].

Acknowledgments The authors gratefully acknowledge the insightful comments of the anonymous reviewers which helped improve the quality and presentation of the paper significantly. This work is partially supported by the US National Science Foundation (NSF) grants 0917089 and 1054935. This work is funded by WiSeMAN Research Lab Department of Computer and Information Science College of Engineering and Computer Science University of Michigan-Dearborn.

References

1. K. Al Agha, M.-H. Bertin, T. Dang, A. Guitton, P. Minet, T. Val, J.-B. Viollet, Which wireless technology for industrial wireless sensor networks? the development of OCARI technology. IEEE Trans. Industr. Electron. **56**(10), 4266–4278 (2009)
2. C. Alcaraz, J. Lopez, A security analysis for wireless sensor mesh networks in highly critical systems. IEEE Trans. Syst. Man Cybern. Part C Appl. Rev. **40**(4), 419–428 (2010)
3. A.S. Altaan, Effects of sensor properties on power consumption in wireless sensor network. in *Computer Research and Development, 2010 Second International Conference on*, pp. 335–339, 7–10 May 2010
4. A. Ashraf, M. Hashmani, B.S. Chowdhry, M. Mussadiq, Q. Gee, A.Q.K. Rajput, Design and analysis of the security assessment framework for achieving discrete security values in wireless sensor networks. in *Electrical and Computer Engineering, 2008. CCECE 2008. Canadian Conference on*, pp. 000855–000860, 4–7 May 2008
5. P. Baronti, P. Pillai, V.W.C. Chook, S. Chessa, A. Gotta, Y. Fun Hu, Wireless sensor networks: A survey on the state of the art and the 802.15.4 and ZigBee standards. Comput. Commun. (ScienceDirect) **30**(7), 1655–1695 (2006)
6. C.-Y. Chang, D. Shiu; Power consumption optimization for information exchange in wireless-relay sensor networks. in *Communications (ICC), 2012 IEEE International Conference on*, pp. 745–750, 10–15 June 2012
7. Y. Chengbo, C. Yanzhe, L. Zhang, Y. Shuqiang, ZigBee wireless sensor network in environmental monitoring applications. WiCom '09. in *5th International Conference on Wireless Communications, Networking and Mobile Computing*, pp. 1–5, 24–26 Sept 2009
8. K. Collins, S. Mangold, G.-M. Muntean, Supporting mobile devices with wireless LAN/MAN in large controlled environments. IEEE Commun. Mag. **48**(12), 36–43 (2010)
9. L. De Nardis, M.-G. Di Benedetto, Overview of the IEEE 802.15.4/4a standards for low data rate Wireless Personal Data Networks. in *Positioning, Navigation and Communication, 2007. WPNC '07. 4th Workshop on*, pp. 285–289, 22–22 March 2007
10. R. Dutta, S. Mukhopadhyay, Improved self-healing key distribution with revocation in wireless sensor network. in *Wireless Communications and Networking Conference, 2007. WCNC 2007. IEEE*, pp. 2963–2968, 11–15 March 2007
11. A.R. Dutta, B.S. Saha, C.A.K. Mukhopadhyay, Efficient clustering techniques to optimize the system lifetime in Wireless Sensor Network. in *Advances in Engineering, Science and Management (ICAESM), 2012 International Conference on*, pp. 679–683, 30–31 March 2012
12. G.K. Ee, C.K. Ng, N.K. Noordin, B.M. Ali, Path recovery mechanism in 6LoWPAN routing. in *International Conference on Computer and Communication Engineering (ICCCE)*, pp. 1–5, 11–12 May 2010
13. P. Ferrari, A. Flammini, D. Marioli, S. Rinaldi, E. Sisinni, On the implementation and performance assessment of a wirelessHART distributed packet analyzer. IEEE Trans. Instrum. Meas. **59**(5), 1342–1352 (2010)
14. H. Forbes, White paper, ISA100 and wireless standards convergence, http://www.isa100wci.org/News-Room/Articles-and-Technical-Papers
15. L. Gang, B. Krishnamachari, C.S. Raghavendra, Performance evaluation of the IEEE 802.15.4 MAC for low-rate low-power wireless networks. in *Performance, Computing, and Communications, 2004 IEEE International Conference on*, pp. 701–706, 2004
16. K. Gill, Y. Shuang-Hua, Y. Fang, L. Xin, A zigbee-based home automation system. IEEE Trans. Consum. Electron. **55**(2), 422–430 (2009)
17. C. Gomez, J. Paradells, Wireless home automation networks: a survey of architectures and technologies. IEEE Commun. Mag. **48**(6), 92–101 (2010)
18. J. Granjal, E. Monteiro, J. Sa Silva, Enabling network-layer security on IPv6 wireless sensor networks. in Global Telecommunications Conference (GLOBECOM 2010), 2010 IEEE, pp. 1–6, 6–10 Dec 2010

19. S. Han, B. Tian, M. He, E. Chang, Efficient threshold self-healing key distribution with sponsorization for infrastructureless wireless networks. IEEE Trans. Wireless Commun. **8**(4), 1876–1887 (2009)
20. T. Hasegawa, H. Hayashi, T. Kitai, H. Sasajima, Industrial wireless standardization scope and implementation of ISA SP100 standard. in *SICE Annual Conference (SICE), 2011 Proceedings of*, pp. 2059–2064, 13–18 Sept 2011
21. M. Healy, T. Newe, E. Lewis, Power management in operating systems for wireless sensor nodes. in *Sensors Applications Symposium, 2007. SAS '07. IEEE*, pp. 1–6, 6–8 Feb 2007
22. A.A.O.A.L. Hester, Y. Huang, Neurfon netform: a self-organizing wireless sensor network. in *11th IEEE ICCCN Conference*, Oct 2002
23. Hitachi global. http://www.hitachi.com/
24. E. Holohan, M. Schukat, Authentication using virtual certificate authorities: a new security paradigm for wireless sensor networks. in *Network Computing and Applications (NCA), 2010 9th IEEE International Symposium on*, pp. 92–99, 15–17 July 2010
25. http://www.zigbee.org/About/AboutAlliance/TheAlliance.aspx
26. http://www.zigbee.org/Specifications.aspx
27. http://www.zigbee.org/Specifications/ZigBee/Overview.aspx
28. J. Hui, M. Devetsikiotis, The use of metamodeling for VoIP over WiFi capacity evaluation (Transactions Letters). IEEE Trans. Wireless Commun. **7**(1), 1–5 (2008)
29. J.W. Hui, D.E. Culler, Extending IP to low power, wireless personal area networks. IEEE Internet Comput. **12**(4), 37–45 (2008)
30. M.Z. Huq, S. Islam, Home area network technology assessment for demand response in smart grid environment. in *Universities Power Engineering Conference (AUPEC), 20th Australasian*, pp. 1–6, 5–8 Dec 2010
31. C. Karlof, D. Wagner, Secure routing in wireless sensor networks: attacks and countermeasures. in *Sensor Network Protocols and Applications, 2003. Proceedings of the First IEEE. 2003 IEEE International Workshop on*, pp. 113–127, 11 May 2003
32. A.N. Kim, F. Hekland, S. Petersen, P. Doyle, When HART goes wireless: Understanding and implementing the wirelessHART standard, emerging technologies and factory automation. in *ETFA 2008. IEEE International Conference on*, pp. 899–907, 15–18 Sept 2008
33. M. Knight, Wireless security How safe is Z-wave? Comput. Control. Eng. J. **17**(6), 18–23 (2006)
34. J.-S. Lee, C.-C. Chuang, C.-C. Shen, Applications of short-range wireless technologies to industrial automation: a ZigBee approach. in *Telecommunications, 2009. AICT '09. Fifth Advanced International Conference on*, pp. 15–20, 24–28 May 2009
35. J.-S. Lee, Y.-W. Su, C.-C. Shen, A comparative study of wireless protocols: Bluetooth, UWB, ZigBee, and Wi-Fi. IECON 2007. in *33rd Annual Conference of the IEEE Industrial Electronics Society*, pp. 46–51, 5–8 Nov 2007
36. J. Li, X. Zhu, N. Tang, J. Sui, Study on ZigBee network architecture and routing algorithm. in *Signal Processing Systems (ICSPS), 2010 2nd International Conference on*, vol. 2, pp. V2-389–V2-393, 5–7 July 2010
37. S. Lin, J. Liu, Y. Fang, ZigBee based wireless sensor networks and its applications in industrial. in *IEEE International Conference on Automation and Logistics*, pp. 1979–1983, 18–21 Aug 2007
38. A. Liu, P. Ning, TinyECC: a configurable library for elliptic curve cryptography in wireless sensor networks. in *Information Processing in Sensor Networks, 2008. IPSN '08. International Conference on*, pp. 245–256, 22–24 April 2008
39. C.-W. Lu, S.-C. Li, Q. Wu, Interconnecting ZigBee and 6LoWPAN wireless sensor networks for smart grid applications. in *2011 Fifth International Conference on Sensing Technology (ICST)*, pp. 267–272, Nov 28 2011–Dec 1 2011
40. J. Ma, M. Gao, Q. Zhang, L.M. Ni, Energy-efficient localized topology control algorithms in IEEE 802.15.4-based sensor networks. IEEE Trans. Parallel Distrib. Syst. **18**(5), 711–720 (2007)

41. L. Mainetti, L. Patrono, A. Vilei, Evolution of wireless sensor networks towards the Internet of Things: A survey. in *Software, Telecommunications and Computer Networks (SoftCOM), 2011 19th International Conference on*, pp. 1–6, 15–17 Sept 2011
42. I. Muller, J.C. Netto, C.E. Pereira, WirelessHART field devices. IEEE Instrum. Meas. Mag. **14**(6), 20–25 (2011)
43. M. Nagajothy, S. Radha, Network lifetime enhancement in wireless sensor network using network coding. in *Control, Automation, Communication and Energy Conservation, 2009. INCACEC 2009. 2009 International Conference on*, pp. 1–4, 4–6 June 2009
44. Official website of HART communication foundation, http://www.hartcomm.org/protocol/wihart/wireless_technology.html
45. T.J. Park, Y.J. Chon, D.K. Park, S.H. Hong, BACnet over ZigBee, a new approach to wireless datalink channel for BACnet. in *Industrial Informatics, 2007 5th IEEE International Conference on*, pp. 33–38, 23–27 June 2007
46. L. Pengfei, L. Jiakun, N. Luhua, W. Bo, Research and application of ZigBee protocol stack. in *2010 International Conference on Measuring Technology and Mechatronics Automation (ICMTMA)*, vol. 2, pp. 1031–1034, 13–14 March 2010
47. C. Perkins, E. Royer, Ad hoc on-demand distance vector routing. in *2nd IEEE Workshop on Mobile Computing Systems and Applications*, 1999, pp. 90–100
48. S. Petersen, S. Carlsen, WirelessHART Versus ISA100.11a: the format war hits the factory floor. IEEE Ind. Electron. Mag. **5**(4), 23–34 (2011)
49. M. Petrova, J. Riihijarvi, P. Mahonen, S. Labella, Performance study of IEEE 802.15.4 using measurements and simulations. in *Wireless Communications and Networking Conference, 2006. WCNC 2006. IEEE*, vol. 1, pp. 487–492, 3–6 April 2006
50. S. Raza, S. Duquennoy, T. Chung, D. Yazar, T. Voigt, U. Roedig, Securing communication in 6LoWPAN with compressed IPsec. in *Distributed Computing in Sensor Systems and Workshops (DCOSS), 2011 International Conference on*, pp. 1–8, 27–29 June 2011
51. S. Raza, T. Voigt, Interconnecting wirelessHART and legacy HART networks. in *2010 6th IEEE International Conference on Distributed Computing in Sensor Systems Workshops (DCOSSW)*, pp. 1–8, 21–23 June 2010
52. Renesas technology europe. http://eu.renesas.com/
53. R. Riaz, K.-H. Kim, H.F. Ahmed, Security analysis survey and framework design for IP connected LoWPANs. ISADS '09. in *International Symposium on Autonomous Decentralized Systems*, pp. 1–6, 23–25 March 2009
54. M. Saleh, I.A. Khatib, Throughput analysis of WEP security in ad hoc sensor networks. in *Proceedings of The Second International Conference on Innovations in Information Technology (IIT'05)*, Dubai, Sept 26–28, 2005
55. Sensinode. http://www.sensinode.com
56. Z. Shelby, C. Bormann, 6LoWPAN: the wireless embedded internet. Wiley Series in Communications Networking and Distributed Systems, 2009
57. E. Shih et al., Physical layer driven protocol and algorithmdesign for energy efficient wireless sensor networks. in *Proceedings of MOBICOM*, 2001, pp. 272–287
58. M.-K. Shin, K. Hyoung-Jun, L3 mobility support in large-scale IP-based sensor networks (6LoWPAN). ICACT 2009. in *11th International Conference on Advanced Communication Technology*, vol. 02, pp. 941–945, 15–18 Feb 2009
59. J. Song, H. Song, A.K. Mok, D. Chen, M. Lucas, M. Nixon, WirelessHART: applying wireless technology in real-time industrial process control. in *IEE Real-Time and Embedded Technology and Applications Symposium, RTAS '08*, pp. 377–386, 22–24 April 2008
60. H. Song, J. Song, X. Zhu, A.K. Mok, D. Chen, M. Nixon, W. Pratt, V. Gondhalekar, WiHTest: compliance test suite for Diagnosing Devices in Real-Time WirelessHART Network. RTAS 2009. in *15th IEEE, Real-Time and Embedded Technology and Applications Symposium*, pp. 327–336, 13–16 April 2009
61. Technical report, Co-existence of wirelessHART with other wireless technologies, http://www.hartcomm.org/protocol/training/resources/wiHART_resources/CoExistence_WirelessHART_LIT122.pdf

62. D.S. Tudose, A. Voinescu, M. Petrareanu, A. Bucur, D. Loghin, A. Bostan, N. Tapus, Home automation design using 6LoWPAN wireless sensor networks. in *Distributed Computing in Sensor Systems and Workshops (DCOSS), 2011 International Conference on*, pp. 1–6, 27–29 June 2011
63. H. Unterassinger, M. Dielacher, M. Flatscher, S. Gruber, G. Kowalczyk, J. Prainsack, T. Herndl, J. Schweighofer, W. Pribyl, A power management unit for ultra-low power wireless sensor networks. in *AFRICON, 2011*, pp. 1–6, 13–15 Sept 2011
64. I. Verhamme, Industrial ethernet book, Issue 64/30, http://www.iebmedia.com/index.php
65. A. Viswanathan, T.E. Boult, Power conservation in ZigBee networks using temporal control. in *Wireless Pervasive Computing, 2007. ISWPC '07. 2nd International Symposium on*, 5–7 Feb 2007
66. A.Y. Wang et al., Energy efficient modulation and MAC for asymmetric RF microsensorsystems. in *IEEE International Symposium on Low Power Electronics and Design*, 2001, pp. 106–111
67. W. Wang, G. He, J. Wan, Research on Zigbee wireless communication technology. in *Electrical and Control Engineering (ICECE), 2011 International Conference on*, pp. 1245–1249, 16–18 Sept 2011
68. Q. Wanzhi, H. Peng, R.J. Evans, An efficient self-healing process for ZigBee sensor networks. in *Communications and Information Technologies, 2007. ISCIT '07. International Symposium on*, pp. 1389–1394, 17–19 Oct 2007
69. G. Wenqi, W.M. Healy, Z. MengChu, ZigBee-wireless mesh networks for building automation and control. in *Networking, Sensing and Control (ICNSC), 2010 International Conference on*, pp. 731–736, 10–12 April 2010
70. White paper, Digi international, http://www.digi.com/pdf/ds_xbeezbmodules.pdf
71. White paper, Ember, http://www.eet-china.com/ARTICLES/2005OCT/PDF/EM250_DATASHEET.PDF
72. White paper, freeScale, http://cache.freescale.com/files/rf_if/doc/BRMC1319192FAM.pdf
73. White paper, Honeywel, https://www.honeywellprocess.com/
74. White paper, Nivis, http://www.nivis.com/resources/WirelessHART
75. White paper, vision for the home ZigBee wireless home automation, https://docs.zigbee.org/zigbee-docs/dcn/06-4720.pdf
76. White paper, ZigBee wireless sensor applications for health, wellness and fitness, https://docs.zigbee.org/zigbee-docs/dcn/09-4962.pdf
77. Wi-Fi CERTIFIED Wi-Fi direct, frequently asked questions, http://www.wi-fi.org/files/faq_20101021_Wi-Fi_Direct_FAQ.pdf
78. Wi-Fi CERTIFIED Wi-Fi Direct: Personal, portable Wi-Fi to connect devices anywhere, any time, http://www.wi-fi.org/register.php?file=wp_Wi-Fi_Direct_20101022_Consumer.pdf
79. J.M. Winter, C. Lima, I. Muller, C.E. Pereira, J.C. Netto, WirelessHART routing analysis software. in *Computing System Engineering (SBESC), 2011 Brazilian Symposium on*, pp. 96–98, 7–11 Nov 2011
80. A.D. Wood, J.A. Stankovic, S.H. Son, in *JAM: A Jammed-Area Mapping Service for Sensor Networks, 24th IEEE Real-Time Systems Symposium, RTSS 2003*, pp. 286–297
81. Y. Xu, J. Heidemann, D. Estrin, Geography-informed energy conservation for ad hoc routing. in *Proceedings of the 7th Annual International Conference on Mobile Computing and Networking. ACM*, 2001, pp. 7084
82. J. Yick, B. Mukherjee, D. Ghosal, Wireless sensor network survey. Comput. Netw. **52**(12), 2292–2330 (2008)
83. S. Yong, Z. Yi, W. Jian, Q. Tinggao, Research and implementation of ZigBee networking. in *Mechatronics and Automation, 2009. ICMA 2009. International Conference on*, pp. 3992–3996, 9–12 Aug 2009
84. X. Zhang, M. Wei, P. Wang, Y. Kim, Research and implementation of security mechanism in ISA100.11a networks. in *Electronic Measurement & Instruments, 2009. ICEMI '09. 9th International Conference on*, pp. 4-716–4-721, 16–19 Aug 2009

85. M. Zhou, Z.-l. Nie, Analysis and design of ZigBee MAC layers protocol. in *2010 International Conference on Future Information Technology and Management Engineering (FITME)*, vol. 2, pp. 211–215, 9–10 Oct 2010
86. Y. Zhou, X. Yang, X. Guo, M. Zhou, L. Wang, A design of greenhouse monitoring & control system based on ZigBee wireless sensor network. in *Wireless Communications, Networking and Mobile Computing, 2007. WiCom 2007. International Conference on*, pp. 2563–2567, 21–25 Sept 2007
87. ZIGBEE sMART eNERGY, http://zigbee.org/Standards/ZigBeeSmartEnergy/

Editor's Biography

Habib M. Ammari is an Associate Professor and the Founding Director of Wireless Sensor and Mobile Ad-hoc Networks (WiSeMAN) Research Lab, in the Department of Computer and Information Science, College of Engineering and Computer Science, University of Michigan- Dearborn, since September 2011. He obtained his second Ph.D. degree in Computer Science and Engineering from the University of Texas at Arlington, in May 2008, and his first Ph.D. in Computer Science from the Faculty of Sciences of Tunis, in December 1996. He has a strong publication record in top-quality journals, such as ACM TOSN, ACM TAAS, IEEE TPDS, IEEE TC, Elsevier COMNET, Elsevier PMC, Elsevier JPDC, Elsevier COMCOM, and high-quality conferences, such as IEEE SECON, IEEE ICDCS, EWSN, and IEEE MASS. He published his first Springer book, "Challenges and Opportunities of Connected *k*-Covered Wireless Sensor Networks: From Sensor Deployment to Data Gathering" in August 2009. Also, he is the author and editor of two Springer books, "The Art of Wireless Sensor Networks: Fundamentals" and "The Art of Wireless Sensor Networks: Advanced Topics and Applications," which will be published in 2014. He has been selected for inclusion in the AcademicKeys Who's Who in Engineering Higher Education in 2012, the AcademicKeys Who's Who in Sciences Higher Education in 2011, Feature Alumnus in the University of Texas at Arlington CSE Department's Newsletter in Spring 2011, Who's Who in America in 2010, and the 2008-2009 Honors Edition of Madison Who's Who Among Executives and Professionals. He received several prestigious awards, including the Certificate of Appreciation Award at ACM MiSeNet 2013, the Certificate of Appreciation Award at the IEEE DCoSS 2013, the Certificate of Appreciation Award at the ACM MobiCom 2011, the Outstanding Leadership Award at the IEEE ICCCN 2011, the Best Symposium Award at the IEEE IWCMC 2011, the Lawrence A. Stessin Prize for Outstanding Scholarly Publication from Hofstra University in May 2010, the Faculty Research and Development Grant Award from Hofstra College of Liberal Arts and Sciences in May 2009, the Best Paper Award at EWSN in 2008, the Best Paper Award at the IEEE PerCom 2008 Google Ph.D. Forum, the Best Graduate Student Paper Award (Nokia Budding Wireless Innovators Awards First Prize) in May 2004, the Best Graduate Student Presentation Award (Ericsson Award First Prize) in February 2004, and Laureate in Physics and Chemistry for academic years 1987 and 1988. Also, he was selected as the ACM Student Research Competition Finalist at the ACM MobiCom 2005. He is the recipient of the Nortel Outstanding CSE Doctoral Dissertation Award in February 2009,

H. M. Ammari (ed.), *The Art of Wireless Sensor Networks*,
Signals and Communication Technology, DOI: 10.1007/978-3-642-40009-4,
© Springer-Verlag Berlin Heidelberg 2014

and the John Steven Schuchman Award for 2006-2007 Outstanding Research by a PhD Student in February 2008. He received a three-year US National Science Foundation (NSF) Research Grant Award, in June 2009, and the US NSF CAREER Award, in January 2011. He is the Founding Coordinator of both of the Research Colloquium Series since September 2011, and the Distinguished Lecture Series since January 2012, in the College of Engineering and Computer Science at the University of Michigan-Dearborn. He has been invited to give invited talks at several reputed universities. He is the Founder of the ACM Annual International Workshop on Mission-Oriented Wireless Sensor Networking (ACM MiSeNet), which has been co-located with ACM MobiCom since 2012. He serves as Associate Editor of several prestigious journals, such as ACM TOSN, IEEE TC, and Elsevier PMC. Also, he has served as Program Chair, Session Chair, Publicity Chair, Web Chair, and Technical Program Committee member of numerous ACM and IEEE conferences, symposia, and workshops.

MIX
Papier aus verantwortungsvollen Quellen
Paper from responsible sources
FSC® C105338

If you have any concerns about our products,
you can contact us on
ProductSafety@springernature.com

In case Publisher is established outside the EU,
the EU authorized representative is:
Springer Nature Customer Service Center GmbH
Europaplatz 3, 69115 Heidelberg, Germany

Printed by Libri Plureos GmbH
in Hamburg, Germany